FUNDAMENTALS OF INTERFACE AND COLLOID SCIENCE

VOLUME II
SOLID-LIQUID INTERFACES

FUNDAMENTALS OF INTERFACE AND COLLOID SCIENCE

VOLUME II
SOLID-LIQUID INTERFACES

J. Lyklema

With special contributions by
A. de Keizer (Chapter 3)
B.H. Bijsterbosch (Chapter 4)
G.J. Fleer and M.A. Cohen Stuart (Chapter 5)

ACADEMIC PRESS

A Harcourt Science and Technology Company

San Diego   San Francisco   New York   Boston
London   Sydney   Tokyo

Academic Press
A Harcourt Science and Technology Company
Harcourt Place, 32 Jamestown Road, London NW1 7BY, UK
http://www.academicpress.com

Academic Press
A Harcourt Science and Technology Company
525 B Street, Suite 1900, San Diego, California 92101-4495, USA
http://www.academicpress.com

ISBN 0-12-460524-9

A catalogue record for this book is available from the British Library

Transferred to digital printing 2005

Printed and bound by Antony Rowe Ltd, Eastbourne

GENERAL PREFACE

Fundamentals of Interface and Colloid Science (FICS) is motivated by three related but partly conflicting observations.

The first is that interface and colloid science is an important and fascinating, though often undervalued, branch of science. It has applications and ramifications in domains as disparate as agriculture, mineral dressing, oil recovery, chemical industry, biotechnology, medical science and many more provinces of the living and non-living world.

The second observation is that proper application and integration of interface and colloid science requires, besides factual knowledge, insight into the many basic laws of physics and chemistry upon which it rests.

In the third place, most textbooks of physics and chemistry pay only limited attention to interface and colloid science.

Between these observations the conflict looms that it is an almost impossible task to master not only some specific domain of application but also interface and colloid science itself, including its foundations. The problem is aggravated by the fact that interface and colloid science has a very wide scope: it uses parts of classical, irreversible and statistical thermodynamics, optics, rheology, electrochemistry and other branches of science. Nobody can be expected to command all of this simultaneously.

The prime goal of FICS is to meet these demands systematically, treating the most important interfacial and colloidal phenomena starting from basic principles of physics and chemistry, whereby these principles are first reviewed. In doing so, it will become clear that common roots often underlie seemingly different phenomena, which is helpful in identifying and recognizing them.

Given these objectives, a deductive approach is indicated. From the beginning to the end, systems of growing complexity are treated gradually, with, as a broad division, in Volume I the fundamentals (F), in Volumes II and III isolated interfaces (I) and in Volumes IV and V interfaces in interaction and colloids (C).

The chosen deductive set-up serves two objectives: the book is not only intended to become a standard reference book, but hopefully parts will be suitable as a textbook for systematic study, either as a self-study guide or as a reference for courses.

In view of these objectives, a certain style is more or less defined and some characteristic style elements are the following:

- Topics are arranged by the main principle(s) and phenomena on which they rest. Since many researchers are not immediately familiar with these principles and as more than one principle may determine a phenomenon met in practice, an extensive and detailed subject index is provided, which in some places has double or triple entries.

- As FICS is a book of principles rather than a book of facts, no attempts are made to give it an encyclopaedic character, although important matter is tabulated for easy reference. For factual information, references are made to the literature, in particular to reviews and books. The philosophy is that experimental observations which illustrate or enforce specific principles are emphasized, rather than given for their own sake. This also implies a certain preference for illustrations with model systems.

- In order to properly formulate physical principles, some mathematics is indispensable and we cannot always avoid complex and abstract formalisms. To that end, specialized techniques that are sometimes particularly suited in solving certain types of problems will be introduced when needed, mostly in the appendices to Volume I. However, the reader is assumed to be familiar with elementary calculus.

- Generally, the starting level of Volume I is such that it can be read without having an advanced command of physics and chemistry. In turn, for the later volumes the physical chemistry of Volume I is the starting point.

- In view of the fact that much space is reserved for the explanation and elaboration of principles, some restriction had to be made with respect to the number of systems to be treated, because otherwise the size would expand out of bounds. In view of the importance of interfacial and colloid science for biology, medicine, pharmacy, agriculture, etc., wet systems, aqueous ones in particular, will be emphasized. "Dry" subjects such as aerosols and solid state physics will be given less attention. Experimental techniques will not be described in great detail except where these techniques have a typical interfacial or colloidal nature.

Considering all these features, FICS is typically a book containing parts that can also be found in more detail in other books but rarely in the present context. Moreover, it stands out by integrating these parts. It is hoped that through this integration many workers will experience the relevance and beauty of interface and colloid science and become fascinated by it.

Hans Lyklema
Wageningen, The Netherlands
1995

PREFACE TO VOLUME II: SOLID-LIQUID INTERFACES

Having laid down the physico-chemical basis of interface and colloid science in Volume I, we can now make a start with the systematic treatment. Solid-gas and, in particular, solid-liquid interfaces are selected as the first topics. The argument for this choice is that such interfaces do not change their area upon adsorption and/or charging, and are therefore the simplest group of model systems. In Volume III, dealing with fluid-fluid interfaces, this restriction will be relaxed.

In line with the general set-up of FICS, attempts are made to keep the treatment systematic and deductive. Recurrent features are that each chapter begins, as much as possible, with the general thermodynamic and/or statistical thermodynamic foundations and the various phenomena are presented in order of increasing complexity. The wish to give the work the nature of both a reference and a textbook is reflected in being comprehensive as far as the fundamentals are concerned and giving the treatment a certain didactic flavour. Completeness always remains a matter of dispute, but it is hoped that many readers will find the information they are looking for, especially regarding basic issues and main principles. No attempts are made to give the book encyclopaedic character as far as the facts are concerned; exceptions are selected data collections, as in appendix 3. Experiments are included on the basis of their power to illustrate certain points; there may be some arbitrariness in the choice. The trend is to shun complex, multivariable systems because for such systems basic features might easily be obscured.

The level is about the same as that of modern journals dealing with the subject. Readers without the required background can hopefully find much of the required information in FICS, going back to earlier chapters or to Volume I where needed. To facilitate this process, extensive back-referencing has been applied, although where appropriate, at the beginning of each chapter the relevant background in Volume I is briefly reviewed, so that the present Volume stands on its own. For the more specialized backgrounds, reference is made to the literature.

In a number of instances a decision had to be taken as to what to call "fundamentals" and what "advanced". In the case of electric double layers, this decision related to the classical Gouy-Stern theory versus modern statistical theories. For pragmatic reasons we decided to emphasize the former: the equations are simple and analytical, and can account for the great majority of situations met in practice. However, a section is included to give an impression of more *a priori* statistical approaches. In the domain of electrokinetics the decision was between simple theories on the level of Helmholtz-Smoluchowski (HS), that may apply to perhaps 30-50% of all systems studied in practice, or on the level of advanced elaborations, accounting for double layer polarization and conduction in the stagnant layer. Here, it was felt that the former did not give enough coverage. In this case the decision was made to treat the subject matter in three stages: first the HS-type cases, then models including polarization and finally models including both polarization and stagnant layer conduction. Although it is beyond the scope of FICS to give all the relevant mathematics in detail, the basic electrokinetic equations are given and explained. Moreover, a simple analytical formula was derived for large κa covering all complications. In this way it is hoped that readers having different levels of interest might all find this chapter useful.

Regarding symbols and units of quantities, IUPAC recommendations have, as a rule, been heeded. However, the scope of FICS leads to a number of clashes that forced us to deviate from these rules now and then. For example, we prefer F for Helmholtz energy instead of A because of the interference with A for area. In chapter 5 we ran into the problem that in polymer adsorption theory the segment adsorption energy parameter χ^s is considered positive for attraction whereas generally attraction energies or forces have negative signs. We decided not to change this unfortunate, though commonly encountered, situation, but where necessary, added appropriate warnings.

As for the spelling of names, we prefer that of the country of origin. So, if referring to people, we would write for instance van der Waals, Deryagin and d'Arcy. However, for phenomena or laws, capitals are used ("Van der Waals equation"). For Slavic names, originally written in the Cyrillic alphabet, we adhere to the *Chemical Abstracts* transcription. However, many Slavic authors have their names transcribed differently in different languages, and as literature citations should be verbatim, some inconsistency is unavoidable. Where appropriate, identifications are made (Deryaguin, Derjagin, Derjaguin, etc. = Deryagin).

In a systematic text the size of FICS, numerous cross-references are unavoidable. To facilitate consultations of previous information, in Volume II we have repeated the numbers referring to subsection headings (**6.5b** instead of just **b**). References to Volume I are preceded by I.

Co-operation and acknowledgements

Although I have written the entire text, except chapter 5, myself, and remain ultimately responsible for any errors, the production of a voluminous project such as FICS is not a one-man show. The same team on which I relied for the writing of Volume I remained available for Volume II, with the sad exception of Dr. J. Scheutjens, who died in a road accident. Several parts of chapter 5 have been inspired by him. Some team members have contributed very substantially to certain chapters; their names are mentioned on the title page. They have also carefully read, and commented on other chapters. In addition I want to acknowledge my appreciation to Prof. B.H. Bijsterbosch (chapters 2 and 3), Dr. M.A. Cohen Stuart (chapter 2), Dr. H.P. van Leeuwen (chapters 3 and 4), Prof. J.Th.G. Overbeek (chapters 3 and 4), and Prof. B. Vincent (chapters 1, 2 and 3). I am particularly indebted to master spy Dr. A. de Keizer, who spotted, and helped to resolve, numerous errors and problems in chapters 1, 2 and 3.

Other co-workers and students from the Department of Physical and Colloid Chemistry of the Wageningen Agricultural University have also greatly contributed, sometimes by considering parts of the text as "study clubs", in other cases by reading entire chapters or parts of them. Of them I acknowledge Dr. L.K. Koopal (chapters 1 and 3), Dr. F.A.M. Leermakers (chapter 2), Dr. N.A.M. Besseling (chapter 1), Dr. J.A.G. Buijs (chapter 1), Mr. A. van der Wal (chapters 3 and 4), Mr. H.P. Huinink (sec. 3.6b), Mr. A.J. van der Linde (chapter 4) and in particular Mr. M. Minor for his expert help with chapter 4.

Outside the department I have also relied on people to check (parts of) chapters. They include Dr. L.G.J. Fokkink (chapter 1 and parts of chapter 3), Dr. J.W.N. Niemantsverdriet (parts of chapter 1), Prof. I. Dékany and Prof. G.H. Findenegg (chapter 2), Dr. A. Breeuwsma (parts of chapter 3), Prof. S.S. Dukhin, Prof. J. Anderson and Dr. R.W. O'Brien (chapter 4).

Preparation of the lay out and the typing were in the dedicated and capable hands of Mrs. Josie Zeevat, with incidental help of Mrs. Yvonne Toussaint and administrative assistance of Mr. A.W. Bouman. For the artwork I am again indebted to Mr. G. Buurman.

All this help and encouragement has been instrumental in completing this Volume.

Hans Lyklema,
Wageningen, The Netherlands
August 1995

CONTENTS OF VOLUME II

List of symbols

LIST OF FREQUENTLY USED SYMBOLS (Volumes I and II)

Symbols representing physical quantities are printed in *italics*.
Thermodynamic functions: capital for macroscopic quantities, small for molecular or subsystem quantities (example: U = total energy, u = pair energy between molecules).

Superscripts

†	standard pressure
o	standard in general
*	pure substance; complex conjugate
σ	interfacial (excess)
E	excess, X (real) – X (ideal)
S, L, G	in solid, liquid or gaseous state
⊥	normal to surface
//	parallel to surface

Subscripts

m	molar (sometimes molecular)
a	areal (per unit area)
g	per unit mass

Recurrent special symbols

$O(10^{20})$	of the order of 10^{20}
ΔX	X (final) – X (initial). Subscript attached to Δ to denote type of process: diss (dissociation), hydr (hydration), mix (mixing), r (reaction), sol (dissolution), solv (solvation), subl (sublimation), trs (transfer), vap (vaporization, evaporation)

Some mathematical signs and operators

Vectors	**bold** face. Example: $\mathbf{F}$ for force, but F_z for z-component of force
Tensors	**bold** face with tilde ($\tilde{\boldsymbol{\tau}}$)

complex quantities bear a circumflex ($\hat{n}$), the corresponding conjugate is $\hat{n}^*$

$|x|$	absolute value of x
$\langle x \rangle$	averaged value of x
$\bar{x}$	Fourier or Laplace transform of x (sometimes this bar is omitted)
$\dot{\gamma}$	time-derivative of γ

For vectorial signs and operators (∇, ∇^2, grad, rot, and ×), see I.appendix **7**.

Latin

a		activity (mol m^{-3})
	$a_\pm$	mean activity of an electrolyte (mol m^{-3})
a		attraction parameter in Van der Waals equation of state (N m^4)
	a^σ	two-dimensional attraction parameter in Van der Waals equation of state (N m^3)
a		radius (m)
	a_g	radius of gyration (m)
	a_m	area per molecule (m^2)
A		area (m^2)
	A_g	specific area (m^2 kg^{-1})
A		Hamaker constant (J)
	$A_{ij(k)}$	Hamaker constant for interaction of materials i and j across material k (J)
b		volume correction parameter in Van der Waals equation of state (m^3)
$\boldsymbol{B}, B$		magnetic induction (T = V m^{-2} s)
$B_2(T)$		second virial coefficient (m^3 mol^{-1} or m^3 molecule^{-1})
	$B_2^\sigma(T)$	interfacial second virial coefficient (m^2 mol^{-1}, m^2 molecule^{-1} or –)
c		velocity of electromagnetic radiation in a vacuum (m s^{-1})
c		concentration (usually mol m^{-3}, sometimes kg m^{-3})
C		(differential) electric capacitance (C V^{-1} or, if per unit area, C m^{-2} V^{-1})
	C^d	(differential) electric capacitance of diffuse double layer (C m^{-2} V^{-1})
	C^i	(differential) electric capacitance of Stern layer (C m^{-2} V^{-1})
C		BET-transformed (-)
C_x		(time-) correlation function of x (dim. x^2)
C_p		molar heat capacity at constant pressure (J K^{-1} mol^{-1})
C_V		molar heat capacity at constant volume (J K^{-1} mol^{-1})
C_A^σ		interfacial excess molar heat capacity at constant area per unit area (J K^{-1} m^{-2} mol^{-1})
C_γ^σ		interfacial excess molar heat capacity at constant interfacial tension per unit area (J K^{-1} m^{-2} mol^{-1})
d		layer thickness (m)
	d^{ek}	electrokinetic thickness (m)
	d^h	hydrodynamic thickness (m)
	d^{ell}	ellipsometric thickness (m)
	d^{st}	steric thickness (polymeric adsorbates) (m)
D		diffusion coefficient (m^2 s^{-1})
	D^σ	surface diffusion coefficient (m^2 s^{-1})
	D_r	rotational diffusion coefficient (s^{-1})

SYMBOL LIST

D_s		self-diffusion coefficient ($m^2\ s^{-1}$)
$\mathbf{D}, D$		dielectric displacement ($C\ m^{-2}$)
De		Deborah number (-)
Du		Dukhin number (-)
e		elementary charge (C)
$\mathbf{E}, E$		electric field strength ($V\ m^{-1}$)
E_{irr}		irradiance ($J\ m^{-2}\ s^{-1} = W\ m^{-2}$)
$\mathbf{E}_{sed}$		sedimentation potential ($V\ m^{-1}$)
E_{str}		streaming potential ($V\ m^2\ N^{-1}$)
f		friction coefficient ($kg\ s^{-1}$)
f		activity coefficient (mol fraction scale) (-)
f_{ij}		Mayer function for interaction between particles i and j (-)
F		Faraday constant ($C\ mol^{-1}$)
F		Helmholtz energy (J)
	F_i, F_{mi}	partial molar Helmholtz energy ($J\ mol^{-1}$)
	F_m	molar Helmholtz energy ($J\ mol^{-1}$)
	F^{σ}	interfacial (excess) Helmholtz energy (J)
	$F_a^{\sigma}, F^{\sigma}/A$	interfacial (excess) Helmholtz energy per unit area ($J\ m^{-2}$)
$\mathbf{F}, F$		force (N)
$g(r)$		radial distribution function (-)
$g^{(h)}$		hth order distribution function (-)
$g(\mathbf{q}, t)$		time correlation function, if real (light scattering usage) (dimensions as C_x)
g		standard acceleration of free fall ($m\ s^{-2}$)
G		Gibbs energy (J)
	G_i, G_{mi}	partial molar Gibbs energy ($J\ mol^{-1}$)
	G_m	molar Gibbs energy ($J\ mol^{-1}$)
	G^{σ}	interfacial (excess) Gibbs energy (J)
	$G_a^{\sigma}, G^{\sigma}/A$	interfacial (excess) Gibbs energy per unit area ($J\ m^{-2}$)
$G(z)$		segment weighting factor in polymer adsorption theory (-)
$G(z, s)$		endpoint distribution in a segment of s segments (polymer adsorption (-)
h		Planck's constant (J s)
$\hbar$		$h/2\pi$ (J s)
h		(shortest) distance between colloidal particles or macrobodies (m)
h		height (m)
$h(r)$		total correlation function (-)
H		enthalpy (J)
	H_i, H_{mi}	partial molar enthalpy ($J\ mol^{-1}$)
	H_m	molar enthalpy ($J\ mol^{-1}$)

H^σ		interfacial (excess) enthalpy (J)
H_a^σ, H^σ/A		interfacial (excess) enthalpy per unit area (J m^{-2})
H, H		magnetic field strength (C m^{-1} s^{-1})
$H(p,q)$		Hamiltonian (J)
i		intensity of radiation (V^2 m^{-2})
	i_i	incident intensity (V^2 m^{-2})
	i_0	intensity in a vacuum (V^2 m^{-2})
	i_s	scattered intensity (V^2 m^{-2})
$\boldsymbol{i}$		unit vector in x-direction (-) (not in chapter I.7)
$\boldsymbol{I}_{str}$		streaming current (C m^2 N^{-1} s^{-1})
I		ionic strength (mol m^{-3})
I		radiant intensity (J s^{-1} sr^{-1} = W sr^{-1})
$I_x(\omega)$		spectral density of x (dim. x^2 s)
$\boldsymbol{j}$		unit vector in y-direction (-) (not in chapter I.7)
$\boldsymbol{j}, j$		(electric) current density (A m^{-2} = C m^{-2} s^{-1})
	$\boldsymbol{j}^\sigma$, j^σ	surface current density (C m^{-1} s^{-1})
$\boldsymbol{J}, J$		flux (mol m^{-2} s^{-1} or kg m^{-2} s^{-1})
	J^σ	surface flux (mol m^{-1} s^{-1} or kg m^{-1} s^{-1})
k		Boltzmann's constant (J K^{-1})
k		rate constant (dimensions depend on order of process)
$\boldsymbol{k}$		unit vector in y-direction (-) (not in chapter I.7)
$\boldsymbol{k}$		wave vector (m^{-1})
$\mathcal{K}$		optical constant (m^2 kg^{-2} or m^2 mol^{-2})
$K(R)$		optical constant (V^4 C^{-2} m^{-6})
$K(\omega)$		absorption index (-)
K		chemical equilibrium constant (general)
	K_p	on pressure basis (-)
	K_c	on concentration basis (-)
K		(integral) electric capacitance (C V^{-1} or C m^{-2} V^{-1})
	K^d	(integral) electric capacitance of diffuse layer (C m^{-2} V^{-1})
	K^i	(integral) electric capacitance of Stern layer (C m^{-2} V^{-1})
K_H		Henry constant (m)
K_i		distribution (partition) coefficient (-)
K_L		Langmuir constant (m^3 mol^{-1})
K		conductivity (S m^{-1} = C V^{-1} m^{-1} s^{-1})
	K^σ	surface conductivity (S = C V^{-1} s^{-1})
L		contour length (polymers) (m)
L_{ik}		cross coefficients in irreversible thermodynamics (varying dimensions)
ℓ		bond length in a polymer chain (m)
m		mass (kg)

SYMBOL LIST

M		(relative) molecular mass (-)
	$\langle M \rangle_w$, M_w	ibid., mass average (-)
	$\langle M \rangle_z$, M_z	ibid., Z-average (-)
	$\langle M \rangle_n$, M_n	ibid., number average (-)
n		refractive index (-)
n		number of moles (- or moles)
	n^σ	number (excess) of moles in interface (- or moles)
$\mathbf{n}_{x,y,z}$		unit vector in x-, y- or z-direction (-) (chapter I.7 only)
N		number of segments in a polymer chain
N		number of molecules (-)
	N_{Av}	Avogadro constant (mol^{-1})
	N_s	number of sites (-)
p		bound fraction (of polymers) (-)
p		pressure ($N\ m^{-2}$)
	Δp	capillary pressure ($N\ m^{-2}$)
p		stiffness (persistence) parameter (polymers) (-)
$\mathbf{p}$, p		dipole moment (C m)
	$\mathbf{p}_{ind}$, p_{ind}	induced dipole moment (C m)
$\mathbf{p}$, p		($= m\ \mathbf{v}$) momentum ($J\ m^{-1}\ s$)
P		probability (-)
$\mathbf{P}$, P		polarization ($C\ m^{-2}$)
Pe		Péclet number (-)
q		heat exchanged (incl. sign) (J)
	$q(isost)$	isosteric heat of adsorption (J)
q		generalized parameter indicating place coordinates in Hamiltonian
q		subsystem canonical partition function (-)
q		electric charge (on ions) (C)
q		persistence length (polymers) (m)
$\mathbf{q}$, q		scattering vector (m^{-1})
Q		electric charge (on colloids, macrobodies) (C)
	$Q(N,V,T)$	canonical partition function (-)
	$Q_{eo,E}$	electro-osmotic volume flow per unit field strength ($m^4\ V^{-1}\ s^{-1}$)
	$Q_{eo,I}$	electro-osmotic volume flow per unit current ($m^3\ C^{-1}$)
$\mathbf{r}$, r		distance (m)
	r_B	Bjerrum length (m)
r		number of segments in a polymer (-)
R		gas constant ($J\ K^{-1}\ mol^{-1}$)
R		(principal) radius of curvature (m)
$\mathbf{R}$		Poynting vector ($W\ m^{-2}$)
	R_θ	Rayleigh ratio (m^{-1}) .

Re		Reynolds number (-)
S		entropy (J K^{-1})
	S_i, S_{mi}	partial molar entropy (J K^{-1} mol^{-1})
	S_m	molar entropy (J K^{-1} mol^{-1})
	S^σ	interfacial (excess) entropy (J K^{-1})
	S_a^σ, S^σ/A	interfacial (excess) entropy per unit area (J K^{-1} m^{-2})
$S(q, R, \Omega)$		spectral density as a function of $\omega_s - \omega_i = \Omega$ (V^2 m^{-2} s)
$S(q, c)$		structure factor (-)
$S(s)$		ordering parameter of s (-)
t		time (s)
t		transport (or transference) number (-)
T		temperature (K)
Ta		Taylor number (-)
u		(internal) energy per subsystem (J)
u		(electric) mobility (m^2 V^{-1} s^{-1})
U		(internal) energy, general (J)
	U_i, U_{mi}	partial molar energy (J mol^{-1})
	U_m	molar energy (J mol^{-1})
	U^σ	interfacial (excess) energy (J)
	U_a^σ, U^σ/A	interfacial (excess) energy per unit area (J m^{-2})
v		excluded volume parameter (polymers) (= $1-2\chi$)
$\boldsymbol{v}$, v		velocity (m s^{-1})
	v_{ef}	electrophoretic velocity (m s^{-1})
	v_{eo}	electro-osmotic velocity (m s^{-1})
	$\boldsymbol{v}_s$	slip velocity (m s^{-1})
V		volume (m^3)
	V_i, V_{mi}	partial molar volume (m^3 mol^{-1})
	V_m	molar volume (m^3 mol^{-1})
w		work (incl. sign) (J)
w		interaction parameter in regular mixture theory (J mol^{-1})
	$w_{(ij)}$	interaction energy between pair of molecules or segments (i and j) (J)
x		mol fraction (-)
x		distance from surface
$\boldsymbol{X}$, X		generalized force in irreversible thermodynamics (varying units)
y		activity coefficient (molar scale) (-)
y		dimensionless potential ($F\psi/RT$) (-)
z		coordination number (-)
z		distance from surface (m)
z		valency (-)

SYMBOL LIST

$Z\ (N,p,T)$		isobaric-isothermal partition function (-)
Z_N		configuration integral for N particles (-)

Greek

α		real potential (V)
α		degree of dissociation (-)
α		contact angle (-)
$\tilde{\alpha},\ \alpha$		polarizability (C V^{-1} m^2 = C^2 J^{-1} m^2)
β_{12}		twice binary cluster integral (-)
β		Van der Waals constant (molecular) (J m^{-6});
	β_D	Debye-Van der Waals constant (molecular) (J m^{-6})
	β_K	Keesom-Van der Waals constant (molecular) (J m^{-6})
	β_L	London-Van der Waals constant (molecular) (J m^{-6})
β		Esin-Markov coefficient (-)
γ		interfacial or surface tension (N m^{-1} or J m^{-2})
γ		activity coefficient (molal scale) (-)
γ		shear strain (-)
	$\dot{\gamma}$	rate of shear (s^{-1})
Γ		surface (excess) concentration (mol m^{-2})
δ		diffusion layer thickness (m)
δx		small variation of x (dim. x)
$\delta(x)$		Dirac delta function of x (dim. x^{-1})
Δ		displacement (m)
ΔX		X (final) $- X$(initial)
$^{\alpha}\Delta^{\beta}X$		X (phase β) $- X$ (phase α)
$^{\alpha}\Delta^{\beta}\phi$		(Galvani) potential difference (V)
$^{\alpha}\Delta^{\beta}\psi$		(Volta) potential difference (V)
	$^{\alpha}\Delta^{\beta}\psi_{diff}$	liquid junction potential (V)
ε		relative dielectric permittivity (dielectric constant) (-)
	ε_o	dielectric permittivity of vacuum (C^2 N^{-1} m^{-2} or C m^{-1} V^{-1})
ε		porosity (-)
ζ		electrokinetic potential (V)
θ		surface coverage $= \Gamma/\Gamma$ (saturated monolayer) (-)
θ		angle, angle of rotation (-)
κ		reciprocal Debye length (m^{-1})
Ξ		grand (canonical) partition function (-)
λ		wavelength (m)
λ		charging parameter (-)
λ		ionic (or molar) conductivity (C V^{-1} m^2 s^{-1} mol^{-1} = S m^2 mol^{-1})
Λ		molar conductivity (S m^2 mol^{-1})
Λ		thermal wavelength (m)

Λ		penetration depth of evanescent waves (m)
μ_o		magnetic permeability in vacuum (V m^{-1} C^{-1} s^2)
μ		magnetic dipole moment (C m^2 s^{-1})
μ		chemical potential (J mol^{-1} or J molecule^{-1})
μ		kinematic viscosity (m^2 s^{-1})
η		dynamic viscosity (N s m^{-2})
ν		frequency (s^{-1} = Hz)
ξ		coupling parameter (Kirkwood) (-)
ξ		grand (canonical) partition function of subsystem (-)
π		surface pressure (N m^{-1} or J m^{-2})
Π		osmotic pressure (N m^{-2})
$\Pi(h)$		disjoining pressure (N m^{-2})
ρ		density (kg m^{-3})
	ρ_N	number density (N/V) (m^{-3})
ρ		space charge density (C m^{-3})
σ		distance of closest approach for hard spheres (m)
σ, σ^o		surface charge density (C m^{-2})
	σ_i	contribution of ionic species i to surface charge (C m^{-2})
	σ^d	surface charge density diffuse layer (C m^{-2})
	σ^i	surface charge density Stern layer (C m^{-2})
σ_x		standard deviation of x (dim. x)
τ		characteristic time (s)
τ		interfacial stress (N m^{-1})
τ		turbidity (m^{-1})
$\tilde{\tau}$		stress tensor (N m^{-2})
	τ_{xy}	flux of x-momentum in y-direction (kg m^{-1} s^{-2}) = shear stress (N m^{-2}), one of the nine components of the stress tensor
	τ_c	rotational correlation time (s)
	τ_r	rotational relaxation time (reorientation time) (s)
ϕ		osmotic coefficient (-)
φ		volume fraction (-)
ϕ		phase (-)
χ		excess interaction energy parameter (-)
	χ_{cr}, φ_{cr}	critical values of χ and φ at phase separation
$\chi(\chi^{\alpha\beta})$		interfacial potential jump (between phases β and α) (V)
χ_e		electric susceptibility (-)
χ^s		adsorption energy parameter (-)
	χ^s_{crit}	critical value of χ^s at the adsorption/desorption point
ψ		electric potential (V)

SYMBOL LIST

ω		angular frequency (rad s^{-1} or s^{-1})
	ω_i	angular frequency of incident radiation (rad s^{-1} or s^{-1})
	ω_s	angular frequency of scattered radiation (rad s^{-1} or s^{-1})
ω		degeneracy of subsystem (-)
Ω		degeneracy of a system or number of realizations (-)
Ω		$\omega_s - \omega_i$ (rad s^{-1} or s^{-1})
Ω		solid angle (sr)
$\Omega(N,V,U)$		number of realizations = microcanonical partition function (-)

1 ADSORPTION AT THE SOLID-GAS INTERFACE

In this first chapter of Volume II we address the least complex interfacial system, the solid-gas interface. As the emphasis of *Fundamentals of Interface and Colloid Science* is on liquid systems, the present topic is not one of direct interest. However, the indirect relevance is considerable. Some of the reasons for this are:

(i) Techniques used to characterize solid-gas and solid-vacuum interfaces are useful, if not indispensable, in establishing the properties of solid surfaces in contact with liquids;

(ii) the solid-gas interface is a model for solid-liquid interfaces. Various phenomena observed with the latter systems are also met in the former ones, but in a quantitatively more accessible way;

(iii) modelling adsorption in terms of adsorption isotherms and surface equations of state is more conveniently introduced with the solid-gas interface.

These three points more or less define the tasks set for the present chapter. However, certain topics typical for solid-gas interfaces, such as heterogeneous catalysis, will not be discussed in detail although they are briefly mentioned in sec. 1.8.

1.1 Selected review of Volume I

In Volume I adsorption phenomena and surface characterization have already been encountered several times, though not in such depth as in the present chapter. By way of introduction to the present, more methodical, treatment we briefly review and integrate some relevant parts of Volume I (Fundamentals)[1].

The phenomenon of adsorption was introduced in sec. I.1.2. There one can find definitions of elementary notions, including those of *adsorbent, adsorbate, adsorptive, desorption, specific surface area, adsorption isotherm (equation)* and *two-dimensional equation of state.* Adsorbed amounts can conveniently be expressed as moles adsorbed (n^σ), moles adsorbed per unit area or *surface concentration* ($\Gamma = n^\sigma/A$) or, if the adsorption in a monolayer $\Gamma(\text{max})$ is known, as

---

[1] Reference to chapters, sections, equations and figures in Volume I are marked by the roman numeral I.

fractional coverage $\theta = \Gamma/\Gamma(\text{max})$. Sometimes adsorbed amounts are expressed in grams or as volumes. When A is not known, adsorbed amounts can only be expressed per unit of weight of the adsorbent.

In sec. I.1.2 some elementary adsorption isotherm equations have been given, viz. [I.1.3.7], [I.1.3.8] and [I.1.3.9]. Rewritten in terms of equilibrium pressures (instead of concentrations) they become

$$\Gamma = K'_H p \qquad (\Gamma \to 0) \tag{1.1.1}$$

$$\Gamma = a\,p^b \tag{1.1.2}$$

$$\theta/(1-\theta) = K'_L p \tag{1.1.3}$$

for the *Henry*, *Freundlich* and *Langmuir isotherm equation*, respectively. The dimensions of K'_H and K'_L on the pressure scale differ from those on the concentration scale, K_H and K_L, respectively. In [I.1.3.7] K_H has the dimensions of a length, in [1.1.1] K'_H is in moles N^{-1}. In [1.1.2] the dimensions of a depend on b, which is not an integer but a fractal and K'_L is a reciprocal pressure because the l.h.s. of [1.1.3] is dimensionless.

The notion of *surface excess* has been discussed in detail in secs. I.2.5 and I.2.22. See particularly fig. I.2.12. Basically, the problem is that thermodynamically Γ can only be defined as an excess with respect to a reference substance. Setting the adsorption of the latter equal to zero implies, in the Gibbs convention, that formally the position of a mathematical plane is defined which separates the two adjoining phases α and β. This plane is known as the *Gibbs dividing plane*. For flat solid-gas interfaces it was demonstrated in sec. I.2.22c that this plane coincides with the physical boundary between solid and gas, the solid being the natural reference. At the same time, the thermodynamic surface excesses occurring in the *Gibbs adsorption equation* [I.2.22.11] simply reduce to the corresponding analytical surface concentrations. Hence, for the adsorption of a mixture of gases we have

$$d\gamma = -S^\sigma_a dT - \sum_i \Gamma_i \, d\mu_i \tag{1.1.4}$$

where γ is the surface tension. For a single gas

$$d\gamma = -S^\sigma_a dT - \Gamma d\mu \tag{1.1.5}$$

In these equations S^σ_a is the *surface excess entropy per unit area*. For adsorption from solution the interpretation of surface excesses requires more scrutiny (chapter 2).

Adsorption models can be tested either by comparison of theoretical isotherm equations, or two-dimensional equations of state, with experimental data. The *two-dimensional* or *surface pressure* has been defined as

$$\pi \equiv \gamma^* - \gamma \qquad\qquad\qquad [1.1.6]$$

where γ^* is the interfacial tension in the absence of adsorbate. Although for solid surfaces the surface tension cannot be readily measured (sec. I.2.24), π is nevertheless obtainable by integration of the Gibbs equation. For the simple case of [1.1.5], setting $d\mu = RT\,d\ln p$,

$$\pi = RT \int_0^\Gamma \Gamma\,d\ln p \left(= RT \int_0^\Gamma \Gamma \frac{d\ln p}{d\Gamma} d\Gamma \right) \qquad\qquad [1.1.7]$$

A plot of Γ as a function of $\ln p$ is a semi-logarithmic adsorption isotherm; hence the surface pressure is accessible as the area under this curve. The difference between isotherm and surface pressure analysis is that the former tests the equilibrium between adsorbate and gas, whereas the latter contains properties of the adsorbate only. Typically, in [1.1.7] the standard chemical potential μ^o does not occur. For an ideal adsorbate we have, according to [I.1.3.6],

$$\pi = RT\,\Gamma \qquad\qquad\qquad [1.1.8]$$

which is the two-dimensional equivalent of the ideal gas law $p = nRT/V$. Equation [1.1.8] follows immediately from [1.1.7] by substitution of the (ideal) Henry isotherm equation [1.1.1].

Adsorption isotherm equations can in principle be derived by first formulating the chemical potential of the adsorbate μ^σ in terms of a model, then equating μ^σ to μ^G. Although it is not impossible to derive expressions for μ^σ by thermodynamic means, statistical approaches are more appropriate because in this way the molecular picture can be made explicit. Moreover, adsorbates are not macroscopic systems, which is a prerequisite for applying thermodynamics, and statistical thermodynamics lends itself very well to the derivation of expressions for the surface pressure. Another approach is based on kinetic considerations: expressions for the rates of adsorption and desorption are formulated; at equilibrium the two are equal.

The required statistical foundations have been outlined in chapter I.3. Basically, two approaches are open, the *canonical* and the *grand canonical*. In the former, the adsorbate is considered as a closed system, characterized by the number of adsorbate molecules N, the area A and the temperature T. By model considerations, the canonical partition function $Q(N, A, T)$ is formulated from which μ^σ is obtained as

$$\mu^\sigma = -kT \left(\frac{\partial \ln Q}{\partial N} \right)_{T,A}$$ [1.1.9]

which is the two-dimensional analogue of [I.A6.12]. In the grand canonical approach, the adsorbate is treated as an open system, the properties of which are defined by the chemical potential of the surroundings, i.e. by μ^G. The grand canonical partition function is $\Xi(\mu, A, T)$. From this, the number of molecules adsorbed is immediately found as

$$\langle N \rangle = kT \left(\frac{\partial \ln \Xi}{\partial \mu} \right)_{T,A}$$ [1.1.10]

which is the two-dimensional analogue of [I.A6.24]. If we want to be precise, we now write $\langle N \rangle$ instead of N because in an open system the number of molecules in the adsorbate can fluctuate around its average value $\langle N \rangle$. In practice this average is measured.

As partition functions contain all the information defining the system, they can also be used to derive the surface pressure. For the canonical and grand canonical ensemble we have

$$\pi = kT \left(\frac{\partial \ln Q}{\partial A} \right)_{T,N}$$ [1.1.11]

and

$$\pi = \frac{kT \ln \Xi}{A} = kT \left(\frac{\partial \ln \Xi}{\partial A} \right)_{T,\mu}$$ [1.1.12]

respectively. These two equations are the two-dimensional analogues of [I.A6.11] and [I.A6.23], respectively.

All of this applies to monocomponent adsorbates. Extension to mixtures is usually not difficult, although formulating Q and Ξ in terms of molecular parameters may become complicated.

In chapter I.3 a number of examples of elaborations have already been given, mostly using lattice statistics. All of them involve a "divide and rule" strategy, in that the system (i.e. the adsorbate) is subdivided into *subsystems* for which subsystem-partition functions can be formulated on the basis of an elementary physical model. For instance, in lattice theories of adsorption one adsorbed atom or molecule on a lattice site on the surface may be such a subsystem. In the simplest case the energy levels, occurring in the subsystem-partition function consist of a potential energy of attraction and a vibrational contribution, the latter of which can be directly obtained quantum mechanically. Having established the partition functions of the subsystems, the partition function of the entire adsorbate is derived by statistical means. For independent subsystems this is easy, but in practice most subsystems are not independent. A compounding complication is that most real surfaces are energetically heterogeneous so that ranges of subsystem partition functions occur. Starting with sec. 1.3, we shall discuss these problems systematically and, where appropriate, review the relevant parts of chapter I.3.

For the physical adsorption of gases on solids the attraction between the molecules and the surface is almost the exclusive driving force. Thermodynamically this means that such gas adsorption is exothermic. Usually the enthalpy of adsorption per molecule depends on θ because of heterogeneity (upon filling an adsorbent with adsorbate the "highest energetic" parts are covered first) and because, with increasing θ, lateral interaction also increases (this contribution may be attractive or repulsive).

Attraction forces may have different origins, but London-Van der Waals contributions are ubiquitous. In chapter I.4 these forces have been discussed at length, for both isolated molecules and condensed media. In [I.4.6.1] we derived the London-Van der Waals part of the interaction energy $u_i(z)$ between an atom i and a flat semi-infinite phase of nature j at distance z

$$u_i(z) = - \pi \beta_{ij} \rho_{N_j} / 6z^3 \qquad [1.1.13]$$

where β_{ij} is the Van der Waals interaction parameter between pairs of atoms (or molecules) i and j (see [I.4.4.19]) and ρ_{N_j} is the number density of the atoms (or molecules) constituting phase j.

For two molecules at very short distance of approach the Born repulsion also has to be accounted for. A common way of doing so is through an r^{-12} term, where r is the distance between the centres of the molecules. For the total interaction between two molecules this leads to the well-known *Lennard-Jones 6-12 pair interaction*. See [I.4.5.1] and fig. I.4.9. The equivalent for an atom at distance z from a flat adsorbent is a *Lennard-Jones 3-9 interaction* which may be generally written as

$$u_i(z) = C_1 z^{-9} - C_2 z^{-3} \qquad [1.1.14]$$

where C_1 and C_2 are constants; C_2 is given by [1.1.13]. There is some inconsistency in the application of [1.1.14] to adsorbates caused by the fact that the integration leading to [1.1.14] presupposes z to be so large that the adsorbent may be considered homogeneous on a molecular scale. This may be correct for large z, but not for the case considered, where z is of the same order of magnitude as the interatomic distances in the adsorbent. Equation [1.1.14] should then rather be replaced by a sum over pair interactions, but these are very sensitive to the position of the adatom. Considering the fact that the repulsive term works only at (very) short distance whereas the Van der Waals attraction has a longer range, a practical solution is to retain the attractive term but replace the repulsive one by a step function (independent of position parallel to the surface); in fact, there is not much difference between this and a z^{-9} decay.

Obviously there are several other types of interactions, including dipole or quadrupole interaction, hydrogen bridges and even chemical bond formation, leading to *chemisorption*.

In the description of adsorption, not only static (equilibrium) features but also kinetic aspects are of importance and we shall consider some of these aspects in sec. 1.5a. Some information is already available in secs. I.6.5d and e, where diffusion transport to and from surfaces has been discussed. One of the more important equations that we derived is [I.6.5.36], describing the increase of Γ with time t due to (semi-infinite) diffusion onto a flat surface. Replacing the concentration c in that equation by the number of moles n per volume V, we have for an ideal gas $c = p/RT$, so that [I.6.5.36] can be written as

$$\Gamma_i(t) = \Gamma_i(t=0) + \frac{2}{RT}\left(\frac{D_i}{\pi}\right)^{1/2}\left[p_i(x=\infty)t^{1/2} - \int_0^{\sqrt{t}} p_i(0,u)\,du^{1/2}\right] \qquad [1.1.15]$$

where D_i is the diffusion coefficient of gas i, and p_i its (partial) pressure, $p_i(x=\infty)$ is the pressure far from the surface, i.e. the analytically determined pressure, and u is the integration variable. This equation is also known as the *Ward-Tordai equation*.

In the initial stages of adsorption, i.e. when a virgin surface is suddenly exposed to a gas of non-zero pressure, the surface starts to act as a sink for adsorptive molecules, then $p_i(0)=0$ and the rate of supply is entirely determined by the second term on the r.h.s. of [1.1.15]:

$$\Gamma_i(t) = 2\left(D_i t\right)^{1/2} p_i / RT\,\pi^{1/2} \qquad\qquad [1.1.16]$$

This equation can, for instance, be used to estimate how fast an adsorbent is covered by gas. Even at very low pressures this is a very rapid process because in gases diffusion coefficients are high ($D = O(10^{-5}\ \mathrm{m^2\ s^{-1}})$[1]), see Table I.6.4). There-

---

[1] Recall that $O(x)$ means "of the order of x".

fore, stringent precautions regarding evacuation have to be taken to keep the surface empty.

In passing, it is noted that differential equations describing diffusion can often be handled by Laplace transformations, the principles of which are laid down in appendix I.10.

Proportionality of Γ_i and $t^{1/2}$ is often (but not always) an indication of a diffusion-controlled process, but such a proportionality does not have to extend over the entire time domain considered. It may happen that diffusion control is realized but that the computed D_i is lower than the corresponding value in the gas phase. One possible explanation for this may be that the supply is followed by a slower surface diffusion process, which is rate-determining. Surface diffusion coefficients D_i^σ tend to be lower than the corresponding bulk values. Such diffusion has been briefly discussed in sec. I.6.5g, under (i). When surface diffusion is zero, the adsorbate is *localized*. In that case equilibration between covered and empty parts of the surface can only take place by desorption and readsorption. For $D_i^\sigma \neq 0$ the adsorbate is *mobile*; it then resembles a two-dimensional gas and we have already given the partition functions for one adsorbed mobile atom in sec. I.3.5d. In sec. 1.5d we shall briefly discuss the transition between localized and mobile adsorption.

Finally it is recalled that in secs. I.7.10 and 11, a number of optical and related techniques have been introduced which are used to characterize surfaces, including reflectometry, ellipsometry and various types of surface spectroscopy. We shall reconsider these in the following section.

1.2 Characterization of solid surfaces

The statement that the surface properties of solids differ from those in the interior of that solid is a truism. The question at issue is now what properties are involved, by how much they differ and how one can characterize them. What atoms are found in the surface, where are they and how are they bound? What do surfaces look like on a molecular scale?

Let us, by way of introduction, consider what may happen when a perfect solid is cleaved under a vacuum. By this process two new surfaces are created. Originally, all atoms or molecules in the solid found themselves in equilibrium positions, mechanically characterized by the fact that the net force on them was zero. Thermodynamically, the Gibbs energy was a minimum at given p and T. Immediately after cleaving, the atoms in the newly-created surfaces are no longer equally surrounded. A net inward force results. Hence, the new situation is not at equilibrium; changes will take place till equilibrium is restored. In the terminology of chapter I.2 we say that the surface has to *relax*. How fast does this relaxation take place and what can be said about the final situation? The answers to these questions vary widely, depending on the system and on the conditions.

Relaxation rates in surfaces depend on such properties as (surface) diffusion coefficients, vibrational degrees of freedom and ductility. For all of these, the temperature is an important parameter. At low temperature all processes are very slow and unrelaxed surfaces may persist over very long times, whereas at high temperatures the readjustment is faster, especially so just below the melting point of the solid. The presence of gases is another factor: they may adsorb and in this way contribute to changing the composition and affect relaxation. It is obvious that relaxation rates strongly depend on the nature of the solid and, for a given solid, they may depend on the crystal planes that have been created. The extent of relaxation will generally also depend on the way in which the new area is created. Cleavage of a monocrystal under vacuum is quite different from, say, ball milling under room conditions. The former process is better controllable than the latter but, in view of the importance in practice of the latter, we cannot ignore ill-defined surfaces. The question is: how reversibly has the surface been formed?

Regarding the direction of the relaxation process, the trend is that, just after cleaving, surfaces respond to the asymmetry of the forces by compaction. In the surface region the atoms are pulled closer to each other, until the short-range Born repulsion, responsible for the r^{-12} term in the Lennard-Jones interaction, inhibits further inward displacement and mechanical equilibrium is re-attained. For some systems, including platinum monocrystals, this compaction has actually been experimentally established; it extends over a few atomic layers. Atoms of different nature can also shift with respect to each other. This happens, for instance, in sodium chloride crystals where the larger, and hence more strongly polarizable, chloride ions occupy positions closer to the boundary than the smaller sodium ions. In this way, an electrical double layer originates in the surface, the resulting charge separation counteracts this process. Yet another possibility is that the surface layer assumes an entirely different crystal habit, say hexagonal when the bulk crystal is cubic. This phenomenon is called *recon-struction*[1]. It may be promoted by adsorption. From all of this it is obvious that for the characterization of solid surfaces we have to consider specific *surface* characterization techniques.

In passing it is noted that in all these respects surfaces of fluids behave very differently. First they have a very short relaxation time or, in the terminology of sec. I.2.3, their *Deborah number De* is low. Secondly, they tend to expand (as

---

[1] A famous example is the so-called Si(111)-(7×7) reconstruction, resolved by G. Binnig, H. Rohrer, Ch. Gerber and E. Weibel, *Phys. Rev. Lett.* **50** (1983) 120.

Table 1.1. Some important surface characteristics of solids and techniques used to study them.

Surface structure. (Morphology, dislocations, porosity, specific surface area, etc.)	Optical microscopy, interferometry, various electron microscopies, atomic force techniques (AFM, STM), surface diffraction, BET (+t-plot), porosimetry.
Chemical composition	Surface spectroscopies, functional adsorption.
Surface tension	No direct experimental methods available.
Surface (Gibbs) energy	No direct methods, but enthalpies of adsorption and/or immersion give some information.
(Energetic) surface heterogeneity	Microcalorimetry. Indirectly by analysis of adsorption isotherms.

compared to the bulk) rather than contract, because entropic factors play a more important role.

In table 1.1 a number of relevant surface characteristics are reviewed with techniques available for their determination.

The first category, surface structure, includes techniques that answer the question "what can we see if looking at the surface with appropriate magnification?"

Obviously, classical microscopy is the first option. Conventional optical microscopy has a resolution down to about 500 nm, depending on the wavelength of the light used. This is usually not enough to establish sizes and shapes of, for instance, colloidal particles. When coherent light waves are used and the interference analyzed by computer, normal resolution down to 1 nm, as compared with a horizontal resolution of about 10 µm, is attainable. However, over the past decades a host of other physical techniques have been developed.

Fast moving electrons have a much shorter wavelength than visible light and features down to about 1 nm can, in principle, be seen. The various electron microscopies are now more or less routine. We introduced them in sec. I.7.11b. They include *transmission electron microscopy* (TEM) and *scanning electron microscopy* (SEM). TEM is often used to obtain the geometry of colloids. However, as electrons can pass through many solids when they have enough energy, TEM is less suited to the study of surface features. SEM is in this respect more appropriate although the resolution is less ($O(10^2$ nm)). In SEM the surface is scanned by an electron beam and the intensity of secondary electrons studied. Figure 1.1 collects a few illustrations clearly showing what information can be

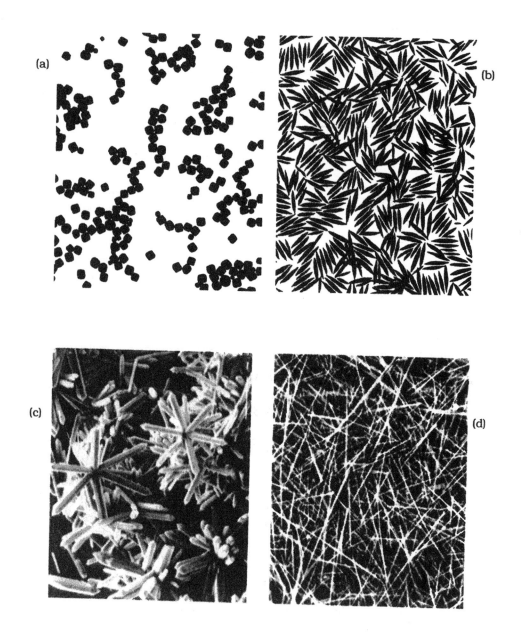

Figure 1.1. Electron micrographs of some model colloids. (a) and (b), TEM of haematite (α-Fe_2O_3), (c) and (d), SEM of zinc oxide. (Courtesy of E. Matijevic. References: (a) S. Hamada, E. Matijevic, *J. Chem. Soc., Faraday Trans.* **78** (1982) 2147; (b) M. Ozaki, S. Kratohvil and E. Matijevic, *J. Colloid Interface Sci.* **102** (1984) 146; (c) and (d) A. Chittofrati, E. Matijevic, *Colloids Surf.* **48** (1990) 65.) (Reprinted with permission of the Royal Society of Chemistry (fig. a), Academic Press (fig. b) and Elsevier Sciences (figs. c and d).) Reprinted with permission of the Royal Society of Chemistry (fig. a), Academic Press (fig. b) and Elsevier Science (figs. c and d).

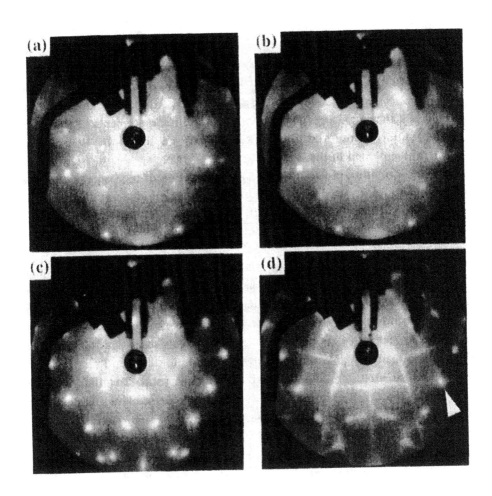

Figure 1.2. LEED picture of W(111) surfaces, covered with Pd. (a), fractional coverage = 0.9; (b), 1.0); (c), 1.1. Pictures at incident energy of 101 eV, taken after annealing for 3 min at 100 K. Figure (d), time exposure while incident energy is increased from 8 to 120 eV; same surface as in (c). After K-J. Song, C-Z. Dong and T.E. Madey, *Langmuir* **7** (1991) 3019. (Reprinted with permission of the American Chemical Society.)

obtained. For the purpose of preparing model colloids it is relevant that very different habits of the same material can be prepared by changing the conditions of nucleation and growth, a matter to which we shall return in Volume IV.

In addition to imaging surfaces to study their morphology, electron beams can also be used to analyze surface structure by monitoring their scattering from surface layers on, or of monocrystals. To that end, the energy of the electrons must be kept low (10-300 eV) to suppress penetration into the solid. Hence the name of the technique, LEED, for *low energy electron diffraction*. The pictures that can be made visible on a fluorescent screen consist of an array of dots, comparable to those obtained from X-ray scattering by crystals; the interpretation is also similar. By varying the incident energy, the depth of the surface layer to be probed can be adjusted. Figure 1.2 gives an example. One of the features that can be seen is that, following monolayer coverage of tungsten by palladium (compare figs. (b) and (c)) dramatic structural changes take place. Figure (d) is a time-exposure, giving information on the movements of the spots. Additional information and interpretation can be found in the original publication, mentioned in the legend.

There are a number of variants of LEED. Diffraction of high energy electrons (≈ 20 keV) can also be used provided that the studies are done at grazing incidence. If the surface is scanned with this technique, information on the structural surface heterogeneity is obtainable.

Atomic force techniques constitute another development. We introduced them in sec. I.7.11b. To this category belong *scanning tunnelling microscopy* (STM) and *atomic force microscopy* (AFM), also known as *scanning force microscopy* (SFM). In STM the (semi-)conducting surface of the substance to be studied is scanned with a needle that has a very sharp tip, preferably only one atom thick. Usually the needle is made of platinum or a platinum-iridium alloy, and sharpened by electrochemical etching. Sometimes it is a monocrystal. Together with the sharpness of the tip, the applied potential difference determines the detail that can be seen. The needle is brought so close to the surface that electron tunnelling takes place under the influence of the applied potential difference. To achieve this the needle is first positioned by eye or with a micrometer screw down to about 0.1 mm above the surface, after which a fine-positioning mechanism is applied, using piezoelectric "walkers". For a number of purposes it is useful to have surface structure pictures at different amplifications, for instance with heteroporous surfaces (surfaces that have pores of different sizes and shapes). In the absence of an applied potential difference the tunnelling current is the same in the two directions, but in its presence one of the directions is favoured. In one of the modes of operation the tunnelling current is recorded as a function of position. As this current depends sensitively (in fact, exponentially) on the distance, surface structures can be determined, in principle down to atomic resolution. A number of variants exist, depending on the way in which the experiment is carried out, including the "constant current" mode (the tip is moved over the surface and the displacement of the tip needed to keep the current constant is recorded) or the "constant height" mode, in which the variation of the current is measured. Needless to say, for such sensitive techniques a number of stringent experimental precautions have to be heeded and great care has to be exercised to avoid artefacts. From all these techniques an atomic scale image of the local electron density is obtained, as it varies along the surface; mostly this corresponds to the topology of the surface (in addition to signals referring to composition). Under the most favourable conditions, the resolution is less than 0.005 nm in the normal (z) direction and 0.01 nm laterally (x, y direction).

STM can also be applied to solids immersed in liquids, with obvious relevance to colloid science. Examples include deposited Langmuir-Blodgett films, electrode surfaces and graphite in oil[1]. However, a Langmuir-Blodgett lipid film can easily be pierced by the tip if the force becomes too high. The force required to displace one lipid molecule in such a layer can in principle be computed from membrane models and is probably of the order of a few tenths of a nanonewton.

When instead of the current or position we measure the repulsive force, experienced by the tip if it is brought very close to the surface, one speaks of *atomic force microscopy* (AFM)[2]. As a trend, AFM is becoming rather a routine technique, with STM as a specialty. The advantage over STM is that the surface does not need to be conducting. AFM forces depend on the distance scale: in the "contact mode" the forces are O(nN) and reflect atom-atom interactions and deformation of the surface may occur; in the "non-contact mode", the forces are O(pN–nN) and rather represent interactions of colloidal origin, particularly Van der Waals forces and, when charges play a role, electrostatic interactions. With regard to the former category, in sec. I.4.6 we have already discussed how the energy can be evaluated from the geometry of the system and Hamaker constants which, in turn, are tabulated in I.App.9. From the energies, the forces are directly obtained by differentiation with respect to the distance, using [I.4.2.3], [I.4.2.4] or variants thereof, depending upon the geometry of the system. Atomic forces are often measured by using soft (cantilever) springs. Obviously these forces must be weaker than the forces keeping the surface together. If "soft" surface features are to be studied, say adsorbed proteins, this must be taken into account.

In fig. 1.3 two examples of an AFM picture are given. Figure 1.3a is a typical AFM image. It shows hexagonal symmetry for the basal plane of graphite measured in the contact mode[3]. In this case the apparent symmetry does not correspond to the presumed trigonal symmetry, because AFM does not always show real atomic scale resolution. Real atomic scale resolution that showed

---

[1] See for instance R. Sonnenfeld, P.K. Hansma, *Science* **232** (1986) 211 and G. Binnig and H. Rohrer's Nobel Prize acceptance paper (Scanning Tunnelling Microscopy - From Birth to Adolescence), *Rev. Mod. Phys.* **59** (1987) 615.
[2] G. Binnig, C.F. Quate and Ch. Gerber. *Phys. Rev. Lett.* **56** (1986) 930.
[3] F. Lin, D.J. Meier, *Langmuir*, **10** (1994) 1660.

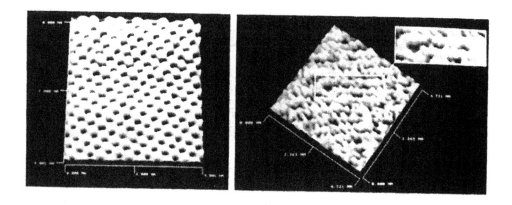

Figure 1.3. (a) AFM image of graphite showing 0.246 nm apparent atomic spacing. (b) Atomic level AFM image of polyethylene chains showing also a hairpin chain fold. Courtesy F. Lin and D. Meier, *Langmuir*, **10** (1994) 1660. (Reprinted with permission of the American Chemical Society.)

STM-like reliability can be achieved e.g. by measuring in the non-contact mode with ultra-low scanning forces ($\approx 10^{-11}$ N)[1]. In fig. 1.3b an atomic-level AFM image of highly oriented chains clearly shows the polyethylene chains as well as the individual methylene groups.

Another, more complicated, illustration is given in fig. 1.4. This sample is an annealed cast film of the copolymer poly(styrene co 2-vinyl pyridine). The two blocks in the polymer are incompatible. Hence, on annealing (keeping the film during a protracted period above the glass transition temperature) the copolymer molecules rearrange themselves into lamellar microdomains; the excess copolymer forms microdomains on the underlying multi-layered film, giving rise to holes in this film. The sharpness of observed features is always the sum-result of tip and sample in the sense that the coarser one determines the definition; for instance, with very fine "super tips", mounted on the regular fine tips, the better defined features can be observed.

Surface structure techniques involving adsorption will be discussed extensively in secs. 1.5-1.7.

---

[1] F. Ohnesorge, G. Binnig, *Science*, **260** (1993) 1451.

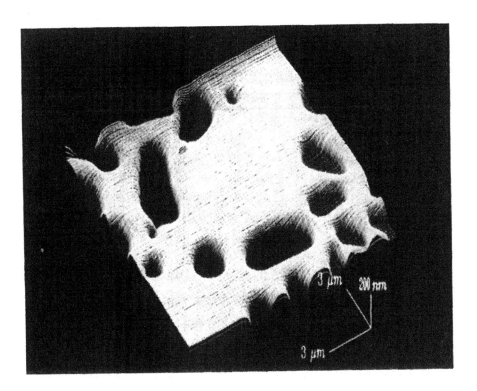

Figure 1.4. Atomic force image of an annealed copolymer film. Explanation in the text. (Courtesy of P.C.M. Grim and G. Hadziioannou, State Univ. of Groningen, The Netherlands).

To determine the chemical composition of surfaces, a number of alternatives present themselves.

In the first place there is a host of surface spectroscopies. Some important ones have been discussed in sec. I.7.11. Adamson gives a list of several tens of approaches and variants[1]. Most widely used are *X-ray photoelectron spectroscopy* (XPS), also called *electron spectroscopy for chemical analysis* (ESCA), *Auger electron spectroscopy* (AES), and *secondary ion mass spectroscopy* (SIMS) and, to a lesser extent, *ion scattering spectroscopy* (ISS). The main characteristics of these techniques have been collected in table I.7.4. Other techniques may be more widely used in certain areas. For example, XPS is very popular in heterogeneous catalysis.

---

[1] A.W. Adamson, *Physical Chemistry of Surfaces*, 5th ed., John Wiley (1990), table VIII-1.

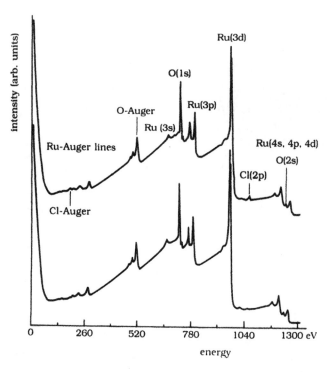

Figure 1.5. XPS (=ESCA) spectrum of ruthenium dioxide; (top) freshly prepared, (bottom) washed with acid, base and water. Explanation in the text. (Courtesy of J.M. Kleijn, Agricultural University Wageningen, The Netherlands; see also J.A. Don, A.P. Pijpers and J.J.F. Scholten, *J. Catal.* **80** (1983) 296.)

As a first illustration, in fig. 1.5 the XPS spectrum of dried colloidal ruthenium dioxide is given. Sols of this oxide can act as a colloidal catalyst for the photoreduction of water. To that end the surface properties must be known and one of the questions is how much chloride is left on the surface (the sample was prepared from ruthenium trichloride by hydrolysis). The spectrum shows peaks for different electrons of ruthenium and oxygen. In addition, some Auger lines emerge; these are due to secondary electrons, i.e. electrons ejected by the X-ray photons act as the incident electrons for the Auger process. The areas under the peaks yield quantitative information on the chemical composition. For instance, the fresh sample contained 5 atomic percent of chlorine and from comparison of the XPS and the Auger peaks it could be concluded that most of this chlorine was on the surface. It is also seen that the washing procedure substantially reduced the chlorine contamination. The maximum at zero energy has a methodical background that does not concern us now. Along the same line, XPS can be used to check the degassing efficiencies of a sample.

Figure 1.6 is an example of a SIMS spectrum. In this case positive ions have been used to bombard the surface. After preparation, (fig. a) this spectrum has

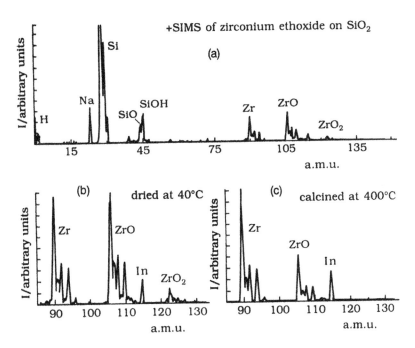

Figure 1.6. Positive SIMS spectrum of a SiO_2 + 9 wt% ZrO_2 catalyst prepared from zirconium ethoxide. a) original sample; b) Zr-Zr-ZrO_2 region for a sample dried at 40°C; c) ibid, calcined at 400°C. (courtesy of J.W. Niemantsverdriet. See A.C.M.Q. Meyers, A.M. de Jong, L.M.P. van Gruijthuijsen and J.W. Niemantsverdriet, *Appl. Catal.* **70** (1991) 53.)

peaks of H^+, C^+, O^+, Na^+, Si^+, K^+, Ca^+, SiO^+, $SiOH^+$, Zr^+, ZrO^+, In^+ and ZrO_2^+. Some of these are artefacts: traces of Na^+, K^+ and Ca^+ always give high yields in $SIMS^+$ and the In^+ peak stems from the indium foil in which the catalyst powder was pressed. The most relevant information regards the Zr-containing ions. Figure (b) indicates small but significant amounts of $ZrOH^+$ ions, see for instance the peaks at 107 atomic mass unit (a.m.u.) ($^{90}ZrOH^+$ and $^{91}ZrO^+$) and 111 a.m.u. ($^{94}ZrOH^+$); this $ZrOH^+$ is probably a debris ion stemming from the zirconium ethoxide. Comparison with fig. (c) shows that calcination in air at 400°C converts the zirconium ethoxide (Zr:O = 1:4) into ZrO_2 (Zr:O = 1:2) which is reflected in the reduction of the ZrO^+/Zr^+ intensity ratio by a factor of about two. SIMS is generally a "dynamic" technique in that, due to the impact of the incident ions, the surface is etched. However, it can also be applied in a more static mode, where damage is a minimum.

These surface spectroscopies can be augmented by surface infrared, NMR, surface-enhanced Raman spectroscopy (SERS), etc. A surface variant of infrared spectroscopy is based on total internal reflection and can be applied for adsorbents which are transparent for the radiation under study. The light beam enters through the sorbent and reflects at the interface. As discussed in sec.

I.7.10a, upon total reflection *evanescent waves* develop in the medium outside the solid. This medium may be gaseous or liquid and may carry adsorbates. The reflected infrared beam provides information on molecular vibrations in the surface region, i.e. on the amount adsorbed, on molecular structure, conformations and interactions. Surface signals can be amplified by letting the beam reflect back and forth many times against two plane parallel surfaces (*multiple reflection*). The abbreviation ATR stands for *attenuated total reflection*, the attenuation of the intensity being caused by absorption.

Functional gas adsorption is the binding of adsorptives, characteristic for special groups on the surface. Specific uptake of certain adsorptives is diagnostic to the presence of certain groups in the surface. This is in distinction to common or *physical adsorption* which is less (or non-) specific. The additional specificity can have a variety of origins, from hydrogen bridges to chemisorption. For instance, the presence of hydrophilic groups on otherwise hydrophobic graphite surfaces will result in specific uptake of water vapour as compared to nitrogen or noble gases. Such specific adsorption can be conveniently followed by infrared spectroscopy (changes in the stretch frequencies of the OH band). When real chemical bonds are formed, spectroscopic detection is particularly useful.

Depending on the nature of the groups to be investigated, a number of "indicator gases" has been proposed. One of the recurrent issues is the determination of the *acidity or basicity of surfaces*, more specifically, the detection of Brønsted (proton donor) or Lewis (electron acceptor) acid sites or groups. Surface OH groups are amphoteric; heavy metal sites (for instance an Al^{3+} site in the surface of alumina) are Lewis acids. Detection of such groups or sites is important for chemisorption and heterogeneous catalysis. Functional gas adsorption can for instance be carried out with CO_2 or NH_3 as (Lewis) bases or with pyridine as an acid adsorptive. Such adsorption measurements can be augmented by spectroscopy and (micro-) calorimetry[1].

The analysis of solids after dispersion in liquids offers a number of advantages because of the availability of a number of titration techniques, which augment the functional adsorption on dry surfaces. For acid-base potentiometric titration of insoluble oxides in aqueous media the principle has already been explained in sec. I.5.6e and we shall return to it in sec. 3.7a. For amphoteric oxides the (pristine) point of zero charge can be measured; it is determined by the difference between pK_{acid} and pK_{base}; with some theoretical analysis these two constants can also be established individually.

Most other techniques mentioned in table 1.1 will be discussed in some detail

---

[1] For a review of the chemical characterization of surfaces, see *Characterization of Powder Surfaces*, G.D. Parfitt, K.S.W. Sing, Eds., Academic Press (1976) chapter 2.

later in this chapter. However, the surface tension of solids is not an operational quantity: there is no unambiguous way of measuring it. Indirect experimental methods have been proposed, for instance measuring the surface tension of the molten material as a function of the temperature and extrapolation to below the melting point, or measuring the solubility of finely divided powders as a function of "the" particle radius, using the Ostwald equation [I.2.23.25]. These methods are very restrictive.

Similar problems arise with the surface excess Gibbs energy G^σ, which is defined in table 1.2 in sec. 1.3. However, a number of enthalpy changes (upon adsorption, immersion, etc.) can be obtained and from them useful thermodynamic information can be deduced, see sec. 1.3. Some of these measurements contribute to the understanding of *surface heterogeneity* (in the energetic sense). In principle such information can also be obtained by isotherm analysis, see sec. 1.7.

All the techniques discussed so far refer to clean surfaces or surfaces with adsorbed molecules. When thicker adsorbed layers are present on the surface, the properties of these layers start to resemble those of the corresponding bulk phases. For instance, for thin water layers on solid surfaces the dielectric permittivity $\varepsilon(\omega)$ and NMR spectra have been measured and compared with those for bulk water. A more or less gradual transition takes place towards wetting films to which we shall return in Volume III and, as far as multilayer adsorption is concerned, in sec. 1.5 g, h.

In this connection it may be added that the characterization of surfaces through contact angle measurements poses a host of problems: pre-phases or wetting films may be formed on the surface: slight adsorption of molecules from the liquid on the "dry" surface may substantially modify the contact angle; hysteresis is usually observed to an extent that may depend on the surface roughness and perhaps even on the way of measurement. In turn, contact angle hysteresis gives some information on the smoothness and homogeneity of the surface. When all these problems are under control information can be obtained from wetting studies regarding the relative *surface hydrophilicity/hydrophobicity*. See further, sec. 1.3f. We intend to return to this in Volume III, chapter 5.

Finally the caveat is repeated that surfaces are not necessarily inert to these various characterization techniques. In order to assess methodical surface modifications, applying more than one technique for a given sample is recommended.

1.3 Thermodynamics of adsorption

In this section we discuss thermodynamic characteristics of adsorbates, emphasizing the surface pressure π, defined in [1.1.6], the enthalpy (or heat) and/or energy and the entropy of adsorption. Regarding the enthalpies and entropy, integral and differential quantities can be distinguished. Moreover, differences between these characteristics as obtained under different experimental conditions will be discussed.

1.3a. Generalities

Our starting point will be the collection of integral and differential thermodynamic functions for flat adsorbates that have been derived in sec. I.2.11 and presented in app. I.5; for convenience these formulas are repeated in table 1.2. Recall that the derivation was based on the Gibbs convention (interfacial excess volume $V^\sigma \equiv 0$) and that other authors may have different definitions for H^σ and G^σ. This last point cannot have any effect on equations connecting operational quantities derived from them. All quantities have their previously defined meanings; n_i^σ is the surface excess of i and the surface concentration Γ_i equals n_i^σ/A. Equilibrium is presupposed, hence the values for the chemical potentials μ_i and the temperature T are those of the surroundings, i.e. the gas phase. Furthermore it is recalled from sec. I.2.22a that for the solid-gas (SG) interface, the Gibbs dividing plane may be identified with the physical SG boundary, which is supposed to be flat. Equations [1.3.5-8] are total differentials, hence, they may be used to obtain relations by cross-differentiations.

As in Volume I, we shall essentially heed IUPAC recommendations for nomenclature, one of the exceptions being that A stands for area and F for

Table 1.2. Integral and differential characteristic functions for flat interfaces. The superscript σ refers to surface excess (Gibbs convention, $V^\sigma \equiv 0$).

$$U^\sigma = T S^\sigma + \gamma A + \Sigma_i \mu_i n_i^\sigma \qquad\qquad [1.3.1]$$

$$H^\sigma = T S^\sigma + \Sigma_i \mu_i n_i^\sigma = U^\sigma - \gamma A \qquad\qquad [1.3.2]$$

$$F^\sigma = \gamma A + \Sigma_i \mu_i n_i^\sigma = U^\sigma - T S^\sigma \qquad\qquad [1.3.3]$$

$$G^\sigma = \Sigma_i \mu_i n_i^\sigma = H^\sigma - T S^\sigma = F^\sigma - \gamma A \qquad\qquad [1.3.4]$$

$$dU^\sigma = T dS^\sigma + \gamma dA + \Sigma_i \mu_i dn_i^\sigma \qquad\qquad [1.3.5]$$

$$dH^\sigma = T dS^\sigma - A d\gamma + \Sigma_i \mu_i dn_i^\sigma \qquad\qquad [1.3.6]$$

$$dF^\sigma = - S^\sigma dT + \gamma dA + \Sigma_i \mu_i dn_i^\sigma \qquad\qquad [1.3.7]$$

$$dG^\sigma = - S^\sigma dT - A d\gamma + \Sigma_i \mu_i dn_i^\sigma \qquad\qquad [1.3.8]$$

Helmholtz energy. The subscripts m and a refer to "per mole" and "per unit area", respectively. Hence, in SI units F_m is in J mol^{-1} and F_a in J m^{-2}. Changes in functions are indicated by a Δ, with a subscript indicating the nature of the change. Thus, $\Delta_{ads}F$ is the change in Helmholtz energy upon adsorption. See further sec. 9a for nomenclature and IUPAC recommendations and the symbol list at the beginning of this book.

1.3b. The surface pressure

This quantity is defined in [1.1.6] and, for SG interfaces, is obtainable from the adsorption isotherm as in [1.1.7]. This pressure is directly related to the change in excess Helmholtz energy of adsorption by using [1.3.3] twice: once with, and once without, adsorbate, followed by subtraction:

$$\pi A = -\Delta_{ads}F^\sigma + \Sigma_i \mu_i n_i^\sigma \qquad [1.3.9]$$

or, per unit area,

$$\pi = -\Delta_{ads}F_a^\sigma + \Sigma_i \mu_i \Gamma_i \qquad [1.3.10]$$

Equation [1.3.10] can, of course, only be applied when the surface area is known.

By the same procedure the corresponding differential expression can be obtained by using [1.3.7] twice. The result is

$$\pi\, d\,A = -d\,\Delta_{ads}F^\sigma - \Delta_{ads}S^\sigma\,dT + \Sigma_i \mu_i dn_i^\sigma \qquad [1.3.11]$$

Here, $\Delta_{ads}S^\sigma$ is the change in entropy of the system upon adsorption. As $d\Delta_{ads}F^\sigma$ is a total differential, we may equate $d\Delta_{ads}F_a^\sigma$ from [1.3.9] with the same in [1.3.11]. The result is

$$A d\pi = \Delta_{ads}S^\sigma\,dT + \Sigma_i n_i^\sigma d\mu_i \qquad [1.3.12]$$

or

$$d\pi = \Delta_{ads}S_a^\sigma\,dT + \Sigma_i \Gamma_i d\mu_i \qquad [1.3.13]$$

For isothermal adsorption and only one component, [1.3.13] reduces to [1.1.5]. It is repeated that the surface pressure is a measurable quantity, also for gas mixtures, provided the area is known.

For a number of purposes it is expedient to consider the surface pressure (and other thermodynamic functions of adsorbates) per mole of adsorbate. One of the reasons is that we are often interested in the difference between adsorbates and corresponding liquids, for instance in considering thin adsorbed liquid films.

For each component this difference is thermodynamically formulated as the difference of chemical potential, $\mu_i^\sigma - \mu_i^L$, which is a difference between partial (molar) Gibbs energies. At the same time this difference is closely related to the *disjoining pressure* of the film. This notion has been introduced in sec. I.4.2. It is very relevant for colloid science because it can be interpreted in terms of the forces determining the stability of colloidal systems: Van der Waals forces (chapter I.4), electrostatic, steric and structural forces. Let us elaborate this for a one-component system.

For the difference between the chemical potentials we may write

$$\mu^\sigma - \mu^L = RT \ln p/p_o \qquad\qquad [1.3.14]$$

where p is the pressure of the vapour with which the adsorbate is in equilibrium and p_o the same for the liquid. Equation [1.3.14] follows immediately from the equality of the chemical potentials in the condensed (or adsorbed) phase and those in the equilibrium phases. If we are dealing with a liquid surface film of thickness h and if the molar volumes of the surface and bulk phases are identical, this expression can be related to the disjoining pressure $\Pi(h)$ through

$$\Pi(h) = -\frac{RT}{V_m} \ln \frac{p(h)}{p_o} \qquad\qquad [1.3.15]$$

but here we shall remain more general. For the liquid $\mu^L = G_m^L = U_m^L - TS_m^L + pV_m^L$, see [I.A3.4]. For the adsorbate, writing, as in [1.3.14], μ^σ for the chemical potential to distinguish it from μ^L, we have from [1.3.9], using [1.3.3], $\mu^\sigma n^\sigma = \pi A + \Delta_{ads} U^\sigma - T \Delta_{ads} S^\sigma$, which may also be written as

$$\mu^\sigma = \frac{\pi}{\Gamma} + \frac{\Delta_{ads} U_a^\sigma}{\Gamma} - \frac{T \Delta_{ads} S_a^\sigma}{\Gamma} \qquad\qquad [1.3.16]$$

Combination with [1.3.14] then leads to

$$\frac{\pi}{\Gamma} = RT \ln \frac{p}{p_o} + \left(\frac{\Delta_{ads} U_a^\sigma}{\Gamma} - U_m^L\right) - T\left(\frac{\Delta_{ads} S_a^\sigma}{\Gamma} - S_m^L\right) - pV_m^L \qquad [1.3.17]$$

in which the pV_m^L term, which is of order of kT, may usually be neglected. The terms in brackets in [1.3.17] represent the required changes in energy and entropy if the material is transferred from the liquid to the adsorbed state. Another advantage of the use of molar quantities is that on the r.h.s. the area does not occur. Using model analyses, these energies and entropies may be interpreted.

Equation [1.3.17] can also be written in terms of partial molar quantities of the adsorbate

$$\frac{\pi}{\Gamma} = RT \ln \frac{p}{p_o} + \Delta_{ads} F^\sigma_m - G^L_m \qquad\qquad [1.3.17a]$$

The thicker the layer, the more $\Delta_{ads} F^\sigma_m$ resembles $\Delta_{cond} F^L_m$; the difference between $\Delta_{ads} F^\sigma_m$ and G^L_m is given by the first term on the r.h.s.; eventually the entire r.h.s. vanishes when adsorbate and bulk liquid are exactly identical.

1.3c. Differential and integral energies and enthalpies of adsorption

Heats of adsorption are important characteristics of adsorbates because they provide information regarding the driving forces for the adsorption. Some scrutiny is desirable, though, since these heats may depend sensitively on the conditions under which they are measured, one issue being what remains constant during the measurement, p, T, V, Γ, π and/or A. In a sense, this is begging the question. Experimentally it is very difficult to keep Γ or π constant, because the heat evolved upon *changing* Γ or π is measured. For the same reason, *isosteric heats* are not measured by establishing how p should vary with T, keeping Γ constant, but by measuring isotherms $\Gamma(p)$ at various T, to make up the differential quotient $(\partial \ln p / \partial T)_\Gamma$. The complexity is compounded if mixtures of gases are studied, but we consider this important theme beyond our present scope.

Several heats, energies and other thermodynamic quantities for adsorption can, and have been, defined, depending on the conditions under which the experiments have been performed. The most rational approach is to consider the way in which the adsorption calorimetry is carried out. We shall do so after defining some relevant parameters[1].

First, we introduce the *integral molar energy of adsorption* as

$$\Delta_{ads} U_m(\text{int}) = U^\sigma_m - U^G_m \qquad\qquad [1.3.18]$$

This is the energy difference between one mole of gas adsorbed and the same free. Integral adsorption energies involve the adsorption of macroscopic amounts of gas; as upon this adsorption Γ changes substantially and most surfaces are energetically heterogeneous, U^σ_m changes during this process. For portions of gas admitted at different Γ, different values for $\Delta_{ads} U_m(\text{int})$ will generally be observed. By the same token, the *integral molar enthalpy of adsorption* is defined as

$$\Delta_{ads} H_m(\text{int}) = H^\sigma_m - H^G_m \qquad\qquad [1.3.19]$$

---

[1] For further reading, see F. Rouquerol, J. Rouquerol and D.H. Everett, *Thermochim. Acta* **41** (1980) 311; S.J. Gregg and K.S.W. Sing, *Adsorption, Surface Area and Porosity*, Academic Press, 2nd ed. (1982).

and the *integral molar entropy of adsorption* as

$$\Delta_{ads}S_m(\text{int}) = S_m^{\sigma} - S_m^{G} \qquad\qquad [1.3.20]$$

For many purposes, including the investigation of surface heterogeneity, the consequences of adsorbing infinitesimal amounts of gas are studied. Then, we obtain the corresponding differential quantities, sometimes also called "derivative" quantities. The required definitions are

$$\Delta_{ads}U_m(\text{diff}) = \left(\frac{\partial U^{\sigma}}{\partial n}\right)_T - U_m^{G} \qquad\qquad [1.3.21]$$

$$\Delta_{ads}H_m(\text{diff}) = \left(\frac{\partial H^{\sigma}}{\partial n}\right)_T - H_m^{G} \qquad\qquad [1.3.22]$$

$$\Delta_{ads}S_m(\text{diff}) = \left(\frac{\partial S^{\sigma}}{\partial n}\right)_T - S_m^{G} \qquad\qquad [1.3.23]$$

for the *differential molar energy, enthalpy and entropy of adsorption*, respectively.

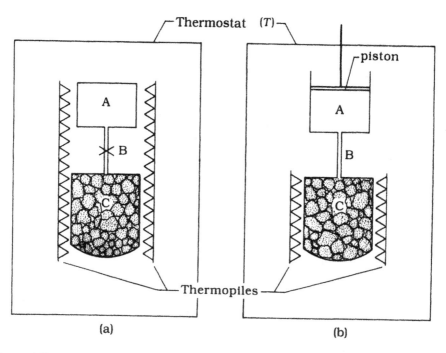

Figure 1.7. Two operational modes for adsorption microcalorimetry (a) the closed mode, (b) the open mode. The gas supply is in A, connected through a small tube (or otherwise) B to the volume C containing the adsorbent. In mode (a) the opening of the obstruction leads to adsorption. The heat evolved in the total system is now determined (A+B+C). In mode (b) the adsorbent is in open connection with the supply. By moving the piston down, the adsorption can be increased in small steps and only the heat evolved in part C is measured.

In these three differentiations the area A is kept constant. For solid surfaces $(\partial U^{\sigma}/\partial n)_T$ and $(\partial H^{\sigma}/\partial n)_T$ are identical. (For liquid surfaces, according to our convention, there is a difference between $(\partial U^{\sigma}/\partial n)_{T,A}$ and $(\partial H^{\sigma}/\partial n)_{T,\gamma}$, see [1.3.5 and 6].) Nevertheless, as $H_m^G = U_m^G + RT$ for an ideal gas we have

$$\Delta_{ads} U_m(\text{diff}) = \Delta_{ads} H_m(\text{diff}) + RT \qquad [1.3.24]$$

With these definitions in mind, let us now consider the measurement of heats of adsorption. In fig. 1.7 two usual modes of measuring this heat are sketched. Figure 1.7a represents the closed mode, fig. 1.7b the open one.

According to Everett[1] the integral energy of adsorption is defined as the experimental thermal effect when the adsorption is carried out by connecting a constant volume gas reservoir with a constant volume adsorbent reservoir, both vessels being immersed in the same calorimetric cell. The process can be carried out in an idealized isothermal calorimeter as shown in fig. 1.7a in which the volume of B is negligible and the volume of C equals the volume of the adsorbent. The total volume remains constant. Hence, there is no volume work and the *absorbed heat* Δq simply equals the integral molar energy of adsorption, defined in [1.3.18], times the amount Δn of gas adsorbed:

$$\Delta q = -\Delta_{ads} U_m(\text{int}) \times \Delta n \qquad [1.3.25]$$

or

$$\Delta_{ads} U_m(\text{int}) = -\left(\frac{\Delta q}{\Delta n}\right)_{T,V} \qquad [1.3.26]$$

For the integral molar enthalpy of adsorption, as above,

$$\Delta_{ads} H_m(\text{int}) = -\left(\frac{\Delta q}{\Delta n}\right)_{T,V} - RT \qquad [1.3.27]$$

In practice the measurement is usually carried out in an open system as represented by part C in fig. 1.7b. The total heat evolved (in A+B+C) follows from the First Law, $\delta q = -\delta U + p\delta V$ where we let the differentials apply to an infinitesimal downward movement of the frictionless piston. When the displacement of the piston results in transport of dn moles from the gas phase to the adsorbate,

$$\delta U = \left(\frac{\partial U^{\sigma}}{\partial n}\right)_T dn - \left(\frac{\partial U^G}{\partial n}\right)_T dn \qquad [1.3.28]$$

where U^{σ} and U^G are the energies of the adsorbate and the gas phase, respect-

---

[1] D.H. Everett, *Pure and Appl. Chem.* **31** (1972) 579.

ively. For small dn, $(\partial U^G/\partial n)_T \approx U_m^G$. Further, in the gas phase, comprising the volumes of A, B and C, both p and V change. Assuming the gas to be ideal, $d(pV) = -RT\,dn$, so that $p\,dV = -RT\,dn - V\,dp$. Combining all of this, using [1.3.21], and realizing that the work $V\,dp$ is completely transformed into heat, but only part of it, $V_c\,dp$ is measured by the thermopiles around vessel C, we arrive at

$$dq = \Delta_{ads}U_m(\text{diff})\,dn + V_c\,dp + RT\,dn \qquad\qquad [1.3.29]$$

or

$$\Delta_{ads}U_m(\text{diff}) = -\left(\frac{\partial q}{\partial n}\right)_T + V_C\left(\frac{\partial q}{\partial n}\right)_T + RT \qquad\qquad [1.3.30]$$

In [1.3.29 and 30] V_C is the volume of the gas in vessel C. Equation [1.3.30] formulates how the differential heat of adsorption can be measured calorimetrically. Because of [1.3.24] we immediately obtain for the differential enthalpy of adsorption

$$\Delta_{ads}H_m(\text{diff}) = -\left(\frac{\partial q}{\partial n}\right)_T + V_C\left(\frac{\partial p}{\partial n}\right)_T \qquad\qquad [1.3.31]$$

Differential and integral molar entropies of adsorption follow immediately from the measured heats as $T^{-1}(\partial q/\partial n)_T$ and $T^{-1}(\Delta q/\partial n)_{T,V}$, respectively. We also have, for equilibrium

$$T\Delta_{ads}S_m(\text{int}) = \Delta_{ads}H_m(\text{int}) \qquad\qquad [1.3.32]$$

and

$$T\Delta_{ads}S_m(\text{diff}) = \Delta_{ads}H_m(\text{diff}) \qquad\qquad [1.3.33]^{1)}$$

1.3d. Isosteric enthalpies of adsorption

Isosteric thermodynamic quantities are in practice not directly measured but follow from the shift of the adsorption equilibrium, caused by changes in temperature and pressure. To formulate this shift, we equate the shifts of the chemical potentials of gas and adsorbate:

$$\delta\mu^\sigma = \delta\mu^G \qquad\qquad [1.3.34]$$

with

$$\delta\mu^G = -S_m^G\,dT + V_m^G\,dp \qquad\qquad [1.3.35]$$

---

1) For further discussion of the measurement of adsorption entropies, see F. Rouquerol, J. Rouquerol, G. Della Gatta and C. Letoquart, *Thermochim. Acta* **39** (1980) 151.

and

$$\delta \mu^\sigma = - S_m^\sigma \, dT + V_m^\sigma \, dp \qquad\qquad [1.3.36]$$

Equation [1.3.35] is the Gibbs-Duhem relation for a homogeneous phase, [I.2.13.4]; eq. [1.3.36] is the equivalent for an adsorbate. As we consider isosteric conditions, dn^G and dn^σ are zero. From [1.3.34-36]

$$\left(\frac{\partial p}{\partial T} \right)_\Gamma = \frac{S_m^G - S_m^\sigma}{V_m^G - V_m^\sigma} \qquad\qquad [1.3.37]$$

As $V_m^G \gg V_m^\sigma$ and $V_m^G = RT/p$, considering [1.3.20 and 33] it follows that

$$\left(\frac{\partial \ln p}{\partial T} \right)_\Gamma = - \frac{\Delta_{ads} S_m (\text{diff})}{RT} = - \frac{\Delta_{ads} H_m (\text{diff})}{RT^2} \qquad\qquad [1.3.38]$$

The enthalpy $\Delta_{ads} H_m(\text{diff})$, obtained in this way is also known as the *isosteric enthalpy of adsorption*. Alternatively, equation [1.3.38] is written as

$$\left(\frac{\partial \ln p}{\partial T} \right)_\Gamma = \frac{q(\text{isost})}{RT^2} \qquad\qquad [1.3.39]$$

where q is counted positive when adsorption occurs, i.e. heat is evolved (*isosteric heat of adsorption*). The term "isosteric" stems of course from the adsorption isosters (relation between p and T at given Γ) from which these heats and enthalpies are obtained. Equation [1.3.39] is the "adsorption" Clapeyron *equation*. Compare its counterparts for the temperature dependence of the solubility (I.2.20.6) and chemical equilibrium (I.2.21.11 and 12).

1.3e. Standard Gibbs energies of adsorption

Although $\Delta_{ads} G_i = 0$ because of equilibrium, the *standard Gibbs energy of adsorption* $\Delta_{ads} G_i^\circ$ is non-zero and, in fact, it is a measure of the *affinity* of the adsorptive for the adsorbate. The situation is reminiscent of that for chemical equilibrium, treated in sec. I.2.21, from which, by analogy to [I.2.21.9 or 10], generally

$$\Delta_{ads} G_i^\circ = - RT \ln K_{ads,i} \qquad\qquad [1.3.40]$$

where K_{ads} is the *adsorption equilibrium constant*.

Establishing these quantities requires consideration of the way in which μ_i^σ and μ_i^G are split into their standard and configurational parts. In [1.3.40] $K_{ads,i}$ should be dimensionless because otherwise its logarithm cannot be taken; this requires the configurational terms to be dimensionless. If this is not so, the resulting $K_{ads,i}$ becomes dependent on the units in which pressures and surface concentrations are expressed and, hence, less well defined. This issue is directly connected to the fact that evaluation of G_i^o requires a model. For instance, for Langmuir-type adsorption, it is assumed that the gas is ideal, i.e. $\mu^G = \mu^{oG} + RT \ln(p/p^\dagger)$, where $p^\dagger$ is a standard pressure and that the adsorbate obeys the statistics discussed in sec. I.3.6d, leading to $\mu^\sigma = \mu^{o\sigma} + RT \ln[\theta/(1-\theta)]$, see [I.2.14.16] and [I.3.6.18], respectively. Equating chemical potentials and introducing $\Delta_{ads}G^o$ as $\mu^{o\sigma} - \mu^{oG}$

$$\frac{\theta}{1-\theta} = \exp\left(\frac{\Delta_{ads}G_i^o}{RT}\right)\frac{p}{p^\dagger} = K_{ads,i}\frac{p}{p^\dagger}$$

[1.3.41]

Often this equation is written as

$$\frac{\theta}{1-\theta} = K_L p$$

[1.3.41a]

or, equivalently,

$$\theta = \frac{K_L p}{1 + K_L p}$$

[1.3.41b]

where K_L is now a parameter having dimensions [pressure]$^{-1}$ and being consequently dependent on the units in which this pressure is expressed.

Standard Gibbs energies of adsorption are often encountered. When $\Delta_{ads}G^o$ is accurately known as a function of temperature, standard enthalpies and entropies of adsorption can also be obtained, using the appropriate Gibbs Helmholtz relations (sec. I.2.15).

1.3f. Experimental evaluation and examples

We shall now briefly review some of the main experimental procedures used to obtain thermodynamic data, and give a few illustrations. Some other examples will follow later in this chapter, for instance where models are tested. For literature reviews see sec. 1.9c.

Surface pressures can be obtained from (semilogarithmic) adsorption isotherms, using [1.1.7] for a monocomponent gas or, for a mixture, by integration of [1.3.13] for any component at fixed temperature.

Four methods are available to obtain heats of adsorption.

(i) The most direct is (micro-)calorimetry which exists in several variants, including titration and flow calorimetry[1]. Heats of immersion in liquids are

---

[1] Dalla Satta, J. Chim. Phys. **70** (1973) 60; D. Furlong, J. Chem. Soc., Faraday Trans I, **76** (1980) 774.

also calorimetrically determined, see under (iv). By calorimetry, integral or differential heats can be obtained, which can be converted into integral and differential enthalpies, see sec. 1.3c.

(ii) From the temperature dependence of the adsorption, the (isosteric) heat of adsorption can be found, using [1.3.27 or 28].

(iii) Isotherm analysis, using models, can also lead to heats (or at least energies) of adsorption. Obviously, the quality of the result depends on the rigour of the theory. A typical example is the BET theory from which an "enthalpic" constant C is obtained, that is related to the difference between the heat of adsorption and the heat of condensation. We will return to this in sec. 1.5f. Alternatively, isotherm models can yield $\Delta_{ads}G$. If this quantity is available as a function of temperature, $\Delta_{ads}H$ and $\Delta_{ads}S$ can be derived, using the appropriate Gibbs-Duhem relationship, see sec. I.2.15. For this purpose very accurate data are needed.

(iv) From *heats of immersion*; $q_{imm} = -\Delta_w H = -\Delta_{imm}H \approx -\Delta_{imm}U$ is minus the *enthalpy of wetting* or of *immersion*. These enthalpy changes are defined as the difference (at constant temperature) between the enthalpy of the solid completely immersed in a wetting liquid and that of solid and liquid taken separately. These heats can be calorimetrically obtained and are very sensitive to the state of the solid surface: is it fully evacuated or is there already some gas adsorbed? In the simplest case q_{imm} is numerically equal to the difference between the integral heat of adsorption at saturation pressure and the heat of vaporization of the liquid. Alternatively, the solid can be precovered with a known amount of adsorbate. In fact, this is the way to determine $\Delta_{imm}H$ as a function of θ.

Enthalpies of wetting can also be obtained from surface tensions, contact angles and their temperature dependencies, basically by using[1]

$$\Delta_w H = \left(\gamma^{SL} - T\,\frac{\partial\gamma^{SL}}{\partial T}\right) - \left(\gamma^{SG} - T\,\frac{\partial\gamma^{SG}}{\partial T}\right) \qquad\qquad [1.3.42]$$

This equation starts from the concept that interfacial tensions are differential Gibbs energies per unit area ($\gamma = (\partial G/\partial A)_{p,T}$ for a monocomponent system, see [I.2.10.7]); subtraction of the corresponding $T(\partial S^\sigma/\partial A)_T$ terms yields the pertaining surface excess enthalpy difference between the solid-liquid and the solid-gas interface, which is the quantity measured. Equation [1.3.42] cannot yet be applied because interfacial tensions involving solids are experimentally inaccessible, but they can be eliminated by using Young's equation for the contact angle (chapter III.5), resulting in

$$\Delta_w H = -\left[\gamma^{LG}\cos\alpha - T\,\partial(\gamma^{LG}\cos\alpha)/\partial T\right] \qquad\qquad [1.3.43]$$

---

[1] W.D. Harkins, G. Jura, *J. Am. Chem. Soc.* **66** (1944) 1362.

The meaning of the contact angle α has been illustrated in fig. I.1.1. Experimentally the procedure is not easy because, in addition to the requirements of surface purity and homogeneity, contact angle measurements may show hysteresis, but in a number of simple systems where such problems did not arise, the identity of $\Delta_w H$ from [1.3.36] and from calorimetry has been ascertained[1].

Differential or *integral entropies of adsorption* can be obtained from calorimetry and together with an adsorption isotherm, or from isotherms at different temperatures (sec. 1.3d).

Adsorption standard Gibbs energies require a model (sec. 1.3e) after which they can be obtained from one isotherm.

Literature abounds with thermodynamic data on adsorption. Some of them are of limited quality because the surface is not sufficiently defined (outgassing? reversibility?) or because of methodical problems (measurements at very low pressures may be difficult). We shall now give a few examples, merely meant as illustrations.

The first example, fig. 1.8, refers to old data on some graphitized "thermal blacks", which are carbons heated to very high temperature (several thousand degrees), after which graphite-like, fairly homogeneous surfaces are formed. The different symbols refer to different blacks which, apparently have very similar surface properties.

Physical adsorption of gases on solids is virtually always enthalpically driven: $\Delta_{ads}H < 0$. Entropically driven adsorption *can* exist but it is very unlikely to find conditions where the (differential) entropy in the adsorbate exceeds that in the gas phase because in the former state the degeneracy is so much lower. (Entropically driven adsorption is more likely with chemisorption, when additional binding entropy may be gained, but this is not being considered here.) The enthalpies reported in fig. 1.8 are indicative of apolar-apolar interactions by Van der Waals forces only. Usually $|\Delta_{ads}H|$ is a decreasing function of coverage because most surfaces are energetically heterogeneous, the parts of the surface with the highest adsorption energy being filled first. In that respect fig. 1.8a is a rare exception in that $|\Delta_{ads}H|$ remains constant over a large range of θ; the horizontal stretch extends from $\theta = 0.1$ to $\theta = 0.9$, with the specific surface area determined by BET (N_2) adsorption. The energetic homogeneity is probably due to the perfect match between the benzene molecules and the graphite-like surface

---

[1] A.W. Neumann, in *Wetting, Spreading and Adhesion*, J.F. Padday, Ed., Academic Press (1978) p.3, especially fig. 4. For further reading, see G.W. Woodbury, L.A. Noll, *Colloids Surf.* **40** (1989) 9; J.M. Douillard, M. Elwafir and S. Partyka, *J. Colloid Interface Sci.* **164** (1994) 238.

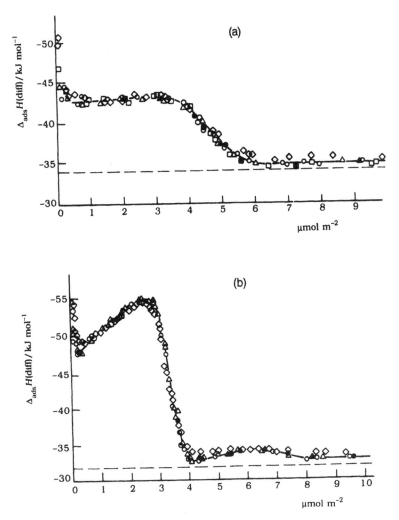

Figure 1.8. Differential heats of adsorption of benzene (a) and *n*-hexane (b) on four different graphitized thermal blacks. Temperature 20°C. (Redrawn from A.A. Isirikyan and A.V. Kiselev, *J. Phys. Chem.* **65** (1961) 601.) Discussion in the text.

structure. At very low θ, $|\Delta_{ads}H|$ may be higher because of attachment of adsorptive molecules to grain boundaries between the single crystals of graphite, whereas after completion of the monolayer, the enthalpy drops to (almost) the enthalpy of liquefaction (dashed line in the figure). In fig. 1.8b there is no such plateau, although from the measurements with benzene it was deduced that the surface is homogeneous. The rise in $|\Delta_{ads}H|$ was attributed to lateral interaction between the hexane molecules.

Figures 1.9a and b refer to the adsorption of water on rutile; fig. (a) gives isosters (plotted as $\ln p$ vs. T^{-1}) and fig. (b) the isosteric heats, derived from

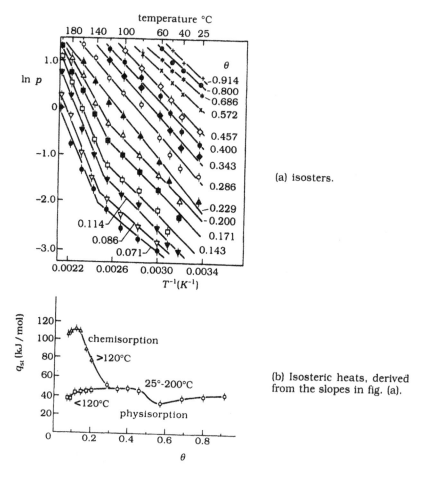

Figure 1.9. Adsorption of water on rutile (TiO_2). (Redrawn from A.C. Zettlemoyer, F.J. Micale and K. Klier, in *Water, a Comprehensive Treatise*, F. Franks, Ed., Plenum Press (1974), ch. 5.)

them. The plots are more or less linear with the striking feature of two distinct slopes at low coverage ($\theta \leq 0.2$). In that region of coverages, the higher slope has been attributed to chemisorption, which in this case is also reversible at the temperatures studied. The chemisorption and physisorption branches are clearly visible in fig. 1.9b. Although the water molecule is smaller than those of benzene and hexane, the heat of adsorption is comparable with that in fig. 1.8; in the present example polar contributions contribute more strongly.

Figure 1.10 is an extreme example of functional adsorption. The strong acid HCl of course has a strong affinity for the basic alumina. The isotherms are of the high-affinity type with a plateau that depends on the modification of the oxide and on its pretreatment, particularly on the temperature of outgassing. Figure (b) demonstrates that the strong affinity for HCl is reflected in the heats of immersion which are much higher in 0.1 M HCl than in water; the latter may

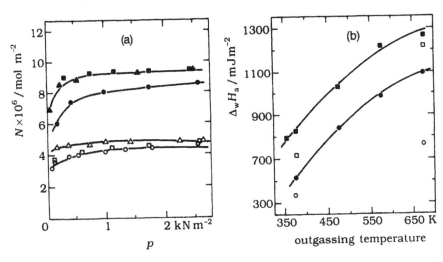

Figure 1.10. Adsorption of gaseous HCl on γ-Al$_2$O$_3$ (open symbols) and α-Al$_2$O$_3$ (filled symbols). Figure (a), isotherms; T = 40°C; outgassing temperature: circles, 80°C, squares, 200°C; triangles, 400°C. Figure (b) heats of immersion. T = 30°C; circles, in water; squares, in 0.1 M HCl. (Redrawn from R.R. Bailey, J.P. Wightman, *J. Colloid Interface Sci.* **70** (1979) 112.)

be compared with the data of table 1.3 in the present subsection.

Entropies of adsorption tend to exhibit substantial variations as a function of the amount adsorbed, even if the data are accurate[1]. Figure 1.11 is an example. Graphon is a graphitized carbon; when its surface contains no or few hydrophilic groups it is hydrophobic and water does not adsorb strongly; indeed the isosteric heat is less than the heat of liquefaction. On the other hand, the entropy of the adsorbate is relatively high, exceeding the molar entropy of the fluid, indicating high mobility of the adsorbed water molecules. Given this insight, it would be appropriate to analyze the entropies by subtracting the translational part and interpret the remainder in terms of rotation and vibration.

In table 1.3 some *enthalpies of wetting* are tabulated, typically per unit area rather than per mole. Hence, independent measurements of the specific surface area must precede these tabulations, with the corresponding proportional uncertainties. The values of $\Delta_w H$ are small, hence the area available for the wetting measurement should preferably be large. Per unit area (for which the symbol is $\Delta_w H_a = \Delta_{imm} H_a$) the values are O(10-10^2) mJ m^{-2}, which is the same order of magnitude as the corresponding interfacial tensions. For the purpose of assessing $\Delta_{ads} H$, wetting at different precoverages tends to be less rewarding than direct determination of the differential heat of adsorption as a function

---

[1] In older papers, entropies are sometimes expressed as "entropy units" (e.u.). One e.u. = 1 cal mol^{-1} K^{-1} = 4.184 J mol^{-1} K^{-1}.

Table 1.3. Enthalpies of wetting (= immersion) for some solid-liquid pairs. Given is $-\Delta_w H_a = -\Delta_{imm} H_a$ in mJ m^{-2} at 25°C.

Solid	Liquid		
	water	n-butanol	toluene
BaSO$_4$	760[1]		
	490[11]		
TiO$_2$ (anatase)	650[1]		
	510[8]		
TiO$_2$ (rutile)	480[1]	290[3]	170[3]
	550[6]	410[6]	
SiO$_2$ (quartz)	610[1]		165[1]
	850[9]		
SiO$_2$ (amorphous)	210[10]		
SiO$_2$ (pyrogenic)	150[1]	115[1]	75[1]
γ-FeOOH (goethite)	465[1]		135[1]
α-Fe$_2$O$_3$ (haematite)	280[1]	140[1]	100[1]
β Cu phtalocyanine (a pigment)[7]	50[1]	95[1]	110[1]
graphon	32[2,5,6]	114[5]	
teflon	6[2,5]	56[5]	
smectites (a type of clay mineral)	55 - 150[4]		
vermiculites (a type of clay mineral)	165 - 300[4]		
kaolinites (a type of clay mineral)	320 - 500[4]		

Data sources: [1] K. Hamann, as reproduced by E. Herrmann in *Characterization of Powder Surfaces* (G.D. Parfitt, K.S.W. Sing, Eds.) Academic Press (1976), p. 212. [2] A.C. Zettlemoyer, *Ind. Eng. Chem.* **57** (1965), 27. [3] R. Sappok, B. Honigmann, as ref. 1), p. 257. [4] H. van Olphen, in *Surface Area Determination*, D.H. Everett, R.H. Ottewill, Eds., Butterworth (1970), p. 255. [5] J.J. Chessick, F.H. Healy and A.C. Zettlemoyer, *J. Phys. Chem.* **60** (1956) 1345. [6] J.J. Chessick, A.C. Zettlemoyer, F.H. Healy and G.J. Young, *Can. J. Chem.* **33** (1955) 251. [7] For a range of other pigments, see J. Schröder, *Progr. Org. Coatings*, **12** (1984) 339; **19** (1991) 227. [8] W.H. Wade, N. Hackermann, *J. Phys. Chem.* **65** (1961) 1682; [9] ibid., *Adv. Chem. Ser.* **43** (1964) 222; [10] J.W. Whalen, *J. Phys. Chem.* **65** (1961) 1676; [11] G.E. Boyd, W.D. Harkins, *J. Am. Chem. Soc.* **64** (1942), 1190, 1195.

of coverage[1]. Nevertheless wetting data have their own interest (chapter III.4) and make a link to adsorption from solution (chapter 2).

The data in table 1.3 give information on the relative affinities (in terms of enthalpies) of the molecules of the liquid for the surface as compared with that between themselves. In spite of the relatively large differences between different

---

[1] C. Létoquart, F. Rouquerol and J. Rouquerol, *J. Chim. Phys.* **70** (1973) 559.

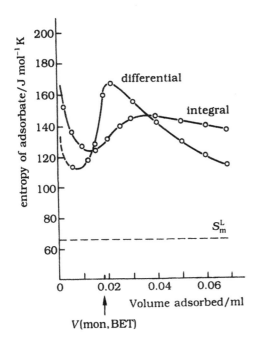

Figure 1.11. Entropy of water, adsorbed on Graphon. The adsorbed volume is in ml standard temperature and pressure and $V(\text{mon})$ is the volume in a saturated monolayer according to BET theory. (Redrawn from A.C. Zettlemoyer, in *Hydrophobic Surfaces*. F.M. Fowkes, Ed., Academic Press (1969) 1.)

authors, some interesting trends emerge. Most striking is the low value for water on hydrophobic surfaces. The almost zero value for teflon may be related to the almost zero enthalpy of hydrophobic bonding in this case and the enthalpic affinity on graphon is reflected in the high mobility of individually adsorbed water molecules (see discussion of fig. 1.11). Heating makes silica less hydrophilic: compare pyrogenic silica with quartz. This is caused by dehydration of silanol groups into siloxanes ($2 \equiv \text{SiOH} \rightarrow \equiv \text{Si} - \text{O} - \text{Si} \equiv + \text{H}_2\text{O}$).

The immersion data for water can also serve as some measure of the hydrophobicity/hydrophilicity of the surface. To that end the data are preferably written as molar quantities ($\Delta_w H_m$). Values below about 40 kJ mol^{-1} indicate physical adsorption, from ~50 - 70 kJ mol^{-1} hydrogen bonds (for instance with surface hydroxyls), from ~70 - 120 kJ mol^{-1} stronger adsorption on Lewis acids and still higher values reflect chemisorption. These data can be compared with the heat of condensation of water, 44 kJ mol^{-1}, which is a measure of the affinity of water for itself[1].

Sometimes surfaces are distinguished between "high energy" and "low energy" on the basis of $\Delta_w H$ for water, but we shall reserve this term for surfaces of high or low U^σ, which is quite different. In fact, for other liquids the sequence is different. For *n*-butanol the surface specificity is much less pronounced because

[1] J. Texter, K. Klier, A.C. Zettlemoyer, *Progr. Surf. Membr. Sci.* **12** (1976) 327.

the molecule can have polar and apolar interactions. The difference between the two rutile samples is related to differences in pretreatment. For toluene the specificity is even less than it is for butanol, because this molecule interacts by apolar interactions only, in this case by Van der Waals forces plus some exchange of π-electrons of the aromatic ring. To a large extent different results between different authors are probably caused by different pretreatments. One of the variables is the *activation temperature* (i.e. the temperature to which the sample is subjected prior to immersion). Depending on the physical process, caused by heating, $\Delta_w H$ may increase or decrease; this is in itself a surface characteristic, but we shall not pursue this feature here.[1]

Enthalpies of wetting are sometimes used to obtain (integral) enthalpies of adsorption by subtracting the enthalpy of condensation. This procedure is not exact because it presupposes a model in which the interaction between the first and the second surface layer is interpreted as purely identical to that in condensation (BET theory assumes the same). However, the heat of adsorption of the second layer is not *exactly* identical to the heat of liquefaction and the configuration of the first layer is affected by the presence of a second. In other words, entropic factors also have to be considered, and, in this connection, the packing in the first layer must be known to convert $\Delta_w H_a$ (in J m^{-2}) into $\Delta_w H_m$ (in J mol^{-2}). Notwithstanding these reservations, a certain similarity may be expected.

Generally speaking, for "simple" adsorption (dominance of short range forces, homogeneous surfaces without pores, monofunctional adsorptives, etc.), the various different approaches (isotherm analysis, calorimetry, isosters, immersion) tend to give similar or identical results, but for more complex situations there may be substantial deviations. See the literature cited in sec. 1.9c.

1.4 Presentation of adsorption data

Adsorbed amounts can be determined volumetrically (depletion from the vapour phase) or gravimetrically (direct measurement of the weight increase of the sample), and augmented by information from other techniques (ellipsometry, spectroscopy, etc.). We assume that such measurements have been carried out over a sufficiently large pressure range, at a number of temperatures, taking the required precautions regarding outgassing, avoidance of impurities etc.[2] and address the issue of the optimal way of presentation.

Experimental adsorption data can be plotted in a variety of ways. By far the

---

[1] A.C. Zettlemoyer, K.S. Narayan, *Heats of Immersion and the Vapor-Solid Interface*, in *The Solid-Gas Interface*, E.A. Flood, Ed., see Vol. I, chapter 6, sec. 1.9c.
[2] For procedures and presentation of data see *Reporting Physisorption Data* (IUPAC report, sec. 1.9a).

most common way is as *adsorption isotherms*, i.e. plots of the amount adsorbed as a function of the equilibrium pressure at constant temperature. One alternative is as *adsorption isosters*, $p(T)$ plots at constant amount adsorbed, and another one is in terms of *two-dimensional equations of state*, i.e. $\pi(\theta)$ plots. However, the primary data are almost exclusively adsorption isotherms and therefore we shall emphasize these in this section.

Isotherms can be plotted in different ways. Linear isotherms are not always the most obvious way of presentation. The information that is relayed (or suggested!) may be very different for different ways of plotting (linear, semilogarithmic, etc.).

1.4a. Plotting adsorption data

What is the best way of plotting the data, and which part of the plot is the most important? That depends on the information wanted. Obviously, when we are interested in the affinity $(\Delta_{ads}G_m^o)$, data at very low coverage are needed, i.e. in the Henry part, where only adsorbent-adsorbate interaction occurs. On the other hand, if there is a plateau (rare in the case of gas adsorption), data in this region give information on the area but not on the adsorbent-adsorbate affinity because of the prevailing lateral interaction.

In figure 1.12 we illustrate the different ways of plotting for the special case of a model Langmuir isotherm. The "measuring points" are chosen such as to obey the Langmuir equation [1.1.3], with $\theta = V/V(mon)$, where $V(mon)$ is the volume adsorbed in a saturated monolayer. It is assumed that the specific surface area (via $V(mon)$) is not known, hence the data are given as V (in ml at standard p and T)as a function of pressure (in Pa or N m^{-2}). The "original" plot is

$$V = V(mon)\frac{K_L' p}{1 + K_L' p} = \frac{k_1 p}{1 + k_2 p} \qquad [1.4.1]$$

with k_1 in m^5 N^{-1} and k_2 in m^2 N^{-1} as the unknowns.

In fig. 1.12a the data are plotted in the direct way for three values of the affinity K_L'. The higher affinity isotherm rises more steeply but attains the same plateau. From these isotherms $V(mon)$ may be obtained by extrapolation; in the present example it is 0.0113 ml g^{-1}. Assuming a certain cross-sectional area for the adsorbate molecules and close packing, the total area can be found. For this procedure no equation is needed. However, once $V(mon)$ is known, θ can be found and then also K_L' as the pressure at which $\theta = 0.5$; for this step an equation is required.

Figure 1.12b gives the same data, now expressed *semilogarithmically*. (As we are only interested in the shapes of the plots here, for the moment we will ignore the problems involved in taking logarithms of quantities that have

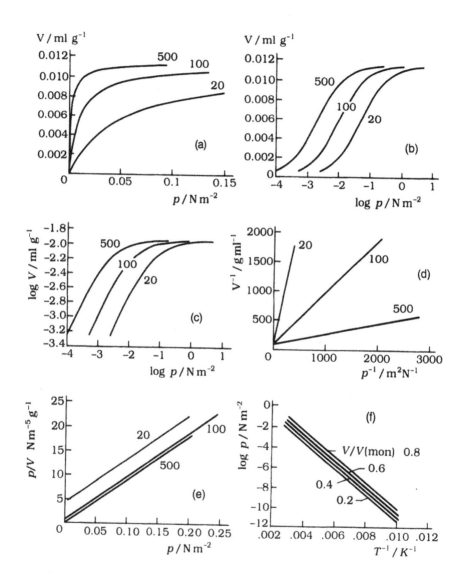

Figure 1.12. Six ways of plotting the same data. Model isotherm according to Langmuir. The values of the Langmuir constant K'_L (in $m^2\ N^{-1}$) are indicated. Explanation in the text.

dimensions.) One of the reasons for such a plot is that in many theories μ is the independent variable (= const + $RT \ln p$), rather than p. Taking logarithms implies giving more weight to the low pressure data. The figure shows that for such plots data over much greater pressure ranges are required, especially in the low p range.

   Double-logarithmic plots are shown in fig. 1.12c. In the low adsorption (Henry) range such plots are linear, but at higher adsorption deviations from linearity occur. This is different for the Freundlich equation [1.1.2] which is linear over the entire range, if plotted double-logarithmically. Therefore, to distinguish data obeying a Langmuir or a Freundlich isotherm, data over a large range of data are required.

   Two ways of plotting $V(p)$ give rise to *linear* isotherms if the Langmuir equation applies, namely

$$\frac{1}{V} = a + \frac{b}{p} = \frac{k_2}{k_1} + \frac{1}{k_1 p} \qquad\qquad [1.4.2]$$

and

$$\frac{p}{V} = a'p + b' = \frac{k_2 p}{k_1} + \frac{1}{k_1} \qquad\qquad [1.4.3]$$

The corresponding plots are given in figs. 1.12d and e. Plots of type [1.4.2] are very sensitive to k_1 (proportional to the reciprocal slope) and hence particularly suited to determine the affinity. For very high pressure ($p^{-1} \to 0$) the intercept is independent of the slope. On the other hand, a plot according to [1.4.3] has a slope that is independent of K'_L, the only effect that K'_L variation has is that it shifts the line parallel to itself. For very high K'_L a limiting line is attained, passing through the origin; here k_1^{-1} in [1.4.3] vanishes. The slope is determined by k_2/k_1, and hence by $V(\text{mon})$, which is a constant. Such slopes are suitable for obtaining the monolayer coverage.

   Finally, fig. 1.12f gives *adsorption isosters*. In this case $\Delta_{ads}H_m(\text{isost})/R$ was fixed at –3000 K (independent of T). From [1.3.39] it is readily derived that a plot of $\log p$ as a function of T^{-1} is linear, the slope being determined by the isosteric enthalpy (or heat) of adsorption.

   Two-dimensional equations of state are plots of π as a function of Γ and T. Recal that surface pressures are experimentally accessible from isotherms; see [1.1.8] and sec. 1.3b. There is a difference of principle between isotherms and isosters, on the one hand, and equations of state on the other, in that the former category describes equilibrium between adsorbate and the gas phase, whereas the latter contains only properties of the adsorbate itself. Typically, isotherms contain constants (K_L, K_V, etc., for a Langmuir, Volmer, etc. adsorbate) describing gas-adsorbent interaction. Factors $\Delta_{ads}U/RT$ dominate these constants. In equations of state such constants are absent; they contain one less parameter. A few illustrations have already been given in Volume I (figs. I.3.3, I.3.6 and I.3.9) and more examples will follow (figs. 1.15b, 1.20b and 1.24b).

These considerations have their consequences regarding the interpretation of experimental data: should it be done in terms of isotherms or in equations of state? Preferably both should be considered, but when specific features are under study a choice may have to be made. For instance, surface heterogeneity shows up very strongly in the shapes of the isotherms (sec. 1.7) but very little in the equation of state; in the model case of local Langmuir isotherms without lateral interaction heterogeneity is not seen at all in the equation of state (because the energy is not considered and the entropy not affected) whereas the isotherm shape is dramatically influenced. On the other hand, for homogeneous model surfaces equations of state may be more suited to observe subtle distinctions in lateral mobility or lateral interaction.

1.4b. *Classification of gas adsorption isotherms*

For physisorption the majority of the isotherms may be grouped into the six types shown in fig. 1.13. The quantity adsorbed may be presented in any suitable unit: moles, grams, volumes s.t.p. (for "standard temperature and pressure"), per gram or per unit area of the adsorbent. The relative pressure is usually written as $p/p(\text{sat})$ where $p(\text{sat})$ is the saturated vapour pressure at that temperature: the equilibrium pressure of the pure adsorbate if it were present as bulk liquid. Types I-V have already been distinguished by Brunauer[1] and are sometimes referred to as BDDT types[2].

Type I is concave and V approaches a limiting value. This type is characteristic for microporous solids[3] with negligible external surface areas (e.g. activated carbons, molecular sieve zeolites and certain porous oxides). Although mathematically type I isotherms are very well described by the Langmuir isotherm equation, calling them "Langmuir isotherms" is not recommended because the limited uptake is governed by the accessible pore volume rather than by the internal surface area.

Types II and III are reversible isotherms for non-porous or macroporous sorbents[3]. The first part of type II resembles that of I, but a plateau is not reached because adsorption in the second and higher layers sets in. Point B, the

---

[1] S. Brunauer, *The Adsorption of Gases and Vapors*, Princeton Univ. Press, Vol. **1** (1945).

[2] After S. Brunauer, L.S. Deming, W.E. Deming and E. Teller, *J. Am. Chem. Soc.* **59** (1937) 2682.

[3] Porosity will be discussed more systematically in sec. 1.6.

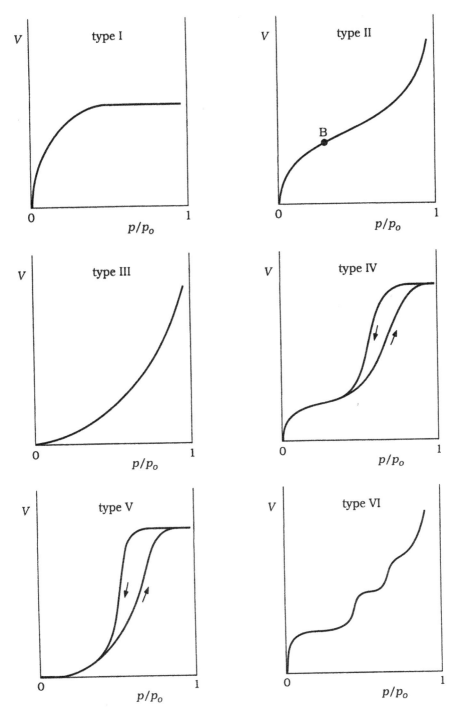

Figure 1.13. Classification of physisorption isotherms for gases on solids.

start of the almost linear middle section of the isotherm, is often interpreted as the stage where monolayer coverage is complete and multilayer adsorption is about to begin. Type III isotherms are observed for gases with a very low affinity to the sorbent: the initial slope is low, but once some molecules have adsorbed, adsorbate-adsorbate attraction induces progressive adsorption. Here adsorption rather resembles *heterogeneous nucleation*; impurities on an otherwise low affinity surface may act as condensation nuclei.

Types IV and V are characterized by *hysteresis loops* and a limiting uptake as $p \rightarrow p(\text{sat})$. Hysteresis is typical for *capillary condensation* in mesopores. Type IV, associated with type II, is more common than type V, which is associated to type III. Many industrial adsorbents give type IV isotherms.

The stepwise isotherms of type VI are only observed under a number of idealized conditions for uniform non-porous surfaces. Ideally, the step-height represents monolayer capacity; in the simplest case, it remains constant for two or three layers. Argon or krypton adsorption on graphitized carbon blacks at liquid nitrogen temperature are amongst the best examples.

In principle, all isotherms approach the linear (Henry) range for $p \rightarrow 0$, but for heterogeneous surfaces it is often difficult to attain this range.

For a description of isotherms for adsorption from solution, see figs. 2.8 and 2.23.

1.5 Adsorption on non-porous, homogeneous surfaces

In this and the following sections we shall, more systematically, derive and discuss a number of important isotherm equations, together with the corresponding two-dimensional equations of state. For easy reference they are collected in app. 1.

Simple systems will be considered first. We shall start with (sub-)monolayer adsorption, that is adsorption in which all the adsorbate molecules are in contact with the adsorbent. It is noted that for gases (sub-) monolayer adsorption is met only at low relative pressures; for relative pressures approaching unity, invariably condensation in more than one layer takes place. Physical adsorption at temperatures above the critical temperature of the gas is also restricted to a maximum of one complete monolayer. For adsorption *from solution* (chapter 2) monolayer adsorption is more usually the rule.

The adsorbents considered here are supposed to be non-porous, inert and energetically homogeneous (except where otherwise stated). In secs. 1.6 and 1.7 these restrictions will be relaxed. Here, only reversible adsorption will be considered.

A recurring distinction is that between *localized* and *mobile* adsorbates, already introduced in sec. I.3.5b, d and I.3.6d, and briefly repeated in sec. 1.1.

For the moment no transitional states between these extremes will be considered. For these two limiting modes of adsorption the establishment of equilibrium proceeds via different paths: in localized adsorption, adsorbed molecules can move to other sites only by desorption and re-attachment, whereas for mobile adsorption the path of tangential transport is also open. Typical representatives of localized isotherms are those of Langmuir, Frumkin-Fowler-Guggenheim (monolayer) and BET (multilayer). For mobile adsorbates examples are the Volmer and the Hill-De Boer (or two-dimensional Van der Waals) equations.

1.5a. Localized monolayer adsorption without lateral interaction (Langmuir)

In the statistical derivation of the Langmuir equation (sec. I.3.6d) the physical premises were specified:

(i) adsorption limited to one monolayer

(ii) surface homogeneity (adsorption on energetically identical sites)

(iii) no lateral interaction between adsorbed molecules

(iv) adsorption is reversible

So far, only monatomic adsorptives have been considered. Conditions (ii) and (iii) imply that the molecular partition function q is only a function of temperature, and not of the adsorbed number of molecules N. Important equations, derived in sec. I.3.6d include the following.

The canonical partition function of the adsorbate is

$$Q(N, N_s, T) = \left[N_s!/N! \ (N_s - N)!\right] q_{loc}^N \qquad [1.5.1]$$

where N_s is the total number of sites available, i.e. $\theta = N/N_s$. We added the subscript loc to q to distinguish it from q_{mob}, to be discussed later.

The grand canonical partition function reads

$$\Xi(\mu, N_s, T) = \left[1 + q_{loc} \exp(\mu/kT)\right]^{N_s} \qquad [1.5.2]$$

The entropy of the adsorbate may be identified with the excess entropy S^σ if this excess is referred to the entropy of gas molecules *at rest*. It consists of additive contributions due to configurations and vibrations (of the adsorbed molecule) around its rest position:

$$S^\sigma = \underbrace{-k N_s[(1 - \theta) \ln(1 - \theta) + \theta \ln \theta]}_{S^\sigma(\text{conf})} + \underbrace{k N\left[T \, d \ln q_{loc}/dT + \ln q_{loc}\right]}_{S^\sigma(\text{vibr})} \qquad [1.5.3]$$

With this equation S^σ can be obtained from the isotherm and the temperature dependence of the adsorption energy, using [1.5.6]. When U^σ is a constant, S^σ has only the configurational term, which is the same as the mixing entropy for gases, see [I.2.17.12]. Typically, for a given adsorbent S^σ(conf) depends only on θ, the degeneracy of the mixing of open and filled sites being solely determined by the *ratio* of the numbers of sites. On the other hand, S^σ(vibr) is proportional to the number N because the adsorbed molecules are independent.

For diatomic molecules the configurational entropy remains the same but the vibrational term has to be modified and now a rotational term, also proportional to N, has to be included. For physisorption the intramolecular vibration is usually not significantly perturbed. Hence, this contribution to S^σ(vibr) may be disregarded. To identify the entropy with the excess entropy, rotation-free molecules must be taken as the reference.

The chemical potential of the adsorbate is given by

$$\mu^\sigma = \mu^{o\sigma} + kT \ln[[\theta/1 - \theta)] \qquad\qquad [1.5.4]$$

with

$$\mu^{o\sigma} = -kT \ln q_{loc} \qquad\qquad [1.5.5]$$

At low temperatures,

$$q_{loc} \approx \exp(-\Delta_{ads}U/RT) = \exp(-U^\sigma/RT) \qquad\qquad [1.5.6]$$

When the temperature is not low, a more complicated equation applies; it should be realized that localized adsorption is especially expected at low temperatures. The energy of the adsorbate consists of a potential energy and a vibration energy contribution, see sec. I.3.5b. If again gas molecules at rest are taken as the reference, $U^\sigma = \Delta_{ads}U$, as assumed in [1.5.6]. The difference with moving molecules is simply $3RT/2$ per mol.

The Langmuir isotherm equation [1.1.3] is obtained either from [1.5.4] by equating μ^σ to μ^G or grand canonically from [1.5.2] using [I.A6.24]. For the Langmuir constant

$$K_L' = \frac{q_{loc}}{(2\pi m/h^2)^{3/2}(kT)^{5/2}} \qquad (\text{m}^2\ \text{N}^{-1}) \qquad\qquad [1.5.7]$$

is obtained, where m is the mass of the atoms of the adsorptive. The denominator stems from the entropy of the gas molecules that is lost; essentially this is a configurational entropy given by the Sackur-Tetrode equation [I.3.1.9]. The configurational entropy that is gained is given by [1.5.3] and reflected in the $\theta/(1-\theta)$ factor in the Langmuir equation via [1.5.4]. From [1.5.6 and 7]

$$K_L' = f(T) \exp(-\Delta_{ads}U/RT) \qquad\qquad [1.5.8]$$

or, using [1.3.40]

$$\Delta_{ads}G^{\circ} = f'(T) + \Delta_{ads}U \qquad [1.5.9]$$

where $f'(T)$ does *not* contain the configurational entropy of the adsorbate and U has a potential and a vibrational contribution.

Finally, for the surface pressure according to [I.3.6.23],

$$\pi a_m = -kT \ln(1-\theta) \qquad [1.5.10]$$

Here, a_m is the area of one site. In the Langmuir picture, the total area is $N_s a_m$. Any part of the surface between the sites are simply not "seen". This has to be considered when interpreting the notion of specific surface area and its determination from monolayer capacity.

Having the statistical framework available, all relevant thermodynamic and mechanical quantities for the adsorbate are readily obtained and the choice between the canonical and grand canonical approach remains a matter of taste and/or convenience. For instance, the surface pressure also follows immediately from the grand canonical partition function, see [1.1.12] with A replaced by $N_s a_m$. The energy of a Langmuir adsorbate follows from the two-dimensional equivalent of [I.A6.8] by substituting [1.5.1]. If this is elaborated, using Stirling's approximation [I.3.6.5] and assuming q_{loc} given by [1.5.6], this energy appears to be U^{σ}, confirming the consistency. This can also be achieved grand canonically, starting from [I.A6.20].

A further feature of the statistical approach is that the fluctuations in the number of adsorbed molecules can be computed. To this end [I.3.7.6] can be used for the *standard deviation* σ_N:

$$\sigma_N = kT(\partial N/\partial \mu)_{N_s,T} = N_s^{-1}(\partial \theta/\partial \ln p)_{N_s,T} \qquad [1.5.11]$$

It is concluded that σ_N is proportional to the slope of the semilogarithmic isotherm plot. According to fig. 1.12b, for a Langmuir adsorbate it passes through a maximum at $\theta = 0.5$; this is expected because then the mixing entropy (first term on the r.h.s. of [1.5.3]) is a maximum. No fluctuations are possible for $\theta = 0$ and $\theta = 1$.

The Langmuir equation can also be derived *kinetically*. In fact, this was the route used by Langmuir himself in his famous seminal paper[1]. Langmuir had chemisorption in mind; obviously the theory is also valid for this, provided the process is reversible.

In the kinetic derivation, adsorption equilibrium is assumed to be established by equating the rate of adsorption ($\sim p$ and $\sim(1-\theta)$) and that of

---

[1] I. Langmuir, *J. Am. Chem. Soc.* **40** (1918) 1316. The statistical derivation is due to R.H. Fowler and E.A. Guggenheim, *Statistical Thermodynamics*, Cambridge Univ. Press (1952).

desorption $(\sim\theta)$. The adsorption energy enters as an activation energy for desorption in the rate constant. For physisorption there is no activation energy in the adsorption step.

Alternatively, the adsorption can be interpreted in terms of the mass action law, considering the equilibrium open site + gas molecule $\rightleftharpoons$ filled site, for which an equilibrium constant similar to that for chemical equilibria can be given. This is a "semi"-thermodynamic method because assumptions have to be made about the division of μ_i^σ into a constant and a configurational part. Mass-action approaches are particularly suited for adsorption from solution, because in this way the replacement of solvent by solute molecules can be accounted for.

It is seen that regarding the derivation of the Langmuir equation, "many roads lead to Rome". The fact that very different physical models all lead to the same result is one of the reasons for its wide application. Considering these disparate approaches quantitatively, the interpretation of K_L' requires more attention. To that end, let us take a closer look at the kinetics, thereby following an idea originally due to Langmuir and elaborated by others[1].

When a gas molecule arrives at a surface onto which it does not adsorb, it will reflect from it. The collision is *elastic*, implying that the angle of incidence equals the angle of exit. Only exchange of momentum normal to the surface takes place. The time that the molecule is "in touch" with the surface is very short, at most of the order of molecular vibrations, say $\gtrsim O(10^{-13}$ s$)$[2]. This time is called the *residence time, τ*.

Now, when the molecule does experience an attractive force from the surface, the time between hitting the surface and leaving it is longer, the stronger the attraction. When τ substantially exceeds 10^{-13} s one may call the molecule adsorbed. Typically, upon desorption such a molecule no longer "remembers" at what angle it had approached the surface. Angle of exit and angle of incidence are uncoupled.

Following Frenkel[1]

$$\tau \approx \tau_0 \exp(-\Delta_{ads}U/RT) \qquad\qquad [1.5.12]$$

where τ_0 may be regarded as a kind of standard residence time, of the order of molecular vibration times. (We have already used such an equation in describing the residence times of water molecules near ions, see [I.5.3.6].) More precisely, $\tau_0 = \nu_0^{-1}$, with ν_0 given by $(f/m)^{1/2}/2\pi$ (see [I.4.5.2]). The standard frequency is

---

[1] J.H. de Boer, *The Dynamical Character of Adsorption*, Clarendon Press, Oxford (1953), chapter 1; Ya.I. Frenkel, *Kineticheskaya Teoriya Zhidkostei*, Engl. transl. *Kinetic Theory of Liquids*, Dover (1955), p.5ff.

[2] Some values have been tabulated by H.J. Kreuzer, Z.W. Gortel, *Physisorption Kinetics*, Springer (1986), table 1.2; the data show considerable scatter.

determined by the mass of the molecule and by the binding strength, which determines f, see [I.4.5.3]. For physical adsorption with $\Delta_{ads}U \approx -25$ kJ mol^{-1} τ exceeds τ_o by about a factor of 10^4 at room temperature. Values of τ_o depend on the nature of the adsorbent and adsorbate, and the temperature. For simple molecules like CH_4, Xe, and N_2 on metals or graphite at temperatures up to 125 K, $\tau_o \sim O(10^{-13}-10^{-12}s)$.

In dynamic equilibrium, the rate of desorption equals $N\tau^{-1}$ because τ^{-1} is the probability that an adsorbed molecule will desorb in one second. The rate of adsorption is determined by the available empty area $a_m(N_s-N)$, the pressure (i.e. by the number of molecules in the gas phase per unit volume) and by the rate at which they move. The result is $a_m p(N_s-N)/(2\pi mkT)^{1/2}$ molecules per second. The factor $(2\pi mkT)^{1/2}$ stems from the kinetic theory of gases and is related to the collision frequency. Equating the two rates gives, after some rearrangements, the Langmuir equation with

$$K'_L = \left(\frac{2\pi}{kT f}\right)^{1/2} a_m \exp(-\Delta_{ads}U / RT) \qquad [1.5.13]$$

which is the classical equivalent of [1.5.7]. The two formulas resemble each other but exact identity cannot be expected because [1.5.7] has a quantum mechanical background. It is noted that in both approaches K'_L can be written as $K'_{L0} \exp(-\Delta_{ads}U / RT)$, in which the major temperature dependence stems from the exponential factor. Letting K'_{L0} be independent of T, as is sometimes done, is not exact, but it is not a bad approximation if the adsorption energy is not too low.

Gas adsorption is a very dynamic process in the sense that the rates of adsorption and desorption are very high, even at low pressures. Contamination of pristine surfaces after exposure to gases is also very rapid.

Regarding the application of the Langmuir equation to physical gas adsorption on homogeneous surfaces, good examples are difficult to find. Those systems that do obey the Langmuir equation appear to be microporous. Of course, the initial parts of multilayer isotherms fulfil the Langmuir equation, and these will be dealt with in the appropriate section. The various ways of plotting and the information obtainable from such plots have been discussed in sec. 1.4.

Given the simple underlying picture and the possibility of obtaining the Langmuir equation in several ways, it is not surprising that variants of it are met in other domains of science. One familiar example stems from biochemistry in formulating the binding of ligands to biomolecules, say oxygen to haemoglobin or small molecules to a protein. Typically, such attachments involve site-binding mechanisms. In the most simple case that all binding sites are independent, and this immediately leads to [1.1.3] with θ now being the fraction of sites occupied, K_L the binding constant and p replaced by the ligand concentration, c_L. The data can be plotted in various ways and then bear specific names. A linear or logarithmic plot of $\theta/(1-\theta)$ as a function of c_L is known as a *Hill plot* and one of θ / c_L versus c_L as a *Scatchard plot*. As in sec. 1.4a, it depends on the information available (is the total number of sites known?) what the preferred way of plotting is.

In Nature, the binding sites are often not independent. Binding of a ligand to one site may promote or inhibit binding to a the second and subsequent site. Biochemists call this (positive or negative) *cooperativity*, but we note that our notion of (positive) cooperativity is restricted to such a strong phenomenon that condensation occurs (vertical parts in the isotherm), corresponding to the notion of "infinite cooperativity" in biochemistry. Although formally dependence of sites can be treated in terms of lateral interactions (sec. 1.5e) the biomechanisms are usually more complex, involving changes of the substrate (corresponding to "non-inertness of the sorbent")[1].

1.5b. *Langmuir adsorption on a binary adsorbent*

By a *binary adsorbent* is meant an adsorbent having two distinct types of sites, each with its own adsorption energy, $\Delta_{ads}U_1$ and $\Delta_{ads}U_2$. Such a system is a first step towards heterogeneous surfaces, exhibiting a range of sites. So long as there is no lateral interaction, the treatment is simple because adsorption on each site type is only determined by the pressure and not influenced by events taking place on other types. For each type i site, the chemical potential of the adsorbed molecules μ_i^σ can be written by analogy with [1.5.4 and 5] as

$$\mu_i^\sigma = kT \ln \frac{\theta_i}{(f_i - \theta_i)q_i} \qquad [1.5.14]$$

if f_i is the fraction of sites of type i and $0 \geq \theta_i \geq f_i$. For a binary adsorbent, in equilibrium with the same gas

$$kT \ln \frac{\theta_1}{(f_1 - \theta_1)q_1} = kT \ln \frac{\theta_2}{(f_2 - \theta_2)q_2} = \mu^G \qquad [1.5.15]$$

Substituting the appropriate pressure dependence in μ^G, an isotherm equation is derived that can be written as

---

[1] For further reading on this matter, D. Freifelder, *Physical Biochemistry*, 2nd ed., Freeman (1982); Ch. Tanford, *Physical Chemistry of Macromolecules* (1961); T.L. Hill, *Cooperativity Theory in Biochemistry*, Springer (1985) chapter 4. (The name "Hill plot" stems from A.V. Hill, *Biochem. J.* **7** (1913) 471.)

$$\frac{\theta_1}{f_1 - \theta_1} = \frac{\theta_2}{f_2 - \theta_2} \cdot \frac{q_1}{q_2} = k_1 p \qquad\qquad [1.5.16]$$

where k_1 is now a constant of dimensions [pressure]$^{-1}$ that does not contain the adsorption energies. Adsorption of molecules on type 2 sites follows that of molecules on type 1 sites with a proportionality factor for the configurational part, determined by $\exp[(\Delta_{ads}U_2 - \Delta_{ads}U_1)/RT]$.

Figure 1.14a gives a plot of $\theta = \theta_1 + \theta_2$ for $f_1 = f_2 = 0.5$ and four values of q_1/q_2. The curve with $q_1/q_2 = 1$ does not correspond to ordinary Langmuir behaviour because each half of the molecules is restricted to half of the sites; this is entropically less favourable than having all sites available to all molecules. Although the curves of fig. 1.14a resemble Langmuir isotherms the fact that the Langmuir premises do not all apply is demonstrated by fig. 1.14b where the curves are replotted as $k_1 p / \theta$ as a function of $k_1 p$; except for $q_1 = q_2$ linearity is not obtained, although in some cases the deviations are small. Typically, fig. 1.14a shows a gradual increase in θ at any adsorption energy difference, rather than steps which would reflect consecutive filling of one type after the other. The curve for $q_1/q_2 = 100$ suggests a plateau at $\theta = 0.5$ because half of the sites are so unattractive as to be virtually inaccessible. However, the fact that θ continues to rise somewhat indicates that this inaccessibility is not absolute.

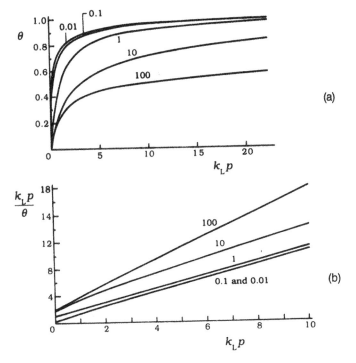

Figure 1.14 Langmuir adsorption on a binary adsorbate. (a) isotherms, (b) "linearized" isotherms. The value of q_1/q_2 in [1.5.16] is given.

1.5c. Mobile monolayer adsorption without lateral interaction (Volmer)

The adsorbate is now visualized as a two-dimensional gas that can move freely tangential to the surface, where the molecules are "caught" in the force field of the surface. They have potential energy (caused by this attraction), vibration energy (normal to the surface) and translational energy. In [I.3.5.15] the following equation was derived for the sub-system partition function of a monatomic adsorbate:

$$q_{mob}(T) = e^{-u_{pot}/kT}\left(\frac{2\pi m k T}{h^2}\right)\frac{e^{-hv/2kT}}{1-e^{-hv/kT}}\,A = C(T)A \qquad\qquad [1.5.17]$$

The factor $(2\pi m k T/h^2)$ is the translational contribution, that with the hv's the vibrational part. The latter also occurs in q_{loc} though in this case only in the normal direction. In the Langmuir treatment the denominator $\exp(-hv/kT)$ was neglected with respect to unity; for mobile adsorption, which is typically encountered at higher temperatures, this approximation is not usually valid. However, for the sake of simplicity, and to maintain a certain analogy to the Langmuir case, we shall assume that the approximation is acceptable and introduce $\Delta_{ads}u = (u_{pot} + \frac{1}{2}hv)$ as the adsorption energy of a (motionless) gas atom to become immobile after adsorption. In other words, the kinetic energy is not counted in $\Delta_{ads}u$. Equation [1.5.17] holds for point-like molecules, each having the entire area A at its disposal. However, we shall now consider the situation where this is not so, i.e. we replace A by $(A - Na_m)$, where a_m is the excluded area per molecule (about twice the real area for disks, depending on the mode of packing). In doing so, the same level of approximation is attained as in a Langmuir adsorbate, where the factor $(1-\theta)$ accounts for the fact that only a fraction of the area is accessible. Considering all of this,

$$q_{mob}(N,T) = \left(\frac{2\pi m k T}{h^2}\right)e^{-\Delta_{ads}U/RT}\left(A - Na_m\right) \qquad\qquad [1.5.18]$$

The canonical partition function of the adsorbate is related to q_{mob} in the same way as it is for a three-dimensional gas [I.3.6.2]

$$Q(N,A,T) = q_{mob}^N/N! \qquad\qquad [1.5.19]$$

The $N!$ accounts for the indistinguishability of the adsorbed atoms. In this case there is no permutation factor $N_s!/(N_s - N)!\,N!$ as in a Langmuir adsorbate.

Substituting [1.5.18] into [1.5.19] and using [1.1.9] for the chemical potential of the adsorbate one finds

$$\mu^\sigma = -kT\ln\left[\left(\frac{2\pi m k T}{h^2}\right)\left(\frac{A - Na_m}{N}\right)e^{-Na_m/(A-Na_m)}e^{-\Delta_{ads}U/RT}\right] \qquad\qquad [1.5.20a]$$

$$\mu^\sigma = -kT \ln\left(\frac{2\pi m k T a_m}{h^2}\right) - \frac{\Delta_{ads} U}{N_{AV}} + kT \ln\frac{\theta}{1-\theta} + kT \frac{\theta}{1-\theta} \qquad [1.5.20b]$$

$$\frac{\mu^\sigma}{kT} = \frac{\mu^{0\sigma}}{kT} + \ln\frac{\theta}{1-\theta} + \frac{\theta}{1-\theta} \qquad [1.5.20c]$$

where θ was introduced for Na_m/A. Equation [1.5.20c] may be compared with the corresponding one for a localized adsorbate, see [1.5.4]; it is seen that the $\mu^{0\sigma}$'s are different (no translational term in [1.5.6]) and that, due to the restriction of the translation with increasing occupancy there is an extra configurational term, $\theta/(1-\theta)$.

Equating [1.5.20b] to μ^G for which we had [I.3.1.10], and introducing the constant

$$K_V = \left[\frac{h^2}{2\pi m(kT)^3}\right]^{1/2} a_m\, e^{-\Delta_{ads} U/RT} \qquad [1.5.21]$$

the following adsorption isotherm equation is obtained:

$$\frac{\theta}{1-\theta}\, e^{\theta/(1-\theta)} = K_V p \qquad [1.5.22]$$

The corresponding two-dimensional equation of state follows immediately from [1.1.11]:

$$\pi a_m = kT \theta/(1-\theta) \qquad [1.5.23]$$

The notion of mobile adsorbates dates back to work by Volmer et al. In 1932 he found that the transfer of gases through pores could, under certain conditions, be up to 1800 times faster than could be accounted for by diffusion only[1]. He suggested the presence of highly mobile gas on the surface. Several years before that, Volmer and Mahnert[2] proposed [1.5.23] on intuitive grounds, based on a resemblance with the three-dimensional Van der Waals equation without interaction, $p(V-nb) = nRT$ for n moles of gas in a volume V (see [I.2.18.26]); this may be compared with its two-dimensional analogue $\pi a_m(1-\theta) = kT\theta$. The referred isotherm equation was proposed by Volmer in a subsequent paper[3]. In agreement with general use, we shall henceforth call [1.5.22] the *Volmer isotherm (equation)* and [1.5.23] the *Volmer equation of state*. The concept of mobility in adsorbates was later advocated by de Boer[4].

[1] M. Volmer, *Trans. Faraday Soc.* **28** (1932) 363.
[2] M. Volmer, P. Mahnert, *Z. Physik. Chem.* **115** (1925) 239, especially p. 251).
[3] M. Volmer, *Z. Physik. Chem.* **115** (1925) 253. In Volmer's original work, and in that of several others, the area correction term is not Na_m but a constant.
[4] J.H. de Boer, *The Dynamical Character of Adsorption*, Clarendon Press, Oxford (1953).

Figures 1.15 and 16 give illustrations and a comparison with the Langmuir model. At low θ, i.e. in the Henry region, mobile isotherms rise more steeply (adsorption is configurationally less restricted) or, equivalently, the pressure is lower. However, at high θ the Volmer isotherm tends to flatten more than the Langmuir one; there it is increasingly difficult to "keep the adsorbate mobile". Correspondingly the pressure rises more steeply. In fig. 1.16 the entropy S^σ of a mobile adsorbate is computed from $Q(N,A,T)$, using the two-dimensional analogue of [I.A6.7]. As expected, for a mobile adsorbate S^σ is higher, asym-

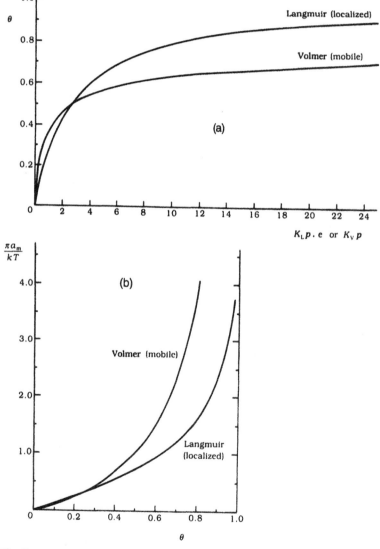

Figure 1.15. Comparison between the Langmuir and Volmer isotherm (a) and equation of state (b). In figure (a) the abscissa axis is scaled by a factor of e to let the two isotherms intersect at $\theta = 0.5$.

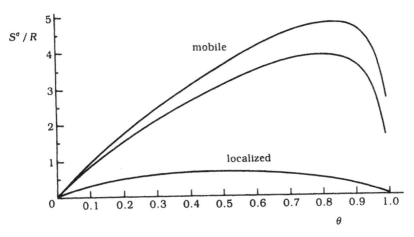

Figure 1.16. Molar entropy (in units of R) of a localized (Langmuir) and a mobile (Volmer) adsorbate. For the former, S^σ is symmetrical in θ, in the latter it is asymmetrical and contains a translational contribution $\theta \ln (2\pi m k T a_m /h^2)$. Parameter values for the mobile isotherms: $a_m = 0.2$ nm^2, lower curve, $M = 28$ ("nitrogen") and top curve, $M = 83$ ("krypton").

metrical with respect to θ and dependent on the nature of the adsorbate because of the translational contribution.

In the Henry region the Langmuir and Volmer isotherms differ only with respect to their affinity constants; this is the only way in which the two modes of adsorption can be distinguished here. In practice the distinction is not always easy because of surface heterogeneity: the most energetical sites are filled first. However, the equations of state are identical (both reduce to $\pi A = NkT$) because such equations do not include adsorbent-adsorbate interactions.

For practical purposes, for a comparison of the curves of figs. 1.15 it is not very easy to decide whether a given set of $V(p)$ data obeys the Langmuir or Volmer model (or neither of them). To that end it is more expedient to choose a mode of plotting that gives linearity for the one but not for the other. By way of illustration fig. 1.17 gives a plot of V^{-1} as a function of p^{-1}. According to fig. 1.12d this mode gives straight lines for the Langmuir case, with a slope determined by K_L and an intercept given by V(mon). The straight line marked "Langmuir" applies to $K_L = 20$ m^{2}N^{-1} and V(mon)$= 0.0113$ dm^{3}g^{-1}. For the Volmer plot the same monolayer capacity was chosen with $K_V = K_L *e = 54.366$. (This is the K_V that in fig. 1.15a leads to isotherm crossing at $\theta = 0.5$.) It is seen that for mobile adsorption linearity is not obtained. Hence, on this basis discrimination is possible. However, in doing so it is noted that the curvature is very strong only in the high pressure end (left) of the isotherm. Suppose that data were available only from $p^{-1} = 20$ m^{2}N^{-1} upwards and subject to some scatter, one could, with high confidence, draw a straight line through the points but with a far too low V(mon) and too high K_V. In conclusion, for discrimination purposes the high pressure region is indispensible.

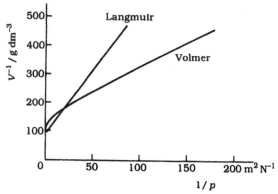

Figure 1.17 Discrimination between Langmuir and Volmer behaviour on the basis of double reciprocal isotherms.

1.5d. Partially mobile adsorbates

As a trend, adsorbates tend to be mobile at high temperature and localized at low temperature. Between these limits there will be transitional states that are difficult to handle because the issue of surface heterogeneity also has to be considered. Whether or not an adsorbed molecule can move laterally depends on the activation energy of such displacements, α. The fraction that can move is proportional to $\exp(-\alpha/RT)$. It is likely that there will be a certain distribution of α over the surface. Because of the exponential dependence, and because the distribution of the parts with low α (patchwise or homogeneous at the site level) also plays an important role, predictions with some general applicability are virtually impossible, as work by Jaroniec et al.[1] has indicated.

Hill[2] idealized lateral transitions by considering the surface as "homogeneous" but with a sinusoidally varying potential energy. At low temperature, the adsorbed molecules can only vibrate in the minima, but with increasing temperature the fraction that can pass the maxima increases. In this model there is a gradual transition between localized and mobile adsorption over the entire (homogeneous) surface.

Another idealization is illustrated in fig. 1.18. Here it is supposed that fractions f_{loc} and f_{mob} of the surface correspond to localized and mobile adsorption, respectively[3]. The adsorptive molecules may then "decide" where they are going. Results depend on the molecular parameters, as accounted for through q_{loc} and q_{mol}. The example given is representative for non-rotating molecules with the mass and area of nitrogen molecules[4]. Figure 1.18a shows that at low θ there is

---

[1] M. Jaroniec, R. Madey, see the references in sec. 1.9c.
[2] T.L. Hill, *J. Chem. Phys.* **14** (1946) 441.
[3] J. Lyklema, *J. Chem. Soc., Faraday Trans. II* **73** (1977) 1646.
[4] C.S. Lee and J.P. O'Connell, *J. Colloid Interface Sci.* **41** (1972) 415, considered three modes of adsorption: localized and mobile over the "mobile" part of the surface, and "mobile" over "sites".

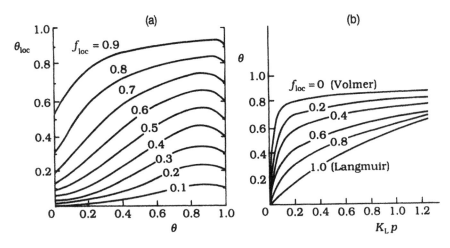

Figure 1.18. Patchwise partially mobile adsorbates. f_{loc} and θ_{loc} are the fraction of the surface on which the adsorption is localized and the fraction of the adsorbate that is localized, respectively. Figure (a) localized fraction, figure (b) isotherms. Parameters in the molecular partition functions correspond to nitrogen atoms, ignoring rotation.

a great preference for the "mobile part" of the surface; fig. (b) illustrates how with increasing f_{loc} a gradual transition from the Volmer to the Langmuir isotherm takes place. There are maxima in the curves of fig. (a) at θ values approaching saturation. The explanation is that under such conditions translation is greatly inhibited whereas localized adsorption still has some configurational degrees of freedom to offer. In other words, these maxima have an entropic background.

Although this analysis is helpful in understanding underlying mechanisms, as an attempt to describe surface heterogeneity it is no more than an embryonic analysis.

1.5e. Monomolecular adsorption with lateral interaction

Adsorbates can become non-ideal for two reasons: non-negligible molecular cross-section and non-negligible lateral interaction. The former non-ideality is automatically accounted for in the Langmuir treatment and this is also the case in our derivation of the Volmer equation. Lateral interaction will now be considered. It is still assumed that the solid surface is ideally flat and homogeneous.

A great deal of this case has already been treated in some detail in sec. I.3.8 and some relevant isotherm expressions and equations of state are included in app. 1. Here we shall briefly review the main results, extending them where appropriate. Most of the treatments are not rigorous, but attempts to improve them tend to follow a law of diminishing returns, because other problems, such as surface heterogeneity, can then no longer be ignored.

At least in lattice statistics, pair interaction energies are usually accounted for by a parameter w, which is positive for repulsion and negative for attraction; for $w = 0$ the situation reduces to the Langmuir case.

For one-dimensional adsorbates exact solutions are possible (sec. I.3.8a) but these are of limited applicability.

Of wider scope is the *Bragg-Williams approximation*, according to which pair interactions do not affect the configurational entropy. It leads to the *Frumkin-Fowler-Guggenheim* (FFG) isotherm equation, derived in sec. I.3.8d, see [A1.5a] with its corresponding two-dimensional equation of state, [A1.5b]. Plots have already been given in figs. I.3.5 and I.3.6. The FFG equation is the simplest one that predicts a two-dimensional phase transition, although inaccurately. Application of the Bragg-Williams approximation to three-dimensional lattices gives rise to an excess Gibbs energy $G_m^E = RT \chi x(1-x)$, [I.3.8.25], where χ is an excess interaction parameter that, in the theory of polymer solutions, is known as the *Flory-Huggins interaction parameter*. Moreover, the formula for G_m^E is almost identical to that in *regular solution theory*. All of this indicates that here one is dealing with models of similar level of sophistication.

Perhaps the best way to apply the FFG equation to experimental data is to plot $\log[\theta/(1-\theta)p]$ against θ. Such plots should be linear with a slope determined by w; the intercept gives K_L. However, this is only possible when $V(mon)$ is known; if this is not so, plots of $V(\log p)$ can be constructed.

As with other equations for monomolecular adsorption, few gas adsorption examples can be found that can be described by the FFG isotherm over the entire pressure range. Applicability over shorter ranges is more likely, and fig. 1.19 is an illustration. These data on the adsorption of krypton on molecular sieves are, according to the authors, a "near-perfect example of localized adsorption on homogeneous surfaces with lateral interaction". To prove this point $\log[p(1-\theta)/\theta]$ is plotted as a function of θ. It is readily verified from [app.1.5a] that this should lead to a straight line whose slope and intercept are determined by zw/kT and $\log K_L$, respectively. The curves are linear up to $\theta \approx 0.5$. However, there is some ambiguity in the establishment of the monolayer capacity because adsorption does not take place on a flat surface, but in cages. In the present example these pores are relatively wide, as could be inferred from the absence of hysteresis. In fig. (b) a lower maximum was assumed ($\theta = 0.75$ corresponds to $\theta_A = 1$) based on a cage filling argument and on data for the isosteric heat of adsorption. If plotted in this way, linearity extends over a longer range of θ_A. The message is that the monolayer capacity is also an adjustable variable. Another possibility that has to be checked is whether perhaps a plot according

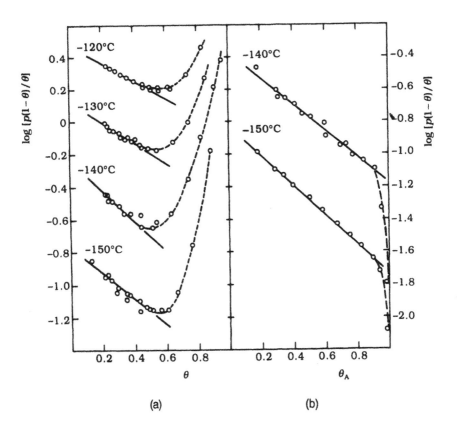

Figure 1.19. Adsorption of krypton on Linde Molecular Sieve 13X. The temperature is indicated. θ_A is a modified degree of occupancy (see text). (Redrawn from L.V.C. Rees and C.J. Williams, *Trans. Faraday Soc.* **60** (1964) 1973.)

to [A1.7a], applicable to *mobile* adsorbates with lateral interaction, would not lead to better linear plots. We found this not to be the case, so that the localized nature appears to be established, at least over part of the θ range. From the slopes it can be derived that the lateral interaction energy term zw is about 0.6 - 1 kT. We return to this issue in connection with fig. 1.20.

A better approach is the *quasi-chemical approximation*, discussed in sec. [I.3.8e]. Here, pairs of nearest neighbours are treated as independent. The isotherm equation [I.3.8.27] is repeated as [app.1.6a] and the two-dimensional equation of state as [A1.6b]. For $w=0$, $\beta=1$ the pair reduces to the Langmuir one, [A1.2a and b]. Figures have already been given (fig. I.3.8 and I.3.9). It is recalled that for $z=4$, the two-dimensional critical point is $w=-1.386$ kT, which is lower than Onsager's "best" value of -1.762 kT but better than the FFG value of -1.000 kT.

Another approach involves *virial expansions*, which have the advantages of being based on some general statistical formalisms (series expansions of grand canonical partition functions) which are applicable to localized and mobile adsorbates. In sec. I.3.8f an elaboration was given for a lattice adsorbate; the results are repeated as [app.1.4a and b]. The factor $B_2^\sigma(T)$ is the *two-dimensional second virial coefficient*. Higher coefficients can in principle be found with the same formalism: they become progressively more complex, implying that the approach is useful only for not too high θ.

For mobile adsorbates, the disappearance of discretization renders the classical statistical approach more appropriate than the quantized one. Now, pair energies are replaced by a continuous function $u(r)$. For $B_2^\sigma(T)$ one has, as the analogue of the three-dimensional variant, (see [I.3.9.12]) the following integral equation:

$$B_2^\sigma(T) = \frac{1}{2} \int\limits_0^\infty [1 - e^{-u(r)/kT}] 2\pi r\, dr \qquad (m^2) \qquad\qquad [1.5.24]$$

For $u(r)$, for instance, the Van der Waals energy could be substituted in the case of non-ideal gases (see [I.4.2.8-11]).

More specifically, let us set ourselves the task of improving the Volmer equation for mobile adsorbates to include lateral interaction. The logic is in the analogy with the three-dimensional Van der Waals equation: [1.5.23] is already a two-dimensional Van der Waals equation of state but without the interaction term. In the three-dimensional case [I.2.18.26] molecular interaction was semi-empirically accounted for by replacing p by $(p + an^2/V^2)$. Let us continue the analogy and introduce the parameter a^σ as the two-dimensional Van der Waals constant. For such an adsorbate, the canonical partition function can be shown to be[1]

$$Q(N, A, T) = \frac{\left[(2\pi m\,k\,T/h^2)(A - Na_m)e^{-\Delta_{ads}U/RT} \right]^N e^{a^\sigma N^2/AkT}}{N!} \qquad\qquad [1.5.25]$$

which is an extension of [1.5.19 and 18]. The SI units of a^σ are [J m^2]. From Q the chemical potential of the adsorbate, the isotherm equation and the two-dimensional equation of state are obtained in the usual way:

---

[1] R.H. Fowler, E.A. Guggenheim, *Statistical Thermodynamics*, Cambridge Univ. Press (1939), 429-43; T.L. Hill, *J. Chem. Phys.* **14** (1946) 441. In *J. Chem. Ed.* **25** (1948) 347, Hill presents a statistical derivation of the three-dimensional Van der Waals equation, which is readily modified to its two-dimensional analogue.

$$\mu^\sigma = - kT(\partial \ln Q / \partial N)_{A,T}$$

$$= - kT \ln\left[\left(\frac{2\pi m k T}{h^2}\right)\left(\frac{A - Na_m}{N}\right)e^{-Na_m/(A-Na_m)}e^{-\Delta_{ads}U/RT}e^{2a^\sigma N/AkT}\right] \qquad [1.5.26]$$

Equating μ^σ to the chemical potential μ^G of the gas [I.3.1.10] leads to the following isotherm equation:

$$\frac{\theta}{1-\theta}e^{\theta/(1-\theta)}e^{-2a^\sigma\theta/a_m kT} = K_V p \qquad [1.5.27]$$

with K_V given by [1.5.21]. The surface pressure $\pi = kT(\partial \ln Q / \partial A)_{T,N}$ is given by

$$\pi = \frac{kTN}{A - Na_m} - \frac{a^\sigma N^2}{A^2} \qquad [1.5.28a]$$

which can also be written as

$$\left(\pi + \frac{a^\sigma N^2}{A^2}\right)\left(A - Na_m\right) = NkT \qquad [1.5.28b]$$

showing the great similarity with the three-dimensional Van der Waals equation of state [I.2.18.26]; hence the name *two-dimensional Van der Waals equations*. Alternatively, [1.5.27 and 28] are called the *Hill-De Boer equation*, after Hill (who laid the statistical foundations) and de Boer (who often used [1.5.28b] in his monograph mentioned in sec. 1.9c). We shall use either name, both for the isotherm and for the pressure.

Equation [1.5.28b] has two parameters that can be adjusted to fit experimental data (a^σ and a_m); [1.5.27] has three, since K_V is in fact also adjustable, although its physical basis is more rigorous. The quality of these equations should not primarily be judged by the ease of fitting data (with three adjustable parameters) but by the reality of parameter values and critical data. Another approach is to consider the analogy to the three-dimensional Van de Waals equation which has been studied in great detail. For instance, in the three-dimensional equation the constant $b = 16\pi a^3/3$, a_m is twice the projected area of an adsorbed molecule, $a_m = 2\pi a^2$, hence $a_m = 2\pi(3b/16\pi)^{2/3}$. Regarding a^σ it is logical to follow the same path of analysis as in the three-dimensional case (compare sec. I.2.18d), i.e. to develop $\pi a_m/kT\theta$ as a series in θ

$$\frac{\pi a_m}{kT\theta} = \frac{1}{1-\theta} - \frac{a^\sigma\theta}{kTa_m} \qquad [1.5.29a]$$

$$\frac{\pi a_m}{kT\theta} = 1 + \left(1 - \frac{a^\sigma}{kTa_m}\right)\theta + \theta^2 + \dots \qquad [1.5.29b]$$

where we used the series expansion $(1-\theta)^{-1} = 1 + \theta + \theta^2 + \dots$. The first term in

[1.5.29b] is the Henry term. It is identical with that for the Langmuir, FFG and Volmer cases. (Note that neither K_L nor K_V enters this term. Hence, in the Henry domain the two-dimensional equation does not discriminate between different mechanisms, although of course the isotherms do.) The factor $(1 - a^\sigma / kT a_m)$ is closely related to $B_2^\sigma(T)$. Specifically the virial expansion can be written as

$$\frac{\pi}{kT} = \frac{\theta}{a_m} + B_2^\sigma(T)\left(\frac{\theta}{a_m}\right)^2 + B_3^\sigma(T)\left(\frac{\theta}{a_m}\right)^3 + \dots \tag{1.5.30}$$

This is the two-dimensional equivalent of [I.2.18.29]. Hence,

$$a_m - a^\sigma / kT = B_2^\sigma(T) \tag{1.5.31}$$

When $u(r)$ is known, integration of [1.5.24] can be carried out. For the common case that $u(r)$ is of the London-Van der Waals type, i.e. $u(r) = -\beta_L / r^6$, the integral can can be split into two parts. The first part goes from $r = 0$ to $r = 2a$, the closest distance of approach of two adsorbed atoms. Here $u(r) = \infty$. The result is $a_m = 2\pi a^2$, that is the excluded area, already used before. When $u(r)$ is so small that only the linear term in the series expansion counts, the second part yields $-\pi\beta_L / 4kT a^4$, which must just equal the second term on the l.h.s. of [1.5.31]. Hence,

$$a^\sigma = -\pi\beta_L / 4 a^4 \tag{1.5.32}$$

relating the two-dimensional interaction parameter in a simple way to the molecular interaction constant.

In fig. 1.20a examples are given for Hill-De Boer adsorption isotherms and in fig. (b) for the surface pressure. The curves for $a^\sigma = 0$ correspond to the Volmer cases. According to this theory, a phase transition can take place and monolayer completion is only slowly attained (but at high θ the theory is not so good any more). At first sight the curves do not differ all that much from their localized counterparts, the FFG isotherms (fig. I.3.5) or the quasi-chemical approximation (fig. I.3.8), so that closer inspection is required to decide which model fits the data best. The location of the critical point offers a better criterion. In the figures it appears to be in the range $6 < 2a^\sigma / a_m kT < 8$. Theoretically it is found as the condition under which the equation of state starts to develop "Van der Waals loops" (compare figs. I.3.6 and I.3.9 for the FFG and quasi-chemical equivalents). From [1.5.29a] after multiplication by θ

$$\frac{d(\pi a_m / kT)}{d\theta} = \frac{1}{(1-\theta)^2} - \frac{2a^\sigma \theta}{kT a_m} = 0 \tag{1.5.33a}$$

$$\frac{d^2(\pi a_m / kT)}{d\theta^2} = \frac{2}{(1-\theta)^3} - \frac{2a^\sigma}{kT \sigma_m} = 0 \tag{1.5.33b}$$

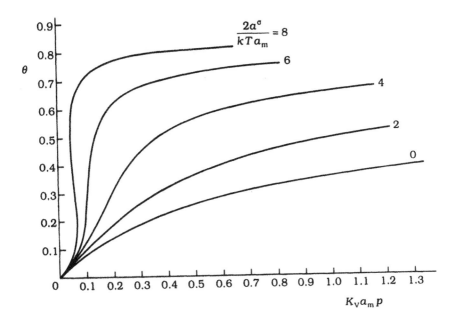

Figure 1.20a. Adsorption isotherms according to Hill and De Boer (two-dimensional Van der Waals isotherms).

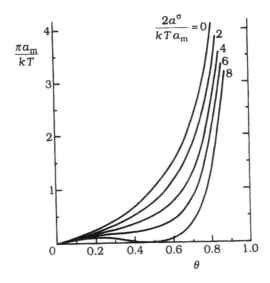

Figure 1.20b. The corresponding equations of state.

from which $(2a^{\sigma}/kTa_m)_{crit} = 6.75$ and $\theta_{crit} = 0.33$ is found. For the FFG case (square lattice) earlier we found $(w/kT)_{crit} = 4$ and $\theta_{crit} = 0.50$, for the quasi-chemical one $(w_{11}/kT)_{crit} = 1.386$ and $\theta_{crit} = 0.50$. Hence some discrimination is possible on this basis. Obviously, this possibility is only viable if the adsorbed molecules attract each other.

How can one in practice discriminate between mobile and localized mono-layer adsorbates with lateral interaction? To that end a comparison of the isotherms of fig. 1.20a with those of fig. I.3.5, or between the corresponding two-dimensional equations of state, figs. 1.20b and I.3.6, is not recommended because the curves resemble each other too much. Rather a plot should be made which shows the discriminating features more strongly.

Figure 1.21 gives such an example. The isotherms (fig. a) are plotted as $\ln[\theta/(1-\theta)]$ as a function of θ. According to [A1.5a] this gives a straight line for localized adsorbates, but not for mobile ones; see [1.5.27] or [A1.7a]. Similar comments can be made about the two-dimensional equations of state (fig. b), where plots of $\pi a_m/kT + \ln(1-\theta)$ as a function of θ^2 are linear for localized adsorbates (see [A1.5b]) but curved for mobile ones ([A1.7c]). Alternatively, one could try $\pi a_m/kT - \theta/\ln(1-\theta)$ which would yield straight lines for mobile adsorbates but curves for localized ones. The plots demonstrate that better discrimination occurs in the region of higher θ's. In this connection the caveat must be made that it is not always easy to establish the monolayer capacity unambiguously, because the plateau is attained very gradually for mobile adsorbates (fig. 1.20a) and for localized ones with repulsive lateral interaction (fig. I.3.5). Uncertainty about this capacity propagates in θ. Also, bi- or multi-

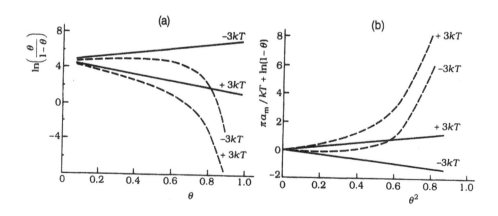

Figure 1.21. Distinguishing between localized (drawn lines) and mobile (dashed curves) adsorbates. Figure (a), adsorption isotherms. Figure (b), two-dimensional equations of state. In figure (a) $K_L = K_V = 100$ m^2 N^{-1}; figure (b) is independent of K. The lateral interaction parameter (in units of kT) is indicated.

layer adsorption should not interfere. At low values of θ all plots become linear; this is readily verified by expanding the ordinate axis unit in a series of θ or θ^2 for figs. 1.21a and b, respectively. Here discrimination is only possible on the basis of numerical values of the constants, but this gives less certainty than the shapes of the curves. Deviations from model behaviour due to surface heterogeneity can only be seen in the isotherms, because two-dimensional equations of state do not involve the adsorbate-vapour equilibrium. Hence to eliminate this possibility it is recommended to study both isotherms and equations of state.

This analysis once again underlines the fact that it is dangerous to use only parts of isotherms under limited conditions as "proof" of the applicability of a given model.

A literature example of a non-ideal mobile adsorbate is presented in fig. 1.22. Qualitatively, the curves resemble some of those in fig. 1.20a. In order to test whether they obey [1.5.27], this equation was linearized as

$$f(\theta, p) = \frac{\theta}{1-\theta} + \ln \frac{\theta}{1-\theta} - \ln p = \ln K_V + \frac{2a^\sigma}{kT a_m} \theta \qquad [1.5.34]$$

and replotted accordingly as fig. 1.23, assuming monolayer coverage to correspond to 39 μmol g^{-1} at 0 and $-19.7°$ and to 40 μmol g^{-1} for the lower temperatures. Straight lines are observed up to $\theta \approx 0.5$. The intercepts reflect K_V and

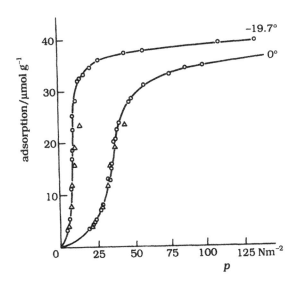

Figure 1.22. Adsorption of carbon tetrachloride on uniform graphite. (Redrawn from C. Pierce, *J. Phys. Chem.* **72** (1968) 1955.)

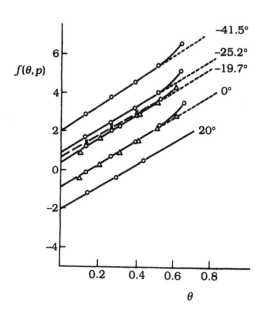

Figure 1.23. Plot according to [1.5.34] for the isotherms of fig. 1.22 including three more temperatures. Open circles are obtained from isotherm data with V(mon) $= 40$ μmol/g for the two low temperatures and 39 μmol/g for the higher temperatures. Open triangles are obtained from isotherm data at 0 and −19.7°C with V(mon) chosen as 50 μmol/g.

depend on $\Delta_{ads}U/RT$, see [1.5.21]. As a function of the temperature, the log of this intercept is linear from which $\Delta_{ads}U$ could be evaluated as −155 J mol^{-1}, well into the realm of physical adsorption. The slopes are virtually temperature independent; they are governed by the kT term in [1.5.34], which, over the temperature range studied, ought to increase by about 15% from the highest to the lowest temperature. Apparently, [1.5.34] works reasonably well, though only over half of the isotherm.

It must be repeated that, regarding the rigour of such analyses, the value of the monolayer capacity is also a critical quantity. If, for instance, it is set at 50 μmol g^{-1}, the plots remain straight till $\theta \approx 0.8$. Given the slow approach to saturation for mobile adsorbates, such a high monolayer capacity is not unreasonable.

Besides finding the best parameter values according to a Hill-De Boer plot, one should of course also try the localized version, i.e. the FFG plot, [A1.5a]. For two values of monolayer capacity (39 and 40 μmol g^{-1}) the data are replotted as $\ln[\theta/(1-\theta)\cdot(p_o/p)]$ as a function of θ, which should yield a straight line. As with the localized case, the isotherms appeared more or less linear if considered over short ranges of θ, although linearity over the entire range is difficult to achieve.

The plots are insensitive for variations in Γ_m. It is concluded that mobility of the adsorbate is likely but that localization cannot be rigorously excluded.

Let us finish with a remark concerning the entropy of a Hill-De Boer adsorbate. As compared with the corresponding situation for the absence of lateral interaction, the log of the canonical partition function has the additional term $a^\sigma N^2/AkT$ (see [1.5.25]). As according to [I.A6.7] $S = k \ln Q + kT(\partial \ln Q/\partial T)_{N,A}$, this term leads to an extra entropy of $ka^\sigma N^2/AkT + kT \, \partial(a^\sigma N^2/AkT)/\partial T$, which is zero if a^σ is independent of temperature. In this respect the Hill-De Boer isotherm relates to the Volmer equivalent in the same fashion as Frumkin-Fowler-Guggenheim relates to Langmuir.

1.5f. Localized multilayer adsorption without lateral interaction (BET)

For gas adsorption on solids, multilayer adsorption is the rule rather than the exception. At the same time the development of models becomes more difficult. One cannot easily see a second layer without adsorbent-adsorbate interaction. Even if the first layer is localized the second and following ones may be mobile. For heterogeneous surfaces the heterogeneity may be reduced, if not completely smoothed out, in the second layer. General, rigorous theories are difficult to develop, but a number of approximations of reasonable practical validity do exist. One of them is a model due to Brunauer, Emmett and Teller, commonly known as the BET theory[1]. Although it is not difficult to identify a number of shortcomings, the model has become popular for practical reasons, including

(i) a number of important isotherm types (types II and III, fig. 1.13) obtain a physical background

(ii) the isotherm equation is simple and can be readily linearized, which is important for practical application

(iii) it is straightforward to obtain from BET theory, the (or at least "a") specific surface area and the (or "a") heat of adsorption

(iv) there are not many competing models that are as "user-friendly".

For multilayer adsorption, the BET model assumes about the same position as that of Langmuir, its monolayer counterpart: adsorption is thought to be localized, the surface is homogeneous and non-porous, lateral interaction is disregarded. The extension is that on one site more than one molecule can adsorb with different affinities, one on top of the other.

---

[1] S. Brunauer, P.H. Emmett and E. Teller, J. Am. Chem. Soc. 60 (1938) 309; S. Brunauer, The Adsorption of Gases and Vapors, Princeton Univ. Press, NJ USA (1943). The abbreviation "BET" was introduced by W.D. Harkins. It is of historical interest that Langmuir himself already anticipated this type of multilayer adsorption, see J. Am. Chem. Soc. 40 (1918) 1316 and D.H. Everett. Langmuir 6 (1990) 1729.

In their original paper, the authors gave a kinetic derivation in which a dynamic equilibrium for each layer was presumed. We shall give a statistical derivation, based on the introduction already given in sec. I.3.6e. Let the adsorbent have N_s sites. On each site 0, 1, 2, 3, ... j, ... ∞ molecules can adsorb. Call the pertaining subsystem partition functions $q(0)$, $q(1)$, $q(2)$, ... $q(j)$, ...$q(\infty)$. It is inherent to the BET picture that for each j only one $q(j)$ is considered; this implies that only one way is allowed by which these j molecules are packed, say as a pile. Otherwise, different q's have to be distinguished for a given j. Let the number of sites covered by 0, 1, 2, ...j, ...∞ molecules be $N_0, N_1, N_2, ... N_j, ... \infty$, with

$$\sum_j N_j = N_s \qquad\qquad [1.5.35a]$$

Also

$$\sum_j j N_j = N \qquad\qquad [1.5.35b]$$

According to sec. I.3.6e the canonical partition function of such an adsorbate is

$$Q(N, N_s, T) = \sum_N N_s! \, \frac{q(0)^{N_0}}{N_0!} \frac{q(1)^{N_1}}{N_1!} \frac{q(2)^{N_2}}{N_2!} \cdots$$

$$= \sum_N N_s! \prod_j \frac{q(j)^{N_j}}{N_j!} \qquad\qquad [1.5.36]$$

where the sum over N stands for the set of all distributions of the adsorbed molecules satisfying [1.5.35a and b]. (In the Langmuir case this sum was absent because if N is fixed so is $(N_s - N)$.) Equations like [1.5.36] can often be conveniently rewritten in terms of a multinominal series of the type [I.3.6.28]

$$\sum_N N_s! \prod_j \frac{(a_j)^{N_j}}{N_j!} = \left(\sum_j a_j\right)^{N_s} \qquad \text{for } \sum_j N_j = N_s \qquad [1.5.37]$$

This reduction is not yet possible with [1.5.36] because of restriction [1.5.35b], but we can get rid of this by considering the system to be open, that is: by changing to the grand canonical partition function:

$$\Xi(\mu, N_s, T) = \sum_N Q(N, N_s, T) e^{\mu N/kT} = \sum_N Q(N, N_s, T) \prod_j e^{\mu j N_j/kT}$$

$$= \sum_N N_s! \prod_j \frac{\left[q(j) e^{\mu/kT}\right]^{N_j}}{N_j!}$$

$$= \left[\sum_j q(j) e^{j\mu/kT}\right]^{N_s} \qquad\qquad [1.5.38]$$

As a digression, recall that in [I.3.5.20] we have introduced the grand canonical equivalent of the subsystem partition function,

$$\xi_j(\mu, T) \equiv \sum_i q(j, i\,T)\,e^{\mu_i/kT}$$

[1.5.39]

for each number adsorbed j. The index i has been used to distinguish the various molecular states from the number of molecules adsorbed. With this definition, [1.5.38] can be written as

$$\Xi(\mu, N_s, T) = \left(\sum_j \xi_j\right)^{N_s}$$

[1.5.40]

which is characteristic for a set of open, but independent subsystems.

Equation [1.5.38] is still fairly general. Except for the premises of localization, surface homogeneity and absence of lateral interaction, we only have the restriction that at given j there is only one type of subsystem, i.e. only one q(j). This generality implies that, for instance, [1.5.38] also applies to the three-dimensional analogue, like water sorption on a non-expanding gel. Furthermore, a number of models are compatible with this expression.

Consider, for instance, the case that each subsystem is "binary" in that it consists of two sites, 1 and 2, with different adsorption energies and, hence, different subsystem partition functions, called q_1 and q_2 if only one molecule adsorbs. Let the maximum number of molecules on the double site be two and let there be an interaction energy between the molecules if there are two of them. The binary sites are independent of each other. This model is a very primitive way to introduce surface heterogeneity and lateral interaction simultaneously. The elaboration runs as follows. In [1.5.38] $q(0) = 1$, $q(1) = q_1 + q_2$ and $q(2) = q_1 q_2 e^{-w/kT}$ where w is the interaction parameter as used in the FFG model. Hence,

$$\ln \Xi = N_s \ln\left[1 + (q_1 + q_2)e^{\mu/kT} + q_1 q_2 e^{-w/kT} e^{2\mu/kT}\right]$$

$$\langle N \rangle = kT \left(\frac{\partial \ln \Xi}{\partial \mu}\right)_{T, N_s} = N_s \frac{(q_1 + q_2)e^{\mu/kT} + 2q_1 q_2 e^{-w/kT} e^{2\mu/kT}}{1 + (q_1 + q_2)e^{\mu/kT} + q_1 q_2 e^{-w/kT} e^{2\mu/kT}}$$

[1.5.41]

By setting $\mu = \mu^G$ in terms of p, the isotherm is obtained, and the surface pressure follows directly as the characteristic function. We shall not elaborate this but now consider the BET model.

In this model, the energetics of the first layer are considered to be different from those in the second and following layers, but the latter are taken as identical. The rationale is that beyond the first layer adsorbate-adsorbate interactions prevail because the influence of the sorbent does not reach so far.

For physical adsorption of apolar molecules this is probably not a bad approximation. The usual picture is that the adsorbate molecules are piled on top of each other, but the model is somewhat less restrictive. The pile does not have to be straight. In fact, the adsorbate-adsorbate enthalpy is commonly identified as the heat of condensation, which does not correspond to a pair interaction but interaction with more molecules. With this interpretation in mind, the necessity of piling molecules on top of each other disappears (molecules can even bridge neighboring sites) but at the expense of losing the connection with the premises of the theory (no lateral interaction) and the rigorous notion of adsorbate-adsorbate pair with the premises of the theory (no lateral interaction). At any rate, mathematically the assumption implies that if we denote the molecular partition functions of molecules in contact with the surface by q_1, those in layer 2, 3, ... are all identical, say q_2. Hence $q(1) = q_1$, $q(2) = q_1 q_2$, $q(3) = q_1 q_2^2$, ... etc. Substitution in [1.5.38] yields

$$\Xi(\mu, N_s, T) = \left[1 + q_1 e^{\mu/kT} (1 + q_2 e^{\mu/kT} + q_2^2 e^{2\mu/kT} + ...)\right]^{N_s}$$

[1.5.42]

$$= \left[1 + \frac{q_1 e^{\mu/kT}}{1 - q_2 e^{\mu/kT}}\right]^{N_s}$$

The following abbreviations are introduced:

$$C = q_1 / q_2$$

[1.5.43]

$$x = q_2 e^{\mu/kT}$$

[1.5.44]

As $q_1 \sim \exp(\Delta_{ads} U/RT)$ and, in the present model, $q_2 \sim \exp(\Delta_{cond} U/RT)$,

$$\ln C \sim \left(\Delta_{ads} U - \Delta_{cond} U\right)/RT$$

[1.5.45]

In passing, [1.5.45] also holds in the kinetic derivation of the BET equation. The quantity x is proportional to the pressure p, hence it is the independent variable in the adsorption isotherm. Introducing C and x in [1.5.42],

$$\Xi = (\mu, N_s, T) = \left[\frac{1 + (C-1)x}{(1-x)}\right]^{N_s}$$

[1.5.46]

$$\langle N \rangle = kT \left(\frac{\partial \ln \Xi}{\partial \mu}\right)_{N_s, T} = x \left(\frac{\partial \ln \Xi}{\partial x}\right)_{N_s, T}$$

$$\theta = \frac{\langle N \rangle}{N_s} = \frac{Cx}{[1 + (C-1)x](1-x)}$$

[1.5.47]

This is the *BET isotherm equation*.

In connection with our earlier remark that Langmuir already anticipated the BET equation it is interesting to note that [1.5.47] can also be written as

$$\theta = \frac{C_L\, x}{1+C_L\, x} + \frac{x}{1-x} \equiv \theta_1 + \sum_j \theta_{j>1} \qquad\qquad [1.5.47a]$$

where $C_L = C-1$ and θ_1 and θ_j represent the degrees of occupancy in the first and j-th layer, respectively. The first term on the r.h.s. is just Langmuir's monolayer isotherm. To this, a term is added formally corresponding to the Raoult equation [I.2.20.15]: the "mole fraction" in the adsorbate beyond monolayer coverage, $N/(N+N_s)$ equals $-\Delta p/p$. Writing for the former $\theta_{j>1}/(1+\theta_{j>1})$ and for the latter $[p(\text{sat}) - p]/p(\text{sat}) = 1-x$, it is immediately found that the fraction adsorbed beyond the monolayer, $\theta_{j>1}$, equals the second term on the r.h.s. of [1.5.47a].

For $x \to 1$, $\theta \to \infty$. This is the *condensation point*, where the pressure reaches the value $p(\text{sat})$. As $x/x(\text{sat}) = \exp\{\mu(p) - \mu[p(\text{sat})]\}/kT = p/p(\text{sat})$,

$$x = p/p(\text{sat}) \qquad\qquad [1.5.48]$$

The equation of state follows from [I.A6.23]

$$\pi a_m = kT \ln\left[\frac{1+(C-1)x}{1-x}\right] \qquad\qquad [1.5.49]$$

In figs. 1.24a and b, BET isotherms and equations of state are plotted, respectively, with varying values of C. The isotherms cover the range between types II and III of fig. 1.13. When C is high enough, the knee-bend B can be located. For high C the part of the isotherm at low x reduces to the Langmuir case. The equations of state differ from all those for monolayer adsorption in that the pressure does not go to infinity for $\theta \to 1$.

For practical purposes [1.5.47] can be linearized as

$$\frac{x}{V(1-x)} = \frac{1}{C V(\text{mon})} + \frac{(C-1)x}{C V(\text{mon})} \qquad\qquad [1.5.50]$$

where, in accordance with common use, the amounts adsorbed are expressed as volumes (stp). (If there is doubt regarding changes in the density of the adsorbate and if a gravimetric method has been used, it may be preferable to work with N and N_s.) The quantity $x/[V(1-x)]$ is known as the *BET transformed*. A plot of this term as a function of x should be linear; from the slope and the intercept one finds C and N_s or $V(\text{mon})$. In practice, situations where $C > 1$ prevail. Then the slope is positive. For very high C the intercept vanishes and the slope becomes equal to $1/V(\text{mon})$; in that case monolayer capacity can be inferred from only one measuring point (the "one-point BET"). For $C < 1$ the slope is negative.

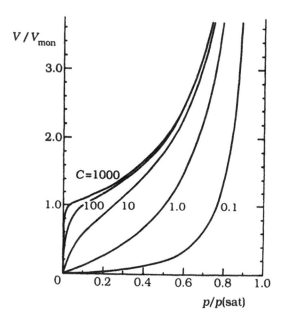

Figure 1.24a. BET adsorption isotherms.

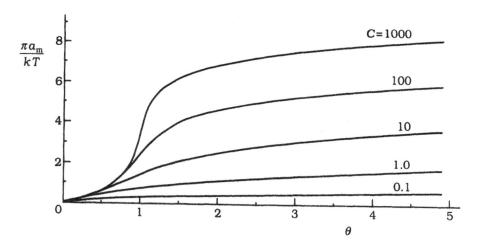

Figure 1.24b. BET equations of state.

The BET equation is widely used to determine specific surface areas A_g of non-porous surfaces. This information is, in turn, used to study porosity, see sec. 1.6. The key is that, despite multilayer adsorption, the monolayer coverage $V(mon)$ can be easily found. Most gas adsorption isotherms on non-microporous sorbents have a linear BET plot in the range $\approx 0.05 < p/p(sat) < \approx 0.3$ including the occupancy $\theta \approx 1$. Assigning realistic values to a_m, from $V(mon)$ one immediately obtains A_g. On closer inspection, establishing a_m is not without ambiguity. According to the model, a_m is the area per site, hence parts of the surface that do not contain sites are not "seen". Positions on a surface, acting as sites for one type of adsorptive, may not be sites for another. Moreover, all imperfections in the model detract in some way from the physical rigour of the notion of monolayer coverage. Because of all these uncertainties, in practice a_m is considered as an adjustable parameter, the value of which is estimated by comparison of the surface area obtained with that from alternative methods. More specifically, experience has shown that for nitrogen at its boiling point (77 K), a frequently used gas and temperature combination, $a_m = 0.162 \ nm^2$ for many adsorbents. This value is somewhere between that of a flat and an upright nitrogen molecule and reproducible within about 20% between a variety of very different adsorbents. On the other hand, for a number of simple and well-defined systems much closer agreements with other techniques are obtained. For the determination of low specific surface areas ($\lesssim 5 \ m^2 \ g^{-1}$) it is, for practical reasons (reduction of the dead space correction), advisable to use adsorptives that at the temperature of the measurement (often ≈ 77 K) have a low vapour pressure, such as krypton or xenon. For these gases the effective molecular cross-sections vary a bit more between different adsorbents; for krypton $a_m = 0.17\text{-}0.23 \ nm^2$ and for xenon $a_m = 0.17\text{-}0.27 \ nm^2$. For argon the measuring range is similar to that for nitrogen and so is the cross-section of the atoms; argon has the advantage that it is more inert. On the other hand, for water a_m varies even more between different sorbents (if A_g is based on a_m for nitrogen), apparently because water adsorption is more specific.

Alternatively, areas can be obtained from isotherms using the "point B method". This point can be read from the isotherm if C is high enough (see fig. 1.13, type II) and interpreted as corresponding to $\theta = 1$. In reality, if the isotherm is twice differentiated to find the inflection point, the latter is found to occur up to 15% above $\theta = 1$.

The conclusion is that the BET method to obtain specific surface areas is semi-empirical. It is always advisable to back up the measurement with an alternative method. Numerous cases are known where such comparisons lead to correspondence; in many others they do not. We shall, henceforth, where appropriate, be specific in indicating areas in terms of the ways they are obtained: *BET(N$_2$) area*, etc.

What has been said about the ambiguity of a_m also applies, *mutatis mutandis*, to the parameter C. According to the model, its interpretation is straightforward (see [1.5.45]), but as in practice the stacking of molecules often does not correspond to that of the model, such deviations reflect in C. Hence, the interpretation becomes somewhat ambiguous; entropic contributions may also come into the picture. For many systems it is found that in the linear BET region the experimental value of C is of the same order as the theoretical value based on $\Delta_{ads}U$ as determined calorimetrically but lower by a factor of up to two.

All these uncertainties are compounded by the heterogeneity of the surface, which is difficult to account for (sec. 1.7), but which is directly reflected in C. Experimentally an average is found. In fact, the empirical observation that [1.5.50] begins to be linear only beyond $x \sim 0.05$ can have its origin in the tendency to occupy very energetical sites below this value.

As with a_m, the conclusion is that C is a semi-empirical parameter.

A host of attempts have been made to improve BET theory, but most approaches have only limited applicability. They include better statistics of the stacking of molecules, account of lateral interaction, or adsorption limited to a certain number of layers to mimic sorption in pores. Regarding this last attempt, isotherms of types IV and V in fig. 1.13 can be predicted, but this is a mathematical rather than physical exercise since limitation is in reality not caused by a spontaneous cessation of adsorption after a given number of layers has been completed. Also in view of the additional complication of surface heterogeneity, the insight gained tends to be proportional to $n^{1/2}$ rather than to n, where n is the number of publications.[1]

1.5g. *Other multilayer models. Semi-macroscopic approaches*

Although over the years the BET model has attained a certain seniority, there has been no lack of alternatives. One automatically thinks of "mobile" equivalents. Multimolecular mobile adsorbates are essentially microscopically thin two-dimensional liquids. The issue is, by how much do these liquids differ from their macroscopic, three-dimensional analogues? It is here that the modelling begins.

The most rigorous approach would be to treat such layers by classical statistical thermodynamics as set forth in sec. I.3.9. An inherent difficulty is that the distribution functions become asymmetrical (different for different

---

[1] For further reading on this and related matters, see e.g. D.M. Young, A.D. Crowell, *Physical Adsorption of Gases*, Butterworth (1962) chapters 5 and 6; S.J. Gregg, K.S.W. Sing, *Adsorption, Surface Area and Porosity*, Academic Press, 2nd ed., (1982), chapter 2, and the references mentioned in sec. 1.9c.

directions). Generally, at equilibrium the chemical potential is the same everywhere, but the way it is composed of the various energetic and entropic contributions changes with position parallel to the surface (say in the x and y directions) and normal to it (in the z-direction). In rigorous statistical treatments this means that formulas for μ have to be modified and supplemented with extra contributions due to the presence of the adsorbent. For instance, in Kirkwood's formula [I.3.9.27] an extra energy term appears; the distribution function and the coupling parameter become anisotropic. Obviously this is not trivial. We shall not pursue this here, except noting that the issue recurs in the interpretation of the surface tensions of liquids (Volume III, chapter 2).

As another approach, one could think in terms of simulations or develop two-dimensional analogues of the semi-empirical equations of state discussed in sec. I.3.9d. Models that fit into the picture of adsorbate mobility ignore variations parallel to the surface, i.e. adopt the mean field approach. Such models have a hierarchy similar to that of the FFG or two-dimensional Van der Waals equations, where any effect that lateral interaction may have on the distribution is also disregarded.

An important category of such models are the so-called *potential theories*, in which near the adsorbate equipotential planes are distinguished as sketched in fig. 1.25. This idea dates back to Polanyi[1]. What is called "potential" is virtually a potential energy, namely that of the adsorbate molecule due to the interaction with the surface. This energy comes on top of the interaction energies it has with its congeners. With increasing z, distances between equipotential planes grow and surface asperities tend to smooth. In this way some account is taken of

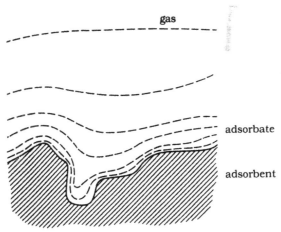

gas

adsorbate

adsorbent

Figure 1.25. So-called equipotential planes near an adsorbent.

[1] M. Polanyi, *Verhandlungen d. Deutsch. Phys. Gesellschaft* **16** (1914) 1012; **18** (1916) 55.

heterogeneity. By modelling the function $u(z)$ and relating u to the amount adsorbed, isotherm equations can be derived.

In the original Polanyi approach the spaces between the various equipotential planes correspond to definite volumes $\phi(z)$, so that there is a function $u = f(\phi)$, called the *characteristic curve*, which is typical for each adsorbent-adsorbate combination and independent of temperature. This picture has, except for the specificity of f, a generic nature and is therefore not only applicable to liquid-like adsorbates but also to the enrichment of gas near the surface, just like nitrogen molecules in the Earth's atmosphere or counterions around a charged colloidal particle.

For a further elaboration we need $u(z)$ and $u(p)$ relations. Regarding the former, it is fairly general to use [1.1.13], here abbreviated to

$$u(z) = -\beta/z^3 \qquad\qquad\qquad [1.5.51]$$

where β is an adsorbent-adsorbate characteristic. Equation [1.5.51] assumes London-Van der Waals interactions to be responsible for the equipotential planes; for not too short distances of approach this is a reasonable approximation. At very short distance the Born repulsion has to be accounted for. In the Lennard-Jones interaction [I.4.5.1] this gives rise to a term $\sim r^{-12}$ for pairs of molecules; for a molecule close to the surface this becomes a z^{-9} term, which is so steep that for all practical purposes the surface may be considered as extending to the minimum in the $u(z)$ curve, with only the attractive part included.

For the relation $u(p)$, one generally writes

$$u(z) = RT \ln p(z)/p(\text{sat}) \qquad\qquad\qquad [1.5.52]$$

where $p(z)$ is the pressure of the vapour in equilibrium with a liquid film of thickness z[1]. The idea behind [1.5.52] is that the energy contribution $u(z)$ to the chemical potential dominates, so that it replaces [1.3.14].

The amount adsorbed follows from the volumes of the equipotential layers that are occupied. Generally, for the mass m_A adsorbed per unit area

$$m_A = \int_0^\infty [\rho(z) - \rho^G] \, dz \qquad\qquad\qquad [1.5.53]$$

For an adsorbed fluid layer of thickness h, $m_A = \rho^L h$ or $V = Ah$ when the density is constant.

The three equations [1.5.51-53], or variants thereof, suffice to formulate

[1] L. Berenyi, *Z. Physik. Chem.* **94** (1920) 628; **105** (1923) 55.

relatively simple isotherm equations. In our case,

$$V = A \beta^{1/3} [RT \ln p(\text{sat}) / p]^{-1/3} \qquad\qquad [1.5.54]$$

Equations of this type appear to fit isotherms of type II (fig. 1.13) quite well, sometimes better than BET theory does. The exponent $-1/3$ stems from [1.5.51]. In practice, values between $-1/3$ and $-1/2$ are usually found. From the viewpoint of dispersion forces this is difficult to account for. Retardation does not play a role and, even if it did, this would further reduce the exponent. Rather, the sum effect of all "hand-waving" approximations (including the assumption of surface homogeneity) leads to a semi-empirical isotherm of the form [1.5.54] in which the constants and exponent are, within certain limits, adjustable. Because of this, the equation is often written in the more general form

$$\left(\frac{V}{V(\text{mon})}\right)^n = \frac{\beta/h(\text{mon})}{RT \ln [p(\text{sat})/p]} \qquad\qquad [1.5.55]$$

where h_{mon} is the thickness of a monolayer and n and β are parameters. This equation is often referred to as the *Frenkel-Halsey-Hill* equation[1]

Another semi-empirical equation, known as the *Dubinin-Radushkevich equation*[2] can be written as

$$\frac{V}{V(\text{mon})} = \exp\left[-b\{RT \ln [p(\text{sat})/p]\}^2\right] \qquad\qquad [1.5.56]$$

where b is an adjustable parameter. This equation also relates the volume adsorbed to a power of $u(z)$. This expression is particularly used in the treatment of porous surfaces (sec. 1.6). Still another similar semi-empirical equation has the form

$$\ln [p(\text{sat})/p] = A - (B/V)^2 \qquad\qquad [1.5.57]$$

It is due to *Harkins and Jura*[3] and was derived on the basis of an empirical two-dimensional equation of state. Here, A and B are constants.

The determination of (specific) surface areas from isotherms of this kind is usually a matter of empiricism because generally the constants cannot be

---

[1] After Ya. I. Frenkel, *Kineticheskaya Teoriya Zhidkostei* (1953), chapter VI. (English transl. J. Frenkel, *Kinetic Theory of Liquids*, Oxford Clarendon Press (1946), also available as a Dover reprint (Dover, 1955)), G.D. Halsey jr., *J. Chem. Phys.* **16** (1948) 931 and T.L. Hill, *Adv. Catal.* **4** (1952) 211.

[2] See M.M. Dubinin, *Adv. Colloid Interface Sci.* **2** (1968) 217, where references to older work can be found.

[3] G. Jura, W.D. Harkins, *J. Chem. Phys.* **11** (1943) 430; W.D. Harkins, G. Jura, *J. Am. Chem. Soc.* **66** (1944) 1366.

rigorously established, and even if that were possible, there is no good explanation for the departure of the power from its theoretical value of $-1/3$ (in [1.5.54]) or $n = 3$ (in [1.5.55]). Empirical areas usually refer to standard surfaces with which the sample under consideration is compared. Adamson[1] gives a table of specific surface areas, and concludes that the correspondence is within about 15% between each other and with the BET area. This is indicative of the accuracy that can be achieved.

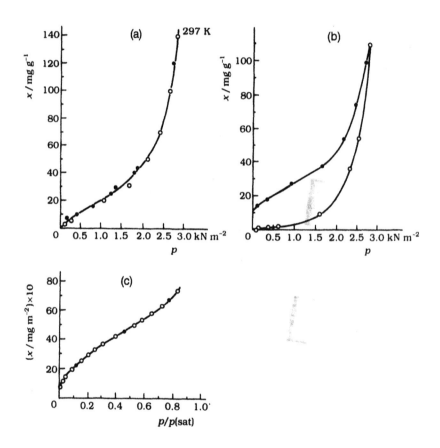

Figure 1.26. Influence of surface treatment on the shape of the isotherm; open circles, adsorption: closed circles, desorption. Adsorbent: Cab-O-Sil. Fig. (a), adsorption of water vapour on the original sample, outgassed at 413 K; fig. (b), the same, after outgassing at 1173 K; fig. (c) adsorption of N_2 at 77.5 K on a sample made hydrophobic by methylation, using chlorosilanes. (Redrawn from P.B. Barraclough, P.G. Hall, *J. Chem. Soc., Faraday Trans. I* **74** (1978) 1360.)

[1] A.W. Adamson, see ref. in sec. 1.9c, table XVI-3.

1.5h. Some examples of multimolecular adsorption

Given the great number of isotherms published in the literature, within the scope of the present chapter we restrict ourselves to a few arbitrary examples, merely meant to illustrate certain points. As shapes of isotherms depend sensitively on the preparation and pretreatment of the sample, especially on the outgassing conditions, these illustrations are not necessarily representative.

In fig. 1.26 the effect of sample pretreatment is illustrated. The original sample is "Cab-O-Sil", a pyrogenic silica. It has a fairly low affinity for water. The isotherm type is between II and III (fig. 1.13). No hysteresis is observed. Stronger outgassing (fig. (b)), further reduces the affinity for water; the curve is now definitely of type II but also shows considerable hysteresis which was attributed to incomplete hydroxylation. In case (c) the surface is made hydrophobic by methylation. The water adsorption isotherm (not shown) remains of type II but as N_2 adsorption is not determined by hydrophilic groups, the corresponding isotherm is of type III. Again, it is hysteresis-free. By application of the theories outlined before, information can be extracted from these isotherms in terms of available areas and enthalpies of adsorption. The authors extended this work with infrared studies.

Figure 1.27 anticipates the discussion of porous materials (see sec. 1.6b and fig. 1.32). One procedure for assessing amounts adsorbed in pores is by comparison with a standard curve on a non-porous sample having otherwise identical surface properties. Two variants exist, the t-plot of De Boer et al. and the α-plot

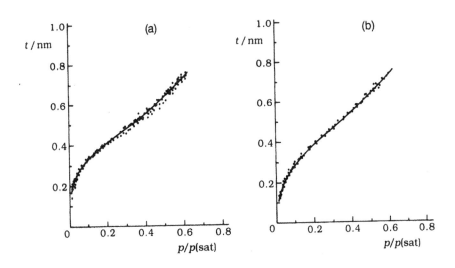

Figure 1.27. t-curve for (a) N_2 isotherms on 22 different solid chlorides and (b) N_2 isotherms on eight different organo-metallic compounds. The BET constant C is between 40 and 70 for cases (a) and between 20 and 30 for cases (b). (Redrawn from A. Lecloux, J.P. Pirard, *J. Colloid Interface Sci.*, **70** (1979) 265.)

of Sing, to which we shall return in sec. 1.6b. Here t is the (average) thickness of the adsorbate layer and $\alpha = V/V_{st}$, where V_{st} is the volume adsorbed at a certain selected standard condition. The problem of selecting the proper t- or α-plot remains under dispute, but various authors have shown that standard isotherms with a wide range of validity do exist. Figure 1.27 illustrates this point. Figure (a) refers to substances such as bulk or supported $AlCl_3$, $CaCl_2$, $MgCl_2$, $CuCl_2$ and $HgCl_2$, fig. (b) to oxygen-containing organo-magnesium and organo-aluminium compounds and some reaction products thereof with oxides or hydroxides. Thicknesses t are computed as $0.354 \, V/V(\text{mon})$ nm. It is seen that there is very little variation between the various samples, although t is slightly higher for the

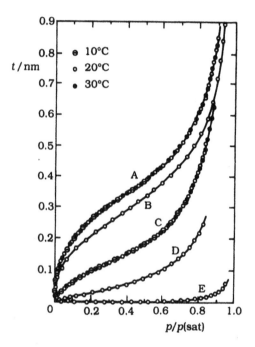

Figure 1.28. Statistical thickness of an adsorbed water film on a number of different non-porous adsorbents. The temperature of the measurements is indicated. Pretreatment: 4h *in vacuo* at 30°C.

Sample	BET (N_2) area/m^2g^{-1}	BET constant C	V(mon) (water) molecules/nm^2
A alumina	125	24-27	7.7-7.9
B alumina	4	28	6.47
C Aerosil 200	201	11	3.52-3.58
D ibid, calcined and hydroxylated	178	5	1.54
E graphitized carbon black	83	5	0.03

(Redrawn from H. Naono, M. Hakuman, *J. Colloid Interface Sci.,* **145** (1991) 405, where further references may be found.)

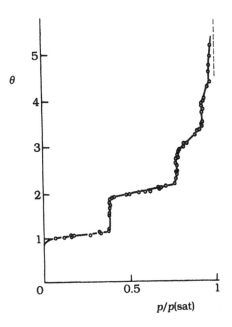

Figure 1.29. Multilayer adsorption of krypton on exfoliated graphite at 77.3 K. (Redrawn from A. Thomy, X. Duval and J. Regnier, *Surface Science Report.* **1** (1981) 1.)

curves of higher C. The corresponding α-curves (not shown) all proved to be identical, within about 5%. This study indicates that working with standard isotherms may be appropriate. The authors used their data to compare the t- and the α-method.

Figure 1.28 is another example of t-isotherms, spanning the entire spectrum of BET-C values. The thicknesses are independent of the temperature and closely follow the $V[p/p(\text{sat})]$ plots of fig. 1.24a.

Figure 1.29 is an example of an isotherm with steps (type VI). Only under specific conditions are such steps observed. As figs. 1.14 and 1.24a demonstrate, they occur neither for monolayer Langmuir adsorption on a binary adsorbent nor for multilayer BET adsorption, not even at extreme differences in the enthalpy of adsorption between the two surfaces, or between the first and the second layer, respectively. However, for monolayer adsorption steps may occur if, due to lateral attraction, two-dimensional phase transitions take place (fig. I.3.7). In the multilayer case, steps may occur if, for instance, the first layer assumes a special stacking or favourable orientation, which grows at the expense of molecules in the second and higher layers. For adsorption from solution, especially with surfactants, such phenomena are not uncommon, but for gas adsorption they are relatively rare, requiring at least very homogeneous surfaces. Apart from giving information on the mode of packing of the adsorbate molecules, and transitions therein, such studies are also relevant for

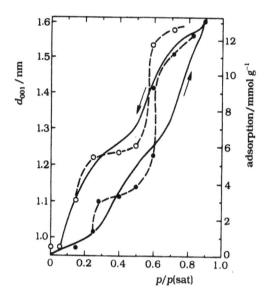

Figure 1.30. Adsorption-desorption of water vapour on a Wyoming montmorillonite (full curves). Also given is the 001 spacing obtained by X-ray diffraction(dashed curves); filled circles, adsorption; open circles, desorption. (Redrawn from J.M. Cases, I. Bérend, G. Besson, M. François, J.P. Uriot, F. Thomas and J.E. Poirier, *Langmuir* **8** (1992) 2730.

wetting phenomena, with the central question of the promotion of second layer formation by the first. If such promotion occurs, the solid is *wetted*, if it is inhibited then wetting is incomplete, the surface plus monolayer is then sometimes called *autophobic*. The authors of fig. 1.29 review a number of such phase transitions especially on lamellar solids.

Figure 1.30 is a typical example of adsorption on a non-inert adsorbent, viz. Na-montmorillonite, a *swelling clay*. Clay minerals consist of stacks of plate-like sheets of aluminosilicates. On exposure to water vapour or aqueous solutions, in some clay minerals, including montmorillonite, penetration of water between these elementary sheets is possible, causing the sample to swell. This is an important phenomenon, for instance in soil science and in drilling muds. Basic colloid-chemical problems are obviously present in that the extent of swelling is determined by the disjoining pressure between the sheets; this pressure is determined by DLVO-type and hydration forces. Figure 1.30 shows the uptake of water vapour, which consists of classical physical adsorption on the external surface of packages of sheets, called *tactoids*, up to about $p/p(\text{sat}) \approx$ 0.16. To that point, the inner layers remain closed, their spacing is 0.96 nm. At higher $p/p(\text{sat})$ massive water uptake occurs in "waves", due to swelling, as evidenced by the parallel lattice expansion. On the basis of this information and other measurements (thermal desorption, heat of immersion) the authors concluded that for $0.16 \leq p/p(\text{sat}) \leq 0.50$ mostly a monolayer of water penetrates, for $0.50 \leq p/p(\text{sat}) \leq 0.90$ mostly a bilayer, etc. As the tactoids are heterogeneous, there are no discontinuities in the adsorbed amounts. Upon desorption, substantial hysteresis occurs but the relation with d persists. To eliminate all swelling water, p/p(sat) has to become as low as 0.05.

Nitrogen adsorption isotherms (not shown) for this sample are of type IV (fig. 1.13); the relatively small hysteresis is now entirely due to capillary condensation between the tactoids and perhaps in pores on their edges. Nitrogen uptake does not lead to swelling.

Generally speaking, a variety of non-inertnesses may be encountered. Besides the type just discussed, in some cases the gas may dissolve in the solid; this is for instance the case with hydrogen and palladium, but this example does not stand on its own.

1.6 Porous surfaces

The presence of pores in a number of adsorbents is at the same time an important physical reality, a nuisance, and a challenge. The nuisance stems from the impossibility to apply the various theories that have been derived for well-defined homogeneous surfaces. The reality is that a host of industrial materials are porous: bricks, tiles, food products, textiles, paper, pigments, zeolites, heterogeneous catalysts, and porous rocks as encountered in enhanced oil recovery. The challenge is to determine sizes and shapes of pores in order to better characterize such adsorbates, to understand the properties of liquids and gases confined to restricted volumes and to explain the selectivities that are often caused by gas uptake in pores. Because of the general complexity of the phenomena, the present-day insight is limited. Some progress is being made by investigating surfaces having "controlled" pores (with respect to sizes and shapes). Such surfaces include synthetic zeolites, molecular sieve carbons, prepared from various polymeric precursors, certain oxides and controlled-pore glasses. They serve as models for those generally encountered in practice which often contain pores of a variety of sizes and shapes.

Gas adsorption is most widely used to assess porosity, especially by analyzing the hysteresis loops appearing in the isotherms due to capillary condensation in pores (fig. 1.13 types IV and V). However, there are a number of alternatives, including mercury porosimetry, neutron and X-ray scattering.

1.6a. Classification

For further discussion it is expedient to systematize pores (with respect to their sizes) and hysteresis curves (with respect to the sizes and shapes of the loops). Regarding the surface, it is often useful to distinguish between the *internal* and *external surface*. Regrettably, there is no unambiguous method of distinguishing between them, because different procedures may yield conflicting results. The external surface is generally regarded as the envelope surrounding the particles or agglomerates thereof; all prominences and wider pores also belong to it; the inner surface then remains reserved for the area in the narrow pores.

In the context of physisorption, the following pore size classification is recommended by IUPAC (see sec. 1.9a):

(i) *macropores*, $\geq$ 50 nm

(ii) *mesopores* $\approx$ 2 nm – $\approx$ 50 nm

(iii) *micropores* $\leq$ 2 nm

These limits are somewhat arbitrary: pore filling mechanisms also depend on the shapes of the pores and on the size of the adsorptive molecule. Despite this inherent vagueness, the classification has its use as a first means of discrimination because it points to different pore filling mechanisms: macropores are so wide that they behave as "virtually flat" surfaces, mesopores are mainly responsible for capillary condensation, whereas micropores are so narrow that one cannot speak of a macroscopic fluid in them. Because in *micropore filling* adsorbates are only a few layers thick, an adsorption plateau is found suggesting monolayer filling and applicability of the Langmuir or Volmer premises. This mechanism is distinct from that in meso- and macropores.

As well as in terms of their sizes, pores can also be classified on the basis of their shapes (cylinders, cones, slits, etc.) and on their *connectivity*: pores may be open on only one side, but they may also run from one side of the adsorbent to the other and thereby be connected to each other.

Experimentally one can study complete hysteresis loops or *scan* them: going back and forth through parts of the loops to obtain information on the pore distribution.

In fig. 1.31 the IUPAC-recommended classification is given for the most common types of hysteresis loops; they are refinements of the general type IV in fig. 1.13. In practice, a wide variety of shapes may be encountered of which types H1 and H4 are the extremes. In the former, the two branches are almost vertical and parallel over an appreciable range of V, whereas in the latter they remain more or less horizontal over a wide range of p/p(sat). Types H2 and H3 are intermediates. Many hysteresis loops have in common that the steep range of the desorption branch leads to a closure point that is almost independent of the nature of the porous sorbent and only depends on the temperature and the nature of the adsorptive. For example, it is at p/p(sat) = 0.42 for nitrogen at its boiling point (77 K) and at p/p(sat) = 0.28 for benzene at 25°C.

Accepting that the effects of various factors on adsorption hysteresis are not fully understood, experience has shown that there is probably a certain correlation between the structures of the pores and the shapes of hysteresis loops. Type H1 has been identified with particles, from other evidence known to consist of *consolidated agglomerates* (rigidly joined particles; loose aggregates of particles could disperse under the influence of uptake of the adsorbate). This type is also found for compacts of approximately uniform spheres in a more or less regular array, i.e. having a fairly narrow pore size distribution. Type H2 is found for many porous adsorbents, including inorganic oxide gels and porous glasses. However, for such systems the pore size and shape distributions are not well defined and such loops are difficult to interpret. In the past this type has been interpreted in terms of so-called 'ink bottle' pores (wide body, narrow neck), but it is now recognized that this is an oversimplified picture: the role of pore-networks also has to be considered. Type H3 loops, which do not exhibit any limiting adsorption at high $p/p(\text{sat})$, are associated with loose aggregates of plate-like particles, giving rise to slit-shaped pores. Similarly, H4-type pores are identified as caused by narrow slit-like pores, but now V remains finite at $p/p(\text{sat}) \rightarrow 1$ (microporosity).

The dashed curves in fig. 1.31 reflect *low pressure hysteresis* which may be

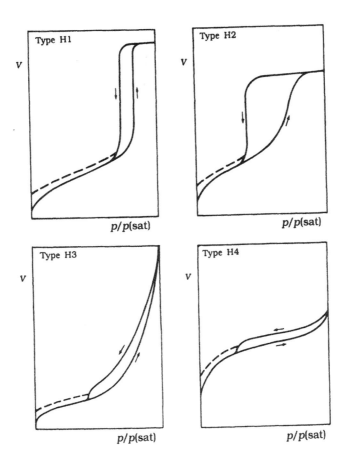

Figure 1.31. Classification of hysteresis loops.

observed down to the lowest attainable pressures. When this occurs, degassing of the adsorbent is possible only at higher temperatures. Such hysteresis may be associated with the swelling of non-rigid pores or with the irreversible uptake of molecules in pores of about the same width as that of the adsorptive molecule. Chemisorption, if irreversible, may also lead to this phenomenon. More generally, substantial hysteresis may point to chemisorption, especially when the loops do not close.

The conclusion is that the loops give preliminary information about the types of pores that may be present. Closer inspection of adsorbate take-up in the various pores is required, if only to establish the 'inner' and 'outer' area.

1.6b. *Assessment of mesoporosity*

The first question to ask is: what are the properties by which the texture of a (meso-)porous surface can be characterized? The next is: how may these properties be estimated?

In the most general case, virtually nothing may be known about the sample. It may have a variety of pores, varying in their sizes, shapes, (cones, cylinders, slits, irregular, with or without blocked entrances, etc.); they may be separate or connected. Porosity depends on the method of preparation. Some adsorbents are well defined, but others, say pigment powders obtained by milling, will generally consist of very irregularly shaped aggregates that may or may not be consolidated. Some preferences for the crystal planes of lower indices may be observed, especially after ageing, because of Ostwald ripening (sec. I.2.23c). Most generally, one wishes to find the complete distributions with respect to sizes and shapes. Clearly this is often an impossible task, and to proceed some empiricism is required. For instance, the problem can be reduced to determining the *pore volume* V_p as a function of some *effective pore radius* a_p regardless of the shapes, or a_p is sought as a function of $p/p(\text{sat})$, or a certain fractal structure may be assigned to the surface, characterized by a fractal dimension, disregarding any information about the history of the sample. Although such reductions may seem Procrustean, they allow simple physical or mathematical models to be applied, which, besides suggesting some insight, function at least as an empirical characterization. We shall now discuss some of these approaches, restricting ourselves to *open pores* (cavities or channels communicating with the surface) and to inert adsorbents (not collapsing under the influence of penetration of a gas or liquid). The *total pore volume* $V_p(\text{tot})$, including all three categories of pore sizes, is often derived from the amount of vapour adsorbed at $p/p(\text{sat}) \rightarrow 1$, assuming that all pores are then filled and the internal area is large as compared with the external one. Macropores are filled near the limit $p \rightarrow p(\text{sat})$, but we now emphasize the mesopores.

In sec. I.6.4f the *hydraulic radius* a_h was introduced for porous bodies as the ratio of the void volume and the void area. We can retain this parameter as a characteristic of pore size. For a cylindrical pore of length ℓ and radius a, $a_h = \pi a^2 \ell / 2\pi a\ell = a/2$, i.e. half the radius; for a slit-shaped pore of area A and width d it is $Ad/2A = d/2$, half the slit width. The *pore size distribution* can now be given as $V(a_h)$ and/or as $a_h(p)$, realizing that a given class of pores will be filled at a given pressure.

Adsorption in mesopores is usually characterized by pore condensation and hysteresis. This can be due to either the occurrence of a metastable state or the presence of a pore network. We will first treat the former mechanism and discuss the network model later.

In mesopores a multilayer film will be adsorbed at the pore wall as the saturation pressure is approached. The stability of this film is determined by the interaction with the wall, e.g. long-range Van der Waals interaction, and by the surface tension and curvature of the liquid-vapour interface. Saam and Cole[1, 2] have advanced a theory, showing how the curved film becomes unstable at a certain critical thickness $t_c = a - r_c$. The adsorption process is shown schematically in fig. 1.32a (1) → (3). During desorption (4) → (6) an asymmetrical state exists in which a partially filled pore is in equilibrium with a film at the metastable thickness $t_m = a - r_m$. As $r_c < r_m$ during adsorption (1) → (2) the adsorbed film will be metastable between r_c and r_m, i.e. the asymmetrical state is energetically more stable than the symmetrical state of a thicker film. Accordingly, emptying a completely filled pore occurs at a lower pressure than pore filling (fig. 1.32b).

As a model[3] we assume an ideal infinite pore with radius a (fig. 1.32a). The liquid in the pore is not identical to that of the bulk because it is influenced by the interaction with the wall. The molar energy U_m depends in some way on the distance to the wall, say $U_m = U_m(r)$, e.g. according to a Lennard-Jones relationship. The pressure difference across the interface is given by the Laplace equation

$$p^L - p^G = \frac{\gamma}{r}$$
[1.6.1]

The chemical potentials in the vapour and liquid phases are given by

$$\mu^G = \mu^{oG}(p^G(\text{sat}), T) + RT \ln\left(\frac{p^G}{p^G(\text{sat})}\right)$$
[1.6.2]

and

1) M.W. Cole and W.F. Saam, *Phys. Rev. Lett.* **32** (1974) 985.
2) W.F. Saam and M.W. Cole, *Phys. Rev. B.* **11** (1975) 1086.
3) A. de Keizer, T. Michalski and C.H. Findenegg, *Pure Appl. Chem.* **63** (1991) 1495.

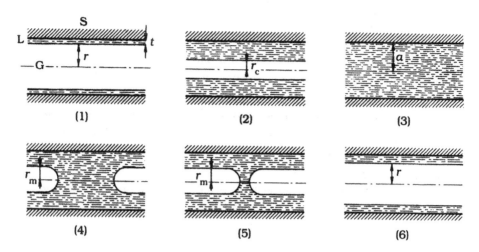

Figure 1.32a. Schematic representation of the adsorption (1) → (3) and desorption (4) → (6) of a fluid from the gaseous phase in a cylindrical pore with radius $a - t$: (1) stable adsorbed film with radius $a - t$; (2) multilayer adsorbed film at the unstability limit r_c; (3) completely filled capillary; (4) unsymmetrical state of a partially filled pore at the metastability limit r_m; (5) further desorption at the metastability limit r_m; (6) stable film with radius r.

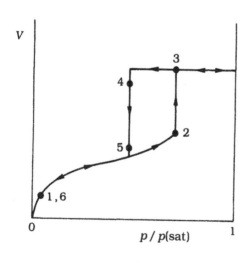

Figure 1.32b. Adsorption and desorption isotherms. The numbers correspond to the adsorption stages in fig. a.

$$\mu^L = \mu^{oL}(p^G(\text{sat}), T) - \frac{\gamma}{\Delta \rho}\left(\frac{1}{R_1} + \frac{1}{R_2}\right) + U_m(r) \qquad [1.6.3]$$

respectively, in which we neglect the interactions between gas and wall and assume ideal behaviour of the vapour phase. Here $p^G(\text{sat})$ is the saturated vapour pressure and $\Delta \rho$ is the difference in the densities of liquid and vapour. R_1 and R_2 are the principal radii of curvature at position r. Let us write $p^G = p$. At equilibrium $\mu^G = \mu^L$ and from [1.6.2 and 3] an "extended" Kelvin equation can be derived:

$$RT \ln \frac{p}{p(\text{sat})} = U_m(r) - \frac{\gamma}{\Delta \rho} \left(\frac{1}{R_1} + \frac{1}{R_2} \right)$$

[1.6.4]

In the symmetrical state $R_1 = \infty$ and $R_2 = r$. The cylindrical interface in the asymmetrical state with $R_1 = \infty$ and $R_2 = r_m$ is in equilibrium with the curved interface, confining the partially filled pore, at every position r. Variations in $U_m(r)$ at the different of the interface are compensated by the Laplace term. The interface between the partially filled pore and the cylindrical surface is not spherical, i.e. $R_1 \neq R_2$.

In the present model, equations for the instability limit r_c and the metastable limit can be derived as follows. According to [1.6.3] μ^L depends on the radii of curvature of the interface. In the symmetrical state, increasing p leads to a thicker adsorption layer. However, at a given thickness $t_c = a - r_c$, $\partial \mu / \partial t$ becomes zero, the layer becomes unstable and the pore fills spontaneously. Upon desorption the process proceeds according to fig. 1.32a (3) → (6). The pore is emptied at a fixed metastable thickness $t_m = a - r_m$. The metastable limit can be obtained by minimizing the free energy for the unsymmetrical state. In the metastable limit the radius is always larger than that in the instability limit r_c. The dependence of r_c and r_m on a dimensionless variable a/a_0, assuming Lennard-Jones interactions between liquid-liquid and liquid-solid, is given in

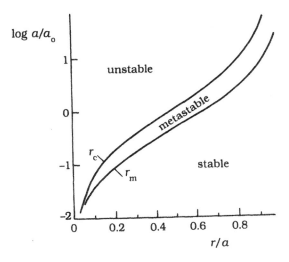

Figure 1.33. Dependence of the unstable limit r_c/a and the metastable limit r_m/a on the dimensionless variable a/a_0 according to the theory by Saam and Cole[1,2].

[1] M.W. Cole and W.F. Saam, *Phys. Rev. Lett.* **32** (1974) 985.
[2] W.F. Saam and M.W. Cole, *Phys. Rev. B.* **11** (1975) 1086.

fig. 1.33. Here $a_0 = (3\pi \alpha \Delta\rho/\gamma)^{0.5}$ where α is the gas-substrate Van der Waals coefficient defined by $\alpha = (\rho^S \beta_{SL} - \rho^L \beta_{LL})/6$, β being the Van der Waals interaction parameter according to [I.4.2.10]. Weak interaction between liquid and substrate or a high surface tension lead to the formation of only thin stable films. On the other hand, for strong interaction and for low surface tension, thick stable films can be formed.

Let us for the moment assume a thin cylindrical adsorbed layer in contact with a hemispherical meniscus, $R_1 = R_2 = R$, see fig. 1.32a (1) neglecting $U_m(r)$. The thickness of the adsorbate layer on the pore walls in this connection is sometimes called t (see below, the "t plot"), although generally we call it h. In the figure it is assumed that the liquid completely wets a thin adsorbed layer of the same molecules. As $R = a - t = 2a_h - t$,

$$a - t = \frac{2\gamma^L V_m^L}{RT \ln(p(\text{sat})/p)} \qquad\qquad [1.6.5]$$

Similarly, for slits $R_1 = d/2 - t = a_h - t$ and $R_2 = \infty$, hence

$$d - 2t = \frac{2\gamma^L V_m^L}{RT \ln(p(\text{sat})/p)} \qquad\qquad [1.6.6]$$

Values of t are obtained from adsorption data of the same adsorptive on a non-porous surface of the same nature, as in fig. 1.28. Substituting bulk values for γ^L and V_m^L, the pore size distribution $a(p)$ or $d(p)$ is obtainable. The occurrence of hysteresis implies that this gives different results for the two branches (ascending and descending) of the curve. In fact, the difference between the two metastable states is a characteristic of the type of pores. For non-connected pores, usually the downward curve is analyzed, because then the menisci have already been formed but for connected networks the ascending one may be more appropriate. After each stepwise change of p, the radius is calculated and from that the exposed pore volume and pore area. This yields a cumulative distribution which, if so desired, can be differentiated.

One empirical procedure to render this more explicit is the *t-plot method* of Lippens and de Boer[1], providing a simple means of analyzing nitrogen isotherms. According to this method, V is plotted against t, obtained from the corresponding reference curve for adsorption on the same, but non-porous, surface. Plotted as $t(p/p(\text{sat}))$ such a t plot is almost identical for the same adsorptive (nitrogen) on a variety of non-porous surfaces; see figs. 1.27 and 1.28.

---

[1] B.C. Lippens, J.H. de Boer, *J. Catalysis* **4** (1965) 319; J.H. de Boer, B.C. Lippens, B.G. Linsen, J.C.P. Broekhoff, A. van den Heuvel and Th. J. Osinga, *J. Colloid Interface Sci.* **21** (1966) 405.

For non-porous surfaces a $V(t)$ plot is of course a straight line through the origin. Upward departures from linearity in the multilayer range are interpreted as caused by capillary condensation in mesopores. Figure 1.34 gives an illustration. The method is empirical because an assumption has to be made regarding the reference curve. Lippens et al[1] set $t = 0.354\ V/V(\text{mon})$, where $V(\text{mon})$ was obtained by the BET equation [1.5.50], which restricts the rigour.

For this reason Sing[2] modified the method by replacing t by $\alpha = V/V_{st}$, where V_{st} is the adsorbed volume at a selected standard pressure. This method is

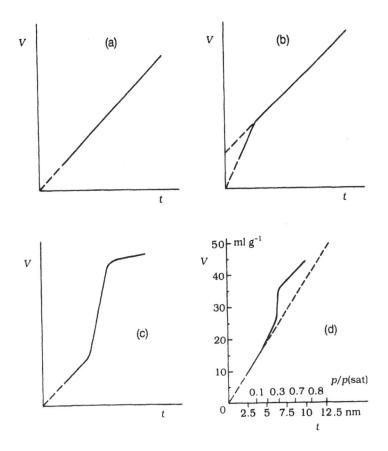

Figure 1.34. Examples of t-plots. (a) Flat homogeneous surface (for nitrogen $t \approx 0.354$ $V/V(\text{mon})$); (b) the same outer surface as in (a) but with some micropores which fill first; (c) ibid, but now with mesopores, giving rise to a sudden upturn. The lower part has the same slope as in figure (a), the upper part with low slope reflects the external surface; (d) practical example: nitrogen on BDH silica (T.F. Tadros, J. Lyklema, *J. Electroanal. Chem.* **17** (1968) 267), indicating two kinds of pore filling before the external surface (which may contain macropores) is occupied. The initial part of the plot is usually not considered.

---

1) B.C. Lippens, B.G. Linsen and J.H. de Boer, *J. Catalysis* **3** (1964) 32.
2) S.J.Gregg, K.S.W. Sing, *Adsorption, Surface Area and Porosity*, Academic Press, 2nd ed. (1982).

now known as the α-method. In this way using the BET monolayer capacity is obviated and replaced by an empirical isotherm. The choice of the standard pressure is not entirely arbitrary; for many systems setting $a = 1$ at $p(\text{sat}) = 0.4$ has been found convenient. Also for the α-method the availability of a well-characterized reference solid is mandatory.

A different, but also widely, used method for obtaining pore size distributions is known as *mercury porosimetry*. Unlike the previous procedures, which are based on the attraction of an adsorptive by the surface, the non-wetting of (most) adsorbents by mercury is now involved. An overpressure is therefore needed to push mercury into the pores; the narrower the pores, the higher the required pressure. By stepwise raising the pressure, and measuring the volume of mercury intruding into the pores, again a pore size distribution is obtained. Automated equipment is now available but unambiguous interpretation is not straight-forward: problems arise with establishing pore shapes, with the mercury-surface contact angle and its hysteresis, the surface tension of the mercury under the prevailing conditions and insufficient mechanical resilience of pore walls against the mercury overpressure. Problems also arise for systems having pores that are connected (forming "networks of pores") and/or may run from one side of the sample to the other; in that case the intrusion will, to a large extent, be determined by the shapes of voids and constrictions ("throats"). In addition, intrusion and extrusion exhibit hysteresis; part of the mercury may be irreversibly trapped. In chapter 5 of Volume III we intend to return to the wetting of surfaces.

In this connection, the determination of the volumes of open pores running from one side of the solid to the other, by gas or liquid flow, may also be recalled. The procedure has been described in sec. I.6.4f. From the flux, the average hydraulic radius or the (internal) surface area can be assessed. This permeability approach is the counterpart of porosimetry, as far as the latter is particularly suited for pores open at one end only.

Other techniques for assessing pore sizes are X-ray and neutron scattering, light scattering (transparant porous glasses or silicas may become opaque during desorption) and NMR. The latter applies to pores filled with liquid. The spin-lattice relaxation time T_1 of, say water, in a confined geometry is increased due to the proximity of the walls. Essentially the ratio between bulk water and water in contact with the pore walls is measured. The advantage that wet pores are studied is partly offset by the difficulty in deconvoluting the signal for heteroporosity.

For the sake of completeness it is noted that many of the surface characterization methods treated in sec. 1.2 also apply for assessing porosity. This applies in the first place for atomic force microscopy (AFM), by which, in

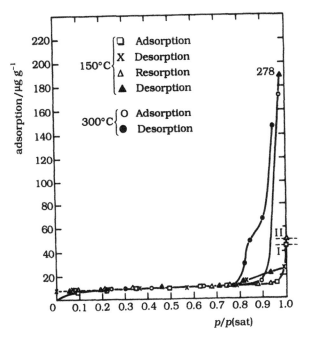

Figure 1.35. Adsorption-desorption-resorption of methanol on a lunar soil sample; outgassing conditions given. I, II, first, second cycle. (Redrawn from R.Sh. Mikhail, D.A. Cadenhead, *J. Colloid Interface Sci.* **61** (1977) 375.)

principle, different classes of pores can be seen if it is operated at different magnifications. As a contrast to physical adsorption methods, AFM is typically applied to small areas. The limitation is that by AFM one cannot "look deeply into the pores". Small angle scattering of X-rays and neutrons is nowadays also used to study porous solids, including gels and ceramic materials; hence porous surfaces can also be investigated. We have already discussed these techniques in sec. I.7.9.

Figure 1.35 is an experimental illustration of adsorption on a porous surface and is, in a sense, a curiosity since it deals with a lunar dust sample, returned to Earth by the Apollo-15 mission. Mineralogically, this sample consists mainly of pyroxene. Lunar dust has a porous structure, formed by tens of millions of years of exposure to a rarified environment, solar wind, solar flare and bombardment by micrometeorites. Exposure to gases readily leads to structural changes; the substrate is not inert. It is particularly susceptible to water, but methanol sorption also gives rise to hysteresis phenomena, as the figure illustrates. The authors interpreted the hysteresis loops at high p(sat) in terms of capillary condensation in slit-shaped pores; this is in agreement with the interpretation of H3 isotherms of fig. 1.31, see sec. 1.6a.

1.6c. *Assessment of microporosity*

Micropores are so small that the close proximity of the pore walls gives rise to an enhanced adsorbent-adsorbate interaction. Because of this, such pores are filled at low p(sat) (*micropore filling*). As stated, the distinction between micro- and mesopores is difficult to specify exactly. Micropore filling is the appropriate term for filling by the primary process of adsorptive-adsorbent interaction, its mesopore equivalent proceeding by the secondary process of capillary condens-ation. It is likely that there are two micropore filling mechanisms. The first involves the entry of individual adsorbate molecules into very narrow pores, it takes place at low p(sat). After this, at somewhat higher p(sat), a cooperative process may take place, involving interaction between adsorbate molecules.

According to present day insight type I isotherms (fig. 1.13) are not represent-ative of a Langmuir-like mechanism, but of micropore filling (in the initial steep part) followed by multilayer adsorption on the small external surface (plateau with a low slope). Therefore, the Langmuir, BET and other equations may not be used to determine the surface area; at best, something can be said about the external area. On the other hand, if the relatively small amount adsorbed on the external surface is subtracted from the total amount (and the absence of mesopores has been ascertained) the remaining amount may be identified as that present in micropores.

For micropores the t-plot method (see sec. 1.6b) does not work quantitatively because thicknesses of multilayers are not suitable for purposes of comparison although the presence of micropores can be inferred, see fig. 1.34b. In this respect the α-method may be somewhat better, but it has the similar drawback that isotherms on non-porous surfaces are not necessarily good references. Con-version of the micropore capacity into a micropore volume requires an assump-tion regarding the density of the adsorbate. As, for micropores, this parameter is elusive, one can at best achieve something like an "effective" micropore volume. The notion of surface area in micropores is a concept that has little real significance and we shall not use it. At this level, shapes of molecules also start to become important. For instance, it has been reported[1] that for a number of microporous carbons the uptake of nitrogen is relatively high, compared to other adsorptives having molecules of similar sizes. Perhaps the best recom-mendation is to compare a number of different adsorbates. *Pre-adsorption* is a variant in which the micropores are first filled with large molecules that are not removed by pumping at ambient temperatures, after which the external area is determined with another adsorptive. Another approach is to perform studies with surfaces of well-defined microporosity.

---

[1] P.J.M. Carrott et al., *Characterization of Porous Surfaces I*, K.K. Unger, J. Rouquerol, K.S.W. Sing and H. Kral, Eds., Elsevier (1987), p. 89.

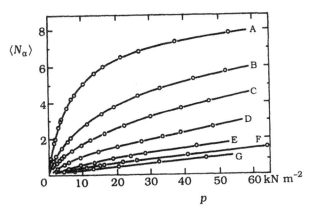

Figure 1.36. Adsorption of methane in the α-cavities of Linde Molecular Sieve 5A. The average number of molecules per cavity $\langle N_\alpha \rangle$ is given for various temperatures: A, 194.65 K; B, 218.15 K; C, 233.15 K; D, 243.15 K; E, 273.15 K; F, 288.15 K and G, 300 K. (Redrawn from H.J.F. Stroud, E. Richard, P. Limcharoen and N.G. Parsonage, *J. Chem. Soc., Faraday Trans.* **72** (1976) 942.)

Typical examples of microporous sorbents are molecular sieves and other well-defined zeolites. See fig. 1.36. Characteristically, sorption is, in the pressure range studied, limited to a few layers, hence isotherms are of the type I category. The authors interpret their data with a variant of the theory presented in sec. 1.5f, starting with an equivalent of [1.5.38] for confined geometries.

1.6d. *Properties of fluids in confined spaces*

In the discussion so far, no molecular picture has been given regarding the fluid (particularly if it is a liquid) inside narrow pores. Because of the proximity of the pore walls, its properties are generally different from those in a macroscopic bulk phase. The issue of the properties of liquids near surfaces is encountered throughout this book; it is, among other places, important in adsorption from solution on uncharged and charged surfaces (chapters 2 and 3), in the interpretation of the surface tension of liquids (Volume III, chapter 2) and in capillarity (Volume III, chapter 5). If, in addition to the practical aspects of pore characterization, progress is to be made in the physical understanding, a study of such fluid properties is necessary. Partly due to the advent of powerful simulation techniques (especially Monte Carlo and Molecular Dynamics) and the preparation of surfaces with controlled porosity, valuable developments have been achieved. Here this matter will be introduced, emphasizing liquids in pores.

The first question is one of definition: how is liquid structure defined? This problem has been addressed in sec. I.5.3c and I.5.4, emphasizing water and solvent structure-originated interactions, including hydrophobic bonding. There it was indicated that statistical methods are the most promising. The notion of bulk liquid structure can be made quantitative in terms of *radial distribution functions* $g(r)$ (for simple liquid), direction-dependent distribution functions $g(r, \theta)$ or higher order distribution functions $g^{(h)}(r_1, r_2, ...)$ (for more complicated liquids) see sec. I.3.9d, e. Such distribution functions are in principle experimentally accessible. For instance, the structure factor $S(q)$, obtainable from scattering data, is the Fourier transform of $g(r)$, see [I.A11.21 - 21b]. Moreover, these distribution functions can in principle be related to macroscopic thermodynamic and mechanical characteristics, as explained in sec. I.3.9d. An alternative is to consider the *correlation length* ξ, introduced in sec. I.7.7c, which is a parameter quantifying the range over which fluctuations are correlated.

For liquids near surfaces the radial distribution functions become asymmetrical and are generally more difficult to handle. The situation is again relatively simple for molecularly flat surfaces when only density variations normal to the surface have to be considered. This is, for instance, the case for spherically symmetrical molecules; for rods, or molecules with asymmetrical interaction (water), the situation is again more complicated. In that case one can introduce a (linear) distribution function $g(z)$ as

$$g(z) = \rho_N(z)/\langle \rho_N \rangle \qquad\qquad [1.6.7]$$

where ρ_N is the number density in moles or molecules per unit volume. Generally, $g(z)$ will show a number of oscillations, damping to zero after a limited number of molecular layers. Figure I.5.6 shows that for water $g(r) \rightarrow 1$ for $r \geq 0.8\ nm$. In the literature long-range influences of surfaces on the structure of adjacent liquids have sometimes been postulated; generally such statements require critical scrutiny.

In connection with quantifying possible changes in the structure of liquids due to confinement in pores, an old rule, known as *Gurvitsch's rule*[1] may be mentioned. It states that for a variety of different liquids the total pore volume is within narrow limits identical, if computed on the basis of their bulk densities. For instance, McKee[2] finds a maximum variation of 4% for a number of organic adsorptives on silica gel with a (Kelvin) radius of 1.7 nm, suggesting that only minor structural deviations take place. However, for more strongly associated liquids like water, and for narrower pores, the deviations may be larger. Then the liquid-solid interaction also starts to play its role.

Some consequences of such structural deviations are that in narrow pores the contact angle and surface tension may differ from their macroscopic values. Also the liquid-vapour phase behaviour differs from that in bulk. Here we shall

[1] L. Gurvitsch, *J. Russ. Phys. Chem. Soc.* **47** (1915) 805.
[2] D.W. McKee, *J. Phys. Chem.* **63** (1959) 1256.

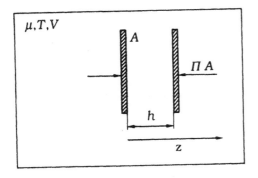

Figure 1.37. Two adsorbing plates in an environment of given μ, T and V. The disjoining pressure is Π.

not yet embark on the statistical modelling but discuss thermodynamic (i.e. phenomenological) aspects, restricting ourselves to the simplest geometry of slit-shaped pores, as in fig. 1.37, and to one-component systems.

As we are dealing with an open system, a grand canonical approach is appropriate. Let us accept that from statistics the grand canonical partition function $\Xi(\mu,V,T)$ has been obtained. Its characteristic function $\Omega = -kT \ln \Xi$ (sometimes called the grand (canonical) potential, especially by physicists) equals the work term, $-pV$ for pure volume work, γA for surface work etc., see [I.3.3.11-12] and [I.A6.23]. For bulk phases $-kT \ln \Xi = -pdV - SdT - Nd\mu$, but this equation has to be extended by surface- and interaction contributions. Generally, there is some problem in separating these. The solid-liquid interfacial tension γ^{SL} depends on h when h is low and so does the interaction force between the two plates. Such a deconvolution is particularly problematic for curved pores. For the present model of slit-shaped pores it seems appropriate to distinguish two types of work. The first involves changes in area that we shall write as $\gamma^f dA$ where $\gamma^f = \gamma^{film}$ is the film tension as introduced in connection with fig. I.2.1. The quantity γ^f depends on h. For large h where the pore surfaces no longer influence each other, $\gamma^f \to 2\gamma^{SL}$. Secondly, changes of h involve work associated with the *disjoining pressure* $\Pi(h)$ across the film. This quantity has been introduced in sec. I.4.2, and was also mentioned in [1.3.15]. It is often used to formulate the interaction between colloidal particles and in thin films. Generally it consists of three components, an electrostatic, a dispersion force and a *solvent structure-originated contribution*. Of these, the last mentioned is the main one under consideration here. Unlike the situation with colloids, where h is a variable, we are now interested in the structure of the confined liquid at fixed h (for rigid pores). With this in mind,

$$-kT \, d \ln \Xi(\mu, V, T) = -p \, dV - S \, dT - N \, d\mu + \gamma^f dA - \Pi A d h \qquad [1.6.8]$$

The volume V is that of the entire system (fig. 1.37); that of the pore is Ah. We repeat that all information on the work is in Ξ and that the division between an area-dependent and a distance-dependent contribution is not always rigorous. However, in the present model the relation between Π and γ^f can immediately be established by cross-differentiation

$$\left[\frac{\partial(\Pi A)}{\partial A}\right]_{\mu,T,h} = \Pi(h) = \left[\frac{\partial \gamma^f}{\partial h}\right]_{\mu,T} \qquad [1.6.9]$$

$$\Pi(h) = \left[\frac{\partial \gamma^f}{\partial h}\right]_{\mu,T} \qquad [1.6.10]$$

The *Gibbs adsorption law* for such a pore can be derived in the same fashion as has been done for a single surface in sec. I.2.13. There we compared differential and integral expressions for the excess Gibbs energy G^σ. In the present case it is more appropriate to apply the procedure to the grand canonical characteristic function $-kT\ln\Xi = -pV + \gamma A$. Taking its differential, and equating it to [1.6.8] yields at constant pressure

$$d\gamma^f = -S_a^f dT - \Gamma^f d\mu - \Pi dh \qquad [1.6.11]$$

Equation [1.6.11] can also be derived by equating the differential dU of $U = TS - pV + \gamma A + \mu N + \Pi Ah$ to $dU = TdS - pdV + \gamma dA + \mu dN + hd(\Pi A)$. For $h \to \infty$, $\Pi \to 0$, $S_a^f \to 2S_a^{SL}$, $\Gamma^f \to 2\Gamma^{SL}$ and $\gamma^f \to 2\gamma^{SL}$, i.e. [1.6.11] reduces to twice the Gibbs equation for a single surface. Equations [1.6.10] and [1.6.11] are consistent. The *amount adsorbed* is related to $g(z)$ via

$$\Gamma^f = \langle \rho_N \rangle \int_0^h [g(z) - 1] dz \qquad [1.6.12]$$

An interesting equation, derived by Henderson[1], relates $g(z)$ to the *solvent structure-originated part of the disjoining pressure*[2] $\Pi_{sol}(h)$ for the case of spherical molecules and hard wall boundaries:

$$\Pi_{sol}(h) = kT\left[\rho_N(0,h) - \rho_N(0,\infty)\right] \qquad [1.6.13]$$

where $\rho_N(0,h)$ is the value of $\rho_N(z,h)$ for $z = 0$, i.e. the local density of molecules at contact with the wall when the two walls are a distance h apart and $\rho_N(0,\infty)$ is the same for $h \to \infty$. Thus, Π_{sol} is determined by the change in the molecular

[1] J.R. Henderson, *Mol. Phys.* **59** (1986) 89. In this paper some more general situations are discussed.
[2] Often also called *solvation force*, or when in water, *hydration force*.

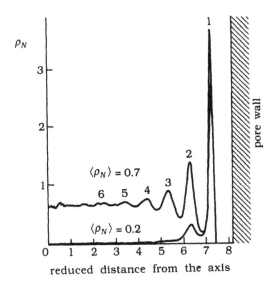

Figure 1.38. Molecular dynamics simulation of the density profiles for spherical molecules in a cylinder, mimicking SF_6 in controlled pore glass (CPG-10). Fluid-fluid and fluid-wall interaction modelled by Lennard-Jones interactions. Reference: A. de Keizer, T. Michalski and G.H. Findenegg, *Pure Appl. Chem.* **63** (1991) 1495.

distribution due to the proximity of the other wall. This force has an oscillatory nature, with maxima (repulsion) if h is about an integral number of molecular diameters and minima when h and the packing are out of register.

This is about as far as thermodynamics takes us. In order to find the profiles statistical theory or simulations are required. By way of illustration a density distribution is given in fig. 1.38. The picture shows the r.h.s. of the distribution. The number density and the distance from the centre of the pore are normalized to become dimensionless; $\langle \rho_N \rangle$ is the density averaged over the pore. For $\langle \rho_N \rangle = 0.2$ most adsorption is in the monolayer, for $\langle \rho_N \rangle = 0.7$ up to six layers can be distinguished. In this case near the centre ρ_N has almost attained $\langle \rho_N \rangle$, implying that now condensation has taken place. If the pores become thinner and/or $\langle \rho_N \rangle$ is further increased, the profiles on opposite walls start to overlap and influence each other, and a solvent structure force develops.[1]

1.6e. Capillary condensation and phase coexistence in pores

It is generally accepted that the phenomenon of adsorption hysteresis has its origin in *capillary condensation*, i.e. condensation of vapour in capillaries under conditions differing from those for bulk phases. The steep rise in the amount adsorbed, that is observed at certain p(sat) below unity, indicates that pores

---

[1] For further reading, see J. Klafter, J.M. Drake, *Molecular Dynamics in Restricted Geometries*, Wiley (1989); R. Kjellander, S. Sarman, *Mol. Phys.* **70** (1990) 215.

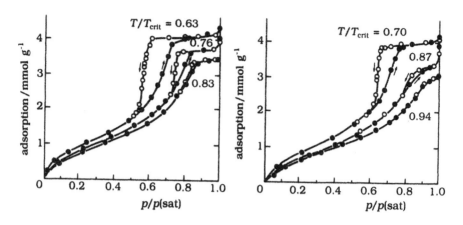

Figure 1.39. Adsorption isotherms of Xenon on Vycor glass. Temperatures relative to the bulk critical temperatures of xenon (289.7 K). Above $T = 0.94$ T_{crit} hysteresis is no longer observed. (Redrawn from S. Nuttall, Ph.D. Thesis, Univ. of Bristol (1974). See also C.G.V. Burgess, D.H. Everett and S. Nuttall, *Pure Appl. Chem.* **61** (1989) 1845 (Courtesy of D.H. Everett).)

with adsorbing walls have the effect of lowering the liquid-gas coexistence curve and the hysteresis implies that capillaries can support metastable liquid.

Regarding the mechanism, of capillary condensation, it is possible that more than one is viable, depending on the nature of the pores. As long ago as 1911 Zsigmundi[1] recognized the phenomenon: he attributed it to contact angle hysteresis.

One way to apprach the phenomenon is by considering the liquid in the pores to have bulk properties and follow the events as a function of increasing and decreasing p/p(sat). We have given such a derivation in connection with fig. 1.32.

For other types of pores other mechanisms may prevail. For instance, for interconnected pores, liquid bridges may form around necks or other constrictions and hence shield the interior from the outside, in this way provoking hysteresis.

However, the influence of the attraction, exerted by the pore wall on the adsorptive, also plays a role and in more detailed interpretations, this should be taken into account. An illustration that liquids in capillaries may behave differently from their bulk counterpart is given in fig. 1.39. With increasing temperature, the hysteresis loops become thinner, disappearing completely at $T = 0.94$ T_{crit}, i.e. the critical point of liquid in pores is lower than that of bulk liquid.

---

[1] R. Zsigmundi, *Z. Anorg. Chem.* **71** (1911) 356.

The thermodynamics presented in the previous subsection is easily extended to account for the coexistence of different phases in pores. In particular, the liquid-vapour transition is relevant here. We shall briefly consider three aspects of this topic.

Let us first discuss the limitation of the LG phase boundary. For the bulk case this is *one curve* in the p, T phase diagram, describing the change of the boiling point with pressure. Such a curve starts at the triple point and runs till the critical point (T_{crit}). Its slope is determined by the heat of evaporation, in accordance with the required Clapeyron equation. With liquids in pores we have *three* state variables, because the size of the pore is now also a parameter. In this way the influence of the wall on the adsorptive enters the discussion. For the case of slits the variables are p, T and h or, more generally, μ, T and h, as in [1.6.11]. From these three variables a three-dimensional coexistence diagram of, say, phases α and β is immediately derived by setting $d\gamma^{f\alpha} = -d\gamma^{f\beta}$ so that from [1.6.11]

$$\left(S_a^{f\alpha} - S_a^{f\beta}\right)dT + \left(\Gamma^{f\alpha} - \Gamma^{f\beta}\right)d\mu + \left(\Pi^\alpha - \Pi^\beta\right)dh = 0 \qquad [1.6.14]$$

This represents a *three-dimensional surface* of which the μ, T-line is only one of the three cross-sections, and now dependent on h. The three cross-sections $(\mu,T)_h$, $(\mu,h)_T$ and $(h,T)_\mu$ are obtained from the appropriate cross-differentiations.

Another option is to derive equations for the pressures at which condensation takes place in narrow capillaries. Let us illustrate this for slits. As before, the characteristic function $-kT\ln\Xi$ of the grand canonical partition function equals $-pV + \gamma^f A$, with $V = Ah$. Let liquid and vapour coexist inside the capillary and assume that we have only these two phases (i.e. the contribution of the (thin) inhomogeneity at the phase boundary to Ξ is ignored). Equilibrium requires $\Xi^L = \Xi^G$ or $p^L Ah + \gamma^{fL} A \approx p^G Ah + \gamma^{fG} A$, so that

$$h(p^L - p^G) = \gamma^{fL} - \gamma^{fG} \qquad [1.6.15]$$

For a given environment, that is, at given μ^G of the surroundings (or, at given pressure p^G of the vapour), the pressure of the liquid differs from p^G due to the capillary pressure, created over the curved LG meniscus. In turn, the shape of the meniscus is determined by the interaction with the wall. Comparing bulk liquid (at μ(sat), i.e. at p(sat)) with liquid in a capillary (at $\mu < \mu$(sat) and pressure p) we have

$$\mu(p(sat)) - \mu(p) = RT\ln\frac{p(sat)}{p} = V_m^L \Delta p = \frac{p^G - p^L}{\rho_N^L} \qquad [1.6.16]$$

where we have introduced the molar volume V_m^L using [I.2.14.15]. Equation [1.6.16] supposes the gas phase to be ideal. Further elaboration of [1.6.16] requires information on the two film tensions. Consider the simple case of fig. 1.40 where in the gas part there is only a very thin adsorbate, so that $\gamma^{fG} = 2\gamma^{SG}$ (= twice the interfacial tension of the solid-vacuum interface minus $RT\Gamma$, see [1.1.8]) and $\gamma^{fL} = 2\gamma^{SL}$. These two surface tensions are connected to each other via the contact angles α through $\gamma^{SG} = \gamma^{SL} + \gamma^{LG} \cos\alpha$. Combining all of this,

$$RT \ln \frac{p(\text{sat})}{p} = \frac{2\gamma^{LG}\cos\alpha}{h\,\rho_N^L} \qquad\qquad\qquad [1.6.17]$$

which is a variant of the *Kelvin equation* [1.6.4] Generally, γ^{LG} and ρ^L differ from their bulk values, due to the interaction with the wall. Especially for narrow slits, this is also so for cos α. All of this is reflected in the difference between the critical points in bulk and in narrow pores, already observed in fig. 1.39. Equation [1.6.17], which could be derived without making any assumption regarding the shape of the meniscus, quantifies how, at pressures below saturation, liquid phases are stabilized provided the solid is wetted. See also fig. I.1.1. When the solid is not wetted, cos α < 0, an *overpressure* is needed to force the liquid into the capillary. This principle is used in mercury porosimetry (sec. 1.6b). Equation [1.6.17] contains only macroscopic quantities but these may differ quantitatively from the corresponding ones for bulk phases.

A third aspect is the close analogy with wetting phenomena. One important issue is whether or not a thin liquid film on a surface is stable or whether it spontaneously (that is, even without a second wall nearby) disproportionates into a droplet in contact with a very thin (usually (sub)-monolayer) film. In wetting language this disproportionation (usually in the other direction, as

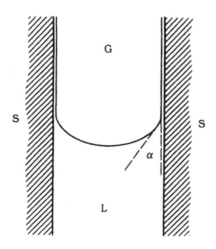

Figure 1.40. Simplified liquid-vapour boundary in a capillary.

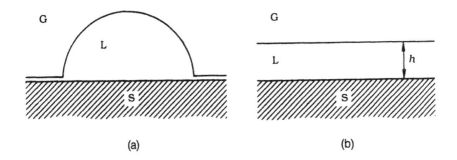

Figure 1.41. The wetting transition.

sketched in fig. 1.41) is called the *wetting transition*. It may be marked by, say, a change of temperature or a change in the material properties. Its occurrence is governed by $\Pi(h)$; as soon as $\Pi(h) > 0$ disproportionation takes place. In the case that only dispersion forces are operative, we have, from chapter I.4, all ingredients available to predict this transition. Recall [I.4.6.6]

$$\Pi(h) = -A^f/6\pi h^3 \qquad\qquad [1.6.18]$$

where A^f, the Hamaker constant of the film, is tabulated for a number of systems in table 7 of I.App.9. A liquid such as that shown in fig. 1.41a, which does not wet a thin layer of itself, is called *autophobic*.

In a narrow capillary, the wetting transition usually takes place under conditions differing somewhat from those in bulk. Spontaneous droplet formation may block pores.

It is beyond the scope of this book to treat modern statistical theories for liquids in an external field, which is a necessary to take wall effects into account. However, one feature that is often revealed by them is an adsorbed amount-pressure graph with a "Van der Waals loop", indicating phase separation and metastable branches. Figure 1.42 gives an example; the system is the same as that which gave rise to the oscillating profile in fig. 1.38. From Molecular Dynamics simulation, the reduction of the hysteresis loops with increasing temperature and the lowering of T_c (fig. 1.39) can also be predicted. Hence, the main physical features are at least semiquantitatively retrieved. On this (or similar) basis, hysteresis loops can be theoretically predicted. The system under study applies to cylinder geometry; similar figures can also be obtained for slits and for interconnected networks of pores[1].

---

[1] For further reading on liquids in capillaries, see the reviews by R. Evans, *J. Phys. Condens. Matter* **2** (1990) 8989; and *Microscopic Theories of Simple Fluids and Their Interfaces*, in *Liquids at Interfaces*, J. Charvolin, J.F. Joanny and J. Zinn-Justin, Eds., Elsevier (1989), and D. Nicholson, N.G. Parsonage, *Computer Simulation and the Statistical Mechanics of Adsorption*, Academic Press (1982).

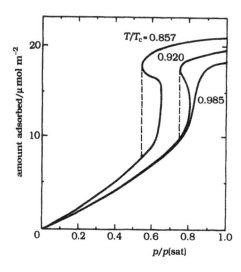

Figure 1.42. Sorption isotherms belonging to the system of fig. 1.39. The numbers at the curves indicate T/T_{crit}.

1.7 Surface heterogeneity

Most real surfaces are energetically and/or structurally heterogeneous. This undeniable fact considerably detracts from the opportunities to apply idealized equations, let alone comparing their qualities on the basis of measurements on "practical" systems. Surface heterogeneity tends to blur such specific features as phase transitions, steps in isotherms and properties of surface entropies.

More than one kind of heterogeneity can be distinguished. Most marked and probably dominant, is *energetic heterogeneity*, meaning that per site, or per patch, the surface energy is not constant. As far as this intrinsic heterogeneity propagates in the adsorption energy, it appears that $\Delta_{ads}u$ is not constant, but a function of coverage, i.e. one is dealing with $\Delta_{ads}u(\theta)$. When an adsorptive is brought in contact with an adsorbate the trend is that the higher-energy spots or patches are filled first. This phenomenon is reflected in the often observed decrease of the (isosteric) heat of adsorption with θ, especially at low θ, as illustrated in fig. 1.9. Heterogeneity can also be *structural* in the sense that the surface is not molecularly flat but exhibits asperities, lattice defects, ridges, dislocations and other irregularities[1]. Often such structural heterogeneities are at the same time energetically different, but there may also be entropical restrictions associated with them. It is difficult to disentangle these two experimentally. Porosity of the surface also involves a kind of heterogeneity but we deem this matter sufficiently discussed in sec. 1.6. We now consider only non-porous adsorbents.

---

[1] For an analysis in terms of fractals, see A. Seri-Levy, D. Avnir, *Langmuir* **9** (1993) 3067.

Regarding the distribution $\Delta_{ads}u(\theta)$, little can be said in general because it depends on the history of the solid. Materials prepared by coagulation of colloids, by controlled growth, by forced precipitation or by milling, will have very different distributions. Moreover, in some systems a certain perfectioning will take place with time (*ageing*) caused by Ostwald ripening, a process to which gas sorption may contribute. Other chemical processes are also known to reduce the heterogeneity, such as thermal graphitization of carbon blacks and silanization of silica gel.

The most direct way of obtaining information on energetic heterogeneity seems by measuring the heat of adsorption, or by establishing the isosteric heat from the temperature dependence of the adsorbed amount (secs. 1.3c and f). However, the enthalpy obtained is the sum effect of adsorbent-adsorbate and adsorbate-adsorbate interactions. Some of the methods discussed in sec. 1.2 are suitable for learning something about structural heterogeneity. Dynamic experiments like adsorption chromatography may also give information on heterogeneity. Temperature-programmed desorption also belongs to this category. In this technique desorption is realized through some time-dependent temperature schedule.

It must be noted that the distribution of the Gibbs energy of adsorption is not exactly the same as that of $\Delta_{ads}u$ because, for entropical reasons, the various sites are not strictly filled in register with decreasing energy. Only at $T \to 0$ is this the case. Another complication is that different adsorptives may "feel" different heterogeneities. For instance, an adsorptive with small molecules might feel the difference between sodium and chloride sites in the surface of a sodium chloride crystal, whereas large molecules, occupying many sites, will see the surface as pseudo-homogeneous.

The question is now whether $\Delta_{ads}u(\theta)$ may be determined from appropriate (adsorption) experiments and analysis. Unfortunately there is no unambiguous way to achieve this. The simplest way is to assume some distribution, then derive an isotherm equation and compare that with experiments to find the distribution parameter. However, such approaches are not generally unique in that different models may predict the same isotherm; in addition, little is known *a priori* about the distributions. The opposite process of finding $\Delta_{ads}u(\theta)$ from experimental isotherms also requires some assumptions on the way in which the sites or patches are distributed and on the way molecules adsorb on spots with identical energy. In addition, great experimental accuracy is demanded. In conclusion, there is no unique solution, although a variety of procedures of limited applicability have been proposed to account for heterogeneity.

A first and basic question is: what is the nature of the distribution is: *patchwise* or *random*? "Patchwise" means that the surface consists of a set of small areas, say crystal faces, each of which is homogeneous and characterized by a certain adsorption energy, say $\Delta_{ads}u_j$ for patch of nature j. For each patch, one of our idealized isotherm equations (app. 1) could be used; lateral interaction between adjacent patches is usually disregarded. As a result, the energy distribution is defined in terms of $\Delta_{ads}u_j$ and the fraction f_j of the surface area, occupied by patches of type j. "Random" means that the individual sites of different adsorption energy are randomly distributed over the surface. Lateral interaction cannot be ignored, but is difficult to account for rigorously because the frequency of each type of pair is not easily established; usually a kind of weighted distribution is assumed. For instance, it could be assumed that the number of nearest neighbours is proportional to the total number of molecules adsorbed. This is the mean field approximation. As stated before, it is likely that the nature of the distribution is co-determined by the history of the sample and by the extent of lateral structural correlation. The two models "patchwise" and "random" are the extremes of a range of intermediates.

Heterogeneity also has its consequences for the distinction between mobile and localized adsorption. The transition between these modes of adsorption is determined by the height of the barrier against lateral flow with respect to kT. On heterogeneous surfaces a distribution of barrier heights may occur; then it is imaginable that part of the adsorption is mobile and the remainder localized. Here, the situation discussed in sec. 1.5d prevails.

For the mathematical description of isotherms on heterogeneous surfaces the following equation may serve as a starting point[1]

$$V(p,T) = \int_0^\infty V_{\ell}(p,T,\Delta_{ads}u) \; f(\Delta_{ads}u) \; d(\Delta_{ads}u/kT) \qquad [1.7.1]$$

Here, $f(\Delta_{ads}u)$ is the fraction of the adsorbent with adsorption energy $\Delta_{ads}u$. This fraction is normalized by

$$\int_0^\infty f(\Delta_{ads}u) \; d(\Delta_{ads}u/kT) = 1 \qquad [1.7.2]$$

The volume V_{ℓ} is that adsorbed on sites or patches having an adsorption energy $\Delta_{ads}u$. The subscript ℓ denotes local: in heterogeneity studies the isotherm, describing the adsorption on such homogeneous parts, is often called the *local iso-therm* for which, as already stated, any isotherm equation of app. 1 can be substituted. Volumes pertaining to different local isotherms have to be added in

---

[1] One of the alternatives is using so-called density-functional theory, which incorporates the topology (spatial distribution) of the various patches. See for instance L. Lajtar, S. Sokolowski, *J. Chem. Soc., Faraday Trans.* **88** (1992) 2545.

order to obtain the total adsorbed volume. In [1.7.1] this sum has been replaced by an integral; this is allowed if the energy distribution is continuous. In reality energies do not run from zero to infinity, but physically the extension of the integration limits does no harm because the parts at the extremes have low (or zero) f, and analytically the range from 0 to ∞ is much easier to handle. The average adsorption energy is

$$\langle \Delta_{ads} u \rangle = \int_0^\infty f(\Delta_{ads} u) \, \Delta_{ads} u \, d(\Delta_{ads} u / kT) \qquad\qquad [1.7.3]$$

From [1.7.1] in principle an isotherm equation can be derived, assuming something about $f(\Delta_{ads} u)$; and $V_{_l}$. Alternatively, from experimental isotherms, information on this distribution is, in principle, obtainable.

Let us, before giving illustrations, discuss some aspects of the statistical foundations. Model assumptions regarding the mode of adsorption (patchwise or random? mobile or localized? mono- or multilayer? with or without lateral interaction?) are reflected in the natures of the local partition functions and in the way they combine to $Q(N, N_s, T)$. Again, no general solution can be given; models of different degrees of sophistication can be developed.

Consider, by way of example, the simple case of localized monolayer adsorption without lateral interaction. Let there be N_{sj} sites of type j. These sites are distinguishable; in the present situation we do not yet have to discriminate between patchwise and random distribution of these sites, but this becomes important as soon as lateral interaction has to be accounted for. We have $N_s = \sum_j N_{sj}$ and $N = \sum_j N_j$ if N_j is the number of molecules adsorbed on sites j in the specific distribution given, i.e. N_j is $N_j(N)$. Many distributions ($N = N_1, N_2, N_3, ...$) are possible. For a given distribution the canonical partition function of the adsorbate $Q(N) = \prod_j Q_j$, where for each class j Q_j is written as [1.5.1], after giving all parameters the subscript j. Summing $Q(N)$ over all distributions N yields $Q(N, N_s, T)$. It is convenient to do this by the maximum term method, explained in sec. I.3.7a, which may be applied if the numbers involved are large. Mathematically it means that the sum $\ln Q(N)$ is replaced by the logarithm of the largest term; that is, we only consider the most probable distribution of the adsorbed molecules over the available sites. To find this maximum, the standard method of Lagrange multipliers is available. We shall not elaborate this here[1] but conclude that in principle all relevant mechanical and thermodynamical parameters of the adsorbate can be derived: N, S, μ and, hence, the isotherm equation. Such a statistical approach adds extra value. For

[1] For such an elaboration, see for example T.L. Hill, *J. Chem. Phys.* **17** (1949) 762.

instance, regarding the entropy, the non-configurational part increases with θ due to the higher vibrational frequencies associated with the lower energies, see [1.5.3]. The configurational entropy is given by

$$S^\sigma(\text{conf}) = k \ln \prod_j \frac{N_{sj}!}{N_j(N_{sj} - N_j)!} \qquad [1.7.4]$$

and, hence, depends on the distribution. The maximum value (in the most probable state) is identical to that for a uniform surface.

In a number of cases, when the heterogeneity is such that groups of sites may be considered as independent subsystems, statistical approaches may also be helpful. This may be so for surfaces with periodical structures like binary crystals or copolymers. Consider, by way of a simple example, a surface where pairs of sites (say A and B) form independent subsystems, and assume that molecules can adsorb on A and/or on B (site partition functions q_A and q_B, respectively) and that there is a lateral interaction energy w if the two sites are both occupied. This leads to isotherm [1.5.41].

Returning to [1.7.1], it is obvious that, unless additional information about V_r and/or f is available, a host of isotherms can be predicted by combining certain assumptions regarding the distribution and the local isotherm, probably enough to "explain" all isotherms ever reported. Suffice it here to give the two illustrations of fig. 1.43(a) and (b). They are taken from Ross and Olivier, who assumed a patchwise heterogeneous Gauss distribution (see [I.3.7.14]):

$$f(\Delta_{ads}u)\, d(\Delta_{ads}u/k\,T) = \frac{e^{-\gamma |\Delta_{ads}u - \langle\Delta_{ads}u\rangle|^2}\, d(\Delta_{ads}u/k\,T)}{\int_0^\infty e^{-\gamma |\Delta_{ads}u - \langle\Delta_{ads}u\rangle|^2}\, d(\Delta_{ads}u/k\,T)} \qquad [1.7.5]$$

where γ is a measure of the width of the distribution: the lower γ, the wider is the distribution, that is the stronger the heterogeneity. For $\gamma \to \infty$ the surface becomes homogeneous, because values of $\Delta_{ads}u$ differing greatly from $\langle\Delta_{ads}u\rangle$ are suppressed. The width can also be expressed in terms of standard deviations, as was done in sec. I.3.7a. For the figure, local isotherms with lateral interaction have been chosen, showing the typical sigmoidal shape for high γ. It is seen that with increasing heterogeneity this sigmoidal shape disappears, so that for $\gamma \sim 1$ the isotherms attain a Langmuir-like shape. They are even linear if plotted as p/V as a function of p, as in fig. 1.12e, illustrating how careful one must be in interpreting such linearity as proof of the viability of the Langmuir premises. In addition, the critical point disappears.

The fact that heterogeneity may so dramatically affect the shapes of adsorption isotherms, whereas the distribution is often not, or only approximately, known, has two practical consequences:

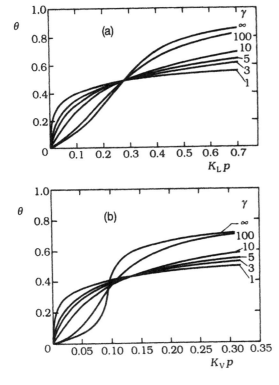

Figure 1.43. Monolayer adsorption isotherms on an adsorbent with a Gauss-type energy distribution. After S. Ross and J. P. Olivier, *On Physical Adsorption*, Interscience Publ. (1964). The local isotherm is FFG in fig. (a) and Hill-de Boer for fig. (b) (see app. 1). The parameter γ is given in [1.7.5]. The horizontal scale is adjustable.

(i) equations, based on the assumption of surface homogeneity, may only be applied to such idealized surfaces.

(ii) for the practical purpose of representing adsorption data *empirical adsorption isotherm equations* remain useful; these are simple equations with a very limited number of adjustable parameters and which are not based on a given model.

One of such empirical equations is that of Freundlich [1.1.2] and another one is the *Temkin equation*[1]

$$V = a \ln b\,p \qquad\qquad [1.7.6]$$

where a, and b are adjustable constants. This latter equation is often suitable for describing the middle part of isotherms. Mathematically, sucn empirical equations can often be derived from [1.7.1] with certain assumptions regarding the local isotherm and the distribution. For instance, the Freundlich equation can be obtained from [1.7.1] assuming the local isotherm to be Langmuir-like

---

[1] Pronounced Tjómkin. Named after M.I. Temkin, *Zhur. Fiz. Khim.* **4** (1933) 573 and M.I. Temkin and W.M. Pyzhev, *Acta Physicochim USSR* **12** (1940) 327.

and the energy distribution exponential[1], and the Temkin isotherm from the same local isotherm and a linear functionality $\Delta_{ads}u = \Delta_{ads}u^o(1 - \text{const. } \theta)$ [2]. However, such a distribution would be more likely if the $\Delta_{ads}u(\theta)$ dependency would be dominated by lateral interaction rather than by adsorbate-adsorbent interaction.

A more practical analytical equaiton for the adsorption on heterogeneous surfaces is

$$\frac{V}{V(\text{mon})} = \frac{(\tilde{k} p)^n}{1 + (\tilde{k} p)^n} \qquad\qquad [1.7.7]$$

which is based on a Langmuir-type local isotherm and a quasi-Gauss distribution. In [1.7.7] n is a measure of the width of the distribution and $\tilde{k}$ its median. It is known as the *generalized Freundlich equation*[3], but may equally wel be called the *generalized Langmuir equation*.

Further discussion of [1.7.7] and other isotherm equations with their derivations can be found. A variety of other empirical isotherms with their "derivations" on the basis of [1.7.1] can be found in the books by Rudzinski and Everett and by Jaroniec and Madey, mentioned in sec. 1.9c.

Let us finally briefly consider the inverse problem: how to find $f(\Delta_{ads}u)$ if $V(p,T)$ is known from experiment, and the local isotherm is known (or assumed). Mathematically this procedure is equivalent to solving so-called *Fredholm integrals of the first kind*; the local isotherm functions serve as the kernel. Computationally speaking, this is an ill-posed problem because minor variations in the data may cause substantial variations in $f(\Delta_{ads}u)$. In other words, the inevitable experimental noise thwarts one obtaining a physically significant distribution. In practice the elaboration may follow one of three paths:

(i) a certain functionality is assumed for the distribution functions, in such a way that spurious oscillations or other instabilities are suppressed.

(ii) an analytical solution is sought.

(iii) oscillations etc. are suppressed by some numerical technique.

The disadvantage of method (i) is that a certain functionality is imposed on an unknown distribution. In approach (ii) usually a simplified local isotherm is selected. The distribution function is obtained as a series of derivatives of the

[1] G.D. Halsey, H.S. Taylor, *J. Chem. Phys.* **15** (1947) 624; G.D. Halsey, *J. Chem. Phys.* **16** (1948) 931.

[2] B.M.W. Trapnell, *Chemisorption*, Butterworth (1955) 124. The derivation does not apply to the entire isotherm.

[3] W. Rudzinsky, K. Nieszporek, Hee Mom and Hyun-ku Ree, *Heterogen. Chem. Revs.* (acc. 1994).

experimental isotherms. The higher the order of the series, the better the distribution function is approximated, but it is difficult to estimate these higher order terms. In this case smoothing of the experimental curve is required; the extent to which this can be done depends on the quality of the data[1].

Method (iii) is most general. Several routes exist to deal with instabilities. Smoothing of the isotherm is needed; this isotherm should be relatively smooth itself.

For a review of these three approaches and a further discussion of method (iii), see ref. [2]. For methods (ii) and (iii) it is recommended to start with many, approximately evenly spaced, data over a large range of pressures.

Heterogeneity also has its consequences for the critical temperatures T_{crit} below which two-dimensional condensation may occur. For some models of monolayer adsorption with lateral attraction, the critical conditions have been established (in sec. I.3.8d for the FFG isotherm, in sec. I.3.8e for the quasi-chemical approximation and in sec. 1.5e for a two-dimensional Van der Waals gas). The value of T_{crit} is a criterion for the validity of an isotherm model, but heterogeneity greatly detracts from it. Heterogeneity inhibits two-dimensional condensation or, in other words, T_{crit} is reduced by an extent that is the greater, the more heterogeneous a surface. General experience confirms this: for instance, for a two-dimensional Van der Waals gas $T_{crit} = 0.5 T_{crit}$ (three-dimensional), but in practice always a factor below 0.5 is found[3].

1.8 Conclusion

The scope and emphasis in this chapter have been dictated mainly by its relevance for the other chapters of the present book and the planned other Volumes. Important purposes in this connection are

(i) characterization of dry surfaces by surface physical techniques

(ii) characterization of these surfaces by gas adsorption

(iii) adsorption features recurring in later chapters and Volumes.

Issue (i) includes aspects of surface structure and composition, (ii) surface area, porosity and heterogeneity. Regarding (iii), there was a certain emphasis on monolayer adsorption (which prevails for adsorption from solution); filling of pores by molecules has its analogies with the stacking of colloidal particles under certain confinements.

[1] M.M. Nederlof, W.H. van Riemsdijk and L.K. Koopal, *Environ. Sc. Tech.* **28** (1994) 1037.

[2] L.K. Koopal, C.H.W. Vos, *Langmuir* **9** (1993) 2593.

[3] J. O'Brien, in *Fundamentals of Adsorption*, A.B. Mersmann, S.E. Scholl, Eds., Am. Inst. Chem. Eng. (1991), 633.

Emphasizing certain aspects implies underexposing others. For instance, we have disregarded the vast and important domains of chemisorption and heterogeneous catalysis, which in our framework are considered as "applications". The relation between physisorption at high relative pressures and wetting is another important issue. A correlation is observed between the way in which the adsorption approaches the $p/p(\text{sat}) \to 1$ limit and the wetting of the solid by the condensed vapour (partial or complete?). We intend to come back to this in chapter 5 of Volume III. Yet other applications in which adsorption plays an important role include the scavenging of gases, adsorption towers, gas chromatography (for the separation of mixtures), heterogeneous nucleation.

Another feature that has not been systematically covered concerns additional means of determining properties of adsorbates. Examples here are the classical spectroscopies, with their surface variants (secs. I.7.10-12), reflection methods, including ellipsometry, reflectometry and evanescent wave studies, NMR, X-ray analysis, neutron diffraction and dielectric spectroscopy. The theory of the last mentioned phenomenon for bulk phases has been discussed in sec. I.4.5f; if applied to adsorbates, the technique can give information on the various degrees of freedom that polar molecules may have, say, for water adsorbed on oxides. For thicker water layers containing ions, measurement of the surface conductivity may yield additional information; see also sec. I.6.6d. The reason for not systematizing these techniques is that we do not consider them typically "surface methods", but rather surface variants of bulk methods.

Surface modification is a domain on its own. To suit certain purposes, dry surfaces can be changed by a number of techniques, such as ion bombardment, molecular beam deposition, *sputtering* (erosion as a result of bombardment with high-energy molecules), doping, coating or chemisorption. Let it be sufficient to note that, after modification, our techniques can be used to detect and analyse the incurred changes[1].

Throughout the entire chapter there has been a certain natural tension between the science of idealized model systems (homogeneous surfaces, pores of well-defined geometries) and the more empirical treatment of practical surfaces. For the former group it is likely that far more penetrating analyses are possible than described here, for instance by considering surface excess heat capacities or by developing the molecular interpretation of the forces responsible for adsorption. On the other hand, for the latter category systems, many of our treatments are definitely overinterpretations.

---

[1] *Chemically Modified Surfaces*, H.A. Mottola, J.R. Steinmetz, Eds. Elsevier (1991). (Proceedings of a symposium, held in the USA (1991).); *Chemically Modified Surfaces*, J.J. Pesek, I.E. Leigh, Eds., Roy. Soc. Chem., Cambridge (UK) (1994).

This contrast between idealized and real systems is a recurring and natural feature and we shall not shun it. A certain emphasis on the model systems is motivated by the consideration that such studies are the basis for further understanding. At the same time a certain prudence against fitting a limited number of equations to a limited number of data is advised. Rarely do such procedures lead to unique solutions. Even a perfect fit does not imply that the underlying model applies, although it may describe the adsorption empirically. The rule remains that for a proper characterization preferably further measurements should be carried out, say adsorption at more temperatures and/or with different adsorptives, or adsorption in conjunction with (micro)-calorimetry. Once such experiments have been done and found to concur, the mathematical confidence limit may become as high as the physical one.

1.9 General references

1.9a. IUPAC recommendations

A Survey of Experimental Techniques in Surface Chemical Physics (prepared for publication by J.H. Block, A.M. Bradshaw, P.C. Gravelle, J. Haber, R.S. Hansen, M.W. Roberts, N. Sheppard and K. Tamaru), *Pure Appl. Chem.* **62** (1990) 2297.

English-derived Abbreviations for Experimental Techniques in Surface Science and Chemical Spectroscopy, Pure Appl. Chem. **63** (1991) 887.

Manual of Symbols and Terminology for Physicochemical Quantities and Units. Definitions, Terminology and Symbols in Colloid and Surface Chemistry (prepared for publication by D.H. Everett), *Pure Appl. Chem.* **31** (1972) 579; ibid, Part II. *Terminology in Heterogeneous Catalysis* (prepared for publication by R.L. Burwell), *Pure Appl. Chem.* **45** (1976) 71.

Manual on Catalyst Characterization (prepared for publication by J. Haber), *Pure Appl. Chem.* **63** (1991) 1227.

Reporting Physisorption Data for Gas/Solid Systems, with Special Reference to the Determination of Surface Area and Porosity (prepared by a committee chaired by K.S.W. Sing), *Pure Appl. Chem.* **57** (1985) 603.

Symmetry, Selection Rules and Nomenclature in Surface Spectroscopies, (in preparation/provisional, 1992).

Recommendations for the Characterization of Porous Solids (prepared by a committee chaired by J. Rouquerol), *Pure Appl. Chem.* **66** (1994) 1739.

1.9b. Characterization of surfaces

A.W. Adamson. *Physical Chemistry of Surfaces*, Wiley, 5th ed. (1990) (chapters VII, VIII and XV).

Surface and Interfacial Aspects of Biomedical Polymers, Vol. 1 *Surface Chemistry and Physics*, J.D. Andrade, Ed., Plenum Press (1985). (Contains chapters on XPS, surface infrared spectroscopy and surface Raman spectroscopy.)

Photons and Low Energy Particles in Surface Processing, C.I.H. Ashby, J.H. Brannon and S.W. Pang, Eds. Materials Res. Soc. Pittsburgh, USA (1992). (Proceedings of a symposium held in Boston, USA, 1991.)

Atomic Scale Imaging of Surfaces and Interfaces, D.K. Biegelsen, D.J. Smith and S.Y. Tong, Eds., Materials Res. Soc., Pittsburgh, USA (1993). (Proceedings of a Symposium held in Boston, USA, 1992.)

Practical Surface Analysis, D. Briggs, M.P. Seah, Eds. Vol. 1, 2nd ed. (1990); *Auger and X-ray Photoelectron Spectroscopy*, Vol. 2 (1992), *Ion and Neutral Spectroscopy*, Wiley. (A standard book with emphasis on techniques and applications, Vol. 1 electron spectroscopy, Vol. 2 ion spectroscopy.)

D. Briggs. *Surface Analysis*, in *Encycl. Polym. Sci. Eng.*, Vol **165** Wiley (1989), 399-442. (Review of XPS and SIMS, principles and application to polymer surfaces.)

Auger Spectroscopy and Electronic Structure, G. Cubiotti, G. Mondio and K. Wandelt, Eds., Springer (1989). (Proceedings of a conference.)

Ion Spectroscopies for Surface Analysis, A.W. Czanderna, D.M. Hercules, Eds., Plenum Press (1991). (Description of electron and ion spectroscopies, emphasizing techniques for which commercial instrumentation is available. Contains appendices on terminology and standard practices.)

Catalysis and Surface Characterization, T.J. Dines, C.H. Rochester and J. Thomson, Eds., Roy. Soc. Chem., U.K. (1992). (Proceedings of a meeting, held in 1992 at Dundee Univ., U.K.)

Surface Area Determination, D.H. Everett, R.H. Ottewill, Eds., Butterworth, 1970. (Proceedings of a symposium and discussion. Various techniques are compared.)

L.C. Feldman, J.W. Mayer, *Fundamentals of Surface and Thin Film Analysis*, North Holland Publ. Cy. (1986). (Very readable introduction, covering all surface techniques involving electrons, ions and X-rays.)

H.-J. Güntherodt, R. Wiesendanger, *Scanning Tunneling Microscopy I. General Principles and Applications to Clean and Adsorbate-Covered Surfaces*, Springer (1992). (First of three Volumes, together intended to offer a comprehensive review.)

P.K. Hansma, V.B. Elings, O. Marti and C.E. Bracker, *Scanning Tunneling Microscopy and Atomic Force Microscopy: Application to Biology and Technology*, *Science* **242** (1988) 209 (very informative review).

M.A. van Hove, W.H. Weinberg and C.-M. Chan, *Low-Energy Electron Diffraction. Experiment, Theory and Surface Structure Determination*, Springer (1986). (Comprehensive overview of LEED).

J.E. Griffith, G.P. Kochanski, *Scanning Tunneling Microscopy*, *Ann. Rev. Mater. Sci.* **20** (1990) 219. (Review.)

Spectroscopic Characterization of Heterogeneous Catalysts. Part A, Methods of Surface Analysis. Part B, Chemisorption of Probe Molecules, J.L.G. Fierro, Ed., Elsevier (1990). (Part A on surface structure methods, surface groups on oxides, X-ray, Mössbauer; Part B on infrared, NMR, EPR, thermal desorption, ...)

The Structure of Surfaces Part I, M.A. van Hove, S.Y. Tong, Eds., Springer Series in Surface Science (1985). (Conference proceedings, rather basic, mono-crystals etc.). Ibid, part II, J.F. van der Veen, M.A. van Hove, Eds. (1988) (emphasis on semiconductors and metals.) Ibid, part III, S.Y. Tong, M.A. van Hove, K. Takayanagi and X. Xie, Eds. (1991) (crystallography, morphology, phase transitions of single crystals, imperfect and amorphous surfaces).

D.A. King, D.P. Woodruff, *The Chemical Physics of Solid Surfaces and Heterogeneous Catalysis*, Elsevier (1981, ...). (A series of the "Advances" type, many experimental results, less emphasis on techniques. Volume **1** (1981) deals with clean solid surfaces, Vol. **2** (1983) with adsorption at solid surfaces, Volume **6** (1993) with co-(chemi-)sorption, promoters and poisons.)

Molecule-Surface Interactions, K.P. Lawley, Ed., *Advances in Chemical Physics*, Vol. **76** John Wiley (1989). (A variety of topics on the characterization of surfaces, adsorbates and surface chemical kinetics.)

Physicochemical Aspects of Polymer Surfaces, K.L. Mittal, Ed., Plenum Press. Vol. **1** and **2** (1983). (Conference proceedings, variety of methods, including "wet" surfaces.)

Surface Analysis Methods in Materials Science, D.J. O'Connor, B.A. Sexton and R.C. Smart, Eds., Springer (1992). (Guide describing major techniques, plus applications in materials science.)

Characterization of Powder Surfaces, G.D. Parfitt and K.S.W. Sing, Eds., Academic Press (1976). (Adsorption and spectroscopic techniques with emphasis on pigments, carbon black, silica, clays.)

X-ray Absorption; Principles, Applications and Techniques of EXAFS, SEXAFS and XANES, R. Prins, D. Koningsberger, Eds., Wiley (1987). (Good general introduction; overview of theory and applications.)

J.C. Rivière, *Surface Analytical Techniques*, Oxford University Press (1990). (Discusses a variety of surface spectroscopies.)

Physical Methods of Chemistry, Vol. **IX**. *Investigations of Surfaces and Interfaces*, B.W. Rossiter, Baetzold, Eds., Wiley-Interscience, 2nd ed. (1993). (Authoritative reviews by various authors; most of the chapters deal with surface spectroscopic techniques.)

D. Sarid, *Scanning Force Microscopy with Applications to Electric, Magnetic and Atomic Forces*, Oxford University Press (1991). (Emphasizing mechanical, technical and constructional features. Thorough presentation.)

G.A. Somorjai, *Chemistry in Two Dimensions: Surfaces*, Cornell University Press (1981). (Composition, structure and reactions of solid surfaces, methods and results, many of them tabulated.)

Catalysts and Related Surfaces, Characterized by X-ray Absorption Fine Structure, G.A. Somorjai, Ed., Baltzer, Basel, Switzerland, 1993. (Proceedings of a symposium held in Tokyo, Japan, 1992.)

G.A. Somorjai, *Introduction to Surface Chemistry and Catalysis*, Wiley (1994). (Adsorption, surface science techniques, with some emphasis on reactions at surfaces and interfaces.)

Chemistry and Physics of Solid Surfaces, R. Vanselow, R. Howe, Eds., Springer Series in Surface Science, VI (1986), VII (1988), VIII (1990). (A variety of contributions, many of them dealing with the physical or chemical characterization of clean surfaces and surfaces with adsorbates.)

D.P. Woodruff, T.A. Delchar. *Modern Techniques of Surface Science*, Cambridge University Press (1986). (Very good, rather basic introduction to all surface techniques.)

Ordering at Surfaces and Interfaces, A. Yoshimori, T. Shinjo and H. Watanabe, Eds., Springer (1992). (Proceedings of a symposium in Hakone, Japan, 1990.)

Solvay Conference on Surface Science, F.W. de Wette, Ed., Springer (1988). (Invited lectures and discussion of a conference.)

Surface X-Ray and Neutron Scattering, H. Zabel, I.K. Robinson, Eds., Springer (1992). (Proceedings of a conference in Germany, 1991.)

A. Zangwill. *Physics at Surfaces*, Cambridge University Press (1988). (Emphasizes physical aspects of clean surfaces and surfaces with adsorbates.)

1.9c. *Adsorption*

Fundamentals of Adsorption. Proceedings of a series of conferences, published by the Am. Inst. Chem. Eng. New York, USA. First Conference, Germany (1983) A.L. Myers, G. Belfort Eds., published (1984); Second Conference, Calif. USA (1986) A.I. Liapis, Ed, published (1987); Third Conference Germany (1989) A.B. Mersmann, S.E. Scholl, Eds., published (1991); Fourth Conference, Kyoto, Japan (1992), M. Suzuki, Ed., published (1993). All books contain a variety of adsorption phenomena.

Dynamics at the Gas/Solid Interface, Faraday Discuss. Roy. Soc. Chem. (London) **96** (1993). (Conference Proceedings.)

A.W. Adamson, *Physical Chemistry of Surfaces*, Wiley, e.g. 5th ed. (1990), chapter XVI. (Standard textbook.)

J.H. de Boer, *The Dynamical Character of Adsorption*, Clarendon Press, Oxford (1953). (Emphasis on the kinetics and dynamics of adsorption.)

Interaction of Atoms and Molecules with Solid Surfaces, V. Bortolani, N. March and M. Tosi, Eds., Plenum (1990). (Surface physics, theory and experiment.)

Carbon Adsorption Handbook, P.N. Cheremisinoff, F. Ellerbusch, Eds., Ann. Arbor Sci. (Michigan, USA), 1978. (Emphasis on the application of activated carbon.)

J.G. Dash, *Statistical Thermodynamics of Physisorption*, in *Progr. Surface Sci.* **5** (1974) 119. (Review of basic principles; some emphasis on thermodynamics.)

Adsorption on Ordered Surfaces of Ionic Solids and Thin Films, H-J. Freund, E. Umbach, Eds. Springer Series in Surface Sci., **33** (1994). (Proceedings of a seminar in Bad Honnef, Germany, 1993.)

The Solid-Gas Interface, E.A. Flood, Ed., M. Dekker (1967). (Older, but comprehensive book, Vol. **1** emphasizes fundamentals. Vol. **2** techniques and case studies.)

S.J. Gregg, *The Physical Adsorption of Gases*. in *MTP International Review of Science, Physical Chemistry*, series 1, Vol. **7**, *Surface Chemistry and Colloids*, M. Kerker, Ed., Butterworth (1972) 189. (Literature review.)

Diffusion at Interfaces: Microscopic Concepts, M. Grunze, H.J. Kreuzer and J.J. Weimer, Eds., Springer (1988). (Proceedings of a workshop; diffusion and nucleation.)

Kinetics of Interface Reactions, M. Grunze, H.J. Kreuzer, Eds., Springer (1986). (Proceedings of a workshop, kinetics, dynamics versus thermodynamics.)

T.L. Hill, *Introduction to Statistical Thermodynamics*, Addison-Wesley (1960). (Chapters 7 and 14 deal with adsorption statistics.)

J. Hobson, *Ultrahigh Vacuum and the Solid-Gas Interface*, Adv. Colloid Interface Sci. **4** (1974) 79. (Discussion of adsorption under conditions where collisions with a solid surface are more likely than those between gas molecules.)

M. Jaroniec, R. Madey, *Physical Adsorption on Heterogeneous Solids*, Elsevier (1988). (Adsorption from gas or liquid phase, emphasis on heterogeneity. Gives a large number of equations.)

M.J. Jaycock, G.D. Parfitt, *Chemistry of Interfaces*, Ellis Horwood (1981). (Introductory, only part on solid-gas interfaces.)

H.J. Kreuzer, Z.W. Gortel, *Physisorption Kinetics*, Springer (1986). (This book contains much information on gas adsorption.)

S.R. Morrison, *The Chemical Physics of Surfaces*, Plenum Press, 2nd ed. (1990). (Adsorption and surface properties from the point of view of solid state physics.)

Interfacial Aspects of Phase Transformations, Nato Advanced Studies, Institute Series, Series C, Vol. **87** (1982), B. Mutaftchiev, Ed., D. Reidel Publ. Cy. (Dordrecht, the Netherlands). (Surface structure, adsorption and heterogeneous nucleation.)

H.C. van Ness, *Adsorption of Gases on Solids. Review of the Role of Thermo-dynamics in Chemistry and Physics of Interfaces*, Am. Chem. Soc. Publication (1971) p. 121. (Review, also covers mixtures of gases.)

D. Nicholson, N.G. Parsonage, *Computer Simulation and the Statistical Mechanics of Adsorption*, Academic Press (1982). (Physical approach, more specialized and more advanced than our treatment.)

A. Patrykiejew, S. Sokolowski, *Statistical Mechanics of Adsorption of Polyatomic Molecules on Solid Surfaces*, Adv. Colloid Interface Sci. **30** (1989) 203. (Review, including lateral interactions and profiles.)

V. Ponec, Z. Knor and S. Cerny, *Adsorption on Solids*, Butterworth, 1974. (Translated from the Czech original and updated. Theory and experiment, also covers chemisorption.)

S. Ross, I.D. Morrison and H.B. Hollinger, *The First Virial Coefficient of an Adsorbed Gas*, Adv. Colloid Interface Sci. **5** (1975) 175. (Comparison between various approaches to derive and interpret $B_2^\sigma(T)$ statistically.)

S. Ross, J.P. Olivier, *On Physical Adsorption*, Interscience Publ. (1964). (Experiments and theory, with some emphasis on the treatment of heterogeneity.)

W. Rudzinski, D.H. Everett, *Adsorption of Gases on Heterogeneous Surfaces*. Acad. Press (1991). (Further reading for most sections of this chapter. Some emphasis on heterogeneous surfaces.)

J.R. Sams, *Application of Statistical Mechanics to Physical Adsorption* in *Progress in Surface and Membrane Science*, D.A. Cadenhead, J.F. Danielli and M.D. Rosenberg, Eds., Vol. **8**, Academic Press (1974). (Review.)

W.A. Steele, *The Interaction of Gases with Solid Surfaces*, Pergamon Press (1974). (Thermodynamical and statistical theories for gas adsorption, molecular beam scattering from surfaces.)

W.A. Steele, *The Physical Adsorption of Gases on Solids* in Adv. Colloid Interface Sci. **1** (1967) 3. (Older but not outdated review; many examples.)

Dynamic Processes on Solid Surfaces, K. Tamaru, Ed., Plenum (1993). (Book by Japanese authors, dealing with catalysis, chemisorption, surface spectroscopies and surface reactions.)

Proceedings of the Third ISSP International Symposium on Dynamical Processes at Solid Surfaces, Tokyo, Japan (1993), K. Terakura, Y. Murata, Eds., North Holland (1993).

J. Texter, K. Klier, and A.C. Zettlemoyer, *Water at Surfaces* in *Progr. Surface Membrane Sci.* **12**, D.A. Cadenhead, J.F. Danielli, Eds. (1978) 327. (Review: adsorption and other methods; some emphasis on biological adsorbents.)

J. Zegenhagen, *Surface Structure Determination with X-ray Standing Waves*, Surface Science Reports **18** (1993) 202. (Review of the X-ray standing wave technique in its application to surface investigations. Discussion of physical principles (interference and scattering of X-rays, generating standing waves, inelastic scattering) and experimental set-up.)

1.9d. Adsorption and porosity

Characterization of Porous Solids (COPS). Proceedings of a series of conferences published by Elsevier. **I** Bad Soden, Germany (1987) K.K. Unger, J. Rouquerol, K.S.W. Sing and H. Kral, Eds., published 1987; **II** Alicante, Spain (1990), F. Rodriquez-Reinosa, J. Rouquerol, K.S.W. Sing and K.K. Unger, Eds., published 1991; **III** Marseille, France (1993) J. Rouquerol, F. Rodriguez-Reinoso, K.S.W. Sing and K.K. Unger, Eds., published 1994.

M.M. Dubinin, *Porous Structure of Adsorbents and Catalysts, Adv. Colloid Interface Sci.* **2** (1968) 217. (Short review.)

R. Evans, *Microscopic Theories of Simple Fluids and Their Interfaces*, in *Liquids at Interfaces*, J. Charvolin, J.F. Joanny and J. Zinn-Justin, Eds., Les Houches XLVIII, Elsevier (1989). (Proceedings of an advanced course, held in France, 1988.)

Characterization of Porous Solids, S.J. Gregg, K.S.W. Sing and H.F. Stoeckli, Eds., Soc. Chem. Ind., London (1979). (Proceedings of a symposium, held in Switzerland, 1978.)

S.J. Gregg, K.S.W. Sing, *Adsorption, Surface Area and Porosity*, Academic Press, 2nd ed. (1982). (Detailed discussion with many examples on the analysis of data obeying various isotherm types.)

S. Lowell, E. Shields, *Powder Surface Area and Porosity*, Chapman and Hall, 3rd ed. (1991). (Adsorption as a tool for the determination of surface area, porosity; porosimetry; hysteresis. Theory and experiment.)

H. Schubert, *Kapillarität in Porösen Feststoffsystemen*, Springer (1982). (Capillarity, porous media, transport phenomena.)

2 ADSORPTION FROM SOLUTION. LOW MOLECULAR MASS, UNCHARGED MOLECULES

For our purposes, adsorption from solution is of more direct relevance than gas adsorption. Most, if not all, topics in the five volumes of FICS involve one or more elements of it. In the present chapter, the basic elements will be introduced, restricting ourselves to low molecular weight, uncharged adsorbates and solid surfaces. Adsorption of charged species leads to the formation of electrical double layers, which will be treated in chapter 3. Adsorption at fluid/fluid interfaces follows in Volume III. Adsorption of macromolecules will be introduced in chapter 5. Between monomers, short oligomers, longer oligomers and polymers there is no sharp transition; in the present chapter we shall go as far as non-ionic surfactants, but omit most of the association and micelle formation features, which will be addressed in a later Volume. There will be some emphasis on aqueous systems.

The present theme follows logically from that discussed in the previous chapter. Further foundation material can be found in Volume I, including adsorption thermodynamics (sec. I.2.20e), interaction forces (secs. I.4.4 and 4.5) and diffusion-determined adsorption/desorption rates (sec. I.6.5).

2.1 Basic features

The main differences between adsorption from the gas and liquid phase are the following.

1. A solution is typically a system of more than one component. In actual cases, there are at least two substances that can adsorb. For a binary fluid mixture, including dilute solutions, adsorption of one type of molecule (say A) involves replacement of the other (B). Thus, adsorption from solution is essentially an *exchange* process. If one molecule of A replaces r molecules of B at the interface, the adsorption equilibrium can be written as

$$n_A^L + r\,n_B^\sigma \leftrightarrows r\,n_B^L + n_A^\sigma \qquad [2.1.1]$$

where n stands for the numbers of moles and the superscripts L and σ denote the solution and adsorbate, respectively[1].

2. As a direct consequence of 1), the adsorption enthalpy and other thermo-dynamic properties are *composite* quantities, containing at least four terms. Some of these terms may dominate, and this is not necessarily the enthalpy of the adsorptive-adsorbent interaction. As a typical illustration, binding of hydrophobic (parts of) molecules on hydrophobic adsorbents from aqueous solution is mainly driven by their dislike for water and not by their attraction to the surface.

3. A second, more analytical, consequence of [2.1.1] is that the change in bulk composition is not only the consequence of the disappearance of, say A from a mixture of A and B, but is also due to the desorption of B. When an isotherm is measured on the basis of depletion of component A in solution, i.e. when Δx_A^L is measured, the resulting isotherm is not an individual isotherm, relating to the interfacial properties of A only, but a *surface excess* (formally called composite) isotherm, relating to the interfacial properties of A *and* B. There is no thermodynamic way to decompose such surface excess isotherms into the two *individual* ones, although there are situations where this can be done with reasonable model assumptions.

One of the typical implications of the excess nature of the measured adsorption is that there are situations where, upon the addition of an adsorbent, A adsorbs positively, although $\Delta x_A^L > 0$, i.e. it seems as if A is expelled. In dilute solutions $(x_A^L \ll x_B^L$, so that $\Delta x_A^L \approx \Delta n_A^L / \Delta n_B^L)$ such a behaviour is not encount-ered. Obviously, such seemingly negative adsorptions are not found when n_A^σ and/or n_B^σ are measured directly, say spectroscopically.

Figure 2.1 gives a numerical example. In the initial situation i (fig. 2.1a) we have a solution consisting of $N_1^L(i) = 360$ white and $N_2^L(i) = 40$ black molecules. The mole fractions are $x_1^L(i) = 0.900$ and $x_2^L(i) = 0.100$, respectively. We do not consider differences with respect to size or shape between molecules 1 and 2. In the final situation f (fig. 2.1b) a piece of adsorbent has been introduced, onto which black molecules have adsorbed preferentially. We see, but ignore, that there is some uncertainty in rigorously discriminating between adsorbed and non-adsorbed molecules. Let us, for the sake of argument, say that 22 black molecules and 14 white molecules are adsorbed. In the adsorbate, $\theta_2 \equiv x_2^\sigma = 0.611$. In the equilibrium solution there remains $N_2^L(f) = 18$ and $N_1^L(f) = 346$, hence $x_2^L(f) = 0.0495$, hence $\Delta x_2^L(f) = -0.0505$. Consider now an initial solution that is very rich in blacks, containing 20 white and 380 black molecules. Assume that again 22 blacks and 14 whites adsorb, i.e. θ_2 is the same

---

[1] In this chapter the superscript S denotes the solid phase. The superscript σ refers to the actual amount present at the interface. Note that some authors use σ for excess with respect to a certain component. In our notation such an excess is indicated by a subscript, giving in parentheses the compound acting as the reference, see sec. 2.3a.

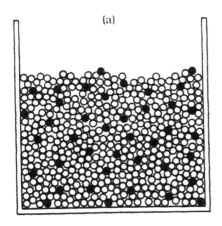

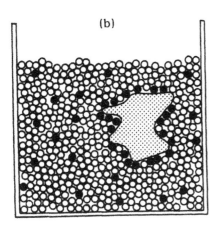

Figure 2.1. Adsorption from solution. An introductory example. (a) is the inital, (b) the final situation. O component 1 ("solvent"), ● component 2 ("solute").

as in the previous example. After adsorption the equilibrium solution consists of 6 whites and 358 blacks, i.e. $\Delta x_2^L(f) = +0.0335$. Hence, the mole fraction of component 2 has increased, although the mole fraction at the surface is identical to that in the previous case. Such observations are typical for surface excess adsorption from binary mixtures.

4. Isotherms from solution may exhibit non-ideality not only because of lateral interaction between adsorbed molecules across the "solvent", but also because of non-ideality in solution (limited solubility, phase transitions, micelle formation of surfactants, etc.).

5. Multilayer adsorption from solution is less common than it is from the gas phase, because of the stronger screening of interaction forces in condensed

fluids. Multilayers may form when demixing conditions are approached. In aqueous systems, containing surfactants, associate structures may develop, caused by hydrophobic bonding.

6. Practical problems, encountered with gas adsorption, like porosity, surface heterogeneity, non-inertness or partial solubility of the solid and the influence of impurities, etc. recur, *mutatis mutandis*. The extent to which porosity manifests itself will in first instance be determined by the sizes of the molecules in the mixtures.

7. As discussed in some detail in chapter 1, for a rigorous test of adsorption models, especially in terms of isotherms, it is necessary to have data for several decades of bulk concentration at one's disposal. Because of analytical limitations, such information is usually restricted to a few decades for adsorption from solution .

A division into "adsorption from dilute solution" and "adsorption from binary (and multicomponent) mixtures covering the entire mole fraction scale" appears to be useful. For simplicity, we shall designate mixtures covering the entire mole fraction scale as *binary mixtures*, as opposed to *dilute solutions*. This distinction is a consequence of issues (1) - (3) above, and reflected in thermodynamic and statistical interpretations. For instance, in dilute solutions locating the Gibbs dividing plane is not a problem, but for a mixture in which one of the components cannot confidently be identified as the solvent, it is.

We recall sec. I.2.22, and in particular fig. I.2.13 where the consequences of basing the Gibbs plane on a major or a minor component are illustrated. Statistically, adsorption from dilute solution is easy when the solvent may be interpreted primitively, i.e. as a structureless continuum. Then, much of chapter 1 may be applied after minor modification. For binary mixtures this becomes more problematic. In practice, adsorption from (dilute) solution is more frequently met than that from binary mixtures.

In this connection, the surface excess Γ_i of a certain component i can always be written as

$$\Gamma_i = \frac{n_i^\sigma}{A} = \frac{1}{N_{Av}} \int_0^\infty \left[\rho_{N_i}(z) - \rho_{N_i}^L \right] dz \qquad\qquad [2.1.2]$$

where ρ_{N_i} is the number density of i and $\rho_{N_i}^L$ is the value of ρ_{N_i} in the bulk reference, that is the mixture with respect to which the excess is counted. A is the area. Equation [2.1.2] is a variant of [1.6.12], which contains the distribution function $g(z)$. The issue of locating the Gibbs dividing plane is embodied in the definition of the lower integration limit, $z \equiv 0$, coinciding with the solid-liquid interface. Alternatively, we can locate the Gibbs dividing plane with respect to a certain reference substance r in solution. In the latter case we have to integrate from $-\infty$ (or the liquid-solid boundary) instead of $z = 0$. Now $\Gamma_r \equiv 0$ and Γ_i becomes $\Gamma_i^{(r)}$, which depends on the functionality $\rho_{N_r}(z)$. When r is the major component, i.e. when the mole fractions of the other components i are very small, the surface excess of i is practically independent of the location of the Gibbs dividing plane. However, when we have to consider the entire range of x_r from 0 to 1 the dependency $\rho_{N_r}(z)$ must be rigorously known. As this function changes with the interfacial composition, the position of the Gibbs plane becomes variable and more difficult to establish.

Summarizing, interpretation of adsorption from binary mixtures poses more problems than adsorption from dilute solutions, but it also exhibits the basic issues more clearly. Therefore, we shall treat these problems first (secs. 2.3, 2.4 and 2.6) and thereafter, starting with sec. 2.7, address dilute solutions.

2.2 The interface between solids and pure liquids

Across vapour, a solid and a macroscopic amount of liquid always attract each other. The Hamaker constant A_{12} for the interaction between the dissimilar macrobodies 1 and 2 is positive, see [I.4.6.3], and recall that to a good approximation $A_{12} \approx (A_{11}A_{22})^{1/2}$ ([I.4.6.5]), where A_{11} and A_{22} are the Hamaker constants across a vacuum for two macrobodies 1 and two macrobodies 2, respectively. In the Lifshits picture (see sec. I.4.7) the same conclusion is obtained. The consequence is that any solid will always exhibit *adhesional wetting* to any liquid, i.e. a solid particle will at least adhere to the surfaces of pure liquids. Whether *immersional wetting* (particles disperse spontaneously in the liquid) also takes place is another matter, to be discussed in chapter 8. For the present purpose the consequence is that a liquid in contact with a solid will never detach itself spontaneously, even if the *cohesion* between the molecules of the liquid greatly exceeds the *adhesion* between liquid and solid. Different situations may arise when the fluid is a binary mixture; due to differences in dispersion forces one of the components may enrich the interface (i.e. adsorb) and, if the two components are not entirely mutually soluble, spontaneous film formation may take place at the solid-liquid boundary.

We now consider the structure of pure (monocomponent) liquids adjacent to solid surfaces, which we assume to be inert. The very idea that liquids vicinal to solids might have different structures is old[1]. Another question is how far such structurally-modified layers extend into the liquid phase. Arguments that such layers are very thick arise regularly. However, considerable restraint is required not to overinterpret excess data, particularly in complex systems. In this respect

[1] W.B. Hardy, *Proc. Roy. Soc. (London)* **A86** (1912) 610.

we shall restrict ourselves to examples that are unambiguously established.

Addressing this problem implies discussing the notion of liquid structure and the influence exerted on it by a nearby, different, phase. The notion of structure of a system in which the molecules are continually changing their positions can only be made rigorously concrete by statistical means, and it is embodied in the notions of radial and angle-dependent distribution functions, $g(r)$ and $g(r, \theta)$, respectively. Distribution functions have been introduced in secs. I.3.9d and e, the structure of solvents, emphasizing water, in sec. I.5.3d. Distribution functions are in principle measurable by scattering techniques, see I.App.11. For liquids near phase boundaries these distribution functions become asymmetrical. However, it is not always possible, and, for that matter, not always necessary to consider the structure in such detail.

Experimental techniques usually only measure certain aspects of the structure, either static (density, number of certain molecular contacts, etc.) or dynamic (local diffusion coefficients, residence times of molecules, etc.). It must be kept in mind that with these techniques only part of the overall structure is recovered. The simplest models are based on *primitive liquids*, defined as those where the liquid is a homogeneous, structureless medium.

2.2a Experiments

A number of experimental techniques are available to obtain information on some aspects of the structure of liquids near surfaces. It should be realized at the onset that measuring the solid-liquid interfacial tension γ^{SL} does *not* belong to this category because it is a thermodynamically inoperable quantity. One cannot isothermally and reversibly extend interfaces with solids (see sec. I.2.24). From contact angles one can obtain $\gamma^{SL} - \gamma^{SG}$ but there are no thermodynamically operational procedures to split this difference into its constituents, although models which do this are available (Volume III).

Enthalpies of wetting or of *immersion* constitute perhaps the most direct thermodynamic information. We refer to table 1.3 (in sec. 1.3f) and the related discussion. Measurements of $\Delta_w H = \Delta_{imm} H$ can be made if a sufficiently large interfacial area is available, that is, with finely dispersed powders. The results depend sensitively on details of the solid surface, especially on the presence of certain surface groups that bind specifically with the liquid into which the solid is to be immersed. For the same reason, inadvertant impurities on the surface may pose a problem. Difficulties may also arise with incomplete wetting, or in establishing the available interfacial area. The quantity $\Delta_w H$ can also be obtained from the temperature dependence of the contact angle, see [1.3.43]. However, it is impossible to find γ^{SL} from $\Delta_w H$.

When the reversibility of the wetting process is assumed, the *entropy of wetting* follows from $\Delta_w S = \Delta_{imm} S = q_{imm} / T$. Reversibility can only be assumed, because the reverse process to give the heat of dewetting is virtually impossible to carry out.

Part of these data on $\Delta_w H$ inform us about the relative preference of SL over LL interactions, which is related to the difference between the Hamaker constants A_{SL} and A_{LL}. Particularly with water as the liquid, the molar enthalpy of wetting $\Delta_w H_m$ can act as a measure of the otherwise vague notion of *surface hydrophilicity/hydrophobicity*. The molar enthalpy of condensation of water is -44 kJ mol^{-1} at room temperature, so one could classify surfaces with higher $|\Delta_w H_m|$ as hydrophilic and those with lower $|\Delta_w H_m|$ as hydrophobic. The higher $|\Delta_w H_m|$, the more hydrophilic the surface is. Here, the hydrophilicity is caused by hydrogen bonds, attachment of water molecules to Lewis sites, or even chemisorption. From hydrophobic surfaces, water molecules are relatively "rejected", in the sense that they prefer to bind to their colleagues in the solution.

In some cases a *functional adsorption* analysis can be carried out by titrating the surface with a reagent for surface groups. For instance, Brönsted acid groups on the surface of a solid can be detected and counted by titration with a Brönsted base. Basically the number of groups and the binding affinity can be obtained. For instance, when this binding obeys Langmuir's law, the solution equivalent of the linear plots [1.4.2 or 3] may be used to find k_1 and k_2 and hence the monolayer coverage and, using [1.3.40], the Gibbs energy of adsorption. For charged particles such titrations are replaced by colloid titrations (sec. I.5.6e and chapter 3).

Other macroscopic properties that in principle can be measured are the *excess density* and the *excess compressibility* of the interfacial liquid. These excess quantities can be positive or negative and follow from a comparison of the corresponding quantities in systems with the liquid and solid separated. Alternatively, liquid behaviour in pores can be studied. An example of this kind has been given by Derjaguin[1] who claims that water in narrow pores of silica gel or Aerosil does not exhibit the typical thermal expansion minimum at 4°C because of structural changes near the surface. Löring and Findenegg[2] studied surface excesses *dilatometrically*.

Spectroscopic techniques give information at a more molecular level. *Neutron reflection*[3] yields information on the liquid density profile $\rho_N(z)$ in the direction normal to the surface. This technique may lead to a non-unique solution, in that a form of $\rho_N(z)$ is found which fits the reflection data; other distributions may fit as well. With ellipsometry, introduced in sec. I.7.10b, the

[1] B.V. Derjaguin, *Colloids Surf.* **38** (1989) 49.
[2] R. Löring. G.H. Findenegg, *J. Colloid Interface Sci.* **84** (1981) 355.
[3] T.L. Crowley, E.M. Lee, E.A. Simister, R.K. Thomas, J. Penfold and A.R. Renie, *Colloids Surf.* **52** (1991) 85.

variation of the refractive index and the depth of the layer over which this variation extends are obtained; this layer is generally interpreted as homogeneous and separated from the bulk through a step function. Law[1] reported that, with Pyrex, surface-induced orientational ordering of water and aniline molecules of up to ≈ 7 and 4.5 molecular diameters respectively occurs. The extent of this ordering is sensitive to the mode of preparation of the adsorbent samples.

Infrared spectra of powders dispersed in liquids may be recorded, but a variety of precautions have to be taken. Obvious problems are absorption by the solvent and by the material of the cell. The former problem can be overcome by working with concentrated dispersions having high interfacial area, and subtracting the spectrum of the solution, or by working with compressed discs of the solid into which the liquid is imbibed. The latter problem can be avoided by choosing cell material other than glass or by using special windows[2]. Working with evanescent waves, especially in the multiple reflection variant, has the advantage that more surface and less bulk is seen.

Nuclear magnetic resonance (NMR) and *electron spin resonance* (ESR) were introduced in sec. I.7.13. They can give information on the dynamics of the molecules and their surroundings. If applied to disperse systems it is, in principle, possible to detect changes in the dynamics for the part of the liquid adjacent to the surface. Broadening of spectral lines can be an indicator of the slowing down of the relevant relaxation process. Obviously, the appropriate isotopes must be available and care has to be taken to identify the physical bases of these relaxations. Most intriguing are systems where the liquid is water. An example is an extensive study by Piculell et al.[3] of aqueous silica dispersions in which ^2H and ^{17}O spin relaxation was studied. One of the problems faced was to discriminate between relaxation of the protons of the water itself and that due to rapid proton or deuteron exchange between silica and water. A second problem was to propose a physical interpretation for an observed, relatively slow, process (longer than 10^{-8} s) for the ^{17}O and ^2H relaxation. The former problem was attacked by systematically comparing the ^2H and ^{17}O relaxation rates, because for pure water these two nuclei relax through intramolecular mechanisms giving virtually the same information on water molecule reorientation. Hence, the bulk process can be distinguished from the bulk-surface exchange process. The observed slow step was assigned to the strongly reduced (by at least two orders of magnitude) translational mobility of

[1] B.M. Law, *J. Colloid Interface Sci.* **134** (1990) 1.
[2] G.D. Parfitt, C.H. Rochester, in *Adsorption from Solution at the Solid/Liquid Interface*, G.D. Parfitt and C.H. Rochester, Eds., Academic Press (1983), sec. 1C.
[3] L. Piculell, *J. Chem. Soc., Faraday Trans. I* **82** (1986) 387; B. Halle, L. Piculell, ibid, 415.

hydrated water; the lateral diffusivity is at least one order of magnitude lower than in bulk. It was also found that the reduction of the translational mobility of the water was caused by many-body correlations. This is not surprising for a liquid as complicated as water[1].

An indirect, but for colloid stability very relevant, technique involves the *surface force apparatus*, mentioned in sec. I.4.8. This apparatus, basically

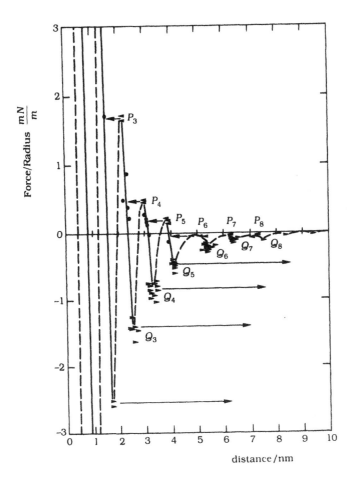

Figure 2.2. Force as a function of distance between two curved mica surfaces of radius 1.65 cm in OMCTS, T=22°C. The arrows indicate inward and outward jumps from unstable to stable points. The dashed parts cannot be measured. (Redrawn from R.G. Horn, J.N. Israelachvili, *J. Chem. Phys.* **75** (1981) 1409.) These results were essentially confirmed by P. Attard and J.L. Parker, *J. Phys. Chem.* **56** (1992) 5086.

---

[1] For reviews of magnetic resonance in colloid and interface science, see *Adv. Colloid Interface Sci.* **23** (1985) and *Colloids Surf.* **45** (1990).

developed by Israelachvili and Adams[1], is more or less routinely used to measure forces between crossed cylindrical mica surfaces with adsorbates on them. The main purpose is the study of colloidal interactions but the method can also be used for liquid structure near surfaces. Any periodicity in the ordering of liquid molecules close to the surface is reflected as a force that undulates as a function of the distance between the surfaces. As such, ordering occurs only over a very limited number of molecular layers, the measurements must be carried out down to very short separations, and extreme precautions have to be taken (smooth surfaces, absence of spurious adsorbates, no electrical double layers, etc.). The phenomenon is the counterpart of liquid ordering in narrow slits which may give rise to phase transitions, see figs. 1.38 and 1.42. Essentially, we are dealing with the colloidal analogue of *solvent structure-mediated interactions*, introduced in sec. I.5.4.

Perhaps the first experimental example observed was that of octamethyl cyclotetrasiloxane (OMCTS) between mica, see fig. 2.2. The molecules of this liquid are spherical with a diameter of about 1 nm. It is seen that at short separations the interaction force has an oscillatory nature with a periodicity corresponding to the molecular diameter. Parts of the curve cannot be measured because upon approach the plates jump into one of the minima, after which they can be pushed closer before jumping into the next minimum and so on. In this example, the magnitudes of the oscillations were insensitive to temperature variation but were reduced by additions of water.

Figure 2.3 shows similar measurements for water. The maxima due to structure-mediated repulsion are very steep and are found at 0 (contact), 0.25, 0.55, 0.72, 0.95, 1.21, 1.45 and 1.67 nm and must reflect the layer-wise orderingof the water near the surface with a periodicity of about 0.24 nm on the average, which is close to the diameter of a water molecule. It is interesting to compare this figure with fig. I.5.6 for the radial distribution functions obtained by X-ray analysis. In both cases the deviations from bulk structure persist to approximately 0.8 nm from the surface, i.e. to about four molecular layers.

Generally, an exponential decay of the height of the maxima, as a function of distance, is found, i.e. when the separation of the two surfaces is h, the solvent structure-mediated contribution to the disjoining pressure can be written as

$$\Pi_{ss}(h) = A\,e^{-h/\xi^{\sigma}} \qquad\qquad\qquad [2.2.1]$$

where A is a factor of dimensions [force/area] and ξ^{σ} is a *surface correlation length*. In sec. I.7.7c the correlation length ξ for bulk phases was defined. The trend is that ξ^{σ} and ξ are of the same order of magnitude, viz. that of

[1] J.N. Israelachvili, G.E. Adams, *J. Chem. Soc., Faraday Trans. I* **74** (1978) 975.

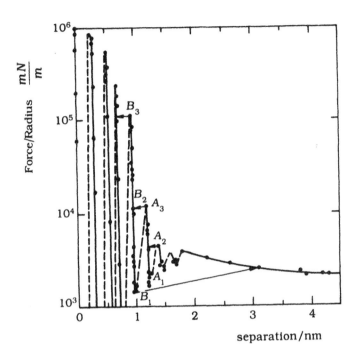

Figure 2.3. Force between mica surfaces in 10^{-3} M KCl. Redrawn from R.M. Pashley and J.N. Israelachvili, *J. Colloid Interface Sci.* **101** (1984) 511. The dashed parts cannot be measured.

molecular size. However, for surfaces that induce long-range structure in the adjacent liquid ξ^σ may exceed ξ.

2.2b Theoretical models

Statistical theories for bulk fluids can be modified to account for the presence of an adjacent external phase. Two routes are available:

(i) Modification of one of the integral equations for the distribution functions, mentioned in sec. I.3.9d (PY, HNC, etc.) or modification of the virial approach (sec. I.3.9c). For instance, Henderson et al.[1] considered a fluid near a wall as the limiting case of a binary fluid where one of the components has an infinitely large molecular size.

(ii) Monte Carlo or Molecular Dynamics simulation.

Generally, a number of model assumptions have to be made regarding the intermolecular interactions. For one thing, the surface has to be modelled; it can attract or repel the molecules of the liquid or be indifferent. When the

---

[1] D. Henderson, F.F. Abraham and J.A. Barker, *Mol. Phys.* **31** (1976) 1291.

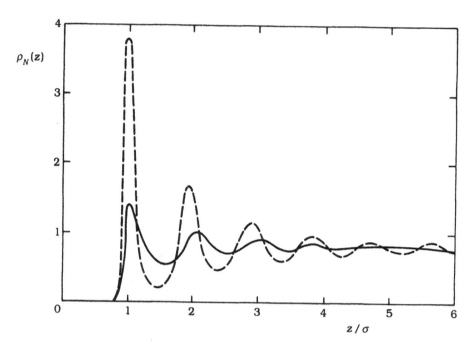

Figure 2.4. Density profile for a Lennard-Jones fluid against a solid wall. The distance z is normalized as z/σ, where $\rho_N \sigma^3 = 0.80$, and ρ_N is the bulk density. Dashed curve, fluid-wall attraction, solid curve, fluid-wall repulsion. (Redrawn from I. Snook and W. van Megen, *J. Chem. Phys.* **70** (1979) 3099.)

liquid molecules are not isotropic, the influence of the surface on the orientation of adjacent molecules also has to be modelled. For the interaction between the liquid molecules, pairwise additive interactions are usually assumed, say of the Lennard-Jones type (sec. I.4.5b). Results are normally presented in terms of *density profiles*, i.e. densities as a function of the distance, $\rho_N(z)$. For asymmetrical molecules, such profiles may differ for different orientations, depending on the interactions of these orientations with the surface.

As a trend, such profiles are similar to those for the SG interface, of which fig. 1.38 was an illustration. Usually there is a maximum near the solid wall, beyond which a few oscillations can be seen. Details of this distribution and the distance over which the undulations damp out are determined by the interaction energies. We shall illustrate some of these features with a few examples.

Figure 2.4 has been taken from a Monte Carlo simulation and applies to simple spherical molecules near a solid wall. The consequences of a number of different types of intermolecular interactions, fluid-wall interactions and bulk densities, ρ_N, were studied. Undulations over a few molecular diameters are obtained, which in this example become more pronounced if there is fluid-wall

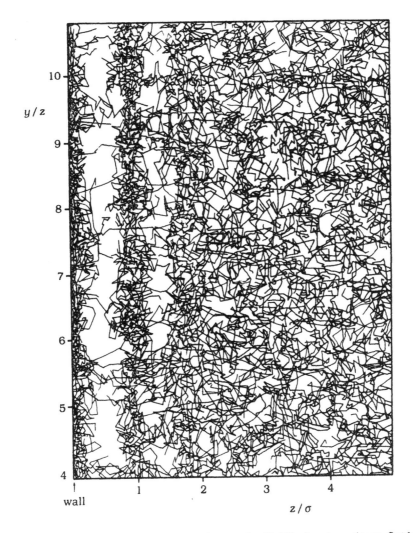

Figure 2.5. Molecular Dynamics simulation of self-diffusion in a dense fluid of "soft" spherical particles near a "hard" solid wall. The "wall" exerts no force on the particles but reverses the z-component of the velocity if a molecule attempts to cross it; σ is the length parameter in the repulsive part of the Lennard-Jones interaction. (Redrawn from J.N. Cape, *J. Chem. Soc., Faraday Trans. II* **78** (1982) 317.)

attraction. The shapes of such curves depend on details of the pair interactions and on ρ_N. Especially for high densities, attractive tails may be "overruled" by the volume constraints, which in itself also leads to periodicity. The repulsive parts of the interactions then dominate.

Figure 2.5 shows a Molecular Dynamics simulation, which is the counterpart of the previous example. Particle trajectories are shown for fluid particles, interacting via an r^{-12} law (as in the Lennard-Jones interaction), with a structureless "hard" wall. A striking feature is that the diffusion between the first layer(s) and the bulk is restricted. In this example, the ratio $D_\perp/D_{//}$ between the perpendicular and parallel self-diffusion coefficients is less than 0.03. In this connection it is interesting to note that in cases where, due to binding with the surface, very low tangential diffusion coefficients are found for molecules in contact with a surface, tangential transport may be faster by desorption to a second or further layer, followed by re-adsorption. The caveat must be made that in fig. 2.5 ideally smooth (Fresnel) surfaces were considered. Real surfaces are usually rough on this scale, leading to smoothing of these contributions.

Not only can simulations predict distributions, dynamic properties can also be obtained using Molecular Dynamics. From I.App.11 it is recalled that the extent to which a molecule is able to retain its original velocity $v(0)$, during a brief period t, is quantified in the *velocity (auto) correlation function* $C_v(t)$, defined as [I.A11.1]

$$C_v(t) \equiv \langle v(0) \cdot v(t) \rangle \qquad\qquad [2.2.2]$$

In I.A11 it was shown that, in bulk liquids for an individual molecule i, C_v is simply related to the self-diffusion coefficient D_s through

$$D_s = \frac{1}{3} \int_0^\infty \langle v_i(0) \cdot v_i(t) \rangle dt \qquad\qquad [2.2.3]$$

For anisotropic fluids, including water, moving parallel to the wall, [2.2.3] has to be replaced by

$$D_{s//} = \frac{1}{2} \int_0^\infty \langle v_x(0) \cdot v_x(t) \rangle + \langle v_y(0) \cdot v_y(t) \rangle dt \qquad\qquad [2.2.4]$$

It is also possible to relate the diffusion coefficients to the corresponding displacements by modifying [I.6.3.2a-b]

$$D_{s//} = \frac{(\Delta x)^2 + (\Delta y)^2}{4t} \qquad\qquad [2.2.5a]$$

$$D_{s\perp} = \frac{\Delta z^2}{2t} \qquad\qquad [2.2.5b]$$

Regarding the surface diffusion coefficient D^σ (and $D^\sigma_{s//}$, $D^\sigma_{s\perp}$) the situation regarding dynamic simulations is that these quantities do not differ by more than an order of magnitude from the corresponding bulk values[1], although the

---

[1] H. Saint-Martin, I. Ortega-Blake, *The Microscopic Structure and Dynamics of Water at a Surface*, in *Biomolecules in Organic Solvents*, A. Gomez-Puyou, Ed., CRC (1992).

question whether $D^\sigma > D$ or $D^\sigma < D$ is under dispute[1]. For electrokinetics it is relevant that even if the surface self-diffusion coefficient does not differ dramatically from that in bulk, the apparent viscosity for tangential flow may substantially exceed that in bulk water. This feature, responsible for the exist- ence of a thin non-slip layer, is well established in electrokinetics: although this layer as a whole appears immobile, ions embedded in it may be quite mobile, leading to a non-negligible surface conductivity (sec. 4.4f). For instance, for silica and haematite sols there are indications that the tangential mobilities of counterions in the stagnant layer are of the same order as, but lower than, those in the bulk, see table 4.3.

2.2c. The solid-water interface

The structure of water near surfaces is of more than average interest and its significance far exceeds the confines of this chapter. For a start, this issue has to be considered in all double layer models of a higher sophistication than the primitive ones. Electrochemists discuss the contribution of the polarization of water near charged surfaces to potential drops and (inner) double layer capacitances (see sec. 3.9). In electrokinetics tangential slip is an issue: why does a thin water layer near the surface remain hydrodynamically immobile, where- as diffusion in this layer is not so strongly reduced? (sec. 4.4). How should one interpret the surface tension of water and the displacement of (or by) water in wetting and dewetting phenomena? In Volume III these issues recur. In later Volumes we intend to deal with the interpretation of water structure-mediated interaction forces between colloids (anticipated in fig. 2.3) and the properties of water near biological surfaces.

Ideally, models of vicinal water should eventually "explain" all established experimental facts. There is a long way to go! However, some general observa- tions have been made. One is that, against a variety of hydrophobic phases (silver iodide, mercury, air) water molecules appear to be oriented with the negative ends of the molecules pointing outward (sec. 3.9). In other words, the polarization of water adjacent to silver iodide and mercury is similar to the spontaneous polarization of water surfaces. The implication is that near such surfaces water-water interactions play at least an important role as water- surface interactions. Another observation, relevant for the interpretation of electrokinetic phenomena, is that tangentially immobile surface layers do occur near both hydrophilic *and* hydrophobic surfaces.

---

[1] For further reading, see *Fundamentals of Inhomogeneous Fluids*. D. Henderson, Ed. Marcel Dekker (1992). (Chapter 5 of this book, by R. Evans, describes the application of density functional theory) *The Liquid-Solid Interface at High Resolution, Faraday Discuss. Roy. Soc. Chem. (London)* (1992).)

Water is a strongly three-dimensionally structured fluid (sec. I.5.3c) with structure-originating interactions reaching several molecular diameters. Considering this, simple models and/or simulations with a limited number of molecules are not really helpful. By "simple" we mean models in which water molecules are represented as point dipoles, point quadrupoles, or as molecules with Lennard-Jones interactions plus an additional dipole, etc., and by "limited" less than, say 10^3 molecules, i.e. 10 molecules in each direction of a cubic box. Admittedly, for a number of simpler problems more embryonic models may suffice. For example, electrochemists often get away with a dipole interpretation when focusing their attention solely on the Stern layer polarization. Helmholtz's equations for the χ-potential [3.9.9] is an illustration.

In this connection it is noted that the water structure problem underlies the much-disputed issue of the range of structure-mediated surface forces.

It is easy to pose the problem; solving it is another matter. All attempts that have been made so far are embryonic, notwithstanding courageous computational efforts.

Regarding water models, some of these recur. In the Rowlinson model[1] which is primarily based upon data for the dipole moment and virial coefficient, the lone-pair negative charges of value $-0.3278e$ are situated directly above and below the oxygen atom at distances 0.02539 nm. The line connecting these charges is normal to the HOH plane. The hydrogen atoms carry the same charges, except for the sign they are situated at 0.096 nm from the centre of the oxygen.

The ST2 model, due to Stillinger and Rahman[2], is also a four-point model, but now the charges are directed at the normal tetrahedral angles. The partial charges are $0.2357e$ and the distances from the positive and negative charges to the centre of the oxygen are 0.10 and 0.08 nm, respectively.

Both in the Rowlinson and the ST2 model there is a Lennard-Jones 6-12 interaction, acting between the oxygens. For computational purposes an attenuation factor is incorporated to reduce the interaction to short distances. Other models include the MCY model[3], in which the molecule is planar, the TIPS 2 picture[4], consisting of a Lennard-Jones interaction plus three point charges and variants of it[5]. In other models hydrogen bonding is included. In a model elaborated by Besseling[6] a water molecule is modelled as having eight

[1] J.S. Rowlinson, *Trans. Faraday Soc.* **47** (1951) 120.
[2] F.H. Stillinger, A. Rahman, *J. Chem. Phys.* **60** (1974) 1545.
[3] O. Matsuoka, E. Clementi and M. Yoshimine, *J. Chem. Phys.* **64** (1976) 1351.
[4] W.L. Jorgensen, *J. Chem. Phys.* **77** (1982) 4156.
[5] W.L. Jorgensen, J. Chandrasekhar and J.D. Madura, *J. Chem. Phys.* **79** (1983), 926; W.L. Jorgenson, J.D. Madura, *Mol. Phys.* **56** (1985) 1381.
[6] N.A.M. Besseling, J. Lyklema, *J. Phys. Chem.* **98** (1994) 11610.

faces. Two of these, the donors, represent the protons, two others, the acceptors, the lone electron pairs and the remaining four are indifferent. The two donors and the two acceptors have a tetrahedral geometry and that is also the case for the four indifferent faces.

In addition to the actual distribution of charges over the water molecule, it is also necessary to consider the extent to which such charges can be displaced under the influence of intermolecular interaction, i.e. the polarizability has to be accounted for. So far, this feature has received little attention.

Before applying such models to vicinal water, they should be checked to account for the properties of bulk water (molar internal energy, pressure, specific heat, singularity at 4°C, etc.), which is sometimes done[1], and for the surface tension as a function of temperature, which is a more critical test but rarely done[2].

Simulation of water adjacent to surfaces is a developing area. So far, a number of publications have appeared, usually differing with respect to the models of the molecules and their interactions, the size of the cell considered, the properties of the surface, treatment of the statistics, etc. A review up to 1991 has been given by Saint-Martin and Ortega-Blake[3]. Obviously the results obtained depend on the specific assumptions made, but a number of features recur, so that they may be considered generic. They include,

(i) The density profile $\rho_N(z)$ has a rapidly damped oscillatory behaviour. Usually the number of oscillations is less than that found with the force apparatus, fig. 2.3. Simulations with much larger numbers of molecules are needed to further analyze this difference.

(ii) After a few molecular layers the properties are essentially bulk-like. Certainly there are no indications of long range (say, beyond 5 nm) structural influences of the surface.

(iii) When there are charges on the surface, their effect on the water structure, notably on the orientation of dipoles, is only noted in the very few first layers adjacent to the surface. If this observation is confirmed then there is an argument for considering water in the diffuse part of a double layer (sec. 3.5) as having bulk properties.

By way of illustration, figs. 2.6 and 2.7 give results of MD simulations of water near silica surfaces[4]. The water molecule exhibits so-called TIP 4P

[1] N.G. Parsonage, in *Trends in Interfacial Electrochemistry*, A.F. Silva, Ed., NATO-ASI Ser., Ser. C (1986) 359.

[2] W.L. Jorgensen, J. Chandrasekhar and J.D. Madura, *J. Chem. Phys.* **79** (1983), 926; W.L. Jorgenson, J.D. Madura, *Mol. Phys.* **56** (1985) 1381.

[3] H. Saint-Martin, I. Ortega-Blake, *The Microscopic Structure and Dynamics of Water at a Surface*, in *Biomolecules in Organic Solvents*, A. Gomez-Puyou, Ed., CRC (1992).

[4] S.H. Lee, P.J. Rossky, *J. Chem. Phys.* **100** (1994) 3334.

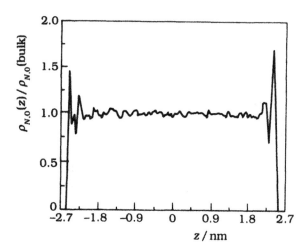

Figure 2.6. Molecular Dynamics simulation of water between two walls of (hydrophilic) silica surfaces. Given is the oxygen number density profile. (Redrawn from S.H. Lee, P.J. Rossky, *J. Chem. Phys.* **100** (1994) 3334.)

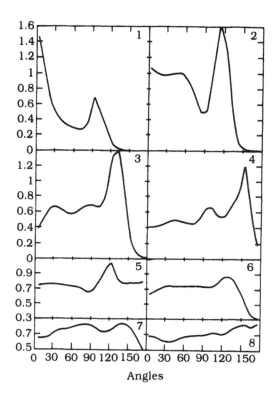

Figure 2.7. Distribution of angles between the OH bond vector and the inward-pointing normal vector. The number of the panel (i) indicates the distance from the surface in that it applies to a layer for which $|z|$ is between $2.55 - 0.05i$ and $2.60 - 0.05i$ nm, with z given in fig. 2.7 (same source as fig. 2.6).

interactions. Simulations have been done with 400 or 584 water molecules. To simulate the nature of the surface, three models were considered: (i) perfectly flat plates with a LJ (9-3) interaction between surface and water molecules; (ii) a close-packed solid of hexagonally packed silicon and oxygen atoms, mimicking hydrophilic silica surfaces and (iii) a surface mimicking hydrophilic silica, having OH groups at the surface and a geometry that was slightly different from that considered in model (ii).

Figure 2.6 gives the oxygen number density, normalized with respect to the same in the bulk, as a function of z, the distance between two "hydrophilic" silica surfaces. As the surfaces are far enough apart, bulk properties are attained in the centre ($z = 0$). Hence, the differences between the distributions near the l.h.s. and the r.h.s. wall illustrate the variability of the simulation results. Near the walls oscillations can be seen; they seem not to extend beyond three layers. For the "hydrophobic" silica density oscillations were also found, but these were less defined and had a wider range (not shown). For bulk water this model predicts an average number of nearest neighbours of about 2.60 of which 2.05 are hydrogen bonded. According to present-day knowledge the former number should in fact be about 4, depending on the temperature, whereas the latter is about right.

Of the various other characteristics that can be obtained, fig. 2.7 gives the angle distribution of the OH bond vector with the normal in layers of increasing distance to the surface. The strong preference for a low angle near the surface (panel 1), caused by the imposed strong attraction with the surface hydroxyl group, is replaced by another with increasing distance (panels 2, 3, ...) to disappear completely at longer distances. For the "hydrophobic" surfaces (not shown) these distributions are very different. From the integrated profile of polar interactions the χ-potential is in principle obtainable.

Dynamic properties can also be obtained. For hydrophilic surfaces the (total) self-diffusion coefficients in the first, the second layer and the bulk are 0.91, 2.08 and 2.87×10^{-9} m^2 s^{-1}. In the first layer, $D_\perp$ is lower than $D_{//}$ by about 16%. These trends are at least semiquantitatively according to expectation; the absolute value for D(bulk) exceeds the experimental value (table I.6.4) by about 30%.

These results give an indication of the achievements of simulations. Further interesting developments may be expected.

2.3 The surface excess isotherm

In this section we consider thermodynamic, statistical thermodynamic (i.e. model-based) and more descriptive ways to analyse surface excesses in binary mixtures.

2.3a. Thermodynamic definition of the surface excess

Starting with the insight that individual surface excesses, i.e. Γ_1 and Γ_2 are inoperational quantities, three types of (accessible) surface excess concentrations are commonly used:

$$\Gamma_2^{(n)} \equiv -n^L\, \Delta_{ads} x_2 / A \tag{2.3.1}$$

$$\Gamma_2^{(1)} \equiv -n^L\, \Delta_{ads} x_2 / A (1 - x_2) \tag{2.3.2}$$

$$\Gamma_2^{(V)} \equiv -V\, \Delta_{ads} c_2 / A \tag{2.3.3}$$

Here, $n^L = n_1^L + n_2^L$ is the total amount (in moles) in the liquid phase, A is the interfacial area, V the liquid volume and $\Delta_{ads} x_2 = x_2(f) - x_2(i)$, where (f) and (i) refer to the final and initial situation, respectively. A plot of one of these Γ_2's as a function of x_2 or c_2 is a *surface excess isotherm*, i.e. there are three kinds of them.

The quantity $\Gamma_2^{(n)}$ is the surface excess of 2 per unit area over the amount of 2 in a reference system containing n^L moles of liquid of uniform composition in which x_2 is the mole fraction of 2. Essentially, this is the excess obtained by the procedures discussed in connection with fig. 2.1. The sum $\Gamma_2^{(n)} + \Gamma_1^{(n)}$ is zero.

Excess $\Gamma_2^{(1)}$ refers to a reference system containing the same amount of 1 as the real system; the composition is again constant up to the surface. It is implicit in this definition that $\Gamma_1^{(1)} \equiv 0$. As

$$\Gamma_2^{(n)} = (1 - x_2)\Gamma_2^{(1)} \tag{2.3.4}$$

$\Gamma_2^{(n)}$ is always smaller than $\Gamma_2^{(1)}$. Physically, this can be understood as follows: as some 1 also adsorbs, the level of the reference goes down, and comparison with a lower reference yields a higher value. Of course, for dilute solutions ($x_2 \ll 1$) the two excesses are identical. For $x_2 \rightarrow 1$ both approach zero.

The excess $\Gamma_2^{(V)}$ is the quantity that is usually measured for adsorption from dilute solutions. Now the reference system has the same volume, in which c_2 is constant up to the phase boundary. V is the total volume minus the volume of the adsorbent which we assumed constant. Working with volumes can have advantages when the fluid density remains constant. For such a constant fluid density (i.e. the molar volumes V_{m1} and V_{m2} not depending on x_2, persisting down to the wall),

$$\Gamma_2^{(V)} = \Gamma_2^{(n)}\left[1 - \left(V_{m2} - V_{m1}\right)c_2^L\right] \tag{2.3.5}$$

Relations between the different types of excess concentrations are discussed in more detail by Király and Dékány[1].

We shall mostly work with $\Gamma_2^{(n)}$.

2.3b. The Ostwald-Kipling isotherm equation

The surface excess $\Gamma_2^{(n)}$ is operational, i.e. it can be measured, but to become more useful it should be related to the amounts n_1^σ and n_2^σ actually present at the surface (although we do not yet have to assume anything about the thickness of the adsorbate layer). Regarding the meanings of the superscripts, see the note in connection with [2.1.1].

We have three material balances. Writing again (i) and (f) for the initial (before adsorption) and final state (after adsorption), respectively:

$$n_1^L(i) + n_2^L(i) = n^L(i) = n \tag{a}$$

$$n_1^L(f) = n_1^L(i) - n_1^\sigma \tag{b}$$

$$n_2^L(f) = n_2^L(i) - n_2^\sigma \tag{c}$$

Realizing that $n_1^L(f)/n_2^L(f) = x_1(f)/x_2(f)$, eq. (b) can also be written as

$$x_1(f)n_2^L(f) = x_2(f)n_1^L(i) - n_1^\sigma x_2(f) \tag{d}$$

Similarly, from (c)

$$x_2(f)n_1^L(f) = x_1(f)n_2^L(i) - n_2^\sigma x_1(f) \tag{e}$$

In (d) and (e) the l.h.s.'s are identical. By subtraction and rearrangement,

$$x_1(f)\, n_2^\sigma - x_2(f)\, n_1^\sigma = x_1(f)\, n_2^L(i) - x_2(f)\, n_1^L(i) = x_1(f)x_2(i)\, n - x_2(f)x_1(i)\, n$$

$$= [1 - x_2(f)]x_2(i)\, n - x_2(f)\, n[1 - x_2(i)]$$

$$= [x_2(i) - x_2(f)]\, n = -\Delta_{ads}x_2\, n \tag{f}$$

Hence, from [2.3.1]

$$\Gamma_2^{(n)} = -\frac{n\,\Delta_{ads}x_2}{A} = \frac{x_1(f)\, n_2^\sigma - x_2(f)\, n_1^\sigma}{A} \tag{2.3.6}$$

We shall further refer to this useful expression as the *Ostwald-Kipling equation*, after Ostwald who was the first to derive it[2] and Kipling who analysed

[1] Z. Király and I. Dékány, Colloid Polym. Sci. 266 (1988) 266.
[2] W. Ostwald, Kolloid-Z. 30 (1922) 279.

it in great detail, paying attention to many of the typical features of adsorption from binary systems[1]. Note that in [2.3.6] a *combination* of n_1^{σ} and n_2^{σ} occurs; there is no thermodynamic way to determine each of them individually. An additional assumption is needed for that. This is not required for dilute solutions, where $x_2 \ll x_1$. The Ostwald-Kipling equation is also consistent with $\Gamma_1^{(n)} + \Gamma_2^{(n)} = 0$.

When the surface area is not known, the following modification can be used:

$$-\frac{n\,\Delta_{ads}x_2}{m} = \frac{x_1(f)\,n_2^{\sigma} - x_2(f)\,n_1^{\sigma}}{m} \qquad [2.3.7]$$

where m is the mass of the adsorbent.

A second variant, in terms of dimensionless quantities, is obtained by introducing the mole fraction θ_2 in the adsorbed state

$$\theta_2 = x_2(f) - n\,\Delta_{ads}x_2/n^{\sigma} \qquad [2.3.8]$$

where n^{σ} stands for $n_1^{\sigma} + n_2^{\sigma}$. Equation [2.3.8] is verified by noting that $A\Gamma_2^{(n)}$, the excess of 2 in moles, is defined as the excess amount of 2 present in the system $(n^{\sigma}\theta_2)$ over the same if the solution composition would persist up to the surface, $n^{\sigma}x_2(f)$. So, $A\Gamma_2^{(n)} = -n\Delta_{ads}x_2 = n^{\sigma}\theta_2 - n^{\sigma}x_2(f)$, from which [2.3.8] directly follows.

2.3c. *Classification of surface excess isotherms*
Before further discussing the thermodynamics and model analyses of surface excess isotherms, let us look at some typical shapes, as shown in fig. 2.8. Types (a) and (b) are the most common. Schay and Nagy[2] distinguish more types by

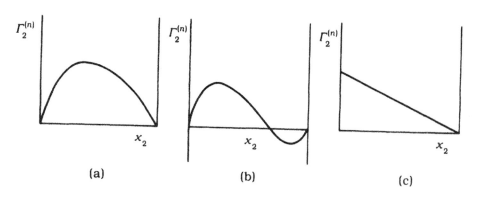

(a)   (b)   (c)

Figure 2.8. Classification of excess isotherms for binary solutions. (a) (inverse) U-shape; (b) (toppled) S-shape; (c) linear.

[1] J.J. Kipling, *Adsorption from Solution of Non-Electrolytes*, Academic Press (1966).
[2] G. Schay, L.G. Nagy, *Periodica Polytech. Budapest* **4** (1960) 45; *Acta Chim. Acad. Sci. Hung.* **50** (1966) 207.

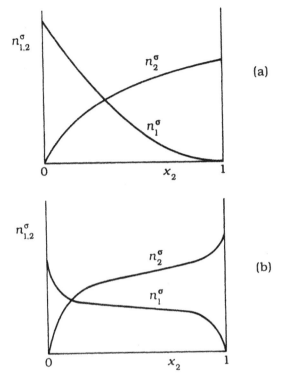

Figure 2.9. Types of partial (or individual) isotherms leading to excess isotherms of the U-type (a) or S-type (b).

recognizing various details in the isotherms of the U- or S-type, such as the occurrence (or not) of linear regions or points of symmetry at $x_2 = 0.5$. Figure 2.9 shows combinations of *individual* or *partial isotherms* which give rise to the two most common types. Type (c) is obtained if one of the components is exclusively sorbed into the pores of, say a molecular sieve, with negligible adsorption on the outer walls of the sieve. This type is rare.

The more complex the individual isotherms, the more complicated is the surface excess isotherm. The simple situation of $n_1^\sigma = 0$ and $n_2^\sigma = $ const. leads to type c. All types of isotherms have in common that $\Gamma_2^{(n)} \to 0$ for $x_2 \to 1$, which follows from the definition: withdrawal of 2 from pure 2 does not affect the bulk composition. Similarly, all isotherms start with $\Gamma_2^{(n)} = 0$ including type c in fig. 2.8, which starts vertically. The additional zero point, observed in S-shaped isotherms, is called an *azeotropic point*: it is characterized by identical composition of adsorbate and solution, $\theta_2 = x_2$, see [2.3.8]·

Models, methods and examples of decomposing surface excess isotherms into their constituent parts will be discussed in sec. 2.4.

2.3d. Enthalpies of immersion (wetting)

In sec. 1.3f the heat of immersion of solids into pure liquids was introduced as one of the thermodynamic means to obtain additional information on the

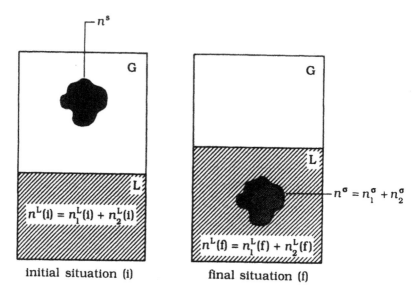

Figure 2.10. Schematic picture of a heat of immersion measurement. $\Delta_{imm}H = H(f) - H(i)$ if p is constant.

energetics of gas adsorption. Recall table 1.3 (sec. 1.3f). Similar analysis can be carried out for the immersion of adsorbents into binary fluid mixtures. This technique does not require non-thermodynamic assumptions.

For convenience we assume that all experimental problems (cleanliness of the sorbent, no vapour pockets upon immersion, inertness of the solid, etc.) have been surmounted. Measurements can be carried out with the mixture, to yield $\Delta_{imm}H$, and also with the pure components, yielding $\Delta_{imm}H_1^*$ and $\Delta_{imm}H_2^*$, which act as references. Generally, $\Delta_{imm}H$ has a number of contributions: adsorption enthalpies of both components and enthalpies of dilution, all of which are, in principle, composition-dependent.

Figure 2.10 illustrates schematically the immersion process. Before immersion the enthalpy of the system is

$$H(i) = n^S H_m^S + n_1^L(i) H_{m1}^L + n_2^L(i) H_{m2}^L$$

$$= n^S H_m^S + n^L(i)\left[x_1(i) H_{m1}^L + x_2(i) H_{m2}^L\right] \qquad [2.3.9]$$

where the H_m's are partial molar enthalpies. (For definitions and "bookkeeping" of partial quantities in mixtures see sec. I.2.16.) Following the procedure discussed in sec. I.2.18b, the enthalpy of the mixture is written as the sum of ideal mixing terms ($x_1 H_{m1}^* + x_2 H_{m2}^*$ where H_m^* is the molar enthalpy of the pure component indicated) and an excess contribution H_m^E accounting for deviations from ideality. Generally, H_m^E is a function of x. We obtain

$$H(i) = n^S H_m^S + n^L(i)\left[x_1(i)H_{m1}^* + x_2(i)\ H_{m2}^* + H_m^E(x)\right] \qquad [2.3.10]$$

Similar reasoning can be applied to the situation after immersion. Then also an adsorbate is present, of which the enthalpy is also written as the sum of two ideal terms and an excess contribution. The result is

$$H(f) = n^S H_m^S + n^L(f)\left[x_1(f)H_{m1}^* + x_2(f)\ H_{m2}^* + H_m^E(x)\right]$$

$$+ n^\sigma\left[\theta_1 H_{m1}^{\sigma*} + \theta_2 H_{m2}^{\sigma*} + H_m^{\sigma E}(\theta)\right] \qquad [2.3.11]$$

Here, H_m^E is a function of $x_2(f)$ and $H_m^{\sigma E}$ is a function of the θ's, accounting for all interadsorbate and adsorbate-solution interactions. Subtraction of [2.3.10] from [2.3.11], using $n^L(f) = n^L(i) - n^\sigma$, yields

$$\Delta_{imm}H = n^L(i)\left[x_1(f)H_{m1}^* + x_2(f)\ H_{m2}^* + H_m^E(x(f)) - x_1(i)H_{m1}^* - x_2(i)\ H_{m2}^* - H_m^E(x(i))\right]$$

$$+ n^\sigma\left[\theta_1 H_{m1}^{\sigma*} + \theta_2\ H_{m2}^{\sigma*} + H_m^{\sigma E}(\theta) - x_1(f)H_{m1}^* - x_2(f)\ H_{m2}^* - H_m^E(x(f))\right]$$

$$[2.3.12]$$

Simplification is possible by realizing that (see [2.3.8])

$$x_1(i) - x_1(f) = \frac{n^\sigma}{n^L(i)}\left[\theta_1 - x_1(f)\right] \qquad [2.3.13]$$

and similarly for component 2. After some algebra, this leads to

$$\Delta_{imm}H = n^\sigma\left[\theta_1(H_{m1}^{\sigma*} - H_{m1}^*) + \theta_2(H_{m2}^{\sigma*} - H_{m2}^*) - H_m^E(x(f)) + H_m^{\sigma E}(\theta)\right]$$

$$+ n^L(i)\left[H_m^E(x(f)) - H_m^E(x(i))\right] \qquad [2.3.14]$$

This rather versatile and general equation has been derived in a slightly different fashion by Everett[1]. It demonstrates the advantage of simultaneously measuring immersional heats and excess adsorption. The terms $(H_{m1}^{\sigma*} - H_{m1}^*)$ and $(H_{m2}^{\sigma*} - H_{m2}^*)$ are nothing else than the immersional enthalpies in the pure liquids, which can be taken from table 1.3 (sec. 1.3f) or from similar data for other substances. When H_m^E is measured from the heat of mixing, it is possible to find $H_m^{\sigma E}$ as a function of θ, which is essentially the information we want to obtain.

A variant of [2.3.14] expresses the H^σ's per unit area, i.e. as H_a^σ, but to relate the sum to contributions of the two components one has to know the molecular cross sections, which are somewhat evasive parameters that may change with composition. Note that the areas of 1 and 2 are also needed to relate the

1) D.H. Everett, *Trans. Faraday Soc.* **61** (1965) 2478.

interfacial excesses θ_1 and θ_2 to the physical fractions of 1 and 2 adsorbed, x_1^σ and x_2^σ.

Equations [2.3.8 and 14] together are not enough to establish n_1^σ and n_2^σ unambiguously, but simultaneous agreement of models with those two expressions of course lends credibility to the assumptions made.

In connection with the temperature dependence of adsorption, an *isosteric heat of adsorption* can also be defined, for instance as

$$\left(\frac{\partial \ln f x_2}{\partial T}\right)_{\Gamma_2^{(n)}} = \frac{q_{ads.st}}{RT^2} = -\frac{\Delta_{ads}H_m}{RT^2} \qquad [2.3.15]$$

where f is an activity coefficient. As compared with gas adsorption, the interpretation of the notion of isostericity is less ambiguous because as a function of temperature both types of molecules may alter their orientation or solvation. Moreover, for x_2 not small, in composite isotherms of, say type 2.8a, where the curves may have different heights for different temperatures, constancy $\Gamma_2^{(n)}$ may require $(\partial x_2/\partial T)$ to have different signs before and after the maximum, although q should not change sign. Alternatively, $(\partial x_2/\partial T)$ can be taken at constant total adsorbed amount. We shall not further discuss this matter.

2.3e. Consistency and congruence tests

There are a few tests of a thermodynamic nature that help to verify the correctness of data and/or theoretical elaborations.

The first of these tests requires adsorption data on the three binary pairs that can be made from the liquids 1, 2 and 3, that must be all miscible in all proportions. Many theoretical pictures (see sec. 2.4, particularly [2.4.32]) lead to exchange constants K_{ij} of the form

$$K_{ij} = \exp\left\{\left[\mu_j^{oL} - \mu_j^{o\sigma} - r\left(\mu_i^{oL} - \mu_i^{o\sigma}\right)\right]/RT\right\} \qquad [2.3.16]$$

When the triad (1+2), (2+3), (3+1) is completely studied and analyzed, the relation

$$\ln K_{12} + \ln K_{23} + \ln K_{31} = 0 \qquad [2.3.17]$$

must hold. When this is not the case there can be defects in the theory or experiments. For non-ideal systems the apparent K's must be split into the ideal and the excess parts; [2.3.17] only applies to the ideal part.

When very accurate data ln on K as a function of T are available, the corresponding standard Gibbs energies $(-RT \ln K_{ij})$ can be split into their enthalpic and entropic parts (sec. I.2.15), of which the former virtually represents the enthalpy of the exchange of i against j under standard conditions, which can be related to the difference $\Delta_{imm}H_j^* - \Delta_{imm}H_i^*$.

A purely thermodynamic variant involves the interfacial tension. The Gibbs adsorption equation at constant p and T [I.2.22.11] may, for a mixture of 1+2, be expressed as

$$-d\gamma = \Gamma_1^{(n)} d\mu_1 + \Gamma_2^{(n)} d\mu_2 = \Gamma_2^{(n)} (d\mu_2 - d\mu_1) \qquad [2.3.18]$$

where we have used $\Gamma_1^{(n)} + \Gamma_2^{(n)} = 0$. Using the Gibbs-Duhem relation $d\mu_1 = -x_2 d\mu_2/x_1$, $d\mu_1$ can be eliminated:

$$-d\gamma = \Gamma_2^{(n)} (1 + x_2/x_1) d\mu_2 = \Gamma_2^{(n)} d\mu_2 /(1-x_2) = \Gamma_2^{(n)} RT d\ln fx_2 /(1-x_2) \quad [2.3.19]$$

of which the r.h.s. is measurable if the surface area and activity coefficient are known. Virtually, this is the differential of the surface pressure. When the r.h.s. has been measured over the entire composition range, integration is possible to yield

$$\gamma_2^* - \gamma_1^* = RT \int_{x_2=0}^{1} \Gamma_2^{(n)} \frac{d\ln f x_2}{(1-x_2)} \qquad [2.3.20]$$

where γ^* refers to the interfacial tension of the pure liquid indicated. Although individual γ^*'s are inoperational, differences between them *are* accessible. The consistency criterion

$$(\gamma_2^* - \gamma_1^*) + (\gamma_1^* - \gamma_3^*) + (\gamma_3^* - \gamma_2^*) = 0 \qquad [2.3.21]$$

is obtained. As $(\gamma_2^* - \gamma_1^*) = \Delta_{imm}H_{a2}^* - \Delta_{imm}H_{a1}^* - T(\Delta_{imm}S_{a2}^* - \Delta_{imm}S_{a1}^*)$, where, as before, the subscript a means "per unit of area", measurements as a function of temperature yield the difference between two heats of immersion, which can be checked experimentally. Hence this method and the previous one are equivalent.

Still another variant consists of predicting the shape of, say, the (1+3) excess isotherm from those for (1+2) and (2+3), and comparing the prediction with experiment.

All these procedures to verify the quality of data and/or elaborations are called *consistency tests*.

Of a slightly different nature is the *temperature congruence* test, which virtually is a test for ideality. Very generally, adsorption isotherms can be written as some function of $x = x_2$, $\theta = \theta_2$ and T,

$$g'(T, \theta) \cdot x = f'(\theta) \qquad [2.3.22a]$$

$$\log g'(T, \theta) + \log x = \log f'(\theta) \qquad [2.3.22b]$$

For ideal systems the standard Gibbs energy of adsorption or, for that matter, K_{ij} is independent of composition and a function of T only. Then, g' may be written as a function of T only.

$$\log g(T) + \log x = \log f(\theta) \qquad [2.3.23]$$

In practice this means that the adsorption isotherm, should shift parallel along the x-axis when the temperature is changed. When this is indeed observed, the isotherms are called *temperature-congruent*; the distance between them is

$$\ln x_2 - \ln x_1 = -\frac{\Delta_{ads}G^\circ}{R}\frac{(T_2 - T_1)}{T_1 T_2} \approx -\Delta_{ads}G^\circ \Delta T / RT^2 \qquad [2.3.24]$$

Temperature congruence is particularly used for individual isotherms.

For further reading on the thermodynamics of adsorption from solution see[1] and the general references at the end of this chapter.

2.4 Adsorption from binary fluid mixtures. Theory

The matter discussed in sec. 2.3 concerned the phenomenology of adsorption from solution. To make further progress, model assumptions have to be made to arrive at isotherm equations for the individual components. These assumptions are similar to those for gas adsorption (secs. 1.4-1.7) and include issues such as: is the adsorption mono- or multimolecular, localized or mobile; is the surface homogeneous or heterogeneous, porous or non-porous; is the adsorbate ideal or non-ideal and is the molecular cross-section a_m constant over the entire composition range? In addition to all of this the solution can be ideal or non-ideal, the molecules may be monomers or oligomers and their interactions simple (as in liquid krypton) or strongly associative (as in water).

Obviously, it makes little sense to review the enormous variety of isotherms that can be conceived on the basis of the multitude of combinations of the above assumptions. Most of these isotherms have limited applicability. Rather, a number of basic principles and a few illustrative elaborations will be given. Following the trends set out before (chapters I.3 and 1), we shall often employ statistical thermodynamical procedures to elaborate models.

---

[1] J.E. Lane, *Adsorption from Mixtures of Miscible Liquids* in *Adsorption from Solution at the Solid/Liquid Interface*, (1983) (see sec. 2.10b) chapter 2; D.H. Everett, *Adsorption from Solution* and D.G. Hall, *Thermodynamics of Adsorption from Mixed Solvents* in *Adsorption from Solution* (1983) (see sec. 2.10b), p.1 and 31, respectively; D.H. Everett, *Trans. Faraday Soc.* **60** (1964) 1803; **61** (1965) 2478; D.H. Everett, *Pure Appl. Chem.* **53** (1981) 2181; D.H. Everett, *Adsorption from Solution* in *Colloidal Dispersions*, J.W. Goodwin, Ed., Roy. Soc. of Chem. (London) (1982) 71; M. Heuchel, P. Brauer, U. Messow and M. Jaroniec, *Chem. Scr.* **29** (1989) 353; G. Schay, *Colloid Polym. Sci.* **260** (1982) 888; *J. Colloid Interface Sci.* **99** (1984) 597.

2.4a. Model assumptions: some generalities

Regarding most assumptions the remarks made in chapter 1 remain valid, *mutatis mutandis*.

Because of the stronger screening in condensed media, monolayer adsorption is much more common than in gas adsorption, where the formation of multilayers is the rule, except at very low pressures. Notable exceptions are strongly associating adsorbates (like surfactants, but even with these adsorption usually remains limited to at most a bilayer) and liquid mixtures close to demixing conditions.

Discrimination between localized and mobile adsorbates must be based on arguments that differ somewhat from those for gas adsorption, where the criterion is whether an adsorbate molecule can at least make one vibration before hopping to the next position. In condensed adsorbates the surface translation of an adsorbed molecule is not an autonomous process because it involves displacement of neighbours. Essentially, this *surface diffusion* process is, as with its bulk analogue, a cooperative process, involving several neighbouring molecules, as Molecular Dynamics has shown. Phenomenologically, the distinction should therefore be made on the basis of the surface diffusion coefficient D^σ with a critical time for surface translation of

$$\tau^\sigma_{\text{crit}} = d^2 / 2D^\sigma \qquad\qquad [2.4.1]$$

where d is the distance of one step. Mechanistically, distinction can also be made between diffusion parallel and normal to the surface and one could define the corresponding residence times, see [2.2.5a and b].

In practice, most models are based on the assumption of localization, the reasoning being that many substances adhere rather strongly to surface sites. As a consequence, frequently used individual isotherms include those of Langmuir or Frumkin-Fowler-Guggenheim (FFG) but not those of Volmer or Hill-De Boer (see app. 1).

Surface heterogeneity involves the variation of the adsorption energy of *two* types of molecules but because of the exchange principle *only one* distribution function is needed. Equations like [1.7.1] can be retained after replacing $f(\Delta_{\text{ads}}u)$ by $f(\Delta_{\text{ads}}u_1 - \Delta_{\text{ads}}u_2)$. As the energy of adsorption has three contributions (adsorptive-surface, adsorptive-solvent and solvent-surface) because of the same exchange principle, and as adsorbed molecules remain partly in contact with the solution, the surface heterogeneity plays a relatively smaller role than in gas adsorption. In a similar vein, porosity is governed by the *difference* between the accessibility of pores to molecules of types 1 and 2.

Non-ideality is another matter, including the issue of constancy of a_{m1} and a_{m2}. All molecules are continually in contact with others, so that non-ideality prevails. Very few systems exist with zero mixing enthalpy and, hence, athermal binary adsorbates are exceptional. More general is the situation where the non-idealities in the bulk and adsorbate are identical or similar. Ideality in bulk also requires additivity of the two partial volumes (no expansion or compression upon mixing). Molar volumes are determined by the interaction between the molecules. Similarly, a_m is determined by the packing of molecules on the surface, which is determined by a number of factors: lateral interaction between identical and dissimilar molecules, interaction with the surface and with the bulk, and the match between molecular sizes and surface site areas. Cases where all of these contributions are independent of coverage are rare. Only in that idealized limiting case is the following expression valid:

$$A = N_1^\sigma a_{m1} + N_2^\sigma a_{m2} \qquad\qquad [2.4.2]$$

where a_{m1} and a_{m2} are constants. In general, however, the a_m's are functions of θ.

The thermodynamics of non-ideal bulk mixtures has been considered in sec. I.2.18. Non-idealities can be expressed in terms of activity coefficients, excess functions, pair interaction energies (as in Regular Solution theory) or through virial expansions. For all these methods surface equivalents can be formulated.

All told, so many aspects are incompletely understood that a rigorous solution is still far away, but for many purposes this is not necessary because a number of approximate methods and expressions have a reasonable range of applicability.

2.4b. Implications of the pair exchange principle

Adsorption from solution is an *exchange process*. Consequences of this "first law" pervade all attempts to define *individual* (or *partial*) isotherms. Any assumption made on the adsorption of component 1 involves an assumption regarding component 2; deriving an equation for 1 implies deriving an equation for 2. This is (or should be) reflected in all models, and all thermodynamics and statistical thermodynamics should be consistent with this principle.

A first step is to write, as in [2.1.1], the exchange in the form of a chemical reaction

$$r N_1^\sigma + N_2^L \rightleftarrows r N_1^L + N_2^\sigma \qquad\qquad [2.4.3]$$

describing the exchange of r molecules 1 with one molecule of type 2. Following the lines of sec. I.2.21, the equilibrium condition

$$\mu_2^\sigma - r\mu_1^\sigma = \mu_2^L - r\mu_1^L \qquad\qquad [2.4.4]$$

is monovariant: $\mu_1^\sigma = \mu_1^L$, $\mu_2^\sigma = \mu_2^L$ and in bulk and adsorbate μ_1 and μ_2 are coupled through Gibbs-Duhem relations (sec. I.2.13; recall that the Gibbs adsorption equation is virtually a surface Gibbs-Duhem relation, defining the number of independent variables, i.e. the degrees of freedom). Monovariancy implies that only one phenomenological isotherm can be derived from this equilibrium, that is: the Ostwald-Kipling equation [2.3.6]. Individual isotherms can be obtained by making assumptions on the relations $\mu_1^\sigma(N_1^\sigma)$, $\mu_1^L(N_1^L)$ etc. In sec. 2.4c-f we shall elaborate this.

The monovariancy also follows from statistical thermodynamics and is reflected in the fact that, if for a binary adsorbate one derives μ_1^σ and μ_2^σ from the canonical partition function $Q^\sigma(N_1^\sigma, N_2^\sigma, T)$, it appears that these two are coupled, the *difference between the two* taking the place of the single μ^σ in gas adsorption. Let us verify this for the simplest case of Langmuir adsorption ($w = 0$, $r = 1$).

For a monolayer lattice gas the canonical partition function is given by [1.5.1]. Realizing that if N is replaced by N_1^σ, $N_2^\sigma = N^\sigma - N_1^\sigma$, so that

$$Q^\sigma(N_1^\sigma, N_2^\sigma, T) = \frac{q_1^{N_1^\sigma} q_2^{N_2^\sigma} N^\sigma!}{N_1^\sigma! \, N_2^\sigma!} \tag{2.4.5}$$

This equation occurred previously as [I.3.6.24]. Proceeding in the usual way (i.e. taking logarithms, using Stirling's approximation ($\ln N! = N \ln N - N$ for large N), μ_1^σ is obtained using the following variant of [I.A6.12]:

$$\mu_1^\sigma = - RT \left(\frac{\partial \ln Q^\sigma}{\partial N_1^\sigma} \right)_{N_2^\sigma, T} = - RT \ln q_1 + RT \ln \frac{N_1^\sigma}{(N_1^\sigma + N_2^\sigma)} \tag{2.4.6}$$

which we write as

$$\mu_1^\sigma = \mu_1^{\sigma o} + RT \ln \theta_1 = \mu_1^{\sigma o} + RT \ln (1 - \theta_2) \tag{2.4.7}$$

By the same token,

$$\mu_2^\sigma = \mu_2^{\sigma o} + RT \ln \theta_2 \tag{2.4.8}$$

Hence,

$$\mu_2^\sigma - \mu_1^\sigma = \mu_2^{\sigma o} - \mu_1^{\sigma o} + RT \ln \left(\frac{\theta_2}{1 - \theta_2} \right) \tag{2.4.9}$$

From a comparison with the gas adsorption equivalent [1.5.4] we see that $\mu_2^\sigma - \mu_1^\sigma$ takes the place of μ^σ and that the standard term now consists of two parts, reflecting the difference in adsorption energy between 1 and 2: from [1.5.6] and [1.5.5] for this case

$$\mu_2^{\sigma o} - \mu_1^{\sigma o} \approx \Delta_{ads} U_{m2} - \Delta_{ads} U_{m1} \qquad\qquad [2.4.10]$$

We can give the process underlying [2.4.6] a physical meaning. The first equality gives us the molar Helmholtz energy if we insert molecules 1 in the adsorbate at constant N_2^σ. As the total number of sites remains fixed, this can only be achieved at the expense of an increase in surface pressure, π. According to [A1.2b] the corresponding work is $-RT\ln(1-\theta_1) = -RT\ln\theta_2$ per mole and now we see this term reflected in the r.h.s. of [2.4.8]: inserting molecules of type 2 involves work against this pressure in addition to the adsorption energy.

This situation is typical for solid-liquid interfaces where, during adsorption, the area A remains constant. In lattice models this is interpreted as the constancy of the total number of sites. With adsorption from solution at liquid-liquid interfaces this restriction is often absent; then one can distinguish between adsorption at given A and that at constant interfacial tension γ or, for that matter, at constant π.

2.4c. The Langmuir case

Beginning with the present sub section we shall give a number of model elaborations. We will start with the Langmuir case, i.e. the simple situation that ideality prevails, all molecules are monatomic, have the same sizes and mix athermally in the solution. The adsorption is localized.

Equilibrium is governed by [2.4.4] with $r = 1$. In line with sec. b, the equilibrium condition can be written as

$$\mu_2^\sigma - \mu_1^\sigma = \mu_2^L - \mu_1^L \qquad\qquad [2.4.11]$$

Ideality of the solution implies that (see [I.2.17.29 and 30])

$$\mu_1^L = \mu_1^{L*} + RT\ln(1-x_2) \qquad\qquad [2.4.12]$$

$$\mu_2^L = \mu_2^{L*} + RT\ln x_2 \qquad\qquad [2.4.13]$$

where μ_1^{L*} and μ_2^{L*} refer to the pure liquids 1 and 2. In this case we may also identify $\mu_1^{\sigma o}$ and $\mu_2^{\sigma o}$ as $\mu_1^{\sigma *}$ and $\mu_2^{\sigma *}$, respectively. We introduce the *exchange constant K* via

$$RT\ln K = -\Delta_{ads}G_m^o = \mu_1^{\sigma*} - \mu_2^{\sigma*} - \mu_1^{L*} + \mu_2^{L*} \qquad\qquad [2.4.14]$$

and arrive at the familiar exchange equation

$$\frac{\theta_2}{\theta_1} = K\,\frac{x_2}{x_1} \qquad\qquad [2.4.15]$$

Substituting $\theta_2 = \theta$, $\theta_1 = (1-\theta)$, etc. this can also be written as

$$\frac{\theta}{1-\theta} = K \frac{x}{1-x} \qquad\qquad\qquad [2.4.16]$$

or as

$$\theta - x = \frac{(K-1)(x-x^2)}{1+(K-1)x} \qquad\qquad\qquad [2.4.17]$$

in which the l.h.s. is the *operational surface excess*, telling us by how much the mole fraction of 2 in the adsorbate exceeds that in the liquid. According to [2.3.8] this is (one of) the measured surface excess isotherms, if plotted as a function of x.

Figure 2.11 gives an illustration of both the individual, $\theta(x)$, and the surface excess, $(\theta-x)(x)$, isotherms. Systems obeying the present premises give rise to type (a) curves (fig. 2.8). With increasing K the maximum increases and moves to the left. At the same time the individual and excess isotherms remain identical till higher θ, which is in line with expectation.

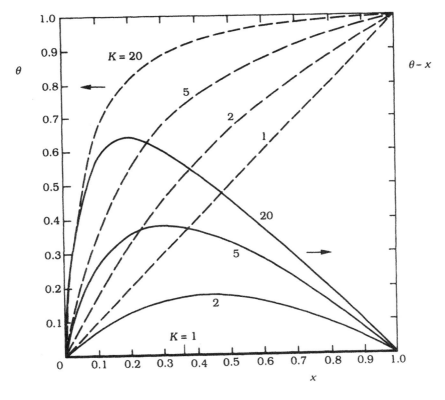

Figure 2.11. Individual (– – –) and excess (——) isotherms for adsorption from a binary solution. K is the exchange constant. Approximations: solution ideal, Langmuir adsorbate. All molecules are the same size.

When $\ln(\theta_2/\theta_1)$ is plotted as a function of $\ln(x_2/x_1)$ at different temperatures the curves are simply shifted by an amount $d \ln K = \Delta_{ads} G_m^{\circ} d T/T^2$. Therefore they are *temperature-congruent* according to the criterion of sec. 2.3e, see [2.3.22].

2.4d. Localized monolayer adsorption with lateral interaction

We now assume the adsorbate and solution to be non-ideal, that is, the molar Gibbs energy differs from its ideal value (corresponding to zero enthalpy of mixing and ideal entropy of mixing (see sec. I.2.17b) by a positive or negative excess G_m^E. Chemical potentials are partial Gibbs energies and for the non-ideal contributions we have derived [I.2.18.15-16] $\mu_1^E = G_m^E - x \partial G_m^E/\partial x$ and $\mu_2^E = G_m^E + (1-x) \partial G_m^E/\partial x$. Making use of the fact that only $\mu_2^E - \mu_1^E$ is needed, it is only necessary to retain $\partial G_m^{LE}/\partial x$ and $\partial G_m^{\sigma E}/\partial \theta$.

Further elaboration requires a model. We shall consider the Bragg-Williams approximation (sec. I.3.8d) in which only the enthalpic part of G_m^E is accounted for, the entropy is assumed to remain ideal. For gas adsorbates this leads to the FFG isotherm [I.3.8.17] and [A1.5a] and in solutions it gives rise to the Regular Solution model, both models being fairly widely applicable. For this approximation, for a binary solution we derived [I.3.8.25]

$$G_m^E = RT \chi x(1-x) \tag{2.4.18}$$

where χ is an interaction parameter containing the pair interaction energies between identical molecules (w_{11} and w_{22}) and those between dissimilar pairs (w_{12}). According to [I.3.8.9], $\chi k T/z = w_{12} - \frac{1}{2}(w_{11} + w_{22})$. In polymer physics, χ is known as the *Flory-Huggins interaction parameter*. When the attraction between dissimilar molecules is stronger than that between the half sum of w_{11} and w_{22}, the system is stable and $\chi < 0$. On the other hand, for $\chi > 0$ the system is, enthalpically speaking, unstable. If χ becomes sufficiently high, demixing (phase separation) occurs. The actual value of χ where this happens depends on the loss in configurational entropy upon demixing. For ideal polymer solutions the critical value is $\chi_{crit} = 0.5$; for shorter molecules it is different, but always positive. Regarding the sign of the critical value of the energy parameter where phase separation occurs, there is a difference of principle between the monocomponent case, as in gas adsorption and adsorption from binary mixtures. In the former case, $w = w_{11}$ should be sufficiently *negative* (see secs. I.3.8d and 1.5e), because attraction between identical molecules is the driving force, whereas in the latter systems χ should be sufficiently *positive*. From [2.4.18], $\partial G_m^E/\partial x = RT \chi (1-2x)$. We use the corresponding expression for the adsorbate. As a result, [2.4.9] is extended to become

$$\mu_2^\sigma - \mu_1^\sigma = \mu_2^{\sigma*} - \mu_1^{\sigma*} + RT \ln\left[\frac{\theta}{1-\theta}\right] - RT\chi^\sigma (1-2\theta) \qquad [2.4.19]$$

and for the bulk

$$\mu_2^L - \mu_1^L = \mu_2^{L*} - \mu_1^{L*} + RT \ln\left[\frac{x}{1-x}\right] - RT\chi^L (1-2x) \qquad [2.4.20]$$

Substitution of [2.4.19 and 20] into the equation for the exchange equilibrium criterion [2.4.11], introducing K according to [2.4.14], leads to

$$\left[\frac{\theta}{1-\theta}\right]e^{\chi^\sigma(1-2\theta)} = K\left[\frac{x}{1-x}\right]e^{\chi^L(1-2x)} \qquad [2.4.21]$$

Illustrations of this behaviour are given in figs 2.12 and 13. The curves may be compared with those in fig. 2.11 for $K = 5$. The individual isotherms are now much steeper. In turn this leads to the near-linear drop of the excess isotherms beyond $x \approx 0.2$. For sufficiently high χ^σ, surface phase separation takes place, as

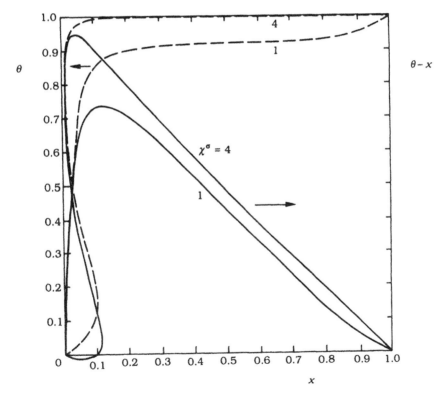

Figure 2.12. Individual (- - -) and excess (———) isotherms for adsorption from a binary solution. Approximations: non-ideal, regular solution and adsorbate. $K = 5$, $\chi^L = 2$. The value of χ^σ is indicated.

observed before (figs. I.3.5 and 7). The individual isotherms intersect at $\theta = 0.5$, independent of χ^σ, but dependent on K and χ^L; this follows immediately from [2.4.21] where on the l.h.s. the exponential with χ^σ vanishes at $\theta = 0.5$. Given the symmetry between adsorbate and bulk, [2.4.21] also accounts for demixing in the liquid; in the example given it is on the verge of doing so, as reflected by the horizontal inflection points in the upper (individual) θ curves. At more positive χ^L the solution would no longer be homogeneous.

In fig. 2.13 χ^L is fixed at -2 (no demixing in the liquid) and χ^σ is varied between -4 (mixed adsorbate stable) and $+2$ (on the verge of demixing). Distinctly different curves are obtained, including some displaying azeotropes. For $\chi^\sigma = -4$ the first molecules adsorb rather strongly, but for high θ's the preference for mixed pairs inhibits completion with pure 2; this gives rise to a negative excess. For $\chi^\sigma = 2$ this trend is reversed. As in the previous case, there is a common intersection point at $\theta = 0.5$; at the corresponding x such a point is also observed in the excess isotherms.

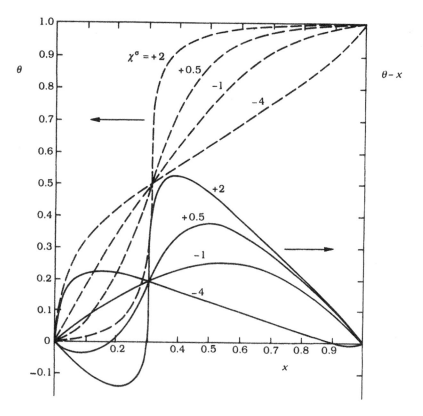

Figure 2.13. Individual (- - -) and excess (———) isotherms for adsorption from a binary solution. Approximation: non-ideal (regular) solution and adsorbate $K = 5$, $\chi^L = -2$. The value of χ^σ is indicated.

2.4e. *Adsorptives of different sizes*

Here we are entering the domain that ultimately leads to polymeric adsorptives. Polymer adsorption will be deferred to chapter 5, although some aspects of it will be discussed in sec. 2.7d on the adsorption of non-ionic surfactants. As a first step, we now consider monolayer adsorption of stiff chains (rods) of r segments from a mixture of r-mers and monomers. The actual conformation of the r-mers is left open. The present theory will approximate relatively short chains; for long chains neglection of the (change of) the conformational entropy of the chains becomes unacceptable. Restriction to monolayer adsorption implies that all the segments of adsorbed r-mers are in contact with the adsorbent. Following Defay, Bellemans and Prigogine, such models will be called *parallel layer models*. Experience with flexible polymers has shown that parallel layers occur only in the very first stage of the adsorption; at higher occupancy loops and tails into the solution start to develop, confining all segments to the first layer being entropically too restrictive.

Equation [2.4.4] is the starting point. The surface fractions are defined as

$$\theta = \theta_2 = r\, N_2^\sigma / N_S^\sigma \qquad\qquad\qquad [2.4.22]$$

$$(1-\theta) = \theta_1 = N_1^\sigma / N_S^\sigma \qquad\qquad\qquad [2.4.23]$$

where $N_S^\sigma = N_1^\sigma + r\, N_2^\sigma$ is the number of sites available, counted in terms of monomers. In the bulk $N^L = N_1^L + r\, N_2^L$; x and $(1-x)$ are defined accordingly. Defining θ as in [2.4.22 and 23] means that the occupancy in the adsorbate is counted as volumes and not as numbers of molecules. Similar reasoning applies to the liquid. Volume statistics is typical for mixtures of molecules with different sizes. The Flory-Huggins theory for polymer solutions is also in terms of volume fractions. When the monomer volumes are invariant, θ may also be interpreted as an area fraction.

To find μ_1^σ and μ_2^σ, let us consider the adsorbent as a two-dimensional lattice, that will now be filled with r-mers in a step by step fashion.

The first segment of the first r-mer has N_S^σ possibilities. If the surface co-ordination number is z^σ, the second segment has $z^\sigma N_S^\sigma$ possibilities. For rigid molecules this automatically fixes the positions of the remaining $(r-2)$ segments. Hence, $z^\sigma N_S^\sigma$ is the number of ways to put a rigid r-mer on a lattice with N_S^σ sites. For the mathematical analysis to follow, it is convenient to replace $z^\sigma N_S^\sigma$ by the following approximate expression:

$$z^\sigma N_S^\sigma \approx \frac{z^\sigma\, N_S^\sigma!}{(N_S^\sigma)^{r-1}(N_S^\sigma - r)!} \qquad\qquad\qquad [2.4.24]$$

which is readily verified by realizing that $N_S^\sigma! / (N_S^\sigma - r)! = N_S^\sigma(N_S^\sigma - 1)(N_S^\sigma - 2) \dots$

$(N_S^\sigma - r + 1)$; the remaining quotient

$$z^\sigma N_S^\sigma N_S^\sigma (N_S^\sigma - 1)(N_S^\sigma - 2) \ldots (N_S^\sigma - r + 1)/(N_S^\sigma)^r$$

reduces to $\approx z^\sigma N_S^\sigma (N_S^\sigma)^r /(N_S^\sigma)^r = z^\sigma N_S^\sigma$ for $N_S^\sigma \gg r$, that is for r-mers that are much smaller than the lattice.

The first segment of the second molecule does not have $z^\sigma N_S^\sigma$ sites available but a fraction $(N_S^\sigma - r)/N_S^\sigma$ less. For the introduction of a rigid r-mer, r sites are needed that are simultaneously empty. Here, the factor $(N_S^\sigma - r)/N_S^\sigma$ is needed r times. In a manner similar to [2.4.24] the number of possibilities of introducing a second molecule is given by

$$z^\sigma N_S^\sigma \left(\frac{N_S^\sigma - r}{N_S^\sigma} \right)^r \approx \frac{z^\sigma (N_S^\sigma - r)!}{(N_S^\sigma)^{r-1}(N_S^\sigma - 2r)!} \qquad [2.4.24a]$$

For the third, approximately,

$$\frac{z^\sigma (N_S^\sigma - 2r)!}{(N_S^\sigma)^{r-1}(N_S^\sigma - 3r)!}$$

and for N_2^σ r-mers, adsorbing simultaneously, all these expressions have to be multiplied to obtain the number of possible configurations as

$$\frac{(z^\sigma)^{N_2^\sigma} N_S^\sigma!}{(N_S^\sigma)^{N_2^\sigma(r-1)}(N_S^\sigma - r N_2^\sigma)!} = \frac{(z^\sigma)^{N_2^\sigma} N_S^\sigma}{(N_S^\sigma)^{N_2^\sigma(r-1)} N_1^\sigma!} \qquad [2.4.24b]$$

Having placed all N_2^σ r-mers on the lattice, the remaining N_1^σ sites are filled with monomers. We further need a factor $N_1^\sigma! N_2^\sigma!$ in the denominator, to account for the indistinguishability of molecules 1 and molecules 2. The factor $N_1^\sigma!$ already occurs in the equation because the way in which the r-mers have been added already accounts for the indistinguishability of the sites. However, the indistinguishability of the r-mers has not yet been accounted for. Considering all of this, we arrive at the following expression for the number of realizations of the mixture:

$$\Omega(N_1^\sigma, N_2^\sigma, r) = \frac{(z^\sigma)^{N_2^\sigma}(N_1^\sigma + r N_2^\sigma)!}{(N_1^\sigma + r N_2^\sigma)^{N_2^\sigma(r-1)} N_1^\sigma! N_2^\sigma!} \qquad [2.4.25]$$

and for the canonical partition function of the adsorbate

$$Q^\sigma(N_1^\sigma, N_2^\sigma, r) = \frac{(z^\sigma)^{N_2^\sigma}(N_1^\sigma + r N_2^\sigma)! \; q_1^{N_1^\sigma} q_2^{N_2^\sigma}}{(N_1^\sigma + r N_2^\sigma)^{N_2^\sigma(r-1)} N_1^\sigma! N_2^\sigma!} \qquad [2.4.26]$$

From this, using [2.4.22 and 23] the two chemical potentials in the adsorbate can be derived in the usual way:

$$\ln Q^{\sigma} = N_1^{\sigma} \ln q_1 + N_2^{\sigma} \ln q_2 + N_2^{\sigma} \ln z^{\sigma} + (N_1^{\sigma} + N_2^{\sigma}) \ln(N_1^{\sigma} + r N_2^{\sigma})$$

$$- N_1^{\sigma} \ln N_1^{\sigma} - N_2^{\sigma} \ln N_2^{\sigma} \qquad [2.4.27]$$

$$\mu_1^{\sigma} = - RT \left(\frac{\partial \ln Q^{\sigma}}{\partial N_1^{\sigma}} \right)_{N_2^{\sigma}, T} = \mu_1^{\sigma o} + RT \ln(1 - \theta) - RT\theta \left(\frac{1-r}{r} \right) \qquad [2.4.28]$$

$$\mu_2^{\sigma} = - RT \left(\frac{\partial \ln Q^{\sigma}}{\partial N_2^{\sigma}} \right)_{N_1^{\sigma}, T} = \mu_2^{\sigma o} + RT \ln \theta - RT \ln(z^{\sigma} r) - RT(r-1)(1-\theta) \qquad [2.4.29]$$

As before, we may replace $\mu_1^{\sigma o}$ by $\mu_1^{\sigma *}$, the standard chemical potential for the interface in contact with pure monomer. However, for μ_2 this is not allowed because the approximations in the derivation of Ω in [2.4.25] break down for $\theta \to 1$.

To formulate the exchange equilibrium condition [2.4.4] we require

$$\mu_2^{\sigma} - r \mu_1^{\sigma} = \mu_2^{\sigma o} - r \mu_1^{\sigma o} + RT \ln \frac{\theta}{(1-\theta)^r} - RT \ln(z^{\sigma} r) + RT(1-r) \qquad [2.4.30]$$

It is logical to apply the same volume statistics to the bulk phase. Apart from the coordination number and the standard terms, the result is the same as [2.4.30]:

$$\mu_2^{L} - r \mu_1^{L} = \mu_2^{Lo} - r \mu_1^{Lo} + RT \ln \frac{\varphi}{(1-\varphi)^r} - RT \ln(z^{L} r) + RT(1-r) \qquad [2.4.31]$$

where φ is now the volume fraction of component 2. Defining the exchange constant now as

$$K(r) = \exp \left[(\mu_2^{Lo} - r \mu_1^{Lo} - \mu_2^{\sigma o} + r \mu_1^{\sigma o})/RT \right] \qquad [2.4.32]$$

the following final individual isotherm equation is obtained:

$$\frac{\theta}{(1-\theta)^r} = \frac{z^{\sigma}}{z^{L}} K(r) \frac{\varphi}{(1-\varphi)^r} \qquad [2.4.33]$$

from which the surface fraction θ, and hence the surface excess, is found as a function of φ.

An illustration is given in fig. 2.14. Here, $z^{\sigma} K(r)/z^{L}$ is kept constant for various r to emphasize the effect of r in the exponents in the two denominators in [2.4.33]; in reality K increases with r. For a given K, an increase of r reduces θ

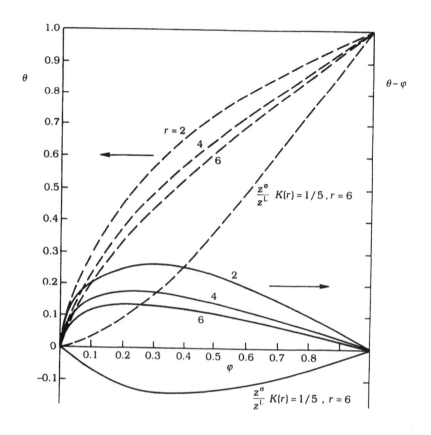

Figure 2.14. Individual (– – –) and excess $(\theta - \varphi)$ (——) isotherms for adsorption of r-mers from a mixture of r-mers and monomers. Lattice theory, no lateral interactions. $z^\sigma K(r)/z^L = 5$ except where otherwise indicated. The value of r is given.

and $\theta - \varphi$ because the number of configurations is reduced. The effect of $K(r)$ is illustrated by the difference between the curves for $r = 6$ between $K = 5$ and $K = 1/5$; this pair of curves is for the individual isotherm mirrored around the bisector, which is the $K = 1$ line.

Variants of [2.4.33] have also been obtained by others, including Everett[1] who subsumed the difference between z^σ and z^L in K and, for rod-like molecules parallel to the surface, by Prigogine and Maréchal[2]. Frisch, Simha and Eirich[3] derived [2.4.33] for $\varphi \ll 1$ in an embryonic polymer adsorption theory. An equation resembling [2.4.33],

[1] D.H. Everett, *Trans. Faraday Soc.* **61** (1965) 2478.
[2] I. Prigogine, J. Maréchal, *J. Colloid Sci.* **7** (1952) 122.
[3] H.L. Frisch, R. Simha and F.R. Eirich, *J. Chem. Phys.* **21** (1953) 365.

$$\frac{\theta}{(1-\theta)^r} = K\,e^{r-1}\varphi \qquad\qquad\qquad\qquad [2.4.34a]$$

has been derived by Conway et al.[1]. These authors assumed the molecules to be flexible, they considered dilute solutions ($\varphi \ll 1$) and defined the coordination numbers and standard states in a slightly different way. Kern and Findenegg[2] introduced a parallel layer model, which is an extension of [2.4.33] including lateral interactions, similar to [2.4.21]

$$\frac{\theta}{(1-\theta)^r}\, e^{\chi^{\sigma}r(1-2\theta)} = K\,\frac{\varphi}{(1-\varphi)^r}\, e^{\chi^{L}r(1-2\varphi)} \qquad\qquad [2.4.34b]$$

Other variants have been discussed by Nikitas[3] and by Torrent and Sanz[4].

Comparing [2.4.33] and [2.4.21] with [2.4.16], it is noted that deviations from ideality can be caused by lateral interactions and/or by size differences between the adsorptives. This is not surprising: even the Van der Waals equation for gases [I.2.18.26] has a molecular interaction and an excluded volume term.

2.4f. Other approaches

The examples discussed in the previous sections illustrate models for deriving isotherms for binary systems. A variety of variants (e.g. mobile adsorbates), alternatives (e.g. models based on computer simulations) and extensions (e.g. multimolecular adsorption, inclusion of surface heterogeneity, can be, and have been, proposed. The extensions usually require more parameters so that agreement with experiment is more readily obtained, but as long as various models are not compared against the evidence, discrimination is impossible. As there are numerous theoretical (e.g. distinction between molecules in the first and second layer) and experimental (presence of minor admixtures, tenaciously adsorbing on part of the surface) variables one tends to enter a domain of diminishing returns. On the other hand, there are detailed models for certain specific, well-defined situations. Here we shall review some approaches for the sake of illustration.

One of the advantages of lattice theories is that interactions between neighbours are easily accounted for. The drawback is that imposing a grid on mobile systems implies an approximation the quality of which is not readily assessed. To avoid this, one can employ a more phenomenological approach at the expense of being more abstract.

[1] B.E. Conway, H.P. Dhar, *Surface Sci.* **44** (1974) 261; H. Dhar, B.E. Conway and K.M. Joshi, *Electrochim. Acta* **18** (1973) 789.
[2] H.E. Kern, G.H. Findenegg, *J. Colloid Interface Sci.* **75** (1980) 346.
[3] P. Nikitas, *J. Chem. Soc., Faraday Trans. I* **80** (1984) 3315.
[4] J. Torrent, F. Sanz, *J. Electroanal. Chem.* **303** (1991) 45.

One of these we shall call the *surface tension method*. Because of its more general interest we shall give the details. Basically it is formulated as follows[1]. The adsorbate is approximated as a thin, homogeneous layer of liquid, bounded on one side by the surface and on the other by the bulk liquid. The Gibbs dividing plane is located between the adsorbate layer and the solution, so that the surface excess Γ_i^σ of component i is identified as the analytical amount of i in the layer, i.e. we write it as Γ_i. To this layer one can assign a Helmholtz energy F^σ, dF^σ satisfying [1.3.7] and a chemical potential

$$\mu_i^\sigma = \left(\frac{\partial F^\sigma}{\partial n_i^\sigma}\right)_{T,A,n_{j\neq i}^\sigma}$$

[2.4.35]

which, using [I.2.14.8], can be rearranged to give

$$\mu_i^\sigma = \left(\frac{\partial F^\sigma}{\partial n_i^\sigma}\right)_{T,\gamma,n_{j\neq i}^\sigma} - \left(\frac{\partial F^\sigma}{\partial A}\right)_{T,n_i^\sigma}\left(\frac{\partial A}{\partial n_i^\sigma}\right)_{T,\gamma,n_{j\neq i}^\sigma}$$

[2.4.36]

Here

$$\left(\frac{\partial F^\sigma}{\partial A}\right)_{T,n_i^\sigma} = \gamma$$

[2.4.37]

and

$$\left(\frac{\partial A}{\partial n_i^\sigma}\right)_{T,\gamma,n_{j\neq i}^\sigma} = a_{mi} = a_i$$

[2.4.38]

is defined as the partial molar area of species i, so that

$$\mu_i^\sigma = \left(\frac{\partial F^\sigma}{\partial n_i^\sigma}\right)_{T,\gamma,n_{j\neq i}^\sigma} - \gamma\, a_i$$

[2.4.39]

The first term on the r.h.s. is a kind of surface chemical potential. It is assumed that this term can be written as a standard term and a configurational part $RT\ln\theta_i$, where $\theta_i = n_i^\sigma/\sum_j n_j^\sigma$. Hence,

$$\mu_i^\sigma = \mu_i^{\sigma o} + RT\ln\theta_i - \gamma\, a_i$$

[2.4.40]

In this approach all deviations from ideality are contained in the γa_i term, that is, in the surface pressure term. Usually, γ (or $\pi = \gamma^* - \gamma$) and a_i both depend on θ_i, so that some authors prefer to replace γa_i by one, composition-dependent

---

[1] D.H. Everett, *Trans Faraday Soc.* **61** (1965) 2478; *Pure Appl. Chem.* **31** (1972) 579.

parameter. Obviously, for γa_i one may write $-RT \ln f_i^\sigma$, where f_i^σ is a surface activity coefficient, or assign a model for it which will ultimately lead to an equivalent of [2.4.19]. Alternatively, one may use both an $RT \ln f_i^\sigma$ and a γa_i term, in this way splitting the non-ideality contribution into a surface pressure and a remainder term. In the last case, the values of the parameters are of course different.

It should also be stressed that there is some arbitrariness in [2.4.40] in that surface excess Helmholtz energies can be defined in ways differing from [1.3.7], as discussed in sec. I.2.10, but that this can never lead to alternative relations between measurable quantities.

For pure i in contact with solids,

$$\mu_i^{\sigma*} = \mu_i^{\sigma o} - \gamma_i^* a_i = \mu_i^{L*} \qquad [2.4.41]$$

where γ_i^* is the interfacial tension of the solid-pure liquid i interface. Hence,

$$\gamma_i^* a_i = \mu_i^{\sigma o} - \mu_i^{\sigma*} = \mu_i^{\sigma o} - \mu_i^{L*} \qquad [2.4.42]$$

is virtually a difference between two standard chemical potentials.

An adsorption isotherm for one component (i) is obtained by equating μ_i^σ to μ_i^L ($= \mu_i^{L*} + RT \ln x_i$ for ideal solutions):

$$\gamma a_i = \gamma_i^* a_i + RT \ln (\theta_i / x_i) \qquad [2.4.43]$$

an equation that has already been derived long ago by Butler[1]. As the surface tension cannot be measured for SL interfaces, [2.4.43] is perhaps more useful for fluid interfaces.

For binary mixtures, assuming $a_1 = a_2 = a = $ constant, setting $\theta_2 = \theta$, $\theta_1 = (1-\theta)$, etc., the r.h.s. of [2.4.43] can be explicited for each of the two components and these are then equated (both have the same γ), leading to the following pair of relationships:

$$\theta = \frac{x \, e^{-\gamma_2^* a/RT}}{(1-x) e^{-\gamma_1^* a/RT} + x e^{-\gamma_2^* a/RT}} \qquad [2.4.44a]$$

$$(1-\theta) = \frac{(1-x) \, e^{-\gamma_1^* a/RT}}{(1-x) e^{-\gamma_1^* a/RT} + x e^{-\gamma_2^* a/RT}} \qquad [2.4.44b]$$

These are individual isotherms. They are inoperational: each r.h.s. contains two undeterminable quantities, γ_1^* and γ_2^*. However, as before, the exchange equation

---

[1] J.A.V. Butler, *Proc. Roy. Soc.* **A135** (1932) 348.

$$\frac{\theta}{1-\theta} = e^{-(\gamma_2^* - \gamma_1^*)a/RT} \frac{x}{1-x} \qquad\qquad [2.4.45]$$

is operational because the difference $\gamma_2^* - \gamma_1^*$ is accessible, for instance via contact angle measurements. This equation may be compared with [2.4.16] and experimental curves can be analysed to check its validity. We further see that [2.4.45] fulfils the consistency test [2.3.21].

It is also possible to eliminate θ between [2.4.44a and b]. Re-introducing γ via [2.4.43], one obtains

$$e^{-\gamma a/RT} = (1-x) e^{-\gamma_1^* a/RT} + x\, e^{-\gamma_2^* a/RT} \qquad\qquad [2.4.46]$$

relating the interfacial tension of the mixture to those of the components (for solid-liquid interfaces this is only of academic interest). Other authors also derived [2.4.46]. For instance, Sircar and Myers[1] arrived at it through a grand canonical approach (recall that the surface work is the characteristic function of the grand canonical ensemble, the equivalent of [I.A6.23]), Schuchowitski[2] derived a variant of it and Defay et al.[3] discussed it in some detail.

Multilayer statistics for simple molecules on a homogeneous surface follows the same lines as for gas adsorption (sec. 1.5f), the main difference being that adsorption in *any* layer is now an exchange process. If derived rigorously, the energy part of the canonical partition function becomes involved because it contains interactions within each layer (i) and also with neighbouring layers (i–1) and (i+1); the contributions of these interactions depend on the composition. Usually certain simplifications are made, for instance, by assuming that, in order to compute the energy it is allowed to assign to each layer its averaged occupancy, i.e. the Bragg-Williams approximation is applied to each layer[4, 5, 6, 7].

Multilayer adsorption on heterogeneous surfaces has been analysed in some detail by Jaroniec et al.[8]. A statistical multilayer theory by Ash et al.[9] also deserves special attention. Here the authors considered mixtures of monomers

---

[1] S. Sircar, A.L. Myers, *J. Phys. Chem.* **74** (1970) 2828, their eq. [19].
[2] A. Schuchowitski, *Acta Physicochim. USSR* **19** (1944) 176.
[3] R. Defay, I. Prigogine, A. Bellemans and D.H. Everett, *Surface Tension and Adsorption*, Longmans (1966), Ch. XII.
[4] S. Ono, *Mem. Fac. Eng. Kyushu Univ.* **12** (1950) 1.
[5] A.R. Altenberger, J. Stecki, *Chem. Phys. Letters* **5** (1970) 29.
[6] J.E. Lane, *Austr. J. Chem.* **21** (1968) 827.
[7] D.H. Everett, *Specialists Periodical Reports Chem. Soc.* (London), *Colloid Science* 1, No. 1 (1973) 70.
[8] M. Borowko, M. Jaroniec, *Adv. Colloid Interface Sci.* **19** (1983) 137; M. Jaroniec, R. Madey, *Physical Adsorption on Heterogeneous Solids*, Elsevier (1988).
[9] S.G. Ash, D.H. Everett and G.H. Findenegg, *Trans. Faraday Soc.* **66** (1970) 708.

plus dimers, flexible trimers or tetramers, all interactions and configurational entropies were all counted and weighted. These authors concluded that isotherms by themselves do not carry enough information to verify the "fine structure" of the adsorbate. This theory is a precursor for subsequent lattice theories of oligomer and polymer adsorption.

Accounting for size differences can also be realized in terms of distribution functions, assuming certain interaction energies. Simply because of size differences between molecules preferential adsorption will take place, i.e. *fractionation* occurs near a phase boundary. In theories where molecular geometries are not constrained by a lattice, this distribution function is virtually determined by the repulsive part of the interaction. An example of this kind has been provided by Chan et al.[1] who considered binary mixtures of adhesive hard spheres in the Percus-Yevick approximation. The theory incorporates a definition of the Gibbs dividing plane in terms of distribution functions. A more formal thermodynamic description for multicomponent mixtures has been given by Schiby and Ruckenstein[2].

Another group of theoretical approaches is based on a comparison with the adsorption of gas mixtures. In principle, one may consider adsorption from a binary fluid mixture as an extrapolation of the adsorption from a binary gas mixture to saturation of the two components. This extrapolation covers the least accessible part of the isotherms, because non-ideality in adsorbate and adsorptive and adsorbate-bulk phase interaction have to be accounted for. General rules are not readily derived, other than those that are so formal as to be of little use in practice. For more simple systems useful limiting laws can be derived. We refer to the literature[3,4,5]. It may be noted that Myers and Sircar[6] developed a consistency test by establishing a relation between surface pressures for the two gases and for the fluid mixture. For some binary mixtures of benzene, cyclohexane, n-heptane and 1,2 dichloroethane on silica gel they found this relation to apply.

Yet another approach is to simulate adsorption by Monte Carlo methods or Molecular Dynamics. In particular with water as the solvent this is an

1) D.Y.C. Chan, B.A. Pailthorpe, J.S. McCaskill, D.J. Mitchell and B.W. Ninham, *J. Colloid Interface Sci.* **72** (1979) 27.
2) D. Schiby, E. Ruckenstein, *Colloids Surf.* **15** (1985) 17.
3) A.L. Myers, S. Sircar, *J. Phys. Chem.* **76** (1972) 3415.
4) B.P. Bering, V.V. Serpinskii, *Izvest. Akad. Nauk S.S.S.R., Otdel Khim. Nauk* (1972), 166 (transl. *Bull. Acad. Sci. U.S.S.R., Chem. Ser.,* (1972) 152).
5) D.H. Everett, in *Adsorption at the Gas-Solid and Liquid-Solid Interface*, J. Rouquerol, K.S.W. Sing, Eds., Elsevier (1982) 1.
6) A.L. Myers, S. Sircar, *J. Phys. Chem.* **76** (1972) 3412.

extension of the similar treatment of pure liquids near surfaces, discussed in sec. 2.2c. This approach awaits further technical developments.

2.5 Experimental techniques

As a preparation to the following sections, we briefly discuss some aspects of measuring adsorption from fluid phases, including dilute solutions. For the sake of systematics, we divide the treatment into two parts: (i) adsorption on disperse systems, sometimes poorly defined, and (ii) the same on well-defined, mostly smooth model surfaces. In case (i) adsorption is almost exclusively determined from solution analysis, i.e. by depletion, so that problems arise with the separation of liquid from solid and the accurate bulk composition determinations. In case (ii), adsorbed amounts can often be determined directly using typical surface analytical techniques.

2.5a. Disperse systems: adsorption

Let us first consider the general case of dispersed particles of unknown or poorly known surface properties. Generally, measuring adsorption in such systems makes little sense unless the sample has been subjected to appropriate cleaning procedures to remove spurious adsorbed impurities, of which some may be held tenaciously. Especially for hydrophilic sorbents in non-aqueous bulk liquids, traces of water may have a dramatic effect. This removal can be achieved by extraction, washing, evaporation or rinsing (for charged impurities in aqueous solution: at various pH's). One criterion of purity is that the isotherms should be reproducible, which is a prerequisite for data analysis anyway. Sometimes the kinetics of adsorption gives a hint: unexpected slow attainment of equilibration may be caused by the slow removal of sorbed impurities or the slow replacement of the adsorbate by impurities. Irreproducibility can also be caused by non-inertness of the adsorbent; for instance it may swell or dissolve. Once reproducibility has been established, the isotherms should preferably be measured at various conditions (temperature), because one isotherm is not usually enough to discern all the subtleties of the adsorbate properties. Isotherms, studied at a variety of pretreatments can also be helpful to identify the conditioning process.

The most common way of measuring surface excesses is by depletion, i.e. to establish $\Delta_{ads}x_2$, needed to obtain $\Gamma_2^{(n)}$, see [2.3.1]. To that end, two composition determinations have to be carried out. We shall not consider such analytical methods in detail here since they are outside the realm of surface science per se. It is noted that, to obtain full coverage ($0 \leq x_2 \leq 1$) sometimes very small composition changes have to be measured in small aliquots, which may render the experiments tedious. Kipling[1] has reviewed some common techniques, including spectroscopy, refractometry, densitometry, gravimetry, for instance with a micro-balance, tracer methods and interferometry. For the application of *in situ* infrared techniques a review by Rochester[2] may be consulted. With the use of proper cells much useful information on surface groups and adsorbed molecules may be obtained for dispersed systems, without the necessity of removing the liquid. As compared with adsorption at the solid-gas interface, one of the main problems with transmission studies is the, sometimes strong, absorption by the liquid, which in cells that have too long a path length makes the system essentially opaque so that absorption bands become invisible. For water, which has a strong infrared absorption, removal of the solvent is virtually unavoidable. To improve precision and speed, Everett and coworkers[3], extending work originated by Kurbanbekov et al.[4], developed a closed-circuit system, containing the thermostatted cell with the adsorbent; the liquid in the cell is circulated by pumping it through the arms of a recording differential refractometer. From the measured change in the refractive index of the liquid in contact with the solid the amount adsorbed is obtained. This procedure is suitable for measuring temperature-dependencies and a great advantage is that it is not necessary to re-immerse the solid sample, or to compare a range of compositions with different samples. Moreover, atmospheric contamination is virtually excluded.

A variant, proposed by Nunn and Everett[5], and briefly called the *null-method* keeps the initial composition $x_2(i)$ constant by adding 2 to the system when this component adsorbs. The amount added is just the surface excess. The advantage is that the original amount $n_2(i)$ need not be known. This variant may also be combined with the circulation method.

These methods are essentially static. Adsorption from solution can also be studied *chromatographically*, which is dynamic. Conversely, information on adsorption from solution is useful to interpret chromatograms.

Various forms of (solid-liquid) chromatography exist, including thin layer chromatography, column chromatography and, more recently, high performance liquid chromatography (HPLC). Column chromatography may be applied in

[1] J.J. Kipling, *Adsorption from Solutions of Non-Electrolytes*, Academic Press (1975).
[2] C.H. Rochester, *Adv. Colloid Interface Sci.* **12** (1980) 43; *Progr. Colloid Polym. Sci.* **67** (1980) 7.
[3] D.H. Everett, *Progr. Colloid Polym. Sci.* **65** (1978) 103; S.G. Ash, D.H. Everett and G.H. Findenegg, *Trans. Faraday Soc.* **64** (1968) 2645; S.G. Ash, R. Bown and D.H. Everett, *J. Chem. Thermodyn.* **5** (1973) 239.
[4] E. Kurbanbekov, O.G. Larionov, K.V. Chmutov and M.D. Yubilevich, *Zhur. Fiz. Khim.* **43** (1969) 1630.
[5] C. Nunn, D.H. Everett, *J. Chem. Soc., Faraday Trans. I* **79** (1983) 2953.

the frontal analysis (or breakthrough) or in the elution mode. In the classical breakthrough approach the column (or plate, etc.) is packed with the adsorbent, originally in a reference liquid, which in the case of adsorption from solution is invariably the solvent, but which in adsorption from binary mixtures may be either one of the pure components or a mixture of given composition. The liquid flows through and equilibrium, or rather establishment of a stationary state, is ascertained by checking constancy of the effluent which is usually called the *eluate*. Then solution containing the adsorptive, or a mixture of different composition is flown through. As (part of) the added adsorptive is withdrawn from the solution it takes a certain volume of liquid to pass over the column before the adsorptive appears in the eluate. This appearance is called *break-through* and the amount adsorbed on the column is proportional to the *retention volume*. The converse process is *elution*, that is desorption by flowing through of solvent, or a mixture containing a displacer. By applying both approaches, an adsorption-desorption consistency check is possible. In *pulse methods* a limited amount of adsorptive is injected only once at the beginning of the column, over which a continuous flow of solvent is maintained. The pulse moves over the column at a certain rate, determined by the *slope* of the adsorption isotherm. As the slope can be detected sensitively, details of the isotherm can also be well established, and hence insight into the adsorption mechanism is obtained, for instance with respect to lateral interaction. For more information on the relation between chromatography and adsorption, see the literature[1].

2.5b. *Disperse systems: calorimetry*

Besides surface excesses, surface excess enthalpies can also be determined directly, especially since the advent of sensitive microcalorimeters. As stated, it is often advisable to study adsorptions and enthalpies together. In calorimetry, in principle two types of measurements can be distinguished.

(i) immersion of the dry adsorbent into a solution, leading to the enthalpy of immersion or enthalpy of wetting, $\Delta_{imm}H = \Delta_{w}H$. This has already been analyzed in sec. 2.3d.

(ii) enthalpies involved in the displacement of a solvent by an adsorptive or of a preadsorbed component by another one. This leads to $\Delta_{ads}H$, but it should be realized that this enthalpy is connected with the exchange between two

---

[1] See for instance, G.H. Findenegg, *Principles of Adsorption at Solid Surfaces and Their Significance in Gas/Solid and Liquid/Solid Chromatography*, in *Theoretical Advancement in Chromatography and Related Separation Techniques*, F. Dondi, G. Guichon, Eds. NATO-ASI Series, Kluwer (1992) where other references can be found, and S.A. Busev, S.I. Zverev, O.G. Larionov and E.S. Yakubov, *Zhur. Fiz. Khim.* **56** (1982), 929 (transl. as *Russ. J. Phys. Chem.* **56** (1982) 563).

adsorptives. It contains at least the difference $\Delta_{ads}H_1 - r\Delta_{ads}H_2$ (if 1 molecule of 1 displaces r molecules of 2) and an enthalpy of mixing in the solution, depending on the way in which the experiment is carried out.

Regarding approach (ii), distinction can be made between *batch* and *flow* techniques, with commercial apparatus available for both.

In the former case, the solid remains suspended in the liquid in the microcalorimeter cell. Then a mother solution is added, either in one step (to obtain an integral heat, $\Delta_{ads}H(int)$) or in several steps, leading to differential heats, $\Delta_{ads}H(diff)$[1]. In the latter case one could also speak of *titration calorimetry*; some commercial microcalorimeters are especially constructed for such titrations. Since, with these techniques, part of the added adsorptive remains in solution, the enthalpy of dilution $\Delta_{dil}H$ must be subtracted; it is dependent on composition and can be determined in a blank without adsorbent. The difference between $\Delta_{ads}H(int)$ and $\Delta_{ads}H(diff)$ has been discussed before, see sec. 1.3c.

In *flow microcalorimetry* the solid is placed in the microcalorimeter cell between, say two filters, and is successively brought in contact with solvent and mixtures (or solutions) of various compositions that flow through the cell or percolate over a small column of the adsorbent[2].

Just as with gas adsorption, an *isosteric enthalpy of adsorption* can also be defined for adsorption from solution. In sec. 2.3d we have refrained from deriving $\Delta_{ads}H_{st}$ from isotherms, measured at different temperatures because a variety of expressions exist, depending on the definition of the notion of "isosteric", see the note following [2.3.15]. For instance, it is unlikely that the composition and the structure of the adsorbate at two different temperatures are identical when the excesses are the same. In this connection, in the differentiation leading to $\Delta_{ads}H_{st}$ from $RT\, \partial \ln x_i / \partial T$ the problem arises as to what has to remain constant. In practice all $\Gamma_j^{(n)}$'s are kept constant. Finally, the process underlying measurement of the isosteric enthalpy (balancing a shift in the adsorption equilibrium by variation of temperature by a change in composition) is different from that in direct adsorption measurements, although a certain analogy may be expected. In fact, this is what is usually found in the

[1] S. Partyka, M. Lindheimer, S. Zaini and B. Brun, *Solid-Liquid Interactions in Porous Media*, Technip, Fr. (1984) 509; S. Partyka, M. Lindheimer, S. Zaini, E. Keh and B. Brun, *Langmuir* **2** (1986) 101.

[2] A.J. Groszek, *Proc. Roy. Soc. (London)* **A314** (1970) 473; G.H. Findenegg, *Carbon* **25** (1987) 119; J. Rouquerol, *Pure Appl. Chem.* **57** (1985) 69; S. Partyka, E. Keh, M. Lindheimer and A. Groszek, *Colloids Surf.* **37** (1989) 309.

literature[1], [2]. Denoyel et al.[3] , and Király et al.[4] and Woodbury[5] have given a systematic analysis of the meanings of the various enthalpies and their experimental determination.

2.5c. Model surfaces

Consider surfaces that are inert and may be made (molecularly) smooth, so that, optically speaking, they may be treated as *Fresnel surfaces*. Mica, certain polished glasses, quartz and silicon wafer surfaces may belong to this category. For such well-defined systems the optical techniques introduced in sec. I.7.10 come to mind: reflectometry, ellipsometry, and (to study the dynamics) fluorescence recovery after photobleaching (FRAP). The principles of these techniques have been outlined in that section.

Ellipsometry and *reflectometry* have in common that information on the adsorbate is extracted from changes induced by this adsorbate in the properties of light after reflection from the surface. At any phase boundary the fractions of light that are reflected and transmitted depend on system properties (such as the refractive indices of solvent, adsorbate and adsorbent, and the thickness of an adsorbed layer) and conditions that can, within some limits, be chosen (such as the angle of incidence θ_i (counted with respect to the normal to the surface), the wavelength λ_i and the polarization). Hence, in principle, adsorbate properties are accessible.

To describe transmission and refraction we have introduced the *transmission* and *reflection coefficients* called t and r, respectively. Generally these are complex quantities, i.e. they are written as $\hat{t}$ and $\hat{r}$, but for non-adsorbing media and Fresnel surfaces they become real, and we recall from [I.7.10.6 and 7] that

$$t_\perp (= t_s) = \frac{2 \sin \theta_t \, \cos \theta_i}{\sin (\theta_t + \theta_i)} \qquad\qquad [2.5.1a]$$

$$t_{//} (= t_p) = \frac{2 \sin \theta_t \, \cos \theta_i}{\sin (\theta_t + \theta_i) \, \cos (\theta_i - \theta_t)} \qquad\qquad [2.5.1b]$$

$$r_\perp (= r_s) = -\frac{\sin (\theta_i - \theta_t)}{\sin (\theta_i + \theta_t)} \qquad\qquad [2.5.2a]$$

$$r_{//} (= r_p) = -\frac{\tan (\theta_i - \theta_t)}{\tan (\theta_i + \theta_t)} \qquad\qquad [2.5.2b]$$

---

[1] G.W. Woodbury, L.A. Noll, *Colloids Surf.* **28** (1987) 233.
[2] I. Johnson, R. Denoyel, J. Rouquerol and D.H. Everett, *Colloids Surf.* **49** (1990) 133.
[3] R. Denoyel, F. Rouquerol and J. Rouquerol, *J. Colloid Interface Sci.* **136** (1990) 375.
[4] Z. Király, I. Dékány and L.G. Nagy, *Colloids Surf. A.* **71** (1993) 287.
[5] G.W. Woodbury, *Colloids Surf.* **8** (1983) 1.

Here θ_t is the angle between the transmitted light and the normal; the angle between the reflected beam and the normal, θ_r is equal to θ_i, see fig. I.7.17. Equations [2.5.1 and 2] are the *Fresnel equations*. The subscripts p and s stand for the parallel and normal to the plane parts of the surface-polarized light (s from the German *senkrecht* = perpendicular). When $\theta_i + \theta_t = \pi/2$, $r_{//} = 0$, i.e. the in-plane reflection vanishes. This situation is met when $\theta_i = n_2/n_1$; the corresponding angle of incidence, θ_B is called the *Brewster angle*:

$$\tan \theta_B = n_2/n_1 \qquad\qquad\qquad\qquad\qquad [2.5.3]$$

Each phase boundary has its own Brewster angle. At the Brewster angle the reflected light is polarized even if the incident beam is not.

The Brewster angle should be distinguished from the angle $\theta_{i,cr} = $ arc $\sin(n_2/n_1)$, see [I.7.10.10], beyond which *total reflection* takes place when the light goes from an optically less dense to a more dense medium; it indicates the angle above which total reflection occurs (vanishing of the transmitted wave). For further reading, see secs. I.7.10a and b and the literature[1]. Ellipsometry and reflectometry both deal with a combination of $r_\perp$ and $r_{//}$, but in a different way. In ellipsometry the *quotient* is measured; it is usually written as

$$\hat{r}_{//}/\hat{r}_\perp = \tan \psi \; e^{i\Delta} \qquad (0 < \psi < \pi/2) \qquad\qquad [2.5.4]$$

where ψ, a positive quantity, is the *modulus* and Δ the *phase*. This is typically a complex equation. In reflectometry one deals with the two complex conjugates, i.e. with the *products*

$$R_{//} = \hat{r}_{//}\, \hat{r}_{//}^* \qquad\qquad R_\perp = \hat{r}_\perp \, \hat{r}_\perp^* \qquad\qquad [2.5.5a,b]$$

which in the case of real reflection coefficients reduce to

$$R_{//} = \hat{r}_{//}^2 \qquad\qquad R_\perp = \hat{r}_\perp^2 \qquad\qquad [2.5.6a,b]$$

In sec. I.7.10b static adsorbates were considered and interpreted as equivalent to thin layers with different refractive indices and/or dielectric permittivities. For such layers one usually introduces the *coefficient of ellipticity* $\bar{\rho}$, which is the imaginary part of $\hat{r}_{//}/\hat{r}_\perp$ at the Brewster angle, where the real part of this quotient is zero. For a thin homogeneous layer of thickness h of a material with relative dielectric permittivity ε between a substrate 2 and a gas or liquid 1 with permittivities ε_2 and ε_1, respectively:

---

[1] P. Drude, *Theory of Optics*, Dover Reprint (1959); R.M.A. Azzam, N.M. Bashara, *Ellipsometry and Polarized Light*, North Holland (1977).

$$\bar{\rho} = \frac{\pi}{\lambda} \frac{(\varepsilon_1 + \varepsilon_2)^{1/2} (\varepsilon - \varepsilon_1) (\varepsilon - \varepsilon_2)}{(\varepsilon_1 - \varepsilon_2) \varepsilon} h \qquad\qquad [2.5.7]$$

Here each ε can be replaced by the corresponding n^2. It is seen that $\bar{\rho}$ is proportional to the thickness. Establishing the value of ε for films which are not macroscopically thick is a problem.

Basically, ellipsometry is a relatively slow but accurate technique because in [2.5.4] the phase Δ and modulus ψ must be simultaneously measured. However, the procedure can be made substantially more rapid by using fast modulators of the polarization plane in so-called Pockels cells. After standardization, ellipsometry can also be very conveniently employed to follow changes in adsorption.

The analysis is carried out using the Drude equations; this leads to a combination of the ellipsometric thickness and the refractive index increment. These characteristics of the adsorbate cannot be unambiguously separated. Conversion of the refractive index increment into the composition of the adsorbate layer is usually done by assuming dn/dx to be the same as in a fluid of composition x; for θ not too high this is usually allowed, but problems may arise when the adsorbate differs substantially from the solution, for instance because of alignment of adsorbed chain molecules. The result obtained is not unique, in the sense that different profiles may lead to the same pair of ellipsometric parameters. Therefore, normally totally adsorbed amounts are presented. For accurate measurements a good optical contrast between adsorbate and solution is mandatory.

So far ellipsometry has been used for adsorption from dilute solution, especially for surfactants, polymers and proteins. Examples will be given where appropriate. For further experimental details, we refer to the literature[1].

Besides ellipsometry, reflectometry has proven its value. By this technique adsorbed masses can conveniently be obtained and, if the measurements are carried out with polarized light, also the orientation of the adsorbed molecules. Experiments are usually done at near-normal incidence, when $R_\perp = R_{//} = R$. Another variant, pertaining to adsorption from solution and sketched in fig. 2.15, can be made fast enough for the kinetics of adsorption to be followed. In the mode shown, fluid is admitted to the surface from bottom to top ("impinging jet"); equations are available for the rate of supply in the stagnation point (the "core" of the fluid flow, which hits the surface perpendicularly). The quotient of the reflected intensities $i_{//}/i_\perp = S$ is obtained by electronic division, it is

[1] S.N. Jasperson, S.E. Schnatterly, *Rev. Sci. Instr.* **40** (1969) 761; B. Heidel, G.H. Findenegg, *J. Phys. Chem.* **88** (1984) 6575; P. Schaaf, Ph. Dejardin and A. Schmitt, *Rev. Phys. Appl.* **20** (1985) 631; **21** (1986) 741; *Langmuir* **3** (1987) 1131; D. Beaglehole, H.K. Christenson, *J. Phys. Chem.* **96** (1992) 3395.

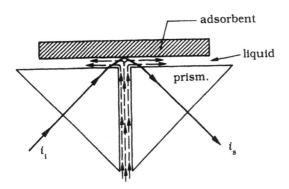

Figure 2.15. Principle of reflectometry under dynamic conditions. The dashed curves indicate fluid flow lines.

proportional to $R_{//}/R_{\perp}$. The proportionality constant is determined by apparatus parameters, by λ, θ_i, the refractive indices of the materials and the polarization angle. This constant has to be standardized, but conditions can be adjusted to attain optimal sensitivity. Particularly around the Brewster angle θ_B the presence of very small adsorbed amounts can be relatively accurately determined. For the adsorption of organic molecules from an aqueous solution the detection limit is as low as 10^{-2} mg m^{-2} under favourable conditions. For adsorbents that have little optical contrast with the liquid the sensitivity can also be enhanced by working with a thin dielectric film, deposited on a strongly reflecting Fresnel surface, say polystyrene on a silicon wafer in aqueous solution. The thickness must be such that the beam reflected from that surface has its phase reversed, leading to an interference that is more strongly dependent on adsorbed amounts (in the example, on the polystyrene[1]).

An example of application to a low molecular mass adsorptive has been given by Yang et al.[2] who measured the adsorption of stearic acid from carbon tetrachloride, which does not absorb much infrared radiation, on to some solids by *Fourier Transform Infrared* spectroscopy (FTIR). To obtain information on the orientation of adsorbed molecules it is necessary to work with polarized light. Fourier transform equipment is commercially available. In this variant, rather than the infrared spectrum being scanned as a function of time, all frequencies are simultaneously passed through the sample, after which the instrument transforms the interferogram obtained to the normal intensity-wavenumber spectrum. The advantages of the Fourier transform mode are greater precision (the sample can be measured several times in succession to improve the signal-

[1] J.C. Dijt, M.A. Cohen Stuart, J.E. Hofman and G.J. Fleer, *Colloids Surf.* **51** (1990) 141; J.C. Dijt, M.A. Cohen Stuart and G.J. Fleer, *Adv. Colloid Interface Sci.* **50** (1994) 79.
[2] R.T. Yang, M.J.D. Low, G.L. Haller and J. Fenn, *J. Colloid Interface Sci.* **44** (1973) 249.

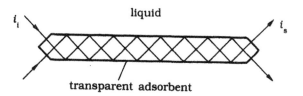

Figure 2.16. Reflectometry in the attenuated total reflection (ATR) mode.

noise ratio) and higher speed (a classical infrared spectrum requires several minutes, a Fourier transform one only a few seconds), so that rates of adsorption can also be studied. After calibration, surface concentrations can be deduced from the appropriate band intensities and rates of adsorption or desorption are inferred from their rates of change with time.

Another variant is sketched in fig. 2.16. For this set-up the adsorbent must be transparent, it looks like a pencil sharpened at both ends, through which the light bundle is led. The angle at which the rays hit the surfaces is such that *total reflection* occurs (sec. I.7.10), hence the *evanescent wave* is measured. The signal is amplified by letting the bundle reflect back and forth several times. Hence the acronym ATR, for *attenuated total reflection*[1]. Such experiments can be carried out in the infrared, visible and ultraviolet part of the spectrum (ATR-infrared, etc.), and they can be executed in the Fourier transform mode. Measurements are often carried out with polarized light to detect the orientation of certain groups near the surface. It is repeated that the depth of the solution "seen" by this method is determined by the decay length of the evanescent waves [I.7.10.12]. For infrared it is typically O (μm). This length usually not only covers the adsorbate, but also part of the bulk liquid (or bulk solution), the contribution of which has to be subtracted. Therefore it is advisable to work in dilute solutions.

Among the other spectroscopies, either in the direct or reflection mode, *fluorescence* spectroscopy may be mentioned. A promising variant is *Total Internal Reflection Fluorescence* spectroscopy (TIRF). The decay rate of an excited fluorescing probe is usually interpreted in terms of the local fluidity and polarity. The technique has been used to estimate the extent of ordering inside adsorbed surfactant layers, but this is not an absolute method because a fluorescent probe has to be inserted, and such probes themselves affect the local fluidity. More rigorous are fluorescence experiments with molecules that possess such a probe as an intrinsic part of their structure, such as tryptophans in

[1] N.Harrick, *Internal Reflection Spectroscopy*, Interscience (1967); *Internal Reflection Spectroscopy, Theory and Applications*, F.M. Mirabella Jr., Ed., Marcel Dekker (1993); J.W. Strojek, J. Mielczarski and P. Novak, *Adv. Colloid Interface Sci.* **19** (1983) 309.

proteins. In addition to the decay, one can study the spatial orientation by working with polarized light.

A group of techniques of growing interest are those based on magnetic resonance, particularly *electron spin resonance* (ESR), also called EPR, and *nuclear magnetic resonance* (NMR). The basic principles have been laid down in sec. I.7.13, together with a brief summary of the kind of information obtainable. Applications to solid-liquid interfaces in the absence of an adsorbate have already been given in sec. 2.2a. Basically, from the widths and positions of peaks conclusions can be drawn regarding the fluidity of the immediate surroundings of the nucleus under investigation and the way in which it is chemically bound. In NMR two relaxation times play important roles, called T_1 and T_2. T_1 is the spin-lattice or longitudinal relaxation time, where "lattice" in NMR jargon means the environment, and T_2 stands for spin-spin or transverse relaxation time. The spin-lattice relaxation is often exponential; the more mobile the surroundings the more rapid this process and, hence, the narrower the peaks. NMR is more generally applicable than ESR because for the latter only molecules can be used that have groups containing an unpaired electron spin, and, hence, are paramagnetic. (The introduction of paramagnetic probes is not a desirable option because the probe itself may change the local structure.) The advent of high resolution NMR apparatus has enabled work with nuclear isotopes in their natural abundances. In addition, "innocuous" replacements like D_2O for H_2O can be helpful; in this way one of the components can be made "invisible".

The nature of the information that can be obtained includes detection of the kind of hydroxyl groups on surfaces, their interaction with water, the mobility of water near surfaces, binding of molecules (via both T_1 and T_2), their orientation and interaction with neighbours and even amounts adsorbed. In principle, a wealth of information can be extracted, but the analysis is not always straightforward because in condensed systems with a substantial background a great number of couplings develop that are not always easy to unravel. However, such interpretations can sometimes be made less ambiguous by supplementing the experiments by classical spectroscopies[1].

If one is interested in the *orientation* of adsorbed molecules one can, as well as using TIRF with polarized light and NMR, also have recourse to non-linear optical techniques, including the analysis of so-called *second harmonics*. The second harmonic is a wave with twice the frequency of the incident one; it is

---

[1] For NMR of solid-liquid interfaces, see K.J. Packer, *Nuclear Spin Relaxation Studies of Molecules Adsorbed on Surfaces*, in *Progress in Nuclear Magnetic Resonance Spectroscopy*, J.W. Emsley, J. Feeney and L.H. Sutcliffe, Eds., Vol. 3, Pergamon (1967), chapter 3, 87; A collection of papers on magnetic resonance in colloid and interface science can be found in *Colloids Surf.* **72** (1993).

sensitive to non-linear phenomena such as occur in the ascending branches of many adsorption isotherms. Even the orientation of small molecules in a monolayer can be measured[1]. *Linear dichroism* has also been proposed to detect chain end-group orientations[2]. Surface plasmon resonance can be used to measure the adsorption on certain metal films (gold, silver, etc.). Essentially this is also a method based on (changes in) refractive indices. For an example see ref. [3]. Neutron reflection may also be mentioned. An example is given in ref. [4].

Most other spectroscopies have a more limited applicability. To this category belong *Raman spectroscopy, Surface-Enhanced Raman Spectroscopy* (SERS), *X-ray standing waves* and *Mössbauer spectroscopy*. We shall not discuss these. Sometimes special techniques, or techniques under special conditions, apply well to certain systems; for instance, surfactant adsorption can be advantageously studied using neutron or X-ray reflection at *grazing incidence* (i.e. at very large angle of incidence).

For reviews on the use of surface spectroscopies to characterize solid-liquid interfaces see[5].

For the sake of completeness it is recalled that AFM techniques (see sec. 1.2) can also be used for adsorption from solution. For this method to work, the adsorbed molecules should be tightly adsorbed, in case the tip displaces them. This requires flat surfaces with molecules fitting in registry, Langmuir-Blodgett layers and similar adsorbates.

An illustration is given in fig. 2.17, which concerns the adsorption of octa-decanol on graphite. A monolayer is formed with a herringbone-like structure, i.e. a real two-dimensional phase with well-defined domain boundaries. In the adsorption isotherms (not shown) this is reflected by an almost vertical ascent (see also fig. 2.28). An interesting aspect of this study is that information on the dynamics of the adsorbate is also obtainable. The adsorbed molecules can flip spontaneously between two different tilt angles and by scanning rapidly this process can be observed: the arrow in fig. 2.17b marks the position of the scan when the flip occurred. From the scan rate the authors concluded that the switching time is much less than 10^{-3} s.

[1] J.H. Hunt, P. Guyot-Sionnest and Y.R. Shen, *Chem. Phys. Lett.* **133** (1987) 189; S.G. Grubb, M.W. Kim, Th. Rasing and Y.R. Shen, *Langmuir* **4** (1988) 452.
[2] R.P. Sperline, Y. Song and H. Freiser, *Langmuir* **8** (1992) 2183.
[3] J.C. Louleague, Y. Lévy, *Macromolecules* **18** (1985) 306.
[4] T.P. Russell, *Mater. Sci. Rept.* **5** (1990) 171.
[5] J. O'M Bockris, M. Gamboa-Aldeco, *Some Recent Spectroscopic Approaches to the Solid-Solution Interface*, in *Spectroscopic and Diffraction Techniques in Interfacial Electrochemistry*, C. Gutiérrez, C. Melendres, Eds., Kluwer (NL) (1990) 55; E. Yeager, A. Homa, B.D. Chan and D. Scherson, *J. Vac. Sci. Technol.* **20** (1982) 628.

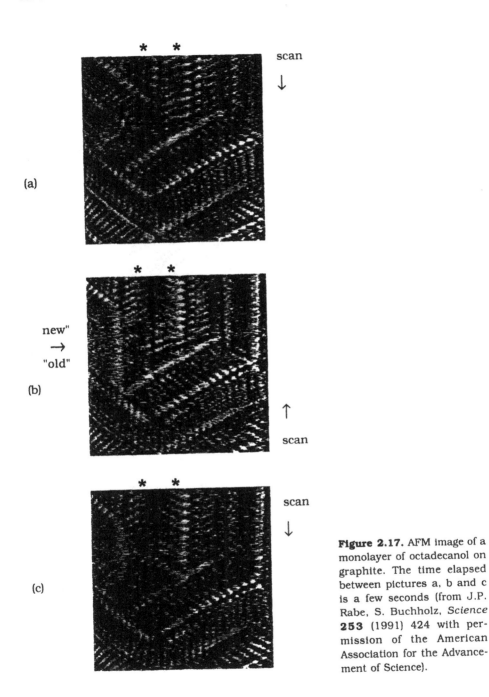

(a)

scan
↓

new"
→
"old"

(b)

scan
↑

(c)

scan
↓

Figure 2.17. AFM image of a monolayer of octadecanol on graphite. The time elapsed between pictures a, b and c is a few seconds (from J.P. Rabe, S. Buchholz, *Science* **253** (1991) 424 with permission of the American Association for the Advancement of Science).

2.6 Binary systems: a few illustrations

To illustrate the theory of secs. 2.3 and 2.4, we now give a few illustrations, obtained with the techniques of sec. 2.5. The selection from the wealth of liter-

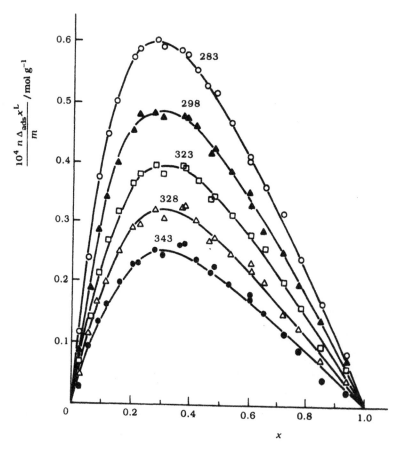

Figure 2.18. Adsorption of *n*-heptane from a cyclohexane-*n*-heptane mixture on Graphon. The temperature (in K) is indicated. (Redrawn from S.G. Ash, R. Bown and D.H. Everett, *J. Chem. Soc., Faraday Trans. I* **71** (1975) 123.)

ature data[1] is arbitrary. Attempts to decompose the surface excess isotherms into its constituents and to obtain further information will be emphasized.

Figure 2.18 is our first example. It concerns a well-defined type (a) isotherm. Graphon is a graphitized carbon black. The measurements were carried out in a closed-circuit system, in which composition changes were recorded by differential refractometry. The data show that in this way great precision can be achieved. Figure 2.18 is a selection from a large set of pairs of organic mixtures collected by the group of Everett; by combining pairs of a triad these data also allow the application of one of the consistency tests of sec. 2.3e. Figure 2.19 gives an example; the self-consistency of this (and other) triads is very

[1] A. Dabrowski and M. Jaroniec list 378 examples in a review, *Adv. Colloid Interface Sci.* **31** (1990) 155.

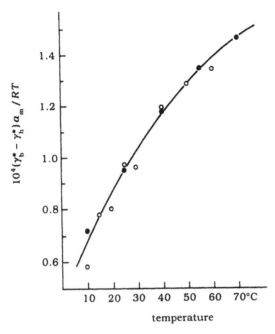

Figure 2.19. Example of a consistency test. Closed points: direct measurements for the system benzene-n-heptane, using [2.3.20]; open circles, calculated from data for the pairs ethanol+benzene and n-heptane + ethanol, using consistency criterion [2.3.21]. (Redrawn from D.H. Everett, *Progr. Colloid Polymer Sci.* **65** (1978) 103.)

satisfactory and supports the reliability of the experiments. Many of these systems exhibit relatively simple behaviour. Among other things, constant values for the area per molecule, a_m could be used and $(\gamma_2^* - \gamma_1^*)/T$ is often linear with T: the difference $\Delta_{imm}H_{a2}^* - \Delta_{imm}H_{a1}^*$ calculated from these slopes agree satisfactorily with the directly measured enthalpies of wetting (fig. 2.20). Equations like [2.4.33], characteristic for the parallel layer model, work relatively well. For a mixture of n-hexane and n-hexadecane a ratio of the r's between 2.15 and 2.25 was found, depending on the temperature; this ratio was the same as it is in bulk mixtures.

Generally speaking, consistency tests in terms of Gibbs energies (i.e. as $\gamma^* a$'s) are less critical than those in terms of enthalpies and/or entropies. For the triad carbontetrachloride, benzene and iso-octane on Aerosil the consistency in terms of G's was excellent, but that in terms of H's and TS's was of somewhat lower quality[1]. It is noted that such tests reflect the consistency of experiments, not the deviations from ideality of the systems studied.

[1] Experiments by S.A. Kazaryan, E. Kurbanbekov, O.G. Larionov and K.V. Chmutov, *Zhur, Fiz. Khim.* **49** (1975) 392, 1243, 1247 (transl. 229, 725, 728), analyzed by D.H. Everett, R.T. Podoll, *Spec. Period. Rept. Chem. Soc. Colloid Science,* Vol. **3** (1978) chapter 2.

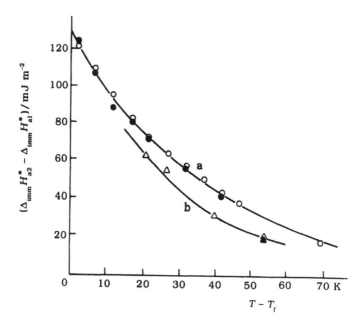

Figure 2.20. Differences between pairs of enthalpies of wetting, obtained from the temperature dependence of the adsorption (open circles) and from immersion calorimetry (closed circles). (a) hexane + hexadecane; (b) pentane + decane. Adsorbent: Graphon. (Source as in fig. 2.19.)

Figures 2.21 and 2.22 refer to the adsorption of low molecular weight aliphatic alcohols from alcohol + benzene mixtures on montmorillonite. This adsorbent is a so-called swelling clay mineral, meaning that it consists of packages of thin (aluminosilicate) layers that, under certain conditions, swell to give ultimately a dispersion of the individual sheets. Upon this swelling the specific surface area increases dramatically, it can readily reach several hundreds of $m^2 g^{-1}$ On adsorption from solution the swelling is determined by the extent to which one or both of the component(s) penetrate(s) between these sheets. In other words, we are dealing here with a non-inert adsorbent. The gas adsorption equivalent has been illustrated in fig. 1.30.

Montmorillonite is a hydrophilic clay mineral. Hence, it is expected, and found, that from their mixtures with benzene the alcohols adsorb preferentially, see fig. 2.21. The surface excess of the alcohol (in mmol g^{-1}) decreases with increasing length of the hydrocarbon tail in the alcohol because of the corresponding increase of the molar area and the hydrophobicity of the alcohols. However, when the clay mineral is made increasingly hydrophobic (hexadecyl-pyridinium cations adsorb with their positively charged pyridinium groups on to the negative sheet surfaces, the hexadecyl tails pointing towards the solution),

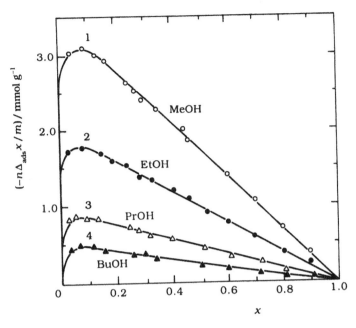

Figure 2.21. Excess adsorption of alcohols from alcohol + benzene mixtures on montmorillonite. (Redrawn from I. Dékány, F. Szántó and L.G. Nagy, *Progr. Colloid Polym. Sci.* **65** (1978) 125.)

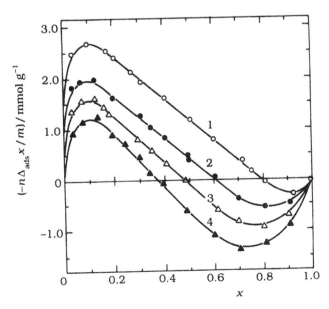

Figure 2.22. Excess adsorption of methanol from methanol + benzene mixtures on montmorillonite treated with increasing amounts (from 1 to 4) of hexadecylpyridinium chloride. (Source as fig. 2.21.)

the preference for the alcohol is reduced and fig. 2.22 shows that over an increasing range of x_{alc} benzene adsorbs preferentially.

These two figures exhibit another feature that is recurrently found, namely that sizable parts of the excess isotherms are linear. Writing the Ostwald-Kipling isotherm [2.3.6] per unit of mass m of the adsorbent as

$$\frac{n\Delta_{ads}x}{m} = \frac{n_2^\sigma(1-x) - xn_1^\sigma}{m} = \frac{n_2^\sigma - (n_1^\sigma + n_2^\sigma)x}{m} \qquad [2.6.1]$$

then it is obvious that linearity may be related to the constancy of n_1^σ and n_2^σ. In other words, the composition of the adsorbate is constant. In fig. 2.9b this constancy would correspond to horizontal parts in the individual isotherms. From [2.6.1] the values of n_1^σ and n_2^σ can be obtained by extrapolation to $x = 1$ and $x = 0$, respectively, see fig. 2.23. Before discussing such extrapolations, the caveat should be made that the inference of constant composition is an assumption, not a thermodynamic fact. Fortuitous compensation between $n_1^\sigma(x)$ and $n_2^\sigma(x)$ may also lead to linearity. It is always good to have additional evidence, say from immersion enthalpies. For example, Dékány et al. [1] found, for the adsorption of methanol from methanol-benzene mixtures on various silicas, isotherms of type b with long linear parts. The enthalpy of displacement was reasonably, but not exactly, constant over this range: it varied by about 10%,

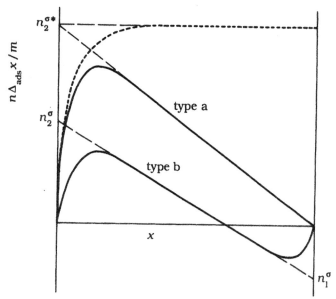

Figure 2.23. Extrapolation of linear parts of surface excess isotherms. Dotted curve: sketch of the individual isotherm for substance 2 in type a.

---

[1] I. Dékány, A. Zsednai, K. László and L.G. Nagy, *Colloids Surf.* **23** (1987) 41.

which gives some feeling for the applicability limits. In another example[1], referring to the displacement of water by n-butanol from a graphitized carbon black, it was found that with increasing x_{BuOH} the mechanism changed: as long as butanol adsorbs in a monomolecular layer, the displacement was enthalpically driven, but at higher x_{BuOH}, where the adsorption becomes multimolecular, it is entropically driven. The latter was attributed to entropy production in the *bulk* solution, i.e. the butanol accumulation was caused by (entropic) rejection from the bulk rather than by attraction to the adsorbent.

For type a curves the intercept n_2^σ may be interpreted as $n_2^{\sigma*}$, that is the real value of n_2^σ in a complete layer from which all 1 is expelled ($n_1^\sigma = 0$). This is the plateau value attained by the individual isotherm of 2, indicated by the dashed curve in fig. 2.23. When an assumption is made about the molecular cross-section a_{m2}, from $n_2^{\sigma*}$ the specific surface area A_s can be obtained. In principle, this method is not different from finding A_s from, say the plateau in a Langmuir isotherm, the only difference being that horizontal Langmuir plateaus (for adsorption from dilute solution) are replaced by linear upper parts, approaching zero at $x \rightarrow 1$ (for excess adsorption from binary mixtures).

For type b curves the intercepts may not be identified with $n_2^{\sigma*}$ and $n_1^{\sigma*}$; they represent the real amount adsorbed in a mixed monolayer over the region where the composition is constant. Two molecular areas, a_{m1} and a_{m2} are now needed to find A_s. Based on the linearity, one could make further assumptions, for instance that a_{m1} and a_{m2} are constant, so that

$$n_1^\sigma a_{m1} + n_2^\sigma a_{m2} = A / N_{Av} \qquad [2.6.2]$$

and hence predict the surface excess isotherm. Alternatively, one could introduce a kind of *mean molecular thickness* t of the adsorbate according to

$$t = \frac{n_1^\sigma}{n_1^{\sigma*}} + \frac{n_2^\sigma}{n_2^{\sigma*}} \qquad [2.6.3]$$

For a number of adsorbents, including porous ones, t has sometimes been found close to unity, indicating that the adsorption is approximately monomolecular[2].

For a more detailed discussion on obtaining A_s under various conditions see the review by Dabrowski and Jaroniec, mentioned at the onset of this section and the references in sec. 2.10.

---

[1] Z. Kiraly, I. Dékány, *Colloids Surf.* **49** (1990) 95.
[2] Systematic analysis of linear isotherm parts dates back to the early 1960s, L.G. Nagy, G. Schay, *Magyar Kém. Folyoirat* **66** (1960) 31; G. Schay, L.G. Nagy and T. Szekrenyesy, *Periodica Polytech. Budapest* **4** (1960) 45; P.V. Kornford, J.J. Kipling and E.H.M. Wright, *Trans. Faraday Soc.* **58** (1962) 74.

2.7 Adsorption from dilute solutions

To judge by the number of papers published annually on adsorption from dilute solution, this subject is more important than adsorption from binary solutions. However, the basic issues can be better illustrated from the latter; so we have emphasized them in the previous sections. Now we shall review some important features of adsorption from dilute solutions. The examples to be given are merely meant to illustrate certain points and do not claim to be a selection based on a "quality test" among the, say, 10^4 isotherms published in the literature.

2.7a. Generalities

Reconsidering the seven characteristics, discussed in sec. 2.1, the following may be said for adsorption from dilute solutions

1. Adsorption from dilute solution is also an exchange process and, hence,

2. molecules do not only adsorb because they are attracted by solids but also because the solution may reject them,

3. analytical problems incurred by composition changes due to desorption of solvent do not occur ($x_{adsorptive} \ll x_{solvent}$).

Points 4)-6) remain valid.

Regarding 7), in practice a sufficient concentration range, if available, allows the establishment of both the initial (Henry) and final (plateau, if any) parts of the isotherm.

It is particularly point 3) that makes adsorption from dilute solution so much easier to deal with. Besides the easier analytical accessibility, it may be added that:

(i) there is no problem with positioning the Gibbs dividing plane because the excess component in the solution (the solvent) is immediately recognized (see sec. I.2.22); it remains the reference over the entire isotherm Consequently, analytical surface excesses differ negligibly from the thermodynamic ones, as occurring in the Gibbs equation.

(ii) plateaus in isotherms do show up as horizontal parts in (or, more often, at the end of) the isotherm because over the plateau region $x_{solvent}$ does not depart markedly from unity. Hence, they are more readily recognized, as contrasted with binary systems where adsorption plateaus show up as slanting straight lines (see fig. 2.21).

(iii) In statistical theories it is often an acceptable approximation to treat the solvent as primitive (structureless). Then all the equations derived for monolayer gas adsorption and collected in app. 1, remain valid after replacing p by c or x and modifying the dimensions of K accordingly if it is understood that the binding energy is now the energy of the displacement and that interaction parameters like w now include contributions from the solvent. That this is indeed so

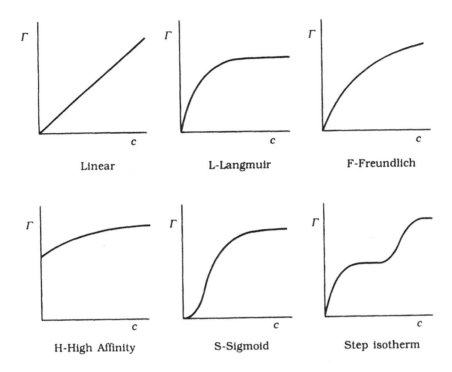

Figure 2.24. Phenomenological classification of adsorption isotherms from dilute solution. The units are arbitrary.

is readily verified by considering the equations derived for adsorption from binary mixtures in the limit $x \ll 1$. For instance, [2.4.16] becomes the Langmuir equation [A1.2a] with x instead of p and [2.4.21] becomes the FFG equation [A1.5a] if the solution is ideal ($\chi^L = 0$) and $\chi^\sigma = z N_{Av} \, w \, \theta/(1-2\theta) \approx z N_{Av} w \, \theta$ for low θ.

Remarks made before on the inertness of the solid, porosity, surface heterogeneity, etc. remain valid.

2.7b. Classification of isotherms

Isotherm shapes being the first diagnostic tool for the nature of a specific adsorption phenomenon, it is expedient to classify the most common types phenomenologically. We shall distinguish the six types of fig. 2.24[1]. For the reasons explained above, these differ from those for adsorption from binary mixtures, see fig. 2.8. There are cases where other ways of plotting are useful, see

---

[1] Some authors prefer a more detailed classification. For instance, C.H. Giles, (J. Colloid Interface Sci. **47** (1974) 755), discriminates between 16 (sub-) types.

for example the semilogarithmic plot in fig. 2.31. The set of fig. 2.23 is not complete. For instance, isotherms increasing rapidly when the solubility in solution is approached are excluded.

Of these shapes, long linear isotherms are uncommon for adsorption on solids (just like type c in fig. 2.8) and occur only if penetration into the solid takes place, leading to a Nernst-type distribution law as in liquid-liquid partitioning, see [I.2.20.1]. Linearity is also found for the (relatively short) initial parts of all isotherms on homogeneous surfaces. We shall call them, as before, *Henry isotherms* or the *Henry parts* of curved isotherms.

The *Langmuir type* (L) occurs often, even when the premises of the Langmuir theory are not satisfied. Perhaps type F is the most common; it is typical for heterogeneous surfaces. *High-affinity isotherms* are characterized by a very steep initial rise, followed by a (pseudo-) plateau. Whether or not the rise is "infinitely" steep is virtually determined by the sensitivity of the technique used to determine the bulk concentration. Hence, there is no difference of principle between types L and H.

Sigmoidal isotherms may result from two-dimensional condensation, as already discussed for gas adsorption, see figs. I.3.5, I.3.7 and 1.20a. For ideally homogeneous adsorbents the steepest rising part is vertical. Isotherms with steps may reflect sudden changes in orientation or, more often, adsorption in more than one layer. To find steps for multilayer formation, which is more rare than in the case of gas adsorption, special orientations are required for the first layer, for instance by having a first-layer structure that is determined by a specific registry with the surface (e.g. hexadecane on graphite from non-aqueous solvents) or by a type of interaction that gives rise to a second layer (as in the adsorption of surfactants). Adsorption in more than two layers is exceptional except close to demixing conditions.

Adsorption isotherms with maxima are indicative of complications, such as formation of new phases, scavenging by micelles, competition of spurious contaminants (including homologues of the adsorptive) or simply analytical artefacts. Such maxima should never be interpreted as a real decrease of Γ with increasing c because this would mean that a chemical potential would decrease with increasing amount of substance, which is thermodynamically impossible. For this reason, such isotherm types are not included in our scheme.

2.7c. *Illustrations of some basic trends*

One important aspect of adsorption studies is to obtain insight into the forces responsible for the process. The notion of force has to be understood in the thermodynamic sense, i.e. in addition to energetic contributions, entropic contributions also have to be included. Another property in which we are interested is the structure of the adsorbate. To achieve all of that, single isotherms do not, as a rule, give enough information. Measurements should preferably be carried out at various temperatures or extended by calorimetry. In addition, independent (spectroscopic) evidence can be very valuable.

For the purpose of fundamental enquiry, the adsorbents used should be well defined. At least the specific surface area should be known. On the other hand, analysis of adsorption from solution data may be one possibility for determining this parameter. The surface should be free of impurities. When these adsorb tenaciously, their presence can often only be established by independent measurements such as spectroscopy or, in the case of charged surfaces, by the shift of the point of zero charge (sec. 3.8). Most of the work reported in the literature has been carried out with convenient but often ill-defined adsorbents, such as different types of carbon, which often have large specific surface areas so that isotherms are easily analytically determined. However, for materials with large specific areas, say $A_g \gtrsim 100$ m^2 g^{-1}, the surfaces are usually porous, so that sorption consists of real adsorption plus pore filling. An indication of this may be that the BET (N$_2$ or Kr) area is much higher than that obtained from adsorption of larger molecules, from either the gas or liquid phase. Non-porous substances, such as single crystals, may be obtained in the laboratory but these have such low areas that contamination (say by polysilicates dissolved from the glass vessel, or present as a result of the preparation of the sample) becomes a serious limiting factor. Low area adsorbents have, in this respect, little buffering capacity. This problem recurs in electrophoresis with dilute sols (sec. 4.5a).

Even in the absence of adsorbed impurities, the structure of the surface of a solid dispersed in a liquid, may differ somewhat from that in the bulk of the solid. Particularly for oxides in aqueous media, superficial hydrolysis takes place, leading to the development of a surface charge, which in itself may alter the adsorption of charged *and uncharged* molecules (sec. 3.12). A check that electrosorption is excluded is that the amount adsorbed is independent of pH. The reverse is not necessarily true: pH-dependence of the adsorption does not prove that electrosorption is involved.

Most surfaces are heterogeneous, as discussed with gas adsorption (chapter 1). The consequence is that thermodynamic parameters like $\Delta_{ads}H_m$ and $\Delta_{ads}G_m^o$ depend on coverage. Only after special surface treatment can a certain degree of homogeneity be achieved; graphitization of carbon black is an example (fig. 1.8).

We begin our illustrations with an old example in which similar adsorptives are compared. See fig. 2.25. The adsorbent, a lignin-based carbon black, is not

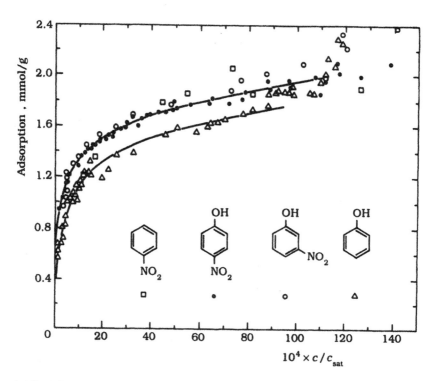

Fig. 2.25. Adsorption isotherms from aqueous solution of four aromatic compounds on active carbon. On the abscissae axis the reduced concentration is plotted. (Redrawn from J.S. Mattson, H.B. Mark Jr., M.D. Malbin, W.J. Weber Jr., and J.C. Crittenden, *J. Colloid Interface Sci.* **31** (1969) 116.)

well-defined. The BET (N_2) area is about 1000 m^2 g^{-1}, which is so high that the surface must have been very porous. Moreover, it must have been very heterogeneous. Nevertheless, this system is an important practical one.

The isotherms are of the H-type without a definitive plateau. The H-type feature is relevant for practice; such coals can "quantitatively" scavenge low amounts of aromatics from aqueous solution and, hence, act as a means for water purification. The absence of well-established plateaus implies that the surface-filling process is more involved than simply packing up to monolayer capacity. Probably pore-filling also takes place. If at the bending point (about 1.4 mmol g^{-1}) a cross-section of ≈0.45 nm^2 is assigned to an adsorbed molecule, an area of ≈380 m^2 g^{-1} would be found, which is lower than the BET (N_2) area but still so high as to suggest contributions from pores.

The isotherms are plotted as a function of the concentration relative to that in a saturated solution. This normalization leads to merging of the isotherms, whereas the original isotherms differ widely. The idea behind this way of plotting is to eliminate differences in "escaping tendency" from the water phase (hydrophobicity), so that the normalized isotherms more truly reflect the affinities for the surface. Thermodynamically speaking, the liquid phase is chosen as the reference point; compare [I.2.20.5]. This indeed eliminates the adsorptive-water interaction, but yet is not identical to the reference states because (minor) differences remain in molar volume and molar entropy of the liquid, etc. However, the fact that after normalization the curves almost co-incide suggests that the binding mechanism is similar. Probably the driving force is π-electron exchange between the aromatic ring and the surface, which can occur if the molecule adsorbs flat. The nitro group enhances the electroneg-ativity; perhaps this is why phenol, which lacks such a group, falls below the others. Aromatic molecules also tend to be adsorbed strongly, and in a flat position, on mercury, the electrochemists' model. Mercury is of course a very polarizable material. Further information regarding fig. 2.25 stems from spec-troscopy (done), studies of the effect of oxidation of the surface (done) and potentiometric titration of the surface (not done).

The strategy of plotting normalized isotherms may be particularly useful if homologues are compared[1].

As stated earlier, adsorption measurements should preferably be amplified by (micro-)calorimetry. A relatively old example where this has been done concerns the adsorption of a homologous series of fatty acids from heptane on hematite (α-Fe_2O_3) using a flow calorimeter[2]. All isotherms are of the L-type and so were the $\Delta_{ads}H$ (c) curves. Over a range of coverages, excluding low θ values, plots of $\Delta_{ads}H$ as a function of the amount adsorbed appeared constant, suggesting little heterogeneity and little lateral interaction. Hence, $\Delta_{ads}H$ can be given per gram of adsorbent, per m^2 (using the BET (N_2) area) or per mole adsorbed (for this the area is not needed). Results obtained by the authors are plotted in fig. 2.26. From the top diagram it is inferred that the molecular cross-section a_m is proportional to n, suggesting that the molecules adsorb flat. How-ever, the slope, about 0.03 nm^2 per CH_2 group, seems a bit too small for that; perhaps the molecules adsorb in a zig-zag conformation. The intercept, 0.15 nm^2, is the cross-section of a carboxyl group. The linearity of the $\Delta_{ads}H_{mi}(n)$ plot suggests negligible lateral interaction. The heat per CH_2 group is –0.9 kJ mol^{-1} which is about –0.38 RT mol^{-1} at room temperature. For adsorption from aqueous solution, where hydrophobic bonding is an important contribution to the driving force, $\Delta_{ads}H_{mi}$ would be about zero. However, from non-aqueous media the adsorption is energetically determined. The intercept, –6.5 kJ mol^{-1},

[1] See for instance R.S. Hansen, R.P. Craig, *J. Phys. Chem.* **58** (1954) 211; E.H.M. Wright, *Chem. Ind.* (1967) 506; G. Altshuler, G. Belfort, *Adv. Chem. Series*, published by the Am. Chem. Soc., **202** (1983) 29.
[2] T. Allen, R.M. Patel, *J. Colloid Interface Sci.* **35** (1971) 647.

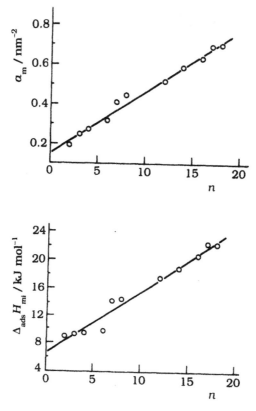

Figure 2.26. Adsorption of fatty acids from heptane on haematite $(\alpha\text{-}Fe_2O_3)$ A_g (BET (N_2)) = 3.45 m^2 g^{-1}. Temperature not stated. Data by Allen and Patel, loc. cit., replotted. n is the number of carbon chains in the chain. Top, molecular cross-section; bottom, molar adsorption enthalpy.

indicates the heat of adsorption of a carboxyl group, which is much higher than that of a CH$_2$ group.

Another illustration of more or less linear $\Delta_{ads}H$ (adsorbed amount) relationships is given in fig. 2.27. The authors found $\Delta_{ads}H_m$ to be proportional to $(\Delta_{ads}\nu_{OH})^{1/2}$, where $\Delta_{ads}\nu$ is the shift of the maximum of the infrared absorption band belonging to the SiOH groups of the "Aerosil" (a non-porous silica). In this way, additional evidence for the mode of adsorption and proof of the correctness of the enthalpies is obtained. In later papers, the authors used this information to establish the fraction of the segments in an adsorbed polymer chain that were in contact with the surface (the so-called *train segments*)[1].

Adsorption enthalpies have also turned out to be useful to investigate the

---

[1] E. Killmann, M. Korn and M. Bergmann, in *Adsorption from Solution*, R.H. Ottewill, C.H. Rochester and A.L. Smith, Eds., Academic Press (1983) 259.

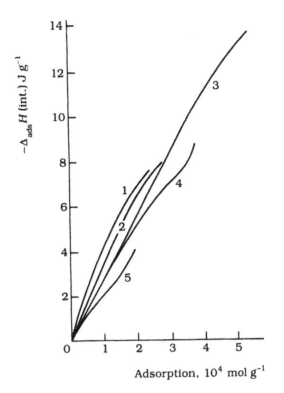

Figure 2.27. Integral enthalpy for the adsorption from carbon tetrachloride on Aerosil 200 of (1) di-*n*-propyl ether; (2) 2-hexanone; (3) 1-hexanol; (4) 2-butanone and (4) acetic acid *n*-butyl ester. (Redrawn from M. Korn, E. Killmann and J. Eisenlauer, *J. Colloid Interface Sci.* **76** (1980) 7.)

nature of the adsorbent. For instance Groszek[1] found that, from cyclohexane, pyrene adsorbed more strongly and with a higher adsorption enthalpy on "polar" graphite than on Graphon ("regular" graphite). The former product was made by oxidizing pure graphite.

In the same vein, fig. 2.28 illustrates the adsorption of long chain *n*-alkanes from *n*-heptane on different grades of graphite, to compare their surface properties. One of these adsorbents, Vulcan 3G, was a standard sample especially prepared under IUPAC auspices. The enthalpies were measured as displacement enthalpies in a flow calorimeter. All enthalpy and adsorption isotherms are S-shaped. In fig. 2.28a, right, $\Delta_{ads}H$ is plotted as a function of $\Delta_{ads}H$ for the standard sample. Straight lines are obtained, implying that the adsorption of *n*-docosane is unaffected by factors such as aggregate size and the fact that the surface must be homogeneous. This linearity also allows one to relate the specific areas of the samples to that of the reference material. In this case, for

[1] A.J. Groszek, *Faraday Discuss. Chem. Soc.* **59** (1975) 109.

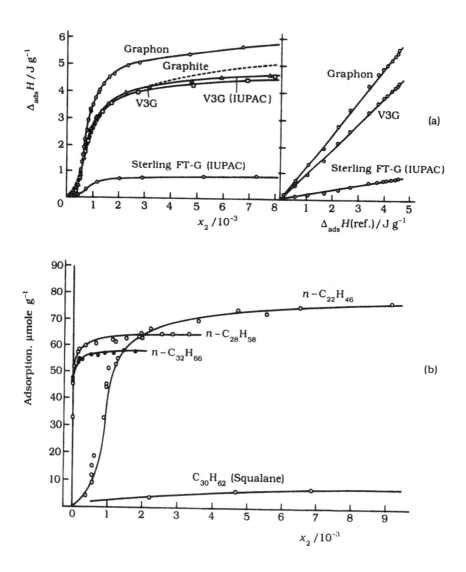

Figure 2.28. Adsorption of long-chain alkanes from *n*-heptane on various grades of graphite. (a, left), Enthalpy isotherms for *n*-docosane (n-$C_{22}H_{46}$); (a, right), correlation with the enthalpy of adsorption for a reference sample; (b) adsorption isotherms for four alkanes on Vulcan 3G. (Redrawn from H.E. Kern, G.H. Findenegg, *J. Colloid Interface Sci.* **75** (1980) 346.)

Graphon, Vulcan 3G and Sterling FTG the following values for A_g were found: 88 ± 3, 69 ± 3 and 13 ± 1 m^2 g^{-1}, respectively (for the reference A_g = 71.3 ± 2.7 m^2 g^{-1}). These data correspond very well with the BET(N$_2$) areas: 87 ± 2, 68 ± 2 and 11.1 ± 0.8 m^2 g^{-1}, respectively (reference 72 ± 2 m^2 g^{-1}). Hence, here we have a calorimetric method to determine specific surface areas. Regrettably, most surfaces are not so homogeneous.

Surface homogeneity is also a prerequisite for obtaining S-shaped isotherms with steeply ascending parts which, for perfect homogeneity, become vertical, as discussed before. From fig. 2.28b it follows that the longer linear alkanes are strongly preferred to n-heptane, indicating attachment in a flat position. For n-$C_{28}H_{58}$ and n-$C_{32}H_{66}$ the plateau is already attained at mole fractions of about 10% of saturation. In line with this, the isotherms become more high-affinity type with increasing length whereas the plateaus (in moles) decrease because the higher members require more space per molecule. Squalane is a branched hydrocarbon; it hardly adsorbs because of packing constraints. That ordering of the alkanes in the adsorbate increases with chain length is also corroborated by the fact that, for long chains, $\Delta_{ads}H$ approaches the melting enthalpy of the alkane under consideration.

The good definition of these systems implies that the models of sec. 2.4 apply. Kern and Findenegg[1] fitted the experimental results for the adsorption of n-docosane from n-heptane on Vulcan 3G graphite at 25, 35 and 45°C with [2.4.34b] assuming $\chi^L = 0$ and $r = 2.67$. A good fit was obtained with a lateral interaction parameter χ^σ of 1.16, 1.10 and 0.91 and equilibrium constants $K^{1/r}$ of 13.9, 10.6 and 8.2 for the three temperatures, respectively. The standard enthalpy of displacement of r moles of n-heptane by 1 mole of n-docosane was calculated from the temperature dependence of K. The value of -54.4 kJ mole^{-1} obtained compared very well with the calorimetric result at monolayer coverage, -58.8 kJ mole^{-1}.

Determination of *specific surface areas* by adsorption from solution is a recurrent issue. Generally speaking there are two prerequisites:

(i) the isotherm should exhibit a plateau

(ii) the molecular area a_m must be known, that is, the area an adsorbed molecule occupies in the system under the prevailing conditions. The problems in defining a_m must not be underestimated: its value may depend on the nature of the sorbent and the solvent; it may even depend on the nature of the crystal plane. In aqueous solution, charge effects may also interfere and make a_m dependent on pH and the salt concentration. As a general recommendation it is advisable to collect additional information to establish the mode of adsorption. We have already given one illustration. Other evidence received concern the orientation of the molecules; for instance, fatty acids orient perpendicularly on hydrophilic surfaces whereas aromatic compounds like nitrobenzene may lay flat.

---

[1] H.E. Kern, G.H. Findenegg, *J. Colloid Interface Sci.* **75** (1980) 346.

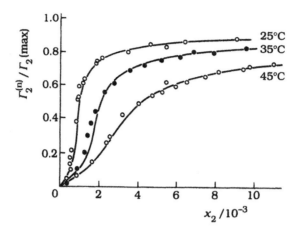

Figure 2.29. Adsorption of n-docosane (n-C$_{22}$H$_{46}$) from heptane on graphite (Vulcan 3G). Influence of temperature. The drawn curves represent a fit by Eq. [2.4.34b]. Γ_2 (max) is the monolayer capacity of n-docosane (= 88.86 μmol g^{-1}) (source as fig. 2.28).

In this respect adsorption of dyes has been popular, especially because their colour allows rapid and accurate spectroscopic detection in the supernatant. However, caveats (i) and (ii) above also apply to these adsorbates. For a discussion the reader may consult the proceedings of a symposium devoted to surface area determination in general[1], containing a contribution by Padday in which the outcome of a comparative survey over 19 different laboratories was reported, dealing with the adsorption of the dye 1,1'- diethyl-2,2' cyanine on silver bromide and a few other adsorbents. Giles et al. have given a list of recommended a_m-values[2].

In addition to calorimetry, information to establish the mode of adsorption is often obtained spectroscopically. Changes in the optical properties of (groups on) the surface or the adsorptive may be monitored. As an illustration of the former, Rochester[3] studied adsorption from the gas and liquid phase on rutile (TiO$_2$) and used infrared spectroscopy to distinguish between attachment at different surface hydroxyls. As an example of an ESR study, McBride[4] investigated the adsorption of fatty acids on amorphous alumina from methanol by labelling them with a spin probe. Relevant information could be

[1] *Surface Area Determination*, D.H. Everett, R.H. Ottewill, Eds. (Proceedings of a Symposium organized by IUPAC and the Soc. Chem. Ind. in 1969.)
[2] C.H. Giles, A.P. d'Silva, *Trans. Faraday Soc.* **65** (1969) 2516; C.H. Giles, A. d'Silva and A.S. Trivedi, in *Surface Area Determination* (loc. cit.), 317.
[3] C.H. Rochester, *Colloids Surf.* **21** (1986) 205.
[4] M.B. McBride, *J. Colloid Interface Sci.* **76** (1980) 393.

obtained by comparing the behaviour of stearic acid with the probe attached to the 4th and to the 15th carbon atom after the carboxyl group, i.e. one near the head group and the other near the tail end.

2.7d. Adsorption of non-ionic surfactants

As a case study we discuss some aspects of the adsorption of non-ionic surfactants, non-ionics for short, from aqueous solution. Such surfactants have invariably long molecules and strongly associate in solution to form micelles. The latter aspect is beyond the confines of the present chapter. Here we shall briefly introduce some main features of the adsorption of non-ionics. Ionic surfactant adsorption belongs to the domain of electrosorption: see sec. 3.12d.

All surfactants have an amphiphilic character, meaning that the molecules have a hydrophilic and a hydrophobic part. For non-ionics the former is commonly a polyoxyethylene (E_m) moiety, where m is the number of $-CH_2 - CH_2 - O$ groups. Experience has shown that chains of this nature dissolve well in water. It is usually stated that the lone electrons on the ether oxygens are responsible for this solubility but this cannot be the complete story because polymethylene oxide, $(-CH_2-O-)_m$ is insoluble in water. So is polypropylene oxide, $(-CH_2 - CH_2 - CH_2 - O-)_m$. The hydrophobic moiety can be a linear hydrocarbon chain (abbreviated C_n, where n is the number of carbons); there can be a phenyl group (ϕ) in the chain (which may assist in the binding when the adsorbent is hydrophobic, and which makes analytical detection straightforward using ultraviolet absorption), some polypropylene segments, etc. Generally the composition can be abbreviated as $C_n E_m$, $C_n \phi E_m$, etc. This code has been proposed by Clunie and Ingram[1]. Non-ionics may also have branched C-blocks. Instead of the E-block there may be polyalcohols or polysaccharides and other variants are sometimes encountered.

When adsorbing from aqueous solution on a hydrophobic surface one expects the molecule to adsorb with its hydrophobic moiety towards the solid. For hydrophilic surfaces adsorption may be the other way around if the affinity of the hydrophilic groups for the surface exceeds that for the water, and if in the adsorbate layer association occurs, reducing extensive exposure of the hydrophobic parts to water. If these auxiliary phenomena do not occur, the outcome is less predictable.

On closer inspection, one encounters here features that are typical for polymer chains. One is that the configurational entropy of the chain starts to exert its influence. Adsorbed polymers are still to a large extent in contact with the solvent. The way in which the chains are folded is important. For instance,

---

[1] J.S. Clunie, B.T. Ingram, Adsorption of non-ionic surfactants, in Adsorption from Solution at the Solid-Liquid Interface (see sec. 2.10b), p. 105.

conditions are imaginable (and do occur), where the balance between attachment of C-blocks or E-blocks is determined by the quality of the solvent, determining the interaction between the chain parts across the solvent.

Another polymer-like feature is heterodispersity, which can lead to fractionation upon adsorption. Most commercial non-ionics are heterodisperse, so it is appropriate to designate them as $C_{\langle m \rangle}E_{\langle n \rangle}$, etc. According to polymer adsorption theory, to be introduced in chapter 5, for a homopolymer longer molecules adsorb preferentially over shorter ones. As non-ionics are not homo-polymers and size-dispersity may occur either or both in the hydrophilic and hydrophobic blocks, closer scrutiny is required. Many reported isotherms deal with heterodisperse non-ionics; such isotherms are difficult to analyse quantit-atively because of the blending of a mixture of causes and effects.

Yet another consequence of the polymer-like nature of the molecules is that establishing the number of segments in contact with the surface solves only part of the problem. One should also find out how far the layer extends into the solution. For stretched chains the layer thickness corresponds to about the linear chain length; for molecules adsorbing as statistical coils this thickness should rather vary with the square root of the chain length, depending on the solvent quality. Adsorbed layer thicknesses can be obtained by a number of methods, including ellipsometry (sec. 2.5c; for a specific example, see[1]), from dynamic light scattering (sec. I.7.8d), small angle neutron scattering (sec. I.7.9b), from reflectometry and electrokinetically from the displacement of the slip plane by the adsorbed layer. (More precisely, the electrokinetically determined layer thickness also depends on the indifferent electrolyte concentration; for sufficiently low ionic strength it approaches the hydrodynamically measured thickness.) These methods do not produce identical information. Ellipsometry is sensitive to the difference between the refractive indices of the chains and the solvent (compare [2.5.7] with n^2's instead of ε's); hence it "sees" especially the inner, denser part of the adsorbed layer. On the other hand, the location of the slip plane is determined by the hydrodynamics of tangential flow, which may be changed dramatically by a few protruding chains. Therefore, the outer layers are detected hydrodynamically and electrokinetically. Invariably, the second group of procedures gives larger thicknesses. In the plateaus of isotherms, the surface concentration Γ is a few mg m^{-2}, depending on chain length and solvent quality, whereas the amount in the first layer is much less than 1 mg m^{-2}. Only in the very initial part of the isotherm may the molecules adsorb in a flat conforma-tion; only then does the parallel layer model (sec. 2.4e) apply. Scattering techniques may provide additional information. To this end refractive index

---

[1] F. Tiberg, M. Landgren, *Langmuir* **9** (1993) 927.

matching may be helpful. For an example, regarding the adsorption of $C_{12}E_5$ and $C_{12}E_6$ on Ludox silica, studied by SANS, see ref.[1].

Although in the literature a large number of isotherms for non-ionics can be found, systematic data for homodisperse samples are not abundant. In particular, isotherms over long concentration ranges (to verify the occurrence of different adsorption regimes, if any) on surfaces of different hydrophobicity are required, preferably in conjunction with auxiliary measurements (calorimetry, spectroscopy, etc.).

Figure 2.30 is the first example we consider. Here, the adsorbent is hydrophilic silica (TK 900, Degussa), containing silanol and siloxane groups on the surface. The pH was not checked; it could have exerted some influence because an electrical double layer will develop at the silica-water interface; the state of hydration of the interface is also pH-dependent. The isotherm clearly exhibits different regimes, with almost vertical parts and shorter or longer plateaus. So much fine structure is not always found; for instance with $C_{12}E_8$ instead of $C_{12}E_5$ the short first plateau is absent (not shown in fig. 2.30).

In order to verify how reproducible the data are and what kind of regimes may generally be distinguished, in fig. 2.31 the data of fig. 2.30 are compared in the semi-logarithmic mode (as in fig. 1.12b) with similar data for $C_{12}E_6$ by another investigator on another type of silica (Ludox A540, DuPont) and largely measured by other techniques. The surfactants are almost identical. (Differences

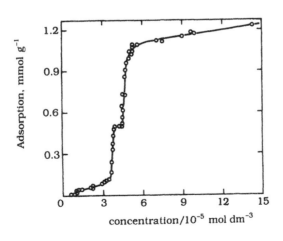

Figure 2.30. Adsorption of homodisperse $C_{12}E_5$ on silica from aqueous solution. Temperature 25°C. (Redrawn from A. Gellan and C.H. Rochester, *J. Chem. Soc., Faraday Trans. I* **81** (1985) 2235.)

---

[1] P.G. Cummins, J. Penfold and E. Staples, *J. Phys. Chem.* **96** (1992) 8092.

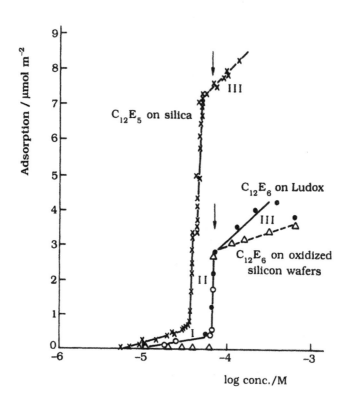

Figure 2.31 Adsorption of non-ionics from aqueous solution. Comparison of data in semilogarithmic plots. X same data as in fig. 2.30, other data obtained by M.R. Böhmer (loc. cit.) for $C_{12}E_6$ on Ludox. Methods of measurement, X, O via surface tension, ● molybdophosphoric acid assay, Δ reflectometry (on oxidized silicon wafers). The arrows indicate c.m.c.'s.

in association behaviour are dominated by the hydrophobic tail length, the c.m.c. of $C_{12}E_5$ is almost identical to that of $C_{12}E_6$. Nevertheless, Cummins et al.[1] concluded from their SANS measurements that there are differences in the pH dependence and temperature dependence between these surfactants on silica.) However, the surfaces are different. In order to establish the adsorption per unit area, specific surface area values were required; differences between them, or methodical differences in assessing these values may in part be responsible for the absolute difference between the two data sets. (For reflectometry there is a standardization step involved.)

Apart from this problem, the occurrence of three regimes I, II and III appears to be recurrent features; first a slow rise, that may or may not be linear (regime

---

[1] P.G. Cummins et. al., loc. cit.

I) then a steep, almost vertical, rise just below the c.m.c. (regime II). The final, slower increase in III would, on a linear plot, almost look like a plateau. Other investigators have essentially confirmed these three regimes, with certain modifications[1]. In isotherms reported in the literature, data in region I are not always available and the rise in II may be less steep, depending on the natures of surfactant and surface, including surfactant heterodispersity and surface homogeneity, respectively.

Before embarking on any interpretation, additional experimental evidence is useful. Böhmer et al.[2] supported their adsorption data by thickness measurements, using dynamic light scattering (sec. I.▼.6) for the Ludox, and streaming potentials (sec. 4.10) for the wafers. He found a sharp increase coinciding with region II. Gellan and Rochester[3] measured enthalpies of adsorption. In region I they found $\Delta_{ads}H_m = -120$ kJ mol^{-1}, which suggests hydrogen bond formation (sec. I.4.5d); for the two rising parts this enthalpy was -1.3 and -1.9 kJ mole^{-1}, before and after the little step, respectively. Such low enthalpies are characteristic for hydrophobic bonding and are also found for micelle formation. The little step in II was reflected in a short segment with very exothermic adsorption enthalpy. In region III the enthalpies are endothermic.

Other evidence that has been invoked stems from contact angles. Gellan and Rochester[3] reported a maximum in the water-air contact angle for $C_{12}E_5$ from water on silica, which coincided with region II. Scales et al.[4] found a similar trend for $C_{(9)}\phi E_{(9)}$ on precipitated silica, but for other systems the maximum is not always very distinct. It is tempting to interpret the increase of the contact angle over region I as caused by a hydrophobization of the surface by surfactant molecules, adsorbing with their C-blocks outwards, but this is an imprudent inference because the molecules in the first layer may have a different *in situ* orientation, which may change as soon as the air bubble is pressed against the adsorbate.

From this evidence the following may be inferred. Region III is not a distinct bilayer of closely packed parallel stretched surfactants, because then the adsorption should be constant. Formation of this layer is entropically driven. Structurally speaking, the layer is rather polymer-like. Region II represents extensive association at the interface, resembling micelle formation in bulk.

[1] K.G. Mathai, R.H. Ottewill, *Trans. Faraday Soc.* **62** (1966) 750, 759; P. Levitz, A. el Miri, D. Keravis and H. van Damme, *J. Colloid Interface Sci.* **99** (1984) 484; B. Kronberg, P. Stenius and G. Igeborn, *ibid.* **102** (1984) 418; S. Partyka, S. Zaini, M. Lindheimer and B. Brun, *Colloids Surf.* **12** (1984) 255.

[2] M.R. Böhmer, L.K. Koopal, R. Janssen, E.M. Lee, R.K. Thomas and A.R. Rennie, *Langmuir* **8** (1992) 2228.

[3] A. Gellan and C.H. Rochester, *J. Chem. Soc., Faraday Trans. I* **81** (1985) 3109.

[4] P.J. Scales, F. Grieser, D.N. Furlong and T.W. Healy, *Colloids Surf.* **21** (1986) 55.

Hydrophobic binding is the driving force; it is largely entropically driven, but there is also a small enthalpic contribution. Generally, the presence of a (hydrophilic) adsorbent facilitates association. In region I the molecules lie rather flat, and at least part of the driving force stems from hydrogen bond formation between surface silanols and the ether oxygens of the polyoxy-ethylene block. Even if this attraction is strong it does not necessarily force all the groups to the surface; this would be entropically unfavourable. However, the hydrocarbon may also be more or less flat on the surface, as it is expelled from water by the hydrophobic effect (sec. I.5.4). Whether association also occurs between molecules adsorbed in the first layer is a matter of discussion (and speculation). It is a regime governed by a subtle balance of forces where different packing possibilities, determined by surface and surfactant properties, play a role, so that the final outcome may be different for different systems, and even for different parts of the same surface. It is not so easy to obtain additional information. Levitz et al.[1] used a fluorescence-decay technique, which can be used when the surfactant contains a phenoxy group. Molecules of $C_8\phi E_8$ and $C_8\phi E_{10}$ show an (additional) slow decay when associated, and this slow decay was observed for these molecules if adsorbed on a hydrophilic silica (Spherosil, from Prolabo), in both regimes I and II, but on a hydrophobic sample it only occurred in II. Further similar information would be helpful to achieve a more detailed understanding of these adsorbates and to verify whether the observed trends may be generalized.

For adsorption from water on hydrophobic surfaces, the hydrophobic parts in the molecule attach to the surface with the hydrophilic part towards the solu-tion. There is no reason for bilayer formation and hence (pseudo-)plateau values are lower than on hydrophilic surfaces.

Systematic studies for homodisperse non-ionics over a large concentration range, where the C- and E-lengths have been independently varied are not yet available. Figure 2.32 shows results for a less-defined sample, but from which some trends may be detected. Here, the molecules contain a polypropylene oxide block P $(-CH_2 - CH(CH_3) - O -)_x$ which is insoluble in water and therefore pro-motes the attachment. All isotherms are of the high-affinity character. The increasing affinity with increasing $\langle x \rangle$ is offset by the simultaneous growth of $\langle m \rangle$; the long ethoxylene blocks have polymer nature, when the hydrophobic moiety of the molecule would adhere strongly and flat on the surface, the E-parts of the coils would become too confined. As expected, the plateau adsorption is lower than for hydrophilic surfaces (fig. 2.31). It decreases with

---

[1] See P. Levitz, H. van Damme, *J. Phys. Chem.* **90** (1986) 1302, where older references can be found.

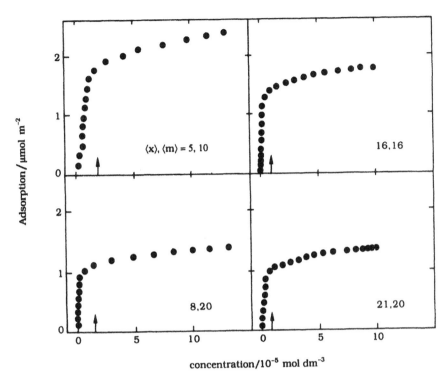

Figure 2.32. Adsorption of $C_9\phi P_{(x)}E_{(m)}$ on emulsifier-free polystyrene latex. The values of $\langle x \rangle$ and $\langle m \rangle$ are indicated; arrows indicate the c.m.c.'s. (Redrawn from B. Kronberg, P. Stenius and Y. Thorssell, *Colloids Surf.* **12** (1984) 113.)

increasing hydrophilic length, i.e. the length of the part that is directed towards the aqueous phase. Corkill et al.[1] had previously observed this trend. It is a polymer feature: water is a good solvent for ethylene oxide, and from good solvents thinner layers adsorb. It is also because of packing constraints that the affinity does not noticeably increase when going from $C_9\phi P_{(8)}E_{(20)}$ to $C_9\phi P_{(21)}E_{(20)}$: it is entropically very unfavourable to have, in addition to the $C_9\phi$ moieties, such long P chains all flat on the surface. Similar trends were found for $C_{12}E_5$ and $C_{12}E_8$ on a graphitized carbon black[2] and for certain commercial non-ionics (Tritons) on sulfur[3].

Enthalpies of adsorption provide keys to establish the driving force but results published in the literature at the time of writing do not yet give a clear picture. Douillard et al. studied $C_8\phi E_x$ (x = 9, 12, 30 or 40) and $C_9\phi E_x$ (x = 9, 11 or 15) adsorption on a carbon black ($A_g = 107$ m^2 g^{-1}) and found the adsorption to

---

[1] J.M. Corkill, J.F. Goodman and J. R. Tate, *Trans. Faraday Soc.* **62** (1966) 750.

[2] A. Gellan, C.H. Rochester, *J. Chem. Soc., Faraday Trans. I* **81** (1985) 1503.

[3] J.M. Douillard, S. Pougnet, B. Faucompre and S. Partyka, *J. Colloid Interface Sci.* **154** (1992) 113.

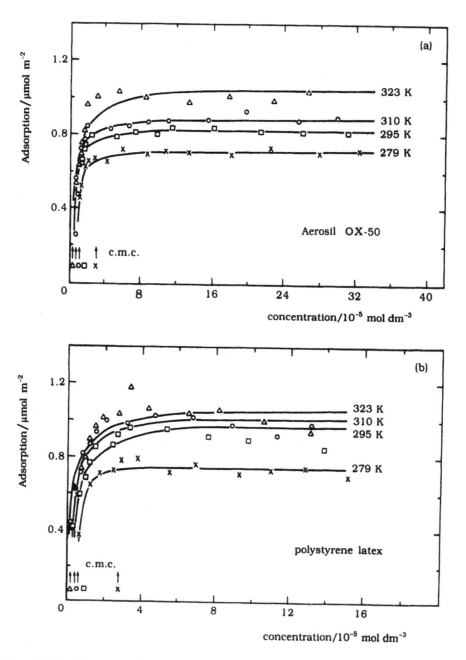

Figure 2.33. Adsorption of $C_9\phi P_{(13)}E_{(27)}$ on a hydrophilic (a) and a hydrophobic adsorbent (b). Influence of the temperature. (Redrawn from Th. van den Boomgaard, Th.F. Tadros and J. Lyklema, *J. Colloid Interface Sci.* **116** (1986) 8.)

be strongly exothermic, almost exponentially decreasing with Γ. Hey et al.[1] found for C_8E_4 a transition from exothermic to endothermic. Findenegg et al.[2], studying C_8E_4 on another carbon black ($A_g = 68 \; m^2 \; g^{-1}$), also found the enthalpy to be negative; region I of the isotherm to become positive, and passing through a maximum at higher Γ's. More systematic studies are required and should prove interesting.

Aqueous solutions of non-ionic surfactants have (at least) one consolute point, meaning that with increasing temperature the solution becomes turbid, due to incipient phase separation. With increasing temperature, water becomes an increasingly poorer solvent for the E-moieties. According to polymer adsorption theory, plateau adsorbed amounts increase the poorer the solvent is. The explanation of this trend is that the poorer the solvent, the denser the polyoxyethylene chains can pack close to the surface. As fig. 2.33 illustrates, this trend is observed for both a hydrophilic and a hydrophobic adsorbent. This is expected because it is a property of the part of the molecule that is in contact with the water. In passing it is noted that the data presented in this diagram, did not emphasize the very initial parts of the isotherms; it is quite possible that here initial slower rises, corresponding to region I in fig. 2.31, are also present.

In summary, this case study demonstrates that adsorption of non-ionic surfactants, an interesting theme in its own right, exhibits not only definite characteristics belonging to the present theme, but also involves aspects of polymers and association colloids.

2.8 Kinetics of adsorption

Adsorption from solution on solid surfaces being an exchange process, its rate is determined not only by transport to the surface and attachment of the component to be adsorbed but also by the rate of detachment of adsorbed other components. Hence, when we speak of a "rate of adsorption", we imply *a combination of an adsorption and a desorption rate*. The slower of the two determines the overall kinetics.

For physical adsorption of small molecules, residence times at the interface are often relatively short. When that is the case, the rate of adsorption of, say component i, is determined by its rate of supply. If this supply is diffusion-controlled we may use [I.6.5.36]

[1] M.J. Hey, J.W. MacTaggart and C.H. Rochester, *J. Chem. Soc., Faraday Trans. I* **80** (1984) 699.
[2] G.H. Findenegg, B. Pasucha and H. Strunk, *Colloids Surf.* **37** (1989) 223.

$$\Gamma_i(t) = \Gamma_i(0) + \frac{2D_i^{1/2}}{\pi^{1/2}}\left[c_it^{1/2} - \int_0^{t^{1/2}} c_i(0,u)\,d(u^{1/2})\right] \qquad\qquad [2.8.1]$$

This equation was first derived by Ward and Tordai, but in sec. I.6.5e we gave a more general derivation without involving the notion of a subsurface. In the integral u is a dummy variable of dimension [time] and $c_i(0,u)$ is the concentration of i on the surface. So, $c_i(0,u) = 0$ at the beginning of adsorption on a pristine adsorbent. Hence, for the initial part of the adsorption on such a clean surface ($\Gamma_i(0) = 0$),

$$\Gamma_i(t) = 2D_i^{1/2}c_it^{1/2}/\pi^{1/2} \qquad (\Gamma_i \ll \Gamma_i(max)) \qquad\qquad [2.8.2]$$

showing the typical $t^{1/2}$ dependence of diffusion-controlled processes. As shown in sec. I.6.5e, [2.8.2] can also be directly obtained from Fick's laws.

Equations [2.8.1 and 2] apply to diffusion-controlled adsorption on flat surfaces from quiescent solutions. Other types of adsorption kinetics have been described in secs. I.6d-g and in the literature[1].

Equation [2.8.2] can be used to estimate the order of magnitude of diffusion-controlled initial adsorption. From [2.8.2] we have $\tau_{diff} = O[(\Gamma/c)^2/D]$. For typical values of $\Gamma = O(10^{-6}$ mol m$^{-2})$, $c = O(10^{-4}$ M $= 0.1$ mol m$^{-3})$, and $D = O(10^{-9}$ m^2 s, see table I.6.4), it is found that $\tau_{diff} = O(0.1$ s). Later in the process the adsorption rate decreases because the surface is no longer a perfect "sink": the gradient in concentration, which drives the process, goes down and the diffusion layer becomes thicker, see [I.6.5.27]. Although in principle it takes an infinite time to attain saturation, in practice monolayer completion may be expected over seconds to tens of seconds, because in practice transport to the surface is enhanced by convection. Maintaining quiescent conditions over long times is experimentally difficult.

When adsorption takes place much faster than according to [2.8.2] the process cannot be purely diffusion-controlled. For instance, when the molecules *and* the adsorbent are oppositely charged, conduction also plays a role. In that case the material flux is given by the *Nernst-Planck* equation, [I.6.7.1 or 3].

Processes slower than predicted also often occur. In these cases sometimes [2.8.2] appears to be obeyed, but with a diffusion coefficient that is (much) lower than its bulk value. In that case there may be two consecutive diffusion processes, of which the second is the slower. This second step may be related to rearrangements on the surface, e.g. involving a slow desorption step. Very slow

---

[1] R.S. Hansen, *J. Colloid Sci.* **16** (1961) 549; K.J. Mysels, H.L. Frisch, *J. Colloid Interface Sci.* **99** (1984) 136; K.J. Mysels, *J. Phys. Chem.* **86** (1982) 4648; I. Ruzič *Croat. Chem. Acta* **60** (1987) 457; D.H. Melik, *J. Colloid Interface Sci.* **138** (1990) 397.

processes may also be the result of minor impurities (being present in low concentrations but with a high affinity for the surface) or, in the case of heterodisperse adsorptives, by slow exchange (when the shorter molecules arrive faster but are eventually displaced by the longer ones).

When *chemisorption* takes place, the rate may be diffusion-controlled or reaction-controlled. The former mode is expected when all arriving molecules are rapidly scavenged by the reaction. Reaction-controlled adsorption has a kinetics typical for chemical processes, with an activation energy and an Arrhenius type of temperature dependence.

When adsorption from solution is monitored by the depletion method, it is very difficult to measure changes in bulk concentration over time intervals down to milliseconds. Perhaps this is the reason that such systematic studies are not abundant in the literature. Fast measurements require stopped-flow, pressure-jump or temperature-jump techniques. The method used to determine concentrations must also be fast; suitable methods include certain spectroscopies and, for charged substances, conductivity[1]. When adsorption on Fresnel surfaces is studied, say by reflectometry, concentration measurements in the solution are not needed.

Since, in first instance, the desorption is proportional to the amount adsorbed and the time, $d\Gamma = -k_{des}\Gamma\,dt$,

$$\Gamma(t) = -\Gamma(t=0)\,e^{-k_{des}t} \qquad\qquad [2.8.3]$$

Such a relationship for desorption rates has, for instance, been found to apply for sodium dodecylsulfate from nylon[2] up to about 0.2 s. This time scale is probably similar for uncharged adsorbates.

Rate studies may also be carried out with model surfaces (sec. 2.5c), with Γ measured *in situ*. With big molecules, like proteins, diffusion is slower, the time scales correspondingly longer and the measurements relatively easier.

At equilibrium the rates of adsorption and desorption are identical. In the kinetic derivation of the Langmuir equation the assumptions are made that the rate of adsorption is proportional to the bulk concentration of the adsorptive and to the empty fraction of the surface,

$$d\,\Gamma(ads)/dt = k_{ads}c(1-\theta) \qquad\qquad [2.8.4]$$

and that the rate of desorption is proportional to the covered fraction of the surface,

1) N. Mikami, M. Sasaki, K. Hachiya, R.D. Atsumian, T. Ikeda and T. Yasunaga, *J. Phys. Chem.* **87** (1983) 1454.
2) S. Miyamoto, M. Tagawa, *Colloid Polym. Sci.* **263** (1985) 597.

$$d\,\Gamma(\text{des})/dt = -k_{\text{des}}\theta \tag{2.8.5}$$

Equating the absolute values of $d\,\Gamma(\text{ads})/dt$ and $d\,\Gamma(\text{des})/dt$ leads to the Langmuir equation in the form

$$\frac{\theta}{1-\theta} = \frac{k_{\text{ads}}}{k_{\text{des}}}\,c \tag{2.8.6}$$

In interface science, rate studies are more routinely performed at liquid-liquid or liquid-gas interfaces. Then the interfacial, or surface, tension can be measured, also under dynamic conditions. Its change with time is a measure of the rate of adsorption, although some interpretational steps are involved.

As both the adsorption and desorption processes are fast, *hysteresis* should be absent. If it does occur, it points to complications, like porosity, surface reactions, non-inertness of the adsorbent, etc. In this sense, its occurrence is a kind of diagnostic tool.

2.9 Conclusions and applications

It was the purpose of this chapter to present a basic understanding of phenomena occurring at solid-liquid interfaces. Restriction was made to uncharged surfaces and small molecules. Difficulties like surface heterogeneity[1] and porosity have not been discussed in detail because the problems are similar to those in chapter 1. One new feature is that heterogeneity with respect to surface hydrophilicity or hydrophobicity may perhaps be approached by studying the adsorption of amphiphilic molecules, which adsorb in a different fashion on these two types of adsorbents (sec. 2.7d). We re-emphasize that, although adsorption from solution is an exchange process, for binary mixtures only one affinity distribution $f(\Delta_{\text{ads}}u)$, as in [1.7.1], is needed and that porosity can best be assessed by comparing the accessibility of the pores to molecules of different sizes. We also avoided the discussion of adsorption from multicomponent mixtures[2]. Basic features, including structural modification of liquids near solids and adsorption, were emphasized.

Adsorption isotherms constitute the first experimental information, though they do not yield the complete story. Typically, adsorptions are excess quantities. For adsorption from binary mixtures, a model is generally needed to

---

[1] For further reading, see M. Jaroniec, R. Madey, *Physical Adsorption on Heterogeneous Surf.*, Elsevier (1988); L. Łajtar, S. Sokołowski, *J. Chem. Soc., Faraday Trans.* **88** (1992) 2545 (via density functionals).

[2] M. Borówko, M. Jaroniec, *Adv. Colloid Interface Sci.* **19** (1983) 137 (review); D. Schiby, E. Ruckenstein, *Colloids Surf.* **15** (1985) 17.

decompose composite into individual isotherms. However, for dilute solutions the excess adsorption may be identified as the amount that is physically present on the surface. In order to obtain further information, adsorption isotherms should be:

(i) measured over a sufficiently large concentration range; sometimes useful insight is obtained from alternative ways of plotting, for instance semi-logarithmically, which is possible if data over large concentration ranges are available.

(ii) measured at several temperatures, to learn more about the driving force.

(iii) augmented by other (calorimetric, spectroscopic, etc.) measurements.

Besides being interesting for their own sake, the insights gained in this way serve several purposes. Some of these purposes recur in this Volume. To start with, in chapter 3 adsorption of charged species (ions) will be discussed. This gives rise to the formation of electrical double layers. Electrosorption will be an important topic. This represents an extension of the present chapter. Another extension is the adsorption of polymers and polyelectrolytes, to be introduced in chapter 5. Adsorption also plays an important role in Volume III, which considers fluid-fluid interfaces. Many experiments to be discussed there deal with Langmuir troughs or similar devices. In such troughs the surface area is often so small that it is impossible to measure adsorbed amounts analytically. This disadvantage is compensated by the advantage that then the interfacial or surface tension or, for that matter, the surface pressure, is accessible. Other chapters also build further on the bricks established in the present chapter, including the chapter planned in Volume III on wetting, which rests on adsorption at three phase boundaries. All these applications involve dynamic, as well as static, phenomena. Electrokinetics (chapter 4) deals with double layers under shear and rates of wetting require insight into the kinetics of adsorption and desorption. More applications follow in the later Volumes.

The direct application of adsorption from solution in practice is abundant. Let us just mention a few examples. Adsorption of obnoxious substances from water on high surface area-carbon blacks is used as a purification step in the *preparation of potable water*. For the assessment of the toxic load in soils it is important to know how *pesticides* and *herbicides* adsorb on soil components (silica, clay minerals, etc.). In fact, many environmental issues involve at least one adsorption step. Adsorption is also involved in the food industry, for instance in decolouring vegetable oils. Dye adsorption is used in colouring various materials, but also in the photographic industry to optically *sensitize silver halides*. Adsorption of so-called "relay molecules" (these are capable of transporting an electron that stems from a photosensitive compound which may be excited by sunlight) on solid catalysts is an important step in the *photolysis of water* by sunlight. It is a potential alternative energy-storing process, which plants do so efficiently in photosynthesis. The last two examples involve a chemical step and can therefore be classified under *chemisorption*, which we shall not treat systematically. It is just noted that, for this and similar reasons, many colloidal dispersions behave as heterogeneous catalysts[1].

Adsorption is one route to *surface modification*, in order to make surfaces more or less hydrophilic, or blood-compatible, or to enrich specific compounds for analytical purposes, as in biosensors or chromatographic columns. For most practical purposes a strong adsorbent-adsorbate attachment is required; when physical adsorption is not strong enough, chemical binding may be appropriate[2]. Surface modification is also a central element in *flotation*, a process in the domain of ore refinement, based upon separation of valuable minerals from gangue by differences in hydrophobicity, so that the more hydrophobic particles may be eliminated by attachment to air bubbles, passed through the slurry. As the components in such a slurry mostly do not differ enough in hydrophobicity to allow such a separation, selective adsorption is needed to enhance the contrast. This is in essence a problem of *selective wetting* to which we shall return in Volume III.

Selective adsorption may also have dramatic consequences for crystal growth if certain crystal planes can be blocked[3]. Trace amounts of impurities may in this way determine the eventual crystal shape. This feature plays an important role in the preparation of (tailor-made) colloids (Volume IV) obviously adsorption kinetics plays an important role here. Certain substances may of course work the other way around in that they "attack" the solid. Examples are some sulfur-containing compounds with silver halides, and chelating agents with haematite (α - Fe_2O_3).

The determination of specific surface areas by adsorption from solution has already been mentioned in secs. 2.6 and 2.7.

Adsorption chromatography and solid-liquid adsorption are mutually coupled, in that adsorption has to be understood to interpret chromatograms but that chromatography can also be used as a tool to study adsorption (sec. 2.5a). As is

---

[1] *Kinetics and Catalysis in Microheterogeneous Systems*, M. Grätzel, K. Kalyana-sundaram, Eds., *Surfactant Series* **38**, Marcel Dekker (1991) (part on micellar catalysis, part on inorganic colloids).
[2] *Chemically Modified Surfaces*, H.A. Mottola, J.R. Steinmetz, Eds. Elsevier (1992); see also *Molecular Monolayers and Films*, a panel report by J.D. Swalen and 11 co-authors, *Langmuir* **3** (1987) 932.
[3] R.F. Strickland-Constable, *Kinetics and Mechanisms of Crystallization*, Academic Press (1968); A.E. Nielsen, *J. Crystal Growth* **67** (1984) 289 (review).

generally known, chromatography is a very powerful analytical separation technique[1].

Adsorption of surfactants has developed into a domain on its own. So far we have only introduced the non-ionic part (sec. 2.7d). Abundant applications are found in detergency, flotation, enhanced oil recovery, drug administration and other pharmaceutical purposes, paints, cosmetics, ceramic materials and the stabilization of suspensions in general.

Most of these applications are rather complicated in that many components interact and that several processes may take place simultaneously and/or consecutively.

2.10 General references

2.10a. IUPAC recommendations

Definitions, Terminology and Symbols in Colloid and Surface Chemistry, Part I. Prepared for publication by D.H. Everett, *Pure Appl. Chem.* **31** (1972) 579 (contains section on adsorption).

Reporting Data on Adsorption from Solution at the Solid/Solution Interface. Prepared for publication by D.H. Everett, *Pure Appl. Chem.* **58** (1986) 967.

G. Schay, *A Comprehensive Presentation of the Thermodynamics of Adsorptive Excess Quantities*, Pure Appl. Chem. **48** (1976) 393.

2.10b. Adsorption from solution

Note: Several references mentioned in sec. 1.9 also contain parts on adsorption from solution; as a rule these are not repeated here.

The Liquid/Solid Interface at High Resolution. Faraday Discuss. Chem. Soc. **94** (1992). (Various contributions, emphasis on STM and interfacial electrochemistry.)

A. Dabrowski, M. Jaroniec, *Adv. Colloid Interface Sci.* **27** (1987) 211, *Theoretical Foundations of Physical Adsorption from Binary Non-electrolytic Liquid Mixtures on Solid Surfaces: Present and Future* (emphasis on adsorption models covering heterogeneity, adsorptives of different sizes and multilayers).

---

[1] F. Helfferich, G. Klein, *Multicomponent Chromatography*, Marcel Dekker, (1970); *Theoretical Advancement in Chromatography and Related Separation Techniques*, F. Dondi, G. Guichon, Eds. NATO-ASI Series, Kluwer, (NL) 1992.

A. Dabrowski, M. Jaroniec, *Excess Adsorption Isotherms for Solid-Liquid Systems and Their Analysis to Determine the Surface Phase Capacity*, *Adv. Colloid Interface Sci.*, **31** (1990) 155. (Essentially an attempt to obtain and define surface areas by adsorption from binary mixtures.)

D.H. Everett, *Adsorption at the Solid-Liquid Interface*, in *Specialist Periodical Report, Colloid Science*. The Chemical Society (London), Vol. **1** (1973) 49 (ready access to the literature up to the year of publication; non-aqueous systems).

D.H. Everett, R.T. Podoll, *Adsorption at the Solid-Liquid Interface: Non-electrolyte systems*, in *Specialist Periodical Report, Colloid Science*. The Chemical Society (London), Vol. **3** (1979) 63. (Literature review up to 1977, especially for homogeneous surfaces.)

D.H. Everett, *Adsorption from Solution*, Spec. Publ. Roy. Soc. Chem. (London), **43** (1982) 71. (Review including colloidal aspects.)

M. Jaroniec, R. Madey, *Physical Adsorption on Heterogeneous Solids*, Elsevier (1988). (Discusses many models of limited validity.)

J.J. Kipling, *Adsorption from Solutions of Non-Electrolytes*, Academic Press (1965). (More or less the root of the story.)

D. Nicholson, N.G. Parsonage, *Computer Simulation and the Statistical Mechanics of Adsorption*, Academic Press (1982). (Contains, besides simulation work, much information on the thermodynamics and structure formation in interfaces.)

Adsorption from Solution, R.H. Ottewill, C.H. Rochester and A.L. Smith, Eds., Academic Press (1983). (Collection of papers dedicated to prof. D.H. Everett, contains several contributions relevant to the present chapter.)

Adsorption from Solution at the Solid/Liquid Interface, G.D. Parfitt, C.H. Rochester, Eds., Academic Press (1983). (Contains chapters on adsorption of small molecules (G.D. Parfitt and C.H. Rochester), adsorption from mixtures of miscible liquids (J.E. Lane) and adsorption of non-ionic surfactants (J.S. Clunie, B.T. Ingram) and many others.)

C.H. Rochester, *Infrared Spectroscopic Studies of Adsorption Behaviour at the Solid/Liquid Interface*, *Adv. Colloid Interface Sci.* **12** (1980) 43. (Emphasis on *in situ* measurements; illustrations of our sections 2.5a and b.)

Adsorption at the Gas-Solid and Liquid-Solid Interface, J. Rouquerol, K.S.W. Sing, Eds. Elsevier (1982). (Proceedings of an international symposium held in Aix-en-Province, France (1981).)

G. Schay, *Adsorption of Solutions of Non-Electrolytes*, in *Surface and Colloid Science*, E. Matijević, Ed., Wiley-Interscience, Vol. **2** (1969) 155. (Review, thermo-dynamics, examples, solid-liquid and liquid-liquid.)

W. Stumm, *Chemistry of the Solid-Water Interface; Processes at the Mineral-Water and Particle-Water Interface in Natural Waters*, Wiley (1992). (Emphasizes chemical reactions and environmental aspects.)

2.10c. *Related references*

Of indirect interest are books on solubilities because the solubility of substances contributes to determining their tendencies to adsorb.

There is a major project, *IUPAC Solubility Data Series*, various volumes published by Pergamon Press.

Extensive data on solubilities can also be found in the *Handbook of Chemistry and Physics*, published by the Chemical Rubber Publishing Co. (Updates appear regularly.)

D.J.W. Grant, T. Higuchi, *Solubility Behavior of Organic Compounds*, Wiley (1990). (Much information on theory, thermodynamics, activities, group contributions: aqueous and non-aqueous solvents.)

J.N. Murrell, A.D. Jenkins, *Properties of Liquids and Solutions*, e.g. 2nd ed., Wiley (1994). (Textbook.)

Chemically Modified Surfaces, H.A. Mottola, J.R. Steinmetz, Eds. Elsevier (1992). (Proceedings of a symposium in the U.S., 1991.)

3 ELECTRIC DOUBLE LAYERS

Electric double layers at phase boundaries pervade the entire realm of interface and colloid science. Especially in aqueous systems, double layers tend to form spontaneously. Hence, special precautions have to be taken to ensure the absence of charges on the surfaces of particles. Insight into the properties of double layers is mandatory, in describing for instance electrosorption, ion exchange, electrokinetics (chapter 4), charged monolayers (Volume III), colloid stability, polyelectrolytes and proteins, and micelle formation of ionic surfactants, topics that are intended to be treated in later Volumes. The present chapter is meant to introduce the basic features.

Given our aim, the approach differs somewhat from that usually taken in electrochemistry. In particular this is manifested in our emphasis on double layers in the absence of an external field. Important electrochemical phenomena and techniques, such as electrocatalysis, conductometry, and electrodeposition in which external fields are imposed, are considered as "applications". Also typical is our choice of model systems, featuring substances that can be dispersed to give a sol or suspension, like silver iodide, oxides and clay minerals, or liquids, to give emulsions. Electrochemists prefer to start with double layers on mercury for reasons of practice and principle. The practical advantage is the easily renewable, virtually flat, conducting surface allowing quantities like the double layer capacitance to be measured with great precision. On the more basic side, the mercury-solution interface is, in the absence of electroactive species, *polarizable* over a wide potential range, i.e. the double layer is the result of an externally applied potential. On the other hand, double layers on silver iodide, oxides, etc., develop spontaneously with the material either as electrodes or as dispersed particles; they are *relaxed* and their Gibbs energy of formation is negative. As a discriminating feature, in polarizable double layers the surface potential is an independent variable, whereas relaxed double layers can phenomenologically (thermodynamically) be described without introducing the notion of surface potential. In secs. I.5.5 and I.5.6 this distinction has already been discussed, see particularly sec. I.5.5b. In addition to these sections, double layers have also been encountered in sec. I.5.2a where we addressed the Debye-Hückel (DH) theory for the spontaneous formation of ionic atmospheres in electrolyte solutions.

It is typical for relaxed double layers in diserse systems that they require ions for their formation. Because of this, the discussion will emphasize aqueous systems; double layers in media of low dielectric permittivity will be treated in sec. 3.11.

Unless stated otherwise, by "double layer" we shall understand "electric double layer".

3.1 Some examples of double layers

Before starting our systematic discussion on the properties of double layers, let us look at the four typical examples sketched in fig. 3.1.

Figure 3.1a is perhaps the most common in colloid science. Particles acquire a surface charge by preferential adsorption or desorption of certain ions. The classical model is silver iodide, which can be negatively or positively charged by excess adsorption of iodide or silver ions, respectively. Similarly, oxides may take up or lose protons from surface hydroxyl groups to acquire a positive or negative surface charge, respectively. The compensating charge on the solution side, the *countercharge*, may be more or less diffusely distributed.

In fig. 3.1b the preference of the hydrophobic tails of the (anionic) surfactant molecules for the oil phase gives rise to the double layer. Such double layers are for instance encountered in some emulsions. They may also occur at the air-water interface; then the driving force for their formation is the expulsion of the hydrocarbon tails from the aqueous phase. We speak of *ionized monolayers* and return to them in Volume III.

Figure 3.1c is an example of a "mixed" double layer as typically encountered with clay minerals. For clarity the countercharge is not drawn. The clay mineral platelet has on the edges a charge that is comparable with that on oxides, in that it is caused by adsorption or desorption of protons. At low pH, as drawn, the edge surface charge is positive. However, the charge on the plates is negative and has a very different origin, viz. *isomorphic substitution in the interior* of the solid. This phenomenon arises from substitution of multivalent cations in the solid by other cations of lower valency ($Al^{3+} \rightarrow Mg^{2+}$, etc.) which has taken place during the genesis of the mineral and is caused by the limited availability of some species. The ensuing frozen-in shortage of positive *space* charge is, for a number of phenomena, felt as a negative *surface* charge, that is manifested on the faces. We shall return to this in sec. 3.10d.

Figure 3.1d illustrates a double layer, created by adsorption of an anionic surfactant on an otherwise uncharged particle, say carbon, in water. This is one of the powerful means of stabilizing hydrophobic materials in aqueous media.

Obviously, combinations may occur in which there is more than one charging mechanism operating. For instance, ionic surfactants may adsorb on silver

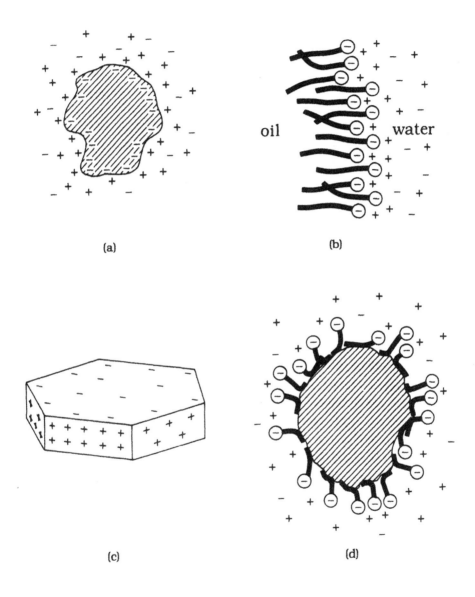

oil   water

(a)

(b)

(c)

(d)

Figure 3.1. Examples of double layers: (a) around a solid particle; (b) at an ionized monolayer of anionic surfactants, adsorbed at the oil-water interface; (c) on a hexagonal clay mineral particle at low pH (only the charge on the particle is drawn); (d) double layer generated by the adsorption of anionic surfactants on a hydrophobic surface. The pictures are schematic.

iodide that already carries a charge. The contributions of silver, iodide and surfactant ions to the total charge are not additive: adsorption of anionic surfactants shifts the iodide-silver ion adsorption balance to the silver side. Surface excesses are coupled by Gibbs' adsorption law (sec. 3.4). Therefore, it is often possible to obtain information on the (variation of the) excess of a given species from that of another. Adsorption of molecules, involving electrostatics is called *electrosorption*; we shall restrict this term to organic molecules and discuss the topic in sec. 3.12.

All the double layers discussed so far, except those generated by isomorphic substitution (Fig. 3.1c), are *relaxed*. They form spontaneously by adsorption of charged species. Hence, the ensuing surface charge depends, according to some isotherm equation, on the concentration of the charge-determining species, i.e. on pAg for silver halide, pH for oxides, on the concentration of surfactants in case (b), etc. On the other hand, the charge due to isomorphic substitution is *non-relaxed*; we consider it as *polarized*; it cannot be varied by changing the composition of the solution. Semiconductors with built-in vacancies or interstitial ions may also belong to this category, depending on the extent to which this charge can relax during an experiment. Sometimes it is convenient to distinguish surfaces *with variable charge*, as in cases (a), (b), (c, edges) and (d), and those *with constant charge* as in case (c, plates).

3.2 Why do ionic double layers form?

Forming polarized double layers is not a real issue. The mercury-solution interface can be polarized by an external source, producing the Gibbs energy required to keep an excess or deficit of electrons on the surface. The depletion of positive charge inside clay platelets is a matter of crystal growth and beyond our topic. Under discussion is now the question why relaxed double layers form, thereby restricting ourselves to ionic layers. (Double layers originating from dipole orientation etc., leading to the so-called χ-potentials introduced in sec. I.5.5a, will be discussed in sec. 3.9).

The basic problem is: do such double layers form by electrostatic forces, by non-electrostatic forces, or by a mixture of these? The answer is simple: *without non-electrostatic forces there is no ionic double layer*.

Some care in defining terms is required. On an atomic level all ionic and molecular interaction can be interpreted as "electric". However, on the colloidal, or mesoscopic, level we may restrict the term "electric" to "Coulombic". Consequently, all other interactions are by definition "non-electric", whatever their origin: the three types of Van der Waals forces, hydrogen bonding, solvent structure-originated or real chemical bond formation.

The same issue occurs in the definition of electrochemical potentials. According to [I.5.1.18]

$$\tilde{\mu}_i \equiv \mu_i + z_i F \psi \qquad\qquad\qquad [3.2.1]$$

relates the *electrochemical potential* $\tilde{\mu}_i$ of an ion i, to its *chemical*, or *thermodynamic*, potential μ_i and the macroscopic electric potential ψ. In [3.2.1] z_i is the valency of i, inclusive of sign, and F the Faraday constant. As discussed at some length in sec. I.4.3c, there is some difficulty in measuring ψ because the averaged macroscopic potential may differ from the potential of mean force. All errors made in interpreting measured values of ψ as the macropotential are automatically assigned to μ_i. Although μ_i is commonly called the "chemical" potential, there is often not much real chemistry in its interpretation: for dissolved species it contains a concentration term and a solvation contribution μ_i^o. Hence the term "chemical" encompasses much more than just chemical bond formation. In our present context we therefore avoid the term "chemical". Returning to the issue of double layer formation we retain the term "non-electrostatic", realizing that in the quantitative evaluation some contributions involved in establishing the electric part may have to be incorporated.

With this in mind, the impossibility of forming a double layer by electric forces only is obvious. Any ion that may attach to a particle will create a potential that keeps out all other ions of the same sign. Accumulation of a number of identical charges on a surface can take place only if the adsorbing ions experience a non-electric affinity for the surface so that they can move *against* the adverse potential. The extent to which this occurs depends on the balance between the attractive non-electrostatic and the repulsive electric forces. In summary: the *reason* for the formation of relaxed double layers is the non-electric affinity of charge-determining ions for a surface; the *extent* to which the double layer develops is determined by the non-electrostatic ⇔ electrostatic interaction balance.

Thermodynamically this issue can be condensed into two equations, previously derived in sec. I.5.7. The electrical part of Gibbs energy for charging a double layer at given p and T is, according to [I.5.7.1],

$$\Delta G_a^\sigma(\text{el}) = \int_0^{\sigma^o} \psi^{o'} d\sigma^{o'} \qquad\qquad\qquad [3.2.2]$$

Here σ^o and ψ^o are the surface charge (density) and surface potential, respectively. The primes indicate the variable values when the double layer is reversibly charged from $\sigma^{o'} = 0$ to its final value, σ^o. As $\psi^{o'}$ and $\sigma^{o'}$ have the same signs, $\Delta G_a^\sigma(\text{el}) > 0$, so a non-electric contribution is needed to make the overall Gibbs energy change negative, as required for a spontaneous process. For relaxed

interfaces ΔG_a^σ (non-el) stems from the non-electrostatic binding of charge-determining ions. For the simple case where there is only one charge-determining
species, for which the non-electrostatic affinity for the surface is independent of
coverage (i.e. for a homogeneous surface) it was derived in sec. I.5.7 that

$$\Delta G_a^\sigma = \Delta G_a^\sigma(\text{el}) + \Delta G_a^\sigma(\text{non-el}) = -\int_0^{\psi^o} \sigma^{o'} \, d\psi^{o'} \qquad\qquad [3.2.3]$$

which is always negative. Here, ΔG_a^σ (non-el) $= -\sigma^o\psi^o = \Gamma_i\Delta_{ads}\mu_i$.

3.3 Some definitions, symbols and general features

The basic quantity is the *charge* on the particle, Q, or the *surface charge density*,
σ^o. For brevity we often call the latter simply "surface charge", but the symbol σ
always stands for a charge per unit area. Following common usage we shall
usually express it in $\mu C \ cm^{-2}$. This charge is not necessarily confined to an
infinitely thin surface layer, but may have some depth, as for the permanent
charge in fig. 3.1c. In fact, for the diffuse part of a double layer the surface
charge σ^d is the total charge in a column of unit area normal to the surface. For
flat symmetry,

$$\sigma^d = \int_0^\infty \rho(x)\,dx \qquad\qquad [3.3.1]$$

where $\rho(x)$ is the space charge density at distance x from where the diffuse layer
starts. For a spherical double layer, with a the particle radius and r the distance
to the centre, if the diffuse part starts at $r = a$,

$$\sigma^d = \frac{1}{a^2}\int_a^\infty r^2 \rho(r)\,dr \qquad\qquad [3.3.2]$$

As a whole, electric double layers are always *electroneutral*. Hence, if the
countercharge is purely diffuse

$$\sigma^o + \sigma^d = 0 \qquad\qquad [3.3.3]$$

When part of the countercharge is bound to the surface by non-electrostatic
forces one speaks of *specific adsorption*. The term derives from the fact that non-
electrostatic binding energies typically depend on the *nature* of the ion, say on
its radius, whereas for purely Coulombic interactions between point charges, as
is the case in diffuse layers, usually the interaction is *generic*: identical for all
ions of the same valency. Generically adsorbing ions are called *indifferent*. Non-

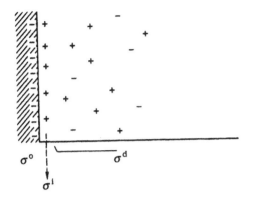

Figure 3.2. Identification of the surface charge σ^o, the specifically adsorbed charge σ^i and the diffuse charge σ^d in a flat double layer.

electrostatic interactions being short-range, specifically adsorbing ions are as a rule found in a thin layer, adjacent to the surface, called the *Stern layer*. Figure 3.2 illustrates such a structure. Calling the specifically adsorbed charge σ^i, the electroneutrality condition now reads

$$\sigma^o + \sigma^i + \sigma^d = 0 \qquad\qquad\qquad [3.3.4]$$

In this case $\sigma^i + \sigma^d$ constitutes the *countercharge*.

There is some arbitrariness in deciding which charge we identify as the surface charge. Typically, this is a non-thermodynamic issue. Let us first consider the case of fig. 3.1a, taking the solid to be silver iodide. In this case there is no ambiguity: silver and iodide ions can adsorb on the particle without modifying the crystal structure, they become part of it and are therefore readily identified as *charge-determining ions*. In such situations one may, as in fig. 3.2, draw the surface charge as residing *in* the solid surface. Even if the adsorbed Ag^+ and I^- ions would not become part of the solid lattice it remains logical to interpret the charge they attribute as *the* surface charge. When Γ_{Ag^+} and Γ_{I^-} are the surface concentrations of Ag^+ and I^- ions, respectively, *by definition*

$$\sigma^o \equiv F(\Gamma_{Ag^+} - \Gamma_{I^-}) \qquad\qquad\qquad [3.3.5]$$

Because double layers are electroneutral, effectively, only adsorption of electroneutral entities takes place; hence operationally the definition is

$$\sigma^o \equiv F(\Gamma_{AgNO_3} - \Gamma_{KI}) \qquad\qquad\qquad [3.3.6]$$

when KNO_3 is the electrolyte. As explained in sec. I.5.6e, σ^o can be determined as a function of pAg by a *potentiometric colloid titration*, a procedure to which we return in sec. 3.7a. At low pAg, $\sigma^o > 0$, at high pAg, $\sigma^o < 0$. There is a certain pAg where $\sigma^o = 0$, the *point of zero charge* (p.z.c.).

For oxides the uniqueness of protons is more or less established. Protons can be adsorbed or removed from the surface by adding acid or base. Therefore, we *define* the surface charge through

$$\sigma^o \equiv F(\Gamma_{HNO_3} - \Gamma_{KOH})$$ [3.3.7]

(in solutions of KNO_3). For low pH $\sigma^o > 0$, for high pH $\sigma^o < 0$. In this case the point of zero charge is a given pH-value. The more acidic the oxide, the lower the p.z.c. We return to the charging mechanism of oxides in sec. 3.6d.

For the sides of clay particles, usually called "edges" (fig. 3.1c), σ^o is defined as in [3.3.7]. One can also assign a point of zero charge to the edges. The face charge is simply determined by the bulk charge density due to isomorphic substitution, using an integral similar to [3.3.1], but now for the solid. There is no way of determining it by a colloid titration but it can in principle be found from the countercharge, using [3.3.4].

The prominent positions of silver and iodide ions for silver iodide and of protons and hydroxyl ions for oxides also follow from the observation that for silver iodide electrodes and for many oxide electrodes the potential is proportional to pAg and pH, respectively, according to *Nernst's law* (sec. I.5.5c). Glass electrodes illustrate this fact. One of the implications is that the potential, generated by the adsorption of these ions is the potential of the solid with respect to the liquid. At the same time, this is the potential needed in [3.2.2 and 3]. Therefore, we call Ag^+ and I^- ions *charge-determining ions* for AgI and we now also identify H^+ and OH^- ions as the *charge-determining ions* for oxides. These ions are also called *potential-determining ions*, but we prefer the former for reasons given in sec. I.5.6.

In fig. 3.1b there may be some difficulty in defining the borderline between the two liquids. Nevertheless, we shall call the charge attributed by the anionic surfactant "surface charge". This choice is obvious if the oil phase is an emulsion droplet. It is a matter of taste whether to denote this charge as σ^o or as σ^i. In fact, there is no difference of principle between surface ions and specifically adsorbed (counter-)ions; intermediate cases exist of very strong specific binding of counterions, like phosphate on haematite or thallium ions on silver iodide. We shall make our choice explicit where necessary.

For the thermodynamics this nomenclature problem must of course be without consequence. If, in the situation of fig. 3.1b or d, the charge attributed by the adsorbed ionics is interpreted as a specifically adsorbed charge σ^i, the Gibbs energy of the layer may be written as

$$\Delta G_a^\sigma = - \int_0^{\psi^i} \sigma^{i'} \, d\psi^{i'} \qquad\qquad [3.3.8]$$

when the non-electrical adsorption Gibbs energy $\Delta_{ads}\mu_i$ is constant. For surfactant adsorption such constancy is unlikely because, especially at high coverages, lateral interaction plays its role. For that case, we shall have to reconsider this complication. In [3.3.8] ψ^i is the potential in the layer where the head groups are located (for a not mathematically thin layer, as drawn, more layers are needed with the resulting potentials). If the charges are interpreted as surface ions, [3.2.3] is appropriate. In the case of both a surface charge and a specifically adsorbed countercharge, as in fig. 3.2, under the proviso of constant non-electrostatic contributions,

$$\Delta G_a^\sigma = - \int_0^{\psi^o} \sigma^{o'} \, d\psi^{o'} - \int_0^{\psi^i} \sigma^{i'} (\sigma^{o'}) \, d\psi^{i'} \qquad\qquad [3.3.9]$$

where $\sigma^{i'}(\sigma^{o'})$ is the relation between σ^i and σ^o during the charging process. This relation is coupled to the Esin-Markov coefficient β, see [3.4.16], and hence obtainable. However, to carry out the integrations in [3.3.9], the charge-potential relationships must also be known.

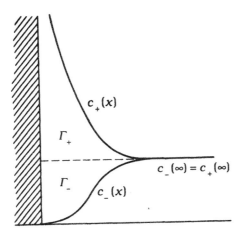

Figure 3.3. Introduction of the notion of negative adsorption in a diffuse double layer. The surface is assumed to be negatively charged. For quantitative pictures, see fig. 3.8.

It is noted that the diffuse part of the double layer does not contribute to ΔG_a^σ because no non-electrostatic interactions are involved: in this part changes in $\tilde{\mu}_i$ due to changes in concentration are exactly balanced by changes in potential, according to Boltzmann's law.

The counthercharge consists of an *excess of counterions* (ions of signs opposite to that of the surface charge) and a *deficit of co-ions* (same charge sign as the surface). For the diffuse part, the distribution may look like that in fig. 3.3. In this case, the excess charge attributed by cations, $\sigma_+^d = z_+ F \Gamma_+$ and that by the co-ions, $\sigma_-^d = z_- F \Gamma_-$, are both positive ($z_- < 0$ and $\Gamma_- < 0$). The quantities σ_+ and σ_- are called the *ionic components of charge*. In the present case

$$\sigma_+^d + \sigma_-^d = \sigma^d$$

[3.3.10]

As will be demonstrated later, at low surface potential $\sigma_+^d = \sigma_-^d = \sigma^d / 2$, but with increasing potential σ_+^d increases, whereas σ_-^d levels off to a constant value. The reason is that, upon making σ^o more negative, $c_+(x)$ can continue to grow, whereas $c_-(x)$ can never become less than zero.

In the language of chapter 2, in a double layer as in fig. 3.3 counterions are *positively*, co-ions *negatively*, *adsorbed*. Because of the longer range of the electrostatic forces, as compared with the short-range specific interactions considered in chapter 2, negative adsorption of ions from charged surfaces is quantitatively more prominent than that of uncharged molecules from an electroneutral surface. As explained in secs. I.5.5f, I.5.6a and b, expulsion of co-ions by a charged particle leads to an increase of the neutral electrolyte concentration in the equilibrium solution around the particle (i.e. the *dialysate*), as compared with the situation for uncharged particles. This phe-nomenon is a manifestation of the *Donnan effect*. It is a general property of charged colloids, resulting in such features as salt-sieving (secs. I.1.1 and I.1.3) and the suspension effect (sec. I.5.5f). We return to this matter in secs. 3.5b, 3.7e and 4.10.

As well as descriptions in terms of charges and potentials, the properties of double layers can also be represented in terms of capacitances. We already defined the *differential (double layer) capacitance* [I.5.1.25]

$$C = d\sigma^o / d\psi^o$$

[3.3.11]

and the *integral capacitance*

$$K = \sigma^o / \psi^o$$

[3.3.12]

For the interconversion $C \Leftrightarrow K$, see [I.5.1.26 and 27].

For completely relaxed interfaces, K cannot be measured because ψ^o is thermodynamically inoperable, but for materials of which electrodes can be made, $d\psi^o$ is accessible, so that C is obtainable. In this respect there is a contrast with the double layer on a polarizable electrode like mercury, where ψ^o is applied and known with respect to a reference. For such a system C is usually determined as the primary variable, from which σ^o is derived by integration with respect to the potential. It may be added that conditions may be devised where the silver iodide-solution interface is polarized (electrodes at high frequency applied a.c. potentials to suppress exchange currents) and that the mercury-solution interface can be "depolarized" by adding electro-active ions (Cd^{2+} at sufficiently negative potential), but we shall not consider these cases (see sec. 3.7d) at this point.

The usefulness of capacitances is that they reflect the extent of *screening* of the surface charge, and hence provide valuable information on the composition of the double layer. The better the surface charge is screened (say by strong accumulation of counterions) the higher σ^o can become.

Although it is logical to discuss *points of zero charge* in the context of electric double layers, these are typical *non-electric* surface properties, because they reflect the difference in non-electrostatic affinity of the two charge-determining species for the uncharged surface. So, for oxides the p.z.c. is related to pK_a and pK_b, the dissociation constants for the acid and base functions of the surface hydroxyl groups.

We return to this notion, and to its electrokinetic counterpart, the *isoelectric point* (i.e.p.) in sec. 3.8.

3.4 Thermodynamics. Application of the Gibbs equation

For many purposes it is conducive to start analyses with thermodynamic considerations. In this way, it is often possible to find laws of general validity and to determine the boundaries between which models can be developed. For the study of (relaxed) double layers the Gibbs adsorption equation is the starting point. Although the interfacial tension of a solid-liquid interface cannot be measured, this equation remains useful because it helps to distinguish measurable and inoperable variables, and because it can be used to correlate surface concentrations of different species (including the surface ions), some of which may not be analytically accessible.

In chapter I.2, especially in sec. I.2.22, the Gibbs equation has been derived and extensively discussed, and in sec. I.5.6 it has been extended to double layers. Here, we briefly repeat and extend the most relevant features.

In its most general form, at given pressure, the Gibbs equation reads (see [I.2.22.9])

$$d\gamma = - S_a^\sigma \, dT - \sum_i \Gamma_i \, d\mu_i \qquad\qquad [3.4.1]$$

where γ is the interfacial tension, S_a^σ the interfacial excess entropy per unit area. The sum extends over all independent components in the system, i.e. all (electroneutral) components except for one because of the phase rule (sec. I.2.13).

and except for those that are dependent because of chemical reactions between them. Surface concentrations are defined by locating the Gibbs dividing plane, sec. I.2.22a. For adsorption from dilute solutions, as is almost exclusively the case in double layer studies, dividing plane, referred to the excess adsorption of solvent (water), coincides in practice with the solid-liquid interface. Hence, all Γ's may be interpreted as the analytically determined surface concentrations. In this way, for silver iodide in an aqueous solution of KI, $AgNO_3$, KNO_3 and an organic substance A,

$$
\begin{aligned}
d\gamma = - S_a^\sigma \, dT - \Gamma_{KI} \, d\mu_{KI} - \Gamma_{AgNO_3} \, d\mu_{AgNO_3} - \Gamma_{KNO_3} \, d\mu_{KNO_3} \\
- \Gamma_A \, d\mu_A - \Gamma_w \, d\mu_w - \Gamma_{AgI} \, d\mu_{AgI}
\end{aligned}
\qquad [3.4.2]
$$

The seven terms on the r.h.s. are not independent. First, there is the Gibbs-Duhem rule [I.2.13.5], which, at constant pressure, for the solution side reads

$$
\begin{aligned}
S_m^L \, dT + x_{KI} \, d\mu_{KI} + x_{AgNO_3} \, d\mu_{AgNO_3} + x_{KNO_3} \, d\mu_{KNO_3} \\
+ x_A \, d\mu_A + x_w \, d\mu_w + x_{AgI} \, d\mu_{AgI} = 0
\end{aligned}
\qquad [3.4.3]
$$

Here, x stands for the mole fraction of the species indicated. For the solid phase

$$
S_m^S \, dT + d\mu_{AgI} = 0
\qquad [3.4.4]
$$

In the second place, because of the equilibrium $KI + AgNO_3 \rightleftarrows AgI + KNO_3$, we have

$$
d\mu_{KI} + d\mu_{AgNO_3} = d\mu_{AgI} + d\mu_{KNO_3}
\qquad [3.4.5]
$$

The three auxiliary conditions [3.4.3 - 5] allow us to eliminate three chemical potentials, for instance μ_w, μ_{AgI} and μ_{KI}, or μ_w, μ_{AgI} and μ_{AgNO_3}. We shall not elaborate this generally[1] but discuss the case of isothermal isobaric systems, where the entropic terms drop out and $d\mu_{AgI} = 0$. Let us eliminate μ_{KI} between [3.4.5] and [3.4.2] by writing $d\mu_{KI} = -d\mu_{AgNO_3} + d\mu_{KNO_3}$

$$
d\gamma = -(\Gamma_{AgNO_3} - \Gamma_{KI}) d\mu_{AgNO_3} - (\Gamma_{KNO_3} + \Gamma_{KI}) d\mu_{KNO_3} - \Gamma_A \, d\mu_A
\qquad [3.4.6]
$$

The surface concentrations in this equation are basically $\Gamma^{(w)}$'s, i.e. they are referred to a Gibbs dividing plane, located on the basis $\Gamma_w \equiv 0$. In sec. I.2.22c it was shown that for dilute solutions $\Gamma^{(w)}$ corresponds to the analytical surface concentration Γ. In [3.4.6], according to [3.3.6], $F(\Gamma_{AgNO_3} - \Gamma_{KI})$ is the surface

---

[1] A more rigorous analysis can be found in B.H. Bijsterbosch, J. Lyklema, *Adv. Colloid Interface Sci.* **9** (1978) 147.

charge σ^o, as obtained by potentiometric colloid titration:

$$\sigma^o \equiv F(\Gamma_{AgNO_3} - \Gamma_{KI})$$ [3.4.7]

Moreover, $F(\Gamma_{KNO_3} + \Gamma_{KI})$ equals all the charge in the double layer, attributed by K^+ ions,

$$\sigma_{K^+} \equiv F(\Gamma_{KNO_3} + \Gamma_{KI})$$ [3.4.8a]

Later we shall also need

$$\sigma_{NO_3^-} \equiv -F(\Gamma_{KNO_3} + \Gamma_{AgNO_3})$$ [3.4.8b]

Taking all of this together, the Gibbs equation reduces to

$$d\gamma = -\sigma^o d\mu_{AgNO_3}/F - \sigma_{K^+} d\mu_{KNO_3}/F - \Gamma_A d\mu_A$$ [3.4.9a]

If we had eliminated $d\mu_{AgNO_3}$ instead of $d\mu_{KI}$, we would have obtained

$$d\gamma = \sigma^o d\mu_{KI}/F + \sigma_{NO_3^-} d\mu_{KNO_3}/F - \Gamma_A d\mu_A$$ [3.4.9b]

Equations [3.4.9a and b] are equivalent because $\sigma_{K^+} + \sigma_{NO_3^-} + \sigma^o = 0$. For oxides the corresponding pair of equations reads

$$d\gamma = -\sigma^o d\mu_{HNO_3}/F - \sigma_{K^+} d\mu_{KNO_3}/F - \Gamma_A d\mu_A$$ [3.4.10a]

$$d\gamma = \sigma^o d\mu_{KOH}/F + \sigma_{NO_3^-} d\mu_{KNO_3}/F - \Gamma_A d\mu_A$$ [3.4.10b]

with $\sigma_{K^+} = F(\Gamma_{KOH} + \Gamma_{KNO_3})$, etc.

In practice, the concentration of the electrolyte containing those ions which become surface ions ($AgNO_3$, KI, HNO_3, KOH) is almost always small as compared with the concentration of the electrolyte not containing such ions (KNO_3). Then $d\mu_{AgNO_3} \approx d\mu_{Ag^+} \approx -2.303\, RT\, dpAg$, etc. Assuming this to be the case, equations [3.4.9 and 10] can be rewritten as

$$d\gamma = 2.303\, RT\, \sigma^o\, dpAg / F - (\sigma^o/2 + \sigma_{K^+}) d\mu_{KNO_3} / F - \Gamma_A d\mu_A$$ [3.4.11a]

$$d\gamma = -2.303\, RT\, \sigma^o\, dpI / F + (\sigma^o/2 + \sigma_{NO_3^-}) d\mu_{KNO_3} / F - \Gamma_A d\mu_A$$ [3.4.11b]

$$d\gamma = 2.303\, RT\, \sigma^o\, dpH / F - (\sigma^o/2 + \sigma_{K^+}) d\mu_{KNO_3} / F - \Gamma_A d\mu_A$$ [3.4.12a]

$$d\gamma = -2.303\, RT\, \sigma^o\, dpOH / F + (\sigma^o/2 + \sigma_{NO_3^-}) d\mu_{KNO_3} / F - \Gamma_A d\mu_A$$ [3.4.12b]

These four equations, or variants thereof, will be used often here. They are rigorous, given the experimental conditions (our identifying of σ^o via [3.4.7] and σ_{K^+} via [3.4.8] is a matter of convention and does not detract from the rigour; other choices could have been made). Upon integrating these equations, care must be taken that dpAg and dpH stand in principle for changes in activities of single ionic entities; we could introduce these only by virtue of the excess of indifferent electrolyte.

Equations [3.4.11 and 12] are useful for a number of purposes, including the following:

(i) The r.h.s.'s only contain measurable variables. The ionic components of charge σ_{K^+} and $\sigma_{NO_3^-}$ are obtainable except for a constant, by integration of the Esin-Markov coefficient with respect to σ^o, see [3.4.16]. Here, no single ionic excesses are counted but sums of electroneutral combinations, including the negative adsorption of electrolyte, Γ_{KNO_3}, see [3.4.8]. Therefore, $d\gamma$ is also operable, although γ itself is not accessible. In fact $d\gamma = -d\pi$, where π is the *surface pressure*. If for π a reference point is chosen, say $\pi \equiv 0$ at the p.z.c. in the absence of electrolyte and organic components, the equation can be integrated to yield a *surface equation of state*. These equations are the electrochemical extensions of [1.1.7].

(ii) For AgI and a number of oxides, where experiments with electrodes could be performed, the applicability of Nernst's law has been ascertained. When that is the case, the first term on the r.h.s. may be replaced by $-\sigma^o \, d\Delta\phi$ where $\Delta\phi$ is the *Galvani potential difference* across the solid-liquid interface (see secs. I.5.5a and c). When no potential drop occurs inside the solid and the χ-potential is constant, this term may also be written as $-\sigma^o \, d\psi^o$. In the Gibbs equation for mercury [I.5.5.16] a similar term also occurs. The distinction is that for mercury $\Delta\phi$ is an independent variable whereas for relaxed double layers this is not the case.

(iii) A number of cross-differentiations can be carried out to interrelate surface concentrations. Especially when one of them is analytically difficult to measure, such relationships may be useful . For example, for the variation of Γ_A with the surface charge, i.e. with pAg for the AgI system

$$\frac{F}{2.303 \, RT} \left(\frac{\partial \Gamma_A}{\partial pAg} \right)_{\mu_{KNO_3}, \, \mu_A} = - \left(\frac{\partial \sigma^o}{\partial \mu_A} \right)_{pAg, \, \mu_{KNO_3}} \qquad [3.4.13]$$

or

$$\frac{F}{2.303} \left(\frac{\partial \Gamma_A}{\partial pAg} \right)_{c_{KNO_3}, \, c_A} = - \left(\frac{\partial \sigma^o}{\partial \ln c_A} \right)_{pAg, \, c_{KNO_3}} \qquad [3.4.13a]$$

where, for the uncharged component, the activity has been replaced by a concentration. In this way, changes in the amount of butanol, adsorbed on silver iodide, too small to be measured, could be inferred from the measured change in the surface charge with the butanol concentration at constant pAg and salt concentration.

(iv) Cross-differentiation also yields *Esin-Markov coefficients* β, introduced in sec. I.5.6d. These coefficients contain information on the relative contributions of the cations and anions to the countercharge, i.e. they help to obtain the composition of the double layer. Experimentally, β is measured as the horizontal spacing between σ^o(pAg) or σ^o(pH) curves, measured at different salt concentrations and defined as

$$\beta \equiv \left(\frac{\partial\, pAg}{\partial \log a_\pm} \right)_{\sigma^o} \qquad\qquad \text{(silver halides)} \qquad\qquad [3.4.14]$$

$$\beta \equiv \left(\frac{\partial\, pH}{\partial \log a_\pm} \right)_{\sigma^o} \qquad\qquad \text{(oxides)} \qquad\qquad [3.4.15]$$

To obtain expressions for β first a change of variables is needed. Say, by way of example, that we are interested in the *lyotropic sequence* for monovalent counterions on AgI. Then we replace in [3.4.11a] K$^+$ by C$^+$ (= Li$^+$, Na$^+$, K$^+$, ...), write μ_s for μ_{salt}, assume organic additives to be absent and introduce the new variable $\xi \equiv \gamma - 2.303\, RT\, \sigma^o\, pAg\, /\, F$ to obtain

$$d\xi = -2.303\, RT\, pAg\, d\sigma^o / F - (\sigma^o / 2 + \sigma_{C^+}) d\mu_s / F$$

from which, by cross-differentiation,

$$2.303\, RT \left(\frac{\partial\, pAg}{\partial \mu_s} \right)_{\sigma^o} = \frac{1}{2} + \left(\frac{\partial \sigma_{C^+}}{\partial \sigma^o} \right)_{\mu_s}$$

or, with $d\mu_s = 4.606\, RT\, d\log a_\pm$, introducing β via [3.4.14],

$$\beta = 1 + 2 \left(\frac{\partial \sigma_{C^+}}{\partial \sigma^o} \right)_{a_\pm} \qquad\qquad [3.4.16]$$

This is a kind of "differential adsorption isotherm", relaying how strongly the counterion charge increases with the surface charge. Figures 3.45 and 3.46 are illustrations of a Esin-Markov analysis for AgI in KNO$_3$. For this system it was also found that β increases from Li$^+$ to Cs$^+$; this is a typical specific effect, caused by the increased non-electric adsorption in this direction. At the same time, the σ^o(pAg) curves for different salt concentrations are wider apart for CsNO$_3$ than for $LiNO_3$. Briefly stated, negative charge on silver iodide is better screened by Cs^+-ions than by Li^+-ions.

For other systems alternatives of [3.4.16] may be derived. We recall [I.5.6.19] for the Esin-Markov coefficient for a $(z_+ - z_-)$ electrolyte

$$\beta = -\frac{1}{|z_+|} - \frac{|z_+| + |z_-|}{|z_+||z_-|}\left(\frac{\partial \sigma_-}{\partial \sigma^\circ}\right)_{a_\pm} = \frac{1}{|z_-|} + \frac{|z_+| + |z_-|}{|z_+||z_-|}\left(\frac{\partial \sigma_+}{\partial \sigma^\circ}\right)_{a_\pm} \qquad [3.4.17a, b]$$

which for symmetrical electrolytes, $|z_+| = |z_-|$, reduces to

$$\beta = -1 - 2\left(\frac{\partial \sigma_-}{\partial \sigma^\circ}\right)_{a_\pm} = 1 + 2\left(\frac{\partial \sigma_+}{\partial \sigma^\circ}\right)_{a_\pm} \qquad [3.4.17c, d]$$

Generally, Esin-Markov coefficients are useful tools to study ionic components of charge and, hence, specific adsorption. They are also required in obtaining the Gibbs energy of a double layer in the case of specific adsorption, see [3.3.9].

In our discussion the Gibbs equation has been formulated in terms of electroneutral components. The purpose was to remain close to thermodynamic rigour. Only electroneutral components are thermodynamically defined and double layers are electroneutral. Hence, their formation must be thermo-dynamically interpretable in terms of adsorption/desorption of electroneutral quantities. On the other hand, when discussing the structures of a double layer one has individual ions in mind. The question rises as to whether the Gibbs equation can also be written *in terms of individual ionic species*. This issue is similar to that of electrolyte solutions, discussed in some detail in sec. I.5.1b and c. Although activities of individual ionic species are inoperable, one can define such quantities provided these activities obey rules that agree with those for the operable electroneutral system, see for instance [I.5.1.6] and [I.5.1.11]. For double layers this means that one may write the Gibbs equation in terms of electrochemical potentials of individual ions, provided (i) the bulk activities satisfy the rules of sec. I.5.1b and (ii) the Gibbs equation is supplemented with the appropriate electroneutrality condition. Thus, instead of [3.4.2] one would write at constant T and setting, as before, $\Gamma^{(w)} = \Gamma$,

$$d\gamma = -\Gamma_{Ag^+}\, d\tilde{\mu}_{Ag^+} - \Gamma_{I^-}\, d\tilde{\mu}_{I^-} - \Gamma_{K^+}\, d\tilde{\mu}_{K^+} - \Gamma_{NO_3^-}\, d\tilde{\mu}_{NO_3^-} - \Gamma_A\, d\tilde{\mu}_A \qquad [3.4.18]$$

with

$$\Gamma_{Ag^+} + \Gamma_{K^+} = \Gamma_{I^-} + \Gamma_{NO_3^-} \qquad [3.4.19]$$

The surface charge is now defined as [3.3.5], $\sigma_{K^+} = F\Gamma_{K^+}$, etc. Instead of [3.4.5] we now have $d\tilde{\mu}_{I^-} + d\tilde{\mu}_{Ag^+} = d\mu_{AgI}$ (= 0 at constant temperature).

Although we prefer the "electroneutral" approach, we shall in a few cases also refer to the "individual ionic" description. Section I.5.6c already contained an example. Obviously, the final outcome in terms of measurable variables must be identical between the two treatments.

The above equations can be extended and/or modified to include other systems and conditions, for instance to mixed or asymmetrical electrolytes and to charged monolayers at LG and LL phase boundaries.

Anticipating sec. 3.7a, it is noted that experimentally only $(\Gamma_{AgNO_3} - \Gamma_{KI})$ or $(\Gamma_{HNO_3} - \Gamma_{KOH})$ can be determined, not the individual terms. By titration it is impossible to establish the condition where $\Gamma_{AgNO_3} = \Gamma_{KI}$ or $\Gamma_{HNO_3} = \Gamma_{KOH}$, i.e. the point of zero charge, without making some model assumption. In line with this, there are no thermodynamic equations containing these individual surface concentrations. The *point of zero charge is thermodynamically inoperable*. Model assumptions are needed to estimate it (sec. 3.8).

3.5 The diffuse part of the double layer

Specific features are most prominent in the part of the countercharge near the surface. For specific adsorption this is so because the non-electrostatic ion-surface interaction has a short range, and for excluded volume effects and ion-ion solvent structure-mediated interactions this is so because these are particularly felt in the most crowded part of the double layer. The most rigorous approach in modelling double layers is to make explicit this gradual decrease of specificity with distance. However, such models tend to become rather involved and are not readily implemented in practice. We shall therefore mostly follow the more pragmatic approach to split the solution side of the double layer into two parts, the *inner part*, or *Stern layer*, and the *outer* or *diffuse part*. The latter is purely generic and ideal. The borderline between the two layers is called the *outer Helmholtz plane, oHp*. This, in principle arbitrary, plane is situated at some molecular diameters from the surface. All specificity is assigned to the inner layer.

Though a bit artificial, the subdivision into a Stern and a diffuse part has proven its value. One reason is that the diffuse part can be described with relatively simple analytical equations that become exact at sufficiently large distance from the surface. These two parts play central roles in electrokinetics, colloid stability and many other phenomena where diffuse double layer theory is found to apply well. From the more theoretical side, the diffuse part is characterized by relatively low potentials, so that deviations from ideality are minor. Charges obeying purely diffuse behaviour seldom exceed a few $\mu C\,cm^{-2}$, whereas σ^o may be as high as 30-50 $\mu C\,cm^2$. Hence, errors made in establishing σ^i from σ^o (experimentally determined) and σ^d (theoretical), using [3.3.4], are negligible. The third reason is that all kinds of problems, like surface heterogeneity and discreteness of charge, are mainly felt in the inner layer and are difficult to account for rigorously, even with the most advanced models. However, from a few molecular layers onwards, many of these 'short-grained' features are smoothed by thermal motion and the rectifying tendency of the potential. Last, but not least, for simple double layers (no hairy layers, no substantial roughnesses etc.) it is more or less proven that the outer Helmholtz plane and the slip plane in electrokinetics are identical or very close (sec. 4.4). Consequently, for practical purposes it is often a good approximation to identify ψ^d with the measurable electrokinetic potential ζ which, under a number of conditions, can be obtained from experiment, invoking certain model assumptions (chapter 4).

In the present section the diffuse part will be treated in some detail.

3.5a. Flat surfaces, symmetrical electrolytes

The theory for flat surfaces and symmetrical electrolytes has been given by Gouy[1] and Chapman[2] and we shall refer to it as *Gouy-Chapman (GC) theory*. The theory antedates Debye-Hückel (DH) theory for ionic atmospheres (sec. I.5.2a) by about two decades and is based on the same premises, viz.

(i) ions are point charges

(ii) the ionic adsorption energy is purely electrostatic ($z_j F\psi(x)$ for an ion j in a place x where the potential is $\psi(x)$).

(iii) the average electric potential (as occurring in Poisson's law) is identified with the potential of mean force (in the Boltzmann equation).

(iv) the solvent is primitive, i.e. a structureless continuum, affecting the distribution only through its macroscopic dielectric permittivity ε, for which the bulk value is taken.

For practical purposes, the difference between GC and DH theory is that the former treats flat surfaces (characteristic for big colloidal particles), as compared to ions in DH theory, and that GC does not impose any restriction on the potential, whereas the DH approximation invokes low potentials. Unlike the GC-case, the DH theory is self-consistent in that the potential of mean force and the mean potential are identical. So the DH equations have the status of exact limiting laws for low potentials and infinite dilution.

---

[1] G. Gouy, *Compt. Rend.* **149** (1909) 654; *J. Phys. (4)* **9** (1910) 457; *Ann. Phys. (9)* **7** (1917) 129.
[2] D.L. Chapman, *Phil. Mag. (6)* **25** (1913) 475.

For flat geometry the Poisson equation relating potential and space charge density [I.5.1.20a] reduces to

$$\nabla^2 \psi(x) = \frac{d^2 \psi(x)}{dx^2} = -\frac{\rho(x)}{\varepsilon_0 \varepsilon} \qquad\qquad [3.5.1]$$

where we have used [I.A7.33]. Here, ε_0 is the permittivity of free space (8.85 × 10^{-12} C V^{-1} m^{-1}, the precise value is given in I.app. 1) and ε the relative dielectric constant, set equal to its bulk value. The distance x is counted from the surface (if the double layer is purely diffuse) or from the outer Helmholtz plane. The space charge density $\rho(x)$ is related to the local electrolyte concentrations. Generally,

$$\rho(x) = F \sum_j z_j c_j(x) \qquad\qquad [3.5.2]$$

where the valencies (also called charge numbers) have the signs included. For a symmetrical electrolyte, for which $z_+ = -z_- = z$, [3.5.2] reduces to

$$\rho(x) = z F \left[c_+(x) - c_-(x) \right] \qquad\qquad [3.5.3]$$

Local concentrations are related to their bulk values $c = c(\infty)$ by Boltzmann's law

$$c_j(x) = c_j e^{-z_j y(x)} \qquad\qquad [3.5.4]$$

where $y(x)$ is a normalized dimensionless potential, defined by

$$y(x) \equiv F \psi(x) / RT \qquad\qquad [3.5.5]^{1)}$$

Later, we shall also use y^o for $F \psi^o / RT$, y^d for $F \psi^d / RT$, etc. Combination of [3.5.1, 3 and 4] leads to the *Poisson-Boltzmann (PB) equation* for the present case

$$\frac{RT}{F} \frac{d^2 y(x)}{dx^2} = -\frac{zFc}{\varepsilon_0 \varepsilon} \left(e^{-zy(x)} - e^{zy(x)} \right) \qquad\qquad [3.5.6]$$

where we have used $c_+(\infty) = c_-(\infty) = c$. The equation can be written in a bit more handy form by introducing the reciprocal Debye length κ, introduced in sec. I.5.2a where also numerical values are given. Generally,

$$\kappa^2 \equiv F^2 \sum_j c_j z_j^2 / \varepsilon_0 \varepsilon RT \qquad\qquad [3.5.7]$$

reducing to

---

1) At 25°C one unit of y corresponds to 25.68 mV.

$$\kappa^2 = 2F^2 c z^2 / \varepsilon_0 \varepsilon RT \qquad\qquad\qquad [3.5.8]$$

for a symmetrical electrolyte. Moreover, the term in brackets in [3.5.6] can be written as $-2\sinh[zy(x)]$[1]. With this in mind, the PB equation becomes

$$\frac{d^2[zy(x)]}{dx^2} = \kappa^2 \sinh[zy(x)] \qquad\qquad\qquad [3.5.9]$$

If desired $d^2y/\kappa^2 dx^2$ may also be written as $d^2y/d(\kappa x)^2$, making the equation entirely dimensionless.

Differential equation [3.5.9] can be integrated after multiplying both sides with $2(dy/dx)$. As

$$2\frac{dy}{dx}\frac{d^2y}{dx^2} = \frac{d(dy/dx)^2}{dx}$$

and

$$2\frac{dy}{dx}\sinh(zy) = \frac{2}{z}\frac{d[\cosh(zy)]}{dx}$$

the equation can be integrated to give

$$\left(\frac{dy}{dx}\right)^2 = \frac{2\kappa^2}{z^2}[\cosh(zy) + \text{const.}]$$

The integration constant is obtained from the boundary condition that at large distance from the surface $dy/dx \to 0$ and $y \to 0$. We find const. $= -1$. Hence,

$$\frac{dy}{dx} = \mp \frac{\kappa\sqrt{2}}{z}\sqrt{[\cosh(zy) - 1]} = \mp \frac{2\kappa}{z}\sinh\left(\frac{zy}{2}\right) \qquad\qquad [3.5.10]$$

where we have used [A2.38]. Introducing the expressions for y and κ into [3.5.10] the *electric field strength* in a diffuse double layer is immediately found as

$$E(x) = -\frac{d\psi(x)}{dx} = \sqrt{\frac{8cRT}{\varepsilon_0\varepsilon}}\sinh\left(\frac{zF\psi(x)}{2RT}\right) \qquad\qquad [3.5.11]$$

which is an uneven function, see fig. A2,1b. We need the minus sign because, for positive potentials ψ decreases with x, whereas for negative potentials it increases. For flat geometry $E = E_x = \mathbf{E}$. The low potential (Debye-Hückel) limit of [3.5.11] is

---

[1] Hyperbolic functions often occur in diffuse double layer theory. Some of their properties are summarized in appendix 2.

$$E(x) = \kappa \psi (x) \tag{3.5.11a}$$

The field strength for $x \downarrow 0$ can be related to the diffuse charge σ^d using Gauss' law [I.4.5.13], expressed as

$$\left(\frac{d\psi}{dx}\right)_{x \downarrow 0} = \frac{\sigma^d}{\varepsilon_0 \varepsilon} = -\frac{\sigma^0}{\varepsilon_0 \varepsilon} \tag{3.5.12}$$

This is a general result, a particular case of [I.A7.50], hence it is not limited to diffuse double layers but can be applied at any position x in any layer: $\sigma(x)$ then means the space charge integrated from $x = 0$ to $x = x$ in a column of unit area normal to the surface at the place where $\sigma(x)$ is to be measured. The sign of this charge to the left is opposite to that to the right. It is useful to realize this in modelling, say of polyelectrolytes. Alternatively, [3.5.12] can be obtained from the Poisson equation, combining [3.3.1] with [3.5.1].

For the *surface charge* we obtain

$$\sigma^d = -\sqrt{8\,\varepsilon_0 \varepsilon \, c \, RT}\, \sinh\left(\frac{z\,y^d}{2}\right) = -\varepsilon_0 \varepsilon \kappa \psi^d\,\frac{\sinh(z\,y^d/2)}{z\,y^d/2} \tag{3.5.13}$$

which for aqueous solutions at 25°C becomes

$$\sigma^d = -11.73\sqrt{c}\,\sinh(0.0195\,z\,\psi^d) \tag{3.5.13a}$$

with σ^d in $\mu C\ cm^{-2}$ for c in M and ψ^d in mV.

The inverse of [3.5.13] can be written as $y^d = (2/z)\sinh^{-1}(-p\sigma^d)$ with $p = (8\varepsilon_0 \varepsilon\, cRT)^{-1/2}$. Using [A2.10] this equation becomes simply

$$y^d = \frac{2}{z}\ln\left[-p\sigma^d + \sqrt{(p\sigma^d)^2 + 1}\,\right] \tag{3.5.14}$$

The negative sign in [3.5.13] means that a negative potential (caused by a negative charge on the surface) leads to a positive charge in the diffuse part. When there is an inner layer of charge σ^i, σ^d may be replaced by $-(\sigma^0 + \sigma^i)$. For a layer which is diffuse right to the solid surface ($\sigma^i = 0$, i.e. volumeless counterions),

$$\sigma^0 = \sqrt{8\,\varepsilon_0 \varepsilon\, c\, RT}\,\sinh(zF\psi^0/2RT) \tag{3.5.15}$$

The trend $\sigma^d(y^d)$ is illustrated by fig. 3.4 for a (1-1) electrolyte. The curves have an inflection point at $y^d = 0$. When one is certain that the double layer is purely diffuse, this is a way to find the point of zero charge (sec. 3.8) from titration. At given y^d the absolute charge increases with increasing ionic strength. This is a consequence of screening of the surface charge by (mainly) counterions, accumulating close to the surface.

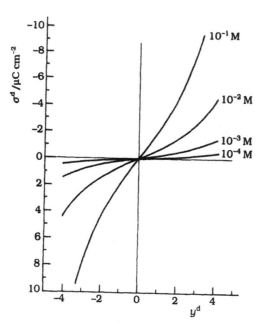

Figure 3.4. Charge as a function of potential in a flat diffuse double layer. Concentration of the (1-1) electrolyte is indicated.

Practice has shown that [3.5.15] applies only close to the point of zero charge: σ^d seldom exceeds a few μC cm^{-2}. Neither do electrokinetic charges, computed from [3.5.14] with ζ replacing ψ^d, exceed these values. The reason is that for higher ψ^d and/or σ^d Stern layers start to develop.

The *differential capacitance* of a diffuse double layer is found directly from differentiation of [3.5.13] with respect to ψ^d. As we are measuring the capacitance for positive charge in a negative potential field, or the other way around, we need a minus sign in

$$C^d = -d\sigma^d / d\psi^d$$

[3.5.16]

The result can conveniently be written as

$$C^d = \varepsilon_0 \varepsilon \, \kappa \, \cosh\left(\frac{z y^d}{2}\right) = \varepsilon_0 \varepsilon \kappa \left[1 + \frac{(z y^d)^2}{8} + O(z y^d)^4\right]$$

[3.5.17]

It is plotted in fig. 3.5. The capacitance is an even function and symmetrical with respect to the point of zero charge. At that point the charge is zero, but the capacitance is finite and equal to

$$C^d = \varepsilon_0 \varepsilon \, \kappa \qquad\qquad (\sigma^d = 0)$$

[3.5.18]

which is nothing else than the formula for a flat condenser with plate distance

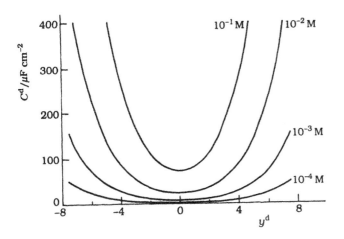

Figure 3.5. Differential capacitance of a flat diffuse double layer; (1-1) electrolyte at 25°C.

κ^{-1}. Just like the charge, C^d increases proportionally to $\sqrt{c}$ because of screening.

With the above information, the Gibbs energy per unit area, G_a^σ can now also be computed. Rephrasing what has been stated in sec. 3.2, for the diffuse countercharge *only* G_a^σ is zero, because for each species j the electrical work involved in carrying ions from the bulk to the layer or conversely, depending on the sign of z_j, ($z_j F \psi$ per mole of ions) is just compensated by the incurred entropy change ($RT \ln[c_j(x)/c_j(\infty)]$), these two quantities being related through the Boltzmann equation [3.5.4]. However, for a diffuse layer *together with a surface charge*, present because of non-electrostatic adsorption of charge-determining ions, the result is finite and given by [3.2.3]. Combination of this equation with [3.5.15] yields for a purely diffuse double layer, i.e. after replacement of σ^o and ψ^o (or y^o) by σ^d and ψ^d (or y^d), respectively.

$$\Delta G_a^\sigma = - \sqrt{8\,\varepsilon_o \varepsilon\, c\, RT} \int_0^{\psi^d} \sinh\left(\frac{zF\psi^{d'}}{2RT}\right) d\psi^{d'} = - \frac{8cRT}{\kappa} \int_0^{zy^d/2} \sinh\left(\frac{z\,y^{d'}}{2}\right) d\left(\frac{z\,y^{d'}}{2}\right)$$

[3.5.19]

from which, using [A2.20]

$$\Delta G_a^\sigma = - \frac{8\,c\,RT}{\kappa}\left[\cosh\left(\frac{z\,y^d}{2}\right) - 1\right]$$

[3.5.20]

This function is plotted in fig. 3.6. Because of the hyperbolic cosine dependency and because both are proportional to $c^{1/2}$ there is some similarity with the differential capacitance, fig. 3.5. The physical background is that both quantities depend in a similar way to screening. Writing ΔG_a^σ as

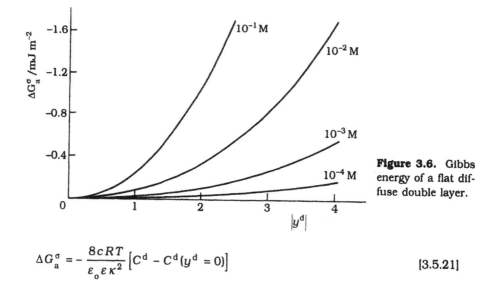

Figure 3.6. Gibbs energy of a flat diffuse double layer.

$$\Delta G_a^\sigma = -\frac{8cRT}{\varepsilon_o \varepsilon \kappa^2}\left[C^d - C^d(y^d = 0)\right]$$

[3.5.21]

it is seen that this quantity is a measure of the screening. However, it is recalled that capacitances are purely electrostatic quantities whereas the Gibbs energy has a non-electrostatical root.

Although the diffuse ions do not contribute to ΔG_a^σ, it is nevertheless possible to use diffuse double layer theory to compute this quantity, because in this model the screening is described, telling us by how much ψ^o rises with increasing σ^o.

Gibbs energies of double layers are of crucial relevance in interpreting colloid stability. Upon overlap of double layers of the same sign screening becomes less efficient, leading to lower capacitances, i.e. to higher electric potentials at given charge, or to lower charges at given potential, or to intermediate situations. All of these cases make ΔG_a^σ less negative, meaning that *work has to be performed to bring the two double layers nearer.* This is the basic reason why two similar double layers repel each other.

The potential distribution in a diffuse double layer is obtained by integration of [3.5.10], which can be carried out analytically, using [A2.5 and 25]. The result is

$$\tanh\left[zy(x)/4\right] = \left\{\tanh\left[zy^d/4\right]\right\}e^{-\kappa x}$$

[3.5.22]

For low potentials (several tens of mV for $z = 1$, the hyperbolic tangent may be replaced by the first, linear, term of its series expansion [A2.9], to give

$$y(x) = y^d e^{-\kappa x} \qquad \text{or} \qquad \psi(x) = \psi^d e^{-\kappa x} \qquad\qquad [3.5.23]$$

According to this familiar limiting law, in a flat diffuse layer the (absolute value of the) potential drops exponentially with distance, reducing to ψ^d/e over a distance κ^{-1}. It is mainly this relation that has led to dubbing κ *the reciprocal double layer thickness*. This notion also follows from [3.5.18] which pictures the double layer as a molecular condenser of capacitance $C^d = \varepsilon_o \varepsilon \kappa$ in which the countercharge is concentrated at a distance κ^{-1} from the surface. However, at higher potentials and for curved interfaces the relation is not so simple, although κ^{-1} remains a useful measure of the extension of the diffuse part of a double layer. When [3.5.23] is valid, the double layer thickness is independent of ψ^d; in all other cases it depends on ψ^d.

Examples of potential-distance curves are sketched in fig. 3.7. The graph is dimensionless; specific situations can be derived by the appropriate scaling. The increase in steepness with decreasing x is caused by the tendency of counterions to enrich the region of high potentials, close to the surface. For higher valency z and/or higher concentration c, κ is higher, i.e. a given value of $y(x)$ is attained at lower x. In the non-dimensionless plot $\psi(x)$, the decay is steeper when κ is higher. We say that due to increase in the electrolyte concentration or in the valency z of the electrolyte the double layer is more *compressed* (thinner). In non-aqueous solvents, which can support very little dissociated electrolyte, the double layers are very extended and the field strength $d\psi/dx$ is very low everywhere. In the limiting case of no screening at all (an academic situation) $\kappa \to 0$, $\psi(x) = \psi^d$ everywhere, and $E = 0$; then the lines of force leave (or go to) the surface parallel to each other and reaching to infinity without changing their density.

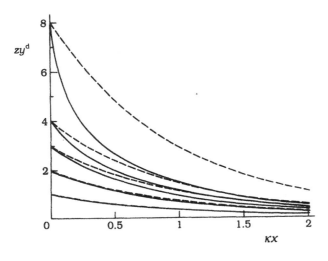

Figure 3.7. Potential distribution in a flat diffuse double layer, for various values of zy^d, that is $zy(x = 0)$. Solid lines: full equation [3.5.22], dashed lines: low potential approximation [3.5.23].

From fig. 3.7 it is also seen that [3.5.23] is a good approximation for $zy \leq 2$ ($\leq$ 50 mV for $z = 1$, $\leq$ 25 mV for $z = 2$, etc.). For higher y^d the approximated equation predicts too slow a decay: $y(x)$ is overestimated. Alternatively stated, the strong rise of y close to the surface is underestimated in the Debye-Hückel approximation. For $x = 0$, considering [3.5.12], at given potential y^d the surface charge σ^d is underestimated or, for that matter, at given σ^d the potential is overestimated.

3.5b. Ionic components of charge. Negative adsorption

In the Gouy-Chapman model the countercharge consists of an excess of counterions and a deficit of co-ions. For a positive surface the corresponding *ionic components of charge* $\sigma_+ = zF\Gamma_+$ and $\sigma_- = -zF\Gamma_-$ are both negative and defined by equations like [3.4.8a and 8b] for silver iodide, or [I.5.6.3c and 3d] for oxides. Let us consider the latter systems and let, as usual, $c_{salt} \gg c_{KOH}, c_{HNO_3}$. In [I.5.6.3c], Γ_{KNO_3} is the *negative adsorption* of neutral electrolyte, leading to the *Donnan effect*, i.e. the expulsion of electrolyte by charged particles. As Γ_{KOH}, which is positive, cannot be measured without simultaneously measuring Γ_{HNO_3} (also positive), σ_{K^+} cannot be measured by titration, but changes of σ_{K^+} are coupled to those in σ^o via the Esin-Markov coefficient, see [3.4.14-16], so that after all they can be assessed except for a constant. The negative adsorption of neutral electrolyte can be measured as an increase in the bulk concentration caused by the addition of the particles. Negative adsorption studies are therefore useful to compute the ionic components of charge for a diffuse double layer.

The two excesses are defined as

$$\sigma_+^d = zF \int_0^\infty \left[c_+(x) - c_+(\infty) \right] dx = zFc \int_0^\infty \left(e^{-zy} - 1 \right) dx \qquad [3.5.24a]$$

$$\sigma_-^d = -zF \int_0^\infty \left[c_-(x) - c_-(\infty) \right] dx = -zFc \int_0^\infty \left(e^{zy} - 1 \right) dx \qquad [3.5.24b]$$

Integrations can be carried out by introducing the field strength[1], for which [3.5.10] is used:

$$\int_0^\infty \left(e^{-zy} - 1 \right) dx = \int_{y^d}^0 \frac{(e^{-zy} - 1) dy}{dy/dx} = \int_{y^d}^0 \frac{(e^{-zy} - 1) dy}{-(\kappa/z)(e^{zy/2} - e^{-zy/2})} = \frac{z}{\kappa} \int_{y^d}^0 e^{-zy/2} dy$$

In this way one obtains

---

[1] D.C. Grahame, *Chem. Rev.* **41** (1947) 441 solved [3.5.24] by taking squares.

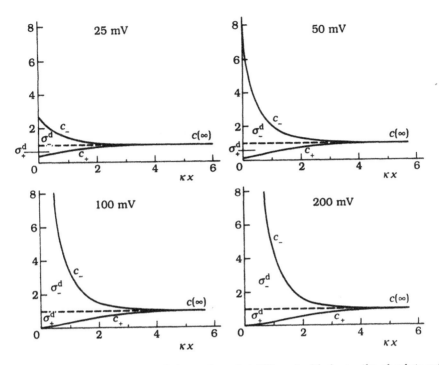

Figure 3.8. Ionic components of charge for a flat diffuse double layer; the absolute values of σ_-^d and σ_+^d are the areas indicated. Symmetrical electrolytes. The (positive) dimensionless potential is also indicated; $z = 1$. These curves are scaled proportionally to the electrolyte concentration.

$$\sigma_+^d = \left(\frac{\varepsilon_0 \, \varepsilon \, RT \, z^2 \, c}{2} \right)^{1/2} \int_{y^d}^{0} e^{-zy/2} \, dy$$

$$\sigma_+^d = \left(2\varepsilon_0 \, \varepsilon RT c \right)^{1/2} \left[e^{-zy^d/2} - 1 \right] = \frac{2c\,z\,F}{\kappa} \left[e^{-zy^d/2} - 1 \right] \qquad [3.5.25a]$$

Similarly,

$$\sigma_-^d = \left(2\varepsilon_0 \, \varepsilon RT c \right)^{1/2} \left[1 - e^{zy^d/2} \right] = \frac{2c\,z\,F}{\kappa} \left[1 - e^{zy^d/2} \right] \qquad [3.5.25b]$$

These components of charge satisfy the requirement $\sigma_-^d + \sigma_+^d = \sigma^d$, and for low potentials reduce to

$$\sigma_+^d = \sigma_-^d = -c\,z^2 \, F \, y^d / \kappa \qquad [3.5.25c]$$

On the other hand, with increasingly positive y^d, σ_-^d increases exponentially, whereas σ_+^d asymptotically approaches the limit

$$\lim_{y^d \to \infty} \sigma_+^d = -(2\varepsilon_0 \varepsilon \, c \, RT)^{1/2} = -2czF/\kappa \qquad\qquad [3.5.26]$$

which for aqueous solutions at 25° is equal to

$$\lim_{y^d \to \infty} \sigma_+^d = -5.866 \sqrt{c} \qquad (\mu C \; cm^{-2} \text{ if } c \text{ in M and } z = 1) \qquad [3.5.26a]$$

Equation [3.5.26] shows that in the limit of maximum expulsion effectively two Debye lengths of the double layer are devoid of co-ions. For very low y^d, where the exponentials may be replaced by their linear terms, $\sigma_+^d = \sigma_-^d$.

These trends are illustrated in fig. 3.8. Equality of σ_+^d and σ_-^d requires potentials below 25 mV. At high y^d, $\sigma_+^d \ll \sigma_-^d$. Plots of the ratios σ_+^d / σ^d and σ_-^d / σ^d as a function of y^d are independent of c_{salt}. For curved surfaces, this is no longer the case, see fig. 3.15.

Negative adsorption is a relatively important phenomenon in concentrated disperse systems and in capillaries. It is responsible for the Donnan effect, for the exclusion of electrolytes from concentrated sols, dispersions and capillaries and the ensuing *salt-sieving effect*, already introduced in chapter I.1. It also plays a role in double layer relaxation as occurs in alternating fields or in particle-particle interaction. As negative adsorption is a purely electrostatic feature and takes place far from the surface, in all these applications its computation from Poisson-Boltzmann statistics is reliable, especially at high $|y^d|$.

3.5c. Asymmetrical electrolytes

When the cation and the anion have different valencies, the screening power for positive and negative surfaces is also dissimilar. For instance, $Ba(NO_3)_2$ screens negatively charged AgI better than KNO_3 does, but for positively charged surfaces it screens as well as KNO_3 with the same NO_3^--concentration. As a result, $\sigma^d(y^d)$ and $C^d(y^d)$ curves become asymmetrical with respect to the zero point $y^d = 0$. Around this zero point a qualitatively new feature emerges.

Suppose one molecule of electrolyte dissociates into v_+ cations of valency z_+ and v_- anions of valency z_- (a negative number), with $v_+ + v_- = v$. Because of electroneutrality,

$$z_+ c_+ = -z_- c_- \qquad\qquad [3.5.27]$$

$$z_+ v_+ = -z_- v_- \qquad\qquad [3.5.28]$$

Further, we have

$$c_+ = v_+ c \qquad\qquad c_- = v_- c \qquad\qquad [3.5.29a, b]$$

where the c's are molar concentrations.

The derivations of sec. 3.5a require some modifications; [3.5.2] may not be replaced by [3.5.3] and κ^2 is maintained in its general form [3.5.7]. Instead of [3.5.6] we now need

$$\frac{RT}{F}\frac{d^2y(x)}{dx^2} = -\frac{F}{\varepsilon_0\varepsilon}\left[z_+c_+ e^{-z_+y(x)} + z_-c_- e^{-z_-y(x)}\right] \qquad [3.5.30]$$

The r.h.s. cannot now be written as a hyperbolic sine, but the integration procedure, starting with the multiplication by $2dy/dx$ remains valid and leads to

$$\frac{dy(x)}{dx} = \mp\left(\frac{2F^2}{\varepsilon_0\varepsilon RT}\right)^{1/2}\left(c_+ e^{-z_+y(x)} + c_- e^{-z_-y(x)} + \text{const.}\right)^{1/2} \qquad [3.5.31]$$

$$\frac{dy(x)}{dx} = \mp\left(\frac{2F^2c}{\varepsilon_0\varepsilon RT}\right)^{1/2}\left(v_+ e^{-z_+y(x)} + v_- e^{-z_-y(x)} - v_+ - v_-\right)^{1/2} \qquad [3.5.32]$$

where we have used the boundary condition $(dy/dx) = 0$ for $y = 0$ and [3.5.29]. From this result, the charge and capacitance are immediately obtained as

$$\sigma^d = -\left(\text{sign } y^d\right)\left(2\varepsilon_0\varepsilon c RT\right)^{1/2}\left(v_+ e^{-z_+y^d} + v_- e^{-z_-y^d} - v_+ - v_-\right)^{1/2} \qquad [3.5.33]$$

and

$$C^d = \left(\frac{\varepsilon_0\varepsilon c F^2}{2RT}\right)^{1/2}\frac{\left|v_+ z_+ e^{-z_+y^d} + v_- z_- e^{-z_-y^d}\right|}{\left(v_+ e^{-z_+y^d} + v_- e^{-z_-y^d} - v_+ - v_-\right)^{1/2}} \qquad [3.5.34a]$$

$$C^d = \left(\frac{\varepsilon_0\varepsilon v_+^2 z_+^2 c F^2}{2RT}\right)^{1/2}\frac{\left|e^{-z_+y^d} - e^{-z_-y^d}\right|}{\left(v_+ e^{-z_+y^d} + v_- e^{-z_-y^d} - v_+ - v_-\right)^{1/2}} \qquad [3.5.34b]$$

$$C^d = \varepsilon_0\varepsilon\kappa\left(\frac{v_+ z_+}{2(z_+ - z_-)}\right)^{1/2}\frac{\left|e^{-z_+y^d} - e^{-z_-y^d}\right|}{\left(v_+ e^{-z_+y^d} + v_- e^{-z_-y^d} - v_+ - v_-\right)^{1/2}} \qquad [3.5.34c]$$

Using [A2.1, A2.34 and A2.40], these equations can be shown to reduce to [3.5.17] for symmetrical electrolytes $(z_+ = z, z_- = -z, v_+ = v_- = 1)$. The absolute bars are needed in [3.5.34] because the numerator is an uneven function, although the capacitance is always positive.

Illustrations are given in figs. 3.9 and 10. Figure 3.9 shows that for negative

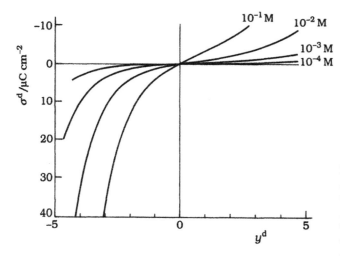

Figure 3.9. Charge as a function of potential in a flat diffuse double layer. The concentration of the (2-1) electrolyte is indicated.

ψ^d, where the bivalent cations are the counterions, the screening is better than for positive ψ^d, where monovalent anions assume this function. In fact, at high negative potential the $\sigma^d(\psi^d)$ relation approaches that for a symmetrical bivalent electrolyte, ([3.5.14] with $z = 2$) because the anions are negatively adsorbed. By the same token, for high positive potential $\sigma^d(\psi^d)$ approaches [3.5.14] for $z = 1$. Around $\psi^d = 0$ there is a transition zone over which the roles of cations and anions as counterions and co-ions are gradually reversed.

Figure 3.10 illustrates the same trends in terms of capacitances. In this example the asymmetry of the electrolyte has been varied at fixed concentration. In this plot the trends are more pronounced than in fig. 3.9. The new feature is that the capacity minimum no longer coincides with the zero point of the diffuse layer potential, but is shifted in the direction where the multivalent ion is the co-ion. (In fig. 3.9 the same can be said of the position of the minimum slope.) The value y^d (min) where the capacitance minimum is located can be obtained by differentiating [3.5.34] with respect to y^d leading to the condition

$$e^{-z_+ y^d(\text{min})}\left[-z_+ z_- e^{-z_+ y^d(\text{min})} - 2z_+(z_+ - z_-) + (z_+^2 + z_-^2 - z_+ z_-)e^{-z_- y^d(\text{min})}\right]$$

$$+ e^{-z_- y^d(\text{min})}\left[-z_+ z_- e^{-z_- y^d(\text{min})} + 2z_-(z_+ - z_-) + (z_+^2 + z_-^2 - z_+ z_-)e^{-z_+ y^d(\text{min})}\right] = 0$$

$$[3.5.35]$$

which requires numerical solution. For (1-1), (2-1), (3-1) and (4-1) electrolytes,

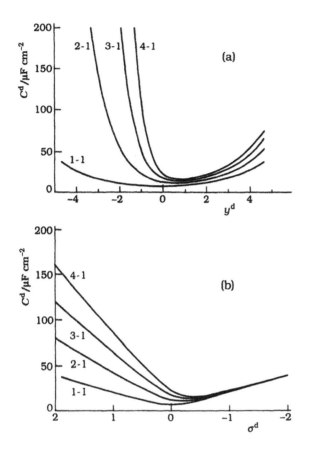

Figure 3.10. Differential capacitance of a flat diffuse double layer. 10^{-3} M (z-1) electrolyte, 25°C. (a) as a function of diffuse layer potential; (b) as a function of diffuse charge.

y^d(min) = 0, 0.631, 0.779 and 0.820, respectively; for (1-2), (1-3) and (1-4) electrolytes, y^d(min) = -0.631, -0.779 and -0.820 respectively.

The minimum in the differential capacitance (at low salt concentration, so that the double layer is mainly diffuse) has been invoked as a method to establish the point of zero charge. It follows from the above that this method only works for symmetrical electrolytes.

For further information, see the literature[1].

3.5d. *Electrolyte mixtures*

For a number of situations, the double layer in mixed electrolytes is of some

[1] Gouy himself already treated the (2-1) and (1-2) cases (G. Gouy, *Compt. Rend.* **149** (1909) 654; *J. Phys.* (4) **9** (1910) 457); D.C. Grahame, (*J. Chem. Phys.* **21** (1953) 1054) and S. Levine and J.E. Jones (*Kolloid-Z* **230** (1969) 306) revisited and extended it, the latter authors included mixtures. R. de Levie (*J. Electroanal. Chem.* **278** (1990) 17) tabulated equations for σ^d, C^d and y^d(min) and gave $y(x)$ profiles. Asymmetrical electrolytes have also been considered in theories involving a more advanced model than Poisson-Boltzmann (see sec. 3.8).

interest. In the treatment of charged dispersions one almost automatically deals with mixtures: two components to regulate the surface charge (say, KOH or HNO_3 for oxides) and one indifferent electrolyte (KNO_3). However, as the double layer charge is only well defined if $c_{KNO_3} \gg c_{HNO_3}$, c_{KOH} (see sec. 3.7a), here one virtually deals with one-component electrolytes. Of more practical interest are ion exchange phenomena, where the displacement of one electrolyte by another is at issue. Also in a number of environmental systems (soils, particles in sea water), we are dealing with mixtures.

It is beyond our scope to analyse all possible cases. In fact, in many multi-component systems with ions of different valencies the Poisson-Boltzmann equation becomes analytically unsolvable. We shall restrict ourselves therefore to the general equations for charge and capacitance and one illustration.

The starting equations are [3.5.1, 2 and 4], as before. They lead to the general expression

$$\frac{d^2 y(x)}{dx^2} = -\frac{F^2}{\varepsilon_0 \varepsilon RT} \sum_j z_j c_j e^{-z_j y(x)}$$

[3.5.36]

where the sum over j covers all ionic species present. This equation can be integrated, using the procedure described before, to give

$$\frac{dy(x)}{dx} = \mp \left(\frac{2F^2}{\varepsilon_0 \varepsilon RT} \right)^{1/2} \left[\sum_j c_j e^{-z_j y(x)} + \text{const.} \right]^{1/2}$$

[3.5.37]

The ionic concentrations c_j can be rewritten in terms of concentrations of electroneutral electrolytes, c_i. For each electrolyte i, dissociating into v_{i+} cations and v_{i-} anions, $c_{i+} = v_{i+} c_i$, $c_{i-} = v_{i-} c_i$; identifying the c_i's as c_{i+} and c_{i-}'s, [3.5.37] becomes generally

$$\frac{dy(x)}{dx} = \mp \left(\frac{2F^2}{\varepsilon_0 \varepsilon RT} \right)^{1/2} \left[\sum_i c_i \left(v_{i+} e^{-z_{i+} y} + v_{i-} e^{-z_{i-} y} \right) - \sum_i c_i \left(v_{i+} + v_{i-} \right) \right]^{1/2}$$

[3.5.38]

When the electrolytes possess common ions, some terms may be grouped together.

The diffuse charge follows immediately from [3.5.12]

$$\sigma^d = -\left(\text{sign } y^d \right) \left(2\varepsilon_0 \varepsilon FRT \right)^{1/2} \left[\sum_i c_i \left(v_{i+} e^{-z_{i+} y^d} + v_{i-} e^{-z_{i-} y^d} \right) - \sum_i c_i \left(v_{i+} + v_{i-} \right) \right]^{1/2}$$

[3.5.39]

and, using [3.5.16] the differential capacitance as

$$C^d = \left(\frac{\varepsilon_0 \varepsilon F^2}{2RT}\right)^{1/2} \frac{\left| -\sum_i c_i \, \nu_{1+} z_{1+} e^{-z_{1+} y^d} + \sum_i c_i \, \nu_{1-} z_{1-} e^{-z_{1-} y^d} \right|}{\left[\sum_i c_i \left(\nu_{1+} e^{-z_{1+} y^d} + \nu_{1-} e^{-z_{1-} y^d} \right) - \sum_i c_i \left(\nu_{1+} + \nu_{1-} \right) \right]^{1/2}}$$

[3.5.40]

No general analytical formula can be given for $y(x)$.

Equations [3.5.39 and 40] can be elaborated for specific cases. For instance, for a mixture of a (2-1) electrolyte like $Cd(NO_3)_2$ (component 1, concentration c_1) and an (1-1) electrolyte with the same anion (component 2, say KNO_3, concentration c_2), we have $z_{1+} = 2$, $z_{1-} = -1$, $z_{2+} = 1$, $z_{2-} = -1$, $\nu_{1+} = 1$, $\nu_{1-} = 2$, $\nu_{2+} = 1$ and $\nu_{2-} = 1$. This is a relevant example because Cd^{2+} ions adsorb specifically on a variety of surfaces. For the charge

$$\sigma^d = -\left(\text{sign } y^d\right)\left(2\varepsilon_0 \varepsilon F RT\right)^{1/2}\left[c_1 e^{-2y^d} + c_2 e^{-y^d} + \left(2c_1 + c_2\right)e^{y^d} - 3c_1 - 2c_2\right]^{1/2}$$

[3.5.41]

is obtained. Expressions for other cases can be formulated accordingly. Integration of [3.5.39] to give $y(x)$ plots is only in special cases analytically possible. Ionic components of charge have been treated by Oldham[1] for a small number of specific cases. Bolt[2] gave rather general solutions in terms of elliptical integrals of the first kind for the situation that the negative adsorption of co-ions is negligible as compared with the positive adsorption of counterions.

By way of illustration, in fig. 3.11 the fractions of the countercharge components NO_3^-, Na^+ and Cd^{2+} are given at constant high concentration of KNO_3 and variable concentration of $Cd(NO_3)_2$. In fig. 3.11a the contribution of the bivalent ion is given. Starting from zero, it increases rapidly with increasing Cd^{2+} concentration, especially at negative y^d, till eventually $Cd(NO_3)_2$ dominates the electrolyte. For $y^d = 0$ this means that then $f_{Cd^{2+}} \to 2/3$ and $f_{Cl^-} \to 1/3$ (fig. 3.11c). This follows from [I.5.6.20] applied to a diffuse double layer

$$\left(\frac{\partial \sigma_\pm}{\partial \sigma}\right)_{a_\pm} (\sigma \to 0) = -\frac{|z_\pm|}{|z_+| + |z_-|}$$

[3.5.42]

but can also be derived from Poisson-Boltzmann theory[3].

For symmetrical electrolytes the ratio is 1/2; compare figs. 3.11b and c for

[1] K.B. Oldham, *J. Electroanal. Chem.* **63** (1975) 139.
[2] G.H. Bolt, in *Soil Science*, Volume B, *Physico-Chemical Models*, G.H. Bolt, Ed., Elsevier (1979), chapter 1.
[3] J. Lyklema, *J. Colloid Interface Sci.* **99** (1984) 109.

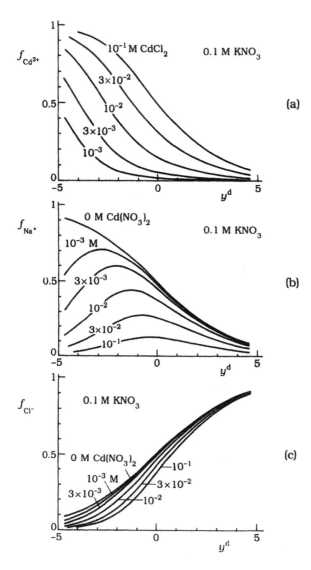

Figure 3.11. Diffuse double layer in a mixture of a 0.1 M (1-1) electrolyte (KNO_3) and the indicated concentration of a (2-1) electrolyte ($Cd(NO_3)_2$). The fractions of the countercharge contributed by the three ionic species in the mixture are shown.

$c_{Cd(NO_3)_2} = 0$. With increasing $c_{Cd(NO_3)_2}$ the negative adsorption of Cd^{2+} (for $y^d > 0$) also rises. The increase of Cd^{2+} uptake goes mainly at the expense of Na^+ (fig. 3.11b), especially at very negative potentials, even leading to decreasing f_{Na^+} with increasingly negative potential. If Na^+ were the sole cation, such a trend could not have occurred. The exchange of Na^+ for Cd^{2+} has a relatively minor effect on f_{Cl^-}, fig. 3.11c. To the right, where Cl^- is positively adsorbed, there is some decrease with increasing $CdCl_2$ concentration because the negative adsorption of cadmium also starts to contribute. To the left, where Cl^- is negatively adsorbed, the reduction is due to the increased positive adsorption of Cd^{2+}.

Pictures like this are the basis of the analysis of *ion exchange phenomena*.

3.5e. *Spherical double layers*

For spherical symmetry the Poisson-Boltzmann equation cannot be solved analytically, but numerical solutions are nowadays available. Analytical solutions exist for low potentials, that is in the Debye-Hückel (DH) approximation, already encountered in the treatment of the ionic atmosphere around ions, sec. I.5.2a. As compared with flat double layers, the low-potential approximation tends to become better, the smaller the particle is. The reason is that, because of the stronger divergence of the lines of force, the potential decays more rapidly; a relatively larger fraction of the countercharge is therefore found in the region of low potentials.

Writing the Laplace operator $\nabla^2 \psi$ in the Poisson equation [I.5.1.20a] in spherical coordinates (first term of the r.h.s. of [I.A7.47], with r instead of ρ) we obtain

$$\frac{1}{r^2}\frac{d}{dr}\left(r^2\frac{d\psi}{dr}\right) = \left(\frac{d}{dr}+\frac{2}{r}\right)\frac{d\psi}{dr} = -\frac{\rho(r)}{\varepsilon_0\varepsilon} \qquad [3.5.43]$$

For $\rho(r)$ we have [3.5.2] with x replaced by r. Relating each local ionic concentration to its bulk value via the Boltzmann equation we arrive at the *Poisson-Boltzmann equation for spheres:*

$$\frac{RT}{F}\frac{1}{r^2}\frac{d}{dr}\left(r^2\frac{dy}{dr}\right) = -\frac{F}{\varepsilon_0\varepsilon}\sum_j z_j c_j e^{-z_j y(r)} \qquad [3.5.44]$$

For one symmetrical electrolyte, this reduces to

$$\frac{1}{r^2}\frac{d}{dr}\left(r^2\frac{dy}{dr}\right) = \frac{\kappa^2}{z}\sinh[z\,y(r)] \qquad [3.5.45]$$

In the DH approximation, the exponentials are replaced by the zeroth- and first-order terms of their series expansions (sec. I.5.2a). Then [3.5.45] reduces to

$$\frac{1}{r^2}\frac{d}{dr}\left(r^2\frac{d\psi}{dr}\right) = \kappa^2\psi \qquad \text{or} \qquad \frac{1}{r^2}\frac{d}{dr}\left(r^2\frac{dy}{dr}\right) = \kappa^2 y \qquad [3.5.46a,b]$$

with κ defined by [3.5.7]. Integration leads to (see [I.5.2.15])

$$\psi(r) = \frac{Q e^{-\kappa(r-a)}}{4\pi\varepsilon_0\varepsilon r(1+\kappa a)} \qquad [3.5.47]$$

where Q is the (total) surface charge and a the radius of the particle (immediately beyond which the diffuse layer is assumed to start). This charge is related to ψ^d according to [I.5.2.16]

$$\psi(a) = \psi^d = \frac{Q}{4\pi\varepsilon_0 \varepsilon a} - \frac{Q\kappa}{4\pi\varepsilon_0 \varepsilon(1+\kappa a)} \qquad\qquad [3.5.48]$$

or

$$Q = 4\pi\varepsilon_0 \varepsilon \psi^d a(1+\kappa a) \qquad\qquad [3.5.49]$$

Combination with [3.5.47] yields

$$\psi(r) = \psi^d \cdot \frac{a}{r} \, e^{-\kappa(r-a)} \qquad\qquad [3.5.50]$$

Because of the r in the denominator, this is not a simple exponential decay. As $Q = -Q^d$ (the diffuse charge) $= -4\pi a^2 \sigma^d$

$$\sigma^d = -\frac{\varepsilon_0 \varepsilon \psi^d (1+\kappa a)}{a} = -\varepsilon_0 \varepsilon \kappa \psi^d \left(1 + \frac{1}{\kappa a}\right) \qquad\qquad [3.5.51]$$

which for large κa reduces to

$$\sigma^d = -\varepsilon_0 \varepsilon \kappa \psi^d \qquad\qquad [3.5.51a]$$

and for small κa to

$$\sigma^d = -\varepsilon_0 \varepsilon \psi^d / a \qquad\qquad [3.5.51b]$$

For the differential capacitance (which in the limit of low potential is equal to the integral capacitance) we obtain

$$C^d = K^d = \frac{\varepsilon_0 \varepsilon(1+\kappa a)}{a} = \varepsilon_0 \varepsilon \kappa \left(\frac{1+\kappa a}{\kappa a}\right) \qquad\qquad [3.5.52]$$

For very low ionic strength, typical for *non-aqueous systems*, screening becomes negligible ($\kappa a \ll 1$) and the expressions acquire the forms for charged spheres without countercharge, that is the Coulomb case:

$$Q = 4\pi\varepsilon_0 \varepsilon \psi^d a \qquad\qquad [3.5.53]$$

$$\psi(r) = a\psi^d / r \qquad\qquad [3.5.54]$$

$$C^d = K^d = \varepsilon_0 \varepsilon / a \qquad\qquad [3.5.55]$$

In the opposite case of very strong screening ($\kappa a \gg 1$), most of the countercharge resides very close to the surface and the double layer assumes the character of a spherical molecular condenser of thickness κ^{-1} ($C^d = \varepsilon_0 \varepsilon / \kappa^{-1}$, etc.)

Another analytical expression of wider applicability than [3.5.51] is

$$\sigma^d = -\frac{2Fcz}{\kappa}\left[2\sinh\left(\frac{zy^d}{2}\right) + \frac{4\tanh(zy^d/4)}{\kappa a}\right] \qquad [3.5.56]$$

for a symmetrical electrolyte. This equation constitutes the first and second term of a series exansion in terms of powers of $(\kappa a)^{-1}$. This equation was proposed by Loeb et al.[1] and analysed in some detail by the Russian School[2] and by Overbeek et al.[3], who also gave a third term. For large κa it leads to the flat plate limit, [3.5.13]. Differentiation gives the capacitance and integration, after replacement of σ^d by $\varepsilon_0\varepsilon\,d\psi/dr$ and of a by r, yields the potential distribution which, incidentally, is not a very good approximation.

Illustrations are given in figs. 3.12 and 3.13 and in tables 3.1 and 3.2. The drawn curves are exact numerical results, obtained by a Runge-Kutta integration[4].

Figure 3.12 illustrates that the charge that a double layer can absorb at given surface potential increases with curvature. The reason is the increasing spatial capacity for countercharge. The DH approximation (dashed curves) is relatively better for low a. Potential-distance curves are given in fig. 3.13. The stronger the curvature, (i.e. the lower κa) the steeper the potential drop is. At a given radius

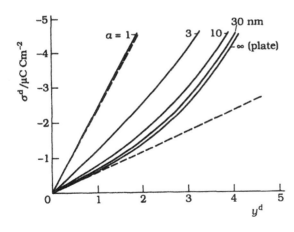

Figure 3.12. Charge in a spherical diffuse double layer. Electrolyte (1-1), 10^{-2} M. Temperature 25°C. The particle radius a is given. Dashed curves: Debye-Hückel approximation.

[1] A.L. Loeb, J.Th.G. Overbeek and P.H. Wiersema, *The Electrical Double Layer around a Spherical Colloid Particle*, M.I.T. Press (1961). (Covers symmetrical and asymmetrical electrolytes, ionic components of charge and Gibbs energies.)
[2] S.S. Dukhin, N.M. Semenikhin and L.M. Shapinskaya, *Doklady Akad. Nauk SSSR* **193** (1970) 385. (Transl. *Dokl. Phys. Chem.* **193** (1970) 540); N.M. Semenikhin, V.L. Sigal, *J. Colloid Interface Sci.* **51** (1975) 215.
[3] J.Th.G. Overbeek, G.J. Verhoekx, P.L. de Bruyn and H.N.W. Lekkerkerker, *J. Colloid Interface Sci.* **119** (1987) 422.
[4] C. Runge, *Math. Ann.* **46** (1895) 167; W. Kutta, *Z. Math. Phys.* **46** (1901) 435. See also J. Stoer, R. Bulisch, *Introduction to Numerical Analysis*, 2nd ed. Springer (1993) 438.

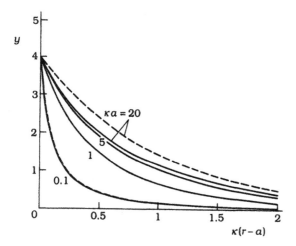

Figure 3.13. Potential distribution in a spherical diffuse double layer. Symmetrical electrolyte. Dashed curves: Debye-Hückel approximation. (Backscaling in terms of concentrations, radii and valencies can be done by using [3.5.8] for κ and a (in nm) = 0.3042 $\kappa a / z\sqrt{c}$ with c in M at 25°C.)

a the decay of $\psi(r)$ is steeper at higher c, but because of the dimensionless abscissa axis, this is not explicit. Tables 3.1 and 3.2 are indicative of the quality of approximations [3.5.51 and 56] and are self-explanatory.

3.5f. Cylindrical double layers

For cylindrical geometry, similar statements can be made as in the first paragraph of sec. 3.5e. We consider infinitely long rods, ignoring end effects for which it is noted that the mean curvature J, defined in [I.2.23.2], is the same as that for a sphere with twice the radius. For such a cylinder, the potential only depends on r. Taking for the Laplacian the first term on the r.h.s. of [I.A7.46], we arrive at

$$\frac{1}{r}\frac{d}{dr}\left(r\frac{dy(r)}{dr}\right) = \left(\frac{d}{dr} + \frac{1}{r}\right)\frac{dy(r)}{dr} = -\frac{F^2}{\varepsilon_0 \varepsilon RT}\sum_j z_j c_j e^{-z_j y(r)} \qquad [3.5.57]$$

for the *Poisson-Boltzmann equation for cylinders*. As before, the sum over j covers all ionic species and $y = F\psi/RT$. For one symmetrical electrolyte

$$\frac{1}{r}\frac{d}{dr}\left(r\frac{dy(r)}{dr}\right) = \frac{\kappa^2}{z}\sinh[zy(r)] \qquad [3.5.58]$$

which in the low-potential limit reduces to

$$\frac{1}{r}\frac{d}{dr}\left(r\frac{dy(r)}{dr}\right) = \left(\frac{d}{dr} + \frac{1}{r}\right)\frac{dy}{dr} = \kappa^2 y(r) \qquad [3.5.59]$$

and which can also be written in the entirely dimensionless form

Table 3.1. Ratio of exact surface charge to charge according to the DH approximation. Spherical double layer, (1-1) electrolyte. (Source: Loeb et al., loc. cit. table 21).

y^d	$\kappa a = 0$	0.1	0.2	0.5	1	2	5	10	20	∞
0	1.0000	1.0000	1.0000	1.0000	1.0000	1.0000	1.0000	1.0000	1.0000	1.0000
1	1.0000	1.0009	1.0026	1.0079	1.0149	1.0230	1.0322	1.0366	1.0392	1.0422
2	1.0000	1.0039	1.0109	1.0336	1.0630	1.0971	1.1347	1.1528	1.1633	1.1752
3	1.0000	1.0097	1.0274	1.0842	1.1563	1.2379	1.3262	1.3681	1.3924	1.4195
4	1.0000	1.0202	1.0573	1.1735	1.3163	1.4739	1.6406	1.7185	1.7635	1.8134
5	1.0000	1.0392	1.1101	1.3250	1.5786	1.8504	2.1322	2.2624	2.3373	2.4200
6	1.0000	1.0736	1.2034	1.5764	1.9970	2.4364	2.8853	3.0911	3.2092	3.3393
7	1.0000	1.1371	1.3666	1.9846	2.6526	3.3373	4.0298	4.3461	4.5271	4.7264
8	1.0000	1.2538	1.6459	2.6338	3.6672	4.7141	5.7668	6.2464	6.5206	6.8225

Table 3.2. Ratio of exact surface charge to charge according to approximation [3.5.56]. Spherical double layer, (1-1) electrolyte. (Source: Loeb et al., loc. cit. table 26)

y^d	$\kappa a = 0.1$	0.2	0.5	1	2	5	10	20	∞
0	1.0000	1.0000	1.0000	1.0000	1.0000	1.0000	1.0000	1.0000	1.0000
1	1.0158	1.0126	1.0073	1.0039	1.0016	1.0004	1.0001	1.0000	1.0000
2	1.0600	1.0464	1.0249	1.0127	1.0051	1.0009	1.0003	1.0001	1.0000
3	1.1232	1.0903	1.0448	1.0203	1.0075	1.0016	1.0004	1.0001	1.0000
4	1.1902	1.1285	1.0550	1.0223	1.0076	1.0015	1.0004	1.0001	1.0000
5	1.2416	1.1458	1.0523	1.0189	1.0059	1.0011	1.0003	1.0001	1.0000
6	1.2599	1.1360	1.0403	1.0130	1.0038	1.0007	1.0002	1.0000	1.0000
7	1.2377	1.1057	1.0261	1.0077	1.0021	1.0003	1.0001	1.0000	1.0000
8	1.1845	1.0696	1.0148	1.0041	1.0011	1.0002	1.0000	1.0000	1.0000
9	1.1218	1.0398	1.0075	1.0019	1.0005	1.0001	1.0000	1.0000	1.0000
10	1.0702	1.0205	1.0036	1.0009	1.0003	1.0000	1.0000	1.0000	1.0000
12	1.0170	1.0043	1.0007	1.0002	1.0000	1.0000	1.0000	1.0000	1.0000
14	1.0032	1.0008	1.0002	1.0000	1.0000	1.0000	1.0000	1.0000	1.0000
16	1.0006	1.0002	1.0000	1.0000	1.0000	1.0000	1.0000	1.0000	1.0000

$$\frac{1}{(\kappa r)}\frac{d}{d(\kappa r)}\left((\kappa r)\frac{dy}{d(\kappa r)}\right) = y \qquad [3.5.59a]$$

This equation can be solved in terms of Bessel functions:

$$y = y^d K_0(\kappa r)/K_0(\kappa r = \kappa a) \qquad [3.5.60]$$

$$dy/d(\kappa r) = -y^d K_1(\kappa r)/K_0(\kappa r = \kappa a) \qquad [3.5.61]$$

where K_0 and K_1 are the zeroth- and first-order Bessel functions of the second kind, respectively[1]. Other quantities, in the appropriate dimensions are immediately derived. For instance, for the surface charge

$$\sigma^d = -\varepsilon_0\,\varepsilon\left(\frac{d\psi}{dr}\right)_{r=a} = -\frac{RT\varepsilon_0\varepsilon\kappa}{F\kappa}\cdot\frac{K_1(\kappa a)}{K_0(\kappa a)}\,y^d \qquad [3.5.62]$$

so that it can be directly obtained from tables of Bessel functions[2]

There is no analytical solution for any of the non-linearized equations but various approximations have been proposed. Often the corresponding derivations start from the flat layer model and correct for curvature in one direction, or they begin with an equation for spherical symmetry which is then "corrected" for the exaggerated curvature. Mathematically, if the Laplacian is written as

$$\nabla^2 \psi(r) = \left(\frac{d}{dr} + \frac{p}{r}\right)\frac{d\psi(r)}{dr} \qquad [3.5.63]$$

the cases $p = 0$, $p = 1$ and $p = 2$ correspond to flat, cylindrical and spherical geometry, respectively, compare [3.5.1, 57 and 43], underlining the intermediate position of cylinders between spheres and plates. This feature is also embodied in the following counterpart of [3.5.56]:

$$\sigma^d = -\frac{2Fcz}{\kappa}\left[2\sinh\left(\frac{z y^d}{2}\right) + \frac{2\tanh(z y^d/4)}{\kappa a}\right] \qquad [3.5.64]$$

For large κa it reduces (as [3.5.56] does) to the flat layer limit. On the other hand, for low κa ($\ll 1$) and low potentials (both hyperbolic functions replaced by their arguments), [3.5.64] reduces to the corresponding equation for spheres with twice the radius, [3.5.51]. For a discussion of this and other approximate

---

[1] F. Bowman, *Introduction to Bessel Functions*, Dover (1958); G.N. Watson, *A Treatise on the Theory of Bessel Functions*, Cambridge Univ. Press, 2nd ed. (1944), reprinted in 1980.
[2] E. Jahnke, F. Emde, *Tables of Functions with Formulae and Curves*, 4th ed., Dover (1945); *C.R.C. Handbook of Chemistry and Physics*, The Chem. Rubber Co.

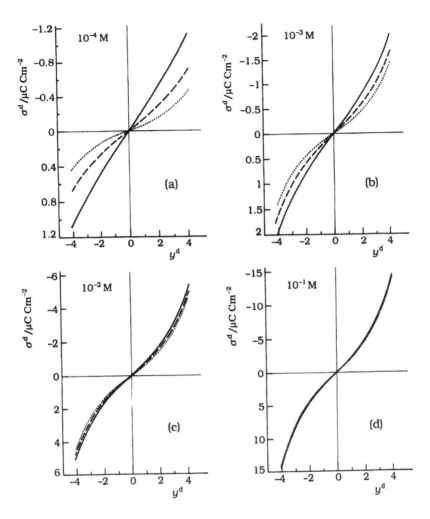

Figure 3.14. Comparison between the diffuse charge on a plate ($\cdots$), a cylinder (- - -) and a sphere (—). The concentration of the (1-1) electrolyte is indicated. Radii of cylinders and spheres 10 nm.

formulas, see the literature[1]. We confine ourselves to an exact numerical illustration, fig. 3.14, which well illustrates the discussed trends. Screening of the

---

[1] V.L. Sigal, N.M. Semenikhin, J. Chem. Phys. **61** (1964) 2170; N.M. Semenikhin, V.L. Sigal, J. Colloid Interface Sci. **51** (1975) 215; D. Stigter, J. Colloid Interface Sci. **53** (1975) 296; H. Ohshima, T.W. Healy and L.R. White, J. Colloid Interface Sci. **90** (1982) 17; D. Bratko, V. Vlachy, Chem. Phys. Lett. **90** (1982) 434; C.J. Benham, J. Chem. Phys. **79** (1983) 1969; R.E. Rice, F.H. Horne, J. Colloid Interface Sci. **105** (1985) 172; R.E. Rice, J. Chem. Phys. **82** (1985) 4337; V.V. Panjukov, J. Colloid Interface Sci. **110** (1986) 556; G.A. van Aken, H.N.W. Lekkerkerker, J.T.G. Overbeek and P.L. de Bruyn, J. Phys. Chem. **94** (1990) 8468.

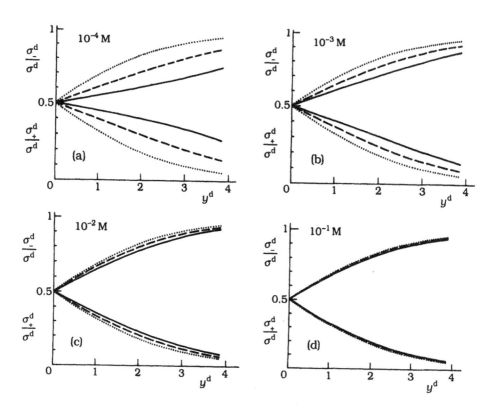

Figure 3.15. Fraction of diffuse charge attributed by negative adsorption of cations (σ_+^d / σ^d) and by positive adsorption of anions (σ_-^d / σ^d) for plates (····), cylinders (- - -) and spheres (——). Radii of cylinders and spheres 10 nm. The concentration of the (1-1) electrolyte is indicated. The curves for the plates are identical for the four concentrations. In this example, $\sigma^o > 0$ and $\sigma^d < 0$.

surface charge increases with increasing curvature (plate < cylinder < sphere), but when the double layer is thin as compared with the particle radius ($\kappa a \gg 1$) the distinction disappears. For all symmetries screening grows rapidly with ionic strength: compare the ordinate axis scales. A similar trend is observed with the ionic components of charge, see fig. 3.15. In this illustration $\sigma^o > 0$, $\sigma^d < 0$. With increasing diffuse layer potential the fraction of charge attributed by negative adsorption of cations decreases, whereas that caused by positive adsorption of anions increases, as in fig. 3.8. Higher ionic strengths tend to suppress the difference between the three geometries.

3.5g. Other geometries; double layers in cavities

Examples of other geometries are surfaces with asperities on them, rough surfaces, and surfaces with an angle between them as is the case for the edges and plates of clay minerals (fig. 3.1c). We shall consider these where necessary.

Double layers around ellipsoids have been considered by a few authors[1,2,3].

Concave double layers, i.e. double layers at the hollow side of curved inter-faces, are of more general interest. Thus double layers inside spherical cavities are relevant for micro-emulsions and vesicles, and those inside cylinders for electrokinetics and membrane transport. Certain contact points exist with double layers in overlap, as occurring in colloid stability.

It follows from the preceding section that the limiting case $\kappa a \gg 1$ (double layer thin as compared with radius of curvature) is simple: then we can simply apply the flat layer theory, discussed extensively in secs 3.5a-d. Beyond this limit, the appropriate Poisson-Boltzmann equation (with p in [3.5.63] depending on the geometry) has to be solved with the appropriate boundary condition, i.e. dy/dr for $r = 0$, so in the centre of a sphere or infinitely long cylinder, the field strength is zero because of symmetry. However, at that location y is not necessarily zero, because double layers from the opposite sides may overlap. This is a new feature as compared with convex double layers around non-interacting particles.

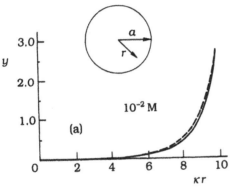

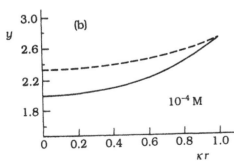

Figure 3.16. Diffuse double layer inside a spherical cavity. (1-1) electrolyte, $y^d = 2.73$ ($\psi^d = 70$ mV), $a = 30$ nm. Solid curves: exact numerical results, dashed curves: linear approximation. The electrolyte concentration is indicated. (Redrawn from J.E. Curry, S.E. Feller and D.A. McQuarrie, *J. Colloid Interface Sci.* **143** (1991) 527.)

[1] N.M. Semenikhin, N.N. Rulev, *Koll. Zhur.* **37** (1975) 907. Transl. *Coll. J. USSR* **37** (1975) 815.
[2] N.N. Rulev, N.M. Semenikhin, in *Poverkhn. Yavleniya Dispersnykh Sist.* (Surface phenomena of disperse systems), N.N. Kruglitskii, Ed., Naukova Dumka, **4** (1975) 197.
[3] M. Teubner, J. Frahm, *J. Colloid Interface Sci.* **82** (1981) 560.

An illustration is given in fig. 3.16 for a spherical cavity. As with convex double layers, the potential drops more rapidly (inward from the surface) with distance in 10^{-2} M than in 10^{-4} M solution; in the latter case the decrease is so weak that in the centre a substantial overlap potential remains. Similar results have been obtained by others[1,2,3], emphasizing other aspects, such as the ion uptake (or exclusion), the Gibbs energy and the disjoining pressure. This information underlies the thermodynamics of vesicle and micro-emulsion formation.

For isolated spheres of water, the value to be taken for the equilibrium electrolyte concentration in a bulk phase, $c(\infty)$, may be a problem. In the situation of fig. 3.16b, c_+ and c_- do not approach this value. When there is equilibrium with an external bulk phase, as with vesicles, this is no problem, because $c_+(r = 0)$ and $c_-(r = 0)$ and $c(\infty)$ are simply related via the Boltzmann equation. When Boltzmann's law applies, the equilibrium concentration $c(\infty)$ in a (virtual) bulk phase can be written as $c(\infty) = [c_+(r = 0) \cdot c_-(r = 0)]^{1/2}$. However, if there is no such equilibrium (say, for microdrops of water formed in a non-conducting oil, under highly dynamic conditions) $c(\infty)$ may differ between one drop and the other and nothing can be said in general. Alternatively, the negative adsorption of electrolyte can be computed if y^d is known (this is the Donnan effect).

The counterpart of fig. 3.16 for cylinders has been elaborated by van Keulen and Smit[4] and was also included in the work by van Aken et al.[3]. Qualitatively, the trends are the same as for spherical cavities. Quantitatively, the potential decay is less steep at given κ. In practice, cylindrical pores are often in equilibrium with a bulk solution, so that the non-equilibrium problem does not arise here.

3.6 Beyond Poisson-Boltzmann. The Stern picture

3.6a. Defects of the Poisson-Boltzmann model

It is not difficult to point to a number of imperfections in the Poisson-Boltzmann theory. Several of these have already been reviewed in discussing the applicability of the Debye-Hückel theory (sec. I.5.2b, c). They include the following:

(i) The finite sizes of the ions are neglected.

---

[1] B.G. Tenchov, R.D. Koynova, B.D. Raytcher, *J. Colloid Interface Sci.* **102** (1984) 337.
[2] M. Caselli, M. Maestro, *J. Electroanal. Chem.* **283** (1990) 67.
[3] G.A. van Aken, H.N.W. Lekkerkerker, J.Th.G. Overbeek and P.L. de Bruyn, *J. Phys. Chem.* **94** (1990) 8468.
[4] H. van Keulen, J.A.M. Smit, *J. Colloid Interface Sci.* **151** (1992) 546.

(ii) Non-Coulombic interaction between counter- and co-ions and surface (*specific adsorption*) is disregarded.

(iii) The permittivity of the medium is assumed to be constant.

(iv) Incomplete dissociation of the electrolyte is ignored.

(v) The average potential and the potential of the mean force are assumed to be identical.

(vi) The solvent is considered to be primitive.

(vii) Polarization of the solvent by the (charged) surface is not taken into account.

(viii) The surface charge is assumed to be homogeneous and smeared-out.

(ix) Surfaces are considered flat on a molecular scale.

(x) Image forces between ions and the surface are neglected.

Considering this long list of iniquities, it is not surprising that conditions are readily found where the picture developed in sec. 3.5 breaks down. A typical illustration is that at high surface potential ($y^o = F\psi^o / RT \gg 1$) the double layer charge and capacitance on mercury and silver iodide remain far below that predicted. On most surfaces and in many electrolytes specific adsorption is observed (different σ^o for different ions of the same valency at given pAg, pH, etc.). The extent of it and the sequence depend on the natures of the surface and electrolyte. On the other hand, perfect applicability of Poisson-Boltzmann equations is observed in other experiments. For instance, interaction forces at not too short distance between two charged surfaces, as measured in the surface force apparatus and the effect of the electrolyte concentration on the thickness of thin liquid films and on the negative adsorption are well described. Hence, it is appropriate to delineate the domain of applicability of Poisson-Boltzmann theory and to consider appropriate corrections.

Let us first discuss the mentioned defects a little more.

Ion size is an obvious feature. According to the Boltzmann equation, $c_i(x) = c_i(\infty) \exp[-y(x)]$. Close to the surface, where potentials of 100-200 mV are not rare, for, say monovalent cations near a negative surface, $\exp[-y(x)]$ can easily become as high as 10^2 - 10^3, predicting local concentrations far exceeding the space physically available. Accounting for finite ion sizes would reduce the high counterion concentrations close to the surface. Non-zero sizes, together with the related non-electric affinity of ions for the surface, are the main reasons for specific adsorption.

Especially for polar solvents the dielectric permeability may be reduced under the influence of the electric field in the double layer. We introduced this *dielectric saturation* in secs. I.5.1d and I.5.3e. The consequence is a reduced screening power of the solvent, especially in the inner part of the double layer where the field is high.

Incomplete dissociation would also reduce the screening power of electrolytes. Ions of higher valency could, as a result of ion association, become entities of lower valency; this has substantial consequences because the valency occurs in the exponentials, see for instance sec. 3.5c. For otherwise strong electrolytes this complication starts when the relative dielectric permittivity is low, say $\varepsilon \lesssim 30$.

Problems (v), (viii), and (ix) are related. Basically, the problem is whether or not an ionic charge, brought from a reference point to a point in the double layer to measure the electric work done, modifies the potential. In reality, a double layer consists of a set of *discrete charges*, with locally very high potentials. When a new ion is introduced, the collection of charges have to be rearranged, producing room for the incoming charge. The isothermal reversible work includes a contribution caused by this rearrangement and hence depends on *ion-ion correlations*. The corresponding potential is called the *potential of mean force*. This potential should appear in the Boltzmann equation. On the other hand, if discreteness of charge is ignored, we obtain a homogeneous field, where in the ideal limit an incoming unit charge is not "seen" by the field, it only acts as a probe. Potentials belonging to this category are called *mean* (or average) *potentials*; such potentials occur in the Poisson equation. In sec. I.4.3c we already introduced and elaborated this distinction. Restating the problem: by identifying the potentials in the Boltzmann and the Poisson equation, an error is made, but it is not straightforward to assess it quantitatively.

Image forces in the solid also play a role when the countercharge is discrete, but not for a smeared-out double layer. For the latter case, all lines of force run parallel (or radially, for spheres), starting at the (positive) surface and ending somewhere in the double layer. However, for discrete charges these lines can bend and, under certain conditions, reach the surface to create image forces. In this context, the nature and discreteness of the surface charge also play a role.

Issues (vi) and (vii) both deal with the nature of the solvent; they are also related to (v). Considering water, the spatial distribution of the molecules is in a very complicated way determined by solvent-solvent, solvent-countercharge and solvent-surface charge interactions. A detailed knowledge of this structure is required to quantify ion-ion correlations, ion-ion and ion-surface solvent structure-originated interactions and the local dielectric permittivity. Polarization of the solvent also contributes to the *interfacial potential jump* or *χ-potential* (secs. I.5.5a and 3.9), which does not occur in Poisson-Boltzmann theory.

Surface heterogeneity means that the chemical contribution to the double layer formation varies from site to site along the surface. This chemical heterogeneity is partly obliterated by the electric field.

3.6b. Statistical thermodynamics of the double layer

Improving the Poisson-Boltzmann model of sec. 3.5 leads to problems similar to those encountered in sec. I.5.2c for the Debye-Hückel theory. Basically it is not all that difficult to account for only one or two of the mentioned defects but such improvements have limited practical relevance because (i) the outcome depends on a number of molecular parameters that are not always known and (ii) several of the defects cannot be repaired without also considering others, especially so because internal compensations occur. By way of examples, accounting for ion correlations leads to an increased concentration in the inner part of the double layer, but considering non-zero ionic volumes leads to a reduction; strong fields lead to dielectric saturation, but lower permittivities lead to weaker fields because such media can carry fewer ions. Hence, for really significant improvement of the theory, all defects should be integrally considered. In considering improvements of the Debye-Hückel theory for strong electrolytes the same conclusion was reached (sec. I.5.2b and c).

The most rigorous approach to improve the Gouy-Chapman theory is by starting from first principles, applying the statistical thermodynamical equations for bulk electrolytes to non-uniform fluids. In this approach all ions (and in general also the solvent molecules) are considered as *individual particles*. The probabilities of finding the individual entities at a given position r and, for asymmetrical molecules, at azimuthal and elevational angles φ and θ are given by distribution functions.

One typical consequence is that we refrain from considering smeared out distributions, as in the Poisson-Boltzmann model. Another implication is that image charges have to be taken into account.

Several excellent reviews on the statistical mechanics of the electrical double layer have been published (Carnie and Torrie, Blum, Blum and Henderson, see sec. 3.15c). In this section we give a summary of the most important elements of the statistical mechanical approach and indicate the improvements with respect to the Gouy-Chapman approach. Our treatment follows the review of Carnie and Torrie[1].

Statistical thermodynamics of the electric double layer starts with modelling the electrolyte and the interface. This can be done by specifying all intermolecular and external interactions in the phase space as a Hamiltonian. The notion of phase space was defined in sec. I.3.9a and the Hamiltonian H was introduced in [I.3.9.1]. As the kinetic part of the Hamiltonian does not contribute to the configuration integrals, we sum only over the potential energies of the ions. In the inhomogeneous system it is customary to separate the interactions with the charged wall (the "external" field) from the interionic ones.

[1] S.L. Carnie, G.M. Torrie, *The Statistical Mechanics of the Electrical Double Layer*, Adv. Chem. Phys. **56** (1984) 142. In sec. 3.6b referred to as Carnie and Torrie, loc. cit.

$$H = \sum_{\alpha=1}^{\nu} \sum_{i_\alpha} u_\alpha \left(r_{i_\alpha} \right) + \frac{1}{2} \sum_{\alpha,\beta=1}^{\nu} \sum_{i_\alpha \neq j_\beta} u_{\alpha\beta} \left(r_{i_\alpha}, r_{j_\beta} \right)$$ [3.6.1]

Here $\alpha, \beta, \ldots, \nu$ denote ionic and non-ionic species and $i, j, \ldots$ is the ith, jth, ... particle of some species located at position $r_i, r_j, \ldots$ The first term on the r.h.s. in [3.6.1] contains the sum of the one-body interactions u_α due to the presence of a (charged) interface on an ion or molecule i of species α at position r_{i_α}. The second term is a sum over two-body interactions $u_{\alpha\beta}$ of a particle i_α at r_{i_α} and a particle j_β at r_{j_β}.

In principle all kind of interactions are contained in [3.6.1]. In the present section we shall consider a solid-liquid interface although the treatment is also valid for liquid-liquid interfaces. Solid and liquid are taken as *primitive*, i.e. as structureless continuums with dielectric permittivities ε^S and $\varepsilon^L = \varepsilon$, respectively. In this model the surface is hard, planar and uniformly charged. Considering the surface charge σ^o as discrete would mean a further improvement. The electrolyte contains hard spherical ions with radii $a_\alpha, a_\beta, \ldots$. The ions have a central charge $q_\alpha, q_\beta, \ldots$. Dipole, multipole, and higher interactions and polarizabilities of the fluid molecule are neglected. It is convenient to separate (non-electric) short-range from (electric) long-range interactions.

It is inherent to treatments involving individual charges to consider *image charges*. In this way, all pair interactions can be accounted for by Coulomb's law. See fig. 3.17 where the positions of ions 1 and 2 (of arbitrary species) are given relative to a surface. For ion 1 the image charge 1* with a magnitude fq_1 is positioned on the opposite side of the wall. From basic electrostatics it follows that

$$f = \frac{\varepsilon - \varepsilon^s}{\varepsilon + \varepsilon^s}$$ [3.6.2]

The image charge is a consequence of the polarization of the semi-infinite solid by ion i and determined by the discontinuity of the dielectric permittivity. It may be considered as a real charge at a position $-z_1$ in a medium of dielectric permittivity ε of such magnitude (fq_1) that the actual field line pattern in the solution is completely reproduced. No images are present if the dielectric permittivities of wall and fluid are equal ($f = 0$). For a very high dielectric permittivity of the solid, as for metals, the image charge is equal in magnitude to the ion charge q, but opposite in sign ($f = -1$). The surface charge and its image, both located at $z = 0$ are also equal in magnitude and opposite in sign. Effectively, the surface charge vanishes and is replaced by an image at $-z_1$. If the dielectric constant of the fluid is much larger than that of the solid charge, q_1 and its image charge are equal in sign and magnitude ($f = 1$) and the charge at the surface becomes twice as high as if there were no image charge. In this case

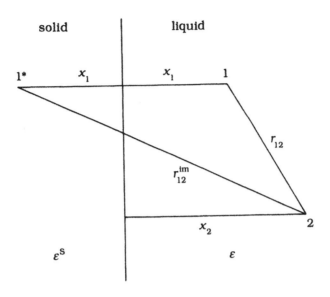

Figure 3.17. Image charges and meanings of symbols. The charge q_1, of ion 1 is "reflected" by the wall to give an image charge fq_1 at position $-x_1$ from the interface. The image charge of ion 2 is not drawn.

we have a *repulsive* image charge, in distinction to the *attractive* image obtained for a solid with a high dielectric permittivity. The occurrence of image charges implies that the interaction between q_1 and a surface charge at x_1 is replaced by an interaction with its image at a larger distance. In all of this the total overall charge remains zero because of electroneutrality.

The one-body interaction in [3.6.1] can now be written as the sum of a Coulomb interaction with the external field, an image term and a short-range non-Coulomb interaction.

$$u_\alpha(x_1) = q_\alpha \psi^{coul}(x_1) + q_\alpha^2 w^{im}(x_1) + u_\alpha^{sr}(x_1)$$ [3.6.3]

The quantities ψ, w and u have dimensions J, J C^{-1} and J C^{-2}, respectively[1]. The one-body interaction only depends on the distance from the wall x_1. In the primitive model the short range contribution can be given by a hard wall repulsion:

$$u_\alpha^{sr}(x_1) = \begin{cases} \infty & x_1 < a_\alpha \\ 0 & x_1 > a_\alpha \end{cases}$$ [3.6.4]

---

[1] Carnie and Torrie, loc. cit. use the same symbol for our ψ, w and u, irrespective of the dimensional differences, and call all of these "potentials".

Here a_α is the radius of particle α. The hard wall repulsion is equivalent to the introduction of a distance of closest approach for the counterions in order to define a Stern layer thickness. In a more restricted primitive model the radius a is set identical for all ionic species. The Coulomb term in [3.6.3] represents the interaction ψ^{coul} of an ion with a charge q_α at x_1 with the surface charge and its image. According to Coulomb's law,

$$\psi^{coul}(x_1) = -\sigma^\circ x_1 \frac{1+f}{2\varepsilon_o \varepsilon}$$ [3.6.5]

The interaction of the ion with its own image also follows from Coulomb's law:

$$w^{im}(x_1) = \frac{f}{16\pi\,\varepsilon_o \varepsilon\, x_1}$$ [3.6.6]

In the absence of a discontinuity in the dielectric permittivity and the wall no image exists. For very high dielectric permittivities of the solid the Coulomb interaction is zero and the ion interacts only with its own image (Table 3.3).

Table 3.3. Ion-image interactions.

	f	$w^{im}(x_1)$	$\psi^{coul}(x_1)$	$w^{im}(r_1, r_2)$
$\varepsilon^S \ll \varepsilon$	1	$1/16\pi\varepsilon_o\varepsilon x_1$	$-\sigma^\circ x_1/\varepsilon_o\varepsilon x_1$	$1/4\pi\varepsilon_o\varepsilon r_{12}^{im}$
$\varepsilon^S = \varepsilon$	0	0	$-\sigma^\circ x_1/2\varepsilon_o\varepsilon x_1$	0
$\varepsilon^S \gg \varepsilon$	-1	$-1/16\pi\varepsilon_o\varepsilon x_1$	0	$-1/4\pi\varepsilon_o\varepsilon r_{12}^{im}$

The interactions between pairs of ions α and β, located at positions 1 and 2 in fig. 3.17 can also be separated in long-range electric interactions and a short-range contribution.

$$u_{\alpha\beta}(r_1, r_2) = q_\alpha q_\beta w^{coul}(r_{12}) + q_\alpha q_\beta w^{im}(r_1, r_2) + u^{sr}(r_{12})$$

$$\equiv q_\alpha q_\beta w^{el}(r_1, r_2) + u^{sr}(r_{12})$$ [3.6.7]

The Coulomb interaction between the ions at 1 and 2 depends, as well as on $q_\alpha q_b$, only on their distance r_{12}, and not on the spatial positions r_1 and r_2:

$$w^{coul}(r_{12}) = \frac{1}{4\pi\varepsilon_o\varepsilon r_{12}}$$ [3.6.8]

The interaction between the charge of ion 2 and the image of ion 1 depends on r_{12}^{im}, i.e. on r_{12}, x_1 and x_2:

$$w^{im}(r_1,r_2) = \frac{f}{4\pi\varepsilon_0\varepsilon\,r_{12}^{im}} = \frac{f}{4\pi\varepsilon_0\varepsilon\,\sqrt{r_{12}^2 + 4x_1x_2}}$$ [3.6.9]

The model electrolyte is now assumed to consist of hard spheres each with a radius a, leading to a hard-core repulsion according to:

$$w_{\alpha\beta}^{sr}(r_{12}) = \begin{cases} \infty & r_{12} < a_\alpha + a_\beta \\ 0 & r_{12} > a_\alpha + a_\beta \end{cases}$$ [3.6.10]

Having defined the different interactions occuring in [3.6.1], we now need to specify the probability of finding an ion α at some position r. The one-particle (singlet) density $\rho_\alpha(r_1)$ is defined[1] in sec. I.3.9d as the number of particles per volume at position r_1. Now we apply the definition to ions. The radial distribution function $g_\alpha(r_1)$ and the *ion-wall total correlation function* $h_\alpha(r_1)$ follow from [I.3.9.22 and 23] as

$$\frac{\rho_\alpha(r_1)}{\rho_\alpha} \equiv g_\alpha(r_1) \equiv 1 + h_\alpha(r_1)$$ [3.6.11]

Here, ρ_α is the bulk density of ion α.

Higher order correlation functions $g_h(r_1,r_2...r_h)$ are defined from higher-order distribution functions $\rho^{(h)}$. For the *pair correlation function* $g_{\alpha\beta}(r_1,r_2)$ we can write

$$\frac{\rho_{\alpha\beta}(r_1,r_2)}{\rho_\alpha(r_1)\rho_\beta(r_2)} \equiv g_{\alpha\beta}(r_1,r_2) \equiv 1 + h_{\alpha\beta}(r_1,r_2)$$ [3.6.12]

The term $\rho_{\alpha\beta}(r_1,r_2)\,dr_1\,dr_2$ expresses the probability of finding simultaneously a molecule α in the infinitesimal volume dr_1 and a molecule β in dr_2.

In sec. I.4.3c we have already introduced the *mean (average) electrostatic potential*. The mean electrostatic potential $\psi(x_1)$ is related to the work necessary to bring an infinitely small probe charge from infinity to x_1 without disturbing the environment. This potential can be defined in terms of the one- and two-body interactions, mentioned above, according to

$$\psi(z_1) = \psi^{coul}(x_1) + \int dr_2\, w^{el}(r_1,r_2)\sum_\alpha q_\alpha \rho_\alpha(x_2)$$ [3.6.13]

The first term on the r.h.s. represents the Coulomb potential due to the wall, the

---

[1] In the present subsection ρ stands for the number density (in m^{-3}). The usual symbol is ρ_N but here we drop the N to avoid cumbersome notation.

second is the sum of all pair interactions between the probe at r_1 with the average charge due to the other ions at all positions z_2. The second term is, of course, zero for a homogeneous ion distribution, but has a finite value in an inhomogeneous system such as an electric double layer. Equation [3.6.13] does not contain the self-image term at x_1 $(q_\alpha w^{im}(x_1))$ as the infinitesimal probe charge (the "spy") does not polarize the surface and correspondingly has no image. From [3.6.13] and the electroneutrality condition we can easily derive the following modification of the Poisson equation [3.5.1] and the (Maxwell) field strength:

$$-\nabla_1 \psi(x_1) = -\frac{d}{dx_1}\psi(x_1) = -\frac{1}{\varepsilon_0 \varepsilon}\int_{x_1}^{\infty} dz_2 \sum_\alpha q_\alpha \rho_\alpha(x_2)$$ [3.6.14]

Integration with respect to x_1 gives the surface potential ψ^0.

Although in [3.6.13] the ψ^{coul} and w^{el} terms are functions of f, in the derivative [3.6.14] the image terms cancel due to the fact that the sum of the image charges is zero. In order to determine the mean potential at some position r_1, we have averaged over all positions of all ions. This mean potential is sometimes called the macropotential.

For some purposes the mean electric potentials at position r_1 is required provided there is an ion α fixed at r_2. We shall call it $\psi_\alpha(r_1;r_2)$. Bringing the probe charge to r_1 leads to a perturbation of the distribution around ion α. However, if we want to determine the potential at some distance from a fixed ion 2, we must realize that the distribution around ion 2 is disturbed and correspondingly the potential at r_1, given a fixed ion at r_2 will be different from the mean potential and is given by

$$\psi_\alpha(r_1;r_2) = \psi^{coul}(r_1) + q_\alpha w^{el}(r_1,r_2) + \int d\, r_3\, w^{el}(r_1,r_3)\sum_\gamma q_\gamma \rho_\gamma(x_3)g_{\gamma\alpha}(r_3,r_2)$$ [3.6.15]

The charges of all other ions are averaged over all positions r_3. The first term on the r.h.s. again accounts for the interaction with the surface charge. The second term is the electrostatic contribution with ion α, fixed at r_2. The third term takes into account the perturbation of the distribution around ion 2. This term is similar to the last term in [3.6.13] except that the distribution function is now multiplied by the pair correlation function $g_{\alpha\beta}(r_3,r_2)$ to account for the probability of finding an ion 3 in the neighbourhood of ion 2. The change in the potential at r_1 as a result of fixing an ion at r_2 is called the *fluctuation potential* (ψ^{fluct}) and is an important parameter for describing deviations from the ideal Poisson-Boltzmann equation. This potential follows directly from the difference between [3.6.15] and [3.6.13], using the definition of $g_{\alpha\beta}(r_3,r_2)$ in [3.6.12];

$$\psi_\alpha^{\text{fluct}}\left(r_1; r_2\right) \equiv \psi_\alpha\left(r_1; r_2\right) - \psi\left(x_1\right)$$

$$= q_\alpha w^{\text{el}}\left(r_1, r_2\right) + \int d\,r_3\, w^{\text{el}}\left(r_1, r_3\right) \sum_\gamma q_\gamma \rho_\gamma\left(x_3\right) h_{\gamma\alpha}\left(r_3, r_2\right) \qquad [3.6.16]$$

Basically, when we know $g_\alpha(x_1)$ or $h_\alpha(x_1)$ and $\rho_{\alpha\beta}(r_1, r_2)$ or $g_{\alpha\beta}(r_1, r_2)$ (and, hence $\psi_1(x_1)$) all required double layer properties can be found. For instance, the surface charge density of each species, i.e. the ionic components of charge, are immediately found by integrating h_α with respect to x_1, considering the appropriate boundary conditions.

The mean potential and the fluctuation potential derived above are expressed in terms of correlation (or distribution) functions. Hence we need next a procedure to calculate these functions. There exist several statistical thermodynamical procedures to express ρ_α into $\rho_{\alpha,\beta}$, $\rho_{\alpha,\beta}$ into $\rho_{\alpha,\beta,\gamma}$, etc. According to the so-called Bogolubov-Born-Green-Kirkwood-Yvon (BBGKY) hierarchy a recurrent expression is derived. The first two terms are[1]

$$kT\,\nabla_1\rho_\alpha\left(r_1\right) = -\rho_\alpha\left(r_1\right)\nabla_1 u_\alpha\left(r_1\right) - \int d\,r_2 \sum_\beta \nabla_1 u_{\alpha\beta}\left(r_1, r_2\right)\rho_{\alpha\beta}\left(r_1, r_2\right) \qquad [3.6.17a]$$

$$kT\,\nabla_1\rho_{\alpha\beta}\left(r_1, r_2\right) = -\rho_{\alpha\beta}\left(r_1, r_2\right)\left[\nabla_1 u_\alpha\left(r_1\right) + \nabla_1 u_{\alpha\beta}\left(r_1, r_2\right)\right]$$

$$- \int d\,r_3 \sum_\gamma \nabla_1 u_{\alpha\gamma}\left(r_1, r_3\right)\rho_{\alpha\beta\gamma}\left(r_1, r_2, r_3\right) \qquad [3.6.17b]$$

In [3.6.17b] the three-particle distribution function $\rho_{\alpha\beta\gamma}\left(r_1, r_2, r_3\right)$ occurs, defined similarly to [3.6.12].

Equation [3.6.17a] provides an expression for the *potential of mean force* on an ion of species α at position x_1. This notion has already been introduced in sec. I.4.3c, but now a more explicit expression can be given. Let this potential be $\psi_\alpha^{\text{mf}}(x_1)$, then[2]

$$q_\alpha\psi_\alpha^{\text{mf}}(x_1) = kT\,\nabla_1 g_\alpha(x_1)$$

$$= -q_\alpha\nabla_1\psi(x_1) - \nabla_1 u_\alpha^{\text{sr}}(x_1) - \int d\,r_2 \sum_\beta \nabla_1 u_{\alpha\beta}^{\text{sr}}\left(r_1, r_2\right)\rho_\beta(x_2) g_\alpha\left(r_1, r_2\right)$$

$$- q_\alpha \lim_{r_1 \to r_2} \nabla_1\left[\psi_\alpha^{\text{fluct}}\left(r_1; r_2\right) - q_\alpha w^{\text{coul}}\left(x_{12}\right)\right] \qquad [3.6.18]$$

The first term on the r.h.s. is the mean potential and after substitution of this term in [3.6.14], ignoring all other terms, the classical Poisson-Boltzmann

---

[1] J.-P. Hansen, I.R. MacDonald, *Theory of Simple Liquids*, Academic Press (1976).
[2] Carnie and Torrie, loc. cit. 152.

equation [3.5.6] is again obtained. The sum of the other terms give the difference between the mean potential and the potential of mean force. The second term on the r.h.s. introduces a charge-free Stern layer near the wall. The third term takes into account the short-distance pair correlations. Together these two terms account for ion specificity, the former between ions and surface, the latter between ions themselves. The last term accounts for the Coulomb pair correlations which depend on the fluctuation potentials in distinction to the mean potential. The potential of mean force at a position x_1 depends on the nature of the species α.

The potential occurring in the Poisson equation is basically the mean potential whereas that in the Boltzmann equation is the potential of mean force. In the Poisson-Boltzmann equation this distinction is not made. The error made in this approximation is quantified by the r.h.s. of [3.6.18], excluding the first term.

With the above, a formal set of equations is given, the elaboration of which requiring a solution for the problem that the recurrent relationships $\rho_\alpha \rightarrow \rho_{\alpha\beta} \rightarrow \rho_{\alpha\beta\gamma}, \ldots$ diverge. Relatively simple densities, or distribution functions, are converted into more complex ones. A "closure" is needed to "stop this explosion". A number of such closures have been proposed, all involving an assumption of which the rigour has to be tested. Most of these write three-body interactions in terms of three two-body interactions, weighted in some way. A well known example is Kirkwood's superposition closure, which reads:

$$\rho_{\alpha\beta\gamma}\left(r_1,r_2,r_3\right) = \rho_\alpha\left(r_1\right)\rho_\beta\left(r_2\right)\rho_\gamma\left(r_3\right)g_{\alpha\beta}\left(r_1,r_2\right)g_{\beta\gamma}\left(r_2,r_3\right)g_{\gamma\alpha}\left(r_3,r_1\right) \qquad [3.6.19]$$

This superposition closure states that the total potential of mean force between three ions can be replaced by the sum of one-body and two-body potentials of mean force. Now [3.6.17b] reduces to

$$kT\,\nabla_1 \ln g_{\alpha\beta}\left(r_1,r_2\right) = -q_\alpha\nabla_1\psi_\beta^{\text{fluct}}\left(r_1,r_2\right) - q_\alpha\int d\,r_3\left[\nabla_1 u^{\text{el}}\left(r_1,r_3\right)\right]$$
$$\cdot \sum_\gamma q_\gamma\rho_\gamma\left(x_3\right)h_{\beta\gamma}\left(r_2,r_3\right)h_{\gamma\alpha}\left(r_3,r_1\right) \qquad [3.6.20]$$

In principle [3.6.20], together with [3.6.18] can be solved for $g_\alpha(x_1)$ and $g_{\alpha\beta}(r_1,r_2)$. In practice these equations are further simplified.

Above we have given the basic strategy for the statistical thermodynamics of the electrical double layer. We shall not discuss the various elaborations in detail, but note that in addition to the BBGYK hierarchy to solve the distribution functions several other methods have been developed. In the Kirkwood hierarchy a coupling constant is introduced to avoid the spatial integration required to obtain distribution functions in the BBGYK hierarchy. Croxton and McQuarrie use the BBGY hierarchy by making approximations for $g_{\alpha\beta}(\mathbf{r}_1,\mathbf{r}_2)$ instead of using the superposition closure. Another class of correlation functions are defined by the Ornstein-Zernike equation (OZ). They are based on physically stronger arguments than the BBGKY recurrent equation. Several closures were proposed in the statistical thermodynamics of liquids. They can also be applied to the statistical thermodynamics of the electrical double layer. We mention the hypernetted chain approximations (HNC) and the mean spherical approximation (MSA). The latter theories were studied in detail by Henderson et al. Density functional theory is another promising approach.

One way to test and compare these various statistical approaches is by computer simulation. Molecular dynamics (MD) simulations are based on the classical equations of motion to be solved for a limited number of molecules. From such simulations information about equilibrium properties as well as the dynamics of the system are obtained. In order to test theories based on primitive models for the solvent, Monte Carlo simulations are more appropriate. In Monte

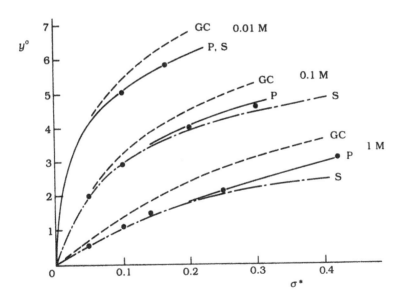

Figure 3.18. Surface charge as a function of surface potential according to some statistical theories. GC = modified Gouy-Chapman, S = HNC/MSA, P = modified "PB 5", $\bullet$ = Monte Carlo simulations. (Redrawn from Carnie and Torrie, loc. cit. 196). The dimensionless surface charge σ^* is scaled in such a way that for aqueous solutions 0.1 unit in σ^* corresponds to 8.8 μC cm^{-2}. The concentration of the (1-1) electrolyte is indicated.

Carlo simulations a random walk through the phase space of the model system is made. In this way a sequence of microscopic states are generated which are either or not accepted based on some criterion. Usually, in double layer problems the chemical potential is kept constant so that the thermodynamic parameters are obtained grand canonically, see I.app. 6.

An illustration is given in fig. 3.18. This diagram gives the (dimensionless) surface charge and potential. As short-range interactions are accounted for, the Stern layer is automatically accounted for, although in the present illustrations specific adsorption is neglected: ions are considered as hard spheres with radius 0.425 nm. The figure may be compared with the GC picture of fig. 3.4. As it is not our intention to compare the various statistical theories we refer to Carnie and Torrie for details, but note that GC stands for Gouy-Chapman theory, allowing for a finite distance of approach from the surface, i.e. zeroth-order Stern theory (sec. 3.6c). It is concluded that all discrete theories have in common that, at given y^o, a higher charge is predicted than by GC theory. This feature is not usually found experimentally. By titration, surface charges on non-porous oxides (systems that can attain relatively high σ^o's) tend to a maximum of about 10, 15 and 30 μC cm^{-2} for $y^o \sim 5$ according to Nernst in 10^{-3} M, 10^{-2} M and 10^{-1} M (1-1) electrolyte, respectively, see fig. 3.63. This corresponds to σ^* values of about 0.11, 0.17 and 0.34, i.e. much lower than predicted by GC theory or one of its extensions. Figure 3.18 also indicates the range where the GC approximation remains accurate. Many more similar comparisons have been published; the literature in sec. 3.15c may serve as a starter.

A qualitatively new feature emerges for (2-2) electrolytes, as illustrated in fig. 3.19. Some of the more sophisticated statistical theories predict a maximum in $y^o(\sigma^*)$. The very strong ion-ion correlations reduce the potentials at high charge; in fact, for extremely high salt concentrations and charges, some theories predict the development of alternating positive and negative ionic layers, leading to oscillations in the $\psi(x)$ dependence. So far, no unambiguous experimental proof of such maxima has been obtained. One of the problems is that (2-2) salts almost invariably contain specifically adsorbing ions, the other is that the maxima are predicted under conditions that are almost inaccessibly high. Some circumstantial indirect evidence may stem from colloid stability studies with highly charged clay minerals. By way of a recent illustration we may refer to a paper by Kékicheff et al.[1], who reported on improved colloid interaction measurements on mica with the surface force apparatus and on Si_3N_4 with atomic force microscopy atomic force microscopy (AFM), in solutions of $Ca(NO_3)_2$. Up to the coagulation concentration ($\sim 10^{-3}$ M) the interaction energy can be satisfactorily

---

[1] P. Kékicheff, S. Marcelja, T.J. Senden and V.E. Shubin, *J. Chem. Phys.* **99** (1993) 6098.

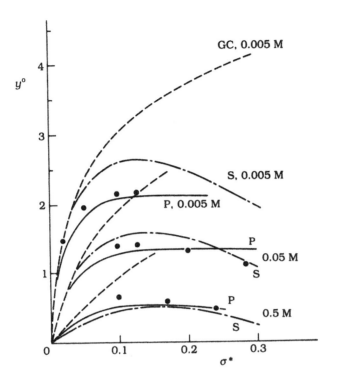

Figure 3.19. As fig. 3.18 but now for a (2-2) electrolyte.

accounted for by classical DLVO stability theory, which is based on a Gouy-Chapman approach. However, for very high concentrations (up to 0.92 M and 4.4 M in the surface force and AFM apparatus, respectively) restabilization was observed, which was attributed to the effect of ion correlations. However, streaming potentials by Scales et al.[1] under the required conditions, did not confirm any reversal in the ζ-potential. Our conclusion is that the phenomena of maxima in $y^o(\sigma^*)$ and sign reversal are open issues.

For further reading on this matter, see[2].

In summary, in this subsection some of the ways to improve GC theory in a fundamental way have been indicated. For the time being the significance of these developments for practice remains somewhat academic because (i) they do not lead to relatively simple analytical formulas that can be easily applied and

[1] P.J. Scales, F. Grieser and T.W. Healy, *Langmuir* **6** (1990) 582.
[2] G.M. Torrie, P.G. Kusalik and G.N. Patey, *J. Chem. Phys.* **88** (1988) 7827; **89** (1988) 3285; **90** (1989) 4513; **91** (1989) 6367; Z.X. Tang, L.E. Scriven and H.T. Davis, *J. Chem. Phys.* **97** (1992) 494.

are valid over a wide range of conditions; (ii) the charge- and potential conditions for which substantial deviations of the GC picture (with charge-free layers) are predicted, are not so often met in real experiments and (iii) to give a really complete picture, the solvent structure should also be accounted for and, hence, solvent structure-originating specific forces and specific ion-surface interaction. In practice, these specific contributions occur widely and sometimes tend to obliterate the electrostatic ionic correlations.

Because of these considerations we shall henceforth base our discussion mainly on the Gouy-Stern picture, which, over recent decades, has shown its power to account adequately for most experimental observations.

For practical purposes the success of the Stern approach has two causes: it adds only one or two parameters (the inner layer capacitance and, sometimes the specific adsorption energy) so that it is relatively easy to handle while remaining physically realistic. The main reason for this last statement is that, if deviations from ideal behaviour occur, this will be particularly the case at high potentials and high local concentrations, that is, close to the surface. As fig. 3.7 shows, there the potential drop is very steep: a little further from the surface, potentials and concentrations have become low enough to consider the remaining part of the layer as ideal (coincidence of the curves in fig. 3.18 at low y^o). A second consideration is that the potential at the location where the double layer becomes diffuse is very close, if not identical, to the electrokinetic potential ($\psi^d \approx \zeta$). Experience has shown that phenomena depending on double layer overlap, such as colloid stability, the thickness of thin liquid films and the electroviscous effect, are essentially determined by ζ, which under certain conditions is obtainable from electrokinetic measurements (chapter 4). As already stated, direct measurement of the interaction between charged surfaces has shown that for not too short distances the Poisson-Boltzmann picture works well.

3.6c. The Stern layer

In the following our discussion will be based on the rather pragmatic, though somewhat artificial, subdivision of the solution side of the double layer into two parts: an *inner part*, or *Stern layer* where all complications regarding finite ion size, specific adsorption, discrete charges, surface heterogeneity, etc., reside and an *outer, Gouy* or *diffuse layer*, that is by definition ideal, i.e. it obeys Poisson-Boltzmann statistics. This model is due to Stern[1], following older ideas of

---

[1] O. Stern, *Z. Elektrochem.* **30** (1924) 508. Note that this improvement of the Gouy-Chapman theory is almost as old as the Debye-Hückel theory (1923). In 1943 Stern received the Nobel Prize in physics for work on the quantization of magnetic moments.

Helmholtz[1], and has over the decades since its inception rendered excellent services, especially in dealing with experimental systems.

Stern layers can be introduced in categories of proficiency of which three are drawn in fig. 3.20a, b and c, and one in fig. 3.21. Figure 3.20a is the most simple picture: only ion size is accounted for, and only in the first layer. We shall refer to this picture as the *zeroth-order Stern layer*. Even in this simple case the double layer is actually a triple layer. The charge distribution remains ideal, meaning that all the relevant equations of sec. 3.5 remain valid after replacing x by $(x-d)$. The borderline between the Stern layer of thickness d and the diffuse layer is called the *outer Helmholtz plane* (oHp). The charge balance is simply

$$\sigma^o + \sigma^d = 0 \qquad\qquad\qquad [3.6.21]$$

but the potential of the diffuse part ψ^d is lower than when the entire double layer were diffuse because then ψ^d would have been equal to ψ^o. In this case, the Stern layer is charge-free. Therefore, the potential decay in it must be linear. The layer acts as a *molecular condenser* (see also fig. I.5.1), to which a *differential capacitance*

$$C^i = \frac{d\sigma^o}{d(\psi^o - \psi^d)} = -\frac{d\sigma^d}{d(\psi^o - \psi^d)} \qquad\qquad [3.6.22a]$$

and an *integral capacitance*

$$K^i = \frac{\sigma^o}{\psi^o - \psi^d} = -\frac{\sigma^d}{\psi^o - \psi^d} \qquad\qquad [3.6.22b]$$

can be assigned. Generally, C^i and K^i depend on σ^o or ψ^o (because the relative dielectric constant in the Stern layer, ε^i depends on the electric field) but only indirectly on the electrolyte concentration (because it affects σ^o at given ψ^o). Differential and integral capacitances can be interconverted, using [I.5.1.26 and 27]; they become identical if independent of σ^o or ψ^o

Alternatively, at a given σ^o Gauss' theorem [I.5.1.22b], written as

$$(d\psi / dx)_{x=0} = -(\psi^o - \psi^d) / d = -\sigma^o / \varepsilon_o \varepsilon^i \qquad\qquad [3.6.23]$$

may be used, from which, considering [3.6.22b],

$$K^i = \varepsilon_o \varepsilon^i / d \qquad\qquad\qquad [3.6.24]$$

When the quotient ε^i / d is independent of σ^o or ψ^o, K^i is also a constant and hence identical to C^i. From [3.6.21 and 22], considering [3.5.16], it further follows that at given σ^o and ψ^o

---

[1] H. Helmholtz, *Pogg. Ann.* **LXXXIX** (1853) 211.

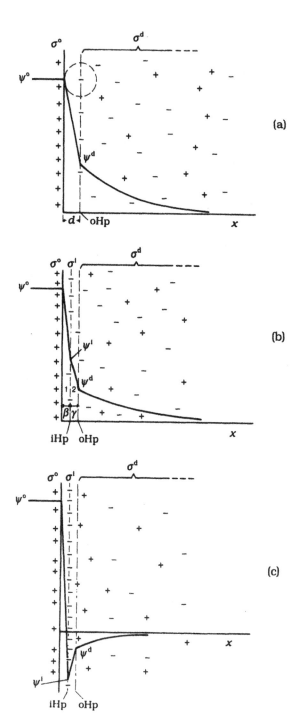

(a)

(b)

(c)

Figure 3.20. Examples of Gouy-Stern layers, considering (a) only finite counterion size, (b) ion size and specific adsorption (c) ion size and superequivalent specific adsorption. All double layers have the same surface potential, but the surface charge increases from (a) to (c).

$$\frac{1}{C} = \frac{1}{C^i} + \frac{1}{C^d} \quad \text{and} \quad \frac{1}{K} = \frac{1}{K^i} + \frac{1}{K^d} \tag{3.6.25}$$

that is, the total double layer capacitance consists of two capacitances in series. The smaller of the two gives the main contribution to the overall capacitance.

In fig. 3.20b *specific adsorption* is also accounted for. The notion of specific adsorption has been defined in sec. 3.3. In disperse systems, its occurrence is *de facto* inferred from the dependence of certain double layer properties on the natures of counter- and co-ions[1]. Generally, ions interacting specifically (non-electrostatically) with the surface approach it to shorter distance ($\beta < d$). The plane where these specifically adsorbed ions reside is called the *inner Helmholtz plane* (iHp)[2]. In colloid science, the model of fig. 3.20b is also known as the *triple layer model*. In this model three charges and three capacitances can be distinguished. For the two inner layer differential capacitances

$$C_1^i = \frac{d\sigma^o}{d(\psi^o - \psi^i)} \tag{3.6.26a}$$

$$C_2^i = \frac{d(\sigma^o + \sigma^i)}{d(\psi^i - \psi^d)} = -\frac{d\sigma^d}{d(\psi^i - \psi^d)} \tag{3.6.26b}$$

The corresponding integral capacitances are

$$K_1^i = \frac{\sigma^o}{\psi^o - \psi^i} = \frac{\varepsilon_o \varepsilon_1^i}{\beta} \tag{3.6.27a}$$

$$K_2^i = \frac{\sigma^o + \sigma^i}{\psi^i - \psi^d} = -\frac{\sigma^d}{\psi^i - \psi^d} = \frac{\varepsilon_o \varepsilon_2^i}{\gamma} \tag{3.6.27b}$$

It must be added that interpretation of capacitances in terms of macroscopic parameters like ε^i's and thicknesses is by no means physically realistic on the scale of one or two molecular diameters.

From [3.6.27a and b], using the charge balance

$$\sigma^o + \sigma^i + \sigma^d = 0 \tag{3.6.28}$$

it follows that

$$\psi^o - \psi^d = \frac{\sigma^o}{K_1^i} - \frac{\sigma^d}{K_2^i} \tag{3.6.29}$$

---

[1] M.A. Habib and J. O'M. Bockris overviewed the issue for polarized interfaces in *Comprehensive Treatise of Electrochemistry*, Vol. 1, J. O'M Bockris, B.E. Conway and E. Yeager, Eds., Plenum Press (1980), 135.
[2] The distinction between iHp and oHp was proposed by R.B. Whitney and D.C. Grahame, *J. Chem. Phys.* **9** (1941) 827.

In this case C^i or K^i cannot be split into the components C_1^i and C_2^i or K_1^i and K_2^i in series. When σ^o and ψ^o are available (from titration) and $\zeta \sim \psi^d$ is known (from electrokinetics) σ^d is also known, and hence ψ^i. One more assumption, or additional piece of information is needed to obtain the two capacitances. As such could serve the shift of the point of zero charge and isoelectric point due to specific adsorption, see the closing remark of sec. 3.8b. Generally, $C_1^i \neq C_2^i$ and $K_1^i \neq K_2^i$, although it is not easy to say by how much they differ because of the problem of assigning a molecular value to the macroscopic parameter ε. The difference between the slopes in layers 1 and 2 is determined by this difference between ε_1^i and ε_2^i and by σ^i

$$\varepsilon_o \varepsilon_1^i \left(\frac{d\psi}{dx} \right)_1 - \varepsilon_o \varepsilon_2^i \left(\frac{d\psi}{dx} \right)_2 = \sigma^i \qquad\qquad [3.6.30]$$

The slope $(d\psi/dx)_1$ in fig. 3.20b is higher than the potential gradient across the Stern layer in fig. 3.20a because at fixed potential ψ^o, σ^o is *higher* than in the situation without specific adsorption. At the same time, ψ^d is *lower*. This is the basis for an otherwise unaccountable observation: at given surface potential (i.e. at fixed pAg, pH, etc.) specific adsorption leads to a higher surface charge but at the same time renders the sol less stable in the colloid sense.

For the triple layer, the counterpart of [3.6.25] is

$$\frac{1}{C} = \frac{1}{C_1^i} - \left(\frac{1}{C_2^i} + \frac{1}{C^d} \right) \frac{d\sigma^d}{d\sigma^o} \qquad\qquad [3.6.31]$$

where C_1^i and C_2^i may be replaced by K_1^i and K_2^i, respectively, and $d\sigma^d/d\sigma^o$ by σ^d/σ^o, if these capacitances are constant (invariant with σ^o). Equation [3.6.31] is readily found by writing in $C^{-1} = d\psi^o/d\sigma^o$ the potential as $(\psi^o - \psi^i) + (\psi^i - \psi^d) + \psi^d$, carrying out the differentiation after changing variables $(\sigma^o \rightarrow \sigma^d)$ in the last two terms. Other relations may be derived by making certain simplifying assumptions, see below. It is repeated that the situation cannot be simply represented by three capacitances in series.

Figure 3.20c is similar to fig. 3.20b, but now the specific counterion adsorption is *superequivalent*: there is more negative charge in the IHP than required to compensate the surface charge. If this occurs, we have definite proof of specific adsorption. Superequivalency leads, in addition to a further increase of σ^o, to a sign reversal of ψ^d, which can be electrokinetically monitored, and which in colloid stability recurs as so-called *irregular coagulation series*: sols that are coagulated by low amounts of specifically adsorbing electrolyte can be restabilized at higher concentrations of that electrolyte. This reversal is loosely called *charge reversal* but the reversal applies only to the diffuse charge, not to the surface charge, which even becomes more positive.

Before discussing the Stern theory further, let us ask what can be measured and how realistic the various models are. First, defining a relative dielectric permittivity for a layer of molecular thickness is precarious. As the thickness of the condenser is also open to discussion it has little sense to go into more detail than just leaving the quotient $\varepsilon_0 \varepsilon / d$, i.e. the capacitance as a variable. The zeroth-order Stern model, fig. 3.20a, requires only one parameter in addition to those of the diffuse layer, viz. C^i or K^i. These capacitances are accessible from measurements at high salt concentration. Then C^d and K^d become very high, see fig. 3.5. On the other hand, C^i and K^i are, by their natures, hardly dependent on c_{salt}, so that $C(\text{total}) \rightarrow C^i$, according to [3.6.25]. For mercury and silver iodide C^i and K^i are now well established at all σ^0. Close to the point of zero charge they amount to about 25-32 μF cm^{-2}, see figs. 3.43 and 3.50.

In the triple-layer model, at high c_{salt}, C_d^{-1} vanishes but in [3.6.31] the two capacitances and $d\sigma^d / d\sigma^o$ remain. In addition, it has to be verified to what extent these capacitances are independent of σ^0. At least three adjustable parameters remain, which seem more than adequate to 'explain' the most recalcitrant surface charge curve. For the mercury system, where very refined capacitance measurements can be made, this is appropriate, but for unruly systems like oxides we easily run the risk of overinterpretation.

A great difficulty is to assess ψ^i, which is required to formulate an adsorption isotherm for counterions. This quantity is rather esoteric. Sometimes ψ^i is identified with ψ^o, but this is a very poor approximation. Somewhat better, simpler models, somewhere between the zeroth and first order have been proposed. Two of them are:

(i) Ignore the break in $d\psi / dx$ at the inner Helmholtz plane. Then ψ^i is simply related to ψ^o and ψ^d as

$$\psi^i = (\gamma \psi^o + \beta \psi^d)/d \qquad\qquad [3.6.32]$$

(ii) Assume that specific adsorption takes place at the outer Helmholtz plane. This means that ψ^i is identified with ψ^d. See fig. 3.21. For this case equation [3.6.31] reduces to

$$\frac{1}{C} = \frac{1}{C^i} - \frac{1}{C^d} \cdot \frac{d\sigma^d}{d\sigma^o} \qquad\qquad [3.6.33]$$

Other assumptions, such as setting $\varepsilon_2^i = \varepsilon$ (bulk) do not lead to simpler formulas.

The approximations involved in simplifications (i) and (ii) are not too far-fetched. Discreteness of charge, with the ensuing multi-imaging, tends to

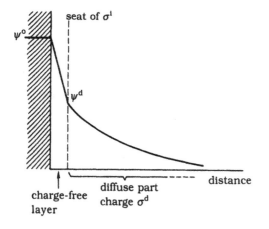

Figure 3.21. Gouy-Stern double layer model with specific adsorption at the outer Helmholtz plane. The inner layer has a constant capacitance C^i.

straighten the potential-distance decay[1, 2]. Assumption (ii) is not unreasonable because in practice ψ^i is closer to ψ^d than to ψ^o. The error made in the Coulomb attraction energy tends to be less than 1 RT per mole whereas for strongly specifically adsorbing counterions the non-electrostatic energy may amount to several RT. On the other hand, for weakly adsorbed ions the relative error is larger, but then the specifically adsorbed amount becomes very small, so that the absolute error remains small. All told, the best way to evaluate ψ^i remains via the establishment of C_1^i and C_2^i from the shifts in the i.e.p. and p.z.c., respectively. Illustrations follow in sec. 3.6d and 3.8b.

3.6d. Specific adsorption of ions

Specific adsorption of ions is their adsorption by non-electrostatic forces. By this mechanism, ions can accumulate on a surface even against electrostatic repulsion. The non-electric Gibbs energy of adsorption generally depends on the natures of the ion and the surface, hence the term "specific". In practice, sometimes situations are met where ions do not specifically adsorb on an uncharged surface (and hence are neither able to shift the point of zero charge nor to adsorb superequivalently), but do so once there are charges on the surface. Alkali ions on silver iodide (fig. 3.41) are examples. Where appropriate, we shall call this type *specific adsorption of the second kind*.

To complete the Stern theory, a model is required to determine the specifically adsorbed charge, σ^i at each σ^o. In other words, one needs an *adsorption isotherm equation*. A number of such isotherms have been derived for uncharged

[1] D.C. Grahame, Z. *Elektrochem.* **62** (1957) 264.
[2] S. Levine, J. Mingins and G.M. Bell, *J. Electroanal. Chem.* **13** (1967) 280.

molecules; they are collected in appendix 1. Adsorption of small ions is now at issue. Larger organic molecules, charged or uncharged, will be considered under *electrosorption*, sec. 3.12.

Theoretically, the extension of the appropriate equations of appendix 1 to ions is straightforward: simply an electrostatic contribution $z_i F\psi^i$ has to be added to the non-electrostatic adsorption Gibbs energy. Multilayer specific adsorption of ions does not have to be considered: ions beyond the Stern layer are (by definition) generically adsorbed. As we mostly consider surface charges, residing on certain sites at the surfaces of the colloidal particles, localized adsorption is the most likely mechanism; in practice this is almost exclusively considered. Lateral interaction is, because of the long range of the electrostatic forces and the usually low degrees of occupancy, dominated by electrostatics and in the mean field treatment accounted for by a $z_i F\psi^i$ term. Under these conditions the *specific* or *non-electrostatic adsorption Gibbs energy* is only determined by the ion-surface interaction. We denote it as $\Delta_{ads}G_{mi}$ per mole of i adsorbed. It is a Gibbs energy, in that the specific binding also has an entropic component, particularly so if the binding is solvent structure-originated. However, the configurational entropy of the ions in the adsorbate is not included in $\Delta_{ads}G_{mi}$. We already encountered this issue in the Langmuir and Frumkin-Fowler-Guggenheim (FFG) treatments of uncharged adsorbates. An *intrinsic binding constant* K_i is introduced as[1]

$$K_i \equiv \exp\left(-\Delta_{ads}G_{mi} / RT\right)$$ [3.6.34]

or

$$pK_i = \Delta_{ads}G_{mi} / 2.303\,RT$$ [3.6.35]

Because of the low coverages, it is often a good approximation to assume K_i to be constant. However, the total Gibbs energy of adsorption is not constant because the electrostatic part changes with σ^o. Equations [3.6.34 and 35] require K_i to be dimensionless. When K_i is not dimensionless and nevertheless these equations are used, the implication is that $\Delta_{ads}G_{mi}$ is referred to an (arbitrary) reference, determined by the concentration units, see sec. I.2.21. There we introduced the symbol K^c for not dimensionless K's, with the concentrations expressed as molarities.

Sometimes the notion of *specific adsorption potential* is used, defined as $\Phi_i = \Delta_{ads}G_{mi} / z_i F$, to obtain a quantity with dimensions of a potential. In the present book this term will be avoided.

---

[1] Note that in some equations this K_i may occur together with K^i, the integral capacitance of the Stern layer.

In formulating adsorption isotherm equations, assumptions have to be made about the kinds of ions that bind, and on the planes where they adsorb. Stern himself considered the specific adsorption of cations and anions, both at the outer Helmholtz plane (approximation (ii) in the previous subsection)[1]. More likely are situations where only one ionic type adsorbs at the inner Helmholtz plane. For that case, the Langmuir equation is readily extended. We start with [A1.2a], which we write as $\theta_i/(1-\theta_i) = K_i x_i$. Here, $\theta_i = N_i/N_s$ is the ratio between the number of specifically adsorbed ions and the number of sites on the surface and x_i is the mole fraction of ions of type i. We prefer this to c_i because now K_i is dimensionless, as [3.6.34] requires. For a charged adsorbate the electrostatic term has to be added, i.e. K_i becomes $K_i \exp(-z_i F\psi^i/RT) = K_i \exp(-z_i y^i)$. Making θ_i explicit and introducing $\sigma^i = z_i e N_i$, we obtain

$$\sigma^i = \frac{z_i e N_s K_i \exp(-z_i y^i) x_i}{1 + K_i \exp(-z_i y^i) x_i}$$

[3.6.36]

The N_s adsorption sites for specifically adsorbing ions are not necessarily identical to those for surface ions. The specifically adsorbed charge can be smaller or larger than the surface charge.

A variant of [3.6.36] applies to the case where, say positive, charges on the surface act as the sites where specific adsorption of anions may occur, i.e. when the specific adsorption is of the second kind. Then ion pairs are formed, held together by electrostatic and non-electrostatic interactions. Defining θ_i as the fraction of σ^o that is compensated by σ^i, $\theta_i < 1$ and obeys

$$\frac{\theta_i}{1-\theta_i} = K_i x_i \exp(-z_i y^i)$$

[3.6.37]

which is the electrostatic equivalent of the FFG equation, [A1.5a]. In this case $\sigma^d = (1 - \theta_i)\sigma^o$.

In summary, it is not difficult to formulate Stern adsorption isotherm equations. The main problem is to determine y^i, for which assumptions have to be made.

First there is the assumption of the mean field, then the localization of the inner Helmholtz plane is at issue. The quality of these models is not easily assessed, but ultimately comparison with the experiment is decisive.

Some elaborations are given in fig. 3.22. Figure 3.22a illustrates, in line with [3.6.25], that for two capacitances in series the lower one plays the greater role in determining the overall value. For $C^i \to \infty$ the purely diffuse limit is recovered, identical to the 10^{-2} M curve of fig. 3.5. Integration gives the $\sigma^o(y^o)$

[1] O. Stern, Z. Elektrochem. 30 (1924) 508.

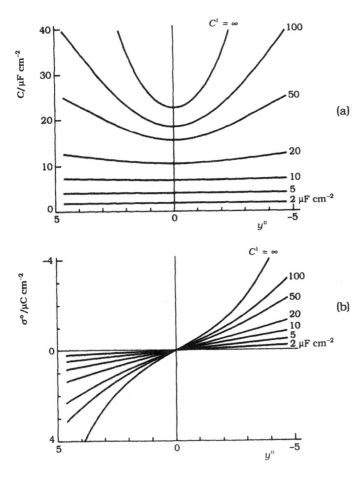

Figure 3.22. Zeroth-order Stern theory. Differential capacitances (a) and surface charges (b) for various values of the (charge-free) inner layer capacitance, C_i. Salt: 10^{-2} M (1-1) electrolyte. Temperature, 25°C.

curves of fig. 3.22b. The curve with $C^i = \infty$ corresponds to the 10^{-2} M curve in fig. 3.4, except for the sign ($\sigma^o = -\sigma^d$ in this case). The lower C^i, the more σ^o is suppressed. For higher electrolyte concentrations (not shown), C^i plays a relatively more important role because then the countercharge tends to accumulate in the inner part. In [3.6.25] $C^d \sim \sqrt{c}$, see [3.5.37], so that, for high c, C_d^{-1} vanishes, hence $C \rightarrow C^i$; in practice this is the means of obtaining C^i.

Cases in which specific adsorption is accounted for require adsorption isotherm equations and hence models. A selection of results is given in figs. 3.23-25. In all these graphs the parameter K_i^c times c_i (in M) equals $K_i x_i$ in [3.6.37], the values $K_i^c = 1$ and 10 dm^3 mol^{-1} corresponding with $\Delta_{ads}G_{mi} = -4.02$ and -6.32 RT, respectively.

Figures 3.23a-d illustrate the properties of the double layer for the simple case of adsorption at the outer Helmholtz plane, as in fig. 3.21. A capacitance of 20 μF cm^{-2} (figs. a and b) is more representative for hydrophobic surfaces, whereas that of 100 μF cm^{-2} (figs. c and d) is more typical for hydrophilic ones.

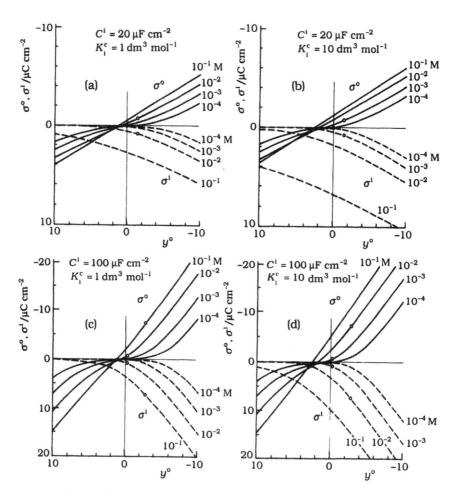

Figure 3.23. Surface charge for a Gouy-Stern layer with specific adsorption of the cation at the outer Helmholtz plane for four combinations of the inner layer capacitance $C^i = K^i$ and specific adsorption energy (via K_i^c). —— surface charge, - - - specifically adsorbed charge. The circles in the curves indicate charge equivalency $\sigma^o = -\sigma^i$. σ^o(max) = 80 μC cm^{-2}, T = 298.16 K, (1-1) electrolyte, concentration given.

In the latter case at given surface potential higher absolute values of the surface charge *and* of the specifically adsorbed charge develop. The latter of course also increases with K_i^c, and therefore, indirectly, the surface charge is also raised.

A new feature is that a common intersection point is no longer found. The *point of zero charge moves to the left* with increasing specific adsorption of cations. This shift is almost independent of the inner layer capacitance. Close to the point of zero charge the specific adsorption is superequivalent, as in fig. 3.20c. Making the surface potential more negative, a point is reached where

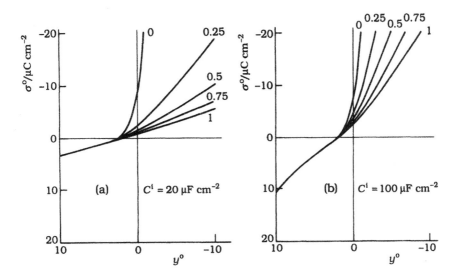

Figure 3.24. Surface charge for a Gouy-Stern layer with specific adsorption at an inner Helmholtz plane of varying positions. The numbers at the curves equal $\beta/(\beta+\gamma)$ in fig. 3.20b. 10^{-2} M (1-1) electrolyte, $K_i^c = 10$ dm^3 mol^{-1}, $T = 298.16$ K, σ^o(max) $= z_i e N_s = 80$ μC cm^{-2}.

$\sigma^o = -\sigma^i$, indicated by circles. At this point $\sigma^d = 0$, i.e. this is virtually the *isoelectric point*. It is concluded that with increasing specific adsorption the p.z.c. and i.e.p. move in different directions. In sec. 3.8b this will be discussed in more detail.

Figure 3.24 shows how, at given specific adsorption energy (via K_i^c) and at given concentration, the surface charge increases when the plane where the specifically adsorbing cations reside (i.e. the inner Helmholtz plane) is moved inward from coinciding with the outer Helmholtz plane to the surface. For lower Stern layer capacitances the position of the inner Helmholtz plane is a more critical parameter than when this capacitance is high.

Figure 3.25 is identical to fig. 3.23, except that the inner Helmholtz plane is now situated at 40% of the outer Helmholtz plane. Qualitatively the same features emerge, but there are quantitative differences with generally more pronounced trends.

The overall conclusion is that even with these relatively simple models a large number of $\sigma^o(y^o)$ curves and accompanying features (like shifts in the zero point of charge and isoelectric point) can be interpreted.

3.6e. Site binding models for the surface charge

In the Poisson-Boltzmann model the surface charge is considered homogeneous and smeared-out. This assumption is retained in many of the more

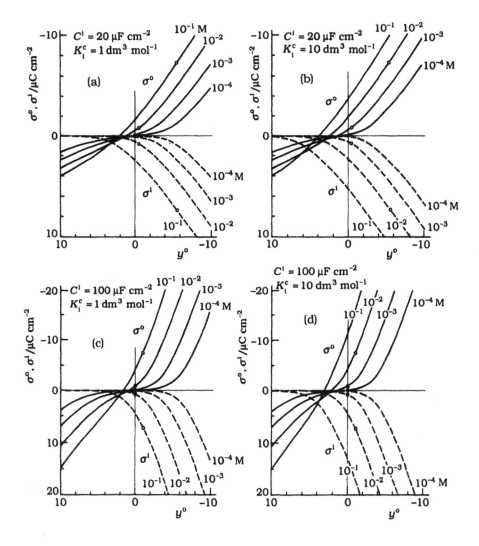

Figure 3.25. As figure 3.23, but with $\beta / (\beta + \gamma) = 0.4$.

advanced treatments of the solution side of the double layer. However, many surface charges have a typically discrete nature. This is so for instance with oxides and latices with covalently bound charged groups. At issue is now how much difference this makes and how one can account for it.

The extent of discreteness depends on the mobility of the surface charges and on their mutual separations. Very rapidly moving charges behave as if the surface charge is smeared-out. This situation is encountered with highly conducting materials like mercury. For less mobile charges, like those on silver iodide and semi-conductors, the assumption of a smeared-out surface charge remains tenable when equilibrium experiments are considered, like titrations. These have a long time scale with respect to that for surface motions (i.e. they have high Deborah numbers).

For the cases of oxides and latices, as mentioned above, there is little or no mobility of the surface charges. In such systems the mutual distance between the charges (ℓ) becomes a characteristic parameter but it is not the absolute value of ℓ that counts, but its relation to the thickness of the diffuse part of the double layer, characterized by κ^{-1}. For $\kappa\ell \ll 1$ the diffuse layer is so much thicker than ℓ that the assumption of a smeared-out surface charge remains tenable, whereas for $\kappa\ell \gg 1$ the charges are relatively so far apart that around each of them a hemispherical diffuse layer is formed. As κ^{-1} is the yardstick, which decreases with c_{salt}, it follows that the surface charge has a more discrete nature at higher indifferent electrolyte concentration.

In the framework of a Stern plus diffuse double layer model the logic is to ignore the discreteness of the surface charge in the diffuse part, but consider it in the Stern layer. However, in practice, usually the Bragg-Williams (mean field) approximation is also made for the surface potential, that is, each charge-determining ion is assumed to experience the same mean potential ψ^o, although for the charges a *site-binding model* is assumed, meaning that the surface charge is modelled as being distributed over a number of surface sites. In other words, this charge is modelled as localized. One of the aims is to account for the non-electrostatic binding (Gibbs) energy and the configurational part of the adsorption entropy. In this way the equivalent of the FFG isotherm for charged systems is obtained. At the same time the influence of neighbouring groups on the dissociation at a given site, also known as the *polyelectrolyte effect*, is accounted for because the potential ψ^o is also determined by the dissociation of adjoining groups. However, the short-range non-electrostatic interaction between charge-determining ions and surface sites is considered independent of the state of dissociation or association of neighbours. One would expect site-binding to be closer to reality for, say latices with covalently bound surface groups, than for mercury.

We shall now elaborate the prototype of the *site-binding model* taking oxides as the examples.

The surface is considered flat and homogeneous as far as the adsorption sites are concerned. By way of illustration, protons and hydroxyl ions are taken as the charge-determining ions. The logic of establishing this couple is purely phenomenological: it may well be that only the proton is charge-determining, but in the practice of potentiometric surface charge determination, protons can only be removed by adding a base so that operationally desorption of a proton is seen as adsorption of a hydroxyl ion. For *insoluble* oxides the high-valency ions in the solid phase (Si^{4+} for silica, Fe^{3+} for haematite, etc.) do not have to be considered; these ions are embedded in their appropriate coordination by oxygens and, at the surface, by hydroxyls.

Surface hydroxyls are written as ROH: the R stands for the solid or for an atom inside the solid, to which the hydroxyl ion is bound. They can become negative by donating a proton (the acid function) or positive by adsorbing one (the base function). The two pertaining dissociation equilibria are characterized by the acid and base dissociation constants K_a and K_b, respectively. Hence we have

$$ROH \rightleftarrows RO^- + H^+ \qquad\qquad K_a = \frac{[RO^-]x^s_{H^+}}{[ROH]} \qquad\qquad [3.6.38a]$$

$$ROH + H_2O \rightleftarrows ROH_2^+ + OH^- \quad K_b = \frac{[ROH_2^+]x^s_{OH^-}}{[ROH]x_w} = \frac{[ROH_2^+]}{[ROH]}\frac{K^s_w}{x^s_{H^+}} \qquad [3.6.38b]$$

Here, the square brackets denote surface concentrations in arbitrary units. The superscript s refers to the surface. The ion product of water, $K^s_w = x^s_{H^+}x^s_{OH^-} = x_{H^+}x_{OH^-}$, see [3.6.41], so we may henceforth drop the superscript s. To keep the K's dimensionless we write them in terms of mole fractions, x, rather than concentrations or activities. Alternatively, when the mole fractions are replaced by concentrations, the K's acquire a dimension. For instance, instead of [3.6.38a and b] we can write

$$K^c_a = \frac{[RO^-]c^s_{H^+}}{[ROH]} \qquad\qquad K^c_b = \frac{[ROH_2^+]c^s_{OH^-}}{[ROH]\,c_w} = \frac{[ROH_2^+]K^c_w}{[ROH]\,c^s_{H^+}} \qquad [3.6.38c,d]$$

In the literature several variants of these three equations can be found, for instance by exchanging the left-hand and right-hand sides of [3.6.38a and b]. Consequently, literature data on K_a and K_b may differ from ours, although their relation to the p.z.c. (see [3.6.43]) should be fully defined. In the practice of potentiometric titration only the ratio between K_a and K_b can be measured. A basically different model is the "one pK model". In this model each site is considered as amphoteric, with the charge equally distributed over the H_2O molecules and the OH^- ions coordinated to a central cation. For instance, in the complex $[Al(OH)_3(H_2O)_3]$ the OH's are assigned a charge of $-\frac{1}{2}$ and each H_2O a charge $+\frac{1}{2}$ unit. For surface groups of metals a similar assumption is made. The basic charging reaction can be written

$$ROH^{-1/2} + H^+ \rightleftarrows ROH_2^{+1/2} \qquad\qquad\qquad [3.6.39]$$

with only one equilibrium constant, which we shall call K_{ab}, given by

$$K_{ab} = \frac{\left[ROH_2^{+1/2}\right]}{\left[ROH^{-1/2}\right]x_{H^+}^s} \qquad\qquad [3.6.40]$$

Mathematically, this is a simpler model at the expense of suppressing the phenomenological equivalence of protons and hydroxyl ions. Later in this section and in sec. 3.6g, we return to this, and other alternatives.

The next assumption is that in [3.6.38a and b] $x_{H^+}^s$ and $x_{OH^-}^s$ differ from their bulk values only because the potentials are different, i.e. according to the Boltzmann rule [3.5.4]. Writing y^o for $F\psi^o/RT$,

$$x_{H^+}^s = x_{H^+} e^{-y^o} \qquad x_{OH^-}^s = x_{OH^-} e^{y^o} \qquad\qquad [3.6.41a, b]$$

or

$$c_{H^+}^s = c_{H^+} e^{-y^o} \qquad c_{OH^-}^s = c_{OH^-} e^{y^o} \qquad\qquad [3.6.41c,d]$$

In passing it is noted that choosing the same y^o for protons and hydroxyls is a good approximation if only the proton, with its negligible size, is charge-determining. The assumption is poorer when the more bulky OH^- ion would adsorb as such. It is implicit in the use of Boltzmann's law in the form [3.5.4] that activity corrections are assumed identical in the bulk and on the surface. Hence, in this approximation the x's in [3.6.41] may be either interpreted as activities or as concentrations. Therefore, within the present approximations,

$$pH = -\log(x_H / V_m) = -\log c_{H^+} \qquad\qquad [3.6.42]$$

where V_m is the molar volume of water. There is no need to discriminate between $-\log c_{H^+}$ and $-\log a_{H^+}$. Experimentally, titrations are usually carried out in excess electrolyte where, say $d\mu_{HNO_3}$, in excess KNO_3 equals $d\mu_{H^+} \approx RT\, d\ln c_{H^+}$ because f_{H^+} is essentially determined by the swamping KNO_3 and is therefore fixed (see sec. 3.7).

Interpreting [RO^-] and [ROH_2^+] as numbers of moles of surface charges per unit area, we find for the surface charge

$$\sigma^o = F\big([ROH_2^+] - [RO^-]\big) = F[ROH]\left\{ \frac{K_b x_H e^{-y^o}}{K_w} - \frac{K_a e^{y^o}}{x_H} \right\} \qquad\qquad [3.6.43]$$

which establishes a relation between σ^o and pH, or rather translates it in terms of y^o.

From [3.6.43] the p.z.c. can be established if under conditions where $\sigma^o = 0$ also $y^o = 0$. As will be discussed in sec. 3.9 this is not generally true for the

Galvani potential difference, although in the present model, where y^o is only that part of the potential that is due to free charges, the step is acceptable.

$$x^o_{H^+} = \left(K_a K_w / K_b\right)^{1/2} \qquad \text{or} \qquad c^o_{H^+} = \left(K^c_a K^c_w / K^c_b\right)^{1/2} \qquad \text{[3.6.44a,b]}$$

or

$$pH^o = \log V_m + \frac{1}{2} p K_a - \frac{1}{2} p K_b + \frac{1}{2} p K_w \qquad \text{[3.6.45a]}$$

$$= \frac{1}{2} p K^c_a - \frac{1}{2} p K^c_b + \frac{1}{2} p K^c_w \qquad \text{[3.6.45b]}$$

where the superscript o indicates the p.z.c. Recall that the K's are dimensionless, but the K^c's are not.

Notwithstanding the assumptions made in its derivation, [3.6.45] is useful. It is seen that the zero point is related to the *difference* ΔpK between pK_a and pK_b, not to their absolute values. For relatively low pK_a and high pK_b, i.e. for acid oxides, pH^o is low, and vice versa. In the one pK model $pH^o = -pK_{ab}$.

The above models further serve as starters for a variety of extensions. One of these is to assume that the surface has two different types of sites, of which one (AH) can be charged negatively, the other (B) positively, say according to

$$AH \leftrightarrows A^- + H^+ \qquad\qquad\qquad B + H_2O \leftrightarrows BH^+ + OH^- \qquad \text{[3.6.46a, b]}$$

Such surfaces may be called *zwitterionic*. Representative systems are proteins or latices with on their surface carboxyl (–COOH) and amino groups (–NH$_2$).

By analogy to the previous analysis, for this case we find

$$\sigma^o = F[B] \frac{K_b x_{H^+} e^{-y^o}}{K^s_w} - F[AH] \frac{K_a e^{y^o}}{x_{H^+}} \qquad \text{[3.6.47a]}$$

or

$$\sigma^o = F[B] \frac{K^c_b c^s_{H^+}}{K^{cs}_w} - F[AH] \frac{K^c_a}{c^s_{H^+}} = F[B] \frac{K^c_b c_{H^+} e^{-y^o}}{K^{cs}_w} - F[AH] \frac{K^c_a e^{y^o}}{c_{H^+}} \qquad \text{[3.6.47b]}$$

For the point of zero charge,

$$pH^o = \log V_m + \frac{1}{2}\left(p K_a - p K_b + p K_w\right) - \frac{1}{2} \log\left([AH]/[B]\right) \qquad \text{[3.6.48a]}$$

or

$$pH^o = \frac{1}{2}\left(p K^c_a - p K^c_b + p K^c_w\right) - \frac{1}{2} \log\left([AH]/[B]\right) \qquad \text{[3.6.48b]}$$

In this case the point of zero charge not only depends on the strength of the two groups but also on their numbers, a result that is intuitively expected and is in line with the known behaviour of proteins.

Other extensions and variants include

(i) consider more surface groups

(ii) include specific adsorption. The binding of an ion to a site can be characterized by [3.6.34]. Recall that our term "specific adsorption" is general and covers all interactions between weak (like solvent structure-originated) to strong (like chemical binding), i.e. it includes ligand exchange and complex formation.

(iii) distinguish between sites for charge-determining and for specifically adsorbing ions, or consider only specific adsorption of the second kind.

(iv) combine site binding models with purely diffuse double layers and/or with Stern layers.

We shall come back to some implications of this modelling in sec. 3.6g. Before finishing this section we mention two elaborations of wider applicability.

The first is that [3.6.45] can be differentiated with respect to the temperature. When V_m is constant, this leads to

$$\frac{dpH^\circ}{dT} = \frac{0.434}{2}\left(\frac{-\Delta_a H + \Delta_b H - \Delta_{diss} H_w}{RT^2}\right) \qquad [3.6.49]$$

where $\Delta_a H$, $\Delta_b H$ and $\Delta_{diss} H_w$ are the enthalpies of reactions [3.6.38a], [3.6.38b] and the dissociation enthalpy of water, respectively.

This equation can be simplified by realizing that $\Delta_a H + \Delta_b H = \Delta_{diss} H_w$ $= -\Delta_{neutr} H_w$: if first a proton is adsorbed and next a hydroxyl ion, the only resulting enthalpy change is that for the dissociation of a water molecule. Hence, one can either eliminate $\Delta_a H$ or $\Delta_b H$. As for many oxides there is spectroscopic evidence that adsorption and desorption of protons is the sole charging mechanism, we eliminate $\Delta_b H$ with the result

$$\frac{dpH^\circ}{dT} = -0.434\,\frac{\Delta_a H}{RT^2} = 0.434\,\frac{\Delta_{ads} H_{H^+}}{RT^2} \qquad [3.6.50]$$

Hence, the binding enthalpy of the proton can be obtained from the shift in the point of zero charge with temperature. The more basic an oxide the higher $\Delta_{ads} H_{H^+}$ is and, hence, the stronger the p.z.c. decreases with temperature. More basic oxidic surfaces also have a relatively high p.z.c., hence we arrive at the simple rule that for oxides (and materials comparable in these respects) *the higher the p.z.c. the more negative its temperature coefficient is.* This rule has been experimentally corroborated, see sec. 3.8c.

The other elaboration applies to monofunctional surfaces (say adsorbents with only carboxyl groups, that may be either dissociated or not). Then the *degree of dissociation* α becomes a logical variable. We define α through

$$\alpha = [RO^-]/([RO^-] + [ROH])$$ [3.6.51]

Using [3.6.38a, 41a and 42] one may write

$$pH = pK_a' + \log \frac{\alpha}{1-\alpha}$$ [3.6.52]

with $pK_a' = -\log(V_m K_a e^{-y^o})$, K_a being the intrinsic dissociation constant. When K_a is written in terms of concentrations, the factor V_m has to be dropped, i.e. then

$$pH = pK_a^{c'} + \log \frac{\alpha}{1-\alpha}$$ [3.6.53]

with $pK_a^{c'} = -\log(K_a e^{-y^o})$. The essential difference between pK_a' and pK_a is in the factor containing y^o, which accounts for the polyelectrolyte effect. Equation [3.6.52 or 53] is the *Henderson-Hasselbalch equation*, already encountered before, see [I.5.2.34]. In this equation pK_a varies with pH because y^o is a function of α. A semi-empirical variant, proposed by Fisher and Kunin and others[1],

$$pH = pK_{app} + n \log\left(\frac{\alpha}{1-\alpha}\right)$$ [3.6.54]

was found to apply to a variety of polyelectrolytes over a relatively wide range of α, with pK_{app} and n constant. Now n is a measure of the polyelectrolyte effect: the more n exceeds unity, the stronger the electrical interaction between the dissociated groups. Because of its empirical nature, no phenomenological meaning can be assigned to pK_{app}. Equation [3.6.54] has proven its use in the detection of, for instance, phase transitions in polyelectrolytes and in the adsorption of proteins and might well be of further value.

3.6f. The Gibbs energy of a Gouy-Stern layer

In practice, double layers are rarely entirely diffuse, and to analyse such features as the lyotropic sequence in, and the influence of adsorbates on, colloid stability, it is necessary to also compute the Gibbs energy ΔG_a^σ for Gouy-Stern layers.

---

[1] S. Fisher, R. Kunin, *J. Phys. Chem.* **60** (1956) 1030; A. Katchalski, P. Spitnik, *J. Polym. Sci.* **2** (1947) 437.

Qualitatively, the zeroth-order Stern layer (fig. 3.20a) differs from the purely diffuse model in that the screening is poorer. Higher potentials are needed to obtain a certain surface charge. When there is specific adsorption (figs. 3.20b and c) a new feature emerges: the non-electrostatic contribution to the (Gibbs) energy of adsorption of counterions leads to an additional contribution to ΔG_a^σ. It is recalled from sec. 3.2 that double layers form only because of non-electrostatic affinities of ions for a surface.

Quantitatively, we have in principle [3.2.3] for the purely diffuse case, which in the zeroth-order Stern model can be modified to account for the fact that ψ can now maximally become ψ^d. The difference between ψ^d and ψ^o is dictated by the Stern capacitance C^i. For the specific adsorption case, [3.3.9] can be used if σ^i is known as a function of σ^o. To this end, one of the Stern equations [3.6.36 or 37] can be used.

However, in practice it is more expedient and transparent to decompose ΔG_a^σ into its four components, of which two are electrical and two non-electrical. To that end it is recalled from sec. I.5.7 that for a purely diffuse counterlayer ΔG_a^σ consists of two contributions

$$\Delta G_a^\sigma = \Delta G_a^\sigma(\text{el}) + \Delta G_a^\sigma(\text{non-el}) = \int_0^{\sigma^o} \psi^{o'} \, d\sigma^{o'} + \Delta G_a^\sigma(\text{non-el}) \qquad [3.6.55]$$

The electrical contribution is always positive. For a clay-plate surface it is the only term and this positive sign reflects the fact that such an interface is not relaxed. The non-electrical term is negative and, for relaxed interfaces, the sole reason why double layers exist. The primes indicate the varying values of ψ^o and σ^o during charging. In practice, the process can be visualized as starting at the point of zero charge, from where the surface is isothermally and reversibly charged by adding electrolyte containing charge-determining ions. For homogeneous surfaces all infinitesimal steps $d\Delta G_a^\sigma(\text{non-el})$ are identical provided in each step the same (infinitesimal) number of ions is transferred. The last step just equals $-\psi^o d\sigma^o$, because electrostatic and non-electrostatic contributions balance. Hence, $\Delta G_a^\sigma(\text{non-el}) = -\psi^o \sigma^o$, combination with $\Delta G_a^\sigma(\text{el})$ and partial integration yields [3.2.3].

Extending this approach to the case of specific adsorption at the outer Helmholtz plane, modelled in fig. 3.21, we arrive at four terms

$$\Delta G_a^\sigma = -\int_0^{-\sigma^d} \psi^{d'} \, d\sigma^{d'} + \int_0^{\sigma^o} (\psi^{o'} - \psi^{d'}) \, d\sigma^{o'} + \Delta G_{a,cd}^\sigma(\text{non-el}) + \Delta G_{a,i}^\sigma(\text{non-el})$$

$$[3.6.56]$$

The first two integrals, representing the purely electric contributions, can be understood by visualizing the charging process to occur in two steps. First, a charge $(\sigma^o + \sigma^i) = -\sigma^d$ is brought to the outer Helmholtz plane; the (positive) electrical work is represented by the first integral. Second, part of this charge, to become the surface charge, is transported from there to the surface, for which the second integral, also positive, is the electrical work involved. The non-electric contributions have been discriminated between those for the charge-determining (cd) and specifically adsorbing ion. As before, $\Delta G^\sigma_{a,cd}(\text{non-el}) = -\sigma^o\psi^o$ but the new item is $\Delta G^\sigma_{a,i}(\text{non-el}) = -\sigma^i\psi^d$, because specific adsorption at the outer Helmholtz plane also proceeds until balanced by the opposing electrical contribution. Hence, generally

$$\Delta G^\sigma_a = -\int_0^{-\sigma^d} \psi^d \, d\sigma^{d'} + \int_0^{\sigma^o} (\psi^{o'} - \psi^{d'}) \, d\sigma^{o'} - \sigma^o\psi^o - \sigma^i\psi^d \qquad [3.6.57]$$

By virtue of charge balance [3.6.28] it is readily verified that this expression is consistent with [3.3.9]. When the differential capacitance of the inner layer (see [3.6.22a]) is known, the second integration can be carried out. For a constant capacitance (implying $K^i = C^i$) the result is

$$\Delta G^\sigma_a = -\int_0^{-\sigma^d} \psi^d \, d\sigma^{d'} + \frac{(\sigma^o)^2}{2C^i} - \sigma^o\psi^o - \sigma^i\psi^d \qquad [3.6.58]$$

In the zeroth-order approximation $\sigma^i = 0$, $\sigma^d = -\sigma^o$ and equation [3.6.57] converts to

$$\Delta G^\sigma_a = -\int_0^{-\sigma^d} \psi^d \, d\sigma^{d'} + \int_0^{\sigma^o} (\psi^{o'} - \psi^{d'}) \, d\sigma^{o'} - \sigma^o\psi^o = -\int_0^{\psi^o} \sigma^o \, d\psi^{o'} \qquad [3.6.59]$$

This result is identical to [3.2.3], for purely diffuse double layers. In words, the Gibbs energy for a double layer with a charge-free inner layer, but with specific adsorption, is the same as that for a purely diffuse layer, the quantitative difference being that, at given ψ^o, σ^o is lower. No additional terms are needed for the charge-free layer because all ions are diffuse and, hence, do not contribute. Equation [3.6.58 or 59] can in this approximation also be written as

$$\Delta G^\sigma_a = \int_0^{\psi^d} \sigma^{d'} \, d\psi^{d'} - \frac{(\sigma^o)^2}{2C^i} \qquad [3.6.60]$$

This is a useful formula because the integral is the equation for a purely diffuse layer ($\sigma^{o'} = -\sigma^{d'}$ in this case); the second term on the r.h.s. modifies it for the charge-free layer. For systems like AgI where ψ^o and σ^o are both known, C^i is also available and ΔG_a^σ can therefore be evaluated.

The purely diffuse limit is obtained for $C^i \rightarrow \infty$ and $\sigma^i = 0$. Writing in [3.6.58] $\psi^{d'} = \psi^{o'} - \psi^{d'} + \psi^{d'}$, it is immediately seen that then also [3.2.3] is recovered.

For specific adsorption at the inner Helmholtz plane (fig. 3.20b or c) we simply extend the charging process leading to [3.6.38]. First, $\sigma^o + \sigma^i = -\sigma^d$ is brought to the outer Helmholtz plane, then $\sigma^o + \sigma^i = -\sigma^d$ is moved to the inner Helmholtz plane and finally σ^o is transported from there to the surface. The result is

$$\Delta G_a^\sigma = -\int_0^{-\sigma^d} \psi^{d'} \, d\sigma^{d'} - \int_0^{-\sigma^d} (\psi^{i'} - \psi^{d'}) \, d\sigma^{d'} + \int_0^{\sigma^o} (\psi^{o'} - \psi^{i'}) \, d\sigma^{o'}$$

$$+ \Delta G_{a,cd}^\sigma(\text{non-el}) + \Delta G_{a,i}^\sigma(\text{non-el}) \tag{3.6.61}$$

or

$$\Delta G_a^\sigma = -\int_0^{-\sigma^d} \psi^{d'} \, d\sigma^{d'} - \int_0^{-\sigma^d} (\psi^{i'} - \psi^{d'}) \, d\sigma^{d'} + \int_0^{\sigma^o} (\psi^{o'} - \psi^{i'}) \, d\sigma^{o'} - \sigma^o \psi^o - \sigma^i \psi^i$$

$$\tag{3.6.62}$$

which is consistent with [3.3.9].

Introducing the differential capacitances of the inner and outer Helmholtz layer (see [3.6.26]), assuming these to be constant, [3.6.61] can be written as

$$\Delta G_a^\sigma = -\int_0^{-\sigma^d} \psi^{d'} \, d\sigma^{d'} + \frac{(\sigma^o + \sigma^i)^2}{2C_2^i} + \frac{(\sigma^o)^2}{2C_1^i} - \sigma^o \psi^o - \sigma^i \psi^i \tag{3.6.63}$$

Equation [3.6.58] is the limit of [3.6.63] for $C_2^i \rightarrow \infty$.

An elegant modification is obtained from [3.6.62] after replacing the first integral by

$$-\sigma^d \psi^d + \int_0^{\psi^d} \sigma^{d'} \, d\psi^{d'},$$

regrouping terms, realizing that differential and integral capacitances are identical. The outcome is

$$\Delta G_a^\sigma = -\frac{(\sigma^o)^2}{2C_1^i} - \frac{(\sigma^d)^2}{2C_2^i} + \int_0^{\psi^d} \sigma^{d'} \, d\psi^{d'} \tag{3.6.64}$$

which is an extension of [3.6.60]. Using [3.5.20], this can be converted into

$$\Delta G_a^\sigma = -\frac{(\sigma^o)^2}{2C_1^i} - \frac{(\sigma^d)^2}{2C_2^i} - \frac{8cRT}{\kappa}\left[\cosh\left(\frac{zy^d}{2}\right) - 1\right]$$ [3.6.65]

The last equation illustrates that ΔG_a^σ has a diffuse contribution plus two additional terms, weighted by the two reciprocal capacitances. When these capacitances are infinitely high, the purely diffuse case is retrieved.

Some illustrations are given in fig. 3.26. They refer to specific adsorption at the outer Helmholtz plane (fig. 3.21). As compared with the situation for a purely diffuse layer (fig. 3.6) two new features emerge. First, the curves become asymmetrical with respect to $y^o = 0$; second, ΔG_a^σ is non-zero at its minimum value. Both phenomena are directly caused by the specifically adsorbed cations, which keep ΔG_a^σ finite even if the (non-electrostatic) adsorption of charge-determining ions is zero.

Figure 3.27 exhibits very little salt concentration sensitivity if ΔG_a^σ is plotted as a function of σ^o. The explanation is that at the rather high values of C^i and

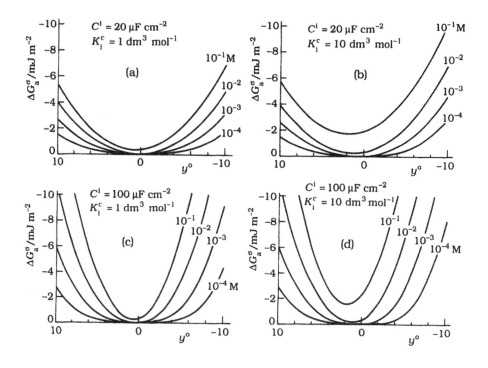

Figure 3.26. Gibbs energy per unit area of double layers with specific adsorption of cations at the outer Helmholtz plane. The combinations of inner layer capacitance C^i and K_i^c are given. Symmetrical (1-1) electrolyte, σ^o(max) = 80 μC cm^{-2}. No distinction is made between C_1^i and C_2^i. The electrolyte concentration is indicated. T = 298.16 K.

K_i^c chosen, a large part of the double layer resides in the inner part, where σ^i and σ^o are closely coupled: ΔG_a^σ is dominated by the spontaneous non-electrostatic adsorption of ions, and, with these adsorptions fixed, ΔG_a^σ is (about) constant. For $\Delta G_a^\sigma(y^o)$ this is not so; now the extent to which the potential rises as the result of this adsorption also plays a role, the extent of which depends on c_{salt}.

Comparison of figs. 3.27a and b also shows that when specific adsorption takes place closer to the surface ΔG_a^σ becomes less negative when the surface is negative. Colloquially stated, this means that it is easier to charge a double layer negative when cations are around in the near neighbourhood.

3.6g. Implementation of models

The models described in secs. 3.6c, d and e and their extensions offer, in principle, a basis for the interpretation of experimental σ^o (pAg, pH, etc.) curves. However, the implementation of this multivariable set of equations may be achieved in a variety of ways, depending on choices made and on analytical or numerical procedures. We shall not discuss these in detail but review some essential elements.

At the outset the criterion for the quality of a certain interpretation should be considered. For the time being this is solely based on the capability to describe titration curves for different electrolytes, different electrolyte concentrations and on different solids. In first instance this task is rewarding because the curves are simple and there are enough parameters. However, a more critical test requires the parameters to be physically realistic and to depend in the proper way on double layer parameters, and also requires the result to be in line with double layer information from other sources. Let us elaborate this.

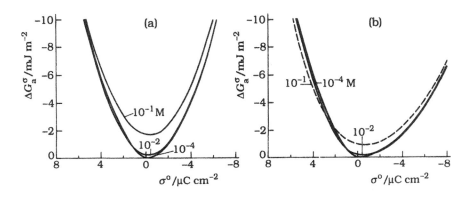

Figure 3.27. As fig. 3.26, but now as a function of the surface charge. $C^i = 20\ \mu\text{F cm}^{-2}$, $K_i^c = 10\ \text{dm}^3\ \text{mol}^{-1}$. (a) iHp = oHp; (b) $\beta\,/\,(\beta + \gamma) = 0.4$.

The most basic problems to solve are (i) dealing with the potential near or in the surface; this is a non-thermodynamic parameter, (ii) matching the smeared-out Gouy-Stern double layer to the localized site binding model of the first layer(s) and (iii) identify the proper binding sites and their numbers.

Regarding issue (i), all theories assume Boltzmann's law to apply, as previously in [3.6.41]. For charge-determining protons on a surface this is a relatively canonical procedure, the more so as all non-electrostatic contributions to the Gibbs energy of adsorption, activity corrections, etc., are convoluted in the relevant pK_a's. Problems arise with establishing ψ^i, the potential at which specifically adsorbing ions adsorb, an issue that is closely related to problem (ii): do we treat the inner Helmholtz layer as discrete or smeared-out? There is no *a priori* way to discriminate. Specific adsorption is caused by non-electrostatic interactions between ion and surface. The ion is usually site-bound, but not necessarily to the same sites that bind charge-determining ions. In fact, when the specific adsorption becomes superequivalent it may well avoid these sites because of the adversary electrostatic repulsion. Binding of hydrophobic ions to hydrophobic surfaces is not site-bound at all. In most site-binding models, specific adsorption is treated similar to that for protons, i.e. with binding constants having an intrinsic and a Boltzmann part. For instance, for a K^+ ion, adsorbing specifically on a RO^- group one would write

$$K_{a,K^+} = \frac{[RO^-]c_{K^+}^i}{[ROH]} = \frac{[RO^-]}{[ROH]} c_{K^+} e^{-y^i} \qquad [3.6.66]$$

The difficulty is then to assign a value for the potential: at the site where the ion binds, it may be very different from the averaged inner Helmholtz layer potential ψ^i and near a charged site ψ^i may differ greatly from its value near an uncharged site. The Gibbs energy of adsorption has two terms, $\Delta_{ads}G^o_{m,K^+} + F\psi^i$. In the various elaborations that have been published, different solutions have been proposed; comparison of their qualities is to some extent possible by judging their capability to account for the screening of the surface charge by electrolytes, that is, to predict the Esin-Markov coefficient. In the case where counterion and proton adsorb at the same potential the issue reduces to that of a common ion exchange process.

Other tests include the requirement of realistic values for the numerous parameters (N_s, pK_a, pK_b, $\Delta_{ads}G^o_{mi}$, C_1^i, and, depending on the model, C_2^i) and realistic dependencies. Sometimes for these parameters independent information is available: after drying, N_s is obtainable from functional gas adsorption studies (sec. 1.3f) or spectroscopically, the enthalpic part of pK_a by calorimetry, C_1 and C_2 from shifts in the point of zero charge (last paragraph of sec. 3.8b). Regarding dependencies, pK_a, pK_b and $\Delta_{ads}G^o_{mi}$ should be independent of pH and salt concentration; C_1^i and C_2^i may depend on σ^0 (because of dielectric satura-tion) but, at given σ^0, not on c_{salt}. Perhaps the ultimate test is to account for electrokinetic potentials, but this suffers from the problem that a model for the slip process is required or, for that matter, that the position of the slip plane is independently established.

The choice of the reactions underlying double layer formation may also vary between different authors. This should be taken into account when comparing tabulated pK's from different sources. For instance, Yates et al.[1], Davis et al.[2], Healy and White[3], James and Parks[4] and Stumm and Morgan[5] write

$$ROH_2^+ \rightleftarrows ROH + H^+ \qquad K_{a1}^s = \frac{[ROH]c_{H^+}^s}{[ROH_2^+]} \qquad\qquad [3.6.67a]$$

$$ROH \rightleftarrows RO^- + H^+ \qquad K_{a2}^s = \frac{[RO^-]c_{H^+}^s}{[ROH]} \qquad\qquad [3.6.67b]$$

Hence, their K_{a1}^s and K_{a2}^s correspond to our K_w^c / K_b^c and K_a^c in [3.6.38a and b], respectively.

Other authors use just the reverses of our [3.6.67a and b]. Van Riemsdijk and Koopal[6] describe the equilibrium in terms of a one-pK model, [3.6.39]; for which Hiemstra and van Riemsdijk[7] have developed an elegant interpretation.

In considering the vast literature on this topic, the reader should also be aware that different authors may define their surface charge differently (some include the specifically adsorbed charge) or define the notion "chemically" differently. (Stumm restricts "chemical" to real chemical bond formation corresponding to our chemisorption.)

This takes us back to the issue of surface heterogeneity. Suspended oxides expose a variety of crystal planes between which the various pK's may vary. Hence, in practice, measurements refer to very heterogeneous surfaces. Accounting for heterogeneity can be done along the same lines as introduced in sec. 1.7 with the potential as an additional parameter but the analysis is very laborious. Perhaps the trend is that the electric potential tends to smooth the

[1] D.E. Yates, S. Levine and T.W. Healy, *J. Chem. Soc. Faraday Trans.* (I) **70** (1974) 1807.

[2] J.A. Davis, R.O. James and J.O. Leckie, *J. Colloid Interface Sci.* **63** (1978) 180.

[3] T.W. Healy, L.R. White, *Adv. Colloid Interface Sci.* **9** (1978) 303.

[4] R.O. James and G.A. Parks, *Characterization of Aqueous Colloids by their Electrical Double-Layer and Intrinsic Surface Chemical Properties*, in *Surface and Colloid Science*, Vol. **12**, E. Matijevic, Ed., Plenum (1982) 119.

[5] W. Stumm, J.J. Morgan, *Aquatic Chemistry*, 2nd ed., Wiley (1981).

[6] See L.K. Koopal, *Adsorption of Ions and Surfactants*, in *Coagulation and Flocculation*, B. Dobias, Ed., Marcel Dekker (1993), chapter 4.

[7] T. Hiemstra, W.H. van Riemsdijk, *Colloids Surf.* **59** (1991) 7.

chemical heterogeneity in the sense that this heterogeneity is less pronounced at some distance from the surface. However, most investigators simply disregard heterogeneity, so that any error created by it is translated into the analysis and the final parameter values that are produced.

Most authors apply some extrapolation procedure to determine the intrinsic (or non-Coulombic or "chemical") part of the standard adsorption Gibbs energy. This may entail extrapolation to zero charge or to infinite salt concentration. James and Parks[1] proposed a double extrapolation procedure.

In secs. 3.8 and 10 a number of experiments and derived parameters will be described. For other reviews etc. on this matter, see [2].

3.7 Measuring double layer properties

Of the several double layer characteristics only certain ones are operational, that is, measurable without making any assumptions. Measurability may depend on the system, or on conditions. In practice, the analysis of most experimental data involves one or more non-thermodynamic arguments or definitions. Many of these assumptions have a certain physical basis. Part of this has been addressed before. We shall now review the main issues, emphasizing double layers on dispersed particles. Experimental details will be discussed only when generally relevant for colloidal systems. Techniques that are typical for special systems or situations will follow in sec. 3.10 if needed.

3.7a. Charges

For disperse systems, charge is the usual primary variable. To determine it, a number of steps have to be made. The first is simply a matter of definition: what kind of charge is to be measured and what is the origin of this charge or, for that matter, what are the charge-determining ions? Here, we shall emphasize the most basic issue, namely the determination of the *surface charge* σ°, as defined in sec. 3.3. For silver iodide and oxides the defining operational equations are [3.3.6 and 7], respectively. These are the best-defined inorganic systems available for which pairs of charge-determining ions have been identified. Such an identification requires a certain knowledge about the system and the process by which the surfaces of the particles may acquire charge. With unknown particles, or known particles with an unknown coating, this is not possible; then one can

[1] R.O. James, G.A. Parks, loc. cit.
[2] *Adsorption of Inorganics at Solid-Liquid Interfaces*, M.A. Anderson, A.J. Rubin, Eds., Ann Arbor Science (1981); T.W. Healy, L.R. White, *Adv. Colloid Interface Sci.* **9** (2978) 303; G.H. Bolt, W.H. van Riemsdijk, in *Soil Chemistry, B. Physico-chemical Models*, G.H. Bolt, Ed., 2nd part, 2nd ed., Elsevier (1982), chapter 13; P.W. Schindler, W. Stumm in *Aquatic Surface Chemistry*, Wiley (1978), chapter 4.

either invoke electrokinetics or try to find out for what kind of ions the surface has a special affinity. Often the structure of the solid gives hints, e.g. for $BaSO_4$ it is logical to consider Ba^{2+} and SO_4^{2-} ions, for calciumhydroxyapatite, the prototype of the main inorganic constituent of bone and teeth, $(Ca_5(PO_4)_3OH)$, Ca^{2+} and phosphate, but H^+ and OH^- may also be tried. For the moment specifically adsorbed charged molecules, such as ionic surfactants, are not considered. When they are present, the titration only measures σ° as defined above, with the caveat that some of these molecules could also strongly bind the charge-determining ions: there is no thermodynamic way to discriminate between ions adsorbed to the surface proper and those strongly bound to adsorbates on that surface.

Even with the "well-behaved" silver halides and oxides, σ° is only defined in not too dilute indifferent electrolyte. For insoluble oxides in pure water, containing only H^+ and OH^- ions, these ions constitute not only the surface charge but also the counter charge. (The situation is a bit academic because carrying out a titration requires the introduction of other ions anyway. Moreover, few oxides are completely insoluble and some ions may be introduced by the wall of the vessel (silicates from the glass).) We shall therefore only consider the realistic cases that c_{HNO_3}, $c_{KOH} \ll c_{KNO_3}$ etc. An additional advantage is that it is easier to establish activity coefficients: in excess KNO_3 f_{H^+} is largely determined by the NO_3^- ions and hence constant. In this way the distinction between dpH $= -d \log c_{H^+}$ and $-d \log a_{H^+}$ fades away.

With the above in mind, σ° can be determined by *colloid titrations*, as described in sec. I.5.6e. To review the experimental ins and outs, consider (insoluble) oxides, subjected to potentiometric acid-base colloid titration. Basically the procedure is that σ° (at say pH', and c_{salt}) is related to σ° at pH" and the same c_{salt} by adding acid or base. The titration is carried out in an electrochemical cell in such a way that not only pH" is obtainable, but also the part of the acid (base) that is not adsorbed and hence remains in solution. Material balance then relates the total amount [of acid minus base] adsorbed, $\sigma^\circ A$ (where A is the interfacial area) at pH" to that at pH'. By repeating this procedure a complete relative isotherm $\sigma^\circ A$ as a function of pH is obtainable. We call such a curve "relative" because it is generally not known what σ° was in the starting position.

The longer the pH range over which $\sigma^\circ A$ is measured, the better it is for subsequent interpretation and characterization. However, this range is limited by experimental constraints. One is that at very low and very high pH it becomes increasingly difficult to determine changes in $\sigma^\circ A$ for analytical reasons ($\Delta\sigma^\circ A$ then becomes a small increment and the relation between the (dominant) amount in solution and measured cell potential is semilogarithmic, i.e. the accuracy drops by a factor of 10 for each unit of pH, pAg, etc., at the lower and higher end.) Another experimental limitation is that oxides are seldom fully inert: some dissolution may take place at extreme pH values.

At this instance it has to be checked whether the titration is reversible. Irreversibility is rarely caused by real hysteresis; more likely it is due either to lack of patience (establishment of equilibrium may take several hours per measuring point, especially so at low c_{salt}) or to chemical problems, like leaching of ions from the solid, dissolution of the solid, adsorption or desorption of inadvertently present ions or molecules, including CO_2, or to ageing of the sample. Spurious ionic adsorptives can often be discarded by double filtration, one at high and one at low pH, pAg, etc., (to desorb anions and cations, respectively). Ageing of the sample prior to titration may render the titrations better reversible. Coagulation is rarely a reason for irreversibility: even in the coagulated state the entire solid-liquid interface usually remains accessible to adsorbing charge-determining ions and the accompanying countercharge, except perhaps in very compact coagulates.

Having established reversibility, the titration should now be repeated at different c_{salt}. Going from a curve at given concentration c'_{salt} to a higher one, c''_{salt}, is simply realized by adding electrolyte and measuring the resulting change in pH. In fact, one can scan all required salt concentrations by titration with electrolyte and, for that matter, this can also be done with organic adsorptives. Establishment of an array of $\sigma^0 A$(pH) curves, relative to each other at different c_{salt} is a prerequisite in establishing the point of zero charge (p.z.c.) to render the set of curves absolute. Basically, when the electrolyte is indifferent and there is a common intersection point (c.i.p.) the pH of this point is identified as the p.z.c., the argument being that for uncharged surfaces there is no salt effect because indifferent ions are generic and do not adsorb. We return to the p.z.c. in sec. 3.8.

The last step (which also may precede the previous one) is to determine the area A. This also is not always obvious. One choice is the BET-area (chapter 1) but a method based on adsorption from solution (chapter 2) may be more appropriate. In a number of cases, however, it was found that charge-determining ions "see" an area that differs from that obtained from adsorption of uncharged compounds. Silver iodide is a notorious example, to which we shall return in sec. 3.10a. In fact, any time that an example of σ^0 is given, the area used to obtain the data set should be indicated.

Historically, the first well-established example of such a set of surface charge curves was not obtained for oxides but for silver iodide. It is reproduced in fig. 3.28. Old as these data may be, they have not been seriously challenged and already illustrate a familiar trend: to the negative side σ^0 becomes more negative, and on the positive side more positive, with increasing c_{salt} because of better screening. However, the increase is far less than predicted by diffuse

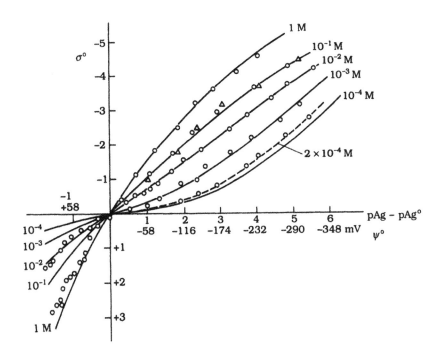

Figure 3.28. Illustration of a seminal colloid titration result obtained after pioneering work by E.J.W. Verwey and H. de Bruyn. Silver iodide in (1-1) electrolytes: drawn curves 7:1 KNO_3 + $NaNO_3$ mixture, O $NaClO_4$, Δ $NaNO_3$. The surface charge could not be exactly established because the surface area was not well known. pAg and ψ^o units are convertible because Nernst's law applies. The (7:1)-KNO_3 + $NaNO_3$ mixture was chosen to suppress the liquid junction potential (sec. F5.5d) with the salt bridge. Source: Redrawn from data by J.A.W. van Laar, PhD Thesis, State Univ. Utrecht (1952); E.L. Mackor, *Rec. Trav. Chim.* **70** (1951) 763, as collated by J.Th.G. Overbeek in *Colloid Science* Vol. **1**, H.R. Kruyt, Ed., Elsevier (1952) 162. Older references include E.J.W. Verwey, H.R. Kruyt, *Z. Phys. Chem.* **A167** (1933) 149; E.J.W. Verwey, *Rec. Trav. Chim.* **60** (1941) 887 and H. De Bruijn, *Rec. Trav. Chim.* **61** (1942) 5, 21.

double layer theory. This can be seen by substituting ψ^o for ψ^d (in this case ψ^o is well established because the interface obeys Nernst's law, [3.7.3a]) in [3.5.15] and fig. 3.4. Obviously there is a Stern layer with a limited capacitance, see [3.6.25] and figs. 3.41-42. Ion specificity (differences between Na^+ and K^+ on negative surfaces and between NO_3^- and ClO_4^- on positive surfaces and shifts of the p.z.c. due to electrolyte addition) is nowadays better known than in the example of fig. 3.28.

Potentiometric colloid titrations can also be carried out on colloids having a fixed number of certain dissociable groups, (say latices with carboxylic or sulfate groups, proteins) or heterogeneous systems such as clay minerals. When there is only one type of group, i.e. in the case of so-called *monofunctional* surfaces, the titration yields the number of those groups. The curve relating the amount of acid or base consumed with the pH is in these cases not gradual but rather has a knee-bend at the equivalence point, which becomes sharper as the groups become stronger. For weak groups the sharpness is increased by adding indifferent electrolyte, because this screens the lateral interaction, i.e. makes the groups effectively stronger. Once the total number of groups has been established, the degree of dissociation α can be computed and hence a Henderson-Hasselbalch plot [3.6.53] constructed, so that pK_a becomes accessible.

An example of such a potentiometric titration curve is presented in fig. 3.29, curve a. The stronger the groups, the steeper the ascending part in curve a, and the better defined the intersection between the two linear parts of curve b. The result is typical for surfaces with strong groups. From the amount of base needed to obtain the point of steepest rise in pH the number of acid groups can be determined. In the present case, where the groups are strong, this yields σ°, which is constant over the entire range. In fact, by the titration the exchange of protons (as counterions) against Na^+ ions is studied, i.e. an *ion exchange capacity* is measured.

In the example, the outcome is confirmed within about 3% by *conductometric titration* (fig. 3.29, curve b). In such a titration the conductivity is followed as a function of the amount of NaOH added. In the first, linearly descending, leg the very mobile protons are replaced by the slower Na^+ ions (see table I.6.5). The ascending linear final part reflects the increase due to addition of NaOH. The

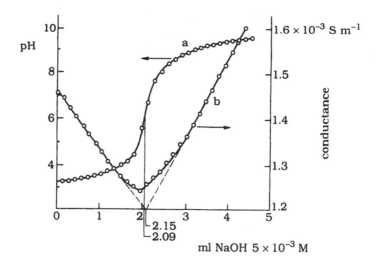

Figure 3.29. Potentiometric (a) and conductometric (b) titration of a latex with sulfonic acid groups. Redrawn from K. Furusawa, W. Norde and J. Lyklema, *Kolloid Z. Z. Polymere* **250** (1972) 908.

two slopes are proportional to $(u_{Na^+} - u_{H^+})$ and to $(u_{Na^+} + u_{OH^-})$, respectively, where the u's stand for limiting ionic mobilities, defined in sec. I.6.6a. Conductometry should preferentially be carried out at low electrolyte content to emphasize the contribution of the colloids to be investigated. The weaker the groups and the lower their number, the larger the angle between the two linear legs.

For monofunctional surfaces with weak surface groups the first leg of a conductometric titration plot is also ascending, though less so than the second leg. Hence the number of groups can still be determined. Different types of groups on one surface can be conductometrically distinguished by titration at different pH's if the various proton affinities are sufficiently apart.

Potentiometric titration has also been widely applied to polyelectrolytes and proteins; conductometric titration to a lesser extent. For colloids with a variety of different groups (proteins), application of the Henderson-Hasselbalch equation [3.6.52 or 53] or its semi-empirical variant [3.6.54] is viable only when the pK_a's of the various groups are sufficiently far apart.

Although colloid titration is the most usual and general method of measuring the surface charge as a function of pH, pAg, c_{salt}, temperature, concentration of organic additives, etc., it is not the only one. A few alternatives are:

(i) For materials that can be made into an electrode, or that can be deposited on an electrode, the differential capacitance can sometimes be measured directly. From this, the surface charge follows by integration. A number of technical problems have to be surmounted, to be discussed in sec. 3.7c. One of these is that *Faradaic currents* (currents across the interface) have to be suppressed or accounted for. Another intrinsic problem is whether the surface properties of the electrode are identical to those of the dispersed particles. For silver iodide and some oxides the capacitance approach has worked well. It is recalled that for polarizable, conducting interfaces, with mercury as the prototype, this is virtually the sole method.

(ii) Adsorption of charge-determining ions can also be measured by radiolabelling or spectroscopically. In these techniques the analysis aims at a certain ion but this does not mean that the adsorption of a single charge-determining ionic species is measurable: when, say labelled I^- adsorbs on AgI, some Ag^+ may co-adsorb and since this co-adsorption is not measured, one cannot say by how much σ^o is changed. Generally, this technique is less accurate than titration.

(iii) A variant of the previous one aims at measurement of the countercharge, for instance by radiolabelling or spectroscopically. As with latices, using this procedure in fact the counterion exchange capacity is measured. Only when negative adsorption of co-ions is negligible (for the conditions, see fig. 3.8) may the counterion charge be identified with minus the surface charge. For very smooth surfaces (mercury, some metals), ion adsorption can also be obtained ellipsometrically, but for dispersed particles this technique is not feasible.

In passing it is recalled that in systems where the surface charge is caused by adsorption of ionic surfactants, the adsorption of these ions can be analytically determined; the ensuing charge can be established if the molecules are fully dissociated. For systems in which both charge-determining ions and ionic surfactants adsorb (sodium dodecyl sulphate on oxides), both charges can be measured to obtain detailed double layer information.

(iv) For the sake of completeness, it is recalled from [I.5.6.17] that for polarizable fluid metals (mercury, gallium) the surface charge can be obtained from *electrocapillary curves*.

Electrokinetic charges (σ^{ek}) can, under a number of conditions, be obtained from one of the electrokinetic techniques, as detailed in chapter 4. For simple surfaces (flat, homogeneous, no hairs) σ^{ek} is probably close to σ^d. Electrokinetics are therefore not alternatives to colloid titrations but rather serve as additional means of establishing the (counter-) charge distribution. Only in media of very low dielectric permittivity, where Stern layers are absent, is $\sigma^{ek} = \sigma^d \approx -\sigma^o$ a good approximation.

All told, situations remain in which the particles are so poorly defined that the charge-determining mechanism is totally unknown. For such systems, electrokinetics remains the sole source of information on (part of the) charge.

3.7b. *Potentials*

Interfacial potentials are more elusive than charges. In sec. I.5.5 various types of potentials have been introduced and discussed. The quantity ψ^o, entering double layer equations, is the potential at the plane of adsorbing charge-determining ions, taken with respect to the solution at infinite distance. When between this plane and the bulk of the solid the composition remains constant and there are no charges, chaninges of ψ^o may be identified with those in the *Galvani potential difference* $\Delta\phi$ between particle and solution. For AgI this appears to be the case, but for oxides and semiconductors further analysis and/or experiments are needed. Galvani potentials are generally unmeasurable because the work of transfer of a single ionic species across a phase boundary involves electric and non-electric contributions that cannot be separated unambiguously.

As with capacitances, potentials with respect to a certain reference can be measured for materials for which electrodes can be made. Then, by changing the composition, $d\Delta\phi$ is obtainable if it is established that the potential of the counterelectrode is constant or if its variation is known and accounted for. For mercury and similar polarizable interfaces changes in $\Delta\phi$ are externally applied. For AgI and a number of oxides it has been experimentally established that *Nernst's law* [I.5.5.9] applies, i.e. that

$$d\,\Delta\phi = \frac{RT}{z_{cd}F}\, d\ln a_{cd} \qquad\qquad [3.7.1]$$

where cd denotes the charge-determining species. When changes in $\Delta\phi$ and ψ^o may be identified, [3.7.1] may also be written as

$$d\,\psi^o = \frac{RT}{z_{cd}F}\, d\ln a_{cd} \qquad\qquad [3.7.1a]$$

In electrochemistry, potentials between two phases, originating from adsorption-desorption processes, i.e. potentials at relaxed interfaces, are sometimes called *open circuit potentials*. As the experiments should be carried out in excess electrolyte, the activity coefficients are determined by the carrier electrolyte, and therefore are independent of the concentration of cd electrolyte, i.e. $d\ln a_{cd} \sim d\ln c_{cd}$, which is actually measured. So,

$$d\psi^o = \frac{RT}{z_{cd}F}\, d\ln c_{cd} \qquad\qquad [3.7.2a, b]$$

For silver halides and oxides

$$d\psi^o = -\frac{2.303\ RT}{F}\, d\,pAg; \qquad d\psi^o = -\frac{2.303\ RT}{F}\, d\,pH \qquad [3.7.3a, b]$$

respectively. At 25°C, $2.303\ RT/F$ equals 0.059 V.

Relative Galvani potentials cannot be made absolute because there is no way to establish the conditions under which there is no potential difference across the boundary between dissimilar phases. For lack of better knowledge, ψ^o is therefore usually referred to the point of zero *charge* which, although not thermodynamically defined, can often be established with some confidence, see below and sec. 3.8. However, the point of zero charge is not necessarily identical to the point of zero potential because even at $\sigma^o = 0$ the *interfacial potential jump* χ, caused by preferential orientation of solvent (water) dipoles and polarization of the particle is non-zero. Anticipating sec. 3.9 it is realized that χ may change with σ^o, so that in equations like

$$\psi^o = -\frac{2.303\ RT}{F}\,(pAg - pAg^o); \qquad \psi^o = -\frac{2.303\ RT}{F}\,(pH - pH^o) \qquad [3.7.4a, b]$$

where the superscript o indicates the point of zero charge, it is not certain to what extent ψ° is due only to the adsorption of charge-determining ions. For model interpretations this presents a problem of principle.

In addition to the above problem there is the question we encountered before, whether the electrode material is identical to that of the dispersed particles in a sol. For AgI and some oxides this identity has been established under certain conditions.

In view of the above uncertainties we shall avoid the use of absolute values of ψ° as much as possible and, for that matter, prefer differential capacitances to integral ones. However, in model interpretations absolute ψ° values are sometimes unavoidable. Equation [3.6.43] is an example: if here [3.7.4b] is used to obtain y° any changes in the pK's due to the change in the polarization of the solvent are ignored.

For materials from which electrodes cannot be made ψ° is virtually inaccessible and only obtainable by inference.

Regarding the *electrokinetic potential* ζ similar remarks can be made as for the electrokinetic charge.

3.7c. *Capacitances*

For polarized interfaces, capacitances can be readily measured directly with great precision. For mercury, capacitance-applied potential curves constitute the basic information for double layer analyses, outweighing that from electro-capillary curves.

Relaxed interfaces cannot be polarized unless special precautions are taken. Capacitances can of course be obtained as *derived* quantities by differentiating the surface charge with respect to the surface potential if changes in the latter are known, which is possible if the Nernst equation applies. We now discuss *direct* capacitance measurements on reversible interfaces. To start with, the response of such an interface to an applied field has to be considered. The basic problem is that not only are double layers built up, but also charge transfer across the interfaces takes place and diffusion of charge-determining ions to or from the surface starts to play a role. With regard to these physical processes only the sum-effect is measured, and this sum has to be divided into its parts to obtain the capacitance. Distinctions can be made because the three constituents mentioned react in a different way to the frequency of the external field.

The physical implication is that a.c. or, more generally dynamic, measurements have to be carried out and the *impedance* spectrum $\hat{Z}(\omega)$ is measured, rather than the resistance. Typically, $\hat{Z}(\omega)$ is a complex quantity. At the same time the static conductance now becomes an *admittance* $\hat{Y}(\omega)$, which is also a complex quantity. Following our custom we generally write

$$\hat{Z}(\omega) = Z'(\omega) - iZ''(\omega)$$

[3.7.5]

$$\hat{Y}(\omega) = Y'(\omega) - iY''(\omega) \qquad\qquad [3.7.6]$$

with a minus sign. Here, the circumflex (^) indicates that the quantity is complex, the prime ' and double prime " indicate the real and imaginary parts, $\mathrm{Re}\hat{Z}$, $\mathrm{Re}\hat{Y}$, $\mathrm{Im}\hat{Z}$ and $\mathrm{Im}\hat{Y}$, respectively. Working with imaginary and complex quantities has been outlined in I. app. 8.

Basically, impedance spectra and admittance spectra contain the same information; it is a matter of experimental and interpretational convenience to decide which one to use. Moreover, the imaginary and real parts of $\hat{Z}$ and $\hat{Y}$ are pairwise coupled by *Kramers-Kronig equations* (variants of [I.4.4.31 and 32]) so that in principle one can be obtained when the other is known[1]. Such a transformation requires very accurate data over a large range of frequencies, which for inorganic solids are not always available, so that for practical purposes these equations have only limited impact. However, mathematical expressions for $\hat{Z}$ and $\hat{Y}$ should always agree with the Kramers-Kronig relations, a fact that can be used to validate electrospectroscopic data[2].

Measurements are carried out in electrochemical cells. They contain at least a reference electrode and a measuring electrode, consisting of (or covered with) the material to be studied. Preparation of such an electrode can be achieved in two ways: (i) deposition of the colloidal particles or (ii) in situ synthesis of the material on a support. Deposition can conveniently be done electrophoretically. However, the deposit formed in this way should also have sufficient mechanical strength. Heating may achieve this, though at the risk of affecting the surface properties. In situ synthesis may be achieved electrolytically, by vapour deposition, sputtering, or a combination of these. Layers formed this way usually have excellent mechanical strength but the identity of the material obtained to that of the dispersed material to be investigated remains to be proven.

Establishing the surface area is the next concern. Electrode surfaces are seldom flat. Instead, they tend to display a set of different crystal faces. Silver iodide electrodes, prepared by amalgamation of silver with mercury, followed by vapour deposition of iodine, look smooth and shiny to the naked eye but reveal crystallites under the electron microscope. Surface irregularities not only complicate the assessment of the real area, they may also interfere in the analysis of impedance spectra in terms of equivalent circuits. After drying, the surface may be studied by the usual optical methods (sec. 1.2) with the familiar caveat that drying may change these properties. Anyway, for a number of oxides and silver iodide it is now established that electrodes can be made which have

[1] For an illustration, see G. Achatz, G.W. Herzog and W.H. Plot, *Surface Technol.* **11** (1980) 431.

[2] G. Láng, L. Cocsis and G. Inzelt, *Electrochim. Acta* **38** (1993) 1047.

properties that are very similar to those of the corresponding sol, although the area has to be adjusted to a certain extent. One illustration regarding silver iodide follows below (fig. 3.32). For haematite (α-Fe_2O_3) the isoelectric point of deposited layers was found to be identical to the point of zero charge of sols under pristine conditions[1] and deposited anatase (TiO_2) films exhibited the same electrochemical and photoelectrochemical activity as semiconducting anatase[2]. On the other hand, in a few other experiments coated materials had isoelectric points or points of zero charge that differed from those of the pristine material. Leaching of ions from the support or incomplete coating could have been the reasons.

Impedance spectra are usually measured with automated frequency-response analysers using single sine wave applied fields. Measured data are resolved into real (in-phase) and imaginary (out of phase) or quadrature constituents. The frequency is usually swept over a certain range, either linearly or logarithmically. The measurements can also be done with sols between the electrodes, to study the dielectric response of colloids. We will return to that in sec. 4.5d. For general information on such techniques and their interpretation, see ref. [3].

Once the required $\hat{Z}(\omega)$ or $\hat{Y}(\omega)$ measurements are completed, the next issue is the data analysis, implying the establishment of the three basic contributions with their appropriate frequency dependencies:

(i) *The capacitance, C.* This quantity can be measured when the frequency of the applied field is low enough for the double layers involved to charge or discharge. In addition to the double layer to be studied, there is always the double layer at the reference electrode (the capacitance of which may be suppressed by giving it a very high area) and inside the solid. Quantitatively, the relaxation time τ_{DL} of the double layer should be small as compared with ω^{-1}. For τ_{DL} we have the Maxwell expression [I.6.6.32]

$$\tau_{DL} = \varepsilon_0 \varepsilon / K << \omega^{-1} \qquad\qquad [3.7.7]$$

where K is the conductivity. For aqueous solutions τ_{DL} varies between 10^{-7} and 10^{-9} s in 10^{-3} - 10^{-1} M solutions, so that periods $>> 10^{-7}$ s, or $\omega << 10^7$ Hz, are safe. With this condition met, the double layer is purely capacitive (that is, it is virtually infinitely rapidly charged) and the capacitance simply appears as an ωC term in Y''.

Deviation from this condition may have different causes. One is that the various double layers have different relaxation times. More troublesome are

[1] N.H.G. Penners, L.K. Koopal and J. Lyklema, *Colloids Surf.* **21** (1986) 459.
[2] L. Kavan, B. O'Regan, A. Kay and M. Grätzel, *J. Electroanal. Chem.* **346** (1993) 291.
[3] A.J. Bard, L.R. Faulkner, *Electrochemical Methods*, Wiley (1980).

non-idealities caused by surface rugosity, which may lead to the impedance behaving as a so-called constant phase element. An inhomogeneous distribution of the current density over the surface is the probable cause. Empirically, the impedance contribution becomes

$$\hat{Z}_c = (i\omega)^{-\alpha} Q \qquad\qquad [3.7.8]$$

where Q is a constant of dimensions $\Omega\ m^2\ s^{-\alpha} = C^{-1}\ V\ m^2\ s^{1-\alpha}$. The exponent α varies between zero and unity and is a measure of the rotation of the $Z''(Z')$ plot as compared with a pure capacitor, see fig. 3.30. A trend like in [3.7.8] is approached by a fractal surface structure. In practice, α and Q are adjustable parameters.

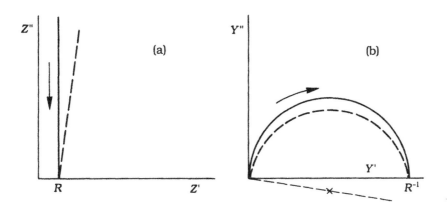

Figure 3.30. Relation between the imaginary and real parts of the impedance (left) and admittance (right). The arrows indicate the direction of the increase of ω. In fig. (a), $Z'' = 0$ and $Z' = R$ for $\omega = \infty$; in fig. (b) $\omega = 0$ in the origin and $\omega = \infty$ for $Y' = R^{-1}$. Figure (b) is circular. Starting from the origin (where $\omega = 0$) the frequency increases. Solid lines: for a resistance and ideal capacitance in series; dashed: for resistance and a constant phase element in series.

(ii) *Ion transfer resistance*, caused by the slowness of the transport of charge-determining ions across the (solid-liquid) interface. This leads to a finite exchange current density j_o and an ion transfer resistance θ given by

$$\theta = RT / z F j_o \qquad\qquad [3.7.9]$$

which has the dimensions of reciprocal capacitance per unit area $\times$ time, $V\,s\,m^2\,C^{-1}$. This equation may be considered as the equivalent of Ohm's law, relating the current through an interface to the corresponding resistance. See for further discussion ref. [1]. The product

$$\tau_{tr} = \theta C \tag{3.7.10}$$

has the dimensions of time and may be interpreted as the characteristic time for interfacial transfer. For $\tau_{tr} \gg \omega^{-1}$, this transfer is suppressed. In electrochemical parlance, this is the case of "irreversible" transfer reactions.

(iii) *Diffusion impedance*, caused by diffusion limitation of the transport of charge-determining ions to or from the interface. (Conduction-limitation by such ions is negligible because the indifferent electrolyte concentration $\gg$ charge-determining ion concentration.) For diffusion to or from a flat surface, the ensuing impedance contribution is

$$\hat{Z}_w = (1 - i)\, \sigma_w \omega^{-1/2} \tag{3.7.11}$$

where

$$\sigma_w = 2^{-1/2}\, RTF^{-2}\, c_{cd}^{-1} D_{cd}^{-1/2} \tag{3.7.12}$$

is the so-called *Warburg coefficient*; c_{cd} and D_{cd} stand for concentration and mean diffusion coefficient of the charge-determining ions, respectively. For a derivation, which is rather involved, see ref.[2]. The original publications from which the name "Warburg coefficient" stems date back to the turn of the century[3].

For the double layer to be studied, the three elements can be grouped into a *Randles circuit* (fig. 3.31), which in turn can be connected in series to other elements of the cell. In passing, a Randles circuit may also be drawn for individual colloidal particles. Then, this circuit is isolated. It stands on its own

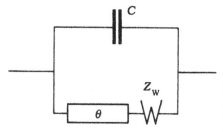

Figure 3.31. Randles circuit, consisting of a pure capacitance C, an ion transfer resistance θ and a diffusion impedance Z_w all counted per unit area.

[1] A.J. Bard, L.R. Falkner, *Electrochemical Methods*, John Wiley, New York (1980).
[2] M. Sluyters-Rehbach and J.H. Sluyters, in *Electroanal. Chem.*, A.J. Bard, Ed., Vol. 4 Marcel Dekker, New York (1970).
[3] E. Warburg, *Wied. Ann.* **67** (1899) 493; *Drud. Ann.* **6** (1901) 125.

and is not series-coupled to the impedances of electrodes or other elements because dispersed particles cannot be charged by external current.

The admittance of the Randles circuit is

$$\hat{Y} = \frac{\theta + \sigma_w \, \omega^{-1/2} - i\sigma_w \, \omega^{-1/2}}{(\theta + \sigma_w \, \omega^{-1/2})^2 + \sigma_w^2 \, \omega^{-1}} - i\omega C \qquad [3.7.13]$$

and can be split into the real and imaginary parts, following the procedures of I.app. 8, leading to

$$\frac{\omega^{1/2}}{Y'} = \theta\,\omega^{1/2} + \sigma_w + \frac{\sigma_w^2 \, \omega^{-1/2}}{\theta + \sigma_w \, \omega^{-1/2}} \qquad [3.7.14]$$

$$\frac{Y''}{\omega} = \frac{\sigma_w \, \omega^{-1/2}}{\omega(\theta + \sigma_w \, \omega^{-1/2})^2 + \sigma_w^2} + C \qquad [3.7.15]$$

In this way, from analysis of the admittance dispersion, the capacitance can be obtained and, if desired, from that the surface charge by integration. An illustration is fig. 3.32. Here a σ^0(pAg) curve for silver iodide, obtained via

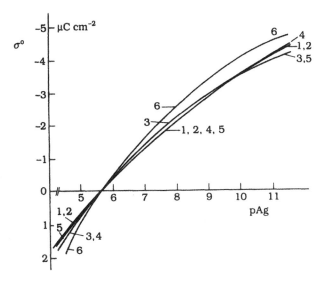

Figure 3.32. Electrical double layer on silver iodide in 10^{-1}M Na$^+$ or K$^+$ (1-1) salts. Comparison of results obtained by different authors, different techniques and different sols. Curves 1-5, potentiometric colloid titration; curve 6, capacitance method for electrodes. References: 1) E.L. Mackor, *Rec. Trav. Chim.* **70** (1951) 763; 2) J.A.W. van Laar, PhD. Thesis, State Univ. of Utrecht, NL (1952); 3) J. Lyklema, *Trans. Faraday Soc.* **59** (1963) 418; 4) B.H. Bijsterbosch, J. Lyklema, *J. Colloid Sci.* **20** (1965) 665; 5) B.H. Bijsterbosch, unpublished; 6) J.H.A. Pieper, D.A. de Vooys, *J. Electroanal. Chem.* **53** (1974) 243. (Redrawn from B.H. Bijsterbosch, J. Lyklema, *Adv. Colloid Interface Sci.* **9** (1978) 147).

capacitance measurements with electrodes, is compared with a variety of curves, obtained on different samples and by different authors, via potentiometric colloid titration. In this case the agreement is very satisfactory; the small difference between curve 6 and the others may be entirely attributed to minor uncertainties in the specific area determination. At 80°C excellent agreement between titration and direct capacitance measurement was also achieved.

3.7d. Other double layer properties

Some important double layer properties will be dealt with elsewhere.

Points of zero charge will be treated in sec. 3.8. Strictly speaking, these are not electrical properties but reflect the difference in affinity of the surface for the two charge-determining species.

Electrokinetics yield information on double layers which are slightly out of equilibrium and also on the distribution of charge and potential. For instance, from the surface conductivity K^σ the product of ionic concentrations and mobilities in (essentially the inner part of) the double layer is obtainable. See further sec. 3.13 and chapter 4.

More indirectly, information on the charge- or potential distribution is also obtainable from studies involving two *double layers in interaction*, as in colloid stability, soap films and wetting films. Overlap is essentially determined by the diffuse parts, the Stern layers defining the boundary conditions. Most of this will be deferred to Volume IV.

The direct determination of the *heat of adsorption* of charge-determining and other ions deserves special attention. The relevance of this thermodynamic information should not be underestimated because the non-electrostatic affinity of ions for a surface is the very driving force for double layer formation (sec. 3.2). Sensitive micro-calorimeters have now been developed with which the required enthalpies can be obtained with sufficient precision, provided the surface area is not too small. The most elegant method is perhaps a *(micro)calorimetric titration*, i.e. a titration in a cell in which adsorbed amounts and heat evolution are simultaneously measured. A good second choice may be an (extensive!) set of heats of immersion of the solid in a range of solutions of different pH and c_{salt}. The obtained enthalpies consist in principle of an electrostatic and non-electrostatic part. The scope includes charge-determining ions, specifically and generically adsorbing ions, although the signal of the last category, which is purely electrostatic, is usually too low to be measurable.

To date, the technique has been successfully applied to a number of oxides. Results will be presented in sec. 3.10c. In principle, the enthalpy of charge formation, whether caused by adsorption/desorption of protons only or by protons and hydroxyls, is also accessible from the temperature dependence of the point of zero charge. For the simple case that only the protons are involved, this follows from [3.6.50]. Hence, a consistency test of data obtained from very different measurements and using very different interpretations, is possible.

Going one step further, the interpretation of the temperature shift of the point of zero charge requires a model for the adsorption equilibrium. From the data and the model, $\Delta_{ads} G^\circ$ is obtainable. If measured as a function of temperature, $\Delta_{ads} H^\circ$ and $T \Delta_{ads} S$ can be derived, using the appropriate Gibbs-Duhem relation (sec. I.2.15). Hence, entropies of double layer formation are also accessible. For the site-binding model the difference between the $\Delta_{ads} G^\circ$'s of protons and hydroxyls follows from $(pK_a - pK_b)$ in [3.6.45]; for a one pK model there is only one such term.

3.7e. Determination of specific surface areas by negative adsorption

Measuring surface charges from depletion of charge-determining ions finds its counterpart in measuring the expulsion of co-ions from the increase in concentration in solution. From this increase, called the *Donnan effect*, information on the double layer can be obtained, but it also offers an original method of determining specific areas, i.e. it is at the same time a technique to measure a double layer property and a double layer application.

Usually, procedures for obtaining specific surface areas (A_g) involving gas adsorption or adsorption from solution have in common that first the monolayer capacity is determined; assigning a cross-section to the close-packed molecules then leads to A_g. Somewhat unexpectedly, A_g can also be determined by ions that are depleted from the double layer, i.e. ions that are absent. Basically, the reason is that the measurable total amount of expelled electrolyte can be written as the product of the amount per unit area times the area. For the former an expression can be derived; for instance if the double layer is diffuse, we have [3.5.25a]. Hence, the area can be found. The negative adsorption method has the advantage that no cross-sectional area is required, but a check of the applicability of the required double layer formula is mandatory. Figure 3.33 depicts the principle. When the surface is uncharged the concentrations of cations and anions, c_{i+} and c_{i-}, respectively, are equal to each other and to the initial concentration in bulk, c_i. Let us now charge the surface negatively (in practice, more systems are negative than positive). In the new distribution c_{i+} increases with decreasing x, whereas c_{i-} decreases. For the sake of simplicity it is assumed that the surface potential is sufficiently negative to let $c_{i-}(x=0)=0$ (compare fig. 3.8 for conditions). In the new equilibrium state the bulk concentration to which both ionic concentrations approach at sufficiently high x, $c_i(f)$ is higher than $c_i(i)$ by an amount Δc_i, because of the Donnan effect. For the uninitiated eye it might seem counterintuitive that cations adsorb positively, but that nevertheless their concentrations *increase* in the solution. The

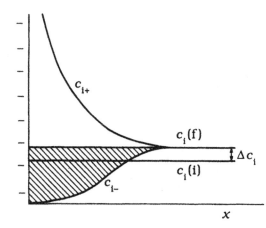

Figure 3.33. Changes in ion concentrations near a surface due to charging of this surface. Symmetrical electrolyte, Δc_i is the Donnan effect.

explanation is that there is indeed a small decrease in the bulk of c_{KI} or c_{KOH}, but the rise of c_{KNO_3} caused by negative adsorption is quantitatively much larger because $c_{KNO_3} \gg c_{KI}$, c_{KOH}. In fig. 3.33 the hatched area gives the surface deficit

$$\Gamma_i = \int_0^\infty \left[c_i(f) - c_{i-}(x) \right] dx \qquad [3.7.16]$$

in moles per unit area. The total expulsion (in moles) is $A \Gamma_i$ which is related to Δc_i by this material balance

$$A \Gamma_i = V \, \Delta c_i \qquad [3.7.17]$$

where V is the total volume of liquid into which the particles, with total area A, are dispersed. For the limiting case of infinitely high negative surface potential [3.5.26] can be used as

$$|\sigma_-^d| = z F \Gamma_i = (2 \varepsilon_0 \varepsilon c_i RT)^{1/2} = 2 c_i z F / \kappa \qquad [3.7.18]$$

Hence,

$$A = \frac{z F V}{(2 \varepsilon_0 \varepsilon RT)^{1/2}} \frac{\Delta c_i}{c_i^{1/2}} = \frac{V \Delta c_i}{c_i} \frac{\kappa}{2} \qquad [3.7.19]$$

with which the problem is solved in principle. Equation [3.7.19] is the simplest available. It can also be obtained by replacing $c_{i-}(x)$ in fig. 3.33 by a step function, where c_{i-} jumps from zero to $c_i(f)$ at a distance $2/\kappa$ from the surface. Hence, the expelled amount is equivalent to the electrolyte depleted from a volume of area A and thickness $2/\kappa$.

Although [3.7.19] allows the determination of A in only one measurement, the validity of the equation should be checked. Over a range of concentrations a plot of $V \Delta c_i / c_i$ as a function of $2/\kappa$ should give a straight line with slope A. Because of the $\Delta c_i / c_i^{1/2}$ dependence, measurements should preferably be done at low concentrations and in small volumes. This is also a prerequisite for making the double layer as diffuse as possible. In fact, negative adsorption is sufficiently large only when the potential ψ^d is high enough but then it has the advantage that, because it concerns the double layer part far from the surface, it works when the Gouy-Chapman theory is optimally valid. Except for the requirement $c_{i-}(0) = 0$ no specific adsorption correction is needed.

The negative adsorption method is nowadays more or less routine for clay minerals, which have highly charged surfaces (to ensure complete depletion) and often large specific areas. It has also found applications for silver halides and oxides. With these systems essentially two electrolytes are needed, one, containing the charge-determining ions to make the surface potential high enough, the other containing the negatively adsorbing ion. Obviously a compromise between the two concentrations has to be found; moreover in these cases an equation for the pertaining electrolyte mixture has to be derived, see sec. 3.5d.

Some illustrations will be given in sec. 3.10a; see in particular fig. 3.40.

3.8 Points of zero charge and isoelectric points

Although points of zero charge (p.z.c.) are typically properties of a surface in the *uncharged* state, they play such an important role in double layers that special attention will be paid to them in the present chapter. This relevance derives from four facts:

(i) Points of zero charge reflect the affinity of the surface under study for charge-determining ions and are therefore a surface characteristic.

(ii) Points of zero charge are the reference points for the surface charge. In particular, potentiometric colloid titrations (sec. 3.7a) only give relative values of σ^0. These can be made absolute when the p.z.c. is known.

(iii) Points of zero charge are sensitive to specific adsorption and can therefore be used to detect and quantify this phenomenon. Impurities on a surface may significantly affect the p.z.c., so that a purity check also offers itself.

(iv) Comparison between p.z.c. and i.e.p. is very helpful to detect and quantify specific adsorption.

Points of zero charge can only be defined for surfaces that can be either positively or negatively charged. For silver halides and oxides this means that at least two charging reactions must be considered. The point of zero charge of an oxide cannot be determined by, say, only adsorption of protons. As a second process the desorption of protons or adsorption of hydroxyl ions presents itself. Thermodynamically the sum effect of these two charging mechanisms, or rather their competition, is observed. Therefore, points of zero charge are related to a difference between two pK's, see [3.6.45], that formally may be modelled into one, using "one-pK" models.

For polarized interfaces (mercury) the point of zero charge is not defined by the composition of the solution; rather, it is a certain applied potential with respect to a reference electrode. It is the potential of the *electrocapillary maximum*, also called the *potential of zero charge*, see [I.5.6.17]. In this case, the two charging processes are supply and withdrawal of electrons, the analogues of desorption and adsorption of protons on oxides.

In relation to the basically dual nature of the charging process, we have two possible ways of identifying the p.z.c.; as pAg$^{\circ}$ or pI$^{\circ}$ for silver iodide, as pH$^{\circ}$ or pOH$^{\circ}$ for oxides, etc. It is customary to choose the first option and we shall follow this usage, but this is not a matter of principle. The two are related by

$$pAg^{\circ} + pI^{\circ} = pL_{Agl} \qquad\qquad [3.8.1]$$

$$pH^{\circ} + pOH^{\circ} = pW \qquad\qquad [3.8.2]$$

where L_{Agl} is the solubility product of AgI and W the ion product of water. At 20 and 25°C pW = 14.17 and 14.00, respectively. Generally,

$$pW = pK_{w}^{c} + \log V_{m,w} \qquad\qquad [3.8.3]$$

Strictly speaking, pAg's, pH's, etc. are negative logarithms of activities. This means that an operation should be agreed to determine them, an issue that is beyond the present topic. However, we shall take the pragmatic stance that we may express them as negative logarithms of concentrations. The arguments for doing so are that (i) P.z.c.'s are typically measured in indifferent electrolytes the concentrations of which are large compared with c_{H^+}, c_{Ag^+}, etc. so that the activity coefficients are more or less fixed. (ii) Uncertainties caused by unknown activity coefficients are almost always smaller than those due to the chemistry of the surface (crystal plane, heterogeneity, trace impurity adsorption, etc.).

Potentials of zero charge are not necessarily identical to "potentials of zero potential" because, as a rule, the interfacial potential jump $\chi \neq 0$ (sec. 3.9). As interfacial potentials across boundaries between dissimilar phases are thermodynamically undefined (sec. I.5.5), the notion of point of zero potential will be avoided. Strictly speaking, the point of zero charge is not operational either; at least from titration, the most common method of measuring them (see sec. 3.8a), one can only find the sum of the adsorptions of two charge-determining species, not their individual absolute values. However, there are methods of determining the condition $\sigma^{\circ} = 0$ with some confidence.

Because of the sensitivity of the p.z.c. to small amounts of adsorbed charged or uncharged molecules ("impurities"), it is for some purposes expedient to introduce the notion of *pristine point of zero charge* (p.p.z.c.) as the point of zero charge of a virgin (uncontaminated) surface[1]. Only the p.p.z.c. is a measure of the intrinsic properties of the dispersed matter. Generally, it depends on the crystal plane, temperature and composition of the solution.

The electrokinetic counterpart of the point of zero charge is the *isoelectric point* (i.e.p.), that is the pAg, pH, etc. where by electrokinetic methods no charge is measured. I.e.p. and p.z.c. are (very) different quantities because the former measures the situation where $\zeta = 0$, that is where essentially the diffuse charge is zero, whereas the latter represents a zero *surface* charge. Only under pristine conditions are the two identical. We shall elaborate this distinction in sec. 3.8b.

In the present section, we shall concentrate on aqueous systems. For non-aqueous fluids, see sec. 3.11.

3.8a. *Experimental determination*

Although the point of zero charge is a property of an uncharged surface, its experimental determination always involves charges because some double layer property is measured as a function of pH, pAg, and interpolated to give pH^o, pAg^o, etc.

The most general and safe procedure to obtain a point of zero charge is by considering the salt effect on σ^o (pAg) or σ^o (pH) curves, obtained by titration. Basically, titrations of the type shown in fig. I.5.17 are performed. When a *common intersection point* (c.i.p.) is found at different indifferent electrolyte concentrations, the pAg or pH, of the c.i.p. is identified as pAg^o or pH^o, reasoning that when there is no double layer, there is no charge to be screened and hence no influence of added indifferent electrolyte. In principle this procedure is correct; the main problem is to ensure that the electrolyte is really indifferent, that is, it contains only generically adsorbing ions.

Let us consider the procedure in some detail, taking oxides as the examples. Let us assume that careful titrations have been carried out, taking all required precautions (sec. 3.7a) and that the pH range around the intersection point(s) of the curves has been studied in some detail. The outcome may look like fig. 3.34a or b.

In fig. 3.34a, a common intersection point is observed. Usually this means that the electrolyte is indifferent, so that c.i.p. = pH^o, but a caveat is needed. A zero salt effect means that the Esin-Markov coefficient β is zero; see the definitions in [3.4.14 and 15]. According to [3.4.16], β is directly related to the fraction of the surface charge that is compensated by cations and anions. For a monovalent electrolyte, at $\beta = 0$,

---

[1] This term has been coined for somewhat different purposes by M.A.F. Pyman, J.W. Bowden and A.M. Posner, *Austr. J. Soil Res.* **17** (1979) 191.

$$\left(\frac{\partial\sigma_+}{\partial\sigma^0}\right)_{a_\pm} = \left(\frac{\partial\sigma_-}{\partial\sigma^0}\right)_{a_\pm} = -\frac{1}{2} \qquad\qquad [3.8.4]$$

For a (2-1) electrolyte the figures are $-\frac{2}{3}$ and $-\frac{1}{3}$, respectively, etc. Equation [3.8.4] means that, upon a very small change in σ^0, cations and anions contribute to the same extent to the compensation, one by positive, the other by negative adsorption. This is entirely in line with a diffuse double layer on a pristine surface; for $y^d \to 0$, $\sigma_+^d \to \sigma_-^d$, as follows from [3.5.25c]. However, a zero value of β may also be found if cation and anion both adsorb specifically to the same extent, or if one may adsorb specifically and the other electrostatically but to the same extent. These are rather fortuitous cases, but they can be confidently excluded if for two different electrolytes (say KF and KNO_3 or KNO_3 and $NaClO_4$) precisely the same c.i.p. is found. At the same time one has then established the pristine point of zero charge and the absence of specific adsorption. A further confirmation stems from the fact that in that case the p.p.z.c. = i.e.p.

Situation b of fig. 3.34 is more common and is indicative of specific adsorption of cations. (For specific adsorption of anions the intersection points

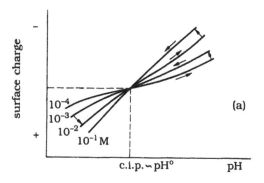

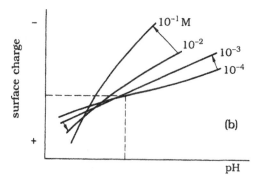

Figure 3.34. Salt effect on the surface charge near the point of zero charge. Top, a common intersection point is found; bottom, with increasing electrolyte concentration, the intersection points shift to lower pH and to more positive surface charge.

move to the right and towards more negative surface charges.) When, starting from the p.p.z.c., salt is added, of which cations adsorb specifically, the originally equal affinity of the surface for H$^+$ and OH$^-$ (or for adsorption and desorption of protons) is unbalanced, in that the surface becomes slightly negative. To re-establish the p.z.c., the pH has to be reduced. The intersection point, and hence the p.z.c., moves to the left, away from its pristine value. In the figure it is assumed that in very dilute electrolyte (crossing of the 10^{-3} M and the 10^{-4} M curves) specific adsorption is negligible so that there p.z.c. $\sim$ p.p.z.c. At the same time, the surface charge where $\beta = 0$ becomes more positive with increasing specific adsorption of the cation. This is also understandable: on an uncharged surface, cations would adsorb more strongly than anions; in order to balance the two, the specific adsorption of the cation must be compensated by rendering the surface more positive. At the cross-over, chemical affinity and electrostatic inhibition balance each other. The trend is that with increasing c_{salt} the specifically adsorbed amount increases, so that the intersection point continues to move to the left and to the more positive side. When a plateau is reached in the specific adsorption, no further change of the intersection point is observed; in that case a second, non-pristine c.i.p. is found which, in contradistinction to the pristine c.i.p., is not situated at $\sigma^o = 0$ and depends strongly on the nature of the electrolyte. See also fig. 3.57, in which connection some literature examples will be cited.

Interpretation of the shifts of intersection points in terms of specific adsorption energies is straightforward but requires a model. Several trends have already been predicted in figs. 3.23 and 25. One could check which combination of K_1 (or K_1^c), C^i (or C_1^i and C_2^i) and position of the inner Helmholtz plane fits the data best.

The titration method described above is the most fool proof and general, but alternatives do exist. Some of these are:

(i) *Ageing.* When the surface area of a suspension decreases by Ostwald ripening charge-determining ions have to desorb, to an extent and in a direction depending on how far away from the p.z.c. the system is. At the p.z.c. there is no measurable preferential desorption. Problems are that (a) it is a slow process and (b) the p.p.z.c. itself may change due to increasing perfection of the crystal faces.

(ii) *Immersion* of the dry powder (say AgI) into solutions of various pAg and noting the direction of the change of pAg. At pAg = pAg^o there is no change. Drawback: it is difficult to ensure extreme purity of the powder.

(iii) *Suspension effect* (sec. I.5.5f). The sign of this effect depends on the sign of the double layer charge; the effect is zero at the p.z.c. The measurements are relatively simple but the interpretation poses problems: the phenomenon depends in a complicated way on the transference numbers of bound ions, and their local concentrations. The problem of whether a p.z.c. or an i.e.p. is found depends on the mobility of the ions behind the slip plane, i.e. on the contribution of the Stern layer to the surface conduction, for which no general rules can be given. See sec. 4.6f.

(iv) *Sol stability*. The minimum in the stability of electrostatically stabilized sols as a function of pH, pAg, etc. does not give the p.z.c. but rather the i.e.p. because particle interaction is governed by the diffuse part of the double layer.

(v) *Capacitance minimum method*. This procedure is based on the minimum of the diffuse double layer capacitance at the zero point of that part of the double layer (fig. 3.5). Experimentally, the minimum is obtainable by differentiation of σ^0(pH) or σ^0(pAg) curves, i.e. one virtually establishes the inflection points in these curves. Alternatively, for electrodes the capacitance can be determined directly but the method requires measurements at various pAg, pH, etc. The capacitance method only works when the double layer is predominantly diffuse (otherwise the minimum is invisible, see fig. 3.22a) in the absence of specific adsorption and with symmetrical electrolytes. For asymmetrical salts the minimum does not coincide with the zero point, see fig. 3.10.

(vi) *Special methods* specific for certain systems[1]. Examples: the electro-capillary maximum for mercury, the flat-band potential for semiconductors. It is noted that some procedures proposed to measure a p.z.c. in fact lead to the i.e.p. Measurement of the adsorption maximum of uncharged organic molecules, identifying this maximum with the p.z.c. is an incorrect procedure because the maxima do not coincide with the p.z.c. (sec. 3.12d).

Isoelectric points require electrokinetic experiments as a function of pAg, pH, etc. which will be discussed in sec. 4.4. Theoretical problems are all but absent because phenomena like surface conduction and relaxation retardation vanish as $\zeta \to 0$. However, experimental problems may arise because the systems become colloidally unstable near the i.e.p. so that, say, no electrophoretic mobility can be measured under conditions where this parameter is required. Another issue is that with micro-electrophoresis the sol is very dilute and hence has a low buffer capacity against inadvertent impurities. For instance, traces of silicate from the vessel wall may adsorb onto the particles and substantially affect the surface properties. To avoid such problems, measurements with plugs of particles (streaming potentials, electro-osmosis) may be recommended. When measurements are carried out with sols, the i.e.p. is obtained by interpolation between data obtained on the positive and negative side. Extrapolation to zero of data

---

[1] For metals, see A.N. Frumkin, O.A. Petrii and B.B. Damaskin, *Potentials of Zero Charge*, in *Comprehensive Treatise of Electrochemistry*, Vol. **1**, *The Double Layer*, chapter 5, J. O'M. Bockris, B.E. Conway and E. Yeager, Eds., Plenum Press (1980).

obtained on one side of the i.e.p. is risky, but in some systems there is no better alternative.

When the electrokinetically found i.e.p. is independent of c_{salt} it is likely that the pristine i.e.p. is obtained, which is equal to the p.p.z.c. (the only caveat being the fortuitous compensation of cation and anion adsorption at the maximum of the isotherm, mentioned in connection with [3.8.1]).

3.8b. The difference between the point of zero charge and the isoelectric point

This difference and experimental methods of discrimination and interpretation will now be briefly summarized for the most general situation where the specifically adsorbing ions are not titratable.

The point of zero charge is the condition where $\sigma^0 = 0$, whereas the isoelectric point is that where $\sigma^{ek} = 0$. The position of the slip plane is not yet fully established (sec. 4.4). However, for all practical purposes it may probably be identified with the outer Helmholtz plane so that for the present discussion the relevant point is that σ^{ek} is about $-(\sigma^0 + \sigma^i)$.

When specific adsorption is absent ($\sigma^i = 0$) p.z.c. = i.e.p. = p.p.z.c. This condition can experimentally be established by the zero salt effect on both p.z.c. and i.e.p. It is illustrated in fig. 3.35a, where for sake of simplicity the χ-potential has been disregarded (only the potentials caused by binding of charge-determining and specifically adsorbing ions are now at issue).

In the presence of specific adsorption the p.z.c. and i.e.p. move in different directions according to the scheme of table 3.4. Figure 3.35 explains these trends pictorially for specific adsorption of anions. Starting from pristine conditions (p.z.c. = i.e.p.), when some anions adsorb specifically (fig. 3.35b), some positive charge appears on the surface: because of electroneutrality, these anions must be accompanied by cations; some (or all) of these are charge-determining and adsorb on the surface. In this situation we are no longer at the p.z.c. To re-establish the p.z.c., charge-determining anions have to be added, i.e. the pH or pAg has to be increased (fig. 3.35c). Situation 3.35b does not represent the i.e.p.

Table 3.4. Direction of the shift of point of zero charge and isoelectric point caused by specific adsorption. "Down" and "up" refer to change in the negative logarithm of the concentration of the charge-determining ions.

specific adsorption of	Direction of shift in pH°, pAg°	
	p.z.c.	i.e.p.
cations	down	up
anions	up	down

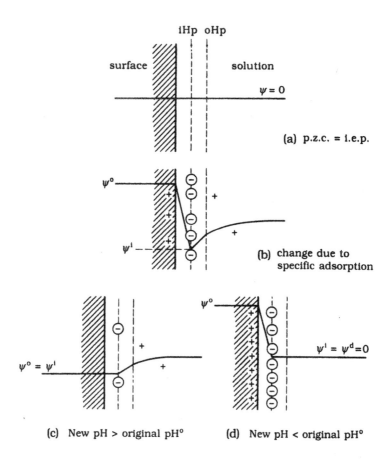

Figure 3.35. Illustrating that due to specific adsorption the p.z.c. and i.e.p. move in opposite directions. Explanation in text.

either since the induced positive (surface) charge is always less than the inducing negative (specifically adsorbed) charge, i.e. $\sigma^o + \sigma^i < 0$. The i.e.p. is re-established by adding charge-determining cations, i.e. the pH or pAg has to be decreased (fig. 3.35d). At the new p.z.c. there is still a small negative charge at the iHp, though less than in situation 3.35b. It is compensated by an equal positive charge in the diffuse part: $\sigma^o = 0$, $\sigma^i + \sigma^d = 0$. At the new i.e.p. there is more negative charge at the iHp than in situation b, now $\sigma^o + \sigma^i = 0$; $\sigma^d = 0$.

For specific adsorption of cations a similar interpretation can be given.

It follows from the above that the shift of the p.z.c. and/or the i.e.p. is direct *proof of specific adsorption*. Together with the observation that these trends are quantitatively different for different specifically adsorbing ions, it is the best criterion. Other means of detection involve the ion specificity of other double layer properties, for instance lyotropic sequences in capacitances or coagulation concentrations.

All of this requires the caveat that the situation becomes different when the specifically adsorbed ion itself is also titrated. This is, for instance, the case with phosphate ions. By potentiometry these ions are then counted as if they were part of the solid. In that case, anions would also reduce the p.z.c. whereas cations increase it. For this reason, SiO_2 dispersions containing traces of alumina have a p.z.c. above that of SiO_2, and alumina dispersions with SiO_2 contamination have a p.z.c. below that of Al_2O_3. Doping of the solid gives rise to similar trends. The difference between the two sets of directions of p.z.c. shifts helps to establish the location and binding of the added ion.

Quantitative interpretations require a model. The various models of sec. 3.6c may be used for that.

3.8c. *Results and interpretation*

For easy reference, a collection of (pristine) points of zero charge is given in appendix 3. These values are fairly well established. Small differences between different experimenters may be real and caused by differences in preparation, pre-treatment, incorporation of traces of foreign substances on the surface or into the solid (doping). Sometimes large differences are found between materials of different origin. For instance, Dumont et al.[1] found for various preparations of TiO_2 the i.e.p. (inferred from coagulation) to vary between 2.9 and 5.9 and Ottewill and Woodbridge[2] observed for specially prepared homodisperse AgI sols an i.e.p. that was about two pAg units lower than that tabulated in app. 3. Honig and Hengst[3] gave an extended compilation for p.z.c. or i.e.p. disparities between different authors and a variety of origins and Somasundaran[4] did so for a number of aluminas, including the minerals corundum and sapphire. Here, pH^o varied between 3.0 and 10.0. Whether such differences are intrinsic properties of the pristine solid surface or caused by contaminants on the surface can be verified by measuring the p.z.c. and i.e.p. simultaneously, but such systematic data are often not available.

Because of these uncertainties, the data in the table are in general restricted to stable, well-reproducible, crystallized materials with the p.z.c. obtained by

[1] F. Dumont, J. Warlus and A. Watillon, *J. Colloid Interface Sci.* **138** (1990) 543.

[2] R.H. Ottewill, and R.F. Woodbridge, *J. Colloid Interface Sci.* **19** (1961) 581.

[3] E.P. Honig, J.H.Th. Hengst, *J. Colloid Interface Sci.* **29** (1969) 510.

[4] P. Somasundaran, in *Clean Surfaces. Their Preparation and Characterization for Interfacial Studies*. Marcel Dekker (1970) 285.

titration. In passing it is noted that surface properties may change upon back-and-forth titration.

Titration of a suspension to establish the point of zero charge invariably means titration of a heterogeneous mixture of different crystal planes that may themselves have different p.z.c.'s. For metal monocrystals this difference has been experimentally established, see appendix 3 table a. Against this background it is perhaps unexpected that many p.z.c.'s are so reproducible. There are a number of possible reasons. One is that points of zero charge are commonly obtained by titrations on rather concentrated dispersions, which have a good buffering capacity with respect to adsorbing impurities, thus reducing the risk of artefacts. A more basic reason is that titrations are usually done on well-aged samples, exhibiting a limited number of low Miller index (i.e. low energy) faces. In titrations with very fresh precipitates the first runs tend to be less reproducible because the higher index surfaces are still disappearing due to Ostwald ripening. A further feature is that some surface groups have such a low or high proton affinity that they are not titrated in the common pH range.

For most solids, points of zero charge, if measured by different authors on different samples, can be reproduced within 0.1 of a pH or pAg unit.

Interpretation of points of zero charge implies establishing and modelling the enthalpy and entropy changes involved in the transfer of charge-determining ions to and from the surfaces. Enthalpies are easier to interpret because they are connected with ion binding for which models can be set up based on the crystal structure of the solid. Entropies are rather related to the organization of adjacent water, which is a more elusive feature. However, entropic contributions may not be ignored because points of zero charge are temperature dependent.

For points of zero charge interpretations, double layer models are not needed.

Thermodynamic interpretation of points of zero charge in terms of Gibbs energies, enthalpies and entropies of the transfer of single charge-determining species from the bulk to the surface is impossible because such a process is inoperable. What *is* possible is to establish thermodynamic functions of the change of the surface charge with pH, pAg, etc. This is the process underlying the determination of σ^o by titration. Let us call this the *charge-formation process*. Its standard Gibbs energy is

$$\frac{\Delta_{cf} G_m^o}{2.303 \, RT} = pK_{cf} \qquad\qquad\qquad [3.8.5]$$

where K_{cf} is the (dimensionless) equilibrium constant of this charge-formation process. The required value at the p.z.c. is

$$\frac{\Delta_{cf}G_m^{\circ}(\text{p.z.c.})}{2.303\ RT} = pK_{cf}(\text{p.z.c.}) \qquad\qquad [3.8.6]$$

that is, at $\sigma^{\circ} \rightarrow 0$.

For the silver iodide system, representing the solid as R, the single ionic adsorption processes

$$RAg^+ + I^- \leftrightharpoons R + AgI \qquad\qquad [3.8.7a]$$

and

$$RI^- + Ag^+ \leftrightharpoons R + AgI \qquad\qquad [3.8.7b]$$

are thermodynamically inaccessible, but (half) their sum

$$\tfrac{1}{2}\left(RI^- + Ag^+ \leftrightharpoons RAg^+ + I^-\right) \qquad\qquad [3.8.7c]$$

is, and the corresponding equilibrium constant is

$$K_{cf} = \left(\frac{[RAg^+]\,x_{I^-}}{[RI^-]\,x_{Ag^+}}\right)^{1/2} = \left(\frac{[RAg^+]\,c_{I^-}}{[RI^-]\,c_{Ag^+}}\right)^{1/2} \qquad\qquad [3.8.8]$$

As the reaction between parenthesis in [3.8.7c] corresponds to a change of σ by *two* charges, we consider half this reaction as the elementary charge-formation step.

At the p.z.c. $[RI^-] = [RAg^+]$, hence

$$K_{cf}(\text{p.z.c.}) = \left(c_{I^-}^{\circ}\,/\,c_{Ag^+}^{\circ}\right)^{1/2} \qquad\qquad [3.8.9]$$

$$\frac{\Delta_{cf}G_m^{\circ}(\text{p.z.c.})}{2.303\ RT} = pK_{cf} = -\tfrac{1}{2}\left(pAg^{\circ} - pI^{\circ}\right) \qquad\qquad [3.8.10]$$

An equation like this was first derived by Bérubé and De Bruyn[1], using potentials and single ion activities in the derivation.

Using [3.8.1], either pAg° or pI° can be eliminated. For the latter case,

$$-pK_{cf} = pAg^{\circ} - \tfrac{1}{2}pL \qquad\qquad [3.8.11]$$

$$\Delta_{cf}G_m^{\circ} = -2.303\ RT\ pAg^{\circ} + 2.303\ RT\ \frac{pL}{2} \qquad\qquad [3.8.12]$$

---

[1] Y.G. Bérubé, P.L. de Bruyn, *J. Colloid Interface Sci.* **27** (1968) 305; see also B. Ball, ibid. **30** (1969) 424.

Expression [3.8.10] shows that it is not the activity of one of the charge-determining species that counts, but rather the ratio between the two. Similarly, [3.8.12] tells us that it is the difference between the activity of one of them and the activity it would have had if there were no preference. For instance, for AgI, $\frac{1}{2}pL \approx 8.1$, $pAg° = 5.6$, meaning that Ag^+ ions have a lower affinity for the uncharged AgI surface than I^- ions. In fact, this is expected in view of the properties of solid AgI and the differences in hydration of Ag^+ and I^- ions. regarding the former, silver halides are built of small Ag^+ ions and big halides (X^-). The (Pauling) crystal ionic radii are $Ag^+ = 0.126$ nm, $I^- = 0.216$ nm, $Br^- = 0.195$ nm and $Cl^- = 0.181$ nm. Locally inside the solid the field is stronger around the Ag^+ ions than around the halides (although part of the binding is covalent). This cannot be the sole reason that on a crystal plane, exposing equal numbers of Ag^+ and X^-, it is more likely that another X^- adsorbs than another Ag^+ because the adsorbing Ag^+-ion also carries a higher field than an adsorbing I^--ion. Packing also plays a role and, last but not least, the Gibbs energies of hydration of the anions, I^-, Br^- and Cl^- tend to be less than that for Ag^+, meaning that the anions are more readily adsorbed, especially if the surface is hydrophobic. In line with this, the asymmetry is most pronounced for AgI. See appendix 3, table c, considering that for AgBr and AgCl $\frac{1}{2}pL \approx 6.1$ and 4.9, respectively.

For oxides the analysis goes along similar lines. The equivalents of the single ionic steps [3.8.7a and b] are $ROH_2^+ + OH^- \rightleftarrows ROH + H_2O$ and $RO^- + H^+ \rightleftarrows ROH$, i.e. the reverse of [3.6.18b and a], respectively. For the overall charge-formation process

$$\frac{1}{2}\left(RO^- + H^+ + H_2O \leftrightarrows ROH_2^+ + OH^- \right)$$ [3.8.13]

for which $K_{cf} = K_b / K_a$

$$K_{cf} = \left(\frac{[ROH_2^+]\, x_{OH^-}^s}{[RO^-]\, x_{H^+}^s \cdot x_w} \right)^{1/2}$$ [3.8.14]

At the p.z.c. $[ROH_2^+] = [RO^-]$, the x^s's become x's and $x_w \approx 1$. Hence, the quotient of mole fractions may also be read as a quotient of concentrations and $K_{cf} \approx K_{cf}^c$. We obtain

$$K_{cf}(\text{p.z.c.}) = \left(x_{OH^-}^o / x_{H^+}^o \right)^{1/2} = \left(c_{OH^-}^o / c_{H^+}^o \right)^{1/2}$$ [3.8.15]

and

$$\frac{\Delta_{cf}G^o_m (\text{p.z.c.})}{2.303\ RT} = pK_{cf} = -\frac{1}{2}\left(pH^o - pOH^o\right) \qquad [3.8.16]$$

Using [3.8.2], pOH^o can be eliminated to give

$$\Delta_{cf}G^o_m (\text{p.z.c.}) = -2.303\ RT\ pH^o + \frac{2.303}{2}RT\ pW \qquad [3.8.17]$$

Equation [3.8.16] shows that oxides for which protons and hydroxyls have the same affinity have $\Delta_{cf}G^o_m$ (p.z.c.) = 0; for these the p.z.c. = 7. For acid oxides, such as silica, $\Delta_{cf}G^o_m$ (p.z.c.) > 0; since $pH^o < pOH^o$, and the p.z.c. is below 7. Again this analysis illustrates that the p.z.c. is not solely determined by the affinity of protons for the surface (a "half-reaction") but by the balance between the affinities for protons and hydroxyls.

In appendix 3b and c values of $\Delta_{cf}G^o_m$ (p.z.c.) are included.

Values for pK_a and pK_b of the half-reactions [3.6.38a and b] can be obtained after making some model assumption. In site binding analyses K_a and K_b, or variants thereof, are usually called *intrinsic dissociation constants* (or binding constants for the reverse reaction), in distinction to the corresponding *apparent dissociation constants* K_a(app.) and K_b(app.) having x_{H^+}, c_{H^+}, etc. in the equation instead of $x^s_{H^+}$, $c^s_{H^+}$, etc. The apparent ones are measured. Differences between $x^s_{H^+}$ and x_{H^+}, etc. are accounted for by Boltzmann factors, as in [3.6.41]. A model is needed to eliminate y^o. In the literature, a number of such approaches have been proposed. They are of different quality (varying from zeroth-order Stern layer to triple layer models (fig. 3.20)), invoking different procedures to eliminate y^o. It is recalled that different authors define their charge-forming mechanisms in different ways (see sec. 3.6f) so that the intrinsic pK's that are eventually obtained may differ widely, although they should be interconvertible.

By way of illustration, in table 3.5 some results are collected, taken from James and Parks[1] who applied a double extrapolation procedure, essentially to go to high ionic strength and low y^o. Other methods tend to be more empirical[2,3]. James and Parks discuss in some detail the choice of their data and parameters needed in the elaboration. The quantity $\frac{1}{2}\left(pK^c_a - pK^c_b + pK^c_w\right)$ corresponds to their choice of the p.z.c. These data may be compared with those in appendix 3.

[1] R.O. James, G.A. Parks, in *Surface and Colloid Science*, Vol. **12**, E. Matijevic, Ed., Plenum (1982) chapter 2.

[2] C.P. Huang, in *Adsorption of Inorganics at Solid-Liquid Interfaces*, M.A. Anderson, A.J. Rubin, Eds., Ann Arbor Sci. (1981) chapter 5.

[3] P.W. Schindler, W. Stumm, in *Aquatic Surface Chemistry*, Wiley (1987) chapter 4.

Table 3.5. Intrinsic pK's for proton uptake or -release for oxides. Defining equations for K_a^c and K_b^c, [3.6.38c and d], respectively. After James and Parks (1982) with the modifications that our pK_a^c is their pK_2^{intr} and our pK_b^c is pK_w^c − their pK_1^{intr}, see sec. 3.6f). Temperature 25°C.

Oxide	Salt	conc./M	pK_a^c	pK_b^c	$\frac{1}{2}(pK_a^c - pK_b^c + pK_w^c)$
TiO_2 (anatase)	NaCl	10^{-3} - 10^{-1}	8.7	10.8	5.95
TiO_2 (rutile)	$NaClO_4$	10^{-3} - 10^{-1}	9.1	11.3	5.90
TiO_2 (rutile)	$NaNO_3$	10^{-3} - 1	9.1	11.2	5.95
TiO_2 (rutile)	KNO_3	10^{-3} - 10^{-1}	9.0	11.4	5.80
α-FeOOH (goethite)	KNO_3	10^{-3} - 10^{-1}	10.5	9.8	7.35
α-Fe_2O_3 (haematite)	KCl	10^{-3} - 1	10.3	7.3	8.50
SiO_2 (Cab-O-Sil)	KCl	10^{-3} - 10^{-1}	7.2		
γ-Al_2O_3 (Alon)	NaCl	10^{-3} - 10^{-1}	11.8	7.8	9.00

The data for pK_a^c and pK_b^c illustrate the relative tendencies of the oxide to become positively or negatively charged: oxides with high pK_a^c and low pK_b^c tend to become positively charged (i.e. they have a high p.z.c.) and conversely. When these pK^c's are converted into ΔG_m^o's of individual charging processes, the reference states have to be defined. They should agree with [3.8.16].

A number of other interpretations of points of zero charge have been offered. Parks and De Bruyn found for oxides a connection between the p.z.c. and the pH of minimum solubility[1]. Some authors claimed a relation between the *heat of immersion* and the p.z.c.[2]. Apart from the fact that $\Delta_{imm}H_m$ is not easy to establish, the relation is at best indirect, as immersion is determined by the interaction between solid and water, the p.z.c. by that between ions and the solid. As far as such a relation exists, it probably involves the structure of the water adjacent to the solid. For a number of metals Trasatti interpreted a certain parallel between the p.z.c. and the electronic work function[3] in terms of the adjacent water structure, particularly its polarization[4]. The work function is a measure of the energy involved in extracting an electron from a metal. Interpretation of the transfer of electrons through polarized solvent layers

[1] G. Parks, P.L. de Bruyn, *J. Phys. Chem.* **66** (1962) 967.

[2] T.W. Healy, D.W. Fuerstenau, *J. Colloid Interface Sci.* **20** (1965) 376; F. Dumont, J. Warlus and A. Watillon, *J. Colloid Interface Sci.* **138** (1990) 543.

[3] This analogy is long established, see for instance A.N. Frumkin and A. Gorodetzkaya, *Z. Phys. Chem.* **136** (1928) 451 and A.N. Frumkin, *J. Colloid Sci.* **1** (1946) 29a. However, the relation is probably even better at negative surface charge, see S. Trasatti, *J. Chem. Soc. Faraday Trans.* I **70** (1974) 1752; A. Frumkin, B. Damaskin, I. Bagotskaya and N. Grigoryev, *Electrochim. Acta* **19** (1974) 75.

[4] S. Trasatti, *Gazz. Chim. Ital.* **106** (1976) 219.

automatically involves the elusive issue of the χ-potential (sec. 3.9). For a number of crystal planes of gold and silver, a relationship was found between the p.z.c. and the surface excess energy of that plane, interpreted in terms of the number of bonds that have to be broken in order to cut a perfect crystal along the plane under consideration[1]. In these cases, the affinity of electrons for the surface will be partly determined by the polarization of the adjacent solvent which, in turn is, at least in part, affected by the surface structure.

More relevant for our purposes are models for oxides in which, in some way, the affinity of charge-determining ions is related to the surface electric field[2]. For example, in the theory of Hiemstra et al. proton affinities are computed in terms of bond distances, coordinations and charge distribution in the solid surface, i.e. it is essentially an energy interpretation. For several oxides there are arguments that this is an acceptable approximation. This model also showed that not all potentially available sites are titrated in the usual pH range.

The temperature dependence of the p.z.c. is particularly relevant because such data offer additional information on the thermodynamics of the charge formation. Reliable data, over a sufficient range of temperatures, to arrive at significant conclusions, are now available for silver iodide (fig. 3.36), haematite and rutile (fig. 3.37). All curves show descending trends with increasing temperature and the p.z.c.'s become more symmetrical in the sense that pAg^o and pI^o, and pH^o and pOH^o approach each other. This is intuitively expected. The p.z.c.'s of rutile and haematite happen to be mirror images of each other which is a consequence of the (fortuitous) fact that pH^o (haematite), within experimental error, is as much above $pW/2$ as pH^o (rutile) is below it. From these data $\Delta_{cf}G^o_m$ can be derived as a function of temperature. For silver iodide $\Delta_{cf}G^o_m/T$ appears to be independent of T over the entire range; for the two oxides this is the case between 5° and $50^\circ C$. From the Gibbs-Helmholtz equation [I.2.15.8a] it then follows that $\Delta_{cf}H^o_m$ is constant. Relevant thermodynamic data are collected in table 3.6. It is concluded that for the hydrophobic silver iodide the entropy contribution to the charge formation is much more important than it is for the more hydrophilic oxides. Transport of a proton from bulk water to water adjacent to hydroxylated oxide surfaces involves much less structural rearrangement than that of silver or iodide ions to the vicinity of water-shunning silver iodide.

---

[1] A. Hamelin, J. Lecoeur, *Surf. Sci.* **57** (1976) 771; A. Hamelin, in *Modern Aspects of Electrochemistry*, J. O'M. Bockris, B.E. Conway, Eds. Plenum, Vol. **16** (1985) 1; R. de Levie, *J. Electroanal. Chem.* **280** (1990) 179.

[2] G. Parks, *Chem. Revs.* **65** (1965) 177; T.W. Healy, A.P. Herring and D.W. Fuerstenau, *J. Colloid Interface Sci.* **21** (1966) 435; R.H. Yoon, T. Salman and G. Donnay, *J. Colloid Interface Sci.* **70** (1979) 483; M.A. Butler, D.S. Ginley, *J. Electrochem. Soc.* **125** (1978) 228; T. Hiemstra, W.H. van Riemsdijk and G.H. Bolt, *J. Colloid Interface Sci.* **133** (1989) 91; W.F. Bleam, *J. Colloid Interface Sci.* **159** (1993) 312.

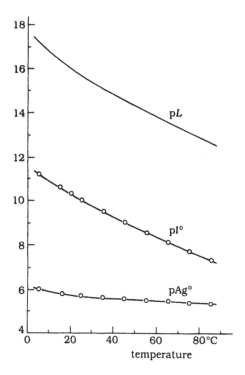

Figure 3.36. Temperature dependence of the point of zero charge of silver iodide. Electrolyte, 10^{-3} M K-biphtalate and 0.099 M KNO_3. After J. Lyklema and A.F.C. Korteweg (partly published *Discuss. Faraday Soc.* **42** (1966) 81).

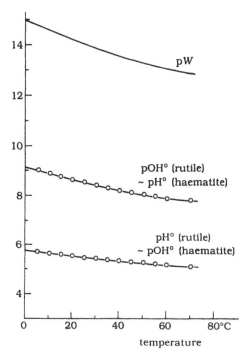

Figure 3.37. Temperature dependence of the points of zero charge of rutile (TiO_2) and haematite (α-Fe_2O_3). The two data are almost exactly mirror images in terms of pH^o and pOH^o. (Data from L.G.J. Fokkink, A. de Keizer and J. Lyklema, *J. Colloid Interface Sci.* **127** (1989) 116.)

Table 3.6. Gibbs energies, enthalpies and entropies of charge formation for silver iodide, rutile and haematite. Electrolyte, 0.02 M KNO_3 $\Delta_{cf}G_m^o$ defined by [3.8.10] or [3.8.16]. All data in kJ mole^{-1}. The assumption is made that $\Delta_{cf}H_m^o$ is independent of temperature.

		$\Delta_{cf}G_m^o$	$\Delta_{cf}H_m^o$	$T\Delta S_{cf}^o$
AgI	10°C	13.34	36.4	23.04
	60°C	9.09	36.4	27.29
TiO_2	10°C	8.86	9.9	1.01
	50°C	8.60	9.9	1.27
α-Fe_2O_3	10°C	-8.86	-9.9	-1.01
	50°C	-8.60	-9.9	-1.27

Dissection of the three parameters given in table 3.6 into their components requires a non-thermodynamic argument. For the two oxides such an assumption may be based on the experimental observation that the directly measured molar enthalpy of charge formation appeared independent of the sign and magnitude of the surface charge[1]. This may mean that the charge-forming process below and above the p.z.c. is the same, viz. uptake or dissociation of a proton by, or from, a certain surface hydroxyl group. Then, it is logical to interpret $\Delta_{cf}H^o = \frac{1}{2}(\Delta_{ads}H_{H^+}^o - \Delta_{ads}H_{OH^-}^o)$ as $\frac{1}{2}(2\Delta_{ads}H_{H^+}^o - \Delta_{neutr}H_w^o)$ and similarly for $T\Delta_{cf}S^o$. As $\Delta_{neutr}H_w^o$ varies almost linearly with temperature from 62 kJ mole^{-1} at 0°C to 49 kJ mole^{-1} at 60°C, $\Delta_{ads}H_{H^+}^o$ can be computed. For 20°C, $\Delta_{ads}H_{H^+}^o =$ -17.6 kJ mole^{-1} for rutile and -36.3 kJ mole^{-1} for haematite. The directly measured values are -22 and -36 kJ mole^{-1}, respectively, indicating satisfactory consistency. The entropies for proton adsorption are almost the same for these oxides and about half the entropy of binding a proton to a hydroxyl in bulk, in line with the suggestion that the water structure near these oxides is not too different from that of bulk water. The expected trend that $\Delta_{ads}H_{H^+}^o$ becomes more negative for oxides with a higher point of zero charge is, with some scatter, confirmed in a review by Blesa et al.[2]. In fig. 3.61 we come back to this rule.

Such a penetrating analysis is not yet possible for silver iodide because direct enthalpy measurements of the charge formation process are not yet available. Systematic studies for oxides with a very high (Al_2O_3) or very low (SiO_2) point of zero charge also deserve attention; these may behave differently.

Regarding mixed oxides, the point of zero charge of the mixture may give a hint about the surface composition (enrichment of one of the components?), a

[1] A. de Keizer, L.G.J. Fokkink and J. Lyklema, *Colloids Surf.* **49** (1990) 149. Similar trends were reported by M.L. Macheski, M.A. Anderson, *Langmuir* **2** (1986) 582.
[2] M.A. Blesa, A.J.G. Maroto and A.E. Regazzoni, *J. Colloid Interface Sci.* **140** (1990) 287.

matter that we shall not pursue here. In addition, deviations of p.z.c.'s from the pristine ones may point to the (inadvertent or intentional) presence of impurities.

For further reading on points of zero charge see [1].

3.9 Interfacial polarization and the χ-potential

Besides the ionic double layers that may be present at phase boundaries there is also a second type of double layer, caused by polarization of the interfacial region, i.e. a double layer not attributable to free ions. An important contribution is the preferential orientation of solvent dipoles and multipoles close to the surface. These molecules may also have induced dipoles. In the surfaces of solids the centres of positive and negative charges are, as a rule, displaced as compared with the situation in the bulk. All these charge displacements together constitute the *interfacial polarization*. The associated potential difference across phase boundaries is called the *interfacial potential (drop)* or *χ-potential*.

In sec. I.5.5a this type of potential was introduced and put in a thermodynamic context. For the interfacial potential of a phase α with respect to a vacuum we write χ^{α}, for that between two phases α and β, $\chi^{\alpha\beta}$ (potential in β with respect to that in α).

Ionic double layers and double layers caused by interfacial polarization occur together but are not independent. Changing the surface charge will affect the polarization of adjacent solvent, so that χ is generally a function of σ^{o}. Specific adsorption of ions in the Stern layer is intimately coupled to the solvent structure and conversely. The inner layer capacitances K_{1}^{i}, K_{2}^{i}, C_{1}^{i} and C_{2}^{i} are also coupled to the interfacial polarization via the local relative dielectric permittivities ε_{1}^{i} and ε_{2}^{i}.

Interfacial polarization is an autonomous phenomenon in the sense that it always occurs, irrespective of the presence of ionic charges. One of the consequences is that at the point of zero charge generally $\chi \neq 0$, so that p.z.c. $\neq$ p.z.p. (point of zero potential). Another manifestation is the existence of a non-zero χ^{w} at the free surface of water.

From all of this it follows that interfacial polarization is an interesting aspect of double layers. At the same time the issue is challenging because the interest in the structure of liquids near phase boundaries clashes with the thermodynamic inoperability of χ-potentials. Attempts to obtain χ from experiment invariably require model assumptions and this is, of course, also true for theoretical interpretations. Therefore, the assessment of χ-potentials has drawn

---

[1] A.N. Frumkin, O.A. Petrii and B.B. Damaskin, *Potentials of Zero Charge*, in *Comprehensive Treatise of Electrochemistry*, J. O'M. Bockris, B.E. Conway and E. Yeager, Eds. Plenum (1980) Vol. 1, chapter 5 (Emphasis on metals).

much attention, also for metal electrodes[1]. In view of the relevance of electrode potentials for a number of phenomena (electrode kinetics, corrosion, electrochemical sensors, etc.) this is not surprising. Much of this work, detailed as it may be, has the limitation that only dipole orientation in one layer is emphasized, i.e. only a part of one of the components of χ is considered. In view of the strongly associated structure of water, this is a simplification. That the structuring of water and other fluids extends over more than one layer is for instance inferred from the solvent structure-induced interaction forces, between surfaces, which extend over a few molecular layers, see figs. 2.2 and 3.

In colloid science the "problem" alluded to the above, can sometimes be avoided by making thermodynamic interpretations in terms of surface charges rather than potentials, but the challenge and relevance remain. We shall therefore now discuss some important attempts to estimate χ experimentally or theoretically, emphasizing water in contact with non-metals.

Before considering model approaches it is necessary to establish the thermodynamic framework. In this way, the limits of the physical models are defined. To that end a few conclusions from Vol. 1 are briefly repeated.

Measuring potential differences across phase boundaries requires determination of the electrical work involved in transferring an ionic species from the bulk of one phase to the other. When the phases are different the total work consists of an electrostatic and a non-electrostatic, or "chemical" contribution. This sum is nothing else than the molar Gibbs energy of transfer, i.e. the difference between the molar Gibbs energies of the ion between the two bulk phases. There is no unambiguous thermodynamic way to divide the total work into these constituents, hence the basic inoperability of the χ-potential. The inoperability of solvation Gibbs energies has been discussed before (sec. I.5.3c).

The (unmeasurable) total, or Galvani, potential difference ${}^{\alpha}\Delta^{\beta}\phi$ consists of the (measurable) Volta potential difference between the two phases and the (unmeasurable) interfacial potential jump $\chi^{\alpha\beta}$ (see [I.5.5.3 and 4]) and is given by

$$ {}^{\alpha}\Delta^{\beta}\phi = \psi^{\beta}(\alpha) - \psi^{\alpha}(\beta) + \chi^{\alpha\beta} \qquad [3.9.1] $$

[1] S. Trasatti, The Electrode Potential, in Comprehensive Treatise of Electrochemistry, Vol. 1, J. O'M. Bockris, B.E. Conway and E. Yeager, Eds. Plenum (1980), chapter 2; B.E. Conway, The State of Water and Hydrated Ions at Interfaces, Adv. Colloid Interface Sci. 8 (1977) 91; W.R. Fawcett, Molecular Models for Solvent Structure at Polarizable Interfaces, Isr. J. Chem. 18 (1979) 3; M.A. Habib, Solvent Dipoles at the Electrode-Solution Interface, in Modern Aspects of Electrochemistry, Vol. 12, J. O'M. Bockris and B.E. Conway, Eds. Plenum (1977) 131; S. Trasatti, Solvent Adsorption and Double Layer Potential Drop at Electrodes, in Modern Aspects of Electrochemistry, B.E. Conway and J. O'M. Bockris, Eds. Vol. 13 Plenum (1979) chapter 2; J. O'M. Bockris, K-T. Jeng, Water Structure at Interfaces: The Present Situation, Adv. Colloid Interface Sci. 33 (1990) 1.

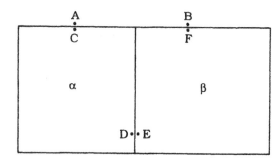

Figure 3.38. Illustration of the meanings of the potentials in [3.9.1 and 2]. The dots indicate positions inside the phase where they are measured, close enough to the adjoining phase to feel its electrical effect, but not so close as to experience its chemical influence. In these positions Volta potentials are measured.

or

$$^{\alpha}\Delta^{\beta}\phi = {^{\alpha}\Delta^{\beta}}\psi + \chi^{\beta} - \chi^{\alpha}$$ [3.9.2]

Volta potentials are potentials just beyond the range of chemical forces so that the work of transfer of a charge from a reference at infinite distance, but in the same phase, to the point where ψ is measured, is purely electrostatic. For a phase α in contact with air, ψ^{α} is obtainable by bringing an electrode just outside the surface, making the gap conducting by adding a radioactive ionic substance. See fig. 3.38, point A and fig. 3.75. Similarly, ψ^{β} is the Volta potential in B, and $^{\alpha}\Delta^{\beta}\psi = \psi^{\beta} - \psi^{\alpha} = \psi(B) - \psi(A)$. Further, $\psi^{\beta}(\alpha)$ is the Volta potential at E, etc. Inside condensed phases, it is not so easy to measure a Volta potential. Perhaps the electrokinetic potential in the absence of specific adsorption may serve as such. Generally, $\chi^{\alpha\beta} \neq \chi^{\beta} - \chi^{\alpha}$, i.e. one cannot find the interfacial potential jump between, say a liquid and a solid, from the difference between the two corresponding χ-potentials with respect to air. Volta potentials vanish at the point of zero charge, because they are caused by free charges. Hence,

$$^{\alpha}\Delta^{\beta}\phi = \chi^{\alpha\beta}$$ (p.z.c.) [3.9.3]

The inoperability of Galvani potential differences between dissimilar phases is closely related to that of single ionic activities. In a homogeneous phase the electrochemical potential of an ionic species i, $\tilde{\mu}_i$ is split into a chemical potential μ_i and an electrical work term (see sec. I.5.1c)[1] according to

$$\tilde{\mu}_i^{\alpha} \equiv \mu_i^{\alpha} + z_i F \phi^{\alpha}$$ [3.9.4]

---

[1] Note that in chapter I.5 and elsewhere in the present volume, we generally use ψ as the symbol for potential. This is current usage and no problems arise as long as no phase boundaries have to be crossed. Here we prefer ϕ to stress that we are dealing with an inner, or Galvani potential, and not with a Volta potential (symbol ψ).

This splitting is also thermodynamically inoperable. For two phases α and β at equilibrium, $\tilde{\mu}_i^\alpha = \tilde{\mu}_i^\beta$. As $^\alpha\Delta^\beta\phi = \phi^\beta - \phi^\alpha$, [3.9.4 and 2] can be combined to give

$$z_i F(\psi^\beta - \psi^\alpha) + z_i F(\chi^\beta - \chi^\alpha) = \mu_i^\alpha - \mu_i^\beta \qquad [3.9.5]$$

Following [I.5.5.6] it is convenient to introduce *real potentials* α_i,

$$\alpha_i^\alpha = \tilde{\mu}_i^\alpha - z_i F\psi^\alpha = \mu_i^\alpha + z_i F\chi^\alpha \qquad [3.9.6]^{1)}$$

Considering the first equality, α_i^α is measurable. It is the isothermal reversible work required to extract ions from phase α; for electrons in a metal it represents the electronic work function. For metals, α_{el} can also be obtained from thermo-emission or the photo-electric effect. Sometimes α_i is called the *real (Gibbs) energy of hydration of ion i*. The logic behind this last definition stems from the second equality in [3.9.6]. The standard molar Gibbs energy of solvation of an ion [I.5.3.1] $\Delta_{solv}G_{mi}^o$ equals μ_i^o when μ_i^o is referred to the gas phase. Assuming the transfer of an ion to proceed under standard conditions, we may, in [3.9.6], replace μ_i by μ_i^o (the concentration term is deleted) and set $\mu^\dagger$ (for the gas phase) equal to zero. Then,

$$\alpha_i^\alpha = \Delta_{solv}G_{mi}^o + z_i F\chi^\alpha \qquad [3.9.7]$$

indicating that α_i^α is the measurable sum of the (inoperable) single ionic Gibbs energy of solvation and the (unmeasurable) potential jump between the liquid and the gas (reference) phase. Obviously for an electroneutral electrolyte $\alpha_+^\alpha + \alpha_-^\alpha = \Delta_{solv}G_{m,salt}^o$, because the χ-terms cancel.

Equation [3.9.7] offers a route to assess χ's: find ionic solvation energies by splitting up $\Delta_{solv}G_m^o$ of salts into its ionic constituents using one of the strategies of sec. I.5.3a, and combine the outcome with measured values of α_i.

The practice of this approach is full of pitfalls. First, values of α are not all that easy to obtain. Measuring Volta potentials is readily disturbed by trace impurities on the surface by static charges, and, for solids, by surface heterogeneity. Already as long ago as 1937, Klein and Lange wrestled with these problems[2]. Randles[3] measured α's for aqueous solutions, descending along the

[1] This quantity has previously been discussed by E. Lange and K. Mischenko, Z. Phys. Chem. **149** (1930) 1.

[2] O. Klein, E. Lange, Z. Elektrochem. **43** (1937) 570; **44** (1938) 562.

[3] J.E.B. Randles, Trans. Faraday Soc. **52** (1956) 1573. For non-aqueous and mixed media such data were obtained by B. Case, N.S. Hush, R. Parsons and M.E. Peover, J. Electroanal. Chem. **10** (1965) 360, see also B. Case, R. Parsons, Trans. Faraday Soc. **63** (1967) 1224. The principle of this method dates back to F.B. Kendrick, Z. Phys. Chem. **19** (1896) 625.

Table 3.7. Real potentials of monovalent ions in aqueous solution, after Randles, ionic Gibbs energies of hydration taken from table I.5.4, and the difference $z_iF\chi^w$ between them. All data in kJ mole^{-1} (25 mV in χ^w and α_i corresponds with 2.4 kJ mole^{-1} in $F\chi^w$ and $F\alpha_i$). The sign convention is such that a positive value of χ^w implies that water is positive with respect to air.

Cation	$z_iF\alpha_i$	$\Delta_{hydr}G^o_{mi}$	$z_iF\chi^w$	Anion	$z_iF\alpha_i$	$\Delta_{hydr}G^o_{mi}$	$z_iF\chi^w$
Li$^+$	-511 ± 2	-481	-30	F$^-$	-415 ± 5	-472	$+57$
Na$^+$	-411 ± 2	-375	-37	Cl$^-$	-296 ± 5	-347	$+57$
K$^+$	-337 ± 2	-304	-33	Br$^-$	-272 ± 5	-321	$+50$
Rb$^+$	-316 ± 2	-281	-35	I$^-$	-239 ± 5	-283	$+44$
Cs$^+$	-284 ± 2	-258	-26				

inside of a hollow vertical tube, in the centre of which a mercury jet acted as the reference electrode. By taking certain precautions he obtained results that were claimed to be reproducible within a few mV. Results are collected in table 3.7 and compared with values of $\Delta_{hydr}G^o_{mi}$, as tabulated in table I.5.4. The values derived in this way for the surface potential of water χ^w illustrate the nagging problem that $z_iF\chi^w$ is only a small difference between two large quantities, of which the absolute value of one (viz. $\Delta_{hydr}G^o_{mi}$) is subject to a basically arbitrary reference point. To establish the individual hydration Gibbs energies, $\Delta_{hydr}H^o_{m,H^+}$ was, on good grounds, taken as -1094 kJ mole^{-1}. The uncertainty of this value is a few percent, say about 30-50 kJ mole^{-1}. To establish absolute values for $\Delta_{hydr}G^o_{mi}$ this is adequate, but it is not good enough to determine χ^w. Even the sign is not certain. Randles[1] tabulated a number of χ^w-values obtained by different authors, making different model assumptions for $\Delta_{hydr}G^o_{mi}$. They vary between +0.5 V and -0.5 V. The highest value is based on a splitting of $\Delta_{hydr}G^o_m$ (salts) by assuming the contributions of F$^-$ and K$^+$ to be equal[2], the lowest on a certain model for the water-ion interaction[3]. In fact, the reverse of our procedure, viz. to obtain χ^w from a simple model, and combine it with α_i-data to obtain individual ionic activities or individual hydration Gibbs energies is less prone to error.

Subsequently, there has been no lack of improved methods to narrow down the uncertainty margin. These approaches invoke information on standard potentials, on improved models to establish individual ionic activities (to determine χ^w, only for one ion is such a datum required) or individual (Gibbs) energies of hydration. For instance, Alfenaar and de Ligny[4] experimented with

[1] J.E.B. Randles, *Phys. Chem. Liq.* **7** (1977) 102.
[2] D.D. Eley, M.G. Evans, *Trans. Faraday Soc.* **34** (1938) 1093.
[3] E.J.W. Verwey, *Rec. Trav. Chim.* **61** (1942) 127, 564.
[4] M. Alfenaar, C.L. de Ligny, *Rec. Trav. Chim.* **86** (1967) 929.

the ferrocene $(C_5H_5)Fe^+$ (C_5H_5) redox couple, reasoning that these ions are so large that the charge has little influence on the water structure around it, so that Born's equation [I.5.3.2] or simple extensions of it, well describe $\Delta_{hydr}G^o_{mi}$. Unfortunately, the cyclopentene groups are rather hydrophobic which leads to additional contributions to the Gibbs energy of solvation.

An important step forward stemmed from measurements of the temperature coefficient $d\chi^w/dT$. Their observation that $d\chi^w/dT$ was negative led Frumkin et al.[1] to the conclusion that χ^w has to be positive. The argument was that temperature increase leads to randomization, so that χ^w and its temperature derivative should have opposite signs. Rigorous analysis requires some scruples, because for instance, the density also changes with temperature, requiring an additional surface entropy term. Frumkin et al. estimated χ^w between 0.1 and 0.2 V. The negative sign of $d\chi^w/dT$ was corroborated by Randles and Schiffrein[2]. Randles[3] later estimated χ^w between +0.08 and +0.13 V. In a detailed study, covering standard electrode potentials and a comparison between solvation Gibbs energies in various media, Trasatti[4] obtained $\chi^w = +0.13 \pm 0.02$ V. The positive sign was also confirmed by Borazio et al.[5] who analysed the EMF of Voltaic cells as it was influenced by the electrolyte concentration. Their value is +0.025 ± 0.010 V.

The conclusion is that, according to present-day insight, $\chi^w > 0$, although its value remains somewhat uncertain. The positive sign means that water is positive with respect to air. In other words, at the free surface of water *the negative sides* of the molecules (the oxygens) are on the average pointed *outward*. This may be another unique property of water. Limited experiments with other liquids gave negative χ-potentials. For instance, Parsons and Rubin[6] found for methanol and ethanol –0.23 and –0.30 V, respectively.

Simulations like those discussed in sec. 2.2c, but now at the water-air interface, support this trend. Using Monte Carlo (MC) or Molecular Dynamics (MD) the preferential orientation of the dipoles can be found, from which the dipolar part of χ^w follows. When p is the value of the dipole moment, $\rho_N(z)$ the number density of the dipoles, and $\langle \cos \theta \rangle$ the average orientation in a position z in the surface layer (z is normal to the liquid-vapour interface), χ follows from

[1] A.N. Frumkin, Z.A. Iofa and M.A. Gerovich, *Zhur. Fiz. Khim.* **30** (1956) 1455.

[2] J.E.B. Randles, D.J. Schiffrin, *J. Electroanal. Chem.* **10** (1965) 480.

[3] J.E.B. Randles, *Phys. Chem. Liq.* **7** (1977) 107.

[4] S. Trasatti, *J. Chem. Soc. Faraday Trans. I* **70** (1974) 1752.

[5] A. Borazio, J.R. Farrell and P. McTigue, *J. Electroanal. Chem.* **193** (1985) 103; see also W. Duncan-Hewitt, *Langmuir* **7** (1991) 1229.

[6] R. Parsons, B.T. Rubin, *J. Chem. Soc. Faraday Trans. I* **70** (1974) 1636.

$$\chi = (\pm)\ \frac{p}{\varepsilon_o} \int\limits_{-\infty}^{+\infty} \frac{\rho_N(z)}{\varepsilon(z)}\ \langle\cos\theta\rangle\ \mathrm{d}z \qquad\qquad\qquad [3.9.8]$$

The signs of χ and $p\langle\cos\theta\rangle$ are uniquely coupled. The direction of p is from the negative to the positive side. For a liquid with the negative side directed towards its vapour, p and χ are both positive. In this case the vapour is the reference. However at solid-liquid interfaces potentials are usually referred to the bulk liquid, in which case a minus sign is required. To remind us of that, the ($\pm$) has been included.

A number of problems present themselves. On the scale of layers only a few molecular cross-sections thick, dipoles may not be modelled as ideal and the value to assign to $\varepsilon(z)$ is difficult to establish. In simulations the dipole issue is translated as the problem of modelling the molecule and, hence, the interaction energy in various orientations. Another limitation of [3.9.8] is that it only gives the dipolar part. Any polarization of the solid (if present) and spill-over of electrons from an adjacent metal electrode are ignored. Nevertheless, MC and MD simulations, involving different numbers of molecules, a variety of molecular interaction models and values for ε differing between 1 and 78.2 all agree about the positive sign of χ^w [1,2,3] and a negative one for methanol[4]. Figure 3.39 illustrates these results and the trend of the absolute value to diminish with increasing temperature. Full comparison with experiments is not (yet) possible because of the lack of studies simultaneously yielding χ^w and $\mathrm{d}\chi^w/\mathrm{d}T$. However, the slopes may be compared with those reported experimentally[5]; in the figure they are indicated as short dashes, more or less arbitrarily fixed at 0.1 V at the temperature of observation. The orders of magnitude agree but further analysis awaits more experimental data.

The tendency of water to direct the negative sides away from the liquid has also been observed for water adjacent to silver iodide. In this case the argument stems from the shift of the points of zero charge upon displacing the water by a monolayer of a number of aliphatic alcohols[6]. Anticipating a conclusion from sec. 3.12, all these alcohols adsorb with their hydrocarbon moieties to the solid,

---

[1] N.I. Christou, J. Whitehouse, D. Nicholson and N.G. Parsonage, *Mol. Phys.* **55** (1985) 39.

[2] M.A. Wilson, A. Pohorille and L.R. Pratt, *J. Phys. Chem.* **91** (1987) 4873.

[3] M. Matsumoto, Y. Kataoka, *J. Chem. Phys.* **88** (1988) 3233.

[4] M. Matsumoto, Y. Kataoka, *J. Chem. Phys.* **90** (1989) 2398.

[5] J.E.B. Randles, D.J. Schiffrin, *J. Electroanal. Chem.* **10** (1965) 480 (averages of their data in 0.054 M NaOH, 0.05 and 0.01 M KCl). In a later paper Schiffrin (*Trans. Faraday Soc.* **66** (1970) 2464), reported $\mathrm{d}\chi^w/\mathrm{d}T = 0.039$ mV deg^{-1} which is lower by about a factor of 10.

[6] B.H. Bijsterbosch, J. Lyklema, *J. Colloid Sci.* **20** (1965) 665.

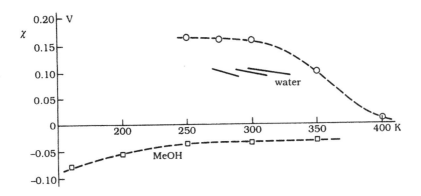

Figure 3.39. Temperature dependence of χ-potentials. Molecular Dynamics of 1000 molecules. O water (Carravetta-Clementi interactions, *J. Chem. Phys.* **81** (1984) 2647) after M. Matsumoto, Y. Kataoka, loc. cit. (1988); □ methanol (TIPS model), same authors (1989). The short dashes are experimental slopes, see text.

thereby "sweeping away" the oriented water dipoles. In all cases a maximum shift of 0.20 ± 0.04 V in the positive direction was found, meaning that originally the silver iodide must have been negative with respect to the water. Assigning a quantitative value involves the uncertainty that the χ-potential after sweeping is not known. It may be slightly negative, as found for methanol and ethanol with respect to vapour, but in the above experiments the alcohols had longer chains. Anyway, the semiquantitative conclusion may be drawn that water polarizes in the same way to silver iodide as it does to air[1]. A similar orientation has also been reported for mercury[2,3].

The overall conclusion that, with respect to all water-shunning or hydrophobic phases studied so far, water turns its oxygen away from the water phase must mean that this is mainly a property of the water and not of water-adjacent phase interaction. For hydrophilic surfaces, containing donors actively promoting the formation of hydrogen bridges, this may be different. Information from inner layer capacitances, suggesting that the relative dielectric permittivity of water adjacent to silver iodide and mercury is much lower than it is for oxides, may be in line with this.

In the literature much attention is paid to Helmholtz's equation

---

[1] For a more detailed analysis, see B.H. Bijsterbosch, J. Lyklema, *Adv. Colloid Interface Sci.* **9** (1978) 147, especially sec. 5.1.
[2] B.E. Conway, H.P. Dhar, *Croat. Chem. Acta,* **45** (1973) 173.
[3] S. Trasatti, *J. Chim. Phys.* **72** (1975) 561.

$$\chi = \sum_i \frac{N_i^\sigma \, p_i \langle \cos \theta_i \rangle}{\varepsilon_o \varepsilon} \qquad\qquad [3.9.9]$$

or simplified versions of it. Here N_i^σ is the number of dipoles i per unit area. In very simple terms, if $p \cos \theta$ is written as the product $\sigma^{dip} d$ of a dipolar charge and the layer thickness over which the polarization has taken place; the quotient $\sigma^{dip}/\chi = \varepsilon_o \varepsilon / d$ may be interpretated as a polarization capacitance. This equation is a simplification of [3.9.8] in that it only considers one layer and ignores interaction with liquid layers further into the liquid. Moreover, the value of ε remains uncertain. Nevertheless, for the relatively simple task of explaining the surface charge and temperature-dependence of the inner layer capacitance such models work reasonably well. See for instance Parsons' four-state model allowing for individual dipoles up or down and water clusters up or down[1]. For further discussions see refs.[2,3,4,5].

More pertinent for our purposes is the problem to what extent ionic double layers and the interfacial polarization are additive. In the absence of specific adsorption the problem is relatively minor: the countercharge resides in the diffuse part, where solvent polarization is negligible (or neglected by definition) and the interfacial polarization more or less coincides with the Stern layer, ignoring any contribution from the solid. To a first rough approximation the two potential drops are spatially separated and hence additive, so that

$$\Delta\phi \sim \chi + \psi^d \qquad\qquad [3.9.10]$$

with $\chi = \psi^o - \psi^d$ and $\Delta\phi \sim \psi^o$ in this approximation. In modelling the inner layer capacitance C^i this is the route often taken. It is acceptable when C^i is independent of the concentration and the nature of the electrolyte. However, C^i does depend on σ^o and on T, and these dependencies can be explained using a molecular model. Statistical theories yield results that are intuitively expected. Aloisi and Guidelli[6] investigated the $\chi(\sigma^o)$ effect for water that was modelled by distinguishing 12 different spatial orientations on a body-centered cubic lattice. Pair interactions were analysed in the quasi-chemical approximation (sec. I.3.8e). It was found that $\chi^{Me,w}$ (Me is a charged metal wall) became more negative with increasing negative σ^o, i.e. in this direction the tendency for water

[1] R. Parsons, *J. Electroanal. Chem.* **59** (1975) 229.
[2] M.A. Habib, in *Modern Aspects of Electrochemistry*, J. O'M. Bockris, B.E. Conway Eds. Vol. **12**, Plenum (1977) 131.
[3] S. Trasatti, ibid. Vol. **13** (1979) 81.
[4] B.E. Conway, *Adv. Colloid Interface Sci.* **8** (1977) 131.
[5] S.K. Rangarajan, *Electrochemistry* **7** (1980) 203.
[6] G. Aloisi, R. Guidelli, *J. Chem. Phys.* **95** (1991) 3679.

molecules to expose their positive sides to the surface grows. However, the absolute value of $\chi^{Me,w}$ was negative at the point of zero charge, so that this model does not recover the situation for mercury. Torrie et al.[1] also studied an ion-free layer, but using one of the integral equations (RHNC, standing for reference hypernetted chain) found the normal density contribution $\rho_N(z)$ "surprisingly resistant" to changes in σ^o. However, as this distribution is to a large extent determined by repulsive interactions, the outcome is not unexpected after all. On the other hand, the polarization did change with σ^o, i.e. at a more or less constant packing the number of field-favourable orientations tends to increase with σ^o.

When specific adsorption does occur, the ionic and solvent polarization parts are no longer additive. Specific adsorption is determined by the state of polarization and conversely. Moreover, the surface charge formed at given pAg, pH, depends on solvent polarization and screening by ions, but it also contributes to the polarization. Hence these three quantities are interrelated. The advanced statistical thermodynamical analysis of this interesting problem is a challenge[2].

3.10 Case studies

In this section the results of a number of double layer studies will be discussed with the dual purpose of reviewing experimental facts and applying the models and theories of the previous sections to real systems. No attempt is made to achieve completeness. Examples are selected because they exhibit characteristic properties or are illustrative for other reasons.

Colloid scientists are more interested in double layers on dispersed materials than on mercury, the classical model system of electrochemists. However, the comparison between those, i.e. between relaxed and polarizable double layers, is of great interest, and therefore a discussion of the mercury-solution interface will be included. Non-aqueous systems will be deferred to sec. 3.11.

Surfaces of disperse materials are less well defined than that of mercury. Problems are incurred with heterogeneity, non-zero solubility, non-inertness on titration, the presence of impurities remaining after the preparation, etc. Some uncertainty as to the specific surface area A_g also presents itself: even if a reproducible value is determined when using an established method such as BET (N_2) gas adsorption some reservation remains about its physical meaning: do ions adsorbing from solution "see" the same surface as N_2 molecules adsorbing from the gas phase at subzero temperatures on evacuated samples? Because of

[1] G.M. Torrie, P.G. Kusalik and G.N. Patey, *J. Chem. Phys.* **88** (1988) 7826.
[2] For a more classical introduction, see B.E. Conway, *The State of Water and Hydrated Ions at Interfaces, Adv. Colloid Interface Sci.* **8** (1977) 91.

these inherent experimental problems, the accuracy and reproducibility of double layer parameters on disperse systems is much less than on mercury. Typically, double layer capacitances on AgI are reproducible to 0.1 μF cm^{-2}, those on mercury to 0.01 μF cm^{-2}. Additional problems of disperse systems are that there may be some arbitrariness in establishing the charge-determining process and that the surface potential range that can be studied is shorter than with mercury. For the latter system the interface can be polarized over a range of about 2 V, whereas for disperse systems to a maximum of about 7 decades in the concentration of charge-determining ions can be covered, corresponding to about 420 mV if Nernst's law applies. These limitations are imposed by the solubility of the material: for instance, oxides become soluble at very high and/or very low pH. The drawback of limited σ^0 or ψ^0 ranges is that certain extrapolations cannot be made. For instance, on AgI the high-potential limiting value of the negative adsorption [3.5.26] can only be reached at low c_{salt}; this limit plays a role in establishing the absolute values of the ionic components of charge (see fig. 3.43). On oxides it is seldom possible to obtain a plateau in σ^0 and hence to establish the number of sites.

To a large extent, these disadvantages are offset by the possibility of obtaining additional information that is difficult to obtain for mercury, such as electrokinetic potentials, colloid stability data and directly measured adsorptions. Even in cases where the evaluation and interpretation of ζ-potentials from electrokinetics are under discussion, the technique is powerful because it informs us about the sign of the diffuse part of the double layer and hence helps to detect (super-)equivalency of specific adsorption. Quantitative information on specific adsorption can also be obtained by comparing p.z.c.'s and i.e.p.'s. Because of all of this, over the past decades systematic studies on a number of well-characterized disperse systems have led to a number of qualitatively new features, whereas the progress in the domain of mercury double layers has been rather quantitative.

Experimental data can be presented per property or per system, both approaches having their merits and demerits. We shall mostly take the second one, to highlight special features of certain systems, but at the end make an exception by comparing lyotropic sequencies on different systems, considering that water structure-mediated interactions can be made most visible if double layers with differing affinities for water are compared. Double layers in the presence of larger ions and/or uncharged adsorbates will be deferred to sec. 3.12.

3.10a. Silver iodide[1]

Silver iodide is the classical model colloid of the Dutch School. It earned its reputation because of a confluence of properties: the charge-determining mechanism is well established, stable positive and negative sols can be made of it for which the stability and electrokinetic properties can be measured, the material is rather inert and insoluble and not particularly sensitive to light. Last, but not least, Ag/AgI electrodes are very stable. Silver bromide and silver chloride are more soluble and light-sensitive, so that only shorter surface potential ranges are accessible and more experimental precautions have to be taken. Historically, AgI sols were first used to study stability phenomena, such as establishing the Schulze-Hardy rule, lyotropic sequences, rates of coagulation, Donnan equilibria and electrokinetics[2]. After pioneering work by Verwey[3] and de Bruyn[4] from about 1950 onwards systematic double layer studies have been conducted. In this way, AgI became the first system for which data on σ^o (or ψ^o), ζ and coagulation concentrations were simultaneously available. Not until then did it become clear that only a (small) fraction of the countercharge, depending on c_{salt}, is relevant for particle interaction and that the inner layer charge σ^i can be estimated with some confidence by subtracting σ^d (from stability and/or electrophoresis) from σ^o (from titration). For the interpretation of colloid stability it meant the replacement of ψ^o by ψ^d in the formula for the *critical coagulation concentration* (c.c.c.). For the present purposes, and anticipating Volume IV, we write this equation as

$$\text{c.c.c.} = c(T)\,\frac{\left[\tanh\left(zy^d/4\right)\right]^4}{z^6 A_{11(w)}^2} \qquad\qquad [3.10.1]^{5)}$$

where $c(T)$ is a known temperature-dependent factor, containing natural constants and the permittivity of the solvent, z is the valency of the counterion and $A_{11(w)}$ the *Hamaker constant* for the attraction between two particles of nature 1 across water, as in sec. I.4.6b, and tabulated in I.app. 9. Originally, the equation contained y^o instead of y^d, the idea being that the double layer was entirely diffuse. For silver iodide sols y^o is high enough to let $\tanh(zy^o/4)$ approach unity, see fig. A2b, hence c.c.c. $\sim z^{-6}$. This finding has been taken as the quantitative validation of the *Schulze-Hardy rule*, stating that the stability of

[1] The double layer on silver iodide has been extensively reviewed by B.H. Bijsterbosch and J. Lyklema, *Adv. Colloid Interface Sci.* **9** (1978) 147, where more details and references can be found.

[2] Several illustrations can be found in *Colloid Science*, Vol. **1**, H.R. Kruyt, Ed., Elsevier (1952).

[3] E.J.W. Verwey, PhD. Thesis, State Univ. of Utrecht, The Netherlands (1934).

[4] H. de Bruyn, *Rec. Trav. Chim.* **61** (1942) 5, 12.

[5] *Colloid Science*, loc. cit. p. 306, where $\gamma = \tanh(zy^d/4)$.

hydrophobic sols decreases very strongly with increasing valency of the counterion. However, as particle interaction is mainly determined by the outer part of the double layers, which is governed by y^d, this potential should be in the equation rather than y^o. Near the c.c.c. y^d is so low that $\tanh(z y^d/4) \approx z y^d/4$, see [A2.9], hence

$$\text{c.c.c.} = c'(T) \frac{(y^d)^4}{z^2 A_{11(w)}^2} \qquad\qquad\qquad [3.10.1a]$$

As y^d tends to decrease with z because multivalent ions compensate more charge in the inner layer, and as this is a strong effect because of the fourth power dependence, the Schulze-Hardy rule remains explainable; in fact, the precise dependence on z and the nature of the counterion now depends on the system, as is experimentally found. This re-interpretation of classical knowledge followed from double layer studies with silver iodide.

Silver iodide is a hydrophobic (or electrocratic) colloid, in the sense that sols are only stable by virtue of electrostatic repulsion between the particles. Ottewill et al.[1] measured the contact angle for water, using the captive air bubble technique. For polished discs around the p.z.c. values of 25-30° were found. Amphipolar molecules like butanol, adsorbing from water, turn their aliphatic moieties to the surface (sec. 3.12). Heats of water vapour adsorption range between 30 and 80 kJ mol^{-1}, which is characteristic of physical adsorption. Impurities in the sample may make the initial adsorption enthalpy higher. These are not so easy to detect and perhaps not fully established; in fact their occurrence will be dependent on the sample. Zettlemoyer et al. found that N_2 and Ar adsorb much better on AgI than water does[2]. The hydrophobicity and the possible presence of a few high-energetic sites on the surface have both been invoked as reasons for the power of AgI crystals to act as a *cloud seeder*. The structural match between the lattices of ice and AgI has also been suggested, but this match is probably fortuitous.

The determination of the specific surface area offered a problem in that electrochemical methods yield areas that systematically exceed the BET (N_2) value by a factor of almost three[3]. This difficulty has not yet been surmounted, although there are good reasons to use the "electrochemical" area for double layers studies: (i) it is the area "seen" by adsorbing ions, (ii) two different

[1] R.H. Ottewill, D.F. Billett, G. Gonzalez, D.B. Hough and V.M. Lovell, in *Wetting, Spreading and Adhesion*, J.F. Padday, Ed., Academic Press (1978) 183.
[2] A.C. Zettlemoyer, N. Tcheurekdjian and J.J. Chessick, *Nature* **192** (1961) 653.
[3] H.J. van den Hul, J. Lyklema, *J. Am. Chem. Soc.* **90** (1968) 3010.

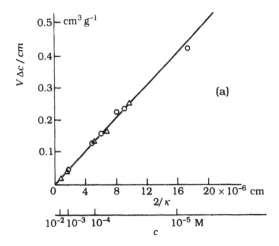

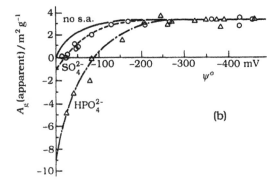

Figure 3.40. Negative adsorption of SO_4^{2-} (o) and HPO_4^{2-} (Δ) ions on AgI, negatively charged by positive adsorption of I^- ions. (a) Dependence on the concentration, (b) on the surface potential. Further discussion in the text. (Redrawn from H.J. van den Hul, J. Lyklema, *J. Colloid Interface Sci.* **23** (1967) 500.)

electrochemical methods confirm each other, viz. negative adsorption and the capacitance method.

Regarding (i): the negative adsorption data are very convincing, as illustrated by fig. 3.40. These results refer to the expulsion of bivalent anions, SO_4^{2-} and HPO_4^{2-}, from silver iodide that was negatively charged by (positive) adsorption of I^- ions. Part of the I^- ions are also negatively adsorbed, as a result of which the equations of sec. 3.7e become a bit more complicated. However, conditions can be found where enough I^- is adsorbed to achieve maximum negative adsorption of bivalent anions although c_{KI} is still so low that the contribution of the I^- ions to the negative adsorption is negligible compared to that of the bivalent ones. In this connection it may be recalled from sec. 3.5d and fig. 3.11 that, because the valencies occur in the exponentials of the Boltzmann equation, multivalent ions tend to dominate the ionic components of charge.

Figure 3.40a confirms [3.7.19]. The "expelled volume" $\Delta V = V\Delta c/c$ is proportional to κ^{-1}, the slope being identical to A_g, because ΔV was taken per mass unit of solid, m. For this sample $A_g = 2.6$ m^2 g^{-1} was found. The proportionality to $c^{1/2}$ and the independence of the nature of the bivalent ion demonstrate that expulsion is complete and that specific adsorption is excluded. In fig. 3.40b the effect of specific adsorption is systematically studied. When the surface potential is sufficiently reduced (i.e. pAg is lowered) the negative adsorption also goes down. The drawn curve indicates what would happen if there were no specific adsorption. The experiment indicates that SO_4^{2-} adsorbs somewhat specifically and HPO_4^{2-} more strongly so. Eventually, with decreasing surface potential the expulsion reverses sign, meaning that the positive adsorption starts to exceed it. The values of ψ^o where the curves cross the abscissa axis can be used to compute $\Delta_{ads}G_m^o$, using one of the models of sec. 3.6d. In fig. 3.40b, the slope of fig. 3.40a is plotted as the ordinate axis. For sufficiently negative surface potential A_g reaches the value of 3.3 m^2 g^{-1}, which is apparently the specific area for this sample.

Regarding (ii), the area obtained by negative adsorption is confirmed by that obtained from capacitance. In 1951 Mackor[1] proposed to equate the differential capacitance in 10^{-3} M non-specifically adsorbing electrolyte around the p.z.c. to that on mercury, which at that time was already well known, thanks to Grahame's work. From C_{AgI} (in μF per gram) and C_{Hg} (in μF cm^{-2}) A_g is then found. The argument was that under those conditions the double layer is mainly diffuse and hence generic, i.e. in [3.6.25] $C \approx C^d$. More detailed experiments more or less confirmed this assumption, see fig. 3.41. Under the given conditions $C^i \sim 30$ μF cm^{-2}, $C^d \sim 6$ μF cm^{-2} ([3.5.17] or fig. 3.5) so that a difference of a factor of two in C^i between mercury and silver iodide gives an uncertainty in the area obtained of about 15% in A_g. Later, the good agreement with directly measured capacitances (sec. 3.7c, fig. 3.32) provided further confirmation.

In summary, the specific surface area to be used in double layer studies appears well-established. As the p.z.c. is also well known, even as a function of the temperature (app. 3, table c), all the prerequisites for obtaining σ^o(pAg) or $\sigma^o(\psi^o)$ curves are available. In fig. 3.28 a historical example has already been given. These data are still valid, with the comment that the specific surface area is now better known. Figure 3.41 gives more modern data, illustrating at the same time the lyotropic sequence for the alkali ions. This lyotropic sequence is also observed in the critical coagulation concentration[2]. In line with the discussion around [3.10.1a] it is typical that these two lyotropic sequencies occur in opposite directions: from Li$^+$ → K$^+$ → Rb$^+$ the surface charge *increases*, whereas the sol stability *decreases*. This lyotropic sequence antagonism is convincing evidence for specific adsorption.

---

[1] E.L. Mackor, *Rec. Trav. Chim.* **70** (1951) 663, 763.
[2] H.R. Kruyt, M.A.M. Klompé, *Koll. Beihefte* **54** (1943) 484.

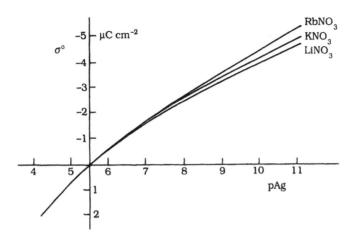

Figure 3.41. Surface charge on silver iodide in 10^{-1} M solutions of three electrolytes. Temperature, 25°C. (Redrawn from Bijsterbosch and Lyklema, loc. cit.)

The adsorption of alkali ions (and of earth alkali ions, not shown) differs from that of the anions SO_4^{2-} and HPO_4^{2-} in that the latter adsorb specifically on uncharged silver iodide, with the concomitant change in p.z.c. (sec. 3.8, fig. 3.23 - 25), whereas the former do not shift the p.z.c. For alkali ions, specificity starts only when there is already I^- on the surface. This is an example of specific adsorption of the second kind, as defined in sec. 3.6d. Apparently, the alkali ions only adsorb on I^- sites, so that there will be some analogy with water structure-originating alkali ion-iodide ion interaction in solution. We will come back to this in sec. 3.10g.

Double layer capacitances in KF, an electrolyte that is indifferent at the p.z.c. are given in fig. 3.42 and Stern layer capacitances in various electrolytes in fig. 3.43. In fig. 3.42 C^d is computed on the basis of [3.5.17], replacing ψ^d by ψ^o. The double layers are diffuse only in low electrolyte ($\leq 10^{-3}$ M) and at low potentials ($\lesssim$ 50 mV). For mercury the same is found. Beyond that range, $\sigma^o(\psi^o)$ curves can be very well interpreted by the series capacitance model [3.6.25] with the Stern capacitance being a function of σ^o *only* (the electrolyte effect is indirect: at given ψ^o, addition of electrolyte increases the absolute value of σ^o). The capacitance C^i in fig. 3.43 has a somewhat hybrid character: it is obtained from [3.6.25] subtracting $(C^d)^{-1}$ from the experimental C^{-1}, choosing for ψ^d in [3.5.17] the value it would have in the absence of specific adsorption. This procedure involves an approximation, but the possibility to reconstruct the basic $\sigma^o(\psi^o)$ curves well indicates that the result is acceptable. Physically $C^i(\sigma^o)$ reflects the sum effect of any change of specific adsorption, dielectric permittivity ε^i and Stern layer thickness d (or, for that matter, β and γ as in fig. 3.20b) with the

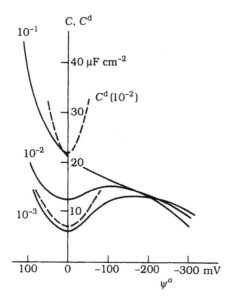

Figure 3.42. Differential capacitance of the double layer on silver iodide in KF. Temperature, 25°C. Solid curves, experiments; dashed curves, diffuse double layer capacitance according to [3.5.17]. (Redrawn from J. Lyklema and J.Th.G. Overbeek, *J. Colloid Sci.* **16** (1961) 595.)

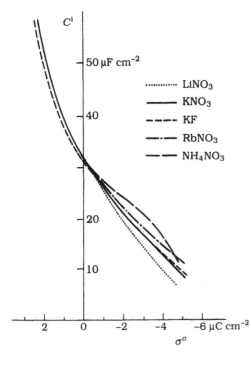

Figure 3.43. Differential capacitance of the Stern layer on silver iodide in various electrolytes (same reference as fig. 3.42).

surface charge, that is, with the electric field across that layer. The steep decrease is not found with mercury. It must therefore be related to surface properties that are specific for silver iodide, probably the asymmetry in the hydration of Ag^+ and I^- ions in the surface[1]. On the other hand, anions as counterions accumulate readily in the iHp. Consequently, on positive AgI, charges and capacitances easily become high, without making positive AgI sols particularly stable. The integral Stern capacitance K^i shows the same features, but decreases less strongly with increasing negative charge.

To further establish specific adsorption and other solvent structure properties, temperature dependent measurements are very helpful. Systematic studies revealed that with increasing temperature the specificity decreases[2]. At 60°C and above, the double layer is entirely generic, at least on the negative side. Randomization of the structure of the adjacent water is a possible candidate to explain this disappearance of specificity. Concomitant sol stability studies showed that in the critical coagulation concentration the lyotropic sequence also vanished. Disappearance of specific adsorption means that then $\sigma^\circ = -\sigma^d$, so that ψ^d can be calculated from the experimental surface charge. Using [3.10.1] this yields the Hamaker constant. (In fact, this is one of the best ways to obtain $A_{11(w)}$ for silver iodide; it is recorded in table I.app. 9.6.) Using this value at room temperature from the c.c.c. values ψ^d can be established, and hence σ^d and σ^i. Using the triple layer model of sec. 3.6d and fig. 3.20b the standard molar specific adsorption Gibbs energies $\Delta_{ads}G^\circ_m$ for Li^+, K^+ and Rb^+ were established at -2, -2.5 and -4 RT, respectively.

Temperature studies provided further insights regarding the state of adjacent water. By way of illustration, in fig. 3.44 the *excess entropy per unit area* is given as a function of the surface charge. This excess was obtained by retaining the $S^\sigma_a dT$ term in [3.4.2] and cross-differentiation with the $\sigma^\circ d\mu_{AgNO_3}$ term[3].

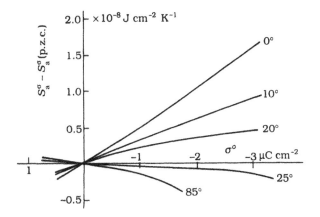

Figure 3.44. Surface excess entropy with respect to the point of zero charge for the double layer on silver iodide. Electrolyte, 10^{-1} M KNO_3. (Redrawn from B.H. Bijsterbosch, J. Lyklema, *J. Colloid Interface Sci.* **28** (1968) 500).

[1] A. de Keizer, J. Lyklema, *Can. J. Chem.* **59** (1981) 1969.
[2] J. Lyklema, *Discuss. Faraday Soc.* **42** (1966) 81.
[3] B.H. Bijsterbosch, J. Lyklema, *J. Colloid Interface Sci.* **28** (1968) 506.

Absolute values cannot be obtained in this way, only the change with σ^o. Given the order of magnitude, it is much more likely that S_a^σ is determined by the structure of adjacent water than by the (very limited) number of specifically adsorbed ions. At high temperatures, where the adjacent water is relatively random, imposing a (negative) field reduces the ordering. At low temperatures it is the other way around. It looks as if around the p.z.c. a certain spontaneous ordering takes place that is partly disrupted by the field. NMR studies confirm this[1]. The data on the positive side do not allow firm conclusions to be drawn although it may be mentioned that at very low temperatures the "pre-freezing" of water leads to the expulsion of NO_3^- ions and a concomitant reduction in the capacitance[2].

The availability of σ^o (ψ^o or pAg) curves at several electrolyte concentrations enables the establishment of the *Esin-Markov coefficient* [3.4.14] and the ensuing determination of the ionic components of charge, integrating [3.4.16][3,4]. Figures 3.45 and 3.46 give results of the former and the latter, respectively.

In fig. 3.45 the bars indicate the spread incurred in taking data obtained by different authors on different samples. To the negative side $\beta \rightarrow -1$, meaning that in the limit of high negative surface charge $(\partial \sigma_+/\partial \sigma^o) \rightarrow -1$ and $(\partial \sigma_-/\partial \sigma^o) \rightarrow 0$. In words, any additionally adsorbed I^- ion is accompanied by a co-adsorbed K^+ ion, not by further increase of the negative adsorption. On positive silver iodide the σ^o range is too short to state anything about the limiting condition.

Figure 3.46 gives the ionic components of charge. The situation around the p.z.c. is not well established and there is the additional problem of finding a reference point for the integration of [3.4.16]. However, it is seen that close to the

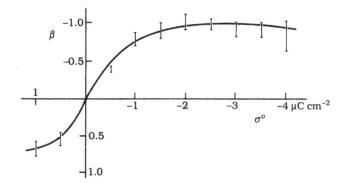

Figure 3.45. The Esin-Markov coefficient for the double layer on silver iodide. (Redrawn from J. Lyklema, J. Electroanal. Chem. **37** (1972) 53.)

[1] A.S. Fawcett, G.D. Parfitt and A.L. Smith, *Nature* **204** (1964) 775.
[2] B. Vincent, J. Lyklema, *Special Discuss. Faraday Soc.* **1** (1970) 148.
[3] B.H. Bijsterbosch, J. Lyklema, *Adv. Colloid Interface Sci.* **9** (1978) 147.
[4] J. Lyklema, *J. Electroanal. Chem.* **37** (1972) 53.

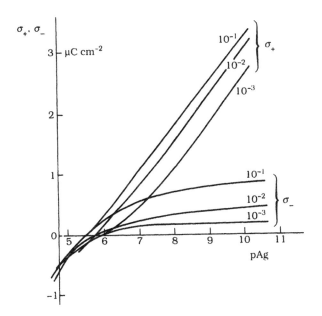

Figure 3.46. Ionic components of charge in the double layer on silver iodide. Temperature 25°C. $\sigma_+ = F\,\Gamma_{K^+}$, $\sigma_- = -F\,\Gamma_{NO_3^-}$. The KNO_3 concentration (in M) is given. (Redrawn from B.H. Bijsterbosch, J. Lyklema, *Adv. Colloid Interface Sci.* **9**, (1978) 147.)

right of the p.z.c. σ_+ and σ_- start being identical but diverge with increasingly negative surface charge. This is in agreement with sec. 3.5b, as illustrated in fig. 3.8. According to [3.5.26] σ_- should eventually attain the limiting value $2cF/\kappa$. For 10^{-3} M electrolyte this is indeed observed, but in 10^{-2} and 10^{-1} M the surface charge range is too short. As stated before, this is a drawback of the silver iodide system as compared with mercury.

We will come back to the lyotropic sequence in sec. 3.10g and to the electro-sorption of organic molecules on silver iodide in sec. 3.12.

3.10b. *Mercury*

The difference of principle with the previous model system is that the mercury-aqueous solution interface can be *polarized* over a large range of potentials, provided *electroactive species* are absent. Ions are said to be electroactive when they are oxidized or reduced at the metal-solution interface and, consequently, transfer charge across it. Whether or not such *depolarization* takes place depends on the potential applied and on the nature of the ion, specifically on its redox potential. Just as a silver iodide electrode can be polarized, suppressing the exchange current by working at sufficiently high frequencies (sec. 3.7c), mercury can be depolarized so that the difference in behaviour between the two electrodes is not absolute. However, in the present section only polarized mercury electrodes will be considered, our purpose being to compare double layers on entirely different surfaces, and studied by entirely different techniques.

In sec. I.5.5b the main difference between polarized and relaxed interfaces was introduced, viz. that for the former the applied potential with respect to a given reference is an independent variable, leading to an additional term in the Gibbs equation (sec. I.5.5c). Measurements can take place in a galvanic cell having the basic nature (see sec. I.5.5e)

$$Cu \mid Hg \mid solution \mid ref.electr. \mid Cu \qquad\qquad [3.10.2]$$

of which the electromotive force (e.m.f.) or cell tension E is given by the Galvani potential difference between the two copper leads:

$$E = \phi_{Cu} \, (\text{r.h.s.}) - \phi_{Cu} \, (\text{l.h.s.}) \qquad\qquad [3.10.3]$$

The elaboration depends on details of the cell, the composition of the solution and the nature of the reference electrode, which can be a (non-reversible) calomel electrode, a reversible (normal) hydrogen electrode or an electrode that is reversible to the cation or anion in the solution. For the last two cases the cell tensions were called E_+ and E_- respectively. For the Gibbs adsorption equations in, say KNO_3, we derived [I.5.6.16]:

$$d\gamma = -S_a^\sigma \, dT - \sigma^\circ \, dE_+ - \Gamma_{NO_3^-} \, d\mu_{KNO_3} \qquad\qquad [3.10.4a]$$

$$d\gamma = -S_a^\sigma \, dT - \sigma^\circ \, dE_- - \Gamma_{K^+} \, d\mu_{KNO_3} \qquad\qquad [3.10.4b]$$

These equations are the counterparts of [3.4.11 and 12] (after omitting the $\Gamma_A d\mu_A$ and adding the $-S_a^\sigma \, dT$ terms). For a Nernstian interface the terms with dpAg or dpH can be written as $-\sigma^\circ \, d\psi^\circ$, where $d\psi^\circ$ can be related to dE, dE_+ or dE_-, depending on the nature of the reference electrode. Hence, there is a close phenomenological similarity between the two different types of charged interfaces. Handling in practice differs, because for mercury the *potential* is externally controllable and hence is the primary variable, whereas for reversible interfaces it is the *charge*.

When surface charges on mercury are required, these can be obtained in two ways: by measuring the surface tension and differentiation with respect to $E_\pm$ or E, according to the *Lippmann equation* [I.5.6.17]

$$\sigma^\circ = -\left(\frac{\partial \gamma}{\partial E_\pm} \right)_{T, \mu_{KNO_3}} \qquad\qquad [3.10.5]$$

or by directly measuring the differential capacitance followed by integration

$$\sigma^o = \int_{E_\pm(p.z.c.)}^{E_\pm} C\,dE_\pm' \qquad\qquad [3.10.6]$$

The integration should start at a reference potential $E_\pm(p.z.c.)$ where $\sigma^o = 0$, which coincides with the *electrocapillary maximum*, e.c.m. Small errors in the establishment of this value lead to an integration constant in [3.10.6]. Generally, the results of the two ways of obtaining σ^o agree with an accuracy that is enviable for those working with disperse systems. Double differentiation of electrocapillary curves and comparison with directly measured differential capacitances is more critical but can be achieved within less than 0.2 μF cm^{-2}, see[1].

In practice, direct measurement of the differential capacitance is preferred. Over the past half century several thousands of publications on the mercury system have appeared in the literature. A seminal review by Grahame[2] deserves special mention. Many of the illustrations in this section stem from his work, which has since then not been materially challenged. For other reviews, see sec. 3.15e and for a compilation of critically evaluated data on capacitances, charges, electrocapillary curves and points of zero charges, see sec. 3.15g.

In fig. 3.47 the principle of Lippmann's *capillary electrometer* is sketched[3]. In such an apparatus the interfacial tension γ of mercury against an aqueous solution, is measured as a function of the applied potential. When the reference R is a calomel electrode, as in Lippmann's design, $\gamma(E)$ is found, otherwise it may be $\gamma(E_+)$ or $\gamma(E_-)$, depending on the nature of R. When R is a calomel electrode, a salt bridge is needed, demanding appropriate precautions against leakage of KCl into the solution. The interfacial tension can in principle also be measured by some alternative technique, say from the shape of a sessile drop or from the drop weight method, using a so-called stalagmometer (Volume III, chapter 1).

Some results are given in fig. 3.48. At first sight the *electrocapillary curves* resemble parabolas but there must be considerable deviations because otherwise C has to be constant, which it is not, see fig. 3.49.

To the far negative side, no difference between the various cations is seen on this scale, but to the left, where the anions adsorb specifically, the curves

[1] D.C. Grahame, *J. Am. Chem. Soc.* **76** (1954) 4819; J. Lawrence, R. Parsons and R. Payne, *J. Electroanal. Chem.* **16** (1968) 193; R. de Levie, S. Sarangapari, P. Czekai and G. Benke, *Anal. Chem.* **50** (1978) 110. A more critical view is taken by H. Vos, J. Wiersma and J.M. Los, *J. Electroanal. Chem.* **52** (1974) 26.
[2] D.C. Grahame, *The Electrical Double Layer and the Theory of Electrocapillarity, Chem. Revs.* **41** (1947) 441. (In sec. 3.10b henceforth called D.C. Grahame, loc. cit.)
[3] G. Lippmann, *Ann. Phys.* **149** (1873) 546; *Ann. Chim. Phys.* (5) **5** (1875) 494; **12** (1877) 265; *Wied. Ann.* **11** (1880) 316.

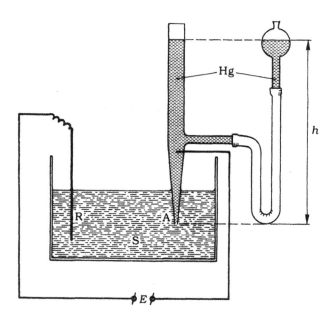

Figure 3.47. Sketch of a Lippmann-type capillary electrometer. The mercury-solution interface resides in the slightly conical capillary A, in which a certain height h is chosen at which the measurements are carried out. The potential is externally applied; R is the reference electrode (in Lippmann's experiments it was a calomel electrode, connected to the solution S via a salt bridge). The interfacial tension between mercury and solution is obtained from the height h, defined in the sketch

diverge. At the same time the *electrocapillary maximum*, i.e. the potential where $\sigma^{\circ} = 0$, shifts to the right, in agreement with the trend for the p.z.c. in table 3.4. In line with this, but not shown, with decreasing electrolyte concentration the curves move upward and the e.c.m. moves to the left (for specifically adsorbing anions) or to the right (when the cation adsorbs specifically). Eventually[1] both approach the *pristine e.c.m.* for which at 25°C $\gamma = 426$ mN m^{-1}. It is noted that the potential range that can be studied is about 1.8 V, i.e. about three times larger than that on silver iodide.

From the electrocapillary curve the surface pressure with respect to the e.c.m. can be derived, using [I.1.3.6]

$$\pi - \pi(\text{e.c.m.}) = \gamma(\text{e.c.m.}) - \gamma = \int_{0}^{\sigma^{\circ}} \sigma^{\circ'} \, dE_{\pm}' \qquad [3.10.7]$$

where we have used [3.10.4] at constant T and μ_{salt}. The result corresponds with

---

[1] C.A. Smolders and E.M. Duyvis, *Rec. Trav. Chim.* **80** (1961) 635, report 426.2 ± 0.2 mN m^{-1} in 0.05 M Na$_2$SO$_4$, obtained by the sessile drop method.

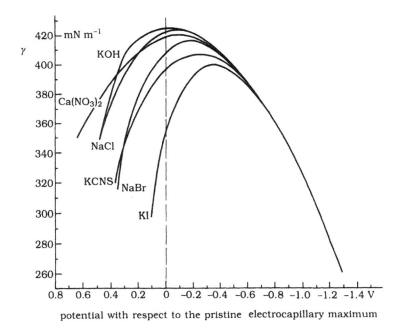

Figure 3.48. Electrocapillary curves after Grahame (loc. cit. p. 448). Concentrations, 1 M. Temperature, 18°C.

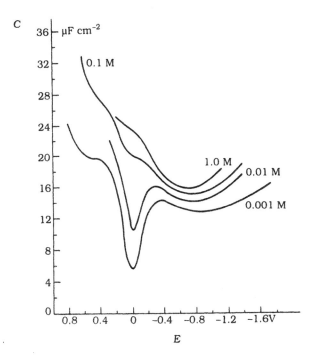

Figure 3.49. Double layer capacitance on mercury in solutions of NaF. Concentration given. Temperature, 25°C. (Redrawn from Grahame, loc. cit. p. 462.)

the Gibbs energy of the double layer, so we may also write

$$\Delta G_a^\sigma = \int_0^{\sigma^0} \sigma^{0\prime} dE_\pm^\prime \qquad\qquad [3.10.8]$$

Unlike the case of relaxed interfaces (see [3.5.20] and [3.6.56]) this is a positive quantity: a polarized double layer does not form spontaneously.

An example of differential capacitance curves is given in fig. 3.49. As with silver iodide there are minima close to the e.c.m. in low concentrations of electrolyte, but beyond that the inner layer capacitances C_1^i and C_2^i dominate. Here the curves show much detail as a function of potential and salt concentration. Numerous attempts have been made to interpret these curves in terms of interfacial polarization. We shall not discuss this here, except to mention that relatively simple models of adjacent water (counting only dipole moments and allowing for only a few orientations) already work relatively well.

From the total capacitances, Stern layer capacitances are obtained in the same way as for silver iodide. Figure 3.50 gives Grahame's differential capacitance for NaF. As with silver iodide, the corresponding integral capacitance K^i exhibits the same features but less pronounced (not shown). Comparison with fig. 3.43 shows that, in the range where comparison is possible, C^i decreases with increasingly negative σ^0, but much less so than with silver iodide. At the p.z.c. the two are comparable, but not exactly identical. The difference determines the error made if the surface area of AgI suspensions is determined by comparison with the capacitance on mercury (Mackor's method, sec. 3.10a). This juxtaposition reveals generic and (substrate-) specific features. Similarity at the

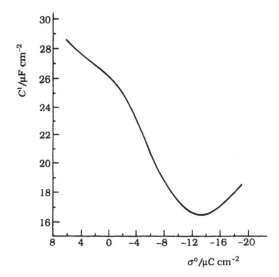

Figure 3.50. Differential capacitance of the Stern layer on mercury obtained from data for 1M aqueous NaF solutions. Temperature, 25°C. (Redrawn from Grahame, loc. cit. 477).

p.z.c. means that there C^i is mainly, if not entirely, determined by the solution sides, i.e. by the structure of the adjacent water. The much higher slope $dC^i/d\sigma^o$ for silver iodide must be attributed either or both to the stronger discrimination between embedding cations and anions as counterions in the Stern layer, or to the difference in hydration of surface Ag^+ and I^- ions. Both features are ultimately related to the water structure. The rise in C^i at very negative σ^o (fig. 3.50) is a new feature, which is not found for silver iodide. It persists at high temperatures, see fig. 3.51. Other "mercury idiosyncrasies" are the *hump* at low temperatures and the *common intersection point* to the positive side. These features are observed for several simple electrolytes so that their interpretation should primarily be based on the structure of adjacent water. Interpreting all these phenomena is a more severe test of the models, and the last word has not yet been said. It may be noted that so far none of the models has explicitly considered the fact that water has a relatively open structure, i.e. that there are holes in it. The number of holes per unit volume will depend on σ^o, p and T. Hills and Payne[1] found the surface excess volume of the inner layer in NaF, NaCl and $NaNO_3$ to be positive and independent of σ^o and the nature of the anion for $\sigma^o < -10\ \mu C\ cm^2$, whereas for positive σ^o it depends strongly on the nature of the anion, even with respect to its sign.

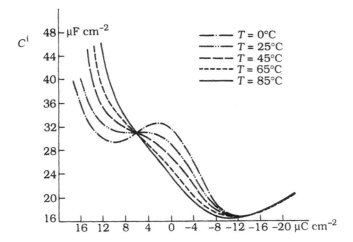

Figure 3.51. Temperature dependence of the inner layer capacitance on mercury, computed from capacitance data in 0.8 M NaF. (Redrawn from D.C. Grahame, *J. Am. Chem. Soc.* **79** (1967) 2093.)

[1] G.J. Hills, R. Payne, *Trans Faraday Soc.* **61** (1965) 326; G.J. Hills, *J. Phys. Chem.* **73** (1969) 3591.

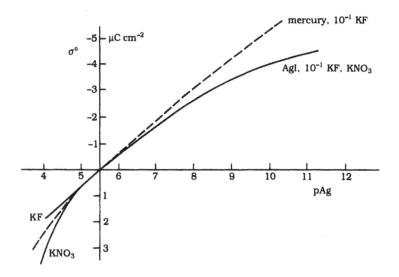

Figure 3.52. Comparison of the electrical double layers on silver iodide and mercury in 10^{-1} M solutions of simple indifferent electrolytes. Temperature, 25°C. One unit of pAg = 0.059 V.

Pursuing the comparison between mercury and silver iodide, the lowering of C^i with increasing T around the p.z.c. is found for both systems. The steeper $C^i(\sigma^o)$ curve for silver iodide corresponds to lower charges on the negative side, see fig. 3.52. On negative silver iodide it is more difficult to accommodate cations near the surface than it is for mercury. This difference is naturally reflected in the lyotropic sequences. On silver iodide, there are clear differences (fig. 3.41) between the alkali ions; on mercury these differences are about an order of magnitude smaller, though in the same direction[1]. In fact, if on silver iodide the differences would have been as subtle as on mercury, they would have escaped detection. Figure 3.53 illustrates these trends. The ordinate axis scales are very different. Based on these observations it may perhaps be predicted that mercury sols, if they could be made without stabilizers, would show a small lyotropic sequence in the critical coagulation concentrations with RbCl slightly lower than LiCl.

With so many detailed data available over large ranges of the surface potential or -charge, ionic components of charges can be readily and accurately obtained with a Esin-Markov type analysis. Figures 3.54 and 3.55 are examples.

Figure 3.54 is the counterpart of fig. 3.46. It shows greater detail and can be

[1] D.C. Grahame, *J. Electrochem. Soc.* **98** (1951) 343.

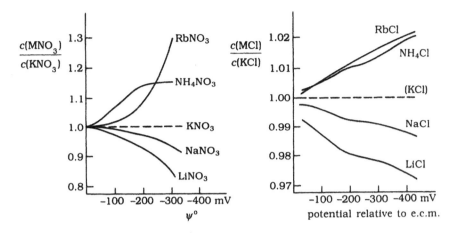

Figure 3.53. Differential capacitance ratios in decimolar solutions of nitrates and chlorides on silver iodide (left) and mercury (right), respectively. Temperature 25°C. (Redrawn from J. Lijklema (= Lyklema), *Kolloid Z.* **175** (1961) 129.)

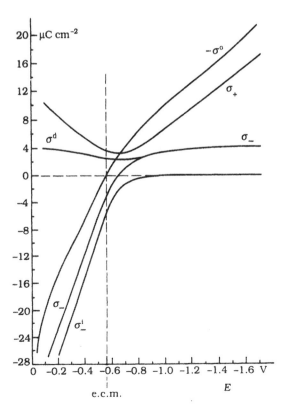

Figure 3.54. Ionic components of charge in the double layer on mercury. Electrolyte, 0.3 M NaCl. Temperature 25°C. Potentials with respect to a 1.0 M NaCl-calomel electrode. Symbols have the same meaning as in secs. 3.5b and 3.6c. (Redrawn from D.C. Grahame, loc. cit. 467.)

analyzed over a wider potential interval, especially on the positive (left of the e.c.m.) side. For the electrolyte, NaCl, the anion adsorbs specifically, and super-equivalently, so that to the positive side the adsorption of cations increases again. This is precisely the picture apparent in fig. 3.20c. In the mercury case there is no difficulty about establishing absolute values because $\sigma_+ = F\Gamma_+$ and $\sigma_- = -F\Gamma_-$ can be directly obtained from [3.10.4], say for NaCl

$$\sigma_{Na^+} = -F\left(\frac{\partial\gamma}{\partial\mu_{NaCl}}\right)_{T,E_-} \qquad \sigma_{Cl^-} = F\left(\frac{\partial\gamma}{\partial\mu_{NaCl}}\right)_{T,E_+} \qquad [3.10.9a, b]$$

Splitting the countercharge into the diffuse (σ^d) and Stern layer (σ^i) contributions is carried out from the observation that the Na^+ ion adsorbs generically so that, from σ_+, ψ^d may be computed, and hence σ^d, σ_-^d and σ_+^d, using [3.5.25 and 13].

At the electrocapillary maximum $\sigma^o = 0$, but this is not a pristine point of charge as defined in sec. 3.8, because of substantial specific adsorption of Cl^- ions. If mercury droplets could be subjected to electrophoresis they would, at the

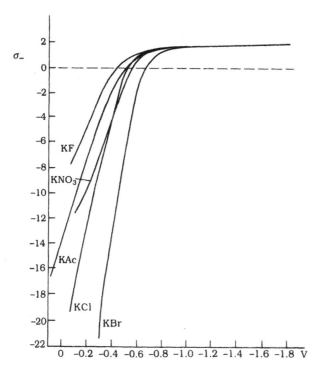

Figure 3.55. Double layer on mercury. Contribution of anions to the countercharge. Temperature 25°C, electrolyte concentration 0.1 M. Potentials with respect to a normal calomel electrode. (Redrawn from D.C. Grahame and B.A. Soderberg, *J. Chem. Phys.* **22** (1954) 449.)

e.c.m., move toward the anode. In fact, as σ^d is positive over the entire range, ψ^d is always negative so that no isoelectric point could be found.

Finally, in fig. 3.55 σ_- is given for a number of potassium salts, indicating increasing specific adsorption from KF (non-specific) to KBr. The plateau at negative potentials is lower than in fig. 3.54 because the electrolyte concentration is less. This figure may be compared with fig. 3.48, where similar lyotropic trends are found, and with fig. 3.40b, which is, apart from the different ordinate axis, the silver iodide counterpart.

In summary, the various models of secs. 3.5 and 3.6 have been successfully, and in detail, applied to the mercury system. Reviews can be found in secs. 3.15b and e. The present section covers only a tiny fraction of this vast literature. It demonstrates the richness of the information that can be obtained, but underlines at the same time the relevance of making a comparison with the AgI system, because in this way some features are observed that are difficult to realize otherwise.

3.10c. *Oxides*

Systematic studies of double layers on insoluble oxides are more recent than those on silver halides and mercury, but for a number of reasons they are not less important.

First, many oxides have great practical relevance. In nature they can be found in, for instance, most minerals, in soils and as finely dispersed matter in lakes and seas. Synthetic oxides are used in paints and ceramics, in magnetic tapes, as supports for heterogeneous catalysts, in batteries, chromatographic columns, etc. Sometimes, such dispersions are non-aqueous. The handling of these systems usually requires insight into the double layer properties. For example, understanding the ion scavenging of oxides in natural waters directly requires such information. The solubility of oxides, and hence their weathering in soils, is also related to their double layer properties. More indirectly, but not less important, is the colloidal stability of these systems, which underlies their sedimentation, packing and settling.

For oxides to become dispersions with relaxed double layers, charge transfer through the interface should take place. Experience has shown that such transport is usually realized via uptake or release of protons, which leads to equilibria such as [3.6.38a and/or b]. For that, some hydration of the surface, leading to surface hydroxyl groups, is needed. Most oxides exhibit this phenomenon. As a consequence, H^+ and OH^- ions may be considered charge-determining. This premise is supported by the observation that several oxides, if made into electrodes demonstrate Nernst or pseudo-Nernst behaviour as a function of pH. Such behaviour has never been observed as a function of the metal ions; apparently these are too deeply embedded in the solid to be liberated without any accompanying chemical process. Having established this charge-determining behaviour, potentiometric acid-base colloid titration (fig. I.5.17 and sec. 3.7a) presents itself as the appropriate technique to obtain σ^{o}. It is also possible to measure the double layer capacitance directly by a procedure as described in sec. 3.7c but only a few charge exchanges across the interface suffice to make the double layer reversible. The implication is that the chemical affinity of the charge-determining ions outweighs the electrostatic interaction caused by an external source. Trasatti et al. report experiments where this is also observed for specifically adsorbing ions like Cl^-, SO_4^{2-} and HPO_4^{2-} on $RuO_2 + TiO_2$ mixtures: these adsorptions are sensitive to pH but not to externally applied potentials[1].

From the double layer point of view oxides constitute a group with a number of specific properties, quite different from those of silver iodide and mercury. Characteristically, on oxides very high positive and negative surface charges can be found, with σ^{o}(pH) curves rather symmetrical with respect to the point of zero charge. These observations can be related to a high inner layer capacitance, which in turn is probably linked to the more hydrophilic nature of most oxide surfaces, and which may mean that many counterions can approach the surface better than on AgI[2]. Figure 3.56 gives an example of the difference. Another

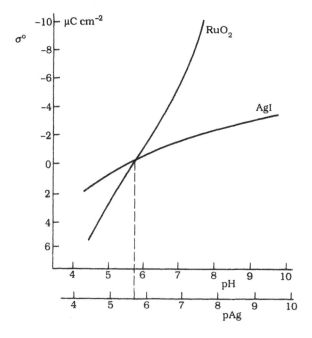

Figure 3.56. Comparison between the electrical double layer on RuO_2 and AgI. Electrolyte, 10^{-1} M KNO_3. Temperature 25°C.

[1] A. Daghetti, G. Lodi and S. Trasatti, *Interfacial properties of oxides*, etc. see sec. 3.15e, especially pages **57** and **58**; S. Trasatti in *Electrochemistry of Novel Materials*, J. Lipkowski, P.N. Ross, Eds., VCH (Weinheim, D) (1994) p. 207.
[2] T. Hiemstra, W.H. van Riemsdijk, *Colloids Surf.* **59** (1991) 7.

interesting feature is that oxides are now available with zero points of charge spanning seven or more pH units (see appendix 3, table b). Chemically speaking, this means that oxides are at our disposal with widely differing surface acidity/basicity. By comparing σ^0(pH) curves on chemically different oxides it is possible to discriminate between oxide-specific and generic double layer features. (For mercury a similar approach involves comparison with other metals.)

The oldest σ^0(pH) determinations, mostly with silica or silica-like oxides, are those carried out by Sears[1], Bolt[2] and Heston et al.[3]. When electrodes can be made and the current through the interface suppressed, capacitances and, hence, charges can be obtained from the impedance. For example, in this way Carr et al.[4] obtained $\sigma^0(E)$ curves for PbO_2, resembling σ^0(pH) curves as obtained by titration on other oxides if Nernst's law relates E to pH. Below, we shall give a few examples where the validity of this law has been verified, but in other cases electrochemists failed to check the pH-dependence of capacitances; the conversion of impedance spectra $Z(\omega)$ into C curves is in itself a problem, requiring assumptions about the double layers[5]. The issue of the value, interpretation and origin of the interfacial potential is not trivial because it focuses on the very issue of reversibility versus polarization. Electrochemists sometimes call the reversible interfacial potential the *open circuit potential*. More systematic studies where σ^0(pH) curves from titration and from impedance measurements are compared, are not known to the author. In applying the titration technique, H^+ and OH^- ions are taken as charge-determining. Given the specific and strong interaction of protons with hydrated oxide surfaces, this is a logical but not exclusive choice: phosphate ions can chemisorb on to iron oxides, imparting a charge, and Mn^{2+} ions contribute to the charge on $\delta\text{-}MnO_2$ particles near the point of zero charge. The arbitrariness reflects the basic tenet that, as a whole, double layers are electroneutral; operationally there is an element of definition in establishing what is called the surface charge. We continue to adhere to the operational definition [3.3.7], thereby realizing that (i) the protons and/or hydroxyls are not necessarily adsorbed on the surface of the oxide but may also be bound to specifically adsorbed ions or even negatively adsorbed and (ii) there are reasons to assume that on the surfaces probably only proton adsorption/desorption takes place but that with the titration technique one cannot distinguish between adsorption of a hydroxyl ion and removal of a proton. In the illustrations below we shall restrict ourselves to non-titratable, specifically or generically adsorbing (counter-) ions.

[1] G.W. Sears, *Anal. Chem.* **28** (1956) 1981.
[2] G.H. Bolt, *J. Phys. Chem.* **61** (1957) 1166.
[3] W.M. Heston jr., R.K. Iler and G.W. Sears, *J. Phys. Chem.* **64** (1960) 147.
[4] J.P. Carr, N.A. Hampson and R. Taylor, *J. Electroanal. Chem.* **27** (1970) 109.
[5] See for instance L. Bousse, P. Bergveld, *J. Electroanal. Chem.* **152** (1983) 25.

A new feature of oxides is that their surfaces are virtually always hydrated. Unhydrated crystal faces are probably inert. In titrations they do not participate but they are included in BET surface areas, and hence show up as a capacity factor. Changes of measured σ^o values with time and/or calcination in the preparatory stage of the sample may in part be caused by this fact. Upon titration, a very thin surface layer is studied, having properties that differ from the bulk oxide. For all applications these are the required parameters, but in modelling the potential some problems are encountered, that are usually formulated in terms of the validity of *Nernst's law*. In order to analyze this issue, the Galvani potential difference $\Delta\phi$ (between the bulk of the oxide and the solution) and ψ^o (at the location of the charged surface groups) must be distinguished.

The applicability of Nernst's law, as judged from the 59 mV change per pH unit at room temperature, can be experimentally verified for oxides that can be made into an electrode. In fact, such a Nernstian dependence has been found for a variety of oxides. Glass electrodes are the most prominent examples; other oxides include Ta_2O_5[1], RuO_2[2], PbO_2[3], ZrO_2[4], TiO_2[5], α-Fe_2O_3[6], β- and γ-MnO_2[7,8], Al_2O_3[9] and Co_3O_4[10]. It may perhaps be concluded that as regards $\Delta\phi$ Nernst behaviour is the rule, rather than the exception.

However, it may *not* be concluded from this that ψ^o also exhibits Nernst-like behaviour. First, there may be some potential drop inside the solid, due to free ions, or to vacancies, electrons or interstitial lattice ions, see sec. 3.10e. This would make ψ^o lower than $\Delta\phi$. Lower values for ψ^o also ensue if the thermodynamic derivation of Nernst's law (sec. I.5.5c) is re-established for a layer of which the composition changes when a charge-determining ion is introduced or removed. The general conclusion is that per pH unit ψ^o may change by less than 59 mV. There is no operational way to confirm this. A number of models have been developed to establish ψ^o quantitatively, but these are beset by a number of uncertainties (the discreteness of the charges being one of them) and experimental verification is only possible by making additional assumptions and then interpreting σ^o(pH) curves, a procedure that is not discriminative. All told, this problem is basically unsolvable.

---

[1] D.A. Vermilyea, *Surface Sci.* **2** (1964) 444.
[2] G. Lodi, G. Zucchini, A. de Battisti, E. Saviera and S. Trasatti, *Materials Chem.* **3** (1978) 179; P. Siviglia, A. Daghetti and S. Trasatti, *Colloids Surf.* **7** (1983) 15.
[3] N. Munichandraiah, *J. Electroanal. Chem.* **266** (1989) 175.
[4] S. Ardizzone, M. Radaelli, *J. Electroanal. Chem.* **269** (1989) 461.
[5] T. Watanabe, A. Fujishima and K. Honda, *Chem. Lett.* **8** (1974) 897.
[6] N.H.G. Penners, L.K. Koopal and J. Lyklema, *Colloids Surf.* **21** (1986) 457.
[7] P. Benson, W.B. Price and F.L. Tye, *Electrochem. Technol.* **5** (1967) 517.
[8] I. Tari, T. Hirai, *Electrochim. Acta,* **27** (1982) 149, 235.
[9] C. Cichos, Th. Geidel, *Colloid Polym. Sci.* **261** (1983) 947.
[10] C. Pirovano, S. Trasatti, *J. Electroanal. Chem.* **180** (1984) 171.

For practical purposes it is good to realize that in site-binding models ψ^o enters the modelling through [3.6.41]. Application of this equation in fact *defines* ψ^o. It may be further recalled that Nernst-type potentials not only arise from static equilibrium but also from stationary states: liquid junction potentials, dominated by one species, can also lead to this behaviour (see equation [I.6.7.8] with $t_+ = t_{H^+} = 1$, $t_- = 0$).

The literature contains a wealth of $\sigma^o(pH)$ curves. All the points of zero charge recorded in appendix 3, table b marked "titr." are based on such curves and may serve as a key to retrieve these data. Between different authors minor or major differences are observed with respect to the zero points (reflecting the surface chemistry), absolute value (reflecting the ratio BET area/electrochemical area) and to the shapes of the curves (reflecting the double layer formation). This takes us automatically to the question of the reproducibility and accuracy of these experimental $\sigma^o(pH)$ curves. Reproducibility depends on such features as inertness of the solid, absence of adsorbing impurities, reversibility of the reference electrode, etc. Inertness of the surface is a real problem. Many oxides slowly dissolve in water, especially at extreme pH's, they may swell or absorb (as well as adsorb) charge-determining and other ions. In practice these phenomena are "betrayed" by slow drifts in the pH-adsorption equilibrium. Incidentally, CO_2 uptake has the same result. Slow steps can also reflect a slow chemisorption process, following a fast electrostatic adsorption (phosphates, arsenates on haematite). For a number of applications these surface reactions are important for their own sake, for example in maintaining equilibria in lakes and river sediments, or in corrosion science. However, for the present purpose they should be avoided. In the titrations this means that a compromise has to be sought between carrying them out slowly enough to establish adsorption equilibrium, but fast enough to suppress the above-mentioned drifts. As a criterion, the curves should be hysteresis-free.

Accuracy is determined by the identity of oxides from different sources. X-ray analysis is not a prime indicator because we are interested in the properties of the surface. A better criterion is the point of zero charge because it reflects the acidities and basicities of the surface groups (sec. 3.8). Moreover, the p.z.c. is affected by specific adsorption, even to the extent that it can be taken as a diagnostic of pristinity. For instance, when a commercial alumina (pristine p.z.c. ≥ 9) has a p.z.c. of 6 this might be caused by the presence on the surface of substantial amounts of silica(tes), intentionally added by the manufacturer to obtain certain dispersion or wetting characteristics. Apart from this, oxides from different sources may have different surface properties and different p.z.c.'s. Synthetic and natural minerals may differ. Differences in the mode of preparation, particularly with respect to calcination, may show up as heating affects the surface hydration and may help to desorb impurities remaining from the preparation stage. By way of illustration, Dumont et al.[1] produced rutile (TiO$_2$) suspensions with varying p.z.c. depending on the way of preparation, and the p.z.c. of ruthenium dioxide shifts with calcination because of release of traces of chlorine (or chloride), left-overs of the RuCl$_4$ from which the sample was prepared[2]. On the other hand, the p.z.c.'s of anatase and rutile, two modifications of TiO$_2$ are very close (app. 3, table c).

For oxides a common intersection point of σ^o(pH) curves at different electrolyte concentrations has also been found in the presence of specifically adsorbed ions[3,4,5,6,7]. The trend is that if such a c.i.p. is observed, it is to the left and to more positive charges for specific adsorption of cations and to the right and a more negative surface charge in the case of specific anion adsorption, see fig. 3.57. The physical explanation of such c.i.p.'s was already anticipated in connection with the interpretation of fig. 3.34b; for the thermodynamics, see[8]. Van Riemsdijk et al. concluded, on theoretical grounds, that surface

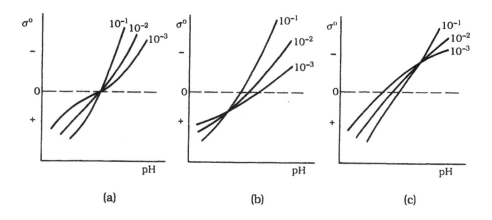

Figure 3.57. Non-pristine common intersection points, as sometimes observed with oxides. (a) No specific adsorption (c.i.p. = p.p.z.c.); (b) specific adsorption of cations; (c) specific adsorption of an ion.

[1] F. Dumont, J. Warlus and A. Watillon, *J. Colloid Interface Sci.* **138** (1990) 543.

[2] J.M. Kleijn, J. Lyklema, *Colloid Polym. Sci.* **265** (1987) 1105; P. Siviglia, P. Daghetti and S. Trasatti, *Colloids Surf.* **7** (1983) 15.

[3] M.A.F. Pyman, J.W. Bowden and A.M. Posner, *Austr. J. Soil Res.* **17** (1979) 191.

[4] S. Ardizzone, L. Formaro and J. Lyklema, *J. Electroanal. Chem.* **133** (1982) 147.

[5] H. Sadek, A.K. Helmy, V.M. Sabet and Th.F. Tadros, *J. Electroanal. Chem.* **27** (1970) 257.

[6] A. Breeuwsma, J. Lyklema, *Discuss. Faraday Soc.* **52** (1971) 324. (In this example, the pH^o was erroneously identified with the c.i.p.)

[7] L. Formaro, R. Giannantonio, C. Pastorelli and R. Carli, *J. Phys. Chem.* **96** (1992) 3197.

[8] J. Lyklema, *J. Colloid Interface Sci.* **99** (1984) 109; D.G. Hall, *J. Chem. Soc. Faraday Trans. I* **84** (1988) 2227.

heterogeneity could also lead to differences between c.i.p.'s and p.z.c.'s[1]. Some care in identifying a c.i.p. as a p.z.c. is therefore indicated; certainly some older sets of data, obtained in the presence of specifically adsorbing ions, must be re-evaluated. On the other hand, most p.z.c.'s in appendix 3, table b are well established and show only minor discrepancies between different authors.

Specific surface areas are usually obtained by BET gas adsorption. In a few cases the specific area obtained in this way was found identical to that from negative adsorption. In modelling the σ^o(pH) curves, the area enters via the inner layer capacitance, which is high and (usually) adjustable. By functional gas adsorption (sec. 1.2) the number and nature (acidity, basicity) of the surface sites can in principle be obtained. This information is relevant for site-binding analyses. For example, Scokart and Rouxhet describe this for a number of oxides[2], determining the adsorbed amounts of Lewis or Brönsted probes by infrared spectroscopy. Surfaces for which all ROH sites are charged carry about one elementary charge per 0.2 nm^2, corresponding to a maximum charge of ≈ 70 - 80 μC cm^{-2}. Titrations usually recover not more than about a third of such values. The maximum number is often extrapolated from data obtained at lower coverage, using a model. Sometimes (fig. 3.64, 65) surface charges far above the maximum are found, indicating penetration of charge-determining ions into the solid. Homodispersity of the sol particles is not directly relevant for the σ^o(pH) curves, but it is of prime importance for a number of derived phenomena (electrophoresis, stability). Indirectly there may be an effect in that homo-disperse sols are often prepared using special recipes, resulting in surface properties that may deviate from more conventional dispersions.

How reproducible are σ^o(pH) curves obtained by different investigators on different samples? Figure 3.58 gives a typical illustration. All three curves deal with synthetic rutile suspensions in concentrated indifferent electrolytes[3]. The points of zero charge are identical within 0.15 of a pH unit, which means that the samples are chemically very much the same. The absolute values differ, implying that the ratios between the electrochemical area (determined by the number of $\equiv$TiOH-groups) and the BET area (as seen by nitrogen adsorption) may differ. In turn, this may result from differences in the method of preparation, especially regarding the temperature regime. It may be significant that the

[1] W.H. van Riemsdijk, G.H. Bolt, L.K. Koopal and J. Blaakmeer, *J. Colloid Interface Sci.* **109** (1986) 219. See also L.K. Koopal, W.H. van Riemsdijk, ibid, **128** (1989) 188.

[2] P.O. Scokart, P.G. Rouxhet, *J. Colloid Interface Sci.* **86** (1982) 96.

[3] Curve (1) data by D.E. Yates, Ph.D. Thesis, Melbourne Australia (1975) as reported by J.A. Davis, R.O. James and J.O. Leckie, *J. Colloid Interface Sci.* **63** (1978) 480; curve (2) H.M. Jang, D.W. Fuerstenau, *Colloids Surf.* **21** (1986) 235; curve (3) data by L.G.J. Fokkink, Ph.D. Thesis, Wageningen, The Netherlands (1987), partly reported in L.G.J. Fokkink, A. de Keizer and J. Lyklema, *J. Colloid Interface Sci.* **127** (1989) 116.

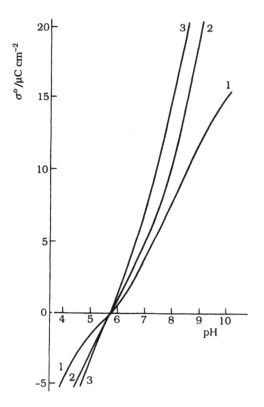

Figure 3.58. Double layer on rutile. Comparison between results from different Schools. (1) Yates, 10^{-1} KNO_3 25°C, (2) Jang and Fuerstenau, 10^{-1} KNO_3, 21°C; (3) Fokkink, 0.2 M KNO_3, 20°C. Further references and discussion in the text.

specific areas reported by the authors increase from 15.4 to 20.6 to 51 m^2 g^{-1} in the direction $1 \rightarrow 2 \rightarrow 3$. The suggestion is that with decreasing particle size *relatively* more surface sites for (de-)protonation are created. When this capacity is considered as an adjustable parameter, the various curves can be multiplied with a factor taking this into account, leading to very close, but not complete congruence. For instance, the inflection point around the p.z.c. in curve (1) is not observed in curves (2) and (3). According to the Gouy-Stern model (fig. 3.22) the visibility of this inflection is related to the difference between C^i and C^d, where the inner capacitance C^i is operationally proportional to the capacity just mentioned. In 10^{-3} M electrolyte all three sources report well-expressed inflection points equal to capacitance minima. The tendency of curve (1) to turn down at high pH is not found for the other two.

Accepting that the relative shapes of the σ^o(pH) curves are well established, the question follows as to what extent the shapes of σ^o(pH) curves vary between *different* oxides. To that end, the sets of curves are displaced horizontally till the zero points coincide. Figure 3.59 gives an example where a striking similarity is observed, which could even be improved by slightly adjusting the specific area.

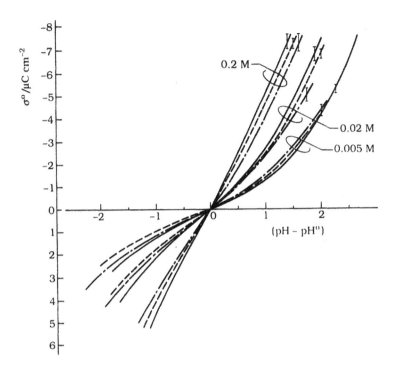

Figure 3.59. Comparison of the surface charge as a function of normalized pH for rutile (—), ruthenium dioxide (– – –) and haematite (–·–). Temperature 20°C. The KNO_3 concentration is given. (Redrawn from L.G.J. Fokkink, A. de Keizer, J.M. Kleijn and J. Lyklema, *J. Electroanal. Chem.* **208** (1986) 401.)

For alumina such curves have also been observed, but not for silica, which is apparently an exception (see below). We say that the curves of fig. 3.59 are *pH-congruent*[1].

Such pH congruence must mean that, starting from the *oxide-specific* point of zero charge, charge formation is, to a first approximation, a *generic* phenomenon. Genericity means that the double layer properties are to a large extent determined by the solution side, which should then be very similar for the various oxides under consideration (and different for silica). In fact, the curves of fig. 3.59 can be well described by a simple Gouy-Stern model with a high C^i, assuming Nernst's law applies.

In this connection fig. 3.52 is recalled, showing the (approximate) congruence between silver iodide and mercury. This pair of curves rather represents the category of hydrophobic surfaces, sometimes called "lower capacitance surfaces"

---

[1] T. Hiemstra et al., *J. Colloid Interface Sci.* **133** (1989) 105 report differences between aluminium and iron oxides but these are to a large extent attributable to different capacity factors.

confirming that it makes sense to distinguish between double layers on hydrophilic and hydrophobic surfaces.

There are strong indications that for rutile and haematite the σ^o(pH) curves are also *temperature-congruent*, i.e. they coincide provided the pH is referred to pH^o at the given temperature[1]. In this way a set of mastercurves is obtained for which at a given electrolyte concentration the integral

$$\int_0^{\sigma^o} (\text{pH} - \text{pH}^o) \, d\sigma^o \qquad\qquad\qquad [3.10.10]$$

is independent of T. When, as is often the case, Nernst's law applies and may be written as

$$d\psi^o = -\frac{RT \, ^{10}\log e}{F} \, d\text{pH} \qquad\qquad\qquad [3.10.11]$$

it can be derived that the electrostatic contribution to the Gibbs energy of double layer formation is proportional to the temperature:

$$\Delta G_a^\sigma(\text{el}) = \int_0^{\sigma^o} \psi^{o'} d\sigma^{o'} = -\frac{RT}{F} \, ^{10}\log e \int_0^{\sigma^o} (\text{pH} - \text{pH}^o) \, d\sigma^o = \text{const}' * T \quad [3.10.12]$$

Hence, according to one of the Gibbs-Helmholtz relations [I.2.15.9a]

$$\Delta H_a^\sigma(\text{el}) = \left\{ \frac{\partial(\Delta G_a^\sigma(\text{el}) / T)}{\partial(1 / T)} \right\}_{c_{\text{salt}}} = 0 \qquad\qquad\qquad [3.10.13]$$

The important implication is that microcalorimetrically determined enthalpies of double layer formation essentially measure the chemical part of the adsorption or desorption of protons, hence their independence of σ^o and c_{salt}. Experimentally this was confirmed for rutile and haematite (straight lines in fig. 3.60). The slope is steeper for the latter, in line with its higher pH^o, see [3.6.50]. For γ-Al$_2$O$_3$ this trend was confirmed at low pH by Machesky and Jacobs[2]. Above pH 6 chemical reactions involving various aluminium hydroxides may also have contributed to $\Delta_{\text{ads}} H$.

Notwithstanding the fact that several literature data on the enthalpy of charge formation are not yet well established (agreement between directly obtained calorimetric data and values derived from dpH^o/dT has not always verified and the absence of an electrostatic contribution has not always been

---

[1] L.G.J. Fokkink, A. de Keizer and J. Lyklema, *J. Colloid Interface Sci.* **127** (1989) 116.
[2] M.L. Machesky, F. Jacobs, *Colloids Surf.* **53** (1991) 297.

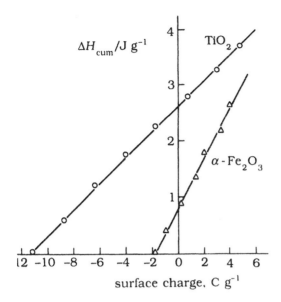

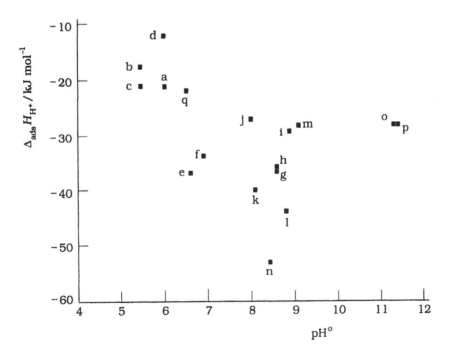

Figure 3.60. Directly measured enthalpy of charge formation (probably $\Delta_{ads}H_{H^+}$) for the double layers on rutile and haematite. The lines are independent of the salt concentration. (Redrawn from J. Lyklema, L.G.J. Fokkink and A. de Keizer, *Progr. Colloid Polym. Sci.* **83** (1990) 46.)

Figure 3.61. Proton adsorption enthalpy, related to the point of zero charge of the oxide. Key: a[32], b,c[20], d[81] = rutile (TiO$_2$); e[25], f[43] = magnetite (Fe$_3$O$_4$); g,h[20], i,j[82] = haematite (α-Fe$_2$O$_3$); k[84] = goethite (α-FeOOH); ℓ[85] = corundum (α-Al$_2$O$_3$); m[25], n[81] = alumina (α-Al$_2$O$_3$); o[86] = NiO; p[86] = Co$_3$O$_4$; q[83] = zirconia (ZrO$_2$). The references correspond to those of appendix 3.

checked or, if this contribution is non-zero, it has not always been accounted for), the data collected in fig. 3.61 deserves attention because they confirm the trend that $\Delta_{ads}H_{H^+}$ becomes more negative with increasing pH°. This rule was predicted near the end of sec. 3.8c. An interesting question is whether extrapolation to the left (silica oxides) leads to endothermic enthalpies. If these would indeed been found experimentally the conclusion must be that for those materials double layer formation is entropically driven. Further studies are welcome.

So far, this picture, with vanishingly low $\Delta_a^\sigma H(el)$ has only been studied for high capacitance double layers. Experiments with silver iodide would be enlightening.

Interpretations of $\sigma^o(pH)$ curves can be classified into several categories. Most general are thermodynamic approaches, because they do not require a model. As far as possible, such interpretations should be dealt with first. The treatment of the temperature dependence and the directly determined enthalpies of charge formation, just given, is one example. Another phenomenological interpretation is in terms of ionic components of charge, of which fig. 3.62 gives an illustration. The relatively small difference between the two sols is probably attributable to minor impurities of chloride ions. Once established, such sets of curves can be subjected to further (model-) analysis. They are the "oxide counterpart" of fig. 3.46. Other examples have been given by Sprycha et al.[1].

Model interpretation in terms of the general Gouy-Stern framework, without any picture for the inner layer capacitance (the counterpart of fig. 3.42) is easy. With C^i or K^i as the adjustable parameters, sets of $\sigma^o(pH)$ curves at various

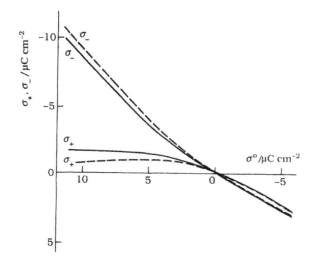

Figure 3.62. Ionic components of charge for the double layer on haematite. Electrolyte, 10^{-1} M KCl. Drawn curves: homodisperse sol; dashed curves: purified heterodisperse sol. (Redrawn from N.H.G. Penners, L.K. Koopal and J. Lyklema, *Colloids Surf.* **21** (1986) 457.)

[1] R. Sprycha, M. Kosmulski and J. Szczypa, *J. Colloid Interface Sci.* **128** (1989) 88.

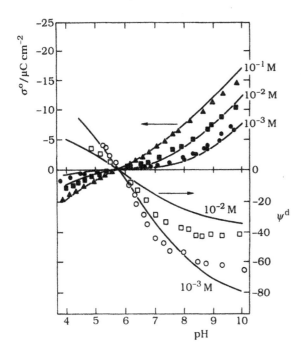

Figure 3.63. Double layer on rutile; example of a site-binding interpretation. Electrolyte, KNO_3, concentrations given. Drawn curves = model analysis. The open symbols are ζ-potentials. (Redrawn from J.A. Davis, R.O. James and J.O. Leckie, *J. Colloid Interface Sci.* **63** (1978) 480.)

ionic strengths are readily reproduced. We shall not give illustrations but note that C^i and K^i are much higher than on AgI or Hg, amounting to several hundreds of μF cm^{-2}. For infinitely high C^i or K^i the double layer attains the diffuse limit.

Somewhat more critical are interpretations in which the inner layer is modelled according to some site-binding model. Figure 3.63 gives an example, which is typical in that the σ^o(pH) curves are well recovered but the electrokinetic potentials are not. The former feature follows simply from the fact that the shapes of the curves are relatively simple and that at least seven parameters can be adjusted: N_s, C_1^i, C_2^i, pK_a, pK_b, $\Delta_{ads}G_{m,+}^o$ and $\Delta_{ads}G_{m,-}^o$ (sec. 3.6e,g). In the present example, N_s = 12 sites per nm^2, C_1^i = 100 μF cm^{-2}, C_2^i = 20 μF cm^{-2}, pK_a^c = 9.0, pK_b^c = 11.4, log $K_{K^+}^c$ = log $K_{NO_3^-}^c$ = 1.9. Some of these values are questionable. Twelve sites per nm^2 seems rather high, the choice $C_2^i < C_1^i$ is not easily supported by a physical picture, and the necessity to assume equal, but slight specific adsorption of K^+ and NO_3^- looks a bit fortuitous. With other sets of parameters, the results can be equally well described[1], hence such analyses are not discriminative, unless independent information supports some of the data (N_s from functional gas adsorption, $\Delta_{ads}H$'s from microcalorimetry).

---

[1] J. Westall, H. Hohl, *Adv. Colloid Interface Sci.* **12** (1980) 265; J.C. Westall, *ACS Symp. Ser.* **323** *Geochem. Processes Miner. Surf.* (1986) 54.

More discriminative is the value of ψ^d that ensues, which may be compared with the electrokinetic potential, ζ. As fig. 3.63 shows, this is a more difficult issue and, in fact, not yet solved because the evaluation of ζ-potentials from, say electrophoretic mobilities, is not easy either. It is likely that the experimental values of ζ, recorded in the figure, have to be increased because surface conduction was neglected in the conversion. We return to this problem in sec. 4.6f.

The experience with oxides, discussed so far is that as a first approximation $\sigma^o(pH-pH^o)$ curves are similar. This is not the case with silica. Figures 3.64 and 65 illustrate this[1]. These curves are fairly representative, although there are some experimental problems that have to be kept in mind. One of them is that the point of zero charge is not readily established; for various silicas no uptake of protons was observed at all, in others sign reversal of the electrokinetic potential was the only way to establish a zero point (i.e.p.). However, all authors agree that pH^o, if it exists, is low: about 2-2.5, depending on the sample, see also appendix 3, table b. Values much higher than that point to admixtures or impurities. For instance, the glass in fig. 3.65 also contained B_2O_3,

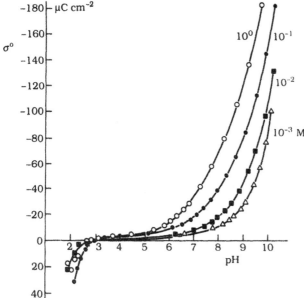

Figure 3.64. Electric double layer on precipitated silica. Electrolyte, KCl, concentration given. Temperature, 25°C. (Redrawn from Th.F. Tadros, J. Lyklema J. Electroanal. Chem. **17** (1968) 267.)

---

[1] Literature sources for fig. 3.65: (Brazilian) quartz, H.C. Li, P.L. de Bruyn, *Surface Sci.* **5** (1966) 203; pyrogenic silica, R.P. Abendroth, *J. Colloid Interface Sci.* **34** (1970) 591; precipitated silica (BDH), see fig. 3.60; glass giving Nernst response to K^+-ions, Th.F. Tadros, J. Lyklema, *J. Electroanal. Chem.* **22** (1969) 9.

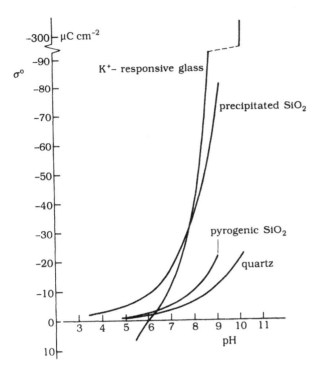

Figure 3.65. Comparison between the electric double layers on various types of silicas in 10^{-1} M (1-1) electrolytes. Provenance of the samples and further discussion in the text. (Redrawn from J. Lyklema, *Croat. Chem. Acta* **43** (1971) 249.)

K_2O, GeO_2 and Al_2O_3. Non-inertness of silica is another problem. At high pH all of them dissolve to some extent, precipitated silica and glass more so than quartz. As the dissolved silicates also consume base, correction for this phenomenon is mandatory to obtain hysteresis-free σ^o(pH) curves. A third problem is the specific surface area. Several silicas are porous for protons, as judged by the absolute values of σ^o, if based on the BET (N_2)-area. The trend is that these increase with porosity, as illustrated by fig. 3.65, with for several samples σ^o exceeding full monolayer coverage.

The most striking difference between the group of silicas and most other oxides is that over several pH units above pH^o the oxide is reluctant to dissociate protons, but beyond that charging becomes very easy. This observation does not stand on its own: for a number of silicas the colloid stability is *inversely related to* σ^o in that uncharged sols are very resistant against coagulation by indifferent electrolytes whereas they become less stable with increasing pH[1,2,3,4]. This

[1] L.H. Allen, E. Matijevic, *J. Colloid Interface Sci.* **31** (1969) 287.
[2] J. Depasse, A. Watillon, *J. Colloid Interface Sci.* **33** (1970) 430.
[3] Yu, M. Chernoberezhskii, E.V. Golikova, *Koll. Zhur.* **36** (1974) 115 (transl. 94).
[4] R.K. Iler, *Colloidal Silica*, in *Surface and Colloid Science*, (E. Matijevic, Ed.), Vol. **VI**, Wiley (1973) 1.

trend is incompatible with an electrostatic stabilization mechanism, and demands a non-electrostatic contribution that is probably related to the tendency of silicas to hydrate strongly around the p.z.c. Further evidence stems from the observation that the trend can be modified by heat treatments[1]. In this connection the mode of $\Delta_{ads}H_{H^+}$ (pH$^\circ$), discussed at the end of sec. 3.8c and in connection with fig. 3.61 may be recalled. Extrapolation of known values for other oxides to pH$^\circ \sim 2$ suggests for $\Delta_{ads}H_{H^+}$ a very low value, perhaps close to zero or even positive, in line with the reluctance of silicas to become positively charged. Recall that double layers are formed by virtue of non-electrostatic or "chemical" ion-surface interactions (sec. 3.2).

The special situation of silicas and the possible role of the water structure at the silica-solution interface have given rise to a number of interesting suggestions.

Properties of σ°(pH) curves are basic elements in the interpretation of more complicated systems involving oxides. One of these is the adsorption of hydrolyzable ions (Cd^{2+}, Al^{3+}, etc.) or anions that themselves can be titrated (HPO_4^{2-}, etc.). In sec. 3.14 some of the relevant applications will be discussed. Another application is that of mixed oxides. The systems include mechanically mixed pure oxides and mixed crystals (such as spinels and ferrites). A number of authors have studied such mixed oxides, thereby reporting the variation of the pH$^\circ$ as a function of the mole fraction of the solid. Sometimes linearity was found, sometimes not. No general rules can be given. The surface composition is not necessarily identical to that of the bulk, molecules of one oxide may leach and adsorb onto the other and lateral interactions of surface groups of the two constituents affect their pK_a's and pK_b's. Mixed oxides are important for a number of technical applications (heterogeneous catalysts with special properties, components of batteries) and also occur in clay minerals, the topic of the following subsection.

3.10d. Clay minerals

For more than one reason clay minerals offer a challenge to colloid scientists. From a practical point of view, the wide occurrence in soils and the relevance for agriculture, fertilization and topsoil mechanical properties may be mentioned. Technical applications are encountered in the paper industry, in ceramics, in brick production, chemical industry and for cleaning purposes ("fuller's earth"). Invariably, these applications involve clay minerals as the adsorbents for polymers, monomers and/or ions. Double layers around clay particles enter the picture in connection with these adsorption phenomena and

---

[1] M. Tschapek, R.M. Torres Sanchez, *J. Colloid Interface Sci.* **54** (1976) 460.

with the interaction of clay particles with each other or with other colloidal or suspended materials.

From the double layer point of view the challenge is that most clay minerals are heterogeneous; part of the surface, the faces or "plates", bears a permanent charge, independent of pH and salt concentration, the other part, the "edges", behaving as a mixed oxide with a pH-dependent charge. Sometimes, these surfaces are distinguished as having a *permanent* and *variable charge*, respectively. More fundamental is their classification as *polarized* versus *relaxed*. The faces are not polarized by the application of an external potential but by *isomorphic substitution* of cations by cations of lower charge, during the genesis of the mineral. This negative (bulk!) charge is frozen on the time scale of the experiment. Thermodynamically, the two types of surfaces require a different approach. Other challenges involve the shapes of the particles (fig. 3.1c gives an elementary sketch) and the fact that some kinds *swell* by lowering the electrolyte concentration. The shape problem, together with the heterogeneous nature of the charge distribution, makes the interpretation of electrophoretic mobilities in terms of ζ-potentials very difficult. Under certain conditions the plate-like structure, with charges of different signs on the edges and plates, can give rise to card house-like coagulates.

Prior to discussing double layer properties, we briefly describe some relevant aspects of clay mineralogy. For more extended treatments, see the textbooks[1,2, 3,4,5,6].

The basic building bricks of clay minerals are *tetrahedrons* with Si^{4+} atoms in the centre which are four-coordinated by oxygens and *octahedrons* with Al^{3+} or Mg^{2+} in the centres, coordinated by oxygens and, especially with Mg^{2+}, hydroxyls. The tetrahedrons can share oxygens to form hexagonal rings, and this pattern can be repeated ad infinitum to obtain a flat tetrahedral sheet. Similarly, by sharing oxygens, the alumina octahedrons can be linked to form an octahedral layer. The tetrahedral and octahedral sheets can be stacked on top of each other, to form a kind of "basis kit" from which nature has created a rich variety of mineral structures.

[1] H. van Olphen, *An Introduction to Clay Colloid Chemistry for Clay Technologists, Geologists and Soil Scientists*, for example the 2nd ed., Wiley (1977).

[2] *Soil Chemistry* (A) *Basic Elements* (G.H. Bolt, M.G.M. Bruggenwert, Eds.), Elsevier (1976); (B) *Physico-Chemical Models* (G.H. Bolt, Ed.), Elsevier (1979).

[3] *Minerals in Soil Environments* (J.B. Dixon, S.B. Weed, J.A. Kittrick, M.H. Milford and J.L. White, editorial committee). *Soil Sci. Soc. Am.*, Madison, USA (1977)

[4] *The Chemistry of Soil Constituents*, D.J. Greenland, M.H.B. Hayes, Eds., Wiley (1978).

[5] N.J. Barrow, *Reactions with Variable Charge Sols*, M. Nijhoff (Dordrecht, The Netherlands) (1987).

[6] *Tonminerale and Tone. Struktur, Eigenschaften, Anwendung und Einsatz in Industrie und Umwelt*, (K. Jasmund, G. Lagaly, Eds.), Steinkopf (1993).

Kaolinite is a clay mineral consisting of a repetition of this double sheet, with the links again realized through oxygens. It belongs to the category of *1:1 clay minerals*. The repeating unit is 0.72 nm. The double sheets are bonded to each other by hydrogen bonds, (involving OH's of the octahedral sheet and oxygens of the adjacent silica sheet) and Van der Waals forces. In view of the number and strength of these links, kaolinite has a rigid crystal structure that cannot be swollen by changing pH or c_{salt}. It is a *non-swelling clay mineral*. Typically, a kaolinite crystal has several tens to over a hundred of such double sheets. The *aspect ratio* (i.e. the area of the plates to that of the edges) is rather high and of the order of 10.

In fig. 3.66 a diagrammatic illustration is given. The upper and lower plates are different: the former exposes Al octahedrons, the latter Si tetrahedrons. The edges therefore contain an alternation of acid and basic surface sites.

The other category, called *2:1 minerals*, has a triple layer as the repeating unit. It consists of an alumina or magnesium oxide layer, flanked on both sides by a silicate layer. It is 1.0 nm thick. The physicochemical behaviour of 2:1 clay minerals depends strongly on the degree of substitution of Si^{4+} by Al^{3+} in the

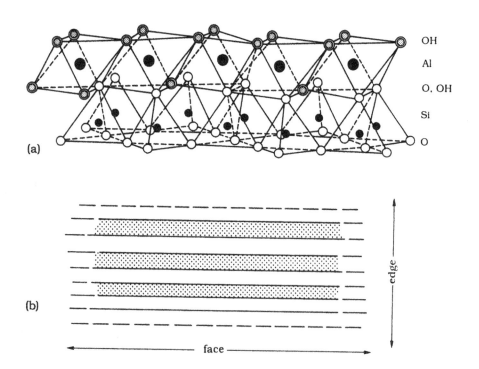

Figure 3.66. Sketch of a kaolinite crystal. (a) the basic double sheet of Al octahedrons and Si tetrahedrons. (b) The crystal structure, as built up of such double sheets. (Redrawn from R.J. Hunter, *Foundations of Colloid Science*, Vol. I, Oxford Univ. Press (1987).)

tetrahedral layers. At high ratios, e.g. $Al^{3+}/Si^{4+} = 1/3$, attraction between the layers is also high and only dehydrated K^+ ions fit between the 2:1 layers. For lower ratios, the attraction is less, allowing K^+ ions to be replaced by hydrated cations such as Na^+ and Ca^{2+}. This can give rise to limited (Ca^{2+}) or unlimited (Na^+) intracrystalline swelling. Important representatives of non-swelling and swelling 2:1 clay minerals (*smectites*) are *illites* (clay micas) and *montmorillonites*, respectively. In fig. 3.67 a sketch is given of the basic structure. With these basic units an enormous variety of minerals can be built, and has been built in nature, depending on variations in the sequences of the basic building kits, or variations in chemical composition (Mg^{2+} for Al^{3+} in the octahedrons, substitutions of Si^{4+} in some tetrahedrons by Al^{3+}, etc.). Mechanically such products may have a wide range of properties, varying between the very soft *talc* (which has Mg^{2+} in the octahedral layer) and *mica* or *muscovite*, which does not swell and tends to form stacks of large smooth layers.

At issue now is how such clay minerals acquire their charge, how this charge can be measured and how it depends on conditions. The first point to make is the electrical non-neutrality of the crystals. In the tetrahedral layer Si^{4+} may be partly substituted by Al^{3+} without altering the lattice (i.e. *isomorphic*). Similarly, in the octahedral layers Al^{3+} may be replaced by Mg^{2+}, Fe^{2+}, Cr^{2+}, Zn^{2+} or other bivalent cations. As a result, a negative volume charge develops inside the solids. It is compensated by an excess of cations on the solution side of the double layer around the particles. This negative charge gives rise to the constant, or *permanent charge*, experimentally seen as essentially residing on the plates. The second type is the *variable charge* at the edges, generated by the uptake or release of protons, just as with oxides. Kaolinite edges possess exposed OH groups of which about half have an acid nature (they stem from the silica tetrahedra), the other half being basic, (from alumina octahedra). The p.z.c. of the edges is probably near neutral. For montmorillonite, pH^o of the edges is perhaps lower than for kaolinite because of the stronger presence of silica.

Edges and plates require different techniques to determine σ^o. For the former, the titration technique as used for oxides is appropriate, whereas on the plates only *counter ion exchange* can be realized, for which the maximum number of exchangeable groups can be established, the *cation exchange capacity* (c.e.c.). The problem is that one cannot easily measure either the plates only or the edges only, but only their summed responses. This is of course also the case with other techniques like conductometry and electrophoresis.

Establishment of the c.e.c. involves converting the plates into a state with only one type of counterions (say NH_4^+, as in the ammonium acetate method) and then desorbing these quantitatively, measuring how many came off (in the example, by exchange with Na^+ and determining the NH_4^+ liberated by a Kjeldahl distillation). Other cationic substances that have been proposed to establish the

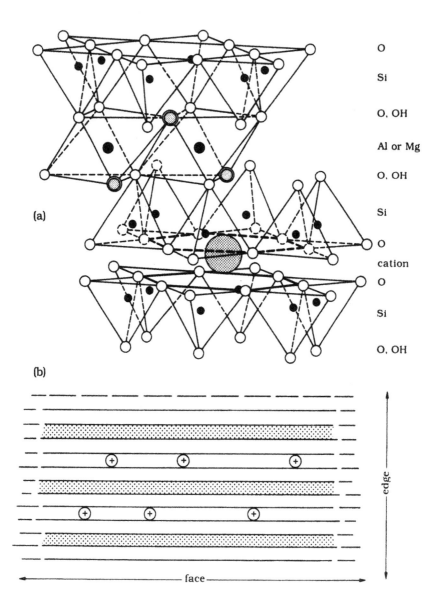

O
Si
O, OH
Al or Mg
O, OH
Si

(a)

O
cation
O
Si
O, OH

(b)

face

edge

Figure 3.67. (a) Sketch of a 2:1 clay mineral. Triple layers are kept together by cations. (b) The ensuing crystal. (The same source as the previous figure.)

c.e.c. include silver-thiourea, butylamine, methylene blue and even cationic surfactants. To exclude adsorption of these cations at the edges, the pH should be well below pH^o (edges). On the other hand, the pH should not become too low because some clay minerals then start to dissolve, a feature that is nowadays well known in connection with acid deposition ("acid rain"). Given the variety in methods and samples, results between different authors and different samples generally agree only within an order of magnitude. For instance, for kaolinite Brady[1] gave 80 μmole g^{-1}, of which a large fraction was pH-dependent. Thomas et al.[2] found 50 μmole g^{-1}, Schofield[3] 100 μmole g^{-1}, Gonzalez et al.[4] 54 μmole g^{-1}, whereas cationic surfactant adsorption rather suggests 36 μmole g^{-1} [5]. As kaolinite is a non-swelling clay, one would expect a relatively larger dependence on the pre-treatment of the sample than, for instance, for montmorillonite. In passing it may be noted that clay scientists have the habit of expressing the c.e.c. in (m)eq. per gram of clay, but the notion of "equivalent" is discouraged by IUPAC. The μmoles refer to monovalent cations. For further information on methods and results, see the textbooks mentioned before and refs.[6,7].

Cation exchange capacities are not yet plate surface charges. First the specific area A_g is needed, which poses at least two problems because of the swelling and the necessity of distinguishing between the plate and edge areas. Regarding the first problem, it is sometimes useful to define an outer and an inner surface, with the latter area being zero for non-swelling clays. Utilizing adsorption of substances that can penetrate between the plates this latter part can be estimated. From studies of adsorption from solution a great deel of interesting information on such intercalation has been obtained, in particular in combination with X-ray determination of the spacings (see Jasmund and Lagaly's book[8], for more information). For the swelling clay montmorillonite we already gave an example of gas adsorption (fig. 1.28) and two examples of alcohol adsorption from binary mixtures (figs 2.20 and 21). Some discrimination between plates and edges can sometimes be achieved from adsorption enthalpies, or by "blocking" one of the two.

Even if the plate area is known together with the c.e.c. the step from the maximum counterion charge density $\sigma_+ = F * \text{c.e.c.}$ (in mole area^{-1}) to σ^o requires conditions where *negative adsorption* may be neglected. For a diffuse double layer in a $(z - z)$ electrolyte, using [3.5.25a and 13]

[1] N.C. Brady, *The Nature and Properties of Soils*, MacMillan (1990).

[2] G.W. Thomas, W.L. Hargrove, in *Soil Acidity and Liming*, F. Adams, Ed., Am. Soc. Agron, Crop Sci. Soc. Am. and Soil Sci. Soc. Am., Madison USA, **12** (1984) 380.

[3] R.K. Schofield, *Soil Sci.* **1** (1949) 1.

[4] L. Gonzalez, P. Marti and L. Ibarra, *Brit. Polym. J.* **21** (1989) 327.

[5] T. Mehrian, A. de Keizer and J. Lyklema, *Colloids Surf.* **A73** (1993) 133.

[6] G.W. Thomas, *Exchangeable Cations*, in *Methods of Soil Analysis*, A.L. Page, R.H. Miller and D.R. Keeney, Eds., Am. Soc. Agron., Soil Sci. Soc. Am., Madison USA (1982) 159.

[7] *A Handbook of Determinative Methods in Clay Mineralogy*, M.J. Wilson, Ed., Blackie (London) (1987).

[8] *Tonminerale und Tone, Struktur, Eigenschaften und Einsatz in Industrie und Umwelt*, K. Jasmund, G. Lagaly, Eds., Steinkopf (1993).

$$\frac{\sigma_+^d}{\sigma^d} = -\frac{1}{2}\frac{\exp(-zy^d/2)-1}{\sinh(zy^d/2)}$$

[3.10.14]

which goes to minus unity for high negative y^d. Fractions of σ^o compensated by co- and counterions are sketched in fig. 3.8. Alternatively, the diffuse layer can be suppressed by working at high electrolyte concentrations, when the cationic countercharge mainly resides in the Stern layer, $\sigma_+ \approx \sigma_+^i \approx -\sigma^o$.

For different kaolinites, in this way σ^o (plates) has been established at a value between -10 and -20 μC cm^{-2}, i.e. it is of comparable order of magnitude as σ^o (edges), depending on pH.

Although for the present purpose the negative adsorption had to be suppressed, for others it is important: Donnan exclusion, pH establishment in soils, and salt-sieving. Traditionally, these topics have greatly benefited from studies with clay colloids.

Electrokinetic studies of clay minerals incur considerable interpretational problems in that, apart from the difficulty to obtain reproducible samples, the conversion of mobilities into ζ-potentials is thwarted by the limited present-day knowledge of the effect of heterogeneity and non-spherical symmetry[1] and the need to correct for surface conduction in such systems[2]. In secs. 4.6f and h we shall come back to these issues. The sole conclusion that can be drawn at present is that ζ tends to be negative over the entire accessible pH range, decreasing in magnitude with decreasing pH and increasing indifferent electrolyte concentration.

With regard to the interpretation, potentiometric σ^o(pH) curves are easier in that they do not require difficult conversion steps. The few literature examples available all tend to exhibit negative charges without intersection points for curves at different electrolyte concentration. Division of the total charge σ^o into its parts σ^o(plates) and σ^o(edges) requires additional information or a model. For instance for the edges one could use a Gouy-Stern model, whereas the proton uptake or release of the plates can be written in terms of an ionic exchange equilibrium

$$H^+(ads.) + K^+(free) \leftrightarrows H^+(free) + K^+(ads.)$$

[3.10.15]

$$K_{HK} = \frac{a_{H^+}}{a_{K^+}}\frac{\sigma_{K^+}}{\sigma_{H^+}}$$

[3.10.16]

For completely ideal (diffuse) layers the exchange is generic and $K_{HK} = 1$ but in

[1] M.C. Fair, J.L. Anderson, *J. Colloid Interface Sci.* **127** (1989) 388.
[2] R.W. O'Brien, W.N. Rowlands, *J. Colloid Interface Sci.* **159** (1993) 471.

most cases Stern contributions are involved, in which case $K_{HK} \neq 1$. In the case that c_{salt} is very high, $a_{H^+} / a_{K^+} \approx c_{H^+} / c_{K^+}$, $\sigma_{H^+} / \sigma_{K^+} \approx \sigma^i_{H^+} / \sigma^i_{K^+}$ and

$$K_{HK} = \exp\left(- \frac{\Delta_{ads}G^o_{mK^+} - \Delta_{ads}G^o_{mH^+}}{RT} \right) \qquad [3.10.17]$$

Additional information may stem from variation of the nature of the cation and/or considering the temperature dependence, to obtain $\Delta_{ads}H_{H^+}$ (isost). See for instance refs.[1,2]. In connection with proving the presence of positive edges at pH < pH° (edge), Thiessen's famous experiments concerned the attachment of small negative Au-sol particles on the edges may be mentioned[3]. Such a diagnostic example of heterocoagulation has found a prominent place as the title page in H. van Olphen's book, mentioned in sec. 3.15e.

3.10e. Semiconductors

As another illustration of the versatility of double layer studies we shall briefly discuss some typically relevant properties of semiconductors. Obviously, the limited space here can neither do justice to the prominence of these materials in transistors and other electronic devices, nor to the ingenuity that has endowed them with special properties.

The notion of conduction in the solid implies that there are charges inside it that can move. In line with the theme of this book, so far we have only considered the solution side of double layers, without troubling ourselves with distribution of charge and potential within the solid. So, we found an analogy between the double layers on Hg (a metallic conductor) and AgI (a solid ionic conductor). Similarly, between the oxides a great similarity was observed, irrespective of their conductivities. Indeed, for many purposes further knowledge of the double layer part "at the other side of the phase boundary" is not needed. Now we shall take a brief look at that side.

With respect to their conductivity, materials can roughly be divided into three categories: *metallic conductors, semiconductors* and *insulators*.

In metallic conductors the metal ions are positively charged and exactly (in solid metals) or more or less precisely (as in liquid metals) fixed by the lattice, but the electrons are free to move, behaving like an "electron gas". Because of their high concentration and high mobility, the conductivity K^S is very high (10^6-10^8 S m^{-1}) and so are the polarizability and the dielectric permittivity, $\varepsilon_o \varepsilon^s$.

[1] G.M. Beene, R. Bryant and D.J.A. Williams, *J. Colloid Interface Sci.* **147** (1991) 358 (illite).
[2] T.M. Herrington, A.Q. Clarke and J.C. Watts, *Colloid Surf.* **68** (1992) 161 (kaolinite).
[3] P.A. Thiessen, *Z. Elektrochem.* **48** (1942) 675.

Supposing that a surface charge σ^o is present at the surface by external polarization, it follows from Gauss' law [3.5.12]

$$\left(\frac{d\psi^S}{dx^S}\right)_{x^S \to 0} = \frac{\sigma^o}{\varepsilon_o \varepsilon^S}$$
[3.10.18]

that $d\psi^S/dx^S$ is very low, almost zero. This means that within the solid the
potential ψ^S is the same everywhere.

On the other side of the conductivity spectrum, in insulators ε and K are
extremely low, say below 10^{-6} S m^{-1}. For our purposes, apolar solvents, like
paraffin oil, toluene, carbon tetrachloride, etc. are important representatives in
connection with colloid stability in apolar media, see sec. 3.11. In such solvents
$d\psi/dx$ is also very low, because the low ε is outweighed by the low σ^o: it is very
difficult to charge particles in such media. By the same argument, field strengths
are also very low inside, say Teflon or most latex particles and inside oil drops
in water (as in oil-water emulsions).

At issue now is the intermediate case of semiconductors. This class is not
primarily defined by the magnitude of K^S but rather by the origin of the few
available charges inside the solid and the possibility of controlling them, e.g. by
chemical means (doping), thermally or optically (illumination). In these respects
semiconductors differ from solid ionic conductors. Also for semiconductors the
charge per unit area is not a real surface charge but stretches into the interior,
over a distance determined by the extentions of the space charge density. For flat
geometry the potential distribution is governed by Poisson's law, [3.5.1], so that
this distribution, just as in the Gouy-Chapman case, can be found if the space
charge density ρ^S is known as a function of position, x^S. This takes us to the
basis of semiconduction. Charges in semiconductors can be ionic or electronic.
Let us consider the second type, which is more typical. For an adequate
discussion quantum mechanics are needed, but that is beyond our scope; we
shall only present some semiquantitative principles.

At the absolute zero point of temperature a typical *intrinsic*, ideal monocrystal
of a semiconductor like germanium, is virtually an insulator. By "intrinsic" is
meant a Ge crystal without any trace of admixture, neither intentionally added,
nor inadvertently present. "Ideal" means without any lattice defects. The
electrons are all bound to the Ge atoms and therefore immobile. When the temperature is raised, some electrons become free, they can move through the
crystal and hence confer a certain conductivity on the crystal. In a semiconductor $dK^S/dT > 0$, this in contrast to metallic conductors, in which the electrons are always present and the randomization due to thermal motion opposes
their directional displacement with increasing temperature. In electrolyte
solutions $dK^L/dT > 0$ because the viscosity decreases with increasing tempera-

ture. At 0°C an intrinsic Ge crystal contains about 3×10^{19} free electrons per m³ and consequently the number of Ge atoms missing an electron is also 3×10^{19} m⁻³, which corresponds to about 5×10^{-8} M, i.e. generally much lower than for electrolyte solutions. On the other hand, the mobilities of electrons are about a factor 10^8 higher than those of ions in solution, so that the conductivities K^S and K^L are of comparable orders of magnitude.

Chemists would describe the liberation of an electron from a Ge atom as an ionization process, Ge $\leftrightarrows$ Ge⁺ + e⁻, but physicists rather recognize the remaining electron-deficient Ge⁺ as a *hole*. A hole h^+ is a site that has a positive charge because it lacks an electron. (In passing, a crystal lattice site lacking an atom or ion is called a *vacancy*.) The above ionization can therefore also be written as

$$0 \leftrightarrows h^+ + e^-$$
[3.10.19]

where the arrow $\leftarrow$ means annihilation of charge by recombination of an electron and a hole.

Although, of course, Ge⁺ atoms are immobile, electrons can move in from a neighbouring Ge atom to the effect that the Ge⁺ atom moves the other way. Phenomenologically speaking, holes are mobile. Their mobilities should not be underestimated; in Ge crystals at room temperature, they are about half as high as those for electrons. The upshot is that in electronic semiconductors, electrons and holes assume the functions that anions and cations have in electrolytes and, consequently, in establishing the space charge density in the double layer.

In an intrinsic semiconductor the space charge density can be written as

$$\rho^S(x^S) = F\left[c_{h^+}^S(x^S) - c_{e^-}^S(x^S)\right]$$
[3.10.20]

entirely equivalent to its ionic counterpart, [3.5.3]. When the distribution of the mobile charges obeys the Boltzmann distribution, i.e. when

$$c_{h^+}^S(x^S) = c_{h^+}^S(\infty)\, e^{-y^S}$$
[3.10.21a]

$$c_{e^-}^S(x^S) = c_{e^-}^S(\infty)\, e^{y^S}$$
[3.10.21b]

with the potentials referred to the heart of the semiconductor ($x^S = \infty$, x^S is counted from the surface inward), the analogy with the solution side in a binary (1-1) electrolyte is complete. For instance, the charge obeys [3.5.13] and the differential capacitance [3.5.17], instead of y^d the normalized potentials y^S at the surface of the semiconductor. The semiconductor-equivalent of the reciprocal Debye length becomes

$$\kappa^S = \left(\frac{2 F^2 \, c_{e^-}^S (\infty)}{\varepsilon_o \varepsilon^S RT} \right)^{1/2}$$ [3.10.22]

which may be compared with [3.5.8]. Substituting $\varepsilon^S = 18$ (typical for Ge) and $c_{e^-}^S (\infty) \approx 5 \times 10^{-8}$ M, one obtains $\kappa^S \approx 50$ nm for the extension of the double layer into the solid, comparable to that of a very dilute aqueous solution.

All of this concerns the limiting case for relatively simple semiconductors. In practice, the situation is more complex and more interesting. These complications involve the quantum mechanics of the system, with the presence of impurities (in particular *donors* and *acceptors*), *crystal defects* and *surface states*. We shall briefly discuss these.

The statistics required in macroscopic physical chemistry at ambient temperatures are invariably *Maxwell-Boltzmann statistics*, the principles and application of which have been laid down in chapter I.3. In a number of situations, especially in solids at not too high temperatures, the number of systems in an ensemble is not small compared to the number of quantum states, leading to degeneracies. Then *Fermi-Dirac statistics* are needed. Equations [3.10.21] remain acceptable at high temperatures, though.

An important achievement of quantum mechanics is the concept of *band structure* for the energies that electrons can assume, see fig. 3.68. Just as in atoms, where electrons are confined to certain orbitals, between which they can "jump" across a forbidden zone, in crystals there are bands of a given width between which there are forbidden zones that under some conditions can be crossed. As is the case with atomic orbitals the number of electrons that can be accommodated in a band is limited by the Pauli principle. As displacement of electrons under the influence of an applied electric field is not possible for a completely filled band, such bands do not contribute to the conductivity. In the

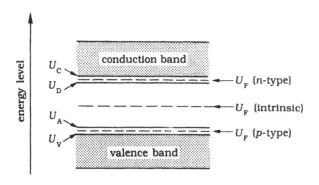

Figure 3.68. Sketch of energy levels in semiconductors. The level of a donor electron is just below (the lower side of) the conduction band, that of an acceptor electron just above (the upper side of) the valence band. Dashed: Fermi levels. The picture applies to a strongly doped semiconductor; for low degrees of doping, the U_D and U_A bands must be replaced by discrete donor-acceptor pairs.

above example of a Ge crystal at $T = 0$ this situation is met. In solid-state language, such electrons are said to be in the *valence band*. Electrons which are freed through uptake of thermal energy appear in the higher, but incompletely filled *conductance band*. For semiconductors the gap between the valence band and the conduction band is relatively small (about 0.75 eV $= 1.2 \times 10^{-19}$ J for Ge), so that thermal excitation is possible. For wider gaps radiation of higher incident energy is required, for instance by irradiation with light of high frequency, leading to *photoconduction*. Insulators are characterized by unsurpassable forbidden zones, i.e. with very high values of $U_c - U_v$ in fig. 3.68.

A number of imperfections may occur in crystals: stacking faults, dislocations, vacancies (*Schottky defects*), movement of an ion or atom to an interstitial position (*Frenkel defects*) and *isomorphic substitution* of "wrong" atoms or ions, as already encountered with clay minerals. The last category is particularly relevant. Solid state scientists have learned from the clay mineral literature that isomorphic *doping* of a semiconductor with minute amounts of ions of a different valency can have drastic consequences. Doping can for instance be achieved by diffusion at elevated temperatures. It makes the crystal non-intrinsic. To be specific: a germanium crystal consisting of tetravalent Ge atoms, can be doped by a pentavalent atom D (for *donor*) like P^{5+} or a trivalent atom A (for *acceptor*), like Al^{3+}. A donor can produce an electron (D $\leftrightarrows$ $D^+ + e^-$); in the case of P in Ge, only four electrons are needed to saturate the chemical bonds; the fifth one becomes a conduction electron. A non-ionized donor has an energy U_D in the forbidden zone, but close to the energy of the conduction band so that ionization of D is easy. Likewise, A is an acceptor according to A $\leftrightarrows$ $A^- + h^+$; the electron it has captured to saturate all required chemical bonds has an energy that is also in the forbidden zone, usually close to the energy of the valence band, see U_A in fig. 3.68. Acceptor atoms create a hole in the valence band, so that this band now also contributes to the conduction. So, donors and acceptors both promote conduction, the former via electrons in the conduction band, the latter via holes in the valence band.

In solid state jargon, a semiconductor with an excess of donors over acceptors is called an *n-type semiconductor*, the other type being a *p-type*. The terms stem from the habit of calling the concentrations of conducting electrons and holes n and p, respectively equivalent to our $c_{e^-}^S(\infty)$ and $c_{h^+}^s$, respectively. The rectifying ability of transistors (containing a junction between an *n*-type and a *p*-type semiconductor) is based on the possibility of electrons to annihilate holes, moving into the opposite direction, but if *n*-type electrons and *p*-type holes move away from each other, a non-conducting layer is created at the interface, inhibiting current transport. Hence, conduction is possible in only one direction.

Further interesting analogies between double layers in semiconductors and in

electrolytes can be mentioned. In the latter, equilibrium demands that for each
ion j the electrochemical potential $\tilde{\mu}_j$ is independent of x, although its constituents $z_j F \psi$ and $RT \ln x_j$ (x_j is the mole fraction) do depend on x, the required
relation between the two being Boltzmann's law. The semiconductor equivalent
of $\tilde{\mu}_j$ is obviously $\tilde{\mu}_{e^-}$, which is equal to $-\tilde{\mu}_{h^+}$ because of reaction [3.10.19]. Solid
state jargon calls this the *Fermi energy* or *Fermi level*, symbol U_F. Hence,

$$U_F \equiv \tilde{\mu}_{e^-} = -\tilde{\mu}_{h^+} \qquad\qquad\qquad\qquad [3.10.23]$$

Relatively simple considerations, based on Fermi statistics, indicate that U_F is
about halfway between U_D and U_C for n-type and halfway between U_V and U_C
for intrinsic semiconductors. These levels are also given in fig. 3.68. The Fermi
level is closely related to the *electronic work function* (the energy required to get
an electron out of a solid).

Near the surface of a semiconductor the components of $\tilde{\mu}_{e^-}$ vary with
position. So do the energy bands. They increase when the potential goes down
and conversely. Figure 3.69 gives a sketch for an intrinsic conductor. The three
upper figures are the semiconductor-equivalent of the ionic components of

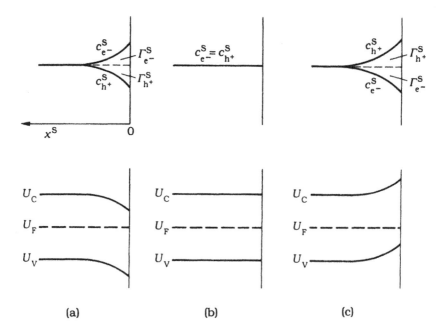

(a)   (b)   (c)

Figure 3.69. Schematic picture for space charges and energy levels in a double layer
inside an intrinsic semiconductor. In case (a) the solid carries an excess of negative
charge ($\sigma^{so} < 0$), in cases (b) and (c) $\sigma^{so} = 0$ and $\sigma^{so} > 0$, respectively. Situation (b)
represents the flatband potential. The compensating charge (positive in case (a), negative
for (c)) is found in the solution to the right of the solid-liquid interface.

charge of sec. 3.5b. Figures 3.69b represent horizontal bands, called *flatbands*. The potential of the flatband is the semiconductor-equivalent of the potential of zero charge of the solid ($\sigma^{so} = 0$).

For doped semiconductors the profiles are somewhat more complicated. Under conditions $c_{e^-}^S(x^S)$ and $c_{h^+}^S(x^S)$ can cross, leading to inversions. The equation for the space charge density [3.10.20] becomes

$$\rho^S(x^S) = F\left[c_{h^+}^S(x^S) - c_{e^-}^S(x^S) + c_A^S - c_D^S\right] \qquad [3.10.24]$$

where c_A^S and c_D^S do not depend on x^S; they are randomly distributed. For elaborations see refs.[1, 2, 3, 4, 5], where further information on semiconductor double layers can be found. Figure 3.70 gives a pictorial illustration and fig. 3.71 presents differential capacitances. They are lower than those for Gouy-Chapman double layers by about two orders of magnitude, but exhibit the same U-shape. For intrinsic semiconductors the minimum coincides with the point of zero potential but for p- and n-type semiconductors it has shifted to the right and to the left, respectively. In Gouy-type double layers such shifts were produced in asymmetrical electrolytes and attributed to the different Boltzmann weighting factors.

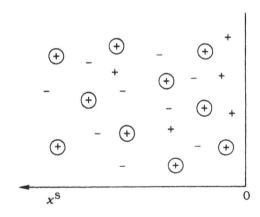

Figure 3.70. Pictorial representation of the space charge near the surface of an *n*-type semiconductor. The encircled positive charges are ionized donor atoms, which have fixed positions.

[1] P.J. Boddy, *J. Electroanal. Chem.* **10** (1965) 199.
[2] M.J. Sparnaay, *The Electrical Double Layer*, Pergamon (1972), chapter 6: *Semiconductor Surfaces*.
[3] Yu.M. Pleskov, *Electric Double Layer on Semiconductor Electrodes*, in *Comprehensive Treatise of Electrochemistry*, Vol. **1**, *The Double Layer*, J. O'M. Bockris, B.E. Conway and E. Yeager, Eds. Plenum (1980) chapter 6.
[4] W.P. Gomes, F. Cardon, *Electron Energy Levels in Semiconductor Electrochemistry*, *Progr. Surf. Sci.* **12** (1982) 155.
[5] S.N. Sze, *Semiconductor Devices, Physics and Chemistry*, Wiley (1985).

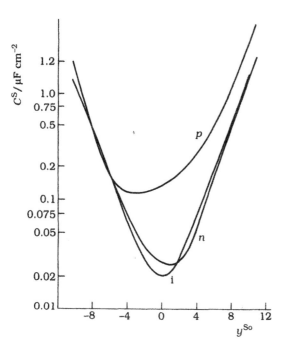

Figure 3.71. Differential capacitance on a p-type, n-type and intrinsic semiconductor as a function of the normalized surface potential. (Redrawn from K. Bohnenkamp, H.J. Engell, *Z. Elektrochem.* **61** (1957) 1184.)

The discussion given above reviews some interesting features of isolated semiconductors. The question now rises how such double layers match with Gouy-Stern layers at the solution side when dispersions or electrodes of semiconductors are considered.

Regarding the charges, at equilibrium their sum must be zero. However, as to the charge and potential distribution not much can be said. The main reason is the existence of so-called *surface states*, semiconductor jargon for spurious charges accumulating at the surface because surface atoms have affinities for holes and/or electrons differing from those in the bulk. Surface states are so called because they may have energies in the forbidden zone; they are not states in the thermodynamic sense. They play a role that is remotely reminiscent of specific adsorption. As such surface states are rather esoteric, the way in which the potential jumps from ψ^S to $\psi = \psi^L$ is very uncertain, the more so since accounting for the χ-potential is not obvious either (sec. 3.9). For the same reason it is not known whether the p.z.c. of the semiconductor ($\sigma^{so} = 0$) or, in semiconductor language, the potential of the flatband, coincides with that in the solution, as reported in appendix 3b. Figure 3.72 gives an "honest" example of this potential distribution. Combined studies of semiconductor and colloidal interfacial electrochemical techniques may help to bridge the gap.

As the capacitances derived from acid-base titration of oxides are several orders of magnitude higher than those for semiconductors (fig. 3.71), such

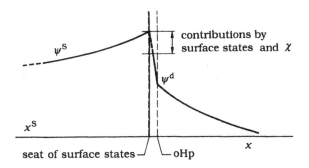

Figure 3.72. Possible junction between a semiconductor and an aqueous solution double layer. The former typically extends a few orders of magnitude further than the latter. The way in which the potential regimes match is highly uncertain. The slope $(d\psi^s/dx^s)$ depends on the sum of all charges starting from the seat of the surface states to the right.

titrations would detect surface hydroxyl groups etc., rather than bulk properties of the solid. However, doping may have its effect on the p.z.c. measured by titration because the admixtures may be surface active. For instance, for AgI pAg^o shifts considerably to the right by doping the solid with sulfur and to the left by doping with lead. Otherwise, besides the caveat that interpretations involving potentials always require a non-thermodynamic step, the double layer inside semiconductors is a challenge on its own and not felt in the properties of the solution side.

3.10f. Water-air

At the pristine water-water vapour interface spontaneous polarization of the water molecules takes place, leading to the χ-potential. Pristinity implies that there are no other ions or dissolved molecules apart from minute amounts of H^+ and OH^- ions, created by spontaneous dissociation of water molecules and which may give rise to a weak superimposed ionic double layer. There is no operational procedure to establish this χ-potential but present-day consensus has it that the air-side is negative, see sec. 3.9. At issue is now the formation of ionic double layers in addition to this when the solution contains simple electrolytes. The more dramatic changes caused by adsorbed or spread surfactants will not be addressed here.

The driving force for this charge formation is again of a non-electrostatic, or chemical nature, as defined in sec. 3.2, and requires preferential enrichment of one of the ionic species. Such a preference is not caused by specific affinity toward the vapour phase, but by a lower reluctance to leave the bulk, i.e. to become dehydrated. Most simple electrolytes have strongly hydrated ions, which avoid the surface so that they are negatively adsorbed.

Adsorbed amounts of electrolyte can be inferred from the change of the surface tension with increasing concentration. According to Gibbs' law [3.4.1] for one electrolyte i

$$d\gamma = -S_a^\sigma dT - \Gamma_i d\mu_i \qquad\qquad [3.10.25a]$$

$$d\gamma = -S_a^\sigma dT - \nu RT \Gamma d\ln y_\pm c \qquad\qquad [3.10.25b]$$

where we have used [I.5.1.17] to relate the chemical potential of the electrolyte to the electrolyte concentration. In [3.10.25b] $\nu = \nu_+ + \nu_-$, ν_+ and ν_- being the numbers of cations and anions, respectively, created by dissociation of one molecule of electrolyte. In [3.10.25b] Γ and c refer to electroneutral electrolyte. For most simple electrolytes $\Gamma < 0$, but for ions with a large radius the expulsion from water can become so strong as to make Γ positive. Surfactants are the extreme examples.

Figure 3.73 gives an illustration. Four out of the five given potassium salts adsorb negatively ($\Delta\gamma > 0$), but KPF_6, which has a bigger anion, enriches the interface ($\Delta\gamma < 0$). Along the same lines, as long ago as 1924, Rebinder[1] already noted that tetramethyl ammonium chloride adsorbs negatively, whereas the tetraethyl compound adsorbs positively. Comparison of figs. (a) and (b) shows that the corresponding acids are more surface active, a matter to which we will return. The surface tension changes are small and not easy to measure; insufficient wetting of solid surfaces by the electrolyte has been a problem in some of the earlier measurements. Using [3.10.25b] Γ can be estimated from the slopes; Γ appears to be $O(10^{-7})$ mol m^{-2}, corresponding with charges of a few µC cm^{-2} if all

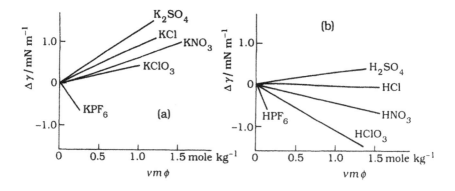

Figure 3.73. Surface tension change caused by potassium salts (a) and the corresponding acids (b); m is the molality and ϕ the osmotic coefficient. (Redrawn from J.E.B. Randles, *Phys. Chem. Liq.* **7** (1977) 107.)

[1] P. Rehbinder (= Rebinder), *Z. Phys. Chem.* **111** (1924) 447.

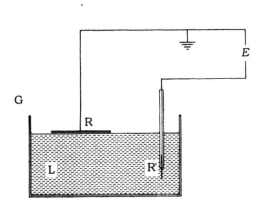

Figure 3.74. Basic set-up for the measurement of Volta potentials of electrolyte surfaces. Explanation in the text.

electrolytes are dissociated in the interfacial region. This charge is of the same magnitude as on silver iodide. However, the extent to which charge separation may take place cannot be deduced from surface tensions.

To this end, *Volta potential* measurements may be helpful. We introduced these as the potential just outside a phase, i.e. just beyond the reach of "chemical" forces, in secs. I.5.5 and 3.9. Figure 3.74 sketches the principle of the measurement. The Volta potential ψ^L of the liquid L is basically measured by the electrode R in the gas phase, which is placed parallel to and very close to the liquid surface. Its potential is kept constant by grounding. The gas in the narrow gap is made conducting by ionization. The potential E is adjusted till no current flows through the circuit containing the reference electrode R'. Then $\psi^R = \psi^L$, and E may be called the *compensation potential*. Now the composition of the solution is changed and the ensuing variation of E recorded. It equals the sum of the change in ψ^L, called $\Delta\psi^L$ and the change in the Galvani potential between solution and reference, $\Delta\phi^{LR'}$. When the latter can be accounted for (for example if R' is reversible to one of the ionic species) or suppressed (if R' is a calomel electrode, connected to the solution through a KCl-containing liquid junction) $\Delta\psi^L$ is obtainable. The method is popular in monolayer studies of surfactants (Volume III) and $\Delta\psi^L$ is then called "surface potential" a term that we shall avoid because it creates confusion. Llopis has reviewed this matter[1]. A variant is the streaming jet method, where R' is another liquid, briefly mentioned in sec. 3.9. When there is only interfacial polarization, $\Delta\psi^L$ may be identified with $\Delta\chi^L$, except for the sign.

Figure 3.75 gives an illustration. The sign convention is thus, that minus signs mean that the gas phase is negative with respect to the liquid. The outcome is in agreement with the expectation that the (bigger) anions tend to accumulate

---

[1] J. Llopis, *Surface Potential at Liquid Interfaces*, in *Modern Aspects of Electrochemistry*, J. O'M. Bockris, B.E. Conway, Eds., Vol. **6** Butterworths (1971) 91.

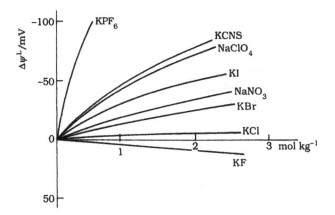

Figure 3.75. Increments in the Volta potential of water due to simple electrolytes. (Redrawn from J.E.B. Randles, *Phys. Chem. Liq.* **7** (1977) 107; his $\Delta\chi$ is our $-\Delta\psi^L$.) $\Delta\psi^L = \psi^L$(solution) $- \psi^L$(water)

at the periphery of the solution. This observation is also in line with the general observation that air bubbles in most electrolytes move electrophoretically towards the anode. For the more strongly hydrated F^- ions this is not the case. Alkali ions remain hydrated and do not contribute to the ionic specificities: between LiCl, NaCl, KCl and CsCl the difference in ψ^L is only a few mV (not shown). Figure 3.73 shows that acids are more surface active than their K^+-salts. It appears that for acids $\Delta\psi^L$ is at least as high for the K^+-salts (not shown) and has the same sign. Therefore, it is unlikely that the acids are surface active because of incomplete dissociation. Perhaps the H_3O^+ ions, orienting with their (three) positive sides toward the water, give rise to a larger contribution to χ than K^+ ions would. Anyway, the structure of the layer invites speculation.

In fig. 3.75 the reference point of $\Delta\psi^L$ at zero concentration is the χ-potential of pure water (sec. 3.9). Starting from this zero point the double layer is created by adding electrolyte, the anion acting as "charge determining". Without model assumptions the anionic charge cannot be established. Randles (loc. cit.) estimated it from $d\gamma/d\ln c$ and an argument derived from double layers on mercury. His result is represented in fig. 3.76. Such curves come as closely as possible to the σ^o-potential curves for other materials. Considering the variety of assumptions that had to be made to arrive at this result it is sufficient to state that the interface of water with air resembles that with mercury or silver iodide rather than that with oxides.

3.10g. Liquid-liquid

For practical purposes this is an extended area, with emulsions as the obvious

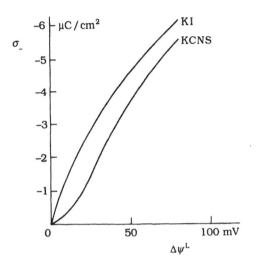

Figure 3.76. Charge-potential curve for the double layer at the air-water interface, computed according to the procedure described in the text (from Randles, see fig. 3.75). Note that in this graph the electrolyte concentration increases from the left to the right.

applications. For two adjacent liquids to be immiscible usually one is aqueous, the other apolar. In common terms one speaks of "water" and "oil". The great majority of emulsions is stabilized by ionic or non-ionic surfactants, polymers or polyelectrolytes. For ionic surfactants and polyelectrolytes the adsorption of such molecules at the liquid-liquid interface leads to the development of an electric double layer at the interface. In the case of ionic surfactants a situation as in fig. 3.1d arises, with on the aqueous side a Gouy-Stern layer, as discussed extensively before, and on the non-aqueous side a double layer of type to be discussed in sec. 3.12. No new features emerge. Double layers created by adsorbed polyelectrolytes constitute a new theme to be introduced in chapter 5 and returned to in Volume V. At issue is now the double layer at oil-water phase boundaries in the presence of low molecular weight electrolytes only.

Double layers created at such interfaces under such conditions resemble those at the air-water interface, but with the additional feature that ions can now also be present in the "oil" phase, though in extremely low concentrations. So, in addition to any polarization of the interface, a (Galvani) potential difference between the two bulk phases may originate from the unequal partition of cations and anions over the two liquids. Usually, anions are larger than cations. Hence, they are better polarizable and therefore enrich the oil phase. Consequently, oil drops in water always tend to be slightly negative, even in the absence of anionic surfactants. Electrophoretically this sign is readily detected; sometimes it is a nuisance when uncharged carrier particles are needed.

Regarding measurements one can either, measure the interfacial tension or, in a circuit, the double layer capacitance, or both. A distinction of principle has to be made in that it has to be decided whether the interface is polarizable or relaxed. In the former case C can be measured as a function of the applied potential difference E and no current should pass the interface. Integration of $C(E)$ yields σ^0. In the latter case preferential adsorption of certain ionic species is exclusively responsible for the surface charge, which in that case should be insensitive to E. In practice mixed situations are often met because the presence of, even minute amounts of, ions is required to render both phases conducting and to suppress spurious electrode polarization: some adsorption of these ions is unavoidable. An illustration of this phenomena has been given by Dupeyrat and Nakache[1]. These authors studied the water-nitrobenzene interface in the presence of KBr (almost insoluble in nitrobenzene) and hexadecyltrimethylammonium picrate (strongly preferentially soluble in nitrobenzene) and found that the interfacial tension became more sensitive to E when these solutes were removed by washing. Girault and Shiffrin[2] found that the change in interfacial tension $\Delta\gamma$, caused by the addition of LiCl, NaCl or KCl was positive for the water-n-heptane, water-1,2 dichloroethane and water-nitrobenzene interfaces, indicating negative adsorption of these electrolytes. Quantitatively $\Delta\gamma$ is for the most apolar "oil", n-heptane, identical to that for the electrolyte-air interface; for the other two, more polar, "oils", $\Delta\gamma$ is smaller. $C(E)$ curves have been given by a number of authors; they invariably exhibit a behaviour reminiscent of diffuse double layers (fig. 3.5). For an example, see Samec et al.[3], who considered tetramethyl (and butyl) ammonium tetraphenyl borate in nitrobenzene as the "oil" phase. Their capacitances were similar to those of fig. 3.5 and increased with the NaBr concentration in the aqueous phase. The extent to which, and the applied potentials at which, current flowed across the interface, was checked by cyclic voltammetry.

3.10h. On lyotropic sequences

Electric double layers are generic only close to the point of zero charge and at low electrolyte concentration. Beyond this range ion specificity starts to show up, as evidenced by the dependence of the surface charge (or capacitance) on the nature of ions of given valency. The direction and extent to which σ^0 and/or C depend on the size and/or nature of the (counter-) ion itself depends on the nature of the surface, i.e. specificity is not simply an ionic size feature but reflects a non-Coulombic ion-surface interaction leading to the specific adsorption Gibbs energy $\Delta_{ads}G_{mi}$ as occurring in Stern theory, see [3.6.34] and elaborations of it. When ions of similar chemistry and the same valence (alkali, earth

[1] M. Dupeyrat, E. Nakache, *J. Colloid Interface Sci.* **73** (1980) 332.
[2] H.H. Girault, D.J. Shiffrin, *J. Electroanal. Chem.* **150** (1983) 43.
[3] Z. Samec, V. Marecek and D. Homolka, *Faraday Discuss, Chem. Soc.* **77** (1984) 197.

alkali or halide ions) are arranged according to their affinity for a surface, their binding to macromolecules, their salting-out efficiency, their coagulation concentration for sols, etc. a so-called *lyotropic series* or *lyotropic sequence* is obtained. For instance, $Li^+ < Na^+ < K^+ < Rb^+ < Cs^+$ is a lyotropic sequence. Especially in protein chemistry such series are also referred to as *Hofmeister series*. We introduced these in sec. I.5.4 in connection with water structure-mediated interactions in electrolyte solutions. Now we consider the binding series as reflected in $\sigma^o(pAg)$, $\sigma^o(pH)$- and related curves. For the sake of argument we shall call a sequence *direct* when it is in the direction of increasing size of the unhydrated ion, as in the example just given, and *inverse* when it is the other way around.

Two types of specific adsorption can be distinguished: that between the (counter-) ion and the uncharged surface and that between the ion and a charge determining ion that is already on the surface. The former is the more general and stronger type. It can give rise to shifts of the point of zero charge (as considered in sec. 3.8b and fig. 3.57) and, because it can become superequivalent, it can lead to sign reversal of ψ^d (as in fig. 3.37c). The second type can neither affect the p.z.c. nor adsorb superequivalently. Specificity shows up only once some charge is already present. Figure 3.41 shows an example. In site-binding pictures it is appropriate to allow the ions to interact chemically with all sites if the first case applies, whereas the second type calls for chemical interaction with the charged sites only. We shall now consider the second category, restricting ourselves to monovalent ions to illustrate some general trends.

In the double layer literature several examples of ion specificity can be found, sometimes covering complete lyotropic series, sometimes only a few ion types. Fuerstenau et al.[1] and De Bruyn[2] tabulated many of these for alkali and alkali earth ions. In table 3.8 we give a selection for the alkali ions on a variety of materials under different conditions. There is consensus about these sequences, the reported trends usually also agree with other observations (coagulation, ion exchange) and are similar for other ions (direct or inverse for alkali ions correspond to the same for alkali earth ions, etc.). With regard to ζ-potentials lyotropic sequences are hardly detectable. This is not surprising because the diffuse part is essentially generic. On the other hand, the difference should show up in the surface conductivity, but systematic experiments are virtually absent. We do not discuss anion binding sequences on positive surfaces because experimentally much shorter ranges are accessible. However, in the more limited number of cases where such data are available there appears to be identity in

---

[1] D.W. Fuerstenau, D. Manmohan and S. Raghavan, in *Adsorption from Solution*, P.H. Tewari, Ed. Plenum (1980) 93.

[2] P.L. de Bruyn, *Phys. Chem. Liq.* **7** (1978) 181.

Table 3.8. Lyotropic sequences for the binding of alkali-ions from double layer measurements unless otherwise stated. c.c.c. = critical coagulation concentration[*].

Material	Sequence	Ref.	Remarks
AgI	Rb > K > Na > Li	1	agrees with c.c.c.
AgI at high T	no detectable specificity	2	" " "
Hg	Rb > K > Na > Li	3	very minor effect
SiO_2 (BDH)	Rb > K > Na > Li	4	
TiO_2 (rutile)	Li > Na > Cs	5, 6	
TiO_2 (low pH°)	Cs > K > Na > Li	6	calcined sample
δ-MnO_2 (low pH°)	Cs > Na	7	manganite IV
β-MnO_2 (high pH°)	Na > Cs	7	pyrolusite
α-Fe_2O_3	Li > Na > K > Cs	8, 9	ref. 8) from c.c.c.
γ-Al_2O_3	Na > K > Cs	10	

*Note: Strong specific binding of a given species is correlated with a low coagulation concentration.

[1] J. Lyklema, J.Th.G. Overbeek, *J. Colloid Sci.* **16** (1961) 595. [2] J. Lyklema, *Discuss. Faraday Soc.* **42** (1966) 81. [3] D.C. Grahame, *J. Electrochem. Soc.* **98** (1951) 343. [4] Th.F. Tadros, J. Lyklema, *J. Electroanal. Chem.* **17** (1968) 267. [5] Y.G. Bérubé, P.L. de Bruyn, *J. Colloid Interface Sci.* **28** (1968) 92. [6] F. Dumont, J. Warlus and A. Watillon, *J. Colloid Interface Sci.* **138** (1990) 543. [7] W. Stumm, C.P. Huang and S.R. Jenkins, *Croat. Chem. Acta.* **42** (1970) 223. [8] F. Dumont, A. Watillon, *Discuss. Faraday Soc.* **52** (1971) 353. [9] A. Breeuwsma, J. Lyklema, *Discuss. Faraday Soc.* **52** (1971) 324. [10] R. Sprycha, *J. Colloid Interface Sci.* **127** (1989) 1.

that "direct" or "inverse" for (earth) alkali ions corresponds with "direct" or "inverse" for halides, etc. In conclusion, the data of table 3.8 are well-established and typical, deserving further discussion.

The first conclusion that may be drawn is that the sequence is direct on well established hydrophobic materials such as AgI and Hg. Old coagulation data by Freundlich[1] confirm this trend for sulfur. Hydrophobicity may be measured in terms of the contact angle for water and, phenomenologically speaking, means that water molecules collectively prefer each other over the solid. With increasing radius, ions also become more hydrophobic, so that the affinity of ions for hydrophobic surfaces also rises, say by increasing hydrophobic bonding. For site adsorption, as is certainly the case for negative AgI, the adsorbing site is an I⁻ ion, which is also large, so that the direct sequence is expected. (In passing, perhaps the absence of such "hydrophobic" surface sites on mercury is responsible

[1] H. Freundlich, *Kapillarchemie, Akad. Verlagsgesellschaft* (Leipzig) Vol. **II** (1932) 387.

for the very low specificity in that case.) Along these lines the affinity for AgI increases in the direction NH_4^+ < tetramethylammonium < tetraethylammonium, etc.

At the other end of the spectrum, inverse sequences are observed for TiO_2, α-Fe_2O_3 and γ-Al_2O_3. Now the sites are the relatively small $\equiv RO^-$ groups which have a relatively strong electric field in their neighbourhood, and therefore prefer the smaller ions. Thus, phenomenologically speaking, for site-adsorption the "like seeks like" rule seems to apply. This rule is also observed for ionic interactions in electrolytes, as expressed in the activity coefficients (sec. I.5.4).

The two opposing trends, with this Gurney-type interpretation, constitute the first, and rather phenomenological way to account for lyotropic sequences. Other authors have tried to interpret them in terms of complexation and co-ordination[1,2], involving "hard" and "soft" Lewis acids and bases. "Hard" corresponds to small ions, "soft" to big ones. "Soft"-"soft" interaction corresponds to our hydrophobic bonding, etc., leading to an interaction scheme known as Pearson's rule[3]. However, the molecular picture is more complicated and involves the water structure, that is, its entropy plays an important, and sometimes even decisive role, as it does in hydrophobic bonding. The increase of entropy stems from the increase in the number of arrangements of water molecules upon association of the cation and anion involved. This (second) type of specific adsorption is typically water structure-mediated.

Returning to table 3.8, a few other remarks can be made. Regarding the oxides, a correlation is observed with the point of zero charge: for those with a low pH^o the sequence is direct, whereas for the high pH^o oxides it is the inverse. This is not unexpected: when the affinity increases with decreasing ion size it is obviously very high for protons. The composition of the solid does not play a significant role: compare the two entries for TiO_2 and for MnO_2. It is rather the surface composition, and the structure of adjacent water that counts. The very low specificity between alkali ions for mercury and air (previous subsection) indicates that there are no sites for adsorption; perhaps this is also the case for AgI at high temperature.

Lyotropic sequences have also been interpreted in terms of the "making" and breaking" of water structure and correlations have been sought with enthalpies of immersion of the solids in water. The first of these is merely paraphrasing the quest for a structure analysis in enthalpy-entropy contributions, the latter considers only the enthalpic part, ignoring the entropic side of the story.

[1] D.J. Barclay, J. Caja, *Croat. Chem. Acta.* **43** (1971) 221.
[2] T.D. Evans, *J. Electroanal. Chem.* **111** (1980) 247.
[3] After R.G. Pearson, *J. Chem. Educ.* **45** (1968) 581, 683.

In conclusion, lyotropic sequences reflect interesting water structure-mediated interactions.

3.11 Double layers in media of low polarity

The present section concerns double layers in liquids with such low dielectric permittivities that dissociated electrolytes are all but absent. Under such conditions, the charge formation and double layer properties obey rules that differ quantitatively, if not qualitatively from their aqueous counterparts. Apart from these academic issues such "apolar" double layers are relevant for a number of applications including several types of paints, emulsions, lacquers and cosmetics.

In broad terms the properties discussed so far do not differ substantially if the water is replaced by, or mixed with, solvents of high ε, such as methanol, ethylene glycol or dimethylformamide. When ε drops below 20-30, electrolytes start to associate, see sec. I.5.2d. For σ^o(pAg,pH) curves this does not have great consequences, although association is felt in the critical coagulation concentrations which are very sensitive to the valency of the counterion. We now go to the other end of the spectrum and discuss media of very low ε, say below 10. This group can again be subdivided into two categories, more or less around $\varepsilon \sim 6$, below which the media are really apolar, and above which a slight solubility of certain ions can be observed. A number of liquid non-ionics belongs to the latter category, whereas paraffin oil, toluene, hexane, etc. constitute the former.

The first question regards the charge-formation process. When there are no ions, no charge can develop, unless electron transfer takes place, This is possible if acceptor and donor molecules are present in the two phases. Fowkes has analyzed the implications in some detail[1]. More typical though, is the situation where the liquid contains *large* molecules which can dissociate. Here, the trend is opposite to that in water, where the smaller ions tend to be more dissociated. As ions cannot be solvated by apolar solvents charges can be created only if large ions are formed, keeping the charges sufficiently apart. For instance, metal sulfosuccinates (including Aerosol OT) and benzene sulfonates, stearic acid, and tetra(isoamyl) ammonium picrate are to a certain extent soluble in heptane, xylene, etc. and can produce a surface charge on dispersed particles, as evidenced by electrophoresis. As the large ions have a stronger preference for the solvent than the smaller ones, the trend is that the smaller one adsorbs on the surface of particles. So, for NaAOT the AOT$^-$ ions remain in solution whereas Na$^+$ ions

[1] F.M. Fowkes, *Interface Acid-base/charge-transfer*, in *Surface and Interfacial Aspects of Biomedical Polymers*, Vol. **1** *Protein Adsorption*, J.D. Andrade, Ed. Plenum (1985) ch. 9.

adsorb on, say dispersed carbon black or rutile, charging them positively; this is in contrast to aqueous systems.

The determination of the surface charges is very difficult. Colloid titrations do not work. Moreover, σ^o is very low, about a factor of 10 below that in water. The best technique is electrophoresis, because zeta potentials are not low and $\sigma^{ek} \sim \sigma^o$ (see below). However, experimentally the technique is replete with problems. First, it is hard to get such systems really water-free; traces of water may accumulate at the interface and dramatically affect the adsorption of the few available ionic species[1]. Parfitt and Peacock[2] list a number of systems studied and conclude that even the sign of ζ can vary between different authors. In addition, measuring mobilities in media of low conductivity is not simple either since electrode polarization and surface conduction have to be suppressed (chapter 4). "Spontaneous" explosions that have sometimes occurred upon pumping oil or gasoline over long distances are a spectacular illustration of the hazards created by accumulating spurious charges that cannot leak away by conduction[3]. In fact, here one is dealing with streaming potentials "out of control". One remedy is to increase the conductivity, for instance by adding tetra-isoamylammoniumpicrate. (The creation of charges by friction is called *triboelectricity*.)

Points of zero charge are, for the same arguments, difficult to define. Only when the system contains both large anions and large cations can a kind of p.z.c. be defined as the concentration ratio of these ions where the particles carry no charge. Experimentally the only way to establish it is electrokinetically, which is a satisfactory technique since in such media p.z.c. $\approx$ i.e.p.

However difficult the measurements may be, the interpretation of the ensuing double layers is very simple. Because of the virtual absence of electrolyte $\kappa \to 0$. There is no screening and the potential decays very slowly with distance. Diffuse double layers extend far from the surface, but are sparsely filled by ions.

The low slope $d\psi/dx$ implies that in this case $\zeta = \psi^o$ is an excellent approximation and, for that matter, $\sigma^{ek} \approx \sigma^o$. In mathematical language, consider a spherical particle. For the total charge we derived [3.5.53], where we now may substitute ψ^o or ζ for ψ^d, i.e.

$$Q = 4\pi \varepsilon_0 \varepsilon \psi^o a \qquad\qquad\qquad [3.11.1]$$

---

[1] A. Kitahara, S. Karasawa and H. Yamada, *J. Colloid Interface Sci.* **25** (1967) 490.

[2] G.D. Parfitt, J. Peacock, *Stability of Colloidal Dispersions in Non-aqueous Media* in *Surface and Colloid Sci.* E. Matijevic, (Ed.), Vol. **10**, Plenum (1978) chapter 4, table 1.

[3] A. Klinkenberg, J.L. van der Minne, *Electrostatics in the Petroleum Industry. The Prevention of Explosion Hazards*, Elsevier (1958).

which is nothing else than the Coulomb expression. The capacitance is that of an isolated sphere without countercharge [3.5.55]

$$C = K = \varepsilon_o \varepsilon / a \qquad\qquad [3.11.2]$$

From [3.11.1] it is inferred that, because of the low ε only a few charges suffice to yield a potential that is of the same order of magnitude as for aqueous systems. Equation [3.11.2] demonstrates that in such systems one deals with *low capacitance double layers*.

Sometimes the question is raised whether it is still permissible to treat double layers with only a few charges on the surface with smeared-out models. The answer is yes, because the thickness of the double layer is proportionally larger and it is the ratio between the distances separating the charges on the surface and the thickness of the double layer which counts.

Double layers in non-polar media recur in colloid stability (Volume IV). The slow decay $d\psi/dr$ (or $d\psi/dx$) means that the field strength is low, and so is the interparticle *force*. On the other hand, the *range* of the interaction is very high, so that even in dilute sols the particles feel each other's presence. Absence of screening means that the pair interaction between particles is completely described by Coulomb's law. In emulsions and at oil-water interfaces a "double diffuse" double layer may be formed, which is more extended in the oil phase[1].

3.12 Electrosorption

Electrosorption is a collective noun for all phenomena involving the influence of double layers on the adsorption of organic molecules and conversely.

Phenomenologically speaking, organic molecules adsorb in the Stern layer, affecting its capacitance and hence the surface charge at given pAg, pH, etc. (Organic molecules remaining in the diffuse layer modify its dielectric permittivity and, in this way, also influence the surface charge, but this is usually a second order effect.) Conversely, charges on the surface modify the structure and solvency of adjacent water and hence the adsorption of the organic molecules.

Analysis generally involves three aspects: measurement, thermodynamic and molecular interpretation. Generally, knowledge of adsorption from solution (chapter 2) is combined with that on double layers (this chapter).

---

[1] For further information see reviews on colloid stability in non-aqueous media: besides the review by Parfitt and Peacock, already mentioned, see J. Lyklema, *Principles of the Stability of Lyophobic Colloidal Dispersions in Non-Aqueous Media*, Adv. Colloid Interface Sci. **2** (1968) 65 and P.C. van der Hoeven, J. Lyklema, *Electrostatic Stabilization in Non-Aqueous Media*, Adv. Colloid Interface Sci. **42** (1992) 205; A. Kitahara, *Non-aqueous Systems* in Electrical Phenomena at Interfaces, (Surfactant Series No. 15). A. Kitahara, A. Watanabe, Eds., Marcel Dekker (1984) 119.

Regarding the forces responsible for adsorption, electrostatic interactions have to be added to the forces that would also operate in uncharged systems. For organic ions they amount to $z_i F \psi^i$ if ψ^i is the potential at the location of adsorption or $\boldsymbol{p}_i \cdot \boldsymbol{E}^i$ for organic dipoles. Electrosorption is complicated because the organic molecule has to compete with the counterion and both may exert either a different or the same influence on the water structure. Hence, the competition may be inhibiting or synergetic, for energetic or entropic reasons, or both.

To judge the competition, the magnitudes of the various interactions may be summarized. Coulombic energies are at room temperature about 1 kT per ion for each 25 mV if $z_i = 1$. As ψ^i is usually not higher than $\approx$ 150 mV, the coulomb energy, being of an energetic nature, may be up to 6 kT. By comparison the, mainly entropic, interaction Gibbs energy per CH_2 group in hydrophobic bonding is also about 1 kT. Hence, hydrophobic bonding can easily overcome electrostatic repulsion if the chain is not too short, and become superequivalent. Dipole terms are about a factor of 5-10 lower than the coulomb interaction but hydrogen bonding is 2.5-10 kT and chemical bonding outweighs all of these. When the solid is easily polarizable, aromatic rings tend to adsorb flat by π-electron exchange.

Measurement of the amount of the organic substance that is adsorbed, Γ_A, can be achieved either directly or indirectly. Direct determination is as described in chapter 2: analytically by depletion, spectroscopically, ellipsometrically, etc., the only additional item being that pAg or pH, etc. should be controlled. Indirectly, Γ_A can be inferred from the change of the surface charge with the concentration of A, see 3.12a, below. The latter approach only gives relative amounts, so a reference is required. Surface charges σ^o are obtained in the usual way, keeping in mind that a few organic adsorptives are also titratable over the pAg or pH, etc. range considered.

Regarding the source of electrosorption information, the situation is the same as with double layers without adsorbed organics: most of it is obtained on mercury. Few common organic molecules have escaped such measurements. Electrosorption on charged colloids as such has also been abundantly investigated, but systematic combination with surface charge measurements is relatively rare; this will be the theme of the present section. Electrokinetic or stability measurements are sometimes enlightening but are insufficient to understand the composition and fine-structure of the adsorbate.

3.12a. Thermodynamic background

The starting point is, as before, the Gibbs equation [3.4.1], which can be elaborated for the system under consideration. Typically, attention is now paid to the $\Gamma_A d\mu_A$ term, expressed as [3.4.2] for the AgI-case. Cross-differentiation between the temperature, the surface charge, or the salt term, leads to useful relationships. We shall, by way of illustration, analyse three cases: adsorption of an uncharged organic molecule (A) on silver iodide and of an ionic surfactant on a constant charge surface and on an oxide. These illustrations recur in the case studies of sec. d. below.

For the first case we have [3.4.11a or 11b] if the temperature is constant. We considered dilute solutions. When the mole fraction x_A of the organic is not low, the equilibrium $d\mu_{AgNO_3} + d\mu_{KI} = d\mu_{KNO_3} + d\mu_{AgI}$, with $d\mu_{AgI} = 0$ at constant temperature, which was needed to sequester the charge-determining terms, remains valid. However, all chemical potentials now become dependent on x_A. Physically this means that σ^o not only depends on the properties of the double layer, but also on the bulk composition, the latter because the chemical potentials of the two charge-determining electrolytes may be affected differently by the organic molecules. The Gibbs equation takes care of this, but in the elaboration of chemical potentials the dependence on x_A should be considered. As the present emphasis is on double layers, most of the analyses will be given for organic molecules that are strongly surface active, i.e. modify the double layer more strongly than the solution.

Returning to [3.4.11a or 11b], cross-differentiation between the first and third terms on the r.h.s. yields [3.4.13 or 13a], which allows the computation of Γ_A as a function of pAg, except for a constant. Sometimes there are ways to find this constant, for instance if there are indications that under certain conditions $\Gamma_A \rightarrow 0$. Alternatively, by introducing the new variable ξ as done four lines above [3.4.16] it is possible to find Γ_A as a function of σ^o.

In the constant surface charge case the term with pHdσ^o drops out. Examples are the plate surfaces of clay minerals and polystyrene latices with sulfate groups. The presence of the surface charge contributes to the excess Gibbs energy of the interface and to the interfacial tension but not to dγ when changes in the adsorption of electrolytes and the organic substances are considered. Let us, for the sake of argument, consider a negative surface onto which the cationic surfactant A^+Cl^- adsorbs from a solution of NaCl. Here A^+Cl^- may stand for dodecylpyridinium or dodecyltrimethylammonium chloride. (Later we shall also consider uncharged molecules of type A.) The Gibbs equation reads

$$d\gamma = -S_a^\sigma\, dT - \Gamma_{NaCl}\, d\mu_{NaCl} - \Gamma_{ACl}\, d\mu_{ACl} \qquad [3.12.1]$$

which can among other things be used to analyse the competition between surfactant and electrolyte. Cross-differentiation between the last two terms gives

$$\left(\frac{\partial \Gamma_{ACl}}{\partial \mu_{NaCl}}\right)_{\mu_{ACl},T} = \left(\frac{\partial \Gamma_{NaCl}}{\partial \mu_{ACl}}\right)_{\mu_{NaCl},T} \qquad [3.12.2]$$

Generally μ_{NaCl} and μ_{ACl} are coupled. Elaboration in terms of activities or concentrations depends on the absolute values of c_{NaCl} and c_{ACl}. Often $c_{NaCl} \gg c_{ACl}$; then $d\mu_{NaCl} = 2RT\,d\ln a_{\pm}$ and $d\mu_{ACl} = d\mu_{A^+} \approx RT\,d\ln c_{surf}$. The last equality is based on the argument that the activity coefficient of the surfactant is mainly determined by the swamping electrolyte NaCl and, hence constant. Using the chain rule [I.2.14.8], the l.h.s. of [3.12.2] can be written as

$$\left(\frac{\partial \Gamma_{ACl}}{\partial \mu_{NaCl}}\right)_{\mu_{ACl},T} = \left(\frac{\partial \Gamma_{ACl}}{\partial \mu_{NaCl}}\right)_{\mu_{A^+},T} - \left(\frac{\partial \Gamma_{ACl}}{\partial \mu_{ACl}}\right)_{\mu_{NaCl},T} \left(\frac{\partial \Gamma_{ACl}}{\partial \mu_{NaCl}}\right)_{\mu_{A^+},T} \qquad [3.12.3a]$$

Changing activities into concentrations, combination of [3.12.2 and 3a] leads to

$$\frac{1}{2}\left(\frac{\partial \Gamma_{surf}}{\partial \ln c_s}\right)_{c_{surf},T} - \frac{1}{2}\left(\frac{\partial \Gamma_{surf}}{\partial \ln c_{surf}}\right)_{c_s,T} \approx \left(\frac{\partial \Gamma_s}{\partial \ln c_{surf}}\right)_{c_s,T} \qquad [3.12.3b]$$

The subscripts s and surf stand for salt and surfactant, respectively. The second term on the l.h.s. is the slope of the semi-logarithmic adsorption isotherm.

When semilogarithmic adsorption isotherms of surfactants are measured at different salt concentrations valuable additional information is obtained that is diagnostic for the mechanism. In some instances the l.h.s. of [3.12.3b] is negative, mainly so in the initial parts of the isotherms. This means that NaCl is expelled. The inference is that the surfactant adsorbs with its cationic charge to the negative surface, displacing Na^+ ions. Under these conditions, the NaCl acts as an *inhibitor*.

Under other conditions, particularly in the upper parts of the isotherms, often $(\partial \Gamma_{NaCl}/\partial \ln c_{surf.}) > 0$. Then, NaCl behaves like a *promotor*. Such an action would concur with the formation of a second layer, with the cationic groups towards the solution. A double layer has to be formed, of which the Gibbs energy would be reduced if Cl^- ions co-adsorb; this co-adsorption is phenomenologically registered as uptake of neutral NaCl.

The conclusion is that the thermodynamic analysis contributes to framing models: changes of mechanism as a function of coverage can be detected and the (positive or negative) adsorption of electrolyte quantified. Similar remarks can be made about the temperature influence, either using [3.12.1] to obtain changes in surface entropy or by obtaining the isosteric enthalpy from adsorption isosters.

Consider now the more complicated case of a double layer with variable surface charge and an ionic surfactant. To be specific, we look at an oxide in a solution containing the anionic surfactant Na^+A^- and NaCl at constant temperature. The Gibbs equations are the following variants of [3.4.10a and 10b]

$$d\gamma = -\sigma^o d\mu_{HCl}/F - (\Gamma_{NaOH} + \Gamma_{NaCl})d\mu_{NaCl} - \Gamma_{NaA}d\mu_{NaA} \qquad [3.12.4a]$$

$$d\gamma = \sigma^o d\mu_{NaOH}/F + (\Gamma_{HCl} + \Gamma_{NaCl})d\mu_{NaCl}/F - \Gamma_{NaA}d\mu_{NaA} \qquad [3.12.4b]$$

In [3.12.4a] it is not allowed to replace $\Gamma_{NaOH} + \Gamma_{NaCl}$ by σ_{Na^+}/F because the surfactant also brings Na^+ ions along. However, $(\Gamma_{HCl} + \Gamma_{NaCl})$ may be replaced by Γ_{Cl^-}. For some purposes σ^o is a more appropriate variable than μ_{HCl} or μ_{KOH}, so by changing variables

$$d\xi = \mu_{HCl}d\sigma^o/F - (\Gamma_{NaOH} + \Gamma_{NaCl})d\mu_{NaCl} - \Gamma_{NaA}d\mu_{NaA} \qquad [3.12.5a]$$

$$d\xi = -\mu_{NaOH}d\sigma^o/F + \sigma_{Cl^-}d\mu_{NaCl}/F - \Gamma_{NaA}d\mu_{NaA} \qquad [3.12.5b]$$

A similar set of four basic equations may be written for the adsorption of a cationic surfactant. For each equation we can write three cross-differential relations, so that in total 24 of these may be formulated. To remain specific, let us consider the variation of the organic adsorption with the surface charge. To that end, the first and third terms on the r.h.s.'s of [3.12.5a and 5b] are cross-differentiated.

$$F\left(\frac{\partial \Gamma_{NaA}}{\partial \sigma^o}\right)_{\mu_{NaCl}, \mu_{NaA}} = -\left(\frac{\partial \mu_{HCl}}{\partial \mu_{NaA}}\right)_{\mu_{NaCl}, \sigma^o} = \left(\frac{\partial \mu_{NaOH}}{\partial \mu_{NaA}}\right)_{\mu_{NaCl}, \sigma^o} \qquad [3.12.6]$$

Elaboration depends on the difference between the magnitudes of c_{HCl}, c_{NaOH}, c_{NaCl}, and c_{NaA}. In surfactant adsorption the limiting cases of excess electrolyte, $c_{NaCl} \gg c_{NaA}$ and no electrolyte are sometimes distinguished. However, complete absence of indifferent electrolyte renders the definition of σ^o ambiguous, as discussed before in this chapter. So we have to ensure that the NaCl concentration is high enough to let $c_{NaCl} \gg c_{HCl}, c_{NaOH}$. Consequently, $d\mu_{HCl} \approx d\mu_{H^+} = 2.303\,RT\,d\log a_{H^+} \approx -2.303\,RT\,dpH$. Activity coefficients, required for the last transition, can be made explicit but for low concentrations they will not change with a_{H^+}. For $d\mu_{KOH}$ the same is found. Hence, [3.12.6] reduces to

$$F\left(\frac{\partial \Gamma_{NaA}}{\partial \sigma^o}\right)_{\mu_{NaCl}, \mu_{NaA}} = 2.303\,RT\left(\frac{\partial pH}{\partial \mu_{NaA}}\right)_{\mu_{NaCl}, \sigma^o} \qquad [3.12.7]$$

Conversion of chemical potentials into concentrations depends on the ratio between c_{NaCl} and c_{NaA}. Assuming constancy of the activity coefficients, the subscripts "μ_{NaCl}, μ_{NaA} constant" for the l.h.s. may be read as c_{NaCl}, c_{NaA}. On the r.h.s. when $c_{NaCl} \ll c_{NaA}$, $d\mu_{NaA} = 2RT\,d\ln c_{NaA}$ but $d\mu_{NaA} = RT\,d\ln c_{NaA}$ for $c_{NaCl} \gg c_{NaA}$. Hence, as the extremes

$$F\left(\frac{\partial \Gamma_{surf}}{\partial \sigma^o}\right)_{c_s, c_{surf}} = \frac{2.303}{2}\left[\left(\frac{\partial pH}{\partial \ln c_{surf}}\right)_{c_s, \sigma^o} - \left(\frac{\partial pH}{\partial \ln c_s}\right)_{c_{surf}, \sigma^o}\right] \quad (c_s \ll c_{surf})$$

[3.12.8a]

$$F\left(\frac{\partial \Gamma_{surf}}{\partial \sigma^o}\right)_{c_s, c_{surf}} = 2.303\left(\frac{\partial pH}{\partial \ln c_{surf}}\right)_{c_s, \sigma^o} \qquad (c_s \gg c_{surf}) \qquad \text{[3.12.8b]}$$

For the intermediate cases $d\mu_{NaA} = d\mu_{Na^+} + d\mu_{A^-} = RT d\ln a_{Na^+} + RT d\ln a_{A^-} \approx RT(d c_{Na^+}/c_{Na^+} + d\ln c_{A^-}) = RT(d c_{Na^+}/(c_{NaCl} + c_{NaA}) + d\ln c_{A^-})$, which varies between $2RT$ and RT, depending on the surfactant/salt concentration ratio. Recall that we consider the case $c_{NaOH} \ll c_{NaCl}, c_{surf}$. Hence in setting up these material balances we may ignore the contribution of the base.

Equations [3.12.8] contain only measurable quantities and may therefore serve to verify consistency (have we forgotten a component?). Alternatively, when Γ_{surf} varies so slightly with σ^o that the change is hardly measurable, it may be assessed from the r.h.s.

Two alternative cross-differentiations are:

(i) Between the first and third terms of [3.12.4a and 4b], which produces the variation of Γ_{surf} with pH from pH-stat cross-sections of titration curves: $(\partial \sigma^o/\partial \ln c_{surf})_{pH}$, the equivalent of [3.4.13].

(ii) Between the second and third terms of [3.12.5b], eventually leading to the salt-dependence of the surfactant adsorption, either at constant pH or at constant surface charge. In the latter case we find

$$\left(\frac{\partial \Gamma_{NaA}}{\partial \mu_{NaCl}}\right)_{\mu_{NaA}, \sigma^o} = -\frac{1}{F}\left(\frac{\partial \sigma_{Cl^-}}{\partial \mu_{NaA}}\right)_{\mu_{NaCl}, \sigma^o} \qquad \text{[3.12.9]}$$

Elaboration in terms of concentrations again depends on the ratio c_{NaA}/c_{NaCl}. The limiting cases are

(a) $c_s \ll c_{surf}$

$$\left(\frac{\partial \Gamma_{surf}}{\partial \ln c_s}\right)_{c_{surf}, \sigma^o} \approx -\frac{1}{2F}\left(\frac{\partial \sigma_{Cl^-}}{\partial \ln c_{surf}}\right)_{c_{Cl^-}, \sigma^o} + \frac{1}{2F}\left(\frac{\partial \sigma_{Cl^-}}{\partial \ln c_{Cl^-}}\right)_{c_{surf}, \sigma^o} \qquad \text{[3.12.10a]}$$

(b) $c_s \gg c_{surf}$

$$\frac{1}{2}\left(\frac{\partial \Gamma_{surf}}{\partial \ln c_s}\right)_{c_{surf}, \sigma^o} - \frac{1}{2}\left(\frac{\partial \Gamma_{surf}}{\partial \ln c_{surf}}\right)_{c_s, \sigma^o} \approx -\frac{1}{F}\left(\frac{\partial \sigma_{Cl^-}}{\partial \ln c_{surf}}\right)_{c_s, \sigma^o} \qquad \text{[3.12.10b]}$$

The co-adsorption of chloride can be positive or negative, as was the case for the salt in [3.12.3]. These equations contain useful mechanistic information.

Thermodynamic analysis can of course also start from directly measured adsorption enthalpies. Detailed and accurate data are required because in the elaboration step often interpolations are needed: for instance, on oxides $\Delta_{ads}H_A(\Gamma_A)$ is required at constant pH or σ^o, but in calorimetry these parameters usually change upon adsorption.

3.12b. Electrosorption isotherms

Electrosorption isotherms $\Gamma_A(c_A)$ can be derived from those for adsorption of uncharged molecules from solution on solids, modifying them by accounting for the charge effect. As we are dealing with solid surfaces, on which sites can usually be identified, it is logical to consider the adsorption as localized. In passing, on mercury localized adsorption isotherm models also appear to work well.

The standard molar Gibbs energy of adsorption now contains electric and non-electric contributions. Formally, we can write

$$\Delta_{ads}G_m^o = \Delta_{ads}G_m^o(\text{non-el}) + \Delta_{ads}G_m^o(\text{el}) \qquad [3.12.11]$$

but there is no unambiguous way to tell which part is which. In handling electrochemical potentials (i.e. molar Gibbs energies) of ions in solution, the same problem was encountered, see sec. I.5.1c. The pragmatic solution was to write $\tilde{\mu}_i(x) = \mu_i + zF\psi(x)$ where $\psi(x)$ is the electric potential at the position x where the electrochemical potential is wanted. Assessing $\psi(x)$ as well as possible, all other contributions, including the errors made in establishing the potential, are subsumed in μ_i and therefore identified as "chemical" or "non-electrical".

For an organic ion the same reasoning can be applied; in fact this can be done in any model. For specifically adsorbed organic ions $\psi(x)$ becomes ψ^i, where it is noted that if inorganic ions also adsorb specifically each type may have its own inner Helmholtz layer potential. Calling this contribution "coulombic",

$$\Delta_{ads}G_m^o(\text{coul}) = z_i F\psi^i \qquad [3.12.12]$$

However, many organic molecules are not ions, but nevertheless react on the presence of double layers. This is a more subtle, second-order effect involving the displacement and re-orientation of water dipoles and quadrupoles, and the breaking or re-formation of hydrogen bridges. Such solvent structure interactions belong to a large extent to the "non-electric" category, as they also occur in bulk solution and at uncharged surfaces (the χ-potential may change). What is needed is the change of this contribution in the field of a double layer.

One way to proceed is to consider the dipole contribution as the leading term, dubbing all remaining parts "non-electric". Assuming the dipoles to be ideal (also a debatable assumption on this scale) at least a first approximation is formulated:

$$\Delta_{ads} G_m^o \text{ (dip)} = \Delta n \, \boldsymbol{p} \cdot \boldsymbol{E}^i \qquad\qquad\qquad [3.12.13]$$

where $\boldsymbol{p}$ is the dipole moment of water molecules and Δn the number of water molecules displaced by one adsorbed organic molecule. Formula [3.12.13] can be extended for organic molecules that themselves carry a dipole. The field strength $\boldsymbol{E}^i$ is taken at the iHp.

Taking all terms together,

$$\Delta_{ads} G_m^o \approx \Delta_{ads} G_m^o \text{ (non-el)} + z_i F \psi^i + \Delta n \, \boldsymbol{p} \cdot \boldsymbol{E}^i \qquad\qquad [3.12.14]$$

as a very first approximation. This term can be substituted in an electrosorption isotherm equation.

Which equation? As already stated, the most appropriate are those for localized adsorption: *Langmuir, Frumkin-Fowler-Guggenheim* (FFG) or *quasi-chemical* (q.c.), see app. 1. Again a number of caveats have to be considered. One is the heterogeneity of the surface, although the electrical double layer tends to smooth it somewhat[1]. Lateral interactions cannot generally be overlooked because electrosorption is rarely restricted to very low degrees of occupancy θ. The FFG and q.c. equations are therefore the only ones on the list that remain. They apply to adsorption in a monolayer. Association into bilayers involves other features that will not be discussed here. However, even for monolayer surfactant adsorbates, where heads and tails interact laterally in a very different way, it is already a very rough approximation to collect all this fine structure in only one interaction parameter w. It should also be realized that most organic molecules are bigger than water molecules. In the statistics this should be accounted for, for instance by using dilute solution variants of equations from sec. 2.4e. There is certainly room for more advanced modelling. We will take the FFG and q.c. expressions as the *starting electrosorption equations* and write them as follows:

$$\frac{\theta}{1-\theta} e^{zw\theta/kT} = K_{el. \, ads} \, x_A \qquad\qquad \text{(FFG)} \qquad [3.12.15]$$

$$\frac{\theta}{1-\theta} \left[\frac{(\beta - 1 + 2\theta)(1-\theta)}{(\beta + 1 - 2\theta)\theta} \right]^{z/2} e^{w/kT} = K_{el. \, ads} \, x_A \qquad \text{(q.c.)} \qquad [3.12.16]$$

Here z is the coordination number in the two-dimensional adsorbate, w the lateral interaction pair parameter,

---

[1] L.K. Koopal, S.S. Dukhin, *Colloids Surf. A* **73** (1993) 201.

$$\beta^2 = 1 - 4\theta(1 - \theta)(1 - e^{-w/kT}) \qquad\qquad\qquad\qquad [3.12.17]$$

x_A in mole dm^{-3} is the mole fraction of A which may be ionic or non-ionic. The logarithm of $K_{el.\ ads}$ may be written

$$\ln K_{el.\ ads} = -\frac{\Delta_{ads} G_m^o}{RT} \qquad\qquad\qquad\qquad [3.12.18]$$

with $\Delta_{ads} G_m^o$ given by [3.12.14] or more elaborate expressions. For lateral repulsion $w > 0$, for attraction $w < 0$; equations [3.12.15 and 16] both reduce to the Langmuir case for $w = 0$:

$$\frac{\theta}{1 - \theta} = K_{el.\ ads}\, x_A \qquad\qquad\qquad\qquad [3.12.19]$$

Expressions [3.12.15 and 16] are both two-parameter formulas. The two unknowns, z, w and $K_{el.\ ads}$, can either be obtained by linearization, for instance, from [3.12.15], using [3.12.14].

$$\ln\left(\frac{\theta}{1 - \theta}\right) + \frac{zw\theta}{kT} = -\frac{\Delta_{ads} G_m^o}{RT} + \ln x_A \qquad\qquad\qquad\qquad [3.12.20]$$

or by analyzing the slope $d\theta/d\ln(c_A\cdot/55.5)$, of which the maximum can be obtained by setting the second derivative equal to zero. For both isotherm equations the maximum is found at $\theta = 0.5$ and

$$\left(\frac{d\theta}{d\ln c_A}\right)_{max} = \frac{1}{4 + zw/kT} \qquad\qquad (FFG) \qquad\qquad [3.12.21]$$

$$\left(\frac{d\theta}{d\ln c_A}\right)_{max} = \frac{1}{4 + 2z(e^{w/2kT} - 1)} \qquad\qquad (q.c.) \qquad\qquad [3.12.22]$$

In the present model these maxima do not depend on $K_{el.\ ads}$. Hence it may be pragmatic to first establish w from the maximum and then $K_{el.\ ads}$ by isotherm linearization. It goes without saying that good data are required.

When w becomes sufficiently negative, *phase separation* takes place. The critical values have been derived for a number of isotherms in sec. I.3.8. They amount to

$$w_{crit.} = -4kT/z \qquad\qquad (FFG) \qquad\qquad [3.12.23]$$

and

$$w_{crit.} = -3kT\ln[z/(z-2)] \qquad\qquad (q.c.) \qquad\qquad [3.12.24]$$

for the two models, respectively, and also do not depend on $K_{el.\ ads}$. It is noted that in the present model w is purely non-electrostatic, because the potential ψ^i

completely accounts for the electrostatics (in the mean field approximation that is) electrosorption w may also have an electrical contribution.

For other models the analysis proceeds along similar lines.

Electrosorption data can also be interpreted in terms of *surface pressures* or, for that matter, *two-dimensional equations of state*. For the above models, the required equations read (see app. 1)

$$\pi a_m = -kT\ln(1-\theta) + \frac{1}{2}zw\theta^2 \qquad\qquad \text{(FFG)} \qquad\qquad [3.12.25]$$

and

$$\pi a_m = -kT\ln(1-\theta) - \frac{zkT}{2}\ln\left[\frac{\beta+1-2\theta}{(\beta+1)(1-\theta)}\right] \qquad \text{(q.c.)} \qquad [3.12.26]$$

No more than the slopes do they contain the electrosorption Gibbs energy. The surface pressure is experimentally accessible by integration of the Gibbs equation $d\gamma = -\Sigma_i \Gamma_i d\mu_i$. The fact that some components are dubbed "charge determining" does not matter provided the sum is taken over electroneutral components. This is our choice, not that of thermodynamics which is model-free. In practice this means that the r.h.s. of, for instance [3.12.4b] is integrated from the pristine surface at the p.z.c. to the final condition, determined by the final values of σ^0, σ_{Cl^-} and $\Gamma_{surf.}$. The surface pressure is a function of state, therefore the order of the three integrations is immaterial and a matter of convenience; moreover one of the terms can be converted (by cross-differentiation) if it might be poorly accessible.

Although interpretation in terms of surface pressures is relatively straightforward and informative, investigators are unaccustomed to apply it to charged solid surfaces. In the mercury double layer field and with monolayers (Volume III) it is more common[1].

---

[1] For further reading, see the following reviews: B.B. Damaskin, O.A. Petrii and V.V. Batrakov, *Adsorption of Organic Compounds on Electrodes*, Plenum (1971; P. Somasundaran, E.D. Goddard, *Electrochem. Aspects of Adsorption on Mineral Surfaces* in *Modern Aspects of Electrochemistry*, B.E. Conway, J. O'M. Bockris, Eds., Plenum, Vol. **13** (1979) 207; B.B. Damaskin, V.E. Kazarinov, *The Adsorption of Organic Molecules* in *Comprehensive Treatise of Electrochemistry*, Vol. **1**, J. O'M. Bockris, B.E. Conway and E. Yeager, Eds., Plenum (1980) Ch. 8, 353; M.V. Sanganarayanan, S.K. Rangarajan, *J. Electroanal. Chem.* **130** (1981) 339; **176** (1984) 1, 29, 45, 65, 99, 119; D.B. Hough, H.M. Rendall, *Adsorption of Ionic Surfactants*, in *Adsorption from Solution at the Solid/Liquid Interface*, G.D. Parfitt, C.H. Rochester, Eds., Academic Press (1983) 247; Ya.A. Shchipunov, *Hydrophobic and Electrostatic Interactions in Adsorption of Surface Active Substances, Tetraalkylammonium Salts*, *Adv. Colloid Interface Sci.* **28** (1988) 135; R. Guidelli, *Molecular Models of Organic Adsorption at Metal-Water Interfaces* in *Adsorption Mol. Met. Electrodes*, J. Lipkowski, P.N. Ross, Eds., VCH New York, (1992) 1.

3.12c. Congruence analysis

By "congruence analysis" is meant a rather formal and general procedure to find out whether $\Delta_{ads}G_m^o$ reacts on the surface potential or on the surface charge. In the field of electrosorption on mercury this procedure is fairly well established[1] to obtain "the primary electric variable".

As discussed in connection with the field dependence of $\Delta_{ads}G_m^o$, the interactions between an adsorbate and a surface are rather involved, so that a sole dependence of $\Delta_{ads}G_m^o$ on either σ^o, or on pAg, pH, etc. must not be anticipated. When an organic molecule is charged its electrical energy is determined by [3.12.12], containing the potential, whereas ΔG_m^o contains the electric field or, by virtue of Gauss's law, the charge when the adsorption only affects the dipole orientations, see [3.12.13]. The problem is of course that ψ^i and E^i are not identical to ψ^o and $E(x=0)$, respectively Hence, the idealized extremes that $\Delta_{ads}G_m^o$(el) is entirely determined by pAg, pH, etc. or entirely by σ^o, are merely indicators for the actual electrosorption mechanism.

Keeping these caveats in mind, the analysis to establish whether $\Delta_{ads}G_m^o$ is a function of (say) pH or σ^o proceeds as follows. The electrosorption isotherm is, quite generally, written as the functionals

$$F'(pAg, \Gamma_A)c_A = f'(\Gamma_A)$$ [3.12.27]

or

$$G'(\sigma^o, \Gamma_A)c_A = g'(\Gamma_A)$$ [3.12.28]

where F', G', f' and g' are mathematical functions of the variables indicated in parentheses. An electrosorption isotherm is now classified as *potential (or pAg, pH, etc.) congruent* provided the function F' is separable into a product of two functions, one depending on pAg, pH, etc. only, the other on Γ_A only, i.e. if

$$F(pAg)c_A = f(\Gamma_A)$$ [3.12.29]

Similarly, it is *charge-congruent* provided

$$G(\sigma^o)c_A = g(\Gamma_A)$$ [3.12.30]

These congruences apply to [3.12.15 and 16] if, via [3.12.18 and 12 or 13], $K_{el.ads}$ only depends on pAg, pH, etc. or only on σ^o, respectively.

Whether the electrosorption is pAg- or charge-congruent is now directly verified by taking logarithms of [3.12.29 and 30]: in the former case plots of $\Gamma_A(\log c_A)$ at different pAg's coincide by shifting along the ln c_A axis; in the latter this is the case for curves at different σ^o. The extent of the shift is log F(pAg)

---

[1] S.K. Rangarajan, *J. Electroanal. chem.* **45** (1973) 279, 283.

or $\log G(\sigma^o)$ per unit of $\log x_A$, with $\Delta G_m^o(el) = -\ln F(pAg)$ or $-\ln G(\sigma^o)$, respectively.

3.12d. Case studies

Systematic electrosorption studies are not abundant in the literature in that rarely are σ^o, Γ_A and the salt effect simultaneously measured over a large range. What *are* available are just millions of adsorption isotherms under a number of conditions, sometimes in connection with electrokinetic and/or stability studies. However useful such studies may be, they do not provide enough information to carry out analyses as intended in the present section. We shall therefore illustrate the subject matter with two case studies, choosing silver iodide and oxides as the, rather representative, examples.

The electrosorption of butanol on silver iodide may be considered as the prototype. The molecule is expected to adsorb with its hydrophobic moiety to the surface and that is indeed found experimentally. Figure 3.77 gives the basic data from which the following conclusions can be drawn:

(i) With increasing c_A the slope, i.e. the capacitance, decreases.

(ii) The point of zero charge moves to the positive side (to the left).

(iii) There is a well-defined common intersection point.

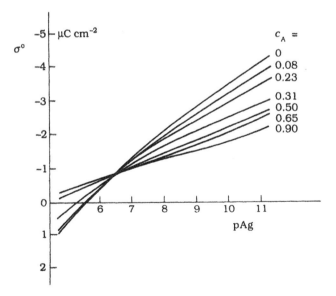

Figure 3.77. Electric double layer on silver iodide in the presence of *n*-butanol (conc. in M given) $c_{KNO_3} = 10^{-1}$ M. Temperature, 20°C. (Redrawn from B.H. Bijsterbosch, J. Lyklema, *J. Colloid Sci.* **20** (1965) 665.)

These data do not stand on their own. Similar sets of curves have been found for other alcohols, for instance for ethylene glycol[1], for tetraalkyl ammonium ions[2], for polymers[3] and also for butanol on mercury[4].

Point (i) is expected: the Stern capacitance decreases because ϵ^i decreases and/or d increases. Considering the main conclusion of sec. 3.9 that water tends to become positively polarized with respect to air and hydrophobic surfaces, point (ii) is also expected: the originally polarized interface is replaced by apolar hydrophobic groups. The extent to which this happens depends somewhat on the nature of the alcohol, in particular on the degree to which the OH group is oriented. Figure 3.78 gives shifts of the isoelectric point for various alcohols. As KNO_3 hardly adsorbs specifically at the zero point, Δi.e.p. $\approx \Delta$p.z.c. The alcohol concentrations are so low that the activity coefficients of Ag^+ and I^- ions are hardly affected; therefore this shift is essentially a shift of the χ-potential: the originally polarized adjacent water dipoles are displaced by the apolar moiety of the alcohol. Extrapolation of the curves gives the maximum shifts collected in table 3.9. On mercury for n-butanol +240 mV is found. The data provide evidence that in the absence of alcohols the χ-potential is about 0.2 V with the negative side of the water molecules towards the AgI or mercury.

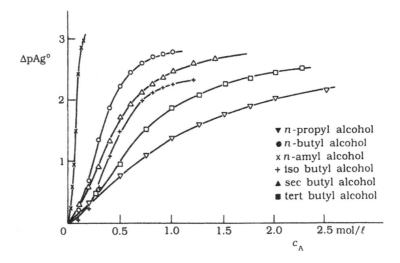

Figure 3.78. Double layer on silver iodide. Shifts of the i.e.p. (from streaming potentials) caused by various alcohols. $c_{KNO_3} = 10^{-1}$ M. Temperature, 20°C. (Same source as fig. 3.77.)

[1] J.N. de Wit, J. Lyklema, *J. Electroanal. Chem.* **41** (1973) 259.
[2] A. de Keizer, J. Lyklema, *J. Colloid Interface Sci.* **75** (1980) 171.
[3] L.K. Koopal, J. Lyklema, *Faraday Discuss. Chem. Soc.* **59** (1975) 230.
[4] E. Blomgren, J. O'M. Bockris and C. Jesch, *J. Phys. Chem.* **65** (1961) 2000.

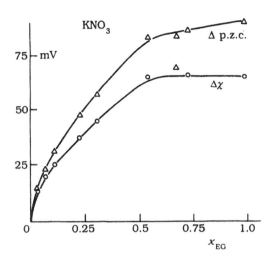

Figure 3.79. Double layer on silver iodide in ethylene glycol. Comparison between the shift in the p.z.c. and in the χ-potential. Electrolyte, 10^{-1} M KNO_3, temperature 20°C. (Redrawn from J.N. de Wit, J. Lyklema, *J. Electroanal. Chem.* **41** (1973) 259.)

Ethylene glycol can be mixed in all proportions with water, so that activity changes in the solution also have to be accounted for. This can be accomplished by using cell EMF measurements. Figure 3.79 shows the result. Only part of Δp.z.c. may now be identified with $\Delta\chi$. This part is much lower than for the alcohols of table 3.9; in view of the structure of the ethylene glycol molecule, this outcome is expected.

Returning to point (iii), considering [3.4.13 or 13a] the common intersection point c.i.p. must coincide with a maximum in the adsorption of butanol as a function of pAg or σ^o, indicated as pAg(max) or σ^o(max), respectively. Even if there is no such c.i.p., at each cross-over of two σ^o(pAg) curves at different c_A, Γ_A is a maximum as a function of σ^o; only in that case does σ^o(max) shift with coverage. A maximum in $\Gamma_A(\sigma^o)$ is expected on the grounds that at very negative and at very positive surface charge water (dipoles) are preferred over (the apolar side of) butanol molecules. The maximum does not coincide with the p.z.c. because of the natural orientation of water in that case. In fact the surface charge at the adsorption maximum rather reflects the natural water polarization, existing at the p.z.c. is more or less compensated by the double layer field. Model interpretations can be, and have been, based on that. As the butanol adsorption is based on the interplay between hydrophobic bonding and dipole displacement, this adsorption is expected to be charge-congruent, but a congruence analysis has not yet been done.

Table 3.9. Maximum shift of the χ-potential (in mV) at the AgI-water interface for a number of alcohols.

n-propyl	189	sec. butyl	240
n-butyl	192	tert. butyl	221
iso-butyl	161		

For the alcohols mentioned in table 3.9 σ^o(max) varies between -1.0 and -1.3 μC cm^{-2}. For butanol on mercury, in 10^{-1} M HCl it is -2.0 μC cm^{-2}. For ethylene glycol σ^o(max) is -3.2 μC cm^{-2} on silver iodide and -3.7 μC cm^{-2} on mercury. Generally, on mercury, this parameter is systematically more negative than (but of the same order of magnitude as) on silver iodide. This is a consideration that ought to be kept in mind in modelling.

The amounts of butanol, desorbed if σ^o is decreased or increased from σ^o(max.) are too small to be measured analytically. More dramatic effects are expected for tetraalkyl ammonium ions that, because of their cationic nature, are more readily desorbed at positive σ^o. Figure 3.80 gives an example. Three features may be mentioned:

(i) The σ^o(pAg)-curves, if extrapolated to the left, seem to merge. Probably this is due to complete desorption. On mercury where much wider σ^o-ranges can be studied, this phenomenon is experimentally retrievable (*desorption peaks* in the capacitance curves). The advantage is that now complete adsorption isotherms $\Gamma_A(c_A)$ at various σ^o or various pAg can be constructed, which can be compared

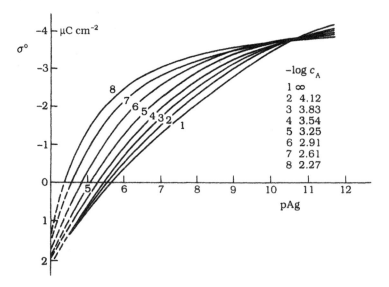

Figure 3.80. Double layer on silver iodide in tetrapropyl ammonium nitrate solutions, of which the (initial) concentration is given. Electrolyte, 10^{-1} M KNO_3, temperature 20°C. (Redrawn from A. de Keizer, J. Lyklema, *J. Colloid Interface Sci.* **75** (1980) 171.)

with direct measurement to verify consistency (this has been done with a satisfactory outcome for a few systems) and to subject the data to a congruence analysis, see under (iii).

(ii) A c.i.p. is observed at σ^o(max.) = -3.7 μC cm^{-2}, implying that if σ^o is made more negative (to the right of the maximum) the tetrapropyl cations *desorb*. It is counter-intuitive that a cation desorbs if the coulomb attraction to the surface is made stronger. The reason must be that in [3.12.14] the dipole term outweighs the Coulomb one. This observation automatically takes us to

(iii) The congruence analysis. Indeed it is found that such isotherms are better charge-congruent than pAg-congruent. Figure 3.81 gives, by way of illustration, the charge congruence for the tetrabutyl compound: the curves almost coincide after a horizontal displacement of -1.31 in $\Delta \ln c$ per μC g^{-1}, so that ΔG^o_{ads} = -1.31 RT μC^{-1} g.

For the AgI system in the presence of n-butanol and several other adsorptives, electrokinetic and stability data are also available. When AgI particles are fully covered by n-butanol they are "hydrophilized"; the lyotropic sequence is then the other way around than on bare AgI.

Much more details can be found in the review by Bijsterbosch and Lyklema, mentioned in sec. 3.15e.

As stated before, systematic electrosorption studies on oxides are much scarcer than those on silver iodide. One of the reasons is that adsorbing molecules do not always affect the (high) surface charge at given pH. For

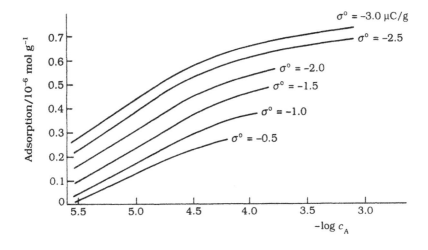

Figure 3.81. Congruence analysis for the electrosorption of tetrabutylammonium nitrate on silver iodide. Electrolyte, 10^{-1}M KNO_3, temperature 20°C. Specific surface area of the sample 0.92 m^2 g^{-1} (reference as fig. 3.80).

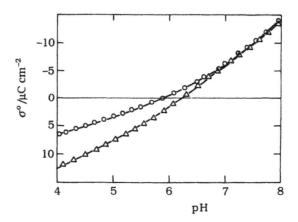

Figure 3.82. Double layer on rutile in 10^{-1} M NaCl; (o) without and (Δ) in the presence of 0.28×10^{-3} M NaNBS. Temperature 20°C. (Redrawn from M.R. Böhmer, L.K. Koopal, *Langmuir* **8** (1992) 2660.)

instance, n-alcohols do not influence σ^o(pH) curves on alumina[1], although they do so on rutile[2]. Best studied is the adsorption of surfactants but, although the influences of c_{salt} and pH are sometimes considered, systematic σ^o plus electrosorption investigations are virtually non-existent.

By way of illustration the adsorption of sodium nonyl-(para)benzene sulfonate (NaNBS) on rutile is considered. For this system the thermodynamics of [3.12.4-10] are applicable, especially when further data become available.

Figure 3.82 shows that NBS⁻ ions adsorb specifically. In fig. 3.83 Γ_{surf} and σ^o/F are plotted as a function of c_{surf} for various combinations of pH and $c_s = c_{KCl}$. At $c_{surf} = 0$, σ^o decreases with increasing pH (in fact at pH 6 the values are negative, which could have been verified had full titrations been done) and increases with c_{salt}. As the surfactant anion adsorbs specifically, it tends to make σ^o more positive; hence σ^o increases with c_{surf}. The KCl concentration has little effect on the plateau attained by σ^o because the double layer capacitance is dominated by the surfactant. The first adsorbing surfactant molecules influence σ^o much more strongly and conversely. This is because the later adsorbing molecules build a second layer. The plateau of the surfactant adsorption increases with the salt concentration because of the mechanism discussed after [3.12.3].

Regarding other illustrations, the work by Iwasaki and De Bruyn on the adsorption of dodecylammonium acetate on silver sulfide may be mentioned as probably the first attempt to measure σ^o and Γ_A simultaneously and analyze

[1] M. Kosmulski, *J. Colloid Interface Sci.* **135** (1990) 590.
[2] J. Szcypa, L. Wasowska and M. Kosmulski, *J. Colloid Interface Sci.* **126** (1988) 592.

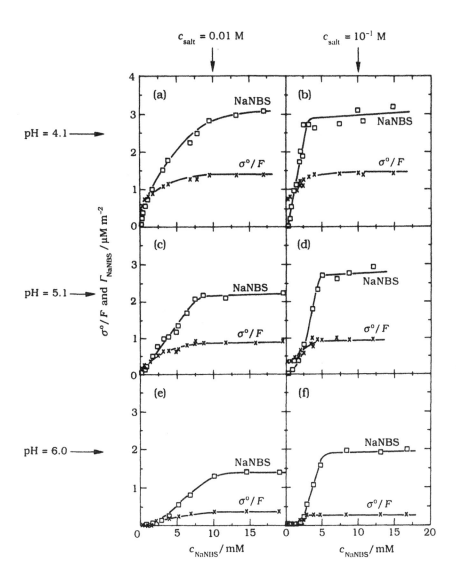

Figure 3.83. Adsorption of NaNBS on rutile, (□), compared with σ^0/F (×) for six combinations of pH and electrolyte (KCl) concentration. (Reference as in fig. 3.82.)

their cross-relation[1]. Temperature studies are also informative, because electrostatic interactions are insensitive to T, whereas hydrophobic bonding typically passes through a maximum. In this way lateral or bilayer association of hydrophobic (parts of) molecules can be distinguished from electrostatic adsorption in

---

[1] I. Iwasaki, P.L. de Bruyn, *Surface Sci.* **3** (1965) 299.

the first layer[1]. Common intersection points of organic and charge-determining ions have been thermodynamically analyzed by de Keizer et al.[2] along the thermodynamic lines of sec. 3.12a. An interesting variant of the sign reversal of the salt effect between the lower and higher parts of surfactant adsorption isotherms, discussed before, was recorded by Bitting and Harwell, who, for sodiumdodecylsulfate on alumina at fixed pH and c_s, observed an inversion of the lyotropic sequence for the alkali ions[3].

3.13 Charged particles in external fields

By way of introduction to electrokinetic phenomena (chapter 4) we shall now briefly discuss the alterations occurring when a particle carrying a double layer is placed in an external electric field. In practice this happens with electro-phoresis and during particle interaction. However, in the present section we shall assume the particle to be at rest, i.e. the electrical force, exerted on it, is compensated by another force that we shall not specify here. The external field may be stationary, as in d.c. electrophoresis, or time-dependent, as in a.c. dielectric dispersion, but here only the former case will be considered. Besides in electric fields, charged colloids can also be placed in a convection, gravitational or centrifugation field, or in a concentration gradient. Each of these gradients can lead to specific electrokinetic phenomena (table 4.1) which are interrelated through Onsager-type cross-relations.

When the applied field E is switched on, a number of phenomena will occur, including:

(i) The double layer becomes *polarized*, meaning that a (small) excess of charge is accumulated on one side, creating a depletion on the other side. This is called *concentration polarization*. (This kind of polarization differs from that of a mercury electrode where an interface is polarized by applying a field across it.)

(ii) An induced dipole moment p_{ind} is created; the distance between the excess positive and negative charges is of order $2a$ if a is the particle radius.

(iii) The electric field is deformed by the particle and its double layer.

(iv) Ion transport, which in the bulk proceeds by conduction, is considerably modified by the presence of the particle. When, as will be assumed here, the particle is kept stationary, the electric force exerted on the countercharge sets this charge into motion. By viscous traction liquid is entrained. This phenomenon is called *electro-osmosis*; it will be treated in sec. 4.3b.

[1] T. Mehrian, A. de Keizer and J. Lyklema, *Langmuir* **7** (1991) 3094.
[2] A. de Keizer, M.R. Böhmer, T. Mehrian and L.K. Koopal, *Colloids Surf.* **51** (1990) 339.
[3] D. Bitting, J.H. Harwell, *Langmuir* **3** (1987) 500.

Thus a complicated interplay of forces and fluxes emerges: diffusion, conduction and hydrodynamic flows. These will be treated in sec. 4.6, but in anticipation of this treatment we shall now emphasize on the (concentration-) polarization, mentioned under (i) and discuss some of its consequences. From the outset it is important to realize that quantities with different length scale interact: the double layer thickness is, and remains, $O(\kappa^{-1})$, but the *polarization field*, caused by the polarization of the particle, has a range $O(a)$.

Hydrodynamic influences are also of order $O(a)$. To emphasize some of the main features we shall restrict the discussion to the situation where $a \gg \kappa^{-1}$.

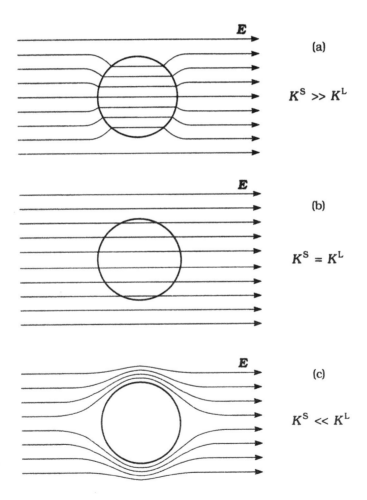

Figure 3.84. Change in the electric field caused by the presence of an uncharged colloid particle.

3.13a. Influence of a particle on the electric field

By way of introduction, consider first an uncharged colloidal particle in an applied homogeneous field. Depending on the conductivity K^S of the particle with respect to that of the solvent, K^L, different situations can be created, as illustrated in fig. 3.84.

For the relatively special case that $K^S = K^L$ (fig. 3.84b) the field lines are straight, as if there were no particle. When there is no barrier for ions to pass the interface, these field lines are identical to the flow lines that ions follow (cations to one side, anions to the other). When the conductivity of the particle exceeds that of the fluid, these lines tend to contract inside the particle (fig. a) but when $K^S \ll K^L$, they avoid the particle (fig. c). In the extreme case of a metallic conductor ($K^S = \infty$) the surface of the sphere is an equipotential plane and the field lines approach the surface perpendicularly from the solution side.

Most colloids rather behave like the *dielectric spheres* of fig. c. The reason is that the ions in the solution can rarely pass the phase boundary. Therefore, the field lines must detour the particles. We shall henceforth emphasize such particles. As these particles are usually charged the extra ions in the solution side of the double layer lead to *surface conduction*, which, in turn, tends to contract the field lines through the double layer. Figure 3.85 illustrates this. Here, we have introduced the dimensionless quantity

$$Du \equiv K^\sigma / a\, K^L \qquad\qquad\qquad [3.13.1]$$

that we shall henceforth call the *Dukhin number*[1]. This number is a measure of the relative contribution of the surface conductance. Recall that K^L is in $S^{-1}m^{-1}$ $= C\ V^{-1}s^{-1}m^{-1}$ and K^σ in $S = C\ V^{-1}s^{-1}$ (sec. I.6.6d). In fig. 3.85 one could visualize Du as the ratio of the surface conduction of a sphere of radius a, $2\pi a K^\sigma$, to the bulk conduction if the particle had the same conductivity as the pure liquid, $\pi a^2 K^L$. This ratio equals Du, except for a factor of 2. If so desired, Du can be split up into contributions of the various double layer parts and/or those of the various ionic types. This will be considered in sec. 4.3f, where explicit formulas for Du will be derived. When $Du \gg 1$ (fig. 3.85b) the double layer "contracts" the field lines around the particle. In the extreme case of $Du = \infty$ and no polarization, the double layer would act as an isopotential envelope around the (dielectric) particle. Then the field lines would approach and leave this envelope perpendicularly. This would also be the case if the particle would be conducting.

---

[1] After S.S. Dukhin who explicitly introduced this parameter in his analyses of electrokinetic phenomena. See e.g. S.S. Dukhin, *Non-equilibrium electric surface phenomena*, *Adv. Colloid Interface Sci.* **44** (1993) 1, where older references can be found. The very idea that the ratio $K^\sigma / a K^L$ plays an important role in electrokinetics is much older and can be found for example in J.J. Bikerman, *Trans. Faraday Soc.* **36** (1940) 154. In sec. 4.5c we shall indicate how K^σ can be measured.

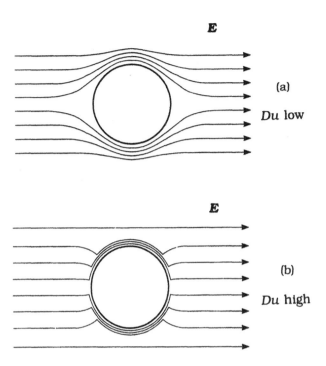

Figure 3.85. Dielectric particle in an external field. (a) non-conducting double layer, (b) conducting double layer.

However, in real situations the angle of incidence depends on position. When Du is small the situation sketched in fig. 3.84a prevails.

Figures 3.84 and 3.85 do not yet suffice to give the electric field at any position in the double layer. On closer inspection, three fields have to be considered: the applied field, just discussed, the field of the double layer and the *polarization field* caused by the concentration polarization. The applied field, characterized by parallel field lines far from the particle, is in most electrokinetic experiments $O(10^3 \text{ V m}^{-1})$, that of the double layer is, at least for an unpolarized system, radially directed and decays from $O(10^6 \text{ V m}^{-1})$ close to the surface to zero for distance $r \to \infty$. The magnitude and extension of the polarization field take intermediate positions. We shall now consider this polarization.

3.13b Double layer polarization. The far field

Figure 3.86 illustrates how double layers around polarized particles and double layers at rest may differ. In fig. 3.86 it is for simplicity assumed that only the countercharge is polarized. In some cases the surface charge can also

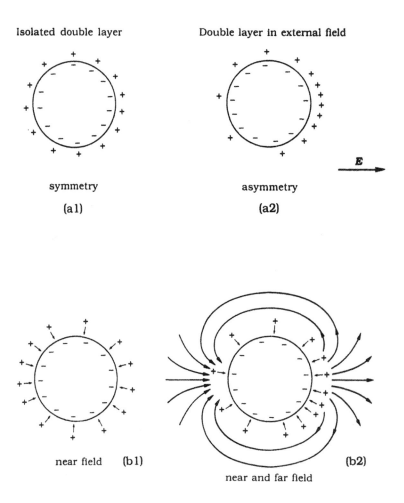

Isolated double layer

Double layer in external field

symmetry

asymmetry

(a1)

(a2)

near field   (b1)

(b2)

near and far field

Figure 3.86. Free and polarized double layers. For simplicity the countercharge is not detailed. (a) charge distribution; (b) electric fields. The lines of force of the field E are not shown: they would overwhelm and skew the far field.

adjust itself at a rate depending on the conductivity of the particles, but in the present introduction we consider the particles to be non-conducting.

The polarization of the double layer has a number of typical consequences. One of these is that an *induced dipole moment* p_{ind} can be assigned to it. It is a vector (sec. I.4.4b). The magnitude of this dipole moment is determined by the applied field, the shape, orientation and volume of the particles, Du and a few solution characteristics. It is the most important characteristic of polarized double layers.

Together with the induction of the dipole, a *polarization field* is developed, as sketched in fig. 3.86. Such a field was already introduced in fig. I.4.5. It has to be superimposed on the double layer field of the undisturbed particle, sketched in fig. 3.86,b1. As a consequence, this double layer field is modified. However, the polarization field extends far beyond the double layer. Therefore, we shall make a distinction between the *far field* and the *near field*, or double layer field. The near field has a range of $O(\kappa^{-1})$, the far field of $O(a)$. The far field contains the polarization field in addition to the applied external field, but no net charges. This field is also the seat of the various hydrodynamic fluxes. Overbeek antici- pated the distinction between these fields in his theory of electrophoresis[1]. He superimposed on the double layer at rest a long-range perturbation field, containing both the polarization field of the double layer and the external field, perturbed by the particle. Later this idea was elaborated by Dukhin and Shilov in a Russian monograph (1972)[2]. The distinction between the two fields is also made in Hunter's book[3]. The name "far field" was probably coined by Fixman[4]. In the present section, only the electrostatic aspects will be considered keeping the particle at rest.

Summarizing, the far and near field differ in three respects. First they do so in range. Common double layer fields extend over distances of order κ^{-1}; in the absence of an external field such fields are radial for a spherical double layer, as shown in fig. 3.86,b1. On the other hand, the range of the far fields is of the order of the particle radius a, which for the case considered, means that they extend far beyond the double layer. In the second place they differ in magnitude, as already stated. Thirdly, the difference is that in the near field there exist local excess charges, whereas in the far field each volume element is electro- neutral. In mathematical language, $\rho^{\mathrm{ff}}(r,\theta) = 0$, where r and θ are defined in fig. 3.87. Consequently, the Laplacian of the potential is also zero in the far field.

$$\nabla^2 \psi^{\mathrm{ff}}(r,\theta) = 0 \tag{3.13.2}$$

For $\nabla^2 \psi$ in spherical coordinates we have [I.A7.47]. For unpolarized double layers, the solution of [3.13.2] is, in such coordinates,

$$\psi^{\mathrm{ff}}(r,\theta) = -E r \cos\theta + \psi^{\mathrm{b}} \frac{a^2}{r^2} \cos\theta \tag{3.13.3}$$

[1] J.Th.G. Overbeek, *Kolloid Beih.* **54** (1943) 287.
[2] English transl. S.S. Dukhin and V.N. Shilov, *Dielectric Phenomena and the Double Layer in Disperse Systems and Polyelectrolytes*, Wiley (1974).
[3] R.J. Hunter, Foundations of Colloid Science, Vol. II, chapter 13 (written in collaboration with R.W. O'Brien).
[4] M. Fixman, *J. Chem. Phys.* **72** (1980) 5177; **78** (1983) 1483.

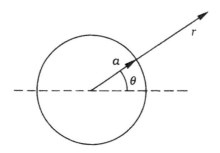

Figure 3.87. Meaning of symbols a, θ and r.

Here, ψ^b is the potential at the border between the near and far field, i.e. it is the potential a few times κ^{-1} from the particle surface. This border is not sharply defined because there is a gradual transition between the near and the far field. However, for further elaborations this graduality has no consequences. In the more general case that the double layer is polarized, the second term on the r.h.s. of [3.13.3] becomes more complicated and dependent on Du, because the polarization field depends on the deformation of the flow lines, see fig. 3.84. Anticipating the derivation leading to [4.6.54], the outcome for large κa is

$$\psi^{ff}(r,\theta) = -E r \cos\theta - \frac{1}{2}\left(\frac{1-Du}{1+2Du}\right)\frac{a^3}{r^2}E\cos\theta \qquad [3.13.4]$$

For other values of κa other expressions are obtained. For $Du = 1$, ψ^{ff} reduces to $-E r\cos\theta$, i.e. only the applied field remains. For $Du = 0$ a more negative value for ψ^{ff} is predicted, as expected. In the extreme case of $Du = \infty$ the double layer conducts so well that no concentration polarization can be created; in that limit $\psi^{ff}(r = a)$ becomes independent of θ. Expression [3.13.4] may also be formally formulated in terms of the induced dipole moment

$$\psi^{ff}(r,\theta) = -E r \cos\theta + \frac{\mathbf{P}_{ind}}{4\pi\varepsilon_0\varepsilon}\frac{\cos\theta}{r^2} \qquad [3.13.5]$$

It follows that in this particular case

$$\mathbf{P}_{ind} = -\frac{1-Du}{1+2Du}\,2\pi\varepsilon_0\,\varepsilon a^3\mathbf{E} \qquad [3.13.6]$$

so that the polarizability of the double layer is

$$\alpha = \frac{Du-1}{1+2Du}\,2\pi\varepsilon_0\,\varepsilon\,a^3 \qquad [3.13.7]$$

where we have used $\boldsymbol{p}_{\text{ind}} = \alpha \boldsymbol{E}$, see [I.4.4.5]. From [3.13.6] it is seen that for large Du the signs of $\boldsymbol{p}_{\text{ind}}$ and $\boldsymbol{E}$ are the same, for $Du < 1$ they are opposite and for $Du = 1$ the induced dipole moment vanishes. Then the double layer is as a whole unpolarized. Finally it is noted that for $r \to \infty$ (far from the particle) only the contribution of the applied field, $-Er\cos\theta$, remains, whereas the $\cos\theta$ term accounts for the angle dependence, that is, for the polarization.

The relative weakness of the far field implies that locally in the double layer the field, and hence the charge and potential distribution, are almost completely determined by the near field, which is almost identical to that at equilibrium. This leads to the concept of *local equilibrium*; at any segment, say between θ and $\theta + d\theta$, the double layer may be considered as being at equilibrium, although over the contour of the double layer a weak gradient in the potential is superimposed. Local equilibrium can also be inferred from the consideration that each double layer segment has intimate contact with the field beyond it, but only little lateral contact with adjacent double layer elements: the contact area between double layer and far field is $O(a^2)$, that between adjacent segments $O(a\kappa^{-1})$; in the case considered (κa large) the latter is much smaller than the former. Consider fig. 3.88. In the field of the counterions (in this case assumed to be all in the Stern layer), i.e. in a contour line around the sphere, the potential and the cation concentration increase slightly from left to right. The fact that the concentration of cations to the right is slightly greater than that to the left also implies (via the local equilibrium theorem) that the bulk concentration in the far field is slightly higher to the right, a phenomenon to which we shall return

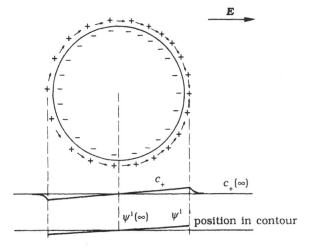

Figure 3.88. Schematic representation of the polarization of a particle with a Stern layer only. The distribution of the cation concentration and the potential in the Stern layer are sketched. Arrows indicate ionic displacements upon polarization.

in sec. 4.6c. This difference is so small that for most purposes it may be ignored. For instance, the value of κ^{-1} is virtually unchanged, so that the thickness of the double layer is hardly affected by the polarization. On the other hand, this concentration difference has to be considered in a.c. studies, requiring the transport of excess salt back and forth around the particle, in electrophoresis and conduction theories and in a number of second-order electrokinetic phenomena like *diffusiophoresis*, to which we shall return in sec. 4.9.

The relatively small deviations from equilibrium allow the polarized double layer to be treated in terms of perturbations from equilibrium. For the potential and the concentrations of each ion we may write

$$\psi(r) = \psi(r,\,eq) + \delta\psi(r) \qquad\qquad\qquad\qquad [3.13.8]$$

and

$$c_i(r) = c_i(r,\,eq) + \delta c_i(r) \qquad\qquad\qquad\qquad [3.13.9]$$

respectively, where r is the general abbreviation for the position, i.e. for r and θ. The equilibrium values follow from static theory, in fact from the Gouy-Stern model. In the far field $\psi^{ff}(r,eq) = 0$; ψ^{ff} is the excess potential there in the presence of the external field, and may therefore be identified to $\delta\psi(r)$. In turn $\delta\psi(r)$ is coupled to the extra ionic concentration, $\delta c_i(r)$, just mentioned. In a.c. fields, and in the initial brief period after applying, or releasing the external field , $\delta\psi(r)$ and $\delta c_i(r)$ become $\delta\psi(r,t)$ and $\delta c_i(r,t)$, respectively. Of course, $\psi(r,\,eq)$ and $c_i(r,\,eq)$ are independent of time. The possibility of applying such a perturbation model considerably facilitates the mathematics, without significant loss of rigour. It will be elaborated in sec. 4.6b.

Returning to the induced dipole moment, this vector is determined with respect to both sign and magnitude by the variety of fluxes in the non-equilibrium double layer. Because of this, and because it is measurable (by dielectric spectroscopy), p_{ind} is the most basic characteristic of non-equilibrium double layers.

In order to establish the sign of p_{ind} it is necessary to have a closer look at the currents in the near and far field. Reconsider fig. 3.86,b2, together with fig. 3.88. Positive current moves from the left to the right through the near field, but in the opposite direction through the far field. As drawn, the former current (the "polarization current") exceeds the latter (the "relaxation current"), hence the directions of p_{ind} and E are the same (the direction of p is counted from the negative to the positive pole, see below [I.4.4.2]). However, in other cases the transport through the far fields may outweigh that through the double layer; then p_{ind} and E have opposite directions. Which of these situations prevails depends on the relative ease by which ions can migrate through the two fields or, more physically, what the value of Du is. The situation, drawn in fig. 3.86,b2, is typical for high Du, i.e. then the directions of $\mathbf{p}_{ind}$ and $\mathbf{E}$ are the same. However, for low Du the bulk (= far field) flux of cations, outweighs that in the double layer. Then $\mathbf{p}_{ind}$ and $\mathbf{E}$ have opposite signs. For the case of high κa, sign reversal occurs when Du becomes lower than unity, as already concluded in connection with [3.13.6].

It follows that there must be a condition where the two fluxes just cancel. At this special value of Du, $\mathbf{p}_{ind} = 0$ and particle and solution are *isoconducting*. Addition of isoconducting particles to an electrolyte solution does not alter the conductivity. The *isoconduction state* plays a similar role in non-equilibrium double layers as the point of zero charge does for equilibrium double layers. For large κa isoconduction is observed for $Du = 1$. In this case $\psi^{ff} = -Er\cos\theta$.

3.13c. *Concentration polarization*

Under stationary conditions, at any position $\mathbf{r}$, the concentrations $c_+(\mathbf{r})$ and $c_-(\mathbf{r})$ of the cations and anions, respectively, remain constant, i.e. they are independent of t. However, this does not mean that no transport (flux) of ions takes place. Just as in an electrolyte solution ions are transported by conduction if a field is applied, ions arrive from one side, pass through and/or around the double layer and disappear into the solution on the other side. In the double layer there are no sources or sinks of ions. On the surface, the normal component of the current density is zero

$$J_{\pm,n}(r=a) = 0 \qquad\qquad\qquad [3.13.10]$$

The conservation of (ionic) mass in each volume element is given by [I.6.1.6] and reads

$$\text{div } \mathbf{j}_\pm(\mathbf{r}) \equiv \nabla \cdot \mathbf{j}_\pm(\mathbf{r}) = 0 \qquad\qquad\qquad [3.13.11]$$

Transport of ions in the double layer is generally caused by diffusion, conduction and convection and the flux $\mathbf{J}_i$ of species i ($= \mathbf{j}_i / z_i F$) is given by the *Nernst-Planck equation* [I.6.7.1] as

$$\mathbf{J}_i(\mathbf{r}) = -D_i \nabla c_i(\mathbf{r}) - \frac{z_i}{|z_i|} c_i(\mathbf{r}) u_i \nabla \psi(\mathbf{r}) + c_i(\mathbf{r}) \mathbf{v}(\mathbf{r}) \qquad\qquad [3.13.12]$$

where u_i is the mobility of species i; $\nabla c_i = \text{grad } c_i$, $\mathbf{J}_i$ is in mol m^{-2} s^{-1} or kg m^{-2} s^{-1}, $\mathbf{j}_i$ in A m^{-2} = C m^{-2} s^{-1}. Equation [3.13.12] is one of the basic equations in the theory of electrokinetic phenomena; we return to it in sec. 4.6a.

Although $\nabla \cdot \mathbf{j}_+ = \nabla \cdot \mathbf{j}_- = 0$, the absolute values of $\mathbf{j}_+$ and $\mathbf{j}_-$ are non-zero and may differ widely. Close to a negative particle there will be an excess of cations, so that $j_+ > j_-$. In the far field the difference between j_+^{ff} and j_-^{ff} is determined by the transport numbers, just as it is in the conduction in electrolyte solutions.

Let us first verify that in the absence of the external field [3.13.12] reduces to known static limits. Consider a flat double layer at rest. Transport in the normal (x-) direction is governed by

$$j_{\pm,n} = - z_\pm F D_\pm \frac{dc_\pm}{dx} - \frac{z_\pm^2}{|z_\pm|} F c_\pm u_\pm \frac{d\psi}{dx} = 0 \qquad [3.13.13]$$

Setting according to the *Nernst-Einstein equation* [I.6.6.15]

$$u_\pm = |z_\pm| F D_\pm / RT \qquad [3.13.14]$$

one obtains

$$\frac{d \ln c_\pm}{dx} = - \frac{z_\pm}{|z_\pm|} \frac{d\psi}{dx} \qquad [3.13.15]$$

which upon integration leads to the Boltzmann equation. So, from a kinetic point of view, this equation can be considered as arise from a balance between diffusion and conduction of ions.

Generally, [3.13.12] has to be solved separately for the various parts inside and beyond the double layer, combining the results at the boundaries between these parts. Let us briefly consider what kind of polarizations and fluxes are expected, working our way "inside out".

(i) *Stern layer.* The concentration is polarized as in fig. 3.88. We may say that $c^i = c^i(\theta)$ or $\sigma^i = \sigma^i(\theta)$. Equations [3.6.16 and 17] for σ^i remain valid but the dimensionless potential y^i must now be replaced by $y^i(\theta)$. Two ion fluxes can be distinguished: $j_t = j_{//}$ in-plane and tangential to the surface, and $j_n = j_\perp$ normal to it. The former is driven by the gradient $\nabla_t \mu$ over the Stern layer, which is parallel to the surface. The normal current is mainly controlled by diffusion and conduction. Stern layers are very thin and have a large contact area with the diffuse part. It is much easier for local ion excesses, created by the tangential field, to relax by transport to the diffuse layer than inside the Stern layer. Because the Stern layer is thin as compared to the particle radius ($d \ll a$) this layer may be mathematically treated as flat.

(ii) *Transition Stern-diffuse layer.* All electrostatic equations (sec. 3.6c, fig. 3.20) remain unaltered after changing all charges and potentials into their respective stationary values. The current passing from the Stern to the diffuse layer is determined by the diffusion coefficient for normal transport, $D_n = D_\perp$. As argued in sec. 2.2c, $D_\perp$ is probably lower than D(bulk) but of the same order. However, special cases are possible, say systems with a very high Stern layer conductivity but inhibited Stern-diffuse layer transport. As in most cases one ionic species dominates the Stern layer charge, the diffusion flux can often be restricted to either cations or anions. The direction of j_n at the border depends on the side of the particle: down or up the field. In fact, the difference between "in" and "out" ionic transports on the two sides is ultimately responsible for the creation of an excess of electroneutral electrolyte in the far field outside the one side of the particle and a deficit outside the other. We shall explain this in detail in sec. 4.6c.

(iii) *Diffuse layer.* The static Boltzmann equation $c_\pm(r) = c_\pm(\infty)\exp[-z_\pm y(r)]$, which refers concentrations and potentials to a bulk phase of zero potential, can now be replaced by

$$c_\pm(r,\theta) = \left[c_\pm(\infty) + \delta c_\pm^{\mathrm{ff}}(\theta)\right] e^{-z_\pm[y(r,\theta)-y^*(r,\theta)]} \qquad [3.13.16]$$

which refers local concentrations to those in the far field. Here y^* is the reference potential in the far field, not including the potential caused by the applied field. For easy reference we may call [3.13.16], or variants of it, the (or a) *dynamic Boltzmann equation.*

As in the Stern layer, ion currents can be normal or tangential, and as in most cases $\kappa a \gg 1$ transport to and from the far fields dominates.

(iv) *Transition diffuse layer - far field.* The distance from the particle surface where this transition is located is $O(\kappa^{-1})$ but not further specified. Its potential is $\psi^b(\theta)$, which occurs in [3.13.3]. In [3.13.13] ψ and ψ^{ff} both equal ψ^b so that

$$c_\pm^b(\theta) = c_\pm(\infty) + \delta c_\pm^b(\theta) \qquad [3.13.17]$$

As $\delta c_\pm \ll c_\pm(\infty)$ the change in the double layer thickness due to polarization is small $(\Delta\kappa^{-1} \ll \kappa^{-1})$, but it may not be neglected because it is responsible for the polarization of the diffuse part.

Regarding the ionic fluxes through the border b the condition

$$\nabla \cdot j_\pm^b = 0 \qquad [3.13.18]$$

must apply because in the stationary state no charge can accumulate or disappear here. Any excess or shortage of charge in the diffuse layer, stemming from lateral transport and from the Stern layer has to disappear into, or be replenished from, the far field. Equation [3.13.18] applies to any volume element in b.

(v) *Far field.* The far potential field and concentration gradient are determined by E and the induced dipole moment p_{ind}, and hence by Du. Each volume element is electroneutral, meaning that $z_+c_+ = -z_-c_-$. Hence, there is no need to distinguish between distributions of cations and anions; considering only $c(r)$ is sufficient. Electroneutrality is also responsible for [3.13.2]. If we combine [3.13.11] with [3.13.12] (with $v = 0$), realizing that $\nabla \cdot \nabla \psi = \nabla^2 \psi$, etc., the fact that $\nabla^2 \psi^{ff} = 0$ implies that also

$$\nabla^2 c^{ff} = 0 \tag{3.13.19}$$

As we may add a constant to c, this can also be written as

$$\nabla^2 \delta c^{ff} = 0 \tag{3.13.20}$$

or as

$$\nabla^2 \mu_{\pm}^{ff} = 0 \tag{3.13.21}$$

where $\mu_{\pm}$ is the chemical potential of the electrolyte. Because of the similarity of [3.13.2 and 19], the general solution for c^{ff} is the same as that for ψ^{ff}. However, the boundary conditions are different: for $r \to \infty$ ψ^{ff} approaches the potential of the applied field, whereas c^{ff} goes to its bulk value. Anyway, the functionalities $c^{ff}(r, \theta)$ and $\psi^{ff}(r, \theta)$ are coupled, and coupled to the induced dipole moment. This coupling can be used in solving electrokinetic equations and, therefore directly coupled to the induced dipole moment, as in [3.13.5]. A similar statement can be made for [3.13.20 and 21].

The equations presented above are the bases for establishing potentials, ion concentrations and fluxes in double layers in external fields. The induced dipole moment is a characteristic and central parameter, related to relevant double layer parameters such as composition and ionic mobilities in the various parts of the double layer. In chapter 4 we shall come back to all of this in more detail.

3.13d. Double layer relaxation

The presence of a near and far field in and around a non-equilibrium double layer leads to the distinction between (at least) *two relaxation times*. Relaxation to the static situation, after switching off the external field, can take place by conduction or by diffusion. Conduction means that ions relax to their equilibrium position by an electric field. Diffusion relaxation implies that a concentration gradient is the driving force. In double layers these two mechanisms cannot be separated because excess ion concentrations that give rise to diffusion, simultaneously produce an electric field, giving rise to conduction. For the same reason, if polarization has taken place under the influence of an external field and this field is switched off, ions return to their equilibrium positions by a mixture of conduction and diffusion.

Conduction relaxation is determined by the excess charge built up (or rather the capacitance C) and the resistance R over which this charge leaks away. See

fig. I.5.11. The product RC has the dimension of time. Introducing the conductivity we find for the *Maxwell relaxation time*, as in [3.7.7] generally

$$\tau \,(\text{cond}) \approx \varepsilon_0 \varepsilon \,/\, K \qquad\qquad\qquad [3.13.22]$$

and for the same of a polarized double layer, relaxing by surface conduction

$$\tau^\sigma \,(\text{cond}) = \varepsilon_0 \varepsilon \, a \,/\, K^\sigma \qquad\qquad\qquad [3.13.23]$$

These two relaxation times differ by a factor of order Du.

Diffusion relaxation is determined by the time needed to cover a certain
displacement. Following [I.6.3.2] the corresponding relaxation time is

$$\tau^{\text{ff}} \,(\text{diff}) \approx (D \,/\, a^2)^{-1} \qquad\qquad\qquad [3.13.24]$$

and

$$\tau^{\text{nf}} \,(\text{diff}) = \tau^{\text{dl}}(\text{diff}) \approx (\kappa^2 D)^{-1} \qquad\qquad\qquad [3.13.25]$$

for the far field and near field, respectively. Usually there is a great difference
between τ^{ff} (diff) and τ^{nf} (diff), except at $\kappa a \approx 1$, but $\tau(\text{cond})$ in [3.13.22 or 23]
and τ^{nf} (diff) are essentially identical, because local charge excesses, created by
diffusion, are opposed by conduction and conversely. For the molar ionic
conductivity, defined in [I.6.6.7] as $\lambda_i = |z_i| F u_i$, [3.13.14] may be used to yield

$$\lambda_i \approx z_i F^2 \,/\, 6\pi \eta \, a_i \qquad\qquad\qquad [3.13.26]$$

which is closely related to

$$D_i \approx k T \,/\, 6\pi \eta \, a_i \qquad\qquad\qquad [3.13.27]$$

where a_i is the ionic radius. Hence,

$$\lambda_i \approx |z_i| F^2 D_i \,/\, kT \qquad\qquad\qquad [3.13.28]$$

For sols of spherical particles the dielectric permittivity, as a function of
frequency, usually has a dispersion with two relaxation times (the *low frequency*
and *high frequency dispersion*), as sketched in fig. 3.89. The first dispersion
results from the fact that the far fields can no longer keep up with the oscillations of the applied field, the second one applies to the double layer itself. The
low frequency dielectric increment can be very large, amounting to 5-20 times ε
for water, depending on sol properties and concentration. Typical values for the
two relaxation frequencies are 10^4 - 10^5 Hz for the first and in the megahertz
range for the second. With decreasing κa the two relaxation times approach
each other.

The above theory is needed in the interpretation of electrokinetic phenomena
and the dynamics of colloid particle interaction. When particles move with

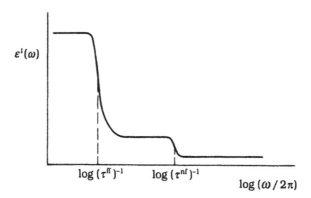

Figure 3.89. The two relaxations for a colloidal particle. The low-frequency relaxation is determined by diffusion in the far field, that at high frequency by *Maxwell conduction* in the double layer. The imaginary part $\varepsilon''(\omega)$ has peaks at the two relaxations. Molecular relaxations, as considered in figs. I.4.7 and I.4.8 come on top of these and are observable at higher frequencies.

respect to the liquid, the problem is compounded because fluid dynamics now also has to be accounted for. Further discussion will follow in sec. 4.6.

For a more rigorous analysis the double layer composition and the fluxes in and around them must be considered in more detail. This requires a substantial amount of mathematics but is also rewarding because more independent information on double layers becomes obtainable. Part of this will recur in chapter 4[1].

3.14 Applications

The present chapter is the longest one in the whole work and it is so for good reasons. Charges at interfaces and the ensuing double layers prevail everywhere in interface and colloid science.

A variety of applications will be presented in the present Volume. To start with, chapter 4 on electrokinetics, builds directly on chapter 3, particularly on sec. 3.13, but also on the sections describing double layer structures at rest, because they differ only slightly from those in applied fields. Double layers also play an inportant role in the adsorption of polyelectrolytes (chapter 5).

Double layers at fluid-fluid interfaces influence the Gibbs energy per unit area ΔG_a^σ and hence the surface or interfacial tension, γ. This recurs in Volume III.

---

[1] For further reading, see S.S. Dukhin, *Non-equilibrium electric surface phenomena* in *Adv. Colloid Interface Sci.* **44** (1993) 1.

For this we need the general expressions derived in sections I.5.7, 3.3 and 3.6f, but for the actual computations a model is again required.

Interaction between double layers, one of the building bricks of colloid stability, is an important theme planned for Volume IV. It has a large number of spin-offs, in, for instance ion exchange, thin wetting films, free films, membranes, association colloids, vesicles, polyelectrolytes, emulsions and rheology. The dramatic influence of electrolytes on these phenomena finds its origin in the changes in the double layer, discussed in this chapter.

Let us now briefly mention a few other applications of double layers.

A topical application of double layers is found in the environment. Top soil essentially consists of coarse structured silicates, oxides, clay minerals and, to a lesser extent, organic colloids, mostly biological debris, like humic acids. Most surfaces bear charges that vary with pH, i.e. with soil acidity. At neutral pH pristine silicates are negatively charged, but iron oxides are positive. On clay minerals the plates are always negative, whereas the edges may be positive or negative, depending on pH. The extent to which humic acids are (negatively) charged also depends on pH. Similar remarks can be made about particles in natural waters, acquifers, estuaries, etc., and, for that matter, sewerage water and industrial effluents.

The variety of charges leads to all kinds of hetero-interactions between the particles, with direct consequences for soil structure. Colloidally speaking, soils should be in the coagulated state, so that a rather open structure is formed, allowing transport of oxygen and nutrients. Of direct relevance to the present theme is the adsorption of ions and organic molecules on the variety of surfaces offered. Let us briefly consider this.

Cation sorption in soils is also of direct interest because a number of cations act as plant nutrients (K^+, Ca^{2+}, Mg^{2+} and, in minor concentrations, Zn^{2+} and Cu^{2+}). Others are poisonous (such as Cd^{2+}, Pb^{2+} and, in higher concentrations, Zn^{2+} and Cu^{2+}). The latter group is usually categorized as *heavy metals*, notorious anthropogenic soil pollutants. The balance between bound and free fractions of these ions determine the nutritional power and the toxicity of soils. Their sorption is electrostatically promoted if the surface charge becomes more negative, i.e. with increasing pH. Soil particles can therefore act as scavengers to an extent which increases with pH. This has transient advantages in that percolation of water, contaminated with heavy metals, is detoxified but creates potential risks because inadvertent pH-lowering (acid rain!) may liberate part of the bound metal. The bound (and not biologically degradable) ions act as a time bomb. Environmentalists have to cope with the same problem when arable land is abandoned and "returned to nature"; acidification may then also occur because calcium-containing fertilizers are no longer administered and may slowly readjust the pH[1].

Similar problems are encountered with anions. Perhaps the *binding of phosphate* to a variety of soil oxides is one of the best-studied systems. The challenge from the double layer side is that phosphate ions strongly bind specifically and remain titratable, i.e. they act as non-constituent charge-determining ions. The adsorption plateau as a function of pH appears to consist of three maxima, coinciding with pH's about equal to pK_{a1}, pK_{a2} and pK_{a3}, superimposed upon a gradually decreasing curve. Environmental relevance stems from fertilization requiring a certain, more or less constant level during the growing season but overdoses are detrimental in that they lead to eutrofication. Nitrogen fixation is another issue. Natural decomposition of a number of nitrogen-containing organics gives first NH_4^+, which is held to clays. Ammonium can be directly used by green plants to (re-)fabricate proteins and other N-containing compounds. However, oxidation by some soil bacteria (*Nitrobacter*, *Nitrosomonas*) leads to nitrate (which is negatively adsorbed and subject to leaching) and additional protons (which may exchange against Na^+, Ca^{2+}, Mg^{2+} and K^+. Nitrate losses can be promoted by deforestation because the uptake by the vegetation is interrupted. It is obvious that all these binding phenomena, with their dependence on pH and c_{salt}, directly reflect double layer mechanisms. The extent to which the adsorption Gibbs energy has a chemical component can be checked by studying the screening by electrolytes. In a model study by Hayes et al.[2] it was thus found that on goethite SeO_3^{2-} ions adsorb specifically, whereas SeO_4^{2-} ions did so electrostatically.

The third category is that of organic molecules for which the general rules of sec. 3.12 are valid. In natural systems several of these react catalytically or under the action of sunlight on, or with, the surface. Photocatalysis can be inferred from diurnal concentration cycles. Photocatalytic decomposition of adsorbed toxic organisms is one of the means nature has for the self-cleaning of lakes. Adsorption on solid particles can sometimes lead to chemical attack and even to dissolution; this is one of the many mechanisms by which nature controls particle size distributions (photo (de-)mineralization). Of the many interesting organic molecules that have been investigated, the bipyridinium cation

$$CH_3 - \,^+N \hspace{-0.3em} \bigcirc \hspace{-0.3em} \bigcirc \hspace{-0.3em} N^+ - CH_3 \hspace{4em} [3.14.1]$$

---

[1] For more information, see *Heavy Metals in the Environment*, J-P. Vernet, Ed., Elsevier (1991); *Chemical Speciation in the Environment*, A.M. Ure and C.M. Davidson, Eds. Chapman and Hall, (1994); W. de Vries, A. Breeuwsma, *Water, Air, Soil Pollution* **28** (1986) 173.
[2] K.F. Hayes, C. Papelis and J.O. Leckie, *J. Colloid Interface Sci.* **125** (1988) 717.

may be mentioned. When used as a herbicide it is called *paraquat*, but if used as an electron-relay in solar energy conversion, it is called *methylviologen* (MV^{2+}). The relay function derives from its ability to pick up an electron, to give the relatively stable MV^{+*} radical; perhaps such radicals also play a role in the so-called plant-protection functioning of the herbicide. The double layer aspects are interesting: does the molecule (electro-)adsorb on surfaces or is it simply a generally adsorbing bivalent counterion? On hydrophobic surfaces like silver iodide MV^{2+} strongly adsorbs specifically, the isotherm obeying FFG statistics[1]. On the other hand, on ruthenium oxide (a catalyst in solar energy conversion) there is hardly any specific adsorption, as can be inferred from the virtual absence of any p.z.c. shift[2]. Here the basis for the application is the extreme sensitivity of pristine surfaces for trace amounts of specific adsorption.

For further reading, see[3, 4], the relevant references in sec. 3.15e and Stumm's books mentioned in sec. 3.15f.

In another vein, double layers play a role in the *salt-sieving* phenomenon, mentioned in the Introduction to Volume I, and already known to Aristotle. When seawater percolates over a compact sediment of silicate-like particles, under some conditions the effluent is potable. Basically the phenomenon is attributable to the negative adsorption of (in this case) anions, leading to the Donnan expulsion of electrolyte, see sec. 3.5b. Over-demand may lead to salt penetration: the screening of the double layers around the silica particles (reduction of κ^{-1}) makes the pores between them effectively wider. For this problem technical solutions had to be found.

Copying these natural processes, a host of membranes having essentially the same propensity have now been synthesized.

Electrosorption has also been applied in electrotechnology to either catalyse or inhibit electrochemical surface reactions. Photoreduction of water is an example of the former, corrosion inhibition of the latter.

Photoreduction of water is one of the methods of absorbing solar energy by converting it to chemical energy, copying in a sense processes that occur during photosynthesis in green plants, algae and some bacteria[5]. The process consists of a number of steps involving transfer of electrons. Light excites a photosen-

---

[1] A. de Keizer, L.G.J. Fokkink, *Colloids Surf.* **57** (1990) 323.

[2] J.M. Kleijn, J. Lyklema, *Colloid Polym. Sci.* **265** (1987) 1105.

[3] G. Sposito, *Surface Complexation of Metals by Natural Colloids*, in *Ion Exchange and Solvent Extraction*, Vol. **11**, J.A. Marinsky, Y. Marcus, Eds. Marcel Dekker (1993).

[4] W. Stumm, J.J. Morgan, *Aquatic Chemistry; an Introduction Emphasizing Chemical Equilibria in Natural Waters*, 2nd ed., Wiley (1981).

[5] *Photochemical Conversion and Storage of Solar Energy*, J. Rabani, Ed., Weizmann Sci. Press, Jerusalem (1981); M. Grätzel, in *Modern Aspects of Electrochemistry* (Eds.) **15** (1983) 83.

sitive compound, the sensitizer, of which the excited electron is transferred by a relay to the catalyst, on the surface of which water is reduced to yield hydrogen. A promising sensitizer is the ruthenium bipyridyl cation; methylviologen [3.14.1] is an often used relay. As colloidal catalysts, oxides of ruthenium and titanium and semiconductors, or combinations of them, have been studied. In semiconductors the radiation can move an electron from the valency to the conduction band; in the presence of, say ruthenium bipyridyl, the captured electron is transferred into the solid, remaining there long enough to effect charge separation.

Corrosion inhibition is a very important but still poorly understood process. The aim is to inhibit the continuous oxidation or reduction of metals. This requires insight into the electrochemistry of the process. A host of interwoven phenomena are involved, including redox reactions on the surface, in the solution and intervention by inorganic or organic molecules. The complexity is illustrated by the fact that there is often no simple correlation between the coverage of a certain inhibitor on the surface and the extent of inhibition.

In connection with the foregoing, double layers of course also play an important role in *electroanalysis*. Transfer of, say, electroactive ions through the polarized mercury-solution interface is preceded by passage through the double layer. Therefore current-potential plots depend in principle on double layer properties. Historically, it was his interest in charge-transfer and corrosion problems that induced Grahame to start his seminal double layer investigations.

In the field of electroanalysis the problem of the operative mechanism of *glass electrodes* may be mentioned. Thin membranes of several types of glass act as perfect Nernst-type electrodes for protons or other ions over many decades of concentration. The basic question is whether a glass electrode reacts to an ion i because i is charge-determining or because a liquid junction potential is set up with the transport dominated by i. In the former case Nernst's law for the electrode potential, [I.5.5.1], applies, in the latter [I.6.7.8] is valid, with $t_+ = t_i = 1$ and $t_- = 0$; the two results are identical in that both give $59/z$ mV per decade of i. On the basis of titrations with ion-specific glasses, it was decided that, at least for a number of glasses, [I.6.7.8] prevails, but not everyone concurs[1, 2].

Double layer studies can also be used to characterize surfaces. Titrations, in combination with electrokinetics, are helpful to identify charge-determining species and hence establish the relative acidity/basicity of a certain surface. These results may be compared with those from functional gas adsorption. In this way heterogeneous catalysts and pigments have been characterized. Trace

[1] H.N. Stein, *The Glass-Electrolyte Interface, Adv. Colloid Interface Sci.* **11** (1979) 67.
[2] K.L. Cheng, *pH-Glass Electrode and its Mechanism*, in *Electrochemistry, Past and Present, ACS Symp. Ser.* **390** (1989), J.T. Stock and M.V. Orna, Eds.

this way heterogeneous catalysts and pigments have been characterized. Trace amounts of impurities can be detected when points of zero charge and isoelectric points are both available (sec. 3.8b). Such techniques have also been attempted for "difficult" minerals of biological origin such as hydroxyapatite, the root material for bones and teeth.

In all these applications knowledge of double layers plays a prominent role, the study of which always leads to promising cross-fertilization between disparate domains.

3.15 General references

3.15a. IUPAC recommendations/data

Solubility Constants of Metal Oxides, Metal Hydroxides and Metal Hydroxide Salts in Aqueous Solution, (prepared for publication by W. Feitknecht, P. Schindler) *Pure Appl. Chem.* **6** (1963) 130.

Interphases in Systems of Conducting Phases, (prepared for publication by S. Trasatti and R. Parsons) *Pure Appl. Chem.* **58** (1986) 438.

The Absolute Electrode Potential: an Explanatory Note, (prepared for publication by S. Trasatti) *Pure Appl. Chem.* **58** (1986) 955.

Electrified Interfaces in Aqueous Dispersions of Electrolytes, (prepared for publication by J. Lyklema) *Pure Appl. Chem.* **63** (1991) 895.

Real Surface Area Measurements in Electrochemistry (prepared for publication by S. Trasatti and O.A. Petrii) *J. Electroanal. Chem.* **327** (1992) 353 (only for electrodes).

Nomenclature, Symbols and Definitions in Electrochemical Engineering, (prepared for publication by G. Gritzner and G. Kreysa) *J. Electroanal. Chem.* **360** (1993) 351.

R.P. Buck, V.V. Cosofret, *Recommended Procedures for Calibration of Ion-Selective Electrodes*, Pure Appl. Chem. **65** (1993) 1849.

Impedances of Electrochemical Systems: Terminology, Nomenclature and Representation. Part I. Cells with Metal Electrodes and Liquid Solutions, (prepared for publication by M. Sluyters-Rehbach) *Pure Appl. Chem.* **66** (1994) 1831.

3.15b. General books and reviews

Adsorption of Inorganics at Solid-Liquid Interfaces. M.A. Anderson, A.J. Rubin, Eds., Ann Arbor Science (1981). (Contains much information on ion adsorption, double layers and site binding in disperse systems.)

A.J. Bard and 11 co-authors, *The Electrode/Electrolyte Interface - A Status Report*, J. Phys. Chem. **97** (1993) 7147. (Review characterization, reactions, chemical materials and microstructure, 346 references.)

Comprehensive Treatise of Electrochemistry, Volume **1**. *The Double Layer*, J. O'M. Bockris, B.E. Conway and E. Yeager, Eds., Plenum Press (1980). (Treats most topics of the present chapter with emphasis on metals and electrodes.)

J. O'M. Bockris, S.U.M. Khan, *Surface Electrochemistry; a Molecular Level Approach*, Plenum (1993). (Quantum features, photo reactions, bio-electro-chemistry, energy storage and other topics.)

B.E. Conway, *The State of Water and Hydrated Ions at Interfaces*, Adv. Colloid Interface Sci. **8** (1977) 91. (Review covering various types of double layers, χ-potentials; models and thermodynamics.)

D.C. Grahame, *The Electrical Double Layer and the Theory of Electrocapillarity*, Chem. Rev. **41** (1947) 441. (A classical review.)

A.T. Hubbard, Electrochemistry of well-defined surfaces, in Acc. Chem. Res. **13** (1980) 177. (Electrode surfaces).

R.J. Hunter, *Zeta Potential in Colloid Science. Principles and Applications*, Academic Press (1981). (Although electrokinetics are emphasized there is also much information on the composition of double layers in disperse systems.)

Electrical Phenomena at Interfaces. Fundamentals, Measurements and Applications, A. Kitahara, A. Watanabe, Eds., Marcel Dekker (1984).

L.K. Koopal, *Adsorption of Ions and Surfactants*, in *Coagulation and Flocculation*, B. Dobias, Ed., Marcel Dekker (1993) 101-207. (Review covering several aspects discussed in this chapter.)

Structure of Electrified Interfaces, J. Lipowski, P.N. Ross, Eds. V.C.H. Publishers (1993). (Double layer structure, including solvent structure; emphasis on metal electrodes.)

J.R. MacDonald, J. Electroanal. Chem. **223** (1987) 1. *Comparison and Discussion of some Theories of the Equilibrium Electrical Double Layer in Liquid Electrolytes*. (Review, emphasis on the mercury-solution interface.)

S.K. Rangarajan, *The Electrical Double Layer*, Electrochemistry **7** (1980) 203. (Review with some emphasis on water structure and electrosorption.)

M.J. Sparnaay, *The Electrical Double Layer*, Pergamon Press (1972). (General properties and application to a variety of systems.)

Adsorption from Aqueous Solution, P.H. Tewari, Ed., Plenum (1980). (Proceedings of a Symposium, contains various chapters on ion adsorption.)

G.R. Wiese, R.O. James, D.E. Yates and T.W. Healy, *Electrochemistry of the Colloid-Water Interface*, International Rev. Sci., Electrochemistry, Physical Chemistry, Series Two, Vol. **6**. A.D. Buckingham, J. O'M. Bockris, Eds., Butterworths (1976) 53. (Review of classical double layers up to about 1975, some emphasis on site-binding interpretation.)

3.15c. Theory: statistics, thermodynamics

L. Blum, *Structure of the Electric Double Layer*, Adv. Chem. Phys. **78** (1990) 171. (Review of various advanced theoretical models.)

L. Blum, D. Henderson, *Statistical Mechanics of Electrolytes at Interfaces*, in *Fundamentals of Inhomogeneous Fluids*, D. Henderson, Ed. Marcel Dekker (1992) chapter 6 (various statistical approaches discussed).

S.L. Carnie, G.M. Torrie, *The Statistical Mechanics of the Electrical Double Layer*, Advan. Chem. Phys. **56** (1984) 141-253. (Gouy-Chapman and more advanced models, including integral equation theories, discrete charges, simulations.)

B.V. Derjaguin, (= Deryagin), S.S. Dukhin and V.N. Shilov, *Kinetic Aspects of Electrochemistry of Disperse Systems*, Adv. Colloid Interface Sci. **13** (1980) 141. (Emphasis on double layers in an external field.)

D.G. Hall, *Thermodynamics of Solutions of Polyelectrolytes, Ionic Surfactants and Other Colloidal Systems*, J. Chem. Soc. Faraday Trans. I **77** (1981) 1121-56. (Paper with a review nature.)

T.W. Healy, L.R. White, *Ionizable Surface Group Models of Aqueous Interfaces*, Adv. Colloid Interface Sci. **9** (1978) 303. (Site-binding).

R. Parsons, *Thermodynamic Methods for the Study of Interfacial Regions in Electrochemical Systems*, in *Comprehensive Treatise of Electrochemistry*, Volume **1**, J. O'M. Bockris, B.E. Conway and E. Yeager, Eds., Plenum Press (1980) 1. (General basic double layer thermodynamics, more emphasis on metals than in the present chapter.)

A. Sanfeld, *Thermodynamics of Charged and Polarized Layers*, in *Monographs in Statistical Physics and Thermodynamics*, Vol. **10**, Wiley (1968). (Based on the application of the method of local thermodynamic balances in continuous, but non-uniform systems.)

3.15d. *Techniques*

Electrochemical Interfaces: Modern Techniques for in-situ Interface Characterisation, H.D. Abruña, Ed., VCH, New York, (1991). (X-ray, SERS, non-linear optics, IR, Mössbauer, etc.)

Spectroelectrochemistry, Theory and practice, R.J. Gale Ed., Plenum (1988). (Includes X-ray, photo-emission, SERS, UV and IR reflectance.)

Spectroscopic and Diffraction Techniques in Interfacial Electrochemistry, C. Gutiérrez, C. Melendres, Eds., Kluwer (NL) (1990).

A. Hamnett, *Ellipsometric Techniques for the Characterization of Electrode Surfaces*, J. Chem. Soc. Faraday Trans. **89** (1993) 1593. (Review-type paper.)

R.J. Hunter, *Zeta Potential in Colloid Science*, Academic Press (1981). (Also contains sections on techniques.)

Techniques for Characterization of Electrodes and electrochemical Processes, R. Varma, J.R. Selman, Eds., Wiley (1991). (Various spectroscopies, ellipsometry, impedance spectroscopy).

3.15e. *Special systems* (see also sec. 3.15b.)
Mercury

D.C. Grahame, *The Electrical Double Layer and the Theory of Electrocapillarity*, Chem. Revs. **41** (1947) 441. (The "classic" for mercury.)

M.A. Habib, J. O'M. Bockris, *Specific Adsorption of Ions*, in *Comprehensive Treatise of Electrochemistry*, J. O'M. Bockris, B.E. Conway and E. Yeager, Eds., Vol. **1**, *The Double Layer*, Plenum (1980) chapter 4. (Review, experiments and models.)

R. Payne, *The Electrical Double Layer: Problems and Recent Progress*, J. Electroanal. Chem. **41** (1973) 277. (Review, almost exclusively on mercury, good entrance to older literature.)

R.M. Reeves, *The Electrical Double Layer; The Current Status of Data and Models, with Particular Emphasis on the Solvent*, in *Modern Aspects of Electrochemistry*, B.E. Conway, J. O'M. Bockris, Eds., Vol. **9**, Plenum (1979). (Review all but limited to mercury.)

R.M. Reeves, *The Double Layer in the Absence of Specific Adsorption*, in *Comprehensive Treatise of Electrochemistry*, J. O'M. Bockris, B.E. Conway and E. Yeager, Eds., Vol. 1 *The Double Layer*, Plenum (1980), chapter 3. (Includes specific adsorption.)

Silver iodide

B.H. Bijsterbosch, J. Lyklema, *Interfacial Electrochemistry of Silver Iodide*, *Adv. Colloid Interface Sci.* **9** (1978) 147. (Extensive review.)

Metals

M.A.V. Devanathan, B.V.K.S.R.A. Tilak, *The Structure of the Electrical Double Layer at the Metal-Solution Interface*, *Chem. Revs.* **65** (1965) 635. (Review with over 300 references to older literature; emphasis on mercury.)

B.E. Conway, *The Solid/Electrolyte Interface*, NATO Conf. Ser. 6, Vol. **5** on *Atomistics of Fracture*, R.M. Latanision, Ed., Plenum (1983) 497. (Review, emphasis on metals; double layers and water structure near charged surfaces.)

Adsorption of Molecules at Metal Electrodes, J. Lipkowski, P.N. Ross, Eds., *Frontiers of Electrochemistry*, Vol. I (1992), VCH Publishers, Weinheim (Germany). (Thermodynamics, kinetics, spectroscopy, electroanalysis.)

Structure of Electrified Interfaces, J. Lipkowski, P.N. Ross, Eds., VCH Publishers (1993). (General double layer work with emphasis on metals; contains contributions on water near (charged) surfaces.)

Oxides

A. Daghetti, G. Lodi and S. Trasatti, *Interfacial Properties of Oxides used as Anodes in the Electrochemical Technology*, *Mater. Chem. Phys.* **8** (1983) 1. (Review, almost 500 references; oxide-gas and oxide-solution interface.)

V.E. Henrich, *The Surfaces of Metal Oxides*, *Rep. Progr. Phys.* **48** (1985) 1481. (Review, structure, especially dry surfaces.)

R.O. James, G.A. Parks, *Characterization of Aqueous Colloids by Their Electrical Double Layer and Intrinsic Surface Chemical Properties*, in *Surface and Colloid Science*, Vol. **12**, E. Matijevic, Ed., Plenum Press (1982) 119. (Review emphasizing methods of determining intrinsic binding constants, mostly for latices and oxides.)

R.O. James, *Characterization of Colloids in Aqueous Systems*, in *Adv. Ceramics*, **21** *Ceramic Powders*, G.L. Messing, J.W. McCauley, Eds., Am. Ceramic Soc. (1987) 349. (Review emphasizing oxides, site binding models.)

D.N. Furlong, D.E. Yates and T.W. Healy, *Fundamental Properties of the Oxide-aqueous Solution Interface*, in *Stud. Phys. Theor. Chem.* **11** (No electrodes conduct. met. oxides. Part B (1981) 367. (Review, structure, double layer.)

The Colloid Chemistry of Silica, P. Somasundaran, Ed., Elsevier, (1992). (Proceedings of a symposium in Washington DC, 1990, dedicated to the memory of R.K. Iler.)

Electrodes of Conductive Metal Oxides, S. Trasatti, Ed., Part A in *Studies in Physical and Theoretical Chemistry*, Vol. **11** Elsevier, (1980). (Emphasis on double layer on oxides.)

Clay minerals
Soil Chemistry, (A) *Basic Elements*, G.H. Bolt, M.G.M. Bruggenwert, Eds., Elsevier (1976); (B) *Physico-Chemical Models*, G.H. Bolt, Ed., Elsevier (1979). (Part B contains much double layer discussion.)

H. van Olphen, *An Introduction to Clay Colloid Chemistry, for Clay Technologists, Geologists and Soil Scientists*, 2nd ed., Wiley (1977). (Explains how colloid characterization and stability can be applied to clay minerals.)

Semiconductors
P.J. Boddy, *The Structure of the Semiconductor-Electrolyte Interface*, J. *Electroanal. Chem.* **10** (1965) 199. (Review)

W.P. Gomes, F. Cardon, *Electron Energy Levels in Semiconductor Electrochemistry*, *Progr. Surf. Sci.* **12** (1982) 155. (Review with 133 references.)

R. Memming, *Charge transfer processes at semiconductor electrodes in Electroanalytical Chemistry*. A.J. Bard. Ed., Vol. **11** (1979), p. 1. Marcel Dekker, New York. (Charge and potential distribution and charge transfer at and through semiconductor-electrolyte solution interfaces.)

W. Mönch, *Semiconductor Surfaces and Interfaces*, Springer (Ser. Surface Sci. No. 26) (1993) (Extension of sec. 3.10e; experiments, clean interfaces and interfaces with adsorbates.)

M.J. Sparnaay, *Semiconductor Surfaces and the Electric Double Layer*, Adv. *Colloid Interface Sci.* **1** (1967) 277. (Review.)

S.N. Sze, *Semiconductor Devices; Physics and Technology*, Wiley (1985). (Physical principles and applications to electronic devices.)

Vapour

J.E.B. Randles, *Structure at the Free Surface of Water and Aqueous Electrolyte Solutions*, Phys. Chem. Liq. **7** (1977) 107. (Review, mainly of the items considered in our secs. 3.9 and 3.10f.)

Carbon

C.A.L.Y. Leon, L.R. Radovic, *Interfacial Chemistry and Electrochemistry of Carbon Surfaces*, Chemistry and Physics of Carbon **24** (1994) 213. (Review, surface characterization by physical and chemical means, double layer, functional groups.)

3.15f. *References emphasizing applications*

Interactions at the Soil Colloid-Soil Solution Interface, G.H. Bolt, M.F. de Boodt, M.H.B. Hayes and M.B. McBride, Eds., NATO-ASI Series E **190**, Kluwer, Dordrecht, The Netherlands, (1991), (Proceedings of an Advanced Research Workshop, Ghent, Belgium, (1986).)

Soil Colloids and Their Associations in Aggregates, M.F. de Boodt, M.H.B. Hayes and A. Herbillon, Eds., Plenum (1990). (korte karakteristiek.)

Minerals in Soil Environment, J.B. Dixon, S.B. Weed, Eds. Soil Sci. Soc. Am. book **#1**, Madison USA (1989) (a standard reference for soil scientists).

The Chemistry of Soil Constituents, D.J. Greenland, M.H.B. Hayes, Eds., Wiley (1978) (includes double layers, electrosorption, water near surfaces); *The Chemistry of Soil Processes*, D.J. Greenland, M.H.B. Hayes, Wiley (1981) (includes transport, precipitation, adsorption).

Electrified Interfaces in Physics, Chemistry and Biology, R. Guidelli, Ed., Kluwer Dordrecht, The Netherlands, 1992. (Proceedings of a NATO-ASI meeting in 1990, dealing with the application of double layers in solid state and surface physics, electrochemistry and biophysics.)

Electrochemistry in Colloids and Dispersions, R.A. Mackay, J. Texter, Eds., VCH-publishers (1992). (Proceedings of a meeting, emphasizing electrochemical (especially redox-) reactions in various colloidal systems.)

Aquatic Surface Chemistry, W. Stumm, Ed., Wiley (1987). (Double layers on oxides, soils, (de-)mineralization; application to natural waters.)

W. Stumm, *Chemistry of the Solid-Water Interface. Processes at the Mineral-Water and Particle-Water Interface in Natural Systems*, Wiley, (1992). (Emphasis on complex formation, kinetics and redox processes especially in relation to environmental issues.)

3.15g. *Data collections*

D.A. Dzombak, F.M.M. Morel, *Surface Complexation Modelling: Hydrous Ferric Oxide*, Wiley (1990). (Compilation of surface complexation constants obtained after critical evaluation of literature data, using the method of the "generalized two-layer model".)

Electrical Properties of Interfaces, Compilation of Data on the Electrical Double Layer on Mercury Electrodes, J. Lyklema, R. Parsons, Eds., U.S Dept. of Commerce, Nat. Bureau Standards, Office of Standard Reference Data (1983).

P.K. Wrona, Z. Galus, *Mercury*, in *Encyclopaedia of Electrochemistry of the Elements*, Vol. **9**, Part A.1, sec. 3. A.J. Bard, Ed. (Contains numerous data, including those in non-aqueous solvents.)

4 ELECTROKINETICS AND RELATED PHENOMENA

As long ago as 1809 Reuss observed that clay particles, dispersed in aqueous media, migrate under the influence of an applied electric field[1]. This was an early example of *electrophoresis*, the most familiar electrokinetic phenomenon.

Electrokinetic phenomena are generally characterized by the tangential motion of liquid with respect to an adjacent charged surface. In the above example the surface was that of a negatively charged clay particle; the particle moved with respect to the stationary liquid. The surface may also be that of a droplet as in emulsions. Alternatively, the particles may be stationary with the liquid moving, as for instance in *electro-osmosis*. For sand this phenomenon was also discovered by Reuss[1].

Electrokinetics have obtained prominent positions in colloid science as additional means of electrical surface characterization. Moreover, there are important practical applications.

By electrophoresis, for example, the sign of the charge on the moving entity can be readily established. However, it is more problematic to quantify this charge or, for that matter, to define what kind of charge is measured. As a whole, electrical double layers are electroneutral; the force exerted by the applied field on the surface charge seems to be exactly balanced by that on the countercharge. Nevertheless, the particles do move. Hence, a closer consideration of forces and flows in the double layer is required to describe these phenomena and to define such notions as *electrokinetic charge* (σ^{ek}) and *electrokinetic* or *zeta potential*, (ζ). Seldom do σ^{ek} and ζ correspond to σ^0 and ψ^0, respectively.

In this chapter these phenomena will be systematically discussed. Related phenomena such as *surface conduction* and *dielectric relaxation* of sols will be included. In view of the fact that the rigorous theory is both physically and mathematically extremely involved, we shall discuss the phenomena in two steps. First, elementary theory will be presented. For electrophoresis this leads to the Hückel-Onsager and Helmholtz-Smoluchowski equations. It gives the leading term for a number of simple conditions regarding potential and particle size and shape. Thereafter, more advanced theoretical treatments will follow.

As before, we shall use SI units throughout, but note that in the literature

---

[1] F.F. Reuss, *Mém. Soc. Impériale Naturalistes de Moscow*, **2** (1809), 327.

papers in the non-rationalized three-quantity or even the c.g.s. system continue to appear. Electrokinetics are quite different from electrode kinetics (reactions at electrodes). The name "cataphoresis" for electrophoresis is obsolete.

4.1 Basic principles

4.1a. Review from Volume I

Electrokinetic phenomena are typically *second-order phenomena*, in that forces of a certain kind create fluxes or flows of another type. For instance, in electrophoresis an *electric* force leads to a *mechanical* motion and in streaming current (see below) an applied *mechanical* force produces an *electric* current. First-order phenomena have been discussed in some detail in chapter I.6, see for instance table I.6.1. Second-order phenomena, as long as they are considered phenomenologically, can be treated advantageously by *irreversible thermodynamics*, of which the basic principles have been outlined in sec. I.6.2, see particularly [I.6.1.19] describing how forces of nature k lead to fluxes of type i. The power of irreversible thermodynamics was convincingly demonstrated in sec. I.6.2c, where *Saxén's rule*[1] was proven without any mechanistic or model assumption. This rule states that for a given system the electro-osmotic flux per unit of electrical current and the streaming potential per unit of pressure are identical (except for the sign) and that the electro-osmotic pressure per unit of field strength equals the streaming current per unit of liquid flow. Here, we are dealing with a typical illustration of Onsager's reciprocal relations. For the case that elementary theory applies, Saxén's rule has also been derived by using models, showing that these models are consistent. However, the Onsager relations remain valid under conditions where elementary theory no longer suffices.

Using irreversible (rather than classical) thermodynamics is also typical in another respect. Classical thermodynamics is suited to describe double layers *at rest*. We did so in secs. I.5.6 and I.5.7 and in chapter 3. In electrokinetics, double layers are not at rest, although in the elementary pictures deviations from the equilibrium distribution of charge and/or potential are disregarded.

Distinction has to be made between the very first moments after the application of the external force and the *stationary state*, developing soon thereafter. Usually, such stationary states are considered, i.e. *inertia* is disregarded. This was also done in sec. 3.13. However, when electrokinetic phenomena are studied in alternating fields, various types of inertia start to play prominent roles, leading to additional *dissipation*. As a consequence, double layer characteristics like the induced dipole moment, the dielectric

---

[1] U. Saxén, *Wied. Ann.* **47** (1892), 46.

permittivity and the conductivity, become frequency-dependent. Mathematically speaking, these quantities become *complex*. Handling complex quantities has been outlined in appendix I.8; resonance and relaxation phenomena have been introduced in sec. I.4.4e.

Flow along uncharged surfaces has been considered in secs. I.6.4f and e, surface conduction in sec. I.6.6d and mixed transport phenomena, simultaneously involving electrical, mechanical and diffusion types of transport in sec. I.6.7. Specifically the *Nernst-Planck equation* ([I.6.7.1 or 2]) is recalled, formulating ion fluxes caused by the sum-effect of diffusion, conduction and convection.

4.1b. *The notions of electrokinetic potential and slip plane*

These notions have already been introduced in Volume I (sec. I.5.5a), but we shall now consider them in more detail. We refer to the problem of the division between ions in the mobile and stagnant layer under conditions of stationary flow.

Consider, by way of example, an immobile, smooth and homogeneous surface, tangential to which liquid moves, as in electro-osmosis or streaming potential studies. (Cases like electrophoresis, in which particles move in a stationary liquid are mirror-images of these.) The tangential component v_t of the fluid velocity $v(r)$ at position r [1] increases from zero at the surface to its bulk value $v_t(\infty)$ far from it. For uncharged surfaces profiles can be computed on the basis of the hydrodynamic equations, introduced in sec. I.6.1 and elaborated in sec. I.6.4. Experience has shown that in such tangential motion usually a very thin layer of fluid remains adhered to the surface. This *stagnant layer* is not more than a few molecular layers thick and for most hydrodynamic purposes its presence is without consequence. However, in electrokinetics it can no longer be ignored because a substantial fraction of the countercharge is located in that layer. Tangential motion of these ions may be inhibited. In other words, in electrokinetics only part of the countercharge is in the hydrodynamically mobile layer.

Which part? Here we encounter one of the basic problems of electrokinetics. Its solution requires insight into the interplay of motions of molecules and ions close to a surface and this information is not easy to obtain. For one thing, a fluid may macroscopically behave as stagnant whereas self-diffusion is hardly impaired, as in many gels. Hence, different techniques may lead to different answers. Macroscopically the problem is why there is a layer of high viscosity close to the surface. It has been suggested that this high viscosity was induced by

---

[1] Bold face symbols refer to vectorial quantities. Working with vectors has been outlined in I.app. 7.

the binding of solvent (particularly water) molecules to the surface, but hydrophobic surfaces have the same propensity.

Surface roughness has also been proposed as the reason. Although it is true that protrusions (like polymeric "hairs") do lead to substantial hydrodynamic immobilization, the stagnant layer has also been observed for smooth surfaces. Hence the phenomenon is more general: whatever its origin, the existence of a stagnant layer is indisputable. Phenomenologically, it appears as if the viscosity of the fluid is for all practical purposes infinitely high adjacent to the wall, dropping to its bulk value over a few molecular layers.

In sec. 4.4 we shall return to this problem. The usual way to solve (or rather avoid) it is by simply postulating a *slip plane* at unspecified distance d^{ek} from the surface. For distances $x < d^{ek}$ ("behind the slip plane") one has the stagnant layer with $\eta = \infty$, beyond it $\eta(x) = \eta(\text{bulk})$. Effectively, the gradual function $\eta(x)$ is replaced by a step function. The space charge for $x > d^{ek}$ is electrokinetically active, the particle (if spherical) behaves as if it had a radius $a + d^{ek}$. Its "electrokinetic surface potential" is the potential at the slip plane and known as the *electrokinetic* or *ζ-potential*.

Experience has shown that this simple picture has advantages. One of these is that colloid interaction appears to be determined by ζ rather than by the surface potential ψ°. In fact, there are indirect, but good reasons to identify ζ with the diffuse layer potential ψ^d (at least for simple surfaces in the absence of polymeric adsorbates). One of the arguments is that the countercharge in the mobile part and the part that gives rise to repulsion upon overlap are both essentially the diffuse part. Experiments with hydrophilic and hydrophobic surfaces have confirmed this. Consequently, d^{ek} corresponds to (about) the thickness of the Stern layer, d. We shall return to this equivalence in sec. 4.4, but already note that in solvents of very low dielectric permittivity the potential decay with distance is so weak that then ζ may be identified with ψ°. Otherwise, ζ and ψ° relate to each other more or less as ψ^d and ψ° do, implying that ζ and ψ° have opposite signs if superequivalent specific adsorption occurs (see sec. 3.6). Electrokinetic charges, computed from ζ, using Gouy-Chapman theory seldom exceed 3-4 µC cm^{-2}. As σ° may readily be ten times as high, it is obvious that only a small fraction of the countercharge is electrokinetically observable.

4.2 Survey of electrokinetic phenomena. Definitions of terms

Here we briefly describe the most familiar electrokinetic phenomena. For easy reference they are tabulated in table 4.1.

Electrophoresis is the movement of colloidal particles or polyelectrolytes, immersed in a liquid, under the influence of an electric field. The *electrophoretic velocity* v (m s^{-1}) is the velocity during electrophoresis and the *electrophoretic*

Table 4.1. Most familiar electrokinetic phenomena

Name	Driving force	Moving phase	Stationary phase	Resulting phenomenon	Quantity usually measured	Symbols	SI units*)
Electrophoresis	d.c. electric field	particles	liquid	particle translation	electrophoretic mobility	$u = v/E$	$m^2V^{-1}s^{-1}$
Sedimentation or centrifugation potential gradient	gravity or centrifugation field	particles	liquid	electric field	potential difference per unit of length	E_{sed}	$V\, m^{-1}$
Electro-osmosis	d.c. electric field	liquid	plug or capillary	liquid displacement	electro-osmotic volume flow per unit field strength	$Q_{eo,E}$	$m^4V^{-1}s^{-1}$
					or per unit current	$Q_{eo,I}$	m^3C^{-1}
				pressure gradient build-up	electro-osmotic (counter-) pressure	Δp_{eo}	$N\, m^{-2}$
Streaming current	pressure gradient	liquid	plug or capillary	electric current	streaming current per unit of pressure difference	I_{str}	$C\, m^2 N^{-1} s^{-1}$ $= A\, m^2 N^{-1}$
Streaming potential	pressure gradient	liquid	plug or capillary	potential difference	streaming potential (difference) per unit of pressure difference	E_{str}	$V\, m^2 N^{-1}$
Colloid vibration potential	ultrasonic field	particles	liquid	alternating electric field	electric field amplitude per unit velocity of ultrasonic field	E_{vibr}, CVP	$V\, s\, m^{-1}$
Electrokinetic sonic amplitude	a.c. electric field	particles	liquid	ultrasonic waves	pressure amplitude per unit field strength	ESA	$N V^{-1}\, m^{-1}$

*) Because of the abundance of electrokinetic phenomena it is sometimes unavoidable to use the same symbol for quantities that have different dimensions.

mobility $(m^2 V^{-1} s^{-1})$ is the velocity per unit field strength, following from $v = uE$, with E (Vm⁻¹) the strength of applied field. The mobility is counted positive if the particle moves toward lower potential and negative in the opposite case. These definitions resemble those for the motion of ions in an electric field (sec. I.6.6a), but the theoretical description of the phenomenon is more involved, except in the first-order approximation. (For electrolytes, electrophoretic retardation and the relaxation effect are usually two smaller corrections of similar magnitude (see I.6.6b), whereas for colloids (with larger κa) the former effect tends to dominate.)

The term *microelectrophoresis*, or even better, *microscopic electrophoresis*, refers to a technique in which the motion of individual particles is followed, usually ultramicroscopically. Other terms refer to alternative ways of measurement, such as *moving boundary electrophoresis*, *paper electrophoresis*, *laser-Doppler electrophoresis*, *gel electrophoresis*, *capillary electrophoresis*, etc. see secs. 4.5 and 10.

Electro-osmosis is the motion of a liquid through an immobilized set of particles, porous plug, capillary or membrane as a consequence of an applied electric field. In the past this phenomenon was also called electro-endosmosis. Electro-osmosis is the result of the force exerted by the applied field on the countercharge in the liquid inside the charged capillaries, pores etc. With their movement, the ions involved entrain the liquid in which they are embedded. The velocity approaches zero at the slip plane. The electro-osmotic velocity is the velocity of the electro-osmotic flow in the liquid beyond the double layer. The *electro-osmotic volume flow* is the volume flow per unit time and either per unit field strength $Q_{eo,E}$ $(m^4 V^{-1} s^{-1})$ or per unit current, $Q_{eo,I}$ $(m^3 C^{-1})$. The latter unit has been used in the derivation of Saxén's rule. Yet other units may be expedient in special cases, for instance the volume flow per unit time and per unit of current density. The flow is is said to be positive if it is in the direction of lower potentials. Alternatively, the *electro-osmotic pressure*, Δp_{eo} or Δp, can be measured as the pressure difference across the plug, membrane, etc., needed to just stop the electro-osmotic volume flow. Δp_{eo} is said to be positive if the higher pressure is on the higher potential side.

A *sedimentation potential gradient*, sometimes simply called *sedimentation potential*, E_{sed}, develops when a charged particle sediments under gravity or in a centrifuge. Upon this movement the centre of the particle charge is shifted with respect to that of the countercharge. In a sol containing many particles, E_{sed} is defined as the potential difference at zero current between two identical electrodes at different levels (or, for the case of a centrifuge, at two different distances from the centre of rotation). E_{sed} is negative if the lower (or peripheral) electrode is positive. The phenomenon is also known as the *Dorn effect*. When the density of the dispersed phase is less than that of the continuous medium (as in many emulsions) creaming rather than sedimentation takes place.

The *sedimentation field strength* E_{sed} is the potential difference per unit length in a sedimentation or centrifugation cell. As the contributions of the interfacial potential differences at the electrodes should not be included in E_{sed}, this quantity can only be measured if the exact identity of the two electrode potentials has been ascertained.

The *streaming potential (difference)*, E_{str} is the potential difference at zero current caused by the flow of liquid under a pressure gradient through a capillary, plug, diaphragm or membrane. The difference is measured across the plug or between the two ends of the capillary. To measure E_{str}, (preferably) identical electrodes must be used on both sides of the capillary. E_{str} is positive if the higher potential is on the high pressure side. Streaming potentials are created by the flow of countercharge inside the capillaries or pores. When the two electrodes are relaxed (see sec. I.5.5b) and short-circuited, a current I_{str} flows through the circuit, which is known as the *streaming current*. I_{str} is counted positive if the current in the capillary, etc. is from the high to the low pressure side (and the other way around in the outside lead). The current per unit area is the corresponding *streaming current density*, j_{str}, which we have assigned vectorial character (sec. I.6.6a).

The *colloid vibration potential (difference)* E_{vibr} or CVP is the a.c. potential difference measured between two identical relaxed electrodes, placed in the dispersion if the latter is subjected to an (ultra)sonic field. CVP is a particular case of the more general phenomenon, *ultrasonic vibration potential* (UVP), applying to any system, whether or not colloids are present. This field sets the particles into a vibrating motion, as a result of which the centres of particle charge and countercharge are periodically displaced with respect to each other. This phenomenon is the a.c. equivalent of that observed in the Dorn effect. Counterpart to this is the *electrokinetic sonic amplitude*, ESA, the amplitude of the (ultra)sonic field created by an a.c. electric field in a dispersion.

The survey in table 4.1 covers most familiar electrokinetic phenomena. Some less frequently encountered alternatives will be discussed in sec. 4.9.

From the discussion it also follows that there is a certain freedom of choice regarding the units into which measured quantities are expressed. Recall, however, the restriction that if cross-relationships are to be derived on the basis of Onsager's reciprocal relationships (secs. I.6.2b and c) the fluxes J, resulting from forces X, must have such dimensions that the product XJ has the dimensions of entropy production per unit time and unit volume (SI units J m^{-3} K^{-1} s^{-1}).

There is also some choice in writing expressions in vector or scalar form. Although the former is more general, the latter is often more convenient, especially when flow or fields in only one direction have to be considered. However, even when writing scalar equations we shall continue to add + or − signs to indicate directions. The reason for doing so is that in electrokinetics two different gradients are always involved (an electric and, for instance, a mechanical one) and distinction has to be made between "upstream" and "downstream" fields or currents, or between "upfield" or "downfield" fluxes or pressure drops. This convention has also been used in chapter I.6.

4.3 Elementary theory

Simple equations for the electrokinetic phenomena discussed in sec. 4.2 can be derived under a number of more or less restrictive conditions, using elementary calculus. Deriving such expressions serves a number of purposes, including:

(i) a leading term is obtained that allows computing ζ quantitatively when a number of limiting conditions are met and that helps to assess ζ semiquantitatively under other conditions.

(ii) a first insight is obtained into the underlying physical picture, together with some feeling for the nature of the simplifications.

For the sake of simplicity, we shall discuss solid-liquid, i.e. inextensible, interfaces. The model assumptions are the following.

1. The liquid is primitive, i.e. it is a continuum, characterized by its bulk dielectric permittivity $\varepsilon_0\varepsilon$ and viscosity η. Liquid properties remain constant down to the slip plane. The potential at this plane is the electrokinetic potential ζ. The solid plus stagnant layer together behave electrokinetically as if the solid persisted up to the slip plane.

2. Surfaces are homogeneous and smooth; when particles are considered, these are spherical and homodisperse.

3. Double layer overlap does not occur.

4. The first premise entails the neglect of surface conduction *behind* the slip plane (except in subsec. 4.3f); surface conduction *beyond* the slip plane is considered where necessary.

5. There is no bulk conduction inside the solid.

6. Liquid flow is laminar; turbulence does not occur.

7. Inertia is disregarded. The system is in a stationary state. (This restriction will be relaxed in subsec. 4.3e.)

8. Only linear electrokinetic phenomena are considered, i.e. cases where the electrophoretic mobility is proportional to the applied field, etc. In practice this means that such problems are not encountered for the field strengths of $\lesssim 10^4$ V m^{-1} under which electrophoresis is usually carried out.

9. Double layer polarization is ignored (except in subsec. 4.3e). This means that polarization fields and induced dipole moments, as introduced in sec. 3.13, are absent.

It is interesting to note that in deriving the most elementary expressions no double layer model is needed; Poisson's equation suffices. In line with this is the fact that Helmholtz derived his equations several decades before the advent of the Gouy-Chapman theory. However, for the interconversion of electrokinetic charges σ^{ek} and ζ-potentials a model is needed. In view of our premises and the closeness of ζ and ψ^d the logical choice for that is the Gouy-Chapman picture or variants of it. For simplicity of language, in this chapter we shall speak of the diffuse (part of the) double layer when referring to the part beyond the slip plane.

More advanced theory will follow in secs. 4.6 and 4.7.

4.3a. Electrophoresis

In sec. I.6.6a the mobility of individual ions was discussed. There, the mobility was introduced as the scalar ratio of the vectors velocity and field strength. This definition also applies to the *electrophoretic mobility u*, defined through

$$v \equiv u\,E \qquad\qquad [4.3.1]$$

Under the simplifying conditions chosen, v and E have the same directions. Our task is to interpret u in terms of double layer and hydrodynamic parameters.

(i) *General considerations.* By way of introduction we start by identifying the various forces acting on a moving colloidal particle, keeping in mind that in reality these forces are not additive, so that it is not always simple to identify each of them. When a particle moves at constant velocity in an electric field, the sum of the forces acting on it must be zero. What are these forces? Double layers as a whole are electroneutral, hence the electric force exerted on the surface charge is exactly opposite to that on the countercharge. If the countercharge would be immobile with respect to the particle the particle would not move. Apparently, the friction felt by the particle differs from that by the countercharge. As a starter for further analysis it is useful to identify the forces involved.

1. F_1, the *driving electrostatic force*, exerted on the surface charge. It simply equals

$$F_1 = Q\,E \qquad\qquad [4.3.2]$$

2. The *frictional force* F_2 caused by the viscous resistance that a moving particle experiences from the fluid in which it moves. For an *uncharged* sphere, according to Stokes

$$F_2 = -6\pi\eta a v \tag{4.3.3}$$

which we derived in sec. I.6.4e. Strictly speaking, [4.3.3] applies to the frictional force of a fluid around a spherical obstacle; the deformation of the flow around the sphere is accounted for and $v = v(\infty)$ is the fluid velocity far from the particle. The problem is, of course, that the particle is not uncharged.

3. The *electrophoretic retardation force* F_3. This is the braking force the moving countercharge exerts on the central particle. The driving force is simply $-QE$, just the opposite of F_1, but the extent to which this force is transmitted to the central particle is rather complicated, and depends on a number of factors. Counterions move in a direction opposite to that of the particle; via the liquid momentum is transported and transmitted to the central particle by viscous traction; this is felt as a braking force. On the other hand, the co-ions move ahead of the central particle, they try to speed it up. The resulting retardation therefore depends on the ionic space charge distribution. Consequently, F_3 depends on the double layer composition (κa, ζ) and on the ionic mobilities (say, via the ionic diffusion coefficients D_+, D_-). The way in which momentum is transferred also depends on the trajectories of the ions, i.e. on the field lines of the electric field. These are determined by Du (see fig. 3.85) so that F_3 also depends on this parameter. The electrophoretic retardation is essentially a hydrodynamic force.

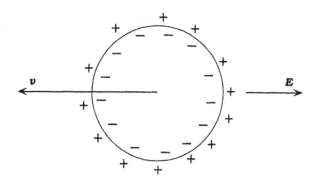

Figure 4.1. Schematic representation of the non-coincidence of the centres of negative (A) and positive charge (B) during electrophoresis. Stationary state. A dipole is induced.

4. A *polarization* or *relaxation force*, F_4, caused by the fact that for a stationary-moving particle the centres of particle charge and countercharge do not coincide. A situation may arise as depicted schematically in fig. 4.1. In this example the particle is negative and moves to the left. The countercharge is represented by positive charges only. Upon this concentration polarization the double layer thickness hardly changes. In sec. 3.13 the resulting induced dipole

moment p_{ind} has been introduced; in fact, fig. 4.1 is similar to fig. 3.87 a2. For the present purpose it is relevant that an additional electric field is generated, that in most cases has the same direction as E (the direction depends on Du). The polarization field, if opposite, further retards the particle motion; this is essentially an electric feature. The name "relaxation" retardation stems from the consideration that a certain time, the relaxation time, is needed, to build up an asymmetric distribution around a moving particle.

Usually F_4 is smaller than F_3, but as F_4 can reduce the mobility by 10-50% it can by no means be neglected. One of the problems is that F_3 and F_4 generally depend on each other so that these, and the other, forces are not additive and, hence, cannot be individually identified. Rigorous solutions for the electrophoretic mobility and other electrokinetic phenomena require the formulation of the hydrodynamic equations, coupling them to the electric field equations and solving the resulting differential equations with the proper boundary conditions. We shall postpone such rigorous approaches to sec. 4.6 and now consider some limiting cases where at least double layer polarization (leading to F_4) is neglected. As stated above, such simplifications are not entirely meaningless, because they lead to relatively simple laws that are valid under limiting conditions and recur in a variety of practical situations.

(ii) *Limiting laws for high and low κa.* Two limiting laws were proposed long ago and that, according to recent insight, they remain valid as limiting laws, one for $\kappa a \gg 1$, the other for $\kappa a \ll 1$. The first is

$$u = \frac{\varepsilon_0 \varepsilon \zeta}{\eta} \qquad (\kappa a \gg 1) \qquad\qquad [4.3.4]$$

It is known as the *Helmholtz-Smoluchowski (limiting) law*, after Helmholtz[1], who derived an expression similar to [4.3.4] but without ε in it, and Smoluchowski[2] who improved Helmholtz's expression.

The other law,

$$u = \frac{2\varepsilon_0 \varepsilon \zeta}{3\eta} \qquad (\kappa a \ll 1) \qquad\qquad [4.3.5]$$

we call the *Hückel-Onsager (limiting) law*, after Hückel[3] and Onsager[4]. That [4.3.4] and [4.3.5] are indeed the limiting laws for low and high κa, respectively,

[1] H. von Helmholtz, *Ann. Phys.* **7** (1879) 337.
[2] M. von Smoluchowski, *Bull. Int. Acad. Sci. Cracovie* (1903) 184; M. von Smoluchowski in *Handbuch der Elektrizität und des Magnetismus*, W. Graetz, Ed., Vol. II, Barth Leipzig (1914) 366; *Z. Physik. Chem.* **92** (1918) 129. Smoluchowski gave his derivation for the inverse case of electro-osmosis.
[3] E. Hückel, *Physik. Z.* **25** (1924) 204.
[4] L. Onsager, *Physik. Z.* **27** (1926) 388.

is confirmed by modern theories. Compare figs. 4.25 and 26.

The fact that u (Hückel-Onsager) $< u$ (Helmholtz-Smoluchowski) must be entirely caused by F_3 mentioned above, and hence by the way the field lines behave around the particle. In the derivation of the Hückel-Onsager formula, the assumption is made that the particle is so small that the field lines are unaffected. Such a situation is also met for ions in an electrolyte solution or for colloids in the special case that the conductivities of particle, K^S and liquid K^L are identical, see fig. 3.84b). On the other hand, a dielectric sphere with a large radius rather generates a field line picture like that in fig. 3.84c. Compare now, by way of example a small particle, radius a, and a large one with radius $100a$, both having the same surface charge density σ^o, and both immersed in the same electrolyte solution (κ is identical). The total charge $Q = 4\pi a^2 \sigma^o$ on the larger particle is 10^4 times that on the smaller one. So, the force acting on the larger one is also 10^4 times that on the smaller, see [4.3.2]. If this were the only difference, the larger one would also move 10^4 times as fast. The Stokes friction [4.3.3] goes with a, rather than with a^2, so this brake is by no means enough to account for the fact that u in [4.3.4] is only 50% higher than that according to [4.3.5]. All the remaining difference must therefore be caused by F_3 : in the high κa case most of the countercharge accumulates relatively close to the surface. The extent to which the ion flow lines are drawn to the particle is determined by the Dukhin number $Du = K^\sigma/aK^L$, see [3.13.1] and fig. 3.85. This countercharge experiences a force that is almost the reverse of that of the particle. Any growth in driving force is therefore compensated by the increase of two brakes. From this qualitative argument it is understood why u (Helmholtz-Smoluchowski) does not excessively exceed u (Hückel-Onsager). Theory is needed to prove that the difference is only a factor of 1.5. For the time being [4.3.4 and 5] can be combined to

$$u = \frac{2}{3}\frac{\varepsilon_o \varepsilon \zeta}{\eta} f\left(\kappa a, \frac{K^S}{K^L}, Du\right) \qquad\qquad [4.3.6]$$

where f is an as yet unknown function, of which we know that $f \rightarrow 1.0$ for $\kappa a \ll 1$ and $f \rightarrow 1.5$ for dielectric spheres of high κa. As Du also depends on ζ (sec. 4.3f), the relation between mobility and electrokinetic potential is in the general case not simple.

(iii) *Early theories.* As an introduction to sec. 4.6, where more advanced approaches will be treated, we now consider some earlier theories, say stemming from before 1935. These approaches ignored the relaxation term, and, in many cases, did not incorporate surface conduction or did so in a primitive way.

Let us first consider the case of low κa. The simplest way to obtain [4.3.5] is to completely ignore the countercharge, that is, treating the particle as an isolated sphere with charge Q and a surface potential identified as ζ, which obeys Coulomb's law ($Q = 4\pi a \varepsilon_0 \varepsilon \zeta$). Using Stokes' law, $v = QE/6\pi\eta a$, eq. [4.3.5] is then immediately obtained. This oversimplified situation applies only when the double layer contains very few ions, as in media of extremely low dielectric permittivity. In sec. 3.11 it was verified that under those conditions screening is negligible so that the Gouy-Chapman relations between charges and potentials reduce to that of Coulomb. Hence, the result is as expected.

Improved theories showed that [4.3.5] remained valid under less restrictive conditions. Onsager[1] gave a derivation in which the electrophoretic retardation is accounted for. This derivation is very similar to his treatment of the electrophoretic correction in the ionic mobilities but contains a number of incorrect steps.

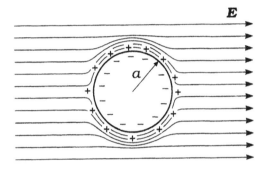

Figure 4.2. Double layer in electric field, $\kappa a \gg 1$.

In the opposite limiting case the double layer looks like fig. 4.2. In this situation most of the lines of force in the double layer are parallel to the surface. The consequence is that the double layer behaves as if it were flat.

Smoluchowski[2] has given a simple derivation of this case. See fig. 4.3. As in the previous case, under stationary conditions the sum of the forces exerted on the particle and the solution around it must be zero. For an infinitesimal slice of thickness dx and area A at distance x from the surface (hatched in fig. 4.3) the force caused by momentum transport, $A\eta(dv_t/dx)_x - A\eta(dv_t/dx)_{x+dx} = -A\eta(d^2 v_t/dx^2)dx$, see [I.6.1.17], is just compensated by the electric force on that element, $A\rho(x)E_t dx$, where E_t is now the tangential component of the field. For $\rho(x)$ we now need the flat geometry variant of the Poisson equation, see [3.5.1]. If this is substituted, the equation can be integrated with the

[1] L. Onsager, loc. cit.
[2] M. von Smoluchowski, *Bull. Int. Acad. Sci. Cracovie* (1903) 184.

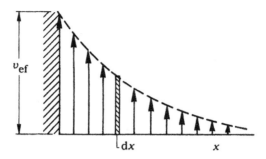

Figure 4.3. Tangential velocity profile $v_t(x)$ when a flat surface moves with respect to the liquid at a velocity $v_{ef} = v_t(x = 0)$.

boundary conditions $(d\psi/dx) \to 0$, and $(dv_t/dx) \to 0$ for $x \to \infty$. The result is

$$\varepsilon_0 \varepsilon \frac{d\psi}{dx} E_t = \eta \frac{dv_t}{dx} \qquad\qquad [4.3.7]$$

Second integration with $\psi(x=0) = \zeta$ and $v_t(x=0) = v_{ef}$, $\psi(x=\infty) = 0$, $v_t(x=\infty) = 0$, as before, leads to [4.3.4]. As in the derivation of the Hückel-Onsager equation, the simplification is made that E_t is the same as the applied field. Although this is not correct, it turns out to be a good approximation for high κa. More rigorous derivations confirm this (sec. 4.6d and elsewhere).

Smoluchowski also presented an elegant theorem showing that [4.3.4] remains valid for any geometry for which all local κa's are $\gg 1$. His argument applied to electro-osmosis and we shall return to it in subsec. 4.3b.

(iv) *Improvement of early theories by consideration of the hydrodynamics.* Equations [4.3.4 and 5] can be derived and extended by explicitly considering the hydrodynamics of the moving particle. We continue to disregard double layer polarization and surface conduction.

The first attempt in this direction dates back to the just-mentioned Smoluchowski theorem. For the development of electrophoresis theory a very important contribution was made by Henry[1] who systematically studied the distortion of the field by spherical and cylindrical particles. It depends on the size and shape of the particle, and on K^S/K^L as illustrated in fig. 3.84. The electrophoretic friction that is created also depends on κ. For this type of theory a double layer picture is needed, but because of mathematical difficulties Henry had to limit himself to the linearized (Debye-Hückel) approximation. As a consequence, Henry's results are only valid for low ζ. Notwithstanding this

---

[1] D.C. Henry, *Proc. Roy. Soc.* **A133** (1931) 106.

limitation, Henry's work was an important step forward because he could prove that, for non-conducting particles, [4.3.4 and 5] are rigorously valid at high and low κa, respectively. Moreover, he was also able to bridge the gap between the two regimes. Henry also considered different values of K^S/K^L.

In fig. 4.4 his results are graphically represented. Here $f(\kappa a, K^S/K^L)$ is given as a function of κa; this function is defined by [4.3.6], but Du is lacking because surface conduction was not taken into account. The figure illustrates all the trends discussed. When κa sufficiently exceeds 0.1, f becomes dependent on K^S/K^L. That f increases with κa if this ratio is low but decreases if it is high, is caused by the difference in electrophoretic retardation, and has been explained above. In practice, the situation with $K^S/K^L = 0$ prevails because charge transport through the SL interface is usually absent because of the absence of interfacial redox reactions or, for that matter, because the overtension needed to force current across the interface is too low. Therefore, we shall continue to emphasize dielectric particles. At high κa the results for spheres and cylinders in any orientation are identical, in line with the independence of κa (i.e. independence of curvature at given κ). At low κa there is a difference of a factor of two between the two orientations.

For cylinders oriented at arbitrary angle to the field, the field and velocity

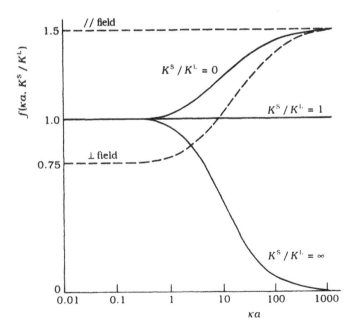

Figure 4.4. The function $f(\kappa a, K^S/K^L)$ according to Henry for three different ratios of K^S/K^L. Also given are the results for dielectric cylinders (dashed); the outcome depends on the orientation. For these systems $K^S/K^L = 0$. (Redrawn from Henry, loc. cit.)

components can be decomposed. Vectorial addition of the two velocity components gives the net velocity; if this is averaged over a random distribution of orientations,

$$\langle u \rangle = u_{//}/3 + 2u_{\perp}/3 \qquad\qquad [4.3.8]$$

This simple outcome has been derived by de Keizer et al.[1]; Stigter[2] proved this result to remain valid if double layer polarization is also included.

For quantitative purposes we give here Henry's series for f for dielectric spheres:

$$f(\kappa a) = 1 + \frac{(\kappa a)^2}{16} - \frac{5(\kappa a)^3}{48} - \frac{(\kappa a)^4}{96} + \frac{(\kappa a)^5}{96} - \cdots$$

$$-\left[\frac{(\kappa a)^4}{8} - \frac{(\kappa a)^6}{96}\right] e^{\kappa a} \int_{\infty}^{\kappa a} \frac{e^{-x}}{x} \, dx \,\ldots \qquad (\kappa a < 1) \qquad [4.3.9a]$$

$$f(\kappa a) = \frac{3}{2} - \frac{9}{2\kappa a} + \frac{75}{2(\kappa a)^2} - \frac{330}{(\kappa a)^3} + \cdots \qquad (\kappa a > 1) \qquad [4.3.9b]$$

As is seen from fig. 4.4 the range of applicability of [4.3.9a] is limited. Ohshima[3] showed that the approximate expression

$$f(\kappa a) = 1 + \frac{1}{2[1 + (2.5/\kappa a)\{1 + 2\exp(-\kappa a)\}]^3} \qquad [4.3.9c]$$

agrees within 1% with [4.3.9a and b] over the entire range of κa provided ζ is not too high, say less than 40 mV. For cylinders he also gives a simple analytical expression.

As a later generalization we present a more rigorous derivation of the Helmholtz-Smoluchowski equation for high κa, in which the curvature of the field is explicitly taken into account. This method, given by Anderson and Prieve[4], may be considered as an elaboration of the Smoluchowski theorem for the case of electrophoresis of large spheres. Surface conduction is still ignored.

See fig. 4.5. Three regions can be distinguished: the (dielectric) particle, the double layer and the far field. As double layer polarization is ignored, there is no polarization field and $\mathbf{E}^{ff} = \mathbf{E}$, the applied field. The border between the

[1] A. de Keizer, W.P.J.T. van der Drift and J.Th.G. Overbeek, *Biophys. Chem.* **3** (1975) 107.
[2] D. Stigter, *J. Phys. Chem.* **82** (1978), 1417, 1424.
[3] H. Ohshima, *J. Colloid Interface Sci.* **168** (1994) 269.
[4] See J.L. Anderson and D.C. Prieve, *Langmuir* **7** (1991) 403, especially app. B; J.L. Anderson, *Ann. Rev. Fluid Mech.* **21** (1989) 61.

double layer and the far field is indicated by the superscript b, as in sec. 3.13. Just outside this border, the electric field $E^b = E$, of which the two components E_n and E_t are indicated in the figure. The velocity in the liquid just outside the double layer v_b equals the vectorial sum of the electrophoretic velocity of the particle v_{ef} and the *slip velocity* v_s, that is the extent by which the liquid velocity at $r = r^b$ lags behind that of the particle. The notion of slip velocity stems from fluid dynamics. This can be denoted as

$$v^b = v_{ef} + v_s \qquad\qquad [4.3.10]$$

For a small double layer segment we may use the argument given in connection with fig. 4.3, leading to

$$v_{s,t} = \frac{\varepsilon_o \varepsilon \zeta}{\eta} E_t i_t \qquad\qquad [4.3.11]$$

where the subscript t refers to the tangential component and i_t is the unit vector tangential to the border and directed opposite to E_t.

Beyond the double layer there is no charge so that the conservation laws are applicable with only the viscous friction and pressure terms.[1] We derived these in sec. I.6.1. Conservation of mass for an incompressible fluid leads to

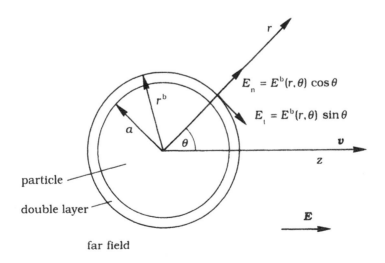

Figure 4.5. Identification of symbols in the derivation of the Helmholtz-Smoluchowski equation for spherical particles at high κa; $r^b - a \approx \kappa^{-1}$.

[1] H. Lamb, *Hydrodynamics*, e.g. 5th ed. Cambridge (1924). H. Brenner, *Chem. Eng. Sci.* **19** (1964) 519, F.A. Morrison, *J. Colloid Interface Sci.* **34** (1970) 210.

$$\text{div } \boldsymbol{v} \equiv \nabla \cdot \boldsymbol{v} = 0 \tag{4.3.12}$$

and conservation of momentum to

$$\eta \nabla^2 \boldsymbol{v} - \nabla p = 0 \tag{4.3.13}$$

which is a variant of the *Navier-Stokes equation*. $\nabla^2 \boldsymbol{v}$ may also be written as rot rot $\boldsymbol{v}$ and equals $(\nabla^2 v_x)\boldsymbol{i} + (\nabla^2 v_y)\boldsymbol{j} + (\nabla^2 v_z)\boldsymbol{k}$, where $\boldsymbol{i}$, $\boldsymbol{j}$ and $\boldsymbol{k}$ are unit vectors in the x, y and z-direction, respectively. Equation [4.3.13] has to be solved under the boundary conditions $\boldsymbol{v}(r = \infty) = 0$, $\boldsymbol{v}(r^b) = \boldsymbol{v}_s + \boldsymbol{v}_{ef}$. Mathematically it appears more convenient to calculate forces than velocities. The exact expression for the force that must be applied to overcome the hydrodynamic (viscous) retarding force at $r = r^b$ is

$$\boldsymbol{F} = 6\pi \eta r^b \, \frac{1}{4\pi} \int_0^{2\pi} \int_0^{\pi} (\boldsymbol{v}_{ef} + \boldsymbol{v}_s) \sin \theta \, d\theta \, d\varphi \tag{4.3.14}$$

Here φ is the azimuthal angle. Because of the spherical symmetry the integration over φ simply yields 2π. We shall not derive [4.3.14] but verify its validity by taking the simple case that $\boldsymbol{v}_s = 0$ (particle of radius r^b without a double layer). As

$$\int_0^{\pi} \sin \theta \, d\theta = 2$$

we obtain

$$\boldsymbol{F} = 6\pi \eta r^b \, \boldsymbol{v}_{ef} \approx 6\pi \eta a \boldsymbol{v}_{ef} \tag{4.3.15}$$

which is *Stokes' law*, already derived in sec. I.6.4e.

Next, $\boldsymbol{v}_s$ has to be substituted in [4.3.14]. At each θ, there is a normal and a tangential component: $v_n^b = v_{ef} \cos \theta$ and $v_t^b = -v_{ef} \sin \theta + \varepsilon_o \varepsilon \zeta E_t^b / \eta$. The value for E_t^b can be obtained by solving Poisson's equation in the far field, $\nabla^2 \psi^{ff} = 0$. The general solution for a dielectric sphere of radius a is

$$\psi^{ff}(r, \theta) = -Er \cos \theta - \frac{1}{2} \frac{a^3}{r^2} E \cos \theta \tag{4.3.16}$$

which we have had before, see [3.13.3]. In the thin double layer case at $r^b \approx a + O(\kappa^{-1})$

$$\psi^b(\theta) = -\frac{3}{2} Er^b \cos \theta \tag{4.3.17}$$

$$E_t^b(\theta) = \frac{3}{2} E \cos \theta \tag{4.3.18}$$

Writing [4.3.11] in vectorial form,

$$v_s = \frac{\varepsilon_0 \varepsilon \zeta}{\eta} \frac{1}{r^b} \left(\frac{\partial \psi}{\partial \theta}\right)_{r,b} i_t = \frac{3}{2} \frac{\varepsilon_0 \varepsilon \zeta}{\eta} E \sin\theta \, i_t$$ [4.3.19]

In passing, Levich's expression[1]

$$v_{ef} = -\frac{2}{3} v_{so,eo}$$ [4.3.19a]

may be mentioned, where v_{so} is the electro-osmotic slip at the equator. This equation also follows from [4.3.19], with $E \sin\theta \, i_t = E$ and $\varepsilon_0 \varepsilon \zeta E / \eta = v_{ef}$. (The sign difference accounts for the opposite directions of solid movement with respect to the liquid and conversely.) Equation [4.3.19a] is useful in deriving equations for the mobility for the case of polarized double layers, as in sec. 4.6d. In the present context the electrophoretic velocity is now obtained by setting the force on the particle, including its (thin) double layer, equal to zero. This unit is electroneutral, and therefore experiences no net electric force. It follows from [4.3.14] that $v_{ef} = -v_s$, integrated over θ and φ. If i_z is the unit vector in the z-direction, using [4.3.18-19]

$$v_{ef} = v_{ef} \cdot i_z = -\frac{1}{4\pi} \int_0^{2\pi}\int_0^{\pi} i_z \cdot v_s \sin\theta \, d\theta \, d\varphi$$ [4.3.20]

$$= \frac{3}{2} \frac{\varepsilon_0 \varepsilon \zeta}{\eta} E \frac{1}{2} \underbrace{\int_0^{\pi} \sin^3\theta \, d\theta}_{4/3} = \frac{\varepsilon_0 \varepsilon \zeta}{\eta} E = [4.3.4]$$

This confirmation of the Smoluchowski derivation also illustrates why it is generally valid at high κa (outside the relatively thin double layer, general hydrodynamics applies with zero electric field) and why the outcome is independent of a (E_t is independent of a). Smoluchowski already anticipated that the equation therefore remains valid for other than spherical geometries (including hollow and irregularly formed surfaces) provided $\kappa a \gg 1$. This was later confirmed by Morrison[2].

(v) *An interim report*

Summarizing the achievements so far, it can be stated that the general validity of [4.3.4 and 5] for large and low κa, respectively, appears established. More advanced theory confirms this. The transition zone remains to be considered because double layer polarization and surface conduction have been ignored. When these approximations are allowed and ζ is low the Henry transition is

[1] V.G. Levich, *Physicochemical Hydrodynamics*, Prentice-Hall (1962), his equation [95.10].
[2] F.A. Morrison, *J. Colloid Interface Sci.* **34** (1970) 210.

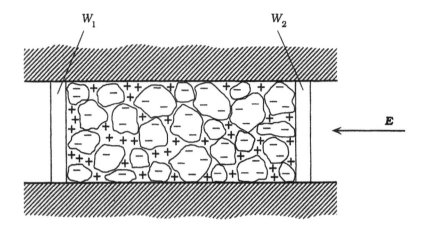

Figure 4.6. Electro-osmosis in a plug, contained in a capillary between two walls W_1 and W_2. The applied electric field exerts a force to the left on the mobile positive charges in the liquid, leading to a liquid flow in that direction. For simplicity the countercharge is only represented by plus signs.

valid. When double layer polarization is properly accounted for, surface conduction beyond the slip plane is automatically included. However, conduction within the slip plane can often not be ignored. Solving these problems rigorously requires consideration of all electric and hydrodynamic fluxes in all parts of the double layer and beyond it. This poses an issue that is physically and mathematically extremely tricky. We shall come back to it in sec. 4.6 after discussing surface conduction (sec. 4.3f) and the essentials of the slip process (sec. 4.4).

4.3b. Electro-osmosis

In fig. 4.6 a sketch is given of the origin of electro-osmosis in a porous plug. The particles are assumed to be negatively charged. Their nature is not relevant for the present purpose but they should be impervious and immobile. Immobilization is for instance, achieved by compressing them between the two walls. Application of an electric field across this plug induces the fluid to move, in the present situation to the left. The force exerted on the mobile charges acts as a body (or volume-) force, because it is transmitted to the liquid via the action = reaction principle, which is valid under stationary conditions. When the capillary walls are also charged, tangential motion with respect to them would also contribute to the osmotic flow. For such irregular plugs it is virtually impossible to derive an equation relating v_{eo} to E; only when everywhere $\kappa R \gg 1$ (R is the local radius of curvature of the particle surfaces) is this possible by virtue of Smoluchowski's theorem, see below.

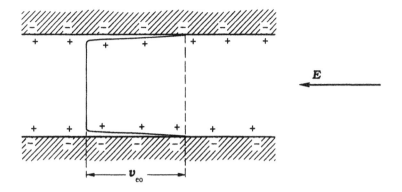

Figure 4.7. Sketch of an electro-osmotic velocity profile in a capillary. For simplicity the countercharge is only represented by plus signs.

For the simpler case of a capillary, as sketched in fig. 4.7 the derivation is straightforward. When the capillary is sufficiently wide, as we are assuming here, no double layer overlap occurs and there is a range of constant velocity in the centre. For wide capillaries (radius $\gg \kappa^{-1}$), the contribution to the volume flow of the layers of variable velocity near the walls may be neglected; in fact their presence is not readily experimentally measurable.

Electro-osmosis is the counterpart of electrophoresis in that the roles of the moving and stationary phases are inverted. Therefore, Smoluchowski's derivation, given in connection with fig. 4.3 and leading to [4.3.4], remains valid, except for the sign and the substitution of v_{ef} by v_{eo}, the velocity of the liquid with respect to the stationary and solid wall

$$\frac{v_{eo}}{E} = -\frac{\varepsilon_0 \varepsilon \zeta}{\eta} \qquad\qquad [4.3.21]$$

The identity of [4.3.21] and (minus) [4.3.4] is an illustration of Onsager's reciprocal relations or, for that matter, of Saxen's relation. The minus sign means that v_{eo} and E are in the same direction when ζ is negative. This is the situation of figs. 4.6 and 4.7, where the cations move from the right to the left. Considering [4.3.10], v_{eo} may also be called the *electro-osmotic slip velocity*, $v_{s, eo}$.

For homogeneous capillaries the field lines are automatically parallel to the wall(s), so that [4.3.21] is valid for any κa provided κa is large. Here, a is the capillary radius.

For a *plug of arbitrary geometry*, Smoluchowski[1] has shown that [4.3.21] and hence the equations that are derived from it, remain valid provided the pore

---
[1] M. von Smoluchowski, *Bull. Intern. Acad. Sci. Cracovie* (1903) 104. See also: J.Th.G. Overbeek in *Colloid Science*, H.R. Kruyt, Ed., Vol. I, Elsevier (1952) p. 202.

diameter is much larger than κ^{-1}. We have dubbed this *Smoluchowski's theorem*. The argument essentially comes down to the decoupling of, and similarity between, the hydrodynamic and electric fluxes in the far field. Close to the surface there is a thin double layer which follows the contours of the solid-liquid interface. Over this layer there is the velocity decay that gives rise to [4.3.21]. The osmotic slip velocity v_{eo} is hydrodynamically interpreted as the liquid velocity just outside the double layer. In the main part of the pore volume, between these borders, there is no net charge, hence $\nabla^2 \psi^{ff} = 0$, $\nabla \cdot E = 0$. Moreover, $\nabla \times E = 0$. The boundary condition is that the normal component of E on the border is zero: there is no net transfer of charge from the double layer to the far field. For the fluid flow the same applies: $\nabla \cdot v = 0$ (incompressibility) and $\eta \nabla^2 v = \nabla p$, as in [4.3.13]; the latter equation reduces to $\eta \nabla^2 v = 0$ if there is no pressure gradient (which, under certain conditions of measurement, can be created by the electro-osmosis). Near the border the electric current (conduction) and the liquid flow are tangential to the border and proportional to each other. There is no reason to suppose that this linearity would break down throughout the whole liquid phase, because the above equations show that the two flows obey identical laws and are independent of each other. So, [4.3.21] should hold for the entire pore. When κa is not large, the de-coupling breaks down: in the most general case, the equations for momentum transport contain both hydrodynamic and electrical terms, see for instance [4.6.5]. For the spherical symmetry case the arguments given recurred in the derivation starting with fig. 4.5. Perhaps the final confirmation of the correctness of this theorem is that all rigorous theories lead to the Helmholtz-Smoluchowski limit at high κa, sec. 4.6d.

In practice, instead of v_{eo} usually the *electro-osmotic volume flow* Q_{eo} is measured. The simplest case concerns capillaries of fixed cross-section A that are so wide that the range close to the surface, where the liquid velocity drops from its bulk value in the core of the lumen to zero at the wall, is so small that the contribution of this surface layer to the flux may be ignored. Then $Q_{eo} = A v_{eo}$, so that for this flow rate per unit of field strength

$$Q_{eo} = -\frac{\varepsilon_0 \varepsilon}{\eta} \zeta A E \qquad\qquad [4.3.22]$$

We have written this as a scalar equation. For the sign, reflecting the direction into which the flow proceeds, the same can be said as after [4.3.21]. When the capillaries are not linear A can be replaced by a vector A, of area A and direction normal to that area. As a result, the inner product $A \cdot E$ appears on the r.h.s., which is a scalar.

When the capillaries are not wide with respect to the surface shear layer, integration is required and the dependency $v_t(x)$ is needed. This relation follows from [4.3.7] and requires a model to obtain $\psi(x)$. The electro-osmotic flux $J_{eo} = Q_{eo}/A$ is again identical to v_{eo}.

From the above, the electro-osmotic volume flow per unit current, $Q_{eo,I}$ can be obtained by relating E to I, which means that the conductivity of the pore has to be considered. Generally it has two terms, a bulk contribution given by Ohm's law [I.6.6.5] and a surface conduction contribution, given by [I.6.6.36]. We therefore write the current as

$$I/E = AK^L + SK^\sigma \tag{4.3.23}$$

Here, A is the cross-section of the capillary, as before, K^L the bulk conductivity (in C V^{-1}s^{-1}m^{-1}), S is the circumference of the capillary and K^σ the *surface conductivity* (in C V^{-1}s^{-1}). Elimination of E between [4.3.22 and 23] leads to

$$Q_{eo} = -\frac{\varepsilon_o \varepsilon \zeta I}{\eta} \frac{A}{(AK^L + SK^\sigma)} \tag{4.3.24}$$

The extent to which surface conduction contributes depends on the geometry and the size of the capillaries: for wider pores, it becomes less important. When surface conduction may be neglected ($SK^\sigma \ll AK^L$), [4.3.24] reduces to

$$Q_{eo} = -\frac{\varepsilon_o \varepsilon \zeta I}{\eta K^L} \tag{4.3.25}$$

and in this case Q_{eo} is independent of the cross-section A.

Expressions for the *electro-osmotic volume flow per unit of field strength* $Q_{eo,E} = Q_{eo}/E$ and *per unit of current*, $Q_{eo,I} = Q_{eo}/I$, mentioned in table 4.1, are now immediately found:

$$Q_{eo,E} = -\frac{\varepsilon_o \varepsilon \zeta A}{\eta} \tag{4.3.26}$$

$$Q_{eo,I} = -\frac{\varepsilon_o \varepsilon \zeta A}{\eta(AK^L + SK^\sigma)} \tag{4.3.27}$$

Finally, the *electro-osmotic (counter-) pressure*, Δp_{eo} is the pressure on the fluid required to create a Poiseuille flow (see [I.6.4.17]) just counterbalancing the electro-osmotic one. For a capillary with $A = \pi a^2$ (a is the radius) and length ℓ

$$Q_{Pois} = \frac{\pi \Delta p \, a^4}{8 \eta \ell} = -Q_{eo} = \frac{\pi a^2 \, \varepsilon_o \varepsilon \zeta E}{\eta} \tag{4.3.28}$$

$$\Delta p_{eo} = \frac{8 \varepsilon_o \varepsilon \zeta E \ell}{a^2} \tag{4.3.29a}$$

or

$$\Delta p_{eo} = \frac{8\varepsilon_0\varepsilon\zeta Il}{\pi a^4 K^L + 2\pi a^3 K^\sigma}$$ [4.3.29b]

where we have used [4.3.23] with $A = \pi a^2$ and $S = 2\pi a$. The positive sign in [4.3.29a] implies that for negative ζ the higher pressure is on the lower potential side of the pore, in agreement with the definition given in sec. 4.2. For porous plugs no simple expressions for Δp_{eo} can be given.

4.3c. Sedimentation potential (Dorn effect)

When a charged colloidal particle sediments under gravity or a centrifugal field, the particle moves ahead of its ionic countercharge. The ensuing inhomogeneous charge distribution in the system gives rise to the sedimentation potential. Suppose the particles are negative. In a sedimenting suspension an electric field is created with an excess negative charge on the bottom side and a positive excess at the top. The precise mechanism depends, as in previous phenomena, on κa. For the case of low κa, in the stationary state the sedimentation field increases the mobilities of the (positive) countercharges till they move as fast as the particle. This countercharge consists of cations and a deficit of anions. As a whole, the countercharge moves faster than the ions in the bulk of the electrolyte. For large κa the double layer rather polarizes as in fig. 4.1. The double layer has about the same thickness everywhere; the dipole moment, related to the concentration polarization is responsible for the sedimentation potential.

When the *sedimentation current* is measured, top and bottom of the cell are externally short-circuited. Then $E_{sed} = 0$, but now a current I_{sed} flows through the external circuit. Under stationary state conditions, at high κa, most of the counterions move with the particle, a small fraction lags behind and is removed from the top side, just as the (small) negative excess is removed from the bottom side of the tube.

For a derivation on the same level as that leading to [4.3.5], consider for κa << 1 the sedimentation of N particles per unit volume, sedimenting with a velocity $\boldsymbol{v}_{sed}$, and each carrying a charge Q. The electric current density caused by this sedimentation equals $N\boldsymbol{v}_{sed}Q$. Electroneutrality is maintained in the system by an ionic current in the opposite direction, $-\boldsymbol{E}_{sed}K^L$, where K^L is again the conductivity of the ionic solution and $\boldsymbol{E}_{sed}$ is the sedimentation field strength, equal to the sedimentation potential difference E_{sed} divided by the distance between the points between which the potential difference is measured. Hence,

$$N\boldsymbol{v}Q = -\boldsymbol{E}_{sed}K^L$$ [4.3.30]

For $\boldsymbol{v}$ we substitute Stokes' law $v = 2a^2 \Delta\rho \boldsymbol{g}/9\eta$ (see [I.6.4.33]) where a is the particle radius, $\Delta\rho$ the density difference between particle and solution and $\boldsymbol{g}$ the standard acceleration of free fall or the corresponding acceleration caused by centrifugation. Electrophoretic retardation is ignored and $Q = 4\pi\varepsilon_0\varepsilon a\zeta$, as for the Coulomb case. We obtain

$$\boldsymbol{E}_{sed} = -\frac{8\pi\varepsilon_0\varepsilon\zeta a^3 \Delta\rho \boldsymbol{g} N}{9\eta K^L} \qquad \text{(low } \kappa a) \qquad\qquad [4.3.31]$$

The arguments are the same as those discussed in the elementary derivation of the Hückel-Onsager equation for electrophoresis, see sec. 4.3a (iii). Smoluchowski (loc. cit) has given a more rigorous derivation for high κa, which led to

$$\boldsymbol{E}_{sed} = -\frac{4\pi\varepsilon_0\varepsilon\zeta a^3 \Delta\rho \boldsymbol{g} N}{3\eta K^L} \qquad \text{(high } \kappa a) \qquad\qquad [4.3.32]$$

Equations [4.3.31 and 32] differ by a factor 3/2, as do the corresponding limiting laws for electrophoresis, [4.3.4 and 5]. The physical reason is that electrophoresis (particle moves under the influence of a potential gradient) and sedimentation potential (moving particle creates a potential gradient) are each other's counterparts, in the sense of the Onsager relations. Stigter[1] pointed this out and later Ohshima et al.[2] proved the identity on the basis of the general equations to be treated in sec. 4.6. These authors also presented results for $\boldsymbol{E}_{sed}$ for intermediate κa.

In [4.3.32] $4\pi a^3 N/3$ may be replaced by the particle volume fraction, φ. Hence,

$$\boldsymbol{E}_{sed} = -\frac{\varepsilon_0\varepsilon\zeta \Delta\rho \boldsymbol{g} \varphi}{\eta K^L} \qquad \text{(high } \kappa a) \qquad\qquad [4.3.33]$$

and similarly for [4.3.31].

By analogy to the other electrokinetic phenomena we could define a scalar sedimentation field strength per unit of acceleration, i.e. $\boldsymbol{E}_{sed}/\boldsymbol{g}$, but this is not customary. We note that the minus signs in [4.3.31-33] are in agreement with the convention of sec. 4.2.

4.3d. Streaming current and streaming potential

Consider a capillary with charged walls as in fig. 4.8, through which liquid is forced under the influence of an applied pressure head Δp. A parabolic

---

[1] D. Stigter, J. Phys. Chem. **84** (1980) 2758.
[2] H. Ohshima, T.W. Healy, L.R. White and R.W. O'Brien, J. Chem. Soc. Faraday Trans. II **80** (1984) 1299.

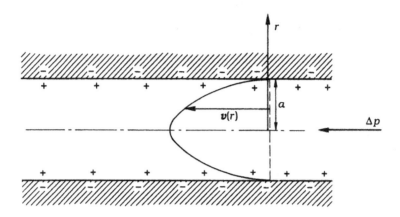

Figure 4.8. Origin of streaming current and streaming potential. For the sake of simplicity the countercharge is only represented by plus signs.

Poiseuille-type of flow ensues, for which we derived [I.6.4.15]

$$v(r) = -\frac{a^2}{4\eta}\left[1 - \left(\frac{r}{a}\right)^2\right]\frac{\Delta p}{\ell} \qquad\qquad [4.3.34]$$

where ℓ is the length of the capillary and a its radius ($\ell \gg a$). In vectorial form,

$$\boldsymbol{v}(r) = -\frac{a^2}{4\eta}\left[1 - \left(\frac{r}{a}\right)^2\right]\text{grad } p \qquad\qquad [4.3.35]$$

With this flow, countercharge is entrained, so that an electric current arises, the *streaming current*, $\boldsymbol{I}_{str}$ given by

$$\boldsymbol{I}_{str} = 2\pi \int_0^a \boldsymbol{v}(r)\,\rho(r)\,r\,\mathrm{d}r \qquad\qquad [4.3.36]$$

so that

$$\boldsymbol{I}_{str} = -\frac{\pi\,\text{grad } p}{2\eta}\int_0^a (a^2 - r^2)\,\rho(r)\,r\,\mathrm{d}r \qquad\qquad [4.3.37]$$

For a general solution of the integral, the cylindrical form of the Poisson equation for $\rho(r)$ is needed, see [I.A7.46] and for the potential distribution [I.3.5.61] is substituted.

We now consider the case of wide capillaries (large κa). For $\rho(r)$ then the one-dimensional variant of the Poisson equation, $\rho(r) = -\varepsilon_o\varepsilon(\mathrm{d}^2\psi/\mathrm{d}z^2)$ may be substituted, where z is the distance from the surface. Further, as only a very thin layer close to the surface contributes to the charge transport (the bulk has no space charge) the velocity profile inside the capillary is immaterial; mathematically this means that only r values very close to $r = a$ have to be considered: $a^2 - r^2 \approx 2a(a-r) = 2az$. Hence [4.3.37] can be rewritten as

$$I_{str} = \frac{\varepsilon_o \varepsilon \pi a}{\eta} \text{ grad } p \int_0^a z \frac{d^2\psi}{dz^2} a \, dz = \frac{\varepsilon_o \varepsilon \pi a^2}{\eta} \text{ grad } p \int_0^a z \frac{d^2\psi}{dz^2} \, dz$$

where the integral can be integrated by parts:

$$\int_0^a z \frac{d^2\psi}{dz^2} \, dz = \int_0^a z \, d\left(\frac{d\psi}{dz}\right) = z \frac{d\psi}{dz} \bigg|_0^a - \int_0^a \frac{d\psi}{dz} \, dz$$

The first term after the second equality sign is zero because for $r = 0$, $d\psi/dz = 0$ and for $r = a$, $z = 0$. The last term equals $-\psi(r = a) + \psi(r = 0)$, which is just the electrokinetic potential ζ. Hence,

$$I_{str} = \frac{\varepsilon_o \varepsilon A \zeta}{\eta} \text{ grad } p \qquad\qquad [4.3.38]$$

where, as before, $A = \pi a^2$ is the cross-section of the capillary. The sign agrees with the convention that I_{str} is counted positive if it runs from the high to the low pressure side.

In sec. I.6.2c irreversible thermodynamics were applied to electrokinetic phenomena. Equation [4.3.38] can be used to present a second illustration. In terms of [I.6.2.6] for the streaming current density one may write

$$\frac{I_{str}}{A} = \left(J_{el}\right)_{\text{grad } \psi = 0} = -L_{21} \text{ grad } p \qquad\qquad [4.3.39]$$

Likewise, for the electro-osmotic velocity

$$v_{eo} = \left(J_r\right)_{\text{grad } p = 0} = -L_{12} \text{ grad } \psi = L_{12} E \qquad\qquad [4.3.40]$$

As, according to Onsager, $L_{12} = L_{21}$ we find $I_{str} = -(v_{eo}/E)A \text{ grad } p$ from which, using [4.3.21], [4.3.38] is immediately retrieved. By the same token, we can use [I.6.2.9], $(E_{str}/\text{grad } p)_{J_{el} = 0} = -L_{21}/L_{22}$, where L_{21} follows from [4.3.38 and 39] as $-\varepsilon_o \varepsilon \zeta/\eta$ and where L_{22} is the conductivity K^L, leading to [4.3.42] for the streaming potential in the absence of surface conduction.

Streaming currents can in principle be measured by placing two identical reversible electrodes at the ends of the capillaries and connecting them by a low-resistance external lead, see sec. 4.4.

When the external circuit has a high resistance, a potential difference builds up across the capillary, the *streaming potential* E_{str}. It leads to a counter-conduction current inside the capillary which, as before, has a bulk and a surface conduction contribution. In the stationary state this counter current equals I_{str}. Using [4.3.23] to eliminate the current,

$$E_{str} = \frac{\varepsilon_o \varepsilon \zeta \, \Delta p}{\eta (K^L + 2K^\sigma/a)} \qquad\qquad [4.3.41]$$

which for the simple case that surface conduction may be neglected reduces to

$$E_{str} = \frac{\varepsilon_o \varepsilon \zeta \, \Delta p}{\eta K^L} \qquad\qquad [4.3.42]$$

In this limiting case the measured signal is independent of the geometry of the capillary (and, hence, the equation is generally applicable, provided the condition $\kappa a \gg 1$ remains valid). This is an advantage of measuring streaming potentials over streaming currents. Streaming potentials are counted positive if the higher potential is on the high pressure side.

In real systems, electro-osmosis and streaming potentials are often measured in porous plugs rather than in capillaries, requiring consideration of hydro-dynamics for other geometries and double layer overlap. However, when κa is high the field closely follows the surface and equations [4.3.26, 27, 41 and 42] remain good approximations. Equations [4.3.41 and 42] demonstrate that in media of low conductivity very high streaming potentials can develop. This feature has been dramatically illustrated by explosions resulting from fuel transport through pipes.

4.3e. Electroacoustics

By this term two phenomena are understood, both referring to the interaction between sound waves and dispersions of charged particles. These methods somewhat transcend the borders set for this section in that the inertia of the particles does play a role; in this respect electroacoustics anticipates a.c. electrokinetics and dielectric dispersion (sec. 4.8).

The first phenomenon is the *colloid vibration potential* CVP or E_{vibr}, already mentioned in table 4.1. Sometimes, the phenomenon is, more generally, called *ultrasound* (or *ultrasonic*) *vibration potential* (UVP). It is obtained if the sol or dispersion is subjected to acoustic waves. The inertia difference between particles and countercharge induces an oscillating dipole, the resulting a.c. electric field across the sol is measurable. The amplitude of this field (in V) per unit velocity amplitude of the (ultra-) sonic field (in m s⁻¹) is the measured quantity called E_{vibr} or CVP, in SI units V s m⁻¹. The phenomenon of a vibration potential has been predicted in 1933 by Debye for electrolyte solutions[1] and was then called *ion vibration potential* (IVP). Later Rutgers[2] and Hermans[3] demonstrated it to occur in colloids and Enderby and Booth[4] gave a first theoretical description. This matter has been reviewed by Zana and Yeager[5]. The order of magnitude of the CVP for colloids is much larger than that for electrolytes, so that in measurements with colloids the latter can usually be neglected.

Electrokinetic sonic amplitude (ESA) is the reverse process: a high-frequency a.c. field is applied to the sol or dispersion, imposing an oscillatory motion on the particles which, in turn, generates a pressure wave with the same frequency, but not necessarily the same phase, as that of the applied field. The pressure wave can conveniently be detected with a piezo-electric transducer. The ESA is the pressure amplitude per unit electric field, SI units m Pa V^{-1} = N V^{-1} m^{-1}. The phenomenon was discovered and patented by Oja et al.[6]. For a paper with some review character see[7].

From measured CVP's and ESA's (sec. 4.5c) the ζ-potential and/or other electrokinetic characteristics must be derived, but the theory is not simple. One of the reasons is that fields and particle polarization are not necessarily in phase so that mobilities, conductivities, etc. become complex quantities. Electrokinetic potentials enter because the inertia difference between the particles and the liquid in which they are embedded leads to shear. In the theories the usual assumption is made that there is a slip plane of which the potential is ζ.

Generally, from CVP or ESA the frequency-dependent mobility of a particle $\hat{u} = \hat{u}(\omega)$ can be obtained. The $\wedge$ indicates that the quantity is complex, i.e. it has a real and an imaginary part that both have to be measured, but then each of these yields information. However, in commercial apparatus measurements are sometimes restricted to only one frequency; then ζ is the sole parameter obtained. Phenomenologically, for non-interacting particles

$$\hat{u} = \frac{CVP(\omega)\hat{K}(\omega)}{\varphi(\Delta\rho/\rho)u_s} \text{ const.}' \qquad [4.3.43]$$

$$\hat{u} = \frac{ESA(\omega)}{\varphi(\Delta\rho/\rho)u_s} \text{ const.}'' \qquad [4.3.44]$$

---

[1] P. Debye, *J. Chem. Phys.* **1** (1933) 13.
[2] A.J. Rutgers, *Physica* **5** (1938) 674.
[3] J.J. Hermans, *Philosoph. Mag.* **25** (1930) 426.
[4] J.A. Enderby, *Proc. Roy. Soc.* **207** (1957) 329; C. Booth, J. Enderby, *Proc. Phys. Soc.* **208** (1957) 357.
[5] R. Zana, E. Yeager, in *Modern Aspects of Electrochemistry.* Vol **14**. J. O'M. Bockris, B.E. Conway and R.E. White, Eds., Plenum (1982) p. 1.
[6] T. Oja, G.L. Peterson and D.W. Cannon. U.S. Patent # 4.497.207 (1985).
[7] R.J. Goetz, M.S. El-Aasser, *J. Colloid Interface Sci.* **150** (1992) 436.

where $\hat{K}(\omega)$ is the (complex) conductivity of the sol or dispersion, φ its volume fraction, $\Delta\rho/\rho$ the relative density difference between particle and solution, giving rise to the inertia, and u_s the sound velocity in the dispersion. We write $\hat{u}$, as usual, in m² V⁻¹ s⁻¹. The volume fraction enters the equation via the mass of all the particles together. The constants stand for cell constants, containing such parameters as shape and size of the electrodes and the properties of the sensor used. These constants are not dimensionless and can either be obtained by a detailed acoustic theory, or, more easily, by standardizing the cell on the basis of measurements with particles of known electrokinetic properties. For dilute sols of spherical particles with high κa a relation between $\hat{u}$ and ζ has been derived by O'Brien[1]

$$\hat{u} = \frac{2\,\varepsilon_o\varepsilon\zeta}{3\eta}\,\hat{G}\,(q)(1+\hat{f}) \qquad\qquad [4.3.45]$$

where $q = \omega a^2/v$ (v is the kinematic viscosity),

$$\hat{G}\,(q) = \left[1 - \frac{iq[3+2\Delta\rho/\rho]}{9\left[1+(1-i)(q/2)^{1/2}\right]}\right]^{-1} \qquad\qquad [4.3.46]$$

$$\hat{f} = \frac{1-i\omega' - \left(2Du - i\omega'\varepsilon_p/\varepsilon_s\right)}{2(1-i\omega') + \left(2Du - i\omega'\varepsilon_p/\varepsilon_s\right)} \qquad\qquad [4.3.47]$$

$$\omega' = \varepsilon_o\varepsilon\omega/K^L \qquad\qquad [4.3.48]$$

$\varepsilon_p/\varepsilon_s$ is the ratio between the dielectric permittivities of particle and solution, K^L is the conductivity of the background electrolyte and $Du = K^\sigma/aK^L$, as before. The factors $\hat{G}$ and $(1 + \hat{f})$ account for the inertia, and for surface conduction and double layer polarization, respectively. Surface conduction will be considered in the next subsection. When inertia does not play a role, that is, when ω is so low that $q \to 0$ and $\omega' \to 0$, $\hat{G}(q) \approx 1$, $\hat{f} \approx 0,5$ so that [4.3.45] reduces to the Helmholtz-Smoluchowski equation. It may be added that in CVP two relaxation frequencies may be distinguished, a hydrodynamic one, determined by $q \approx 1$, and a concentration polarization one, for which $\omega' \approx 1$. Mostly the former frequency exceeds the latter by a factor of about 10^3.

Using this set of equations, from $\hat{u}(\omega)$ not only can ζ be evaluated but also the surface conductivity, if the particle size is known. Alternatively, if Du is known independently, the radius can in principle be established.

[1] R.W. O'Brien, J. Fluid Mech. **190** (1988) 71. See also A.J. Babchin, R.S. Chow and R.P. Sawatzky, Adv. Colloid Interface Sci. **30** (1989) 111.

CVP and ESA are techniques that appear to remain essentially valid for concentrated systems, with obvious advantages for applications. However, many systems are neither homodisperse nor do they contain spherical particles. Then, the computed ζ-potentials are rather semiquantitative (but useful) electrokinetic characteristics. Loewenberg and O'Brien have extended the theory to non-spherical particles[1].

For reviews of this matter see ref.[2].

4.3f. Surface conduction and the Dukhin number

Surface conduction is the excess conduction tangential to a charged surface. The phenomenon originates in the excess concentrations of the ions constituting the countercharge, and has been known for a long time: Smoluchowski recognized it at the beginning of the 20th century[3]. Later, Rutgers[4] used equations like [4.3.41] to estimate K^σ from the radius dependence of the streaming potential. In this way he solved the then open problem that ζ appeared spuriously dependent on the capillary radius if surface conduction was not accounted for. Surface conduction is most easily observed with macroscopic surfaces but also plays an, often underestimated, role in disperse systems. Below the case of large κa is considered, i.e. the double layer is thin as compared to the radius of curvature of the surface and may therefore be considered as virtually flat.

The phenomenon has been introduced in sec. I.6.6d and is sketched in fig. 4.9. The basic equation is

$$\boldsymbol{j}^\sigma = K^\sigma \boldsymbol{E} \tag{4.3.49}$$

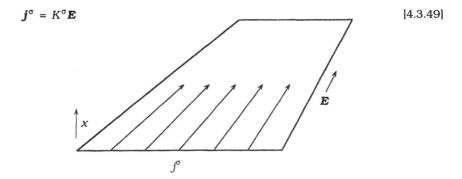

Figure 4.9. (= fig. I.6.23). Sketch to illustrate the excess current density parallel to a surface.

[1] M. Loewenberg, R.W. O'Brien, *J. Colloid Interface Sci.* **150** (1992) 158.
[2] B.J. Marlow, D. Fairhurst and H.P. Pendse, *Langmuir* **4** (1988) 611; A.J. Babchin, R.S. Chow and R.P. Sawatzky, *Adv. Colloid Interface Sci.* **30** (1989) 111.
[3] M. von Smoluchowski, *Physik. Z.* **6** (1905) 529.
[4] A.J. Rutgers, *Trans. Faraday Soc.* **36** (1940) 69

where j^σ is the (excess) current density tangential to the surface (in $C\,s^{-1}\,m^{-1} = A\,m^{-1}$), E the applied field causing this current (in $V\,m^{-1}$) and K^σ the (excess) *surface conductivity* (in $A\,V^{-1} = S$). Equation [4.3.49] may be interpreted as the two-dimensional analogue of Ohm's law.

In sec. I.6.6d we derived the equation

$$j^\sigma = \int_0^\infty \left[j(x) - j(\infty) \right] dx \qquad\qquad [4.3.50]$$

where $j(x)$ is the bulk current density in an infinitesimally thin layer of thickness dx parallel to the surface. Expressing $j(x)$ and $j(\infty)$ in terms of concentrations and mobilities

$$j(x) - j(\infty) = \sum_i \left[c_i(x) - c_i(\infty) \right] |z_i| F u_i(x) E \qquad\qquad [4.3.51]$$

Hence,

$$K^\sigma = F \sum_i |z_i| \int_0^\infty \left[c_i(x) - c_i(\infty) \right] u_i(x)\, dx \qquad\qquad [4.3.52]$$

which relates the surface conductivities to local concentrations and local mobilities. Further analysis requires more information on the double layer composition and these mobilities. Moreover, convection also has to be considered.

In view of the success of Gouy-Stern theory in interpreting static double layer properties it is consistent to do the same for the surface conductivity, i.e.

$$K^\sigma = K^{\sigma i} + K^{\sigma d} \qquad\qquad [4.3.53]$$

with, to a first approximation, $K^{\sigma i}$ determined by the ions in the inner Helmholtz layer and $K^{\sigma d}$ by the diffuse part; these two parts being distinguishable as "behind" and "beyond" the slip plane[1]. The assumption is made that the surface charge is immobile.

For the general case of a number of different ionic species i in the Stern layer

$$K^{\sigma i} = \sum_i K_i^{\sigma i} = \sum_i \sigma_i^i u_i^i \qquad\qquad [4.3.54]$$

which, using the *Nernst-Einstein equation*

$$u_i = \frac{|z_i| F D_i}{RT} \qquad\qquad [4.3.55]$$

---

[1] Sometimes $K^{\sigma i}$ is called "anomalous surface conductivity", and the phenomenon of conduction behind the slip plane as "anomalous surface conduction". We shall not use this term because there is nothing anomalous about it.

see [I.6.6.15], may be formally written as

$$K^{\sigma i} = \frac{F}{RT}\sum_i |z_i| D_i^\sigma \sigma_i^i \qquad [4.3.56]$$

Where, as before, σ_i^i is the surface charge density in the Stern layer, attributed by ions i. Often there is only one mobile ionic species in the Stern layer. Then the expressions reduce to

$$K^{\sigma i} = \sigma^i u^i = \frac{|z_i| F D_i^\sigma \sigma^i}{RT} \qquad [4.3.57]$$

Values for σ^i can be obtained with one of the strategies outlined in sec. 3.6c-e. Needless to say, lateral mobilities and diffusivities are interesting and informative parameters.

The diffuse part can be, and has been, analyzed in detail, mainly thanks to Bikerman[1]. The advantages are that in this part the mobilities may be equated to those in bulk, whereas the local concentrations obey Gouy-Chapman statistics.

Bikerman realized that besides conduction, the applied field also creates an electro-osmotic fluid transport. In his equation for $j(x)$, applied to one electrolyte,

$$j(x) = F\left[z_- c_-(x) - z_+ c_+(x)\right] \frac{\varepsilon_0 \varepsilon}{\eta} \left[\zeta - \psi(x)\right] E$$

$$+ \frac{F^2}{RT}\left(z_+^2 D_+ c_+(x) + z_-^2 D_- c_-(x)\right) E \qquad [4.3.58]$$

the first term is the electro-osmotic current density, whereas the second covers the conductive transfer of diffuse layer ions relative to the liquid. The ionic diffusion coefficients are assumed to be the same as in the bulk; they enter through equation [4.3.55]. Equation [4.3.58] can be substituted in [4.3.50], taking into account that all concentrations follow from Gouy theory. Replacing the integration over x by one over y, Bikerman eventually derived for a symmetrical electrolyte

$$K^{\sigma d} = \sqrt{8\varepsilon_0 \varepsilon c\, RT}\, \left\{\frac{u_+}{A-1} - \frac{u_-}{A+1} + \frac{4\varepsilon_0 \varepsilon c\, RT}{\eta z F}\, \frac{1}{A^2-1}\right\} \qquad [4.3.59]$$

where $A = \coth(-z F \zeta / 4RT)$. After some algebra this can be rewritten as[2]

---

[1] J.J. Bikerman, Z. Physik. Chem. **A163** (1933) 378; Kolloid-Z. **72** (1935) 100.
[2] B.V. Deryagin and S.S. Dukhin, Koll. Zhur. **31** (1969) 350 (transl. 277) extended this equation to polarized double layers.

$$K^{\sigma d} = \frac{2F^2 z^2 c}{RT \kappa} \left[D_+ \left(e^{-zF\zeta/2RT} - 1 \right) \left(1 + \frac{3m_+}{z^2} \right) + D_- \left(e^{zF\zeta/2RT} - 1 \right) \left(1 + \frac{3m_-}{z^2} \right) \right]$$

[4.3.60]

where

$$m_\pm = \left(\frac{RT}{F} \right)^2 \frac{2\varepsilon_0 \varepsilon}{3\eta D_\pm}$$

[4.3.61]

is a dimensionless parameter indicating the relative contribution of electro-osmosis to surface conductance. For aqueous solutions at room temperature $m \approx$ 0.15. Expressions for $K^{\sigma d}$ that do not contain the terms with m are simplifications in which electro-osmosis is neglected. Counterions contribute relatively more to the electro-osmosis than co-ions do, to an extent determined by ζ. When cation and anion have the same diffusion coefficients, the m's are also identical; in that case the equation can be simplified to

$$K^{\sigma d} = \frac{4F^2 c z^2 D[1 + 3m/z^2]}{RT \kappa} \left[\cosh \left(\frac{z F \zeta}{2RT} \right) - 1 \right]$$

[4.3.62]

Figure 4.10 gives a graphical representation of [4.3.62] for (1-1) electrolytes. Experimental values of ζ vary between about 150 mV in 10^{-3} M electrolyte to less than 25 mV in 10^{-1} M. Consequently, experimental values of $K^{\sigma d}$ are between 0 and $\approx 2 \times 10^{-9}$ S. (There is some circularity in the arguments because many reported ζ-potentials have to be upgraded because in the computation surface conduction was neglected; however, this upgrading does not greatly affect the above order of magnitude.)

When ζ is high enough to let $zF\zeta/2RT \gg 1$ only one ionic species has to be considered, viz. the counterions. Then [4.3.60] reduces to

$$K^{\sigma d} = \frac{4F^2 z^2 c}{RT \kappa} D e^{zF\zeta/2RT} \left(1 + \frac{3m}{z^2} \right)$$

[4.3.63]

Diffuse layer surface conductivities may be compared with those in the inner layer. For the case of a triple layer with all Stern ions in the inner Helmholtz plane we have $K^{\sigma i} = \sigma_i^i u_i^i$ for ions of type i, see [4.3.54]. Stern charges on non-porous surfaces can be anything between 0 and ≈ 50 μC cm^{-2}. High values are, for example, encountered on some oxides where the titration charge σ° attains values of several tens of μC cm^{-2}, whereas $\sigma^d \approx \sigma^{ek}$ may be lower by a factor of ten. On silver iodide, on the other hand, σ^i remains as low as ≈ 3-4 μC cm^{-2}. The (translational) mobilities are unknown but, considering the arguments given in secs. 2.2c and 4.4, they are probably of the same order as those in bulk, though somewhat lower. Bulk limiting ionic mobilities range, with a few exceptions from 4-9 $\times 10^{-8}$ m^2 V^{-1} s^{-1} (table I.6.5). Therefore, maximum values for $K^{\sigma i}$ are

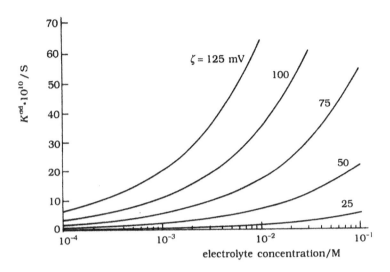

Figure 4.10. Surface conduction in the diffuse part of the double layer according to [4.3.62]. Monovalent symmetrical electrolyte with $D_+ = D_-$.

$\approx 10^{-8}$ S, i.e. they may exceed $K^{\sigma d}$ by a factor of about 5-10, but they may also be lower. Assuming an adsorption model, σ^i can be related to c_i and the specific Gibbs binding energy $\Delta_{ads}G_{mi}$. Examples are [3.6.16 and 17]; figs. 3.20-22 give illustrations. Making ΔG_{mi} more negative increases σ^i_i, but does not necessarily enhance $K^{\sigma i}$ because tighter bound ions may have a lower lateral mobility. Very high values for $K^{\sigma i}$ may be expected for porous surfaces, containing thick, hydrodynamically stagnant layers with mobile ions in them, as with porous glass and bacterial cells. In summary, $K^{\sigma i}$ can be of the same order as $K^{\sigma d}$ but may also exceed $K^{\sigma d}$ by a factor of 10. The ratio $K^{\sigma d}/K^{\sigma i}$ provides interesting information on the distribution of mobile charges in the double layer. We will return to this matter in sec. 4.6e.

With the above information the ratio between K^{σ} and the bulk conductivity K^L can now be made explicit. An elegant way of doing so is by introducing a dimensionless ratio $K^{\sigma}/\ell K^L$, where ℓ is a characteristic length. In sec. 3.13a we called this ratio the *Dukhin number, Du*. For spherical particles ℓ becomes a, the particle radius. Then

$$Du \equiv \frac{K^{\sigma}}{a\,K^L} = \frac{K^{\sigma d}}{a\,K^L} + \frac{K^{\sigma i}}{a\,K^L} = Du^d + Du^i \qquad\qquad [4.3.64]$$

For hollow capillaries, a is the inner capillary radius. The denominator of [4.3.41] can also be written as $\eta K^L(1 + 2\,Du)$.

In the electrokinetic literature Du, or closely related ratios, are represented by different symbols. Dukhin himself used the symbol Rel[1]. In the Australian literature[2] a "surface conduction parameter" λ is introduced, which is identical to our Du and a quantity β which is equal to $\lambda(D_+ + D_-)/D_+$ for a z-z electrolyte, i.e. for ions of equal diffusivity $\beta = 2Du$. The very idea that the ratio K^σ/aK^L is important in discriminating various electrokinetic regimes is older, and has for instance been put forward by Bikerman[3].

On the basis of [4.3.60 or 62] Du^d can be made explicit. To that end $K^{\sigma d}$ is divided by $a K^L$, with $K^L = 2z F c u^L$ (this follows from [I.6.6.8] for a symmetrical electrolyte, using $u_+^L = u_-^L = z F D/RT$, as in [4.3.55]). The result is

$$Du^d = \frac{2}{\kappa a}\left(1 + \frac{3m}{z^2}\right)\left[\cosh\left(\frac{z F \zeta}{2RT}\right) - 1\right] \qquad [4.3.65]$$

The low ζ variant is, using [A2.8],

$$Du^d \approx \left(\frac{z^2 + 3m}{2\kappa a}\right)\left(\frac{F \zeta}{RT}\right)^2 \qquad [4.3.66]$$

and the simplification for not too low ζ and only one ion type, from [4.3.63] becomes

$$Du^d \approx \frac{(1 + 3m/z^2)}{\kappa a}\exp\left(\frac{z F |\zeta|}{2RT}\right) \qquad [4.3.67]$$

Equations for Du^d, not containing the $3m/z^2$ term, can also be found in the literature. These are approximations in which electro-osmosis is ignored. The error made is about 45% and 11% for mono- and bivalent electrolytes, respectively, i.e. the order of magnitude is correct. When $K^{\sigma i} \gg K^{\sigma d}$ it is of course not necessary to estimate Du^d so accurately and conversely for $K^{\sigma i} \ll K^{\sigma d}$. An equation for Du which contains $K^{\sigma i}/K^{\sigma d}$ explicitly will be given later: [4.6.56].

For the Stern layer, containing only one species i

$$Du^i = \frac{\sigma_i^i u_i^i}{2a z_i F c_i u_i^L} \qquad [4.3.68]$$

This quantity depends on the ratios u_i^i/u_i^L and σ_i^i/c_i (i.e. on the adsorption isotherm for ions in the Stern layer).

---

[1] S.S. Dukhin, Developments of Notions as to the Mechanism of Electrokinetic Phenomena and the Structure of the Colloid Micelle, in Surface and Colloid Science, E. Matijevic, Ed., Vol. 7, Wiley (1974), ch. 1.
[2] R.J. Hunter, Zeta Potential in Colloid Science. Principles and Applications, Academic Press (1981)
[3] J.J. Bikerman, Trans. Faraday Soc. 36 (1940) 154.

For large κa or a both Du's go to zero, meaning that then surface conduction may be neglected. For the electrophoretic mobility this means that the Helmholtz-Smoluchowski equation [4.3.4] also remains the correct limit for high κa if surface conduction does occur.

We return to more complicated situations and/or more advanced theories for electrokinetics in sec. 4.6. As a part of this, surface conduction will be further considered in secs. 4.6e and f.

4.4 Interpretation of electrokinetic potentials

Although electrokinetic potentials belong to the most common characteristics of charged surfaces, their interpretation in terms of double layer potentials is not yet straightforward. Rarely is ζ identical to ψ°; mostly it is lower, if not much lower. When superequivalent counterion adsorption occurs even the signs of ζ and ψ° differ. In these respects ζ rather resembles ψ^d, a correspondence that is supported by the frequently observed correlation between electrokinetic potentials and the stability of electrocratic colloids.

How close are ζ and ψ^d ? A reply to this question involves the comparison of results from disparate measurements and models. Electrokinetic potentials can be obtained from electrokinetics after making some assumption on the slip process, i.e. about the hydrodynamics of tangential flow, whereas ψ^d requires information on double layers and/or stability and some model about the double layer under static conditions. There is no *a priori* reason to expect identity between the results. However, for practical purposes the outcome might be that the two are close enough to consider them as the same. Let us therefore now consider the slip process and the charge distribution in a double layer in more detail than in sec. 4.1b.

On a molecular scale there is no sharp boundary between hydrodynamically stagnant and movable solvent molecules. As discussed in sec. 2.2, the, say tangential, diffusion coefficient of water near many surfaces may be somewhat lower than in bulk, but it is not zero. The very existence of ionic conduction in the layer(s) adjacent to surfaces also points to non-zero mobility. Yet, phenomenologically such layers behave as immobilized. This looks like a paradox, but the phenomenon is encountered in other places as well. For instance, a few percent of gelatin added to water may hydrodynamically immobilize the liquid completely, without markedly impairing ionic conduction or self-diffusion of dissolved ions. Macroscopic immobilization of a fluid is not in conflict with mobility on a molecular scale.

Consider in this connection the normal (perpendicular) distribution of ions and solvent molecules. In the liquid adjacent to the solid the number density of the solvent molecules may exhibit some oscillations that damp out over a few molecular layers. The distribution of ions in these first layers has to obey a similar pattern. In primitive double layer models any solvent-mediated oscillations in the distribution of (counter-) ions are ignored, although it is generally acknowledged that the positions of the inner and outer Helmholtz planes are determined by the "hydration" of these ions. In turn, this hydration is controlled by the way in which these ions are embedded in water. The most direct information available for the positions of the two Helmholtz planes stems from capacitances, but these only give the quotients ε_1^i / β and ε_2^i / γ (see [3.6.7]), which defy the determination of β and $\beta + \gamma$ because of the undefinedness of the relative dielectric permittivities. Anyhow, on a molecular level, there is no sharp distinction between molecules and ions "under influence of the surface" and those "having bulk properties". Nevertheless, in double layer modelling the abstraction is that an outer Helmholtz plane exists, within which all deviations from the purely diffuse behaviour are concentrated, and beyond which the distributions of charge and potential obey the Gouy-Chapman model.

The abstraction made in electrokinetics is that for practical purposes the gradual transition from hydrodynamically stagnant to free solvent is replaced by a discrete slip plane at which a sudden transition takes place. The basic question of how close ζ and ψ^d are, i.e. the problem of assessing the distance between the positions of the slip plane and the outer Helmholtz plane, comes down to comparing the qualities of disparate approximations.

Given this issue, three roads are open to establish the relation between ζ and ψ^d:

(i)   Molecular modelling and simulations.

(ii)   Introduction of theories for the slip process.

(iii) Establishment of ζ and ψ^d by experiment as well as possible and comparing the outcomes.

These three approaches should also answer the question of the dependency on the nature of the material, including its surface roughness. In this respect, it is noted that Stern and stagnant layers are found for both hydrophobic and hydrophilic surfaces, i.e. we are dealing with a very general phenomenon. The presence of hairs on the surface can lead to a substantial increase of the stagnant layer thickness, an aspect that we shall not consider here. Below we shall restrict ourselves to aqueous systems.

Regarding (i), the most academic approach, at the time of writing only embryonic attempts have been made. In sec. 2.2c the structure of water near interfaces has been discussed and in sec. 3.9 the same has been done for charged surfaces. As to the dynamics, fig. 2.5 may be reconsidered. Although this figure does not specifically apply to water, for this liquid the dynamics may be similar. It may be inferred that residence times of fluid molecules are relatively long at the crests in the density oscillations $\rho_N(z)$ but shorter in the minima. Otherwise stated, molecules can move relatively easily parallel to the surface, but have more trouble in moving normally to or away from it. This is in line with the observation that diffusion coefficients parallel to the surface $D_{//}^{\sigma}$ tend to be higher than those normal to it, $D_{\perp}^{\sigma}$. It further means that for ionic transport similar trends apply, which has its consequences for the interpretation of the surface conductance K^{σ}. This picture also offers a ground for resolving the apparent paradox of observing a relatively high K^{σ} in a liquid that is hydrodynamically stagnant. Ever since the pioneering work of Bernal and Fowler it has been known that viscous flow in water is a very co-operative process[1], in that the viscous displacement of molecules at a given position requires the motion of other molecules, several molecular distances away. Restriction of the latter by the presence of a domain of long residence times is therefore macroscopically experienced as an increase in viscosity. This is why high local viscosities and high ionic mobilities in stagnant layers are not in conflict with each other.

Elaborations along these lines are the most fundamental, and in that respect the most satisfactory. They require simulations with large samples under lateral shear, with ions embedded in the water and charges on the surface, and with realistic interactions between the water molecules. Surface roughness on a molecular scale should also be accounted for: the presence of material obstacles on a surface may drastically affect the tangential motion of molecules and ions that have to negotiate such asperities. There is still a long way to go.

(ii) As far as theories for the slip process are concerned, there are only two models. According to the first, proposed by Bikerman[2], the slip phenomenon is entirely attributed to surface roughness. Although roughness is expected to play a role, it is unlikely to be the whole story because slip planes also have to be invoked in the interpretation of electrokinetic phenomena with perfectly smooth surfaces. Another drawback is that the roughness effect is difficult to quantify.

A more quantitative model has been proposed by Lyklema and Overbeek[3]. In their picture the notion of slip plane is replaced by a slip layer, characterized by a viscosity for tangential fluid motion that gradually decreases from a very high value close to the surface ($z = 0$) to its bulk value far away ($z = \infty$). This idea is illustrated in fig. 4.11. As a next step it was assumed that, at any z, $\eta(z)$ was

[1] J.D. Bernal and R.H. Fowler, *J. Chem. Phys.* **1** (1933) 515.

[2] J.J. Bikerman, *Surface Chemistry for Industrial Research*, Academic Press (1948) 1386; *Physical Surfaces*, Academic Press (1970) 175.

[3] J. Lijklema (= Lyklema), J.Th.G. Overbeek, *J. Colloid Interface Sci.*, **16** (1961) 507; J. Lyklema, *Colloids Surf.* **A92** (1994) 41.

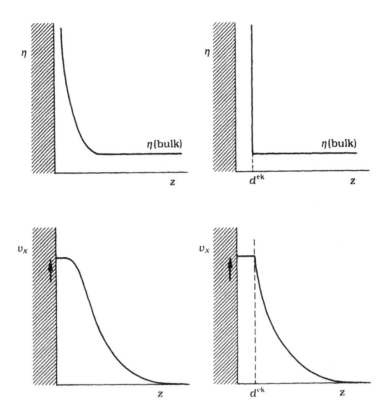

Figure 4.11. Viscosity (top) and tangential velocity profile (bottom) for a real double layer (left) and a layer containing a discrete slip plane at distance d^{ek} from the surface. The surface is supposed to move upward in the x-direction with respect to a stationary liquid.

determined by the electric double layer field strength at that place, $E(z)$. The underlying idea was that the passage of molecules along each other is impeded if the dipoles of the molecules of the liquid are oriented. Experimentally, this *viscoelectric effect* has been established for chloroform, amylacetate and chlorobenzene as[1]

$$\frac{\eta(E) - \eta(0)}{\eta(0)} = \frac{\Delta\eta}{\eta} = f\,E^2 \qquad\qquad [4.4.1]$$

where the *viscoelectric coefficient* f is 1.89×10^{-16}, 2.74×10^{-16} and 2.12×10^{-16} m^2 V^{-2}, for these three liquids, respectively. Here, $\eta(0) = \eta$ is the viscosity in the absence of a field. For water f_w is not known, but it was estimated at $f_w = 10.2 \times 10^{-16}$ m^2 V^{-2} on the basis of a molecular model for the influence of dipole orientation

---

[1] E.N.Da.C. Andrade, C. Dodd, *Proc. Roy. Soc. (London)* **A187** (1946) 296; **A204** (1951) 449.

on the activation energy for viscous flow; this theory could account for the shape of [4.4.1]. This model is too simple, because it only considers pair interactions and ignores the idiosyncrasies of the water structure. Hence, the value of f_w is not more than a semi-quantitative estimate. However, the form of [4.4.1] is probably correct. In any case, η must be an even function of E.

On the basis of this model, the derivation in connection with fig. 4.3 leading to [4.3.4] can now be improved by replacing $\eta = \eta$(bulk) by $\eta(E)$, with the double layer field strength given by [3.5.11] for a flat double layer. There is no problem in using diffuse theory for E because most of the charge transport by hydrodynamic traction takes place in the mobile outer part of the double layer. Strictly speaking, the relative dielectric permittivity ε is also field-dependent, but provisional models for the dielectric saturation of water[1] indicate this phenomenon to take place at field strengths as high as $O(10^8$ V m$^{-1})$, which is about a factor of 10 higher than those where the viscoelectric effect becomes prominent. Hence, for the present purpose the assumption $\varepsilon(z) = \varepsilon$(bulk) is justified.

Instead of [4.3.4]

$$u = \frac{\varepsilon_0 \varepsilon R T}{\eta F} \int\limits_0^{y^d} \frac{dy}{1 + A_{ve}\,\sinh^2{(zy/2)}}$$

[4.4.2]

is now obtained[2], with

$$A_{ve} = 8cRTf/\varepsilon_0\varepsilon$$

[4.4.3]

For $A_{ve} = 0$ and substitution of $F\zeta/RT$ for y^d the Helmholtz-Smoluchowski limit is retrieved. In [4.4.2] y^d may be replaced by y^o if tangential fluid motion also occurs in the Stern layer, but for the present purpose this is assumed to be negligible.

The integration in [4.4.2] can be carried out analytically, but the results are somewhat complicated and depend on A_{ve}. They are graphically presented in fig. 4.11. Here ζ(app) is the electrokinetic potential calculated from

$$\zeta(\text{app}) \equiv u\eta/\varepsilon_0\varepsilon$$

[4.4.4]

i.e. it is the potential obtained if the Helmholtz-Smoluchowski equation [4.3.4] were applicable. For $A_{ve} = 0$, ζ(app) $\rightarrow \zeta = \psi^d$. The higher c is at given f_w (as shown) or, the higher f_w is at given c (not shown), the lower the observed electrokinetic potential is.

---

[1] F. Booth, J. Chem. Phys. **19** (1951) 391, **23** (1955) 453; see also J.A. Schellman, J. Chem. Phys. **26** (1957) 1225.
[2] J. Lyklema, J.Th.G. Overbeek (loc. cit. 1961).

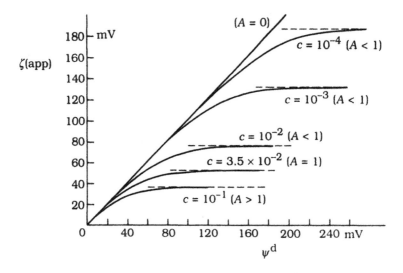

Figure 4.12. Apparent electrokinetic potential as a function of the outer Helmholtz plane potential, when the slip process is determined by the viscoelectric effect. $f_w = 10.2 \times 10^{-16}$ V^{-2} m^{-2}; c is the concentration of the (1-1) electrolyte in M. (Redrawn from J. Lyklema, *Colloids Surf.* **A92** (1994) 41.)

The qualitatively new feature is that ζ(app) does not increase continually but attains a plateau, which is lower for higher values of $c f_w$. The physical reason is that, with increasing y^d, the electric field in the double layer also rises and hence the part of the double layer that is immobilized. If translated in terms of the familiar slip picture, one could say that the slip plane moves outward if y^d increases. In fig. 4.12 the nature of the surface (hydrophilic or hydrophobic) does not arise, apart from possible differences with respect to f_w. However, this is perhaps a small effect because the packing of liquids near a surface is more determined by liquid-liquid than by liquid-surface interactions.

Figure 4.13 gives an experimental illustration for a few systems for which both titration and electrokinetic data as a function of c_{salt} and pH or pAg are available[1]. Data are plotted in terms of charges because the surface charge σ^o is experimentally accessible; the electrokinetic charge σ^{ek} is obtained from [3.5.13] after replacing σ^d and ψ^d by σ^{ek} and ζ_{obs}, respectively. Electrokinetic potentials have been computed from mobilities taking surface conduction into account, see sec. 4.6f. The levelling of σ^{ek}, predicted by the theory is

---

[1] Older comparisons have been discussed by R.J. Hunter, *Zeta Potential in Colloid Science. Principles and Applications*, Academic Press 1981).

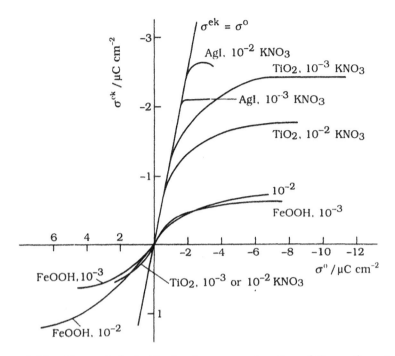

Figure 4.13. Observed electrokinetic charge as a function of the surface charge for a number of systems (same source as fig. 4.12).

experimentally retrieved. Plateaus are only a few μC cm^{-2}. They depend neither on c_{salt} nor on the nature of the surface in an obvious way, suggesting that here one is dealing with a typical property of the solvent water. Moreover, the plateau values can be well accounted for by viscoelectric constants $f_w \approx 10 \times 10^{-16} m^2 V^{-2}$, as estimated.

The satisfactory agreement between theory and experiment does not validate the former because the levelling of σ^{ek} can also be explained by other mechanisms. One is ion condensation [1], but according to this phenomenon the transition between the ascending and the horizontal parts of the σ^{ek} (σ^o) plot should be sharp and this has never been observed. More realistic is that σ^d lags behind σ^o because of counterion binding in the Stern layer and/or finite Stern layer capacitances. This effect comes in addition to the viscoelectric effect. Discrimination is possible in that specific adsorption would be rather system-specific and would not yield a plateau but a continually rising σ^{ek}. There is certainly need for more extended, high quality data.

(iii) The third approach is to assess ψ^d from colloid stability studies and

[1] G.S. Manning, *Quart. Rev. Biochem.* **11** (1978) 179; *Ann. Rev. Phys. Chem.* **23** (1972) 117.

compare the result with ζ. Such a comparison requires several interpretational steps and involves the estimation of additional parameters. For instance, critical coagulation concentrations or rates of coagulation should be interpreted; after that a Hamaker constant has to be substituted. Therefore, the results of such comparisons are not unambiguous, although for some well-studied model systems, like the hydrophobic silver iodide and the hydrophilic aqueous side of anionic surfactant monolayers the difference between ψ^d and ζ appears to be small[1]. In passing, it may be noted that estimation of the stagnant layer thickness d^{ek} from Gouy theory, (i.e. using [3.5.22] with $y(x) = F\zeta/RT$, $y^d = y^o$ and $x = d^{ek}$, leads to erroneously high results.

Given the uncertainties that are still encountered in the interpretation of electrokinetic potentials or charges, we shall generally interpret electrokinetic phenomena in terms of ζ-potentials, assuming ζ to be the potential of a slip plane. Its position is close to, or for practical reasons identical to, the outer Helmholtz plane.

4.5 Experimental techniques

Electrokinetics play important roles in colloid and surface science but also beyond these domains. The purpose of the present section is to consider some principles behind their measurements, emphasizing methods relevant for characterizing colloids and interfaces. The large number of semiquantitative methods, including paper or gel electrophoresis for diagnostic purposes and isoelectric focusing in protein chemistry, are considered "applications" (sec. 4.10). For more details, see the literature of sec. 4.11, especially the books by Hunter, and Righetti et al. Unless where explicitly mentioned otherwise, the considerations apply to aqueous systems.

4.5a. Electrophoresis

We are interested in measuring the electrophoretic mobility u of individual colloidal particles. This requires the preparation of a sol, applying an external electric field **E** and determining the velocity (or velocities) **v** of the particle(s).

Preparing the sol is basically a matter beyond electrokinetics, but it should be noted that, particularly for microelectrophoresis (see below), the sols have to be stable and very dilute. Strong dilution implies a low area to liquid volume ratio and hence a low buffering capacity against impurities that may be introduced inadvertently upon diluting the sol. Obviously, pH and electrolyte concentration should be controlled. The pH may not only affect the surface

---

[1] J. Lyklema, *J. Colloid Interface Sci.* **58** (1977) 242.

charge on the particles but also that on the wall of the cell, which is often made of glass or quartz. In turn, this surface charge gives rise to additional electro-osmosis and surface conduction, which, in some cases, have to be controlled. In electrophoresis measurements surface conduction of the cell wall poses no problem, because in that technique it is automatically accounted for. Electrolytes determine the zeta potential, but also the colloid stability of the particles, unless they are sterically stabilized. At high electrolyte concentrations electrocratic sols coagulate, and settle on the walls of the capillary so that then no mobility measurements can be made. Streaming potential and electro-osmosis measurements are then still possible. Efficient temperature control is another prerequisite because the sol tends to be heated by the passing electric current and the resulting convection should be suppressed. We assume the particle to be solid, although experience has shown that much of the following also applies to moving droplets in emulsions.

Regarding the determination of particle velocities, two approaches are possible. In the first, known as *microelectrophoresis*, individual particles are made visible and their translation is monitored. This is the most basic procedure because the variation of the velocities over the particles can be measured. When microelectrophoresis is not feasible, the collective displacement of a large number of particles can be studied. This technique is known as *moving boundary electrophoresis*. We shall now discuss these modes of operation.

(i) *Microelectrophoresis.* All methods for the visualization of individual particles are in some way or other based on light scattering. In the oldest version, particles are observed *ultramicroscopically*, i.e. their presence is established by the strong lateral scattering known as the *Tyndall effect* (sec. I.7.6). A thumbnail of the optics is given in fig. 4.14. Ultramicroscopically, the particles are seen as bright spots on a dark background, and their velocities can be measured by using a standardized grid. More recently this technique has

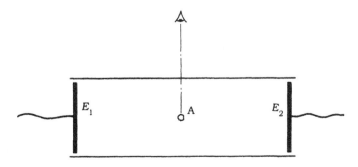

Figure 4.14. Ultramicroscopic observation of the motion of a particle A, moving in the field between the electrodes E_1 and E_2. The incident light is normal to the plane of the sketch. The cell may be flat or cylindrical.

largely been replaced by *electrophoretic light scattering*, which is based on *quasi-elastic*, or *dynamic, light scattering (QELS)*, also known as *photon correlation spectroscopy* PCS, and/or *laser-Doppler microscopy*. The optical principles have been discussed in sec. I.7.8. According to [I.7.3.8] the relative Doppler shift of the frequency $\Delta\omega/\omega = -\boldsymbol{v} \cdot \boldsymbol{n}/c$, where $\boldsymbol{v}$ is the required velocity, $\boldsymbol{n}$ a unit vector in the direction of the incident light and c the velocity of electromagnetic radiation in vacuum. Hence, $\boldsymbol{v}$ can be obtained[1,2]. Optical visibility requires sufficient contrast between particles and background. In practice, this means that the particles should not be too small and have a refractive index that differs sufficiently from that of the solvent. Static and dynamic light scattering has already been discussed in detail in chapter I.7.

From QELS the diffusion coefficient, and hence the particle radius, can be found, provided the particles are spherical and homodisperse and the sol is dilute. Deconvolution is difficult when one of these premises does not apply. Dilution of the sol is a prerequisite to avoid multiple scattering and any hydrodynamic interaction between the particles. A variety of apparatus is nowadays commercially available, based on one of the above-mentioned techniques, or a combination of them, so that velocity and size are both obtained. With some of these instruments size and/or velocity distributions can be derived, but the caveat must be made that these tend to be based on software programs in which a certain type of distribution is presupposed.

In order to guarantee well-defined hydrodynamic conditions, usually closed cells are used, thus avoiding inadvertent variations of the external pressure. However, contrary to what one might expect, closing the cell does not imply immobilization of the continuous phase. The reason is that the (transparent) cell walls usually carry a charge so that an electro-osmotic fluid displacement is created, as in fig. 4.7. The electrophoretic displacement is superimposed on this electro-osmotic flow and to subtract the latter the hydrodynamics have to be considered in some detail.

Electro-osmosis in a closed cell leads to a hydrodynamic pressure which, in turn, causes a Poiseuille-type back flow (sec. I.6.4d and fig. I.6.10), leading to a velocity profile as in fig. 4.8. For the, most common, cylindrical cell, the resulting velocity profile is as in fig. 4.15. The mathematical elaboration is as follows. Let z be the axial direction in the cylinder and r the radial one, then the fluid velocity in the z-direction at a distance r from the axis can be written as

$$v_z(r) = b\left[1 - (r/a)^2\right] + v_{eo} \qquad\qquad [4.5.1]$$

---

[1] J.R. Goff, P. Luner, *J. Colloid Interface Sci.* **99** (1984) 468.
[2] J.F. Miller, K. Schätzel and B. Vincent, *J. Colloid Interface Sci.* **143** (1991) 532. See further the reviews mentioned in sec. 4.11c.

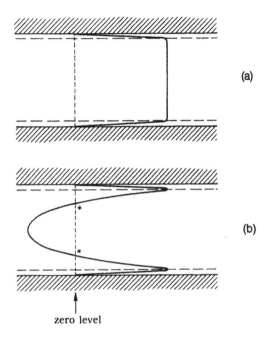

Figure 4.15. Sketch of a velocity profile in a closed electrophoresis capillary. (a) electro-osmotic slip only, as in fig. 4.7. (b) electro-osmosis plus Poiseuille counterflow. Asterisks indicate stationary levels. (Not to scale, the layers close to the surfaces are only a few times κ^{-1} thick.)

The first term on the r.h.s. is the Poiseuille term, which has been derived before, see [I.6.4.15]; $b = \Delta pa^2/4\eta\ell$ where $\Delta p/\ell$ is the pressure drop over the cell length ℓ, a the inner cylinder radius and η the (dynamic) viscosity. The second term is the electro-osmotic contribution. The two terms act in opposite directions because of the back-flow nature of the Poiseuille flow. When the direction of the applied field is reversed, both terms obtain the opposite sign. The cell being closed, the net flow is zero. Hence,

$$2\pi \int_0^a r\upsilon_z(r)\,dr = 0 \qquad\qquad [4.5.2]$$

Substitution of [4.5.1] and carrying out the integration leads to

$$(b + \upsilon_{eo})a^2 - ba^2/2 = 0 \qquad\qquad [4.5.3]$$

or $2\upsilon_{eo} = -b$, which can be substituted into [4.5.1] to yield

$$\upsilon_z(r) = \upsilon_{eo}\left[2(r/a)^2 - 1\right] \qquad\qquad [4.5.4]$$

According to [4.5.4] in the centre of the cell ($r = 0$), the velocity is equal but opposite to v_{eo}. This provides a means to determine the latter, using sol particles of known electrophoretic mobility as "tracers" to make the flow visible. However, [4.5.4] is also suitable for visualizing the entire velocity profile in the cell. The distance where the liquid is fully at rest ($v_z(r) = 0$) is given by $r = a/\sqrt{2} = 0.707\ a$. If measured at this depth the velocity of a particle is just the electrophoretic velocity. Most, if not all, modern equipment provide for such a measurement. In practice this requires very precise positioning of the optical beam which, in turn, is usually achieved by working with thin-walled capillaries, narrow light beams and/or microscopes with restricted focal length. As at the depth of zero flow the velocity gradient dv_z/dr is large, mispositioning can lead to relatively large errors. A practical check is the identity of the electrophoretic mobility at the stationary level $r = 0.707\ a$ in a cross-section through a capillary: two of such stationary levels are encountered (asterisks in fig. 4.15b), which should be symmetrical with respect to the axis of the capillary. Differences between these two observations may have different causes, such as optical mispositioning, sedimentation or creaming of the sol particles, the latter two leading to different values of the ζ-potential of the upper and lower cell wall, and hence to an asymmetric electro-osmotic profile. In fact, spurious sedimentation or creaming can also lead to particle trajectories that are no longer fully in the direction of the field. For this reason, to obtain enhanced experimental control it is advisable (though more time consuming) to measure the complete profile, which should consist of two straight lines of the same slope, intersecting at the axis, if $v_z(r)$ is plotted as a function of $(r/a)^2$. In practice the check is usually done with tracer particles. Preferably, such a tracer particle should be sterically stable and have zero ζ-potential. For an example (latex covered by a dense layer of hydroxypropyl cellulose) see[1].

To distinguish between, and establish, the real electrophoretic and the electro-osmotic velocity, the following analysis may be helpful. Just near the cylinder wall (positions at the dashed lines in fig. 4.15) the mobility of the particle is $v_p = v_{ef} + v_{eo}$. At other positions the liquid velocity according to [4.5.4] has to be subtracted. Introducing X as the relative distance from the wall into the tube, relative to the diameter

$$X = \frac{1}{2}\left(1 - \frac{r}{a}\right)$$ [4.5.5]

the velocity of the particle may be written as

$$v_p(X) = v_{ef} + v_{eo} - 8v_{eo}\left(X - X^2\right)$$ [4.5.6]

[1] K. Furusawa, Q. Chen and N. Tobori, J. Colloid Interface Sci. 137 (1990) 456.

Similar things can be said for cells of other geometries. For rectangular cells the result is rather complicated[1]. The profile generally depends on the thickness $2h$, the width and the length of the cell. When the width is at least twenty times the thickness, the result is relatively simple:

$$v_z(y) = \left(v_{eo}/2\right)\left[3y^2/h^2 - 1\right] \tag{4.5.7}$$

where y indicates the vertical position ($y = 0$ in the interior of the cell and varies from $y = -h$ to $y = +h$). In cells where the profile is again parabolic, the stationary levels are at $y = \pm 0.577\, h$. In this case, the velocity of a particle is found to be [2]

$$v_p(X) = v_{ef} + v_{eo} - 6v_{eo}\left(X - X^2\right) \tag{4.5.8}$$

with X defined through

$$X = \frac{1}{2}\left(1 - \frac{y}{h}\right) \tag{4.5.9}$$

In the literature several examples of parabolic profiles can be found [2,3,4]. In fig. 4.16 we give a more recent example[5], in which the mobility of latex particles was studied in a flat cell with mica surfaces. The stationary levels were found at $X = 0.2113$ and 0.7887. That the profile in fig. 4.16a is parabolic is

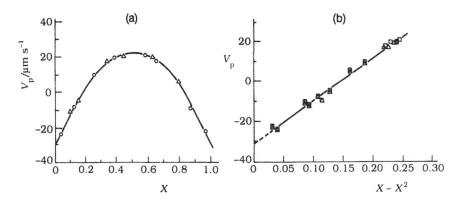

Figure 4.16. Velocity profile for negatively charged polystyrene latex particles in a rectangular cell with mica surfaces. □ or O and △ refer to different field directions. (a) $v(X)$; (b) $v(X - X^2)$. (Redrawn from Debacher and Ottewill, loc. cit.)

[1] H.A. Abramson, L.S. Moyer and M.H. Gorin, *Electrophoresis of Proteins*, Rheinhold (1942) or Hafner (New York) (1964).
[2] For more details about the computation, see Hunter, *Zeta Potential in Colloid Science*, Academic Press (1981), his sec. 4.13.
[3] G.E. van Gils, H.R. Kruyt, *Koll. Beihefte* **45** (1936) 60.
[4] J.Th.G. Overbeek, in *Colloid Science*, Vol. **I**, H.R. Kruyt, Ed., Elsevier (1952) p. 218, 225.
[5] N. Debacher, R.H. Ottewill, *Coll. Surf.* **65** (1992) 51.

illustrated by the linearity of $v_p(X - X^2)$, see fig. 4.16b, in agreement with
[4.5.8]. From slope and intercept v_{ef} and v_{eo} can be obtained. The theoretical
zero point value in fig. 4.16b is $\frac{1}{6} = 0.16667$; the experimental data come very
close to this value. A faster but less accurate way is to take measurements at two
positions r only, for instance near the wall and in the centre of the tube. In that
case there is no way to verify the correctness of the shape of the profile.

Measures to reduce the electro-osmosis include reduction of the surface charge
on the inner side of the capillary wall. Entirely uncharged materials are
virtually non-existent. Coating the surface may help, provided the adhered
molecules bind strongly and permanently; otherwise they may desorb and re-
adsorb on the particles, which may have a dramatic effect on their mobilities,
considering the low adsorption buffer capacity of the dilute sol, mentioned
before. Leaching of (oligo-) silicates from (particularly soft) glasses may have the
same effect. One way to verify the absence of such contaminants might be to
establish that measured mobilities are independent of the sol concentration.
Contaminants, already present in the sol, cannot be detected in this way.

The electrodes should be reversible and non-gassing and products of electrode
reactions should not contaminate the system, which can be achieved by
positioning them far from the place where measurements are done. For simple
geometries the electric field E acting on the particles is readily found, otherwise
it can be obtained from the (measured) current density and the conductivity of
the dispersion, using Ohm's law. In practice, the applied field is usually kept
below 10^4 V m^{-1} to avoid heating of the cell. Under those field strengths velocities and field strengths remain linear.

Over the years a variety of improvements have been suggested to facilitate the
measurements, to increase the optical contrast, to suppress external influences,
to extend measurements to more difficult systems, etc. A number of these
improvements have been implemented in commercial apparatus. Perhaps the
caveat is appropriate that sophisticated equipment can never compensate ill-
defined (surface) chemistry.

Problems on their own are posed by sols in media of very low polarity.
Because such systems contain very few ions, they have extremely low conductivity. In addition, because of the low dielectric permittivity the electrophoretic
mobility is much lower than in aqueous media. Therefore, the applied electric
field should be higher by about a factor of 10 as compared to aqueous systems,
say $\approx 100 - 300$ V cm^{-1}. The combination of high fields and low conductivity is
particularly troublesome, and provisions have to be taken to prepare electrodes
that do not polarize under such conditions and cells that do not allow the field
to leak away through or outside the walls. A number of solutions have been

proposed, starting with pioneering work by van der Minne and Hermanie[1]. Kornbrekke et al.[2] have reviewed this matter. When the measurements have been finished, the theoretical problem of *dielectrophoresis* has to be solved, that is the spontaneous motion of uncharged particles under the influence of a strong heterogeneous applied field if the difference between the dielectric permittivities of particle and medium is large[3]. Only after this phenomenon has been accounted for, is the remaining task simple because in view of the low κa value the Hückel-Onsager equation [4.3.5] will usually apply.

In summary, micro-electrophoresis is a powerful and basic colloidal technique but in order to obtain reliable results a number of precautions have to be taken[4].

(ii) *Moving boundary electrophoresis.* The moving boundary method to measure electrophoretic mobilities differs from the previous one in that no attempt is made to observe individual particles. Rather the collective motion of all the colloids in a sol is studied. If the sol adjoins an equilibrium colloid-free solution the boundary between sol and solution will move under the influence of a field applied across this border, hence the name. The rate at which this boundary is displaced gives the electrophoretic mobility.

Moving boundary electrophoresis is easy to carry out if the boundaries are (and remain) sharp and visible. For this situation the technique dates back to Burton[5] and the principle is sketched in fig. 4.17. The initial situation, with the two levels even, is usually obtained by first filling the U-tube partly with the equilibrium liquid, and then introducing the sol from beneath. This can be achieved through an external source, e.g. through an additional tube (not shown) under the influence of a hydrostatic head. (Most sols have a higher density than the equilibrium liquid.) For the colloid-free liquid preferably the dialysate or pressure (or gravity-) filtrate should be used so that moving particles always experience the same composition of the liquid carrier. The electrodes should be non-gassing and electrode reaction products should not reach the sol. To achieve this, additional measures can be taken, such as placing electrodes into a special side-compartment that is connected to the sol but does not allow convection of those products. To prevent convection the cell can be closed, at least on one side. When the potential is applied, the boundary starts moving in a direction dictated by the direction of the applied field and the (electrokinetic) sign of the

---

[1] J.L. van der Minne, P.H. Hermanie, *J. Colloid Sci.* **7** (1952) 600; **8** (1953) 38.

[2] R.H. Kornbrekke, I.D. Morrison and T.Oja, *Langmuir* **8** (1992) 1211.

[3] H. Pohl, *Dielectrophoresis*, Cambridge Univ. Press (1978).

[4] For further reading on the micro-electrophoresis technique, see the review by A.M. James, cited in sec. 4.11c.

[5] E.F. Burton, *Phil. Mag. (1)* **11** (1906) 425; a more primitive mode was used earlier by H. Picton and S.E. Linder, *J. Chem. Soc.* **61** (1892) 148.

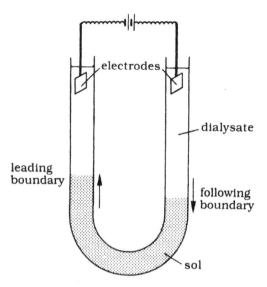

Figure 4.17. Basic principle of Burton-type moving boundary electrophoresis.

colloids. When both tubes are tightly capped, electro-osmosis, induced by the charge on the tube's inner surface, has to be accounted for. The rising and descending boundaries are also known as *leading* and *following* boundary, respectively.

The sharpness of the boundaries depends on a number of factors. Conductivities of sols are often lower than those of the dialysates. Hence, the applied field strength in the sol is higher than in the dialysate, so that colloids, moving fortuitously ahead at the rising boundary, are soon overtaken by the others. This is a self-regulating mechanism, contributing to sharpen this boundary. By the same token, the descending boundary tends to become more diffuse. For systems where the sol conductivity is lower than that of the dialysate, the situation is the other way around. Besides this, gravity and heterogeneity also affect the definiteness of the borders. In practice, all these situations have been observed.

Given the non-homogeneous distribution of the field over the tube, there is also some uncertainty about the actual field at the boundaries, and hence the question has to be posed which of the two moving boundaries, if any, and if observable, gives the right mobility. Dilution of the sol to homogenize the conductivities as much as possible has the drawback that the boundary becomes less visible. Tison[1] has experimentally studied the anomalous behaviour of the two boundaries by measuring the conductivity at various positions in the tube for colloidal silica. He concluded that during the electrophoresis (in his experiments at constant current) the electrolyte composition, and hence the conductivity and local field, changed with time in a rather complicated way. These changes may account for the diffuseness of the boundaries, or at least contribute to it. At the same time they thwart the establishment of the driving force. The extent to which local composition changes occur will be determined by the way in which the field lines run around the particles and on their sizes, shapes, volume fractions, conductivities and on the conductivity of the interparticle solution. Mathematically, these changes are determined by κa, φ, K^s and Du. Marmur investigated this issue theoretically[2]; he advised doing the experiments at constant current. Only when the boundaries are sharp can well-defined velocities be estimated, but the establishment of the mobility is not without ambiguity.

The conclusion is that the moving boundary technique is simple but not rigorous. For rapid semi-quantitative purposes it may be useful.

Many colloids are not visible because they are neither opaque nor coloured. In particular, this applies to hydrophilic colloids like polyelectrolytes and proteins. For this category, moving boundary electrophoresis can still be carried out by using a Schlieren optical system, that is a method based on the difference in refractive index between sol and dialysate. Such a system is also applied in ultracentrifuges. In the 1930s this method was perfected by Tiselius[3] who also devised procedures to obtain sharp initial boundaries and to thermostat the contents. Working with rectangular cells with displaceable segments was one of the improvements. Alternatively, measurements can be carried out at 4°C where water has its maximum density, so that temperature fluctuations and ensuing convection have only minor effect. One of Tiselius' feats was the separation of human serum proteins into albumin and the four globulin fractions, α_1, α_2, β and γ. By this elegant technique Tiselius earned a Nobel prize.

Tiselius' work has given rise to a steady development of newer techniques with increasing resolution. In modern practice the method is almost exclusively used for separation and identification purposes. Quantitative determination of mobilities, let alone ζ-potentials, is virtually impossible because the complications discussed above are compounded by those caused by interaction between

---

[1] R.P. Tison, *J. Colloid Interface Sci.* **60** (1977) 519.

[2] A. Marmur, *J. Colloid Interface Sci.* **85** (1982) 556.

[3] A. Tiselius, Ph.D. Thesis, Univ. Uppsala (S) (1930); *Svensk. Kem. Tid.* **50** (1938) 58; *Kolloid Z.* **85** (1938) 129.

the molecules and the additional problem of transforming mobilities into ζ-potentials for partly draining coiled molecules.

For further information see the reviews by Longsworth[1].

4.5b. *Streaming potentials, electro-osmosis and other measurements with plugs*

Electro-osmosis is the counterpart of electrophoresis in that now the materials to be studied are stationary, whereas the liquid moves at a given velocity, driven by an applied field. In streaming potential measurements an applied pressure difference is the driving force; in that case a potential difference is measured. In practice, two ways are open: working with *capillaries* or with *plugs*.

(i) In the case of capillaries, it is the ζ-potential and the surface conductivity of the inner wall that are the purposes of the measurement. Perhaps this is the most ideal model because the geometry, and hence the flow pattern, is well-defined. Streaming potentials, electro-osmosis, and surface conduction can be readily measured and interpreted. The method is limited to materials that can be forged into capillaries, such as glasses and quartz. Somewhat more general are measurements in flat cells of which the plates are made of the material to be investigated.

(ii) Porous plugs of the materials to be studied provide a more general alternative because the material that can be studied can be selected more widely. It is sufficient that it can be made into a plug which can support mechanical stress. In order to stick together there is no need for the particles to be in the coagulated state because the plug can be kept in place by permeable membranes. In some implementations porous electrodes are used to that end. Some materials are easier to pack than others, some are so difficult to manage that centrifugation onto one of the porous membranes is required, after which the second membrane is put in position. Upon permeation of the liquid, the plug should not leach particles. However, if a plug can be made, there is no further restriction on the material, be it colloidally stable or unstable, transparent or opaque, having identical or very different particles, both with respect to size or shape. So, in these respects, electrokinetic studies with plugs have advantages over electrophoresis. Drawbacks are the interpretational problems: the usually irregular stacking of the particles prohibits the development of simple flow patterns, there is double layer overlap and the particles touch each other.

Generally, for plugs where the local radii of curvature are $>> \kappa^{-1}$, analysis and interpretation become simpler. (Smoluchowski's theorem, sec. 4.3b). Hetero-

---

[1] L.G. Longsworth, *Moving Boundary Electrophoresis*, in *Electrophoresis, Theory, Methods and Applications*, M. Bier, Ed., Academic Press (1959) p. 91 *Theory* and p. 137 *Practice*.

geneous plugs containing a wide distribution of pore sizes offer great interpretational difficulties because the hydrodynamic volume flow $(\sim a^4)$, the
conduction current $(\sim a^2)$ and the electrokinetic quantities $(\sim a^2)$ depend in a
different way on a.

We shall emphasize plug experiments but sometimes refer to capillaries,
which will be considered model systems.

With plugs and capillaries a number of electrokinetic (streaming potentials,
electro-osmosis, streaming currents) and related phenomena (conductivity,
permeability) can be measured, all of these requiring a different mode of
operation.

– *Streaming potentials* (E_{str}), see [4.3.41 and 42], require an applied pressure
head, a pair of electrodes to measure E_{str}, and another (or the same, see below)
pair to apply a potential difference, in order to measure the total conductivity
(bulk plus surface conductivity).,

– *Streaming currents* (I_{str}), see [4.3.38]), do not require a conductivity
measurement but suffer from the drawback of electrode polarization; this
polarization leads to a decrease of I with time, so that it is often impossible to
obtain a realistic value for the streaming current. We shall not extensively
discuss this type of measurement.

– *Electro-osmotic flux*, Q_{eo}, see [4.3.24-27], requires the application of a field or
forcing a controlled current I through the plug. For $Q_{eo,I}$ the conductivity has to
be known, but for $Q_{eo,E}$ this is not necessary. Similar things can be said about
the *electro-osmotic counterpressure*, Δp_{eo}, see [4.3.29].

– *D.c. conductivity measurements* require current-carrying electrodes. If these
electrodes polarize, K can still be measured if I is kept constant. However, it is
mostly more convenient to measure the a.c. conductivity, see sec. 4.5e. For
capillaries the total conductivity equals the sum of the contributions of the
bulk, K^L and that of the surface, $2K^\sigma/a$ if a is the capillary radius (see the
denominator of [4.3.41]):

$$K = K^L + 2K^\sigma/a = K^L(1 + 2Du) \tag{4.5.10}$$

We return to this in subsec. 4.5c.

– *Liquid permeabilities*, i.e. the flux of fluid passing through the plug under the
influence of an applied pressure gradient, have already been discussed in sec.
I.6.4f for the case of uncharged systems. It is recalled that this flow is governed
by d'Arcy's law [I.6.4.35 or 36] and that, under appropriate conditions, from the
liquid flux the specific surface area A_g of the plug may be estimated, for
instance by using the *Kozeny-Carman* equation, [I.6.4.41]. For charged plugs
these conditions are best met by working at swamping electrolyte concentration,

or by short-circuiting the electrodes at the two sides of the plug. When electrokinetic effects are not suppressed, apparent viscosities are measured exceeding the bulk values of the liquid. We shall return to this type of secondary electrokinetic phenomenon in sec. 4.9.

Besides these phenomena, plug experiments can also be carried out to measure *salt-sieving*, the retention of electrolytes caused by the negative adsorption of co-ions from the (overlapping) double layers, also known as the Donnan exclusion. The phenomenon was already met on p. 1 of Volume I and further analysed in sec. 3.5b.

All told, a variety of interesting experiments can be carried out with a plug, each type requiring its specific set-up. To implement this in practice a number of cells have been described in the literature. Some of these have been mentioned by Hunter[1]. Figure 4.18 shows another versatile example. It is an illustration of a four-electrode system: the porous electrodes 2,2' also act as the membranes to contain the plug; electrodes 1,1' can be used to apply a potential gradient. Electrolyte solution is in the reservoirs G. It can be forced through the plug (in the sketch from left to right) by using nitrogen under pressure (A), a valve B and regulator C. D is a pressure-buffering vessel with outlets, including one for a manometer (E) and the others to be connected to independent cells. Aircock F allows rapid depressurization, if needed. Three-way cocks H allow reversal of the flow direction. Such a device is a must since it is always necessary to measure $E_{str}(\Delta p)$ in two directions to verify absence of polarization, either of the plug or of the electrodes. The flow-measuring device, among other things needed for electro-osmosis, is a calibrated tube J, with a

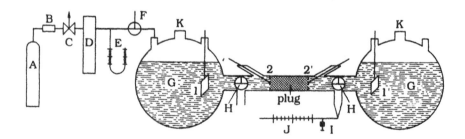

Figure 4.18. Experimental set-up for a variety of electrokinetic and other experiments with a porous plug. (Redrawn from A.G. van der Put and B.H. Bijsterbosch, *Acta Polymerica* **32** (1981) 311, with minor modifications by Th.J. van der Hoven (Ph.D. thesis. Agricultural University Wageningen, The Netherlands, 1984).)

---

[1] R.J. Hunter, *Zeta potential in Colloid Science*, Academic Press (1981) his sec. 4.2. See also H.-J. Jacobasch, G. Bauböck and J. Schurz, *Monatsh. Chem.* **117** (1986) 1133; H.-J. Jacobasch, J. Schurz, *Österr. Chem. Z.* **88** (1987) 104.

piston I to adjust the initial meniscus position in this tube. The caps K allow for changing the contents of the reservoir vessels.

A critical test of this, and other, apparatus involves at least two checks: one on the absence of electrode polarization and the other on reversibility upon reversal of the applied pressure gradient (in streaming potential measurements) or electric field (in electro-osmosis). Electrode polarization leads to drifts, whereas electrode reaction products, including gas bubbles, may contaminate or destabilize the system. For this reason, it is not recommended to use electrodes 2,2' as charge-carrying and avoid streaming current measurements. Such polarization may also be detected by the impossibility of obtaining exact sign reversal (asymmetries) or spurious concentration dependencies. Van der Linde and Bijsterbosch[1] demonstrated that the maximum in the ζ-potential as a function of the electrolyte concentration that is sometimes observed, is in some cases an experimental artefact caused by electrode polarization. (In other cases the phenomenon persists even if polarization is absent; then it has to be attributed to unjustified neglect of surface conduction in the conversion leading to ζ-potentials. We return to this interpretational aspect in sec. 4.6e.)

In figure 4.19 an example is given of a result meeting all the imposed requirements: $E_{str}(\Delta p)$ is linear, passes through the origin and reverses perfectly

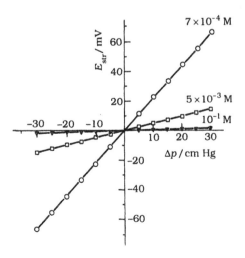

Figure 4.19. Streaming potentials obtained with the cell of fig. 4.18. Negatively charged polystyrene latex. Concentration of electrolyte (KCl) indicated.

[1] A.J. van der Linde, B.H. Bijsterbosch, *Colloids Surf.* **41** (1989) 345.

upon reversal of the sign of Δp, in the figure indicated as negative values. In this case the sensing electrodes were made of platinum and aged for over a day.

It may finally be noted that, because in plugs the area/liquid volume ratio is not small, the danger of contamination is much less than in micro-electrophoresis.

Having executed the appropriate measurement, an interpretational step is needed to obtain ζ and/or K^σ or, for that matter, Du. As with electrophoresis, this conversion is straightforward only under idealized conditions, in the present case for capillaries. Then one of the equations of sec. 4.3 can be used. With capillaries it is also simple to obtain K^σ from [4.5.10] by first measuring K in swamping electrolyte ($K \gg 2K^\sigma/a$). This gives the cell constant, which may then be used to obtain K and K^L at lower c_{salt}, using the bulk relationship between K^L and c_{salt}.

4.5c. Surface conductivities

Surface conductivities are excess quantities. They can only be established under conditions where the bulk conductivity can be subtracted from the total conductivity in an unambiguous way. In practice this means in systems with well-defined geometry, i.e. for cylindrical pores. The bulk conductance of a pore of radius a is $\pi a^2 K^L$, the surface conductance $2\pi a K^\sigma$, so that the latter can be found by subtraction of the bulk term from the measured total conductance. Equation [4.5.10] anticipated this. The procedure is not very accurate. An improvement is consideration of suitable electrokinetic phenomena for a range of capillaries of the same material, but having different radii. The premise is that the electrokinetic potential is determined by the nature of the material, its surface charge, and the nature and concentration of the electrolyte, but not by the radius of curvature. Capillaries of the same material under the same conditions but having different radii must therefore have the same ζ-potential. We shall call this requirement *electrokinetic consistency*. Anticipating further discussion (sec. 4.6e) electrokinetic consistency should also apply to different types of electrokinetic measurements. In the 1930s, failure to achieve electro-kinetic consistency led to the insight that surface conduction has to be considered. Compare for instance [4.3.24 and 25] for the electro-osmotic flow rate: neglect of surface conduction renders the equation independent of the cross-section, where several measurements indicated that Q_{eo} depends on a.

Making a virtue of necessity, the appropriate electrokinetic equations can be written explicitly in terms of a and then subjected to experimental verification. For instance, [4.3.24] can be transformed to give

$$-\frac{\varepsilon_0 \varepsilon \zeta I}{\eta \mathcal{Q}_{eo}} = K^L + \frac{2K^\sigma}{a} \qquad\qquad [4.5.11]$$

so that a plot of the l.h.s. as a function of the reciprocal radius should be linear with intercept K^L and slope $2K^\sigma$. In this way Du can also be measured. Similar procedures can be developed for other appropriate electrokinetic phenomena.

Generally speaking, the establishment of K^σ has not yet obtained the attention it deserves. We come back to some results and implications in sec. 4.6e.

4.5d. Electroacoustics

The principles of UVP, or its special variant CVP, in which we are now interested, and ESA have been introduced in sec. 4.3e. Under discussion is now how the measurements are carried out in practice. As a consequence of the relatively complicated experiments, involving wave generation (acoustic or electric), transducers, response analysis, etc., measurements are mostly done with commercial apparatus, including a lot of electronics and software. Some of these apparatus allow for additional options, such as measuring the particle size distribution, conductivity or titrating the sample to change the surface charge.

The principle of the technique is sketched in fig. 4.20[1] . In the CVP mode, an electric field E_2 is applied; through the piezo-electric transducer this feed signal is converted into a sound wave, which migrates through the (non-conducting) spacer towards the sol or suspension. The spacer is needed to separate the circuits E_2 and E_1, to avoid induction. Inside the sample, the particles start to oscillate at the frequency of the sound wave. Amplitude and phase lag are determined by the density difference between particles and solution and by the "braking power" of the double layer, hence the occurrence of $\Delta \rho / \rho$ in equations [4.3.43-44] and the possibility of saying something about ζ and K^σ. These oscillations are measured as an electrical field E_1. So, although incident and emerging fields are both electrical, the physical process involves sound waves (although the apparatus is not "noisy"). In the ESA mode, the process is the other way around: now an electric field E_1 is applied to set the particles into an oscillatory motion; the sum of all these motions produces a sound wave that, via the spacer and transducer, is measured as an a.c. field E_2. The two modes require different electronics. For instance, in the CVP mode the impedance of the external circuit should be high with respect to that of the sample, whereas this is the other way around for the ESA mode.

Depending on the design of the cell, a variety of measurements can be carried

---

[1] B.J. Marlow, D. Fairhurst and H.P. Pendse, *Langmuir* **4** (1988) 611; See also A.J. Babchin, R.S. Chow and R.P. Sawatzky, *Adv. Colloid Interface Sci.* **30** (1989), 111 and, for a more recent development, R.W. O'Brien, D.W. Cannon and W.N. Rowlands, *J. Colloid Interface Sci.* **173** (1995) 406.

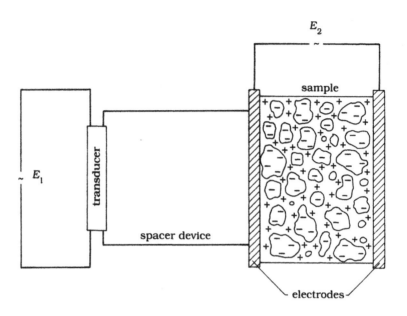

Figure 4.20. Principle of electroacoustics. Discussion in the text.

out, including CVP or ESA as functions of ω, pH, addition of the same colloid (to check the effect of the volume fraction φ) or of another (to study heterocoagulation), or of polymeric flocculants. Isoelectric points are readily established. There is little restriction on the properties of the sample, except that a certain minimum volume is required (about 200 ml). The volume fraction up to which the signal remains proportional to φ depends on the nature of the system, in particular on the (hydrodynamic and electrical) particle interaction. For instance, Marlow et al.[1] found for an Al_2O_3-clad TiO_2 sol proportionality to φ up to about $\varphi \approx 0.3$ for the conductivity and till $\varphi \approx 0.15$ for the mobility. Klingbiel et al.[2] found linearity of the ESA signal with φ up to 0.06 or 0.07 for poly(methyl-methacrylate) latex and a "Ludox" silica, respectively. When linearity is established, extrapolation to $\varphi \to 0$ will yield the required interaction-independent limiting value. In very concentrated samples, measurements can still be carried out, provided the stirrer in the system can keep the system homogeneous, but then the interpretation becomes cumbersome.

From equations [4.3.43 and 44] it follows that ESA measured in a given cell, and the to CVP of the same sample but measured in the same or a different cell, should be related according to

[1] B.J. Marlow, D. Fairhurst and H.P. Pendse, *Langmuir* **4** (1988) 611.
[2] R.W. Klingbiel, H. Coll, R.O. James and J. Texter, *Colloids Surf.* **68** (1992) 103.

$$CVP(\omega)\,\hat{K}(\omega) = \text{const. } ESA(\omega) \qquad\qquad [4.5.12]$$

The correctness of this identity has been verified by O'Brien et al. for a number of colloidal systems[1] (and so was the corresponding relation between UVP and ESA for electrolyte solutions). This is support for the consistency which, in view of the Onsager relations, had to be anticipated. It means that a physically real phenomenon has been properly measured.

The second step, the interpretation in terms of ζ-potentials and conductivities, requires theory. For non-dilute dispersions this implies consideration of hydrodynamic and electrostatic particle interaction. James et al.[2], working with poly(styrene) and poly(methyl methacrylate) latices, alumina and silica sols confirmed that u obtained from ESA agreed with the (static) values, obtained micro-electrophoretically, if the theory by O'Brien (see [4.3.45-48]) was applied in the analysis. Marlow et al.[3] already noted the same for dilute rutile dispersions; their mobility (or ζ) curves as a function of pH agreed with those in fig. 3.63.

In view of these, and several other, experimental observations, it may be concluded that ESA and CVP are full-grown electrokinetic techniques.

4.5e. Dielectric dispersion and a.c. conductivity

The dielectric dispersion $\hat{\varepsilon}(\omega)$ and the a.c. conductivity $\hat{K}(\omega)$ are counterparts of one another. Basically the one can be converted into the other if data are available over the entire frequency range, using the Kramers-Kronig relations [I.4.4.31-32]. However, this requires extremely precise data to carry out the required integrations; in practice measurements of $\hat{\varepsilon}(\omega)$ and $\hat{K}(\omega)$ are carried out at different parts of the ω-spectrum. See the discussion in connection with [3.7.5 and 6].

Because of the relaxation of double layers, the resistive and capacitive components of the impedance $\hat{Z}(\omega)$ vary with frequency; the former leads to $\hat{K}(\omega)$, the latter yields $\hat{\varepsilon}(\omega)$. The quantity $\hat{Z}(\omega)$ or, for that matter, the admittance $\hat{Y}(\omega)$ is usually measured in an impedance spectrometer. Quite generally, the impedance may be written as

$$\hat{Z}(\omega) = \frac{1}{\hat{Y}(\omega)} = \frac{1}{K(\omega) - i\omega\,\varepsilon_0\,\varepsilon(\omega)} \qquad\qquad [4.5.13]$$

where the complex denominator consists of a real part, the conductivity, and an imaginary or capacitative part $(C = \varepsilon_0\varepsilon$ if C in F m$^{-1})$. Alternatively, the denominator may be formulated as a complex dielectric permittivity, consisting

---

[1] R.W. O'Brien, P. Garside and R.J. Hunter, *Langmuir* **10** (1994) 931.
[2] R.O. James, J. Texter and P.J. Scales, *Langmuir* **7** (1991) 1993.
[3] B.J. Marlow, D. Fairhurst and H.P. Pendse, loc. cit.

of a real dielectric part and an imaginary conductive contribution. These two modes of formulation are equivalent. The ratio between the real component $K(\omega)$ of $\hat{Y}(\omega)$ and the imaginary component $i\omega C(\omega)$ is often represented in terms of the phase angle, δ. For pure electrolytes K and C are frequency-independent. For low volume fractions φ, changes in K and C are caused by the particles proportional to φ[1] but for high φ this linearity will break down. Similar trends were observed with electroacoustics (sec. 4.5d).

We shall now discuss some practical aspects of such measurements.

One of the most characteristic problems in the measurement of low frequency dielectric dispersion is that the ratio between the resistive and capacitive component of $\hat{Z}(\omega)$ is very large. In other words, the phase angle is high. Therefore, accurate measurements of the latter component require an extremely powerful phase-sensitive detector. A second, nagging, problem, is electrode polarization, which may become particularly troublesome at low ω. There are two options to avoid, or suppress, electrode polarization:

(i) construction of a cell with variable electrode distance, so that the bulk impedance can be separated from that of the electrodes

(ii) working with four-electrode systems, in which the sensor electrodes carry no current, so that they are not polarized. We already met this strategy in connection with fig. 4.18.

Discussions of measuring cells and electronics can be found in the literature[2,3]. An example of a measuring cell is given in figure 4.21. The cell

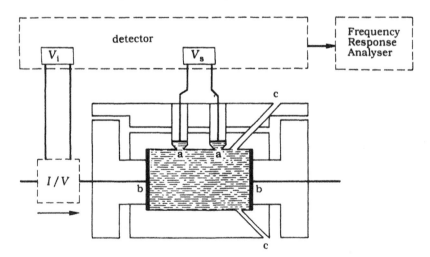

Figure 4.21. Example of a cell to measure $\varepsilon(\omega)$ and $K(\omega)$ of sols. Explanation in the text. (Redrawn from Kijlstra loc. cit.).

[1] E.H.B. de Lacey, L.R. White, *J. Chem. Soc. Faraday Trans. II* **77** (1981) 2007.
[2] D.F. Myers, D.A. Saville, *J. Colloid Interface Sci.* **131** (1989) 448, 461.
[3] J. Kijlstra, R.A.J. Wegh and H.P. van Leeuwen, *J. Electroanal. Chem.* **366** (1994) 37.

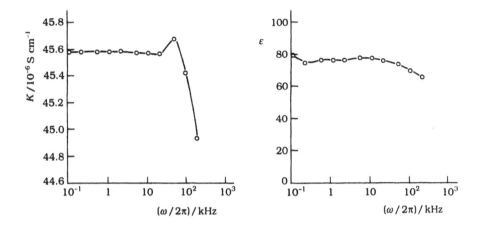

Figure 4.22. Permittivities (left) and conductivities (right) of a 10^{-4} M HCl solution. (Redrawn from D.F. Myers, D.A. Saville, loc. cit.).

contains two large current-carrying electrodes b, usually made of platinum. An a.c. voltage is supplied from a generator and the current j through the cell is transformed into a voltage V_j by a current-to-voltage converter. Both V_j and the voltage drop between the sensor electrodes V_s are fed into a detecting unit. This unit generates a potential difference $V_s - V_j$ suitable for analysis by a frequency response analyzer which yields the real and imaginary components of the admittance. These can then be interpreted according to [4.5.13].

It is convenient to supply the cell with openings for filling via a pump. Thus, the cell can be alternately filled with suspension and with blank electrolyte solution without changing the set-up otherwise. Precise thermostatting is necessary.

Accurate measurement of the dielectric spectrum of a colloidal dispersion requires careful calibration. Obviously, the most pertinent tests are done with the background electrolyte solution. Figure 4.22 gives an illustration. It is seen that K is constant up to over 10^4 Hz. As stated, the determination of ε is less accurate, but the results are acceptable because the most unfavourable case of lowest ε is shown. For colloidal dispersions ε can attain very high values (> 10^3), so that for these systems the measurements are relatively more reliable. By way of illustration, fig. 4.23 gives a result for the dielectric and conductivity increment, measured with the cell of fig. 4.21. There is some scatter in $\Delta\varepsilon$ in the range of low ω; this is the most cumbersome range to measure this quantity because δ increases with decreasing frequency. The accuracy drops further when

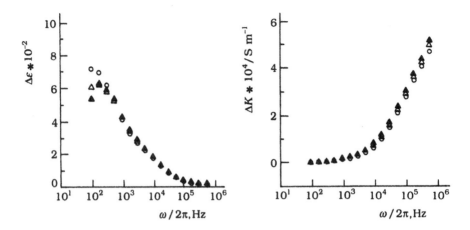

Figure 4.23. Dielectric (left) and conductivity (right) dispersion of a haematite sol in 1.3×10^{-3} M KCl, measured in the cell of fig. 4.20. Volume fractions: O, $\varphi = 4\%$,; Δ, 3% and $\blacktriangle$, 2%; the last two are converted into values for 4% assuming linearity. (Redrawn from Kijlstra, et al.)

the electrolyte concentration is increased. Master curves, independent of volume fraction, can be obtained by converting the dielectric and conductivity increments to those at higher φ, assuming linearity.

Interpretation of such curves will follow in sec. 4.8.

4.6 Generalized theory of electrokinetic phenomena. Application to electrophoresis and experimental verification

In the elementary theory of sec. 4.3 a number of simplifying assumptions were made. One of the most restrictive of these was the approximation that the structure of the double layer was not affected by the applied field.

Other simplifications included the disregarding of surface conduction (i.e. only the case $Du = 0$ was considered) and the limitation to very simple geometries (spheres, capillaries, etc.) without double layer overlap. Inclusion of all these features is physically and mathematically extremely difficult and as yet only rigorously solved under limiting conditions. In order to identify the various problems we shall, in the present section, retain the restrictions to simple geometries (emphasizing electrophoresis of spheres) and absence of double layer overlap but do automatically consider double layer polarization (i.e. the relaxation retardation, force F_4 in sec. 4.3a(i)) and always take surface conduction into account (i.e. the contributions of Du, in [4.3.53 and 62]). The treatment of a.c. electrokinetics will be deferred to sec. 4.8; this implies that for the moment only stationary states will be considered.

Regarding the physics of this issue, the task is to formulate the appropriate equations, describing convection, diffusion and conduction. The equations of motion are coupled to electric equations, like Poisson's law. The mathematical task is to solve the ensuing set of differential vector equations, some of which are non-linear, with the appropriate boundary conditions. General analytical solutions do not exist, but there are numerical solutions and good approximate equations for a number of limiting situations. Although the full mathematical analysis is beyond the confines of the present chapter, we shall present their main elements because these are needed to understand the physics of the phenomena.

When these differential equations are properly solved, surface conduction is automatically taken into account, because then all the ionic currents and osmotic flows have been considered. However, most elaborations start from the oversimplified model of a discrete slip plane without surface conduction within the stagnant layer. In such models $K^{\sigma i} = 0$. We shall call models with $K^{\sigma i} = 0$ *rigid particle* theories. The rigid particles are therefore dielectric spheres, characterized by an outer ("surface") potential ζ. The most *a priori* approach is to introduce non-zero $K^{\sigma i}$'s from the very onset, but regarding literature elaborations, we shall follow the historical development and first discuss models with $K^{\sigma i} = 0$, thereafter (sec. 4.6e) consider experimental evidence for non-zero $K^{\sigma i}$ and eventually introduce this phenomenon in the theory.

The foundations of this approach to electrokinetics date back to Overbeek[1] and Booth[2]. Important improvements include work by Dukhin[3], Wiersema, Loeb and Overbeek,[4], O'Brien and White[5], Fixman[6] and others. Excellent background reading is found in ref. [7]. See further the references in sec. 4.11. Although the mathematical elaborations of the recent theories[3,5], which have been developed independently of each other, may differ, their basic premises are identical, viz. those of our sections 4.6a and b.

[1] J.Th.G. Overbeek, *Kolloid Beih.* **54** (1943) 287.

[2] F. Booth, *Nature* **161** (1948) 83; *Proc. Roy. Soc. (London)* **A203** (1950) 514.

[3] S.S. Dukhin, in *Research in Surface Forces*, B.V. Deryaguin, Ed. Consult. Bureau (1966) 54; ibid, Vol. **3** (1971).

[4] P.H. Wiersema, A.L. Loeb and J.Th.G. Overbeek, *J. Colloid Sci.* **22** (1966) 78.

[5] R.W. O'Brien, L.R. White, *J. Chem. Soc. Faraday Trans. (II)* **74** (1978) 1607.

[6] M. Fixman, *J. Chem. Phys.* **72** (1980) 5177; **78** (1983) 1483.

[7] *The Electrokinetic Effects*, (prepared by R.W. O'Brien) chapter 13, in R.J. Hunter, *Foundations of Colloid Science*, Vol. **2**, their 4.12. In sec. 4.6 further referred to as O'Brien in Hunter, loc. cit.

4.6a. Basic electrokinetic equations

Describing the various flows in electrokinetics requires vector field formalism of which the basics have been explained in I.app. 7. Here we shall use the shorthand notation $\nabla\psi$, $\nabla \cdot \mathbf{a}$ and $\nabla^2\psi$, standing for grad ψ, div $\mathbf{a}$ and the Laplacian of ψ, respectively. Background hydrodynamics can be found in chapter I.5 and the electrostatics in I.6. Some basic aspects of double layers in an external field have already been introduced in sec. 3.13.

Below, the assumption will be made that no chemical reactions take place, i.e. no ions are created or annihilated, that turbulence does not occur (i.e. the Reynolds number is below 500 (see table I.6.2 in sec. I.6.4b) and that the liquid is incompressible (i.e. $\nabla \cdot \mathbf{v} = 0$). Besides these, not very restrictive, conditions the equations below are general, although sometimes further assumptions are made.

We begin with the *conservation of mass* law, stating that all ions flowing into an infinitesimal volume minus those that go out contribute to the charge accumulation in that volume. For a derivation, see sec. I.6.1a. For each ionic species i at location $\mathbf{r}$

$$\frac{\partial c_i(\mathbf{r},t)}{\partial t} = -\frac{\nabla \cdot \mathbf{j}_i(\mathbf{r},t)}{z_i F} = -\nabla \cdot \mathbf{J}_i(\mathbf{r},t) \qquad \text{(mole m}^{-3}\text{ s}^{-1}\text{)} \qquad [4.6.1]$$

where $\mathbf{j}_i$ is the contribution from type i ions to the current density (C m^{-2}s^{-1}) and $\mathbf{J}_i$ the flux of ions of type i. The sign of the ion is included in z_i. Under stationary conditions the dependency on t disappears.

Generally, ions can be transported by three mechanisms: diffusion, conduction and convection. To $\mathbf{j}_i/z_i F$ these transport components contribute by $-D_i\nabla c_i$, $-D_i z_i F c_i \nabla\psi / RT$ and $c_i \mathbf{v}$, respectively, where according to the *Nernst-Einstein relationship*, see [4.3.55], the ion mobility u_i is written as $|z_i| FD_i / RT$. In Volume I we derived the *Nernst-Planck equation* for the corresponding fluxes, see [I.6.7.1 and 2] and [3.13.12]. In the present context we write the Nernst-Planck equation as

$$\frac{\mathbf{j}_i(\mathbf{r},t)}{z_i F} = -D_i(\mathbf{r})\left[\nabla c_i(\mathbf{r},t) + \frac{z_i F c_i(\mathbf{r},t)}{RT}\nabla\psi(\mathbf{r},t)\right] + c_i(\mathbf{r},t)\mathbf{v}(\mathbf{r},t) \qquad [4.6.2]$$

In the usual electrokinetic double layer picture there is a slip plane. Beyond it the fluid has bulk properties, i.e. we can replace $D_i(\mathbf{r})$ by D_i, the bulk diffusion constant

$$\frac{\mathbf{j}_i(\mathbf{r})}{z_i F} = -D_i\left[\nabla c_i(\mathbf{r},t) + \frac{z_i F c_i(\mathbf{r},t)}{RT}\nabla\psi(\mathbf{r},t)\right] + c_i(\mathbf{r},t)\mathbf{v}(\mathbf{r},t) \qquad [4.6.2a]$$

In the part within the slip plane no convection takes place, i.e. $v(r,t) = 0$ if counted with respect to the particle (or $v(r) = v_{ef}$, etc., if counted with respect to the solution.) This layer is usually thin with respect to the particle radius, meaning that only the tangential components $\nabla_t c_i$ and $\nabla_t \psi$ remain. For a flat surface, where y denotes the direction parallel to the surface, these gradients could be written as dc_i/dy and $d\psi/dy$, respectively; for a spherical surface of radius a these would become $dc_i/ad\theta$ and $d\psi/ad\theta$, respectively. Furthermore, making the common assumption that all the Stern charge is concentrated in the inner Helmholtz plane, it makes sense to transform [4.6.2] into its two-dimensional analogue, i.e. j_i becomes j_i^s in C m^{-1} s^{-1} and c_i is replaced by Γ_i^l, the surface concentration of i in mole m^{-2}. We obtain

$$\frac{j_i^s}{z_i F} = - D_i^s \left[\nabla_t \Gamma_i^l + \frac{z_i F \Gamma_i^l}{RT} \nabla_t \psi^l \right] \qquad [4.6.2b]$$

$$j_i^s = - D_i^s \left[\nabla_t \sigma_i^l + \frac{z_i F \sigma_i^l}{RT} \nabla_t \psi^l \right] \qquad [4.6.2c]$$

where σ_i^l is the Stern charge contributed by ion species i, as before. In [4.6.2b and c] j_i^s, Γ_i^l, σ_i^l and ψ^l generally depend on the position and on t but these dependencies have not been made explicit to facilitate the readability of the equations. Equation [4.6.2c] can elegantly be written as

$$j_i^s = - D_i^s \nabla_t \sigma_i^l - K_i^{ls} \nabla_t \psi^l \qquad [4.6.2d]$$

where

$$K_i^{ls} = D_i^s z_i F \sigma_i^l / RT = u_i^s \sigma_i^l \qquad [4.6.3]$$

is the contribution of ions i to the *surface conductance* in the Stern layer and u_i^s is the tangential mobility of ions i in that layer. Often there is only one ionic species in that layer, so that $K_i^{ls} = K^{ls}$. In that case K^{ls} is identical to $K_t^{\sigma l}$, as in [4.3.57]. Stern layer conduction is often ignored, leading to an overestimation of the electrophoretic mobility at given ζ. We return to this matter in sec. 4.6e.

It must be added that eqs. [4.6.2-3] apply for the case that the Stern layer only carries a current tangential to the surface. However, as discussed in sec. 3.13c, there are cases in which both tangential and normal fluxes occur in that layer. The ratio between j_t and j_n depends not only on the two conductivities $K_t^{\sigma l}$ and $K_n^{\sigma l}$ or diffusion coefficients D_t and D_n, but also on the area/cross-sectional ratio of the Stern layer segment under consideration. Usually this ratio is large, so that it is much easier for incoming ions ("pumped" into the Stern layer by the applied field) to leave this layer towards the diffuse part than to continue their migration parallel to the surface. The sum of all fluxes j_n leads to the concentration polarization of electroneutral electrolyte beyond the double layer, introduced in sec. 3.13 and to which we return in sec. 4.6c. As, however, the overall current of these consecutive processes is determined by the slower one, in the final equations for the potential distribution, the polarization and the electrophoretic mobility under stationary conditions eventually $K_t^{\sigma i}$ appears in the final equations rather than $K_n^{\sigma i}$.

For a flat diffuse double layer at rest the term in square brackets in [4.6.2a] vanishes for each ion, so that

$$\frac{d c_i(x)}{dx} = - \frac{z_i F c_i(x)}{RT} \frac{d \psi(x)}{dx} \qquad\qquad [4.6.4]$$

which, after integration, leads to Boltzmann's law.

Conservation of momentum leads to Newton's second law, derived in sec. I.6.1b, see [I.6.1.14], and variants of it, including the *Navier-Stokes equation*. For the present case it is written in the following general form

$$\rho \frac{\partial v(r,t)}{\partial t} = \eta(r)\nabla^2 v(r,t) - \nabla p(r,t) - \sum_i z_i F c_i(r,t) \nabla \psi(r,t) \qquad [4.6.5]$$

The l.h.s. is the acceleration (ρ is the density of the fluid[1]), the three terms on the r.h.s. account for the frictional, the pressure and the electric force, respectively. Because of the presupposed low Reynolds number only the linear term is needed on the l.h.s. The last term on the r.h.s. is the electrical force, exerted on the ions in the double layer, but as under stationary conditions this force is equal but opposite to the frictional force of the fluid in which they are embedded, we may use the action = reaction principle and consider this force as exerted on the liquid.

To familiarize with this equation some special cases can be mentioned. For stationary states the time drops out of the equation and $\partial v / \partial t = 0$. (In fact, this equality remains valid for a.c. fields where the frequency is low enough to let $\omega \ll a^2 \eta/\rho$.) Hence,

$$0 = \eta(r) \nabla^2 v(r) - \nabla p(r) - \sum_i z_i F c_i(r)\nabla \psi(r) \qquad\qquad [4.6.6]$$

Stationary flow of fluids in the absence of electric fields is governed by $\eta \nabla^2 v = \nabla p$. Several examples of this have been treated in sec. I.6.4. Poiseuille flow through a cylindrical capillary is one of the typical illustrations.

---

[1] The symbol ρ is also used for space charge density. When confusion is possible the two will be distinguished.

For tangential flow parallel to a charged flat surface under the influence of an applied field, that is, the high κa case, transport is determined by the first and the third terms on the r.h.s. of [4.6.6]. This leads to Smoluchowski's derivation of electro-osmotic slip, see the text leading to [4.3.7].

Finally, the last two terms on the r.h.s. are needed if the potential and the pressure gradients in the normal direction are considered, when no fluid flow occurs in that direction. This is often the case because the surface cannot act as a source of liquid ($v_n = 0$ at $x = 0$). Formation of a diffuse layer leads to local excesses of ions exerting a certain osmotic pressure, which is just equal and opposite to ∇p, because otherwise the liquid would start to flow. Equation [4.6.6] then reduces to

$$\frac{dp}{dx} = -\sum_i z_i F c_i(x) \frac{d\psi}{dx} \qquad [4.6.7]$$

which is readily integrated after substituting the Boltzmann equation for $c_i(x)$ to give

$$p(x) = p^{ff} + RT \sum_i \left[c_i(x) - c_i^{ff} \right] \qquad [4.6.8]$$

Here, the far field has been taken as the reference.

These examples illustrate the richness of [4.6.5].

As before, we can write [4.6.5] for the two parts of the double layer. Beyond the slip plane the only change is that $\eta(r) \approx \eta$ (= the bulk viscosity)

$$\rho \frac{\partial v(r,t)}{\partial t} = \eta \nabla^2 v(r,t) - \nabla p(r,t) - \sum_i z_i F c_i(r,t) \nabla \psi(r,t) \qquad [4.6.9]$$

Within the slip plane the viscosity is infinitely high and all velocities are zero with respect to the solid. As in [4.6.7], in the Navier-Stokes equation, the ∇p and the $\nabla \psi$ terms are the only ones remaining. Consider a Stern layer, containing only one type of ions, i, around a spherical particle of radius a, then [4.6.6] reduces to

$$\frac{dp}{a\,d\theta} = -z_i F c_i^i \frac{d\psi^i}{a\,d\theta} \qquad [4.6.10]$$

where θ has the same meaning as in fig. 3.87. The pressure p acts on a Stern layer segment of thickness d^{ek} which is, according to sec. 4.4, close to the thickness of the Stern layer, d. Replacing the volume concentration in that segment by a surface charge σ^i on the inner Helmholtz plane, $z_i F c_i = \sigma^i / d^{ek}$ and from [4.6.10]

$$\frac{dp}{a\,d\theta} = -\frac{\sigma^i}{d^{ek}}\frac{d\psi^i}{a\,d\theta}$$ [4.6.11]

The last expression needed is the *Poisson equation*. In a field of homogeneous dielectric constant ε,

$$\nabla^2\psi\,(r,t) = -\frac{\sum_i z_i F\,c_i(r,t)}{\varepsilon_o \varepsilon}$$ [4.6.12]

When ε is not constant, the equation must be written as

$$\nabla\cdot(\varepsilon_o \varepsilon(r)\nabla\psi) = -\sum_i z_i Fc_i(r,t)$$ [4.6.13]

As we are not considering relativistic rates of transport, the potential can adjust itself instantaneously to the space charge density. Therefore, although the Poisson equation is essentially of a static nature, it is also valid under the dynamic conditions considered here.

We have already given expressions for the Laplacian $\nabla^2\psi$ for unpolarized flat, spherical and cylindrical geometries, see [I.app.7.46-49]. Let us again make the subdivision into the two electrokinetically distinguishable double layer parts. Beyond the slip plane [4.6.12] applies with $\varepsilon = \varepsilon$ (bulk). In the far field the space charge density is zero. Hence

$$\nabla^2\psi\,(r) = 0 \qquad\qquad \text{(any } t) \qquad\qquad\qquad [4.6.14]$$

The general solution of [4.6.14] is a series of Legendre polynomials, each next term representing a higher moment. For our purposes two terms suffice:

$$\psi(r,\theta) = -E\,r\cos\theta - f(Du)\frac{a^3}{r^2}\,E\cos\theta$$ [4.6.15]

where $f(Du)$ is a dimensionless function of Du, depending on ζ and κa. The precise functionality depends on conditions: high or low κa, zero or finite conduction in the Stern layer, zero or finite charge exchange between the Stern layer and the diffuse layer, etc. When $Du = 0$, $f(Du) = \frac{1}{2}$ and [4.6.15] reduces to [3.13.3]. The first term on the r.h.s. of [4.6.15] is the potential caused by the applied field, the second term represents the polarization field, which decays to zero over distances of order a. Equation [4.6.15] can also be expressed in terms of the polarizability of particle plus double layer or of its induced dipole moment. We already gave the solutions for the large κa case, see [3.13.5-7] and return to this in [4.6.54].

Within the stagnant layer the integration of [4.6.13]cannot be carried out unless it is known how ε depends on distance and how the charges are distributed. For the Stern layer picture we have the charge-free inner and outer

Helmholtz layer where $\nabla^2 \psi$ is zero, meaning that $d\psi/dx$ is constant. From sec. 3.6c we know these field strengths to be

$$\left(\frac{d\psi}{dx}\right)_1 = -\frac{\sigma^o}{\varepsilon_o \varepsilon_1^i}$$

[4.6.16a]

$$\left(\frac{d\psi}{dx}\right)_2 = -\frac{(\sigma^o + \sigma^i)}{\varepsilon_o \varepsilon_2^i} = \frac{\sigma^d}{\varepsilon_o \varepsilon_2^i} \approx \frac{\sigma^{ek}}{\varepsilon_o \varepsilon_2^i}$$

[4.6.16b]

where the subscripts 1 and 2 refer to the inner and outer Helmholtz layer, respectively. The difference between these two slopes is governed by

$$\varepsilon_o \varepsilon_1^i \left(\frac{d\psi}{dx}\right)_1 - \varepsilon_o \varepsilon_2^i \left(\frac{d\psi}{dx}\right)_2 = \sigma^i$$

[4.6.17]

which can also be written as

$$D_1 - D_2 = -\sigma^i$$

[4.6.18]

where D_1 and D_2 are the dielectric displacements in the inner and outer Helmholtz layer, respectively.

Let us, at this stage, recapitulate the equations to be solved. Most generally, these are [4.6.1, 2, 5 and 13]. These apply for any κa, any ζ and any frequency of the applied field.

For the rigid particle model and stationary states, we have instead [4.6.1, 2a, 6 and 12]. In this approximation equations for the Stern layer [4.6.2b-2d, 3, 10 and 11] are needed.

This set of differential equations constitute the starting points for the theory of all electrokinetic phenomena. Differences between these phenomena are reflected in the *boundary conditions*.

For all phenomena the assumption has been made that there is no charge transfer through the particle surface. In fact, often the particles are taken as dielectric solids through which the lines of electric flow do not pass. For fluid drops, as in emulsions, this may be a poor approximation because internal flow inside the drops may occur. However, practice has shown that, due to Marangoni effects, the liquid-liquid interface also often behaves as if it were inextensible. Absence of ion transfer across the phase boundary means that there the normal component of j is zero

$$\left(j_n\right)_{surf} = 0$$

[4.6.19]

The presence of a stagnant layer means that in electrophoresis

$$\left(v_t\right)_{sp} = v_{ef}(\theta) \qquad\qquad [4.6.20]$$

where sp stands for slip plane, whereas in the bulk

$$v(\infty) = 0 \qquad\qquad [4.6.20a]$$

For electro-osmosis

$$\left(v_t\right)_{sp} = 0 \qquad\qquad v(\infty) = v_{eo} \qquad\qquad [4.6.21, 21a]$$

and for sedimentation

$$v(\infty) = -v_{sed} \qquad\qquad [4.6.22]$$

Regarding the potentials, in electrophoresis and electro-osmosis

$$\nabla\psi(\infty) = -E \qquad\qquad [4.6.23]$$

or

$$\psi(\infty) = -Er\cos\theta \qquad\qquad [4.6.23a]$$

whereas for sedimenting particles this field is just the sedimentation field E_{sed} required to balance the current flow created by the sedimenting particles.

Other boundary conditions will be introduced, where appropriate.

We consider the physics of this set of basic equations sufficiently explained and now turn to the mathematical elaboration.

4.6b. Small deviations from equilibrium

The mathematical complexity of the equations of sec. 4.6a is caused not only by the large number of coupled differential equations that have to be simultaneously solved, but also by the fact that some of these equations are nonlinear (i.e. contain products of the unknown quantities, like $c_i\nabla\psi$ and $c_i\nabla v$ in [4.6.2]). One simplification that can be made with little loss of rigour, and which leads to linearization, is to consider only small deviations from equilibrium, for which we may write

$$c_i(r,t) = c_i(r, \text{eq}) + \delta c_i(r,t) \qquad\qquad [4.6.24]$$

$$\psi(r,t) = \psi(r, \text{eq}) + \delta\psi(r,t) \qquad\qquad [4.6.25]$$

with $\delta c_i \ll c_i(\text{eq})$ and $\delta\psi \ll \psi(\text{eq})$. Here, (eq) means: in the absence of the applied field. Formally, v may be replaced by δv, because $v(\text{eq}) = 0$. [4.6.24 and 25] are good approximations for all electrokinetic phenomena that do not take place under extreme conditions, such as extremely high applied fields. The increments may be positive or negative.

For the part beyond the slip plane [4.6.1 and 2a] are combined, and these substitutions carried out. The $\partial c_i(eq)/\partial t$ term drops out, the terms $\nabla c_i(eq)$ and $z_i F c_i(eq)\nabla \psi(eq)/RT$ in the expression between the square brackets cancel because of [4.6.4]. Products of perturbations, like $\delta c_i \, \delta \psi$ are neglected. Regarding the convection contribution we obtain $-\nabla \cdot \left[c_i(eq) + \delta c_i \right] v$, which can be written as $- v \cdot \nabla \left[c_i(eq) + \delta c_i \right] - \left[c_i(eq) + \delta c_i \right] \nabla \cdot v$ where we have used $\nabla \cdot (\psi v) = \psi \nabla \cdot v + v \cdot \nabla \psi$, see [I.app.7.35]. Because of the incompressibility of the fluid $\nabla \cdot v = 0$. Considering all of this,

$$\frac{\partial \delta c_i(r,t)}{\partial t} = D_i \, \nabla \cdot \left[\nabla \delta c_i(r,t) + \delta c_i(r,t) \frac{z_i F}{RT} \nabla \psi(r,eq) + c_i(r,eq) \frac{z_i F}{RT} \nabla \delta \psi(r,t) \right]$$

$$- \nabla c_i(r,eq) \cdot v(r,t) \qquad\qquad [4.6.26]$$

Similar transformations can be carried out for the other relevant equations. For the Stern layer, from [4.6.1 and 2c], considering that $\nabla_t \sigma_i^l(eq) = 0$ and $\nabla_t \psi^l(eq) = 0$,

$$z_i F \frac{\partial \delta c_i^l(\theta,t)}{\partial t} = D_i^s \, \nabla \cdot \left[\nabla_t \delta \sigma_i^l(\theta,t) + \delta \sigma_i^l(\theta,t) \frac{z_i F}{RT} \nabla_t \psi(eq) + \sigma_i^l(eq) \frac{z_i F}{RT} \nabla \delta \psi(\theta,t) \right]$$

$$[4.6.27]$$

where θ, defined in fig. 4.5, refers to the position of the increment considered parallel to the surface. For a non-spherical contour a more detailed specification may be needed. Equation [4.6.27] can also be written as

$$\frac{\partial \delta \sigma_i^l(\theta,t)}{d^{ek} \partial t} = D_i^s \, \nabla \cdot \left[\nabla_t \delta \sigma_i^l(\theta,t) + \delta K_i^{ls}(\theta,t) \nabla_t \psi(eq) + K_i^{ls} \nabla \delta \psi(\theta,t) \right] \qquad [4.6.28]$$

where, as in [4.6.11], d^{ek} is the Stern layer ($\approx$ stagnant layer) thickness and

$$\delta K_i^{ls}(\theta) = D_i^s z_i F \delta \sigma_i^l(\theta) / RT \approx u_i^s \delta \sigma_i^l(\theta) \qquad\qquad [4.6.29]$$

is the contribution of ions i to the Stern layer conductivity increment.

From [4.6.6], by the same procedure

$$\rho \frac{\partial v(r,t)}{\partial t} = \eta \nabla^2 v(r,t) - \nabla \delta p(r,t)$$

$$- \sum_i \left\{ z_i F \delta c_i(r,t)\nabla \psi(r,eq) + z_i F c_i(r,eq)\nabla \delta \psi(r,t) \right\} \qquad [4.6.30]$$

To obtain this result, we used $\nabla p = \sum_i z_i F \delta c_i \nabla \psi$. As before, the inertia term

$\partial v / \partial t$ in [4.6.30] will be zero for most electrokinetic phenomena. Regarding the electrostatic terms, [4.6.17] becomes

$$\varepsilon_0 \varepsilon_1^i \left(\frac{\partial \delta \psi (y,t)}{\partial x} \right)_1 - \varepsilon_0 \varepsilon_2^i \left(\frac{\partial \delta \psi (y,t)}{\partial x} \right)_2 = \delta \sigma^i(y,t) \qquad [4.6.31a]$$

When the flow lines can pass through the particle, a similar equation can be given for the boundary between the particle (p) and the inner Helmholtz layer

$$\varepsilon_0 \varepsilon^p \left(\frac{\partial \delta \psi (y,t)}{\partial x} \right)^p - \varepsilon_0 \varepsilon_1^i \left(\frac{\partial \delta \psi (y,t)}{\partial x} \right)_1^i = \delta \sigma^o(y,t) \qquad [4.6.31b]$$

where $\delta \sigma^o$ is the dynamic surface charge increment. However, we do not generally consider this case. At the slip plane

$$\varepsilon \left(\frac{\partial \delta \psi (y,t)}{\partial x} \right) = \varepsilon_2^i \left(\frac{\partial \delta \psi (y,t)}{\partial x} \right)_2^i \qquad [4.6.31c]$$

Because of the many symbols needed to characterize positions and states of the various parameters, these equations look more complicated than those of sec. 4.6a. In fact they are simpler because the non-linear terms have disappeared by, so to say "subtracting" the equilibrium situation. The equations deal essentially with perturbations only. Linearity implies that a sum of solutions of the set is also a solution. We shall henceforth only consider situations where linearization is allowed, except where explicitly stated.

Let us finally check that our equations do lead to classical results if the assumption is made that the double layers are not perturbed. Consider the situation of thin double layers. In [4.6.30] only two terms remain, $\eta \nabla^2 v = \eta d^2 v_t / dx^2$ and $\sum_i z_i F c_i(x) \nabla \delta \psi(x,t)$. The latter accounts for the difference between the potential in the presence and absence of the outer field, $\nabla \delta \psi(x,t)$, which is just the applied field E. For $\sum_i z_i F c_i$ we write the space charge density which, according to the Poisson equation, equals $\varepsilon_0 \varepsilon d^2 \psi / dx^2$. Equating the frictional drag and electric terms and integrating, the Helmholtz-Smoluchowski equation is recovered.

4.6c. *Double layer relaxation processes*

So far, we have encountered relaxation times of diffuse double layers a few times, for instance in sec. 3.13. The considerations given were based on diffusion and/or conduction of ions. With the equations of sec. 4.6b these phenomena can be put on a more general and quantitative footing.

In electrokinetics ions move around the particles. For the present purpose we assume the particles to be dielectric and large $(\kappa a \gg 1)$, so that the applied field is essentially parallel to the surface, as in fig. 4.2. Application of an external field polarizes the double layer as in figs. 3.86a2, 3.88 and 4.1, meaning that, as compared with the equilibrium situation, the double layer contains more positive charge on the one side (on the right side in these figures) and less to the other (left). When the field is stationary, ions are continually replenished, maintaining the polarization. The composition of the excess charge depends on ψ^0 and c_{salt}. When both are low, about 50% of the countercharge is caused by an excess of counterions (cations) and 50% by a deficit of co-ions (anions). For higher potentials the relative contribution of the cations grows, and so does the contribution of the cations to Du, till eventually all the tangential transport is solely governed by cations. For that situation one might say that the transference number of cations in the double layer is unity: $t_+^\sigma = 1$; $t_-^\sigma = 0$. This situation differs from that in the bulk. How do the two match? Let us assume that we are dealing with one symmetrical electrolyte ($z_+ = z$, $z_- = -z$) of which the bulk ionic diffusion coefficients are the same: $D_+ = D_- = D$ and $t_+ = t_- = 0.5$, so that in the solution 50% of the conduction is carried by the cations and 50% by the anions. Figure 4.24 illustrates what happens. Here, a small section of the double layer is drawn. By the action of the external field there is an excess positive charge in the solution side of the double layer. As the double layer is thin as compared with the particle radius, this excess will almost exclusively leak away into the solution. In the figure this is indicated by long radial arrows,

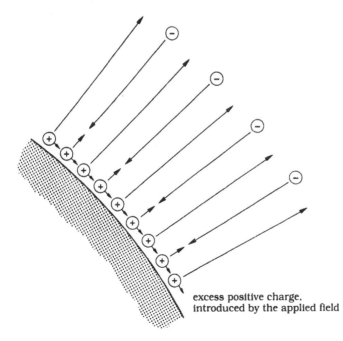

excess positive charge, introduced by the applied field

Figure 4.24. Explanation of the creation of electroneutral electrolyte near a particle surface as a consequence of the transition between conduction mainly by cations near the surface to that by cations and anions in the bulk. Further discussion in the text.

as compared to short tangential ones. A fraction of the cations will make it to the solution, others will meet an oncoming anion to form an ion pair. In this way, electroneutral electrolyte is created in the double layer. The fraction of cations that is neutralized by anions is determined by the bulk transport number; in our example it is 0.5. The main conclusion is that *electroneutral electrolyte is created near the surface*, which has to diffuse away into the solution, beyond the double layer, i.e. into the region of the far field. In passing it may be mentioned that analysis of excesses or deficits of electrolytes near electrodes in electrical cells, created when a current passes, also leads to excesses or deficits of electroneutral electrolyte; this is one of the traditional ways to determine transport numbers.

By a similar mechanism, in other double layer parts where there is a deficit of positive charge (to the left in figs. 3.86a2 and 3.88), a local shortage of electrolyte is built up, which has to be replenished by diffusion from beyond the double layer. In fig. 4.25 these salt fluxes are sketched. Recall that for equilibrated double layers the equilibrium electrolyte concentration is higher than in the absence of a double layer because of the negative adsorption of electrolyte; the present asymmetrical disturbance develops in addition to this. The excess of electrolyte on the one side and the deficit to the other try to annihilate each other by diffusion in the far field.

So it is seen that the movement of a charged particle under a potential gradient $(\nabla \psi)$ creates a concentration gradient $(\nabla c_{\pm})$. According to the Onsager

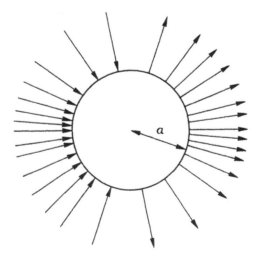

Figure 4.25. Schematic illustration of the creation of excess electroneutral electrolyte to the right, and a corresponding deficit to the left of the particle. Same situation as in fig. 4.24. Beyond the particle, (not shown), electroneutral electrolyte diffuses from the right to the left, around the particle. (Not to scale. The length of the arrows is $O(\kappa^{-1})$; in reality mostly $\kappa a > 1$.)

principles, the inverse phenomenon must also exist: movement of a charged particle under the influence of a concentration gradient, thereby polarizing the particle. Such reciprocal behaviour is observed in *diffusiophoresis*.

The main point of this discussion is that *two* transfer processes have to be distinguished, each with its inherent relaxation time:

(i) (re-) formation of an electrical double layer. Range: κ^{-1}; mechanisms: conduction and diffusion;

(ii) transport of salt beyond the double layer, through the solution around the particle, and back on the other side. Range: a; mechanism: diffusion.

In sec. 3.13 only process (i) has been considered. Process (ii) is obviously much slower.

Quantitative elaboration of these mechanisms is very complicated and beyond the scope of the present chapter. However, on the basis of the equations of sec. 4.6b it can be indicated how solutions can be obtained. To that end, consider the, somewhat academic, case of a flat uncharged surface on the outside of which at $t = 0$ charge is created. One could think of an electrode that is charged by switching on a potential difference, but the mechanism is not important for the present reasoning. We want to apply [4.6.26]. At $t = 0$ $\nabla \psi(x, eq)$ and $\nabla c_i(x, eq) = 0$. Moreover, $v = 0$: fluid will neither move tangentially for symmetry reasons nor normally because the presumed incompressibility of the fluid would resist this. With this in mind [4.6.26] reduces to

$$\frac{\partial \delta c_i(x,t)}{\partial t} = D_i\left[\nabla^2 \delta c_i(x,t) + c_i(eq)\frac{z_i F}{RT}\nabla^2 \delta \psi(x,t)\right]$$

[4.6.32]

For a binary electrolyte with $D_+ = D_- = D$, eliminating $\nabla^2 \delta \psi$ by using the Poisson equation, the following pair is obtained:

$$\frac{\partial \delta c_+(x,t)}{\partial t} = D\left[\nabla^2 \delta c_+(x,t) + \frac{\kappa^2}{2}\left(\delta c_+(x,t) - \delta c_-(x,t)\right)\right]$$

[4.6.33a]

$$\frac{\partial \delta c_-(x,t)}{\partial t} = D\left[\nabla^2 \delta c_-(x,t) - \frac{\kappa^2}{2}\left(\delta c_+(x,t) - \delta c_-(x,t)\right)\right]$$

[4.6.33b]

Considering our discussion, we are interested in the time dependency of the excess space charge density $\delta \rho(x,t) = z F\left(\delta c_+(x,t) - \delta c_-(x,t)\right)$ and in that of the concentration of the electroneutral electrolyte, $\delta c_s(x,t) = \frac{1}{2}\left(\delta c_+(x,t) + \delta c_-(x,t)\right)$. In other words, from [4.6.33a and b] the time derivatives of the differences between, and the sums of, δc_+ and δc_- must be obtained. To that end, the equations are subtracted and added, yielding

$$\frac{\partial \delta \rho(x,t)}{\partial t} = D\left[\nabla^2 \delta \rho(x,t) + \kappa^2 \delta \rho(x,t)\right]$$

[4.6.34a]

$$\frac{\partial \delta c_s(x,t)}{\partial t} = D\nabla^2 \delta c_s(x,t)$$ [4.6.34b]

respectively. As expected, [4.6.34b] contains no double layer parameter, whereas [4.6.34a] does.

Equations like these are typically solved by Laplace transforms, of which the principles have been laid down in I.app.10. Elaborations for a number of diffusion problems can be found in sec. I.6.5. In fact, [4.6.34b] is nothing else than Fick's second law. The final result depends on the boundary conditions which are different for the different electrokinetic phenomena, see [4.6.19-23 and 31a-c]. In more complicated cases not considered here (asymmetrical electrolytes, electrolytes with different D_i's, κa not large, interacting particles, etc.) often only numerical solutions are available.

Analytical solutions have been given for special situations by O'Brien[1]. The case discussed so far has been analyzed by him to solve [4.6.34a]. The boundary conditions are $\delta\rho \to 0$ for $x \to \infty$ and $\partial\delta\rho/\partial x \to -\kappa^2\sigma^i/2zF$ for $x \to 0$ if σ^i is the surface charge density, pumped into the Stern layer at $t = 0$. The latter condition stems from $\varepsilon_0\varepsilon(\partial\delta\psi) = -\sigma^i$, compare [4.6.31a]. Integration then gives

$$\delta\rho(x,t) = \kappa\sigma^i e^{-\kappa x}\left\{\text{erf}\left[(\kappa^2 Dt)^{1/2} + \frac{x}{2(Dt)^{1/2}}\right] - \text{erf}\left[-(\kappa^2 Dt)^{1/2} + \frac{x}{2(Dt)^{1/2}}\right]\right\}$$

[4.6.35]

This equation describes how rapidly a surface charge spreads.

Equation [4.6.34b] has been solved for the case that ions are continually pumped into the Stern layer, from where they exit radially with a current density j_n. The boundary conditions are $\delta c_s \to 0$ for $x \to \infty$ and $\partial\delta c_s/\partial x \to -j_n/2zFD$ for $x \to 0$. The result is

$$\delta c_s(x,t) = \frac{j_n}{2zFD^{1/2}}\left\{2\left(\frac{t}{\pi}\right)^{1/2}e^{-x^2/4Dt} - \frac{x}{D^{1/2}}\left[1 - \text{erf}\left(\frac{x}{2(Dt)^{1/2}}\right)\right]\right\}$$

[4.6.36]

This equation describes how rapidly (positive or negative) excess neutral electrolyte is created. As discussed before, this model is representative of ion transport in double layers as occurs in electrophoresis and electro-osmosis. It is recalled that the error function erf b is defined as

$$\text{erf}\, b = \frac{2}{\pi^{1/2}}\int_0^b e^{-u^2}du$$ [4.6.37]

---

[1] O'Brien in Hunter, loc. cit. sec. 13.5.

It is a function that rises monotonically from zero at $b = 0$ to 1 for $b \to \infty$. For high b it can be approximated as

$$\text{erf } b \approx 1 - \frac{e^{-b^2}}{b\,\pi^{1/2}} \left(1 - \frac{1}{2b^2}\right) \qquad (x \to \infty) \qquad [4.6.37a]$$

From [4.6.35] it follows that in the limit of long times

$$\delta\rho(x) \approx \kappa\,\sigma^i\,e^{-\kappa x} \qquad (t \to \infty) \qquad [4.6.38]$$

Then the charge distribution is, at given influx σ^i, independent of t and only a function of distance. So, a situation is attained in which all introduced charge, accounted for by σ^i, can leak away by conduction. The time, required to attain this equilibrium situation is a few multiples of $(\kappa^2 D)^{-1}$. Hence,

$$\tau^{dl}\,(= \tau^{nf}) \approx (\kappa^2 D)^{-1} \qquad [4.6.39]$$

can be taken as a measure of the *(diffuse) double layer relaxation time*. In this way a footing has been given for [3.13.25]. This relaxation time is $O(10^{-7}-10^{-6})$s in $10^{-3} - 10^{-1}$ electrolytes.

In d.c. electrokinetics the time scale of the applied disturbance is usually much larger than that. The implication is that locally, in each segment of the double layer, the double layer may be considered at equilibrium, although over the particle as a whole, there is a tangential gradient in the ionic concentration. In sec. 3.13 we have referred to this feature as the *local equilibrium principle*. This principle starts to fail when the time scale of the applied disturbance is shorter than 10^{-7} s. For instance, this may be the case in a.c. fields of frequencies $\gtrsim$ 20 MHz.

In contradistinction to $\delta\rho$, the excess electrolyte δc_s continues to grow even for $t \to \infty$, as long as the charges in the double layer continue to be replenished. Substituting, for large t, [4.6.37a] in [4.6.36] one obtains

$$\delta c_s(x,t) \approx \frac{2j_n\,t^{3/2}D^{1/2}}{z\,F\,x^2}\,e^{-x^2/4Dt} \qquad (4Dt \gg x^2) \qquad [4.6.40]$$

whereas the limiting value for $t \to \infty$ follows from [4.6.36] as

$$\delta c_s(x,t) \approx \frac{j_n\,t^{1/2}}{z\,F\,D^{1/2}\,\pi^{1/2}} \qquad (t \to \infty) \qquad [4.6.41]$$

which is independent of x but still depends on t, meaning that the ionic concentrations in the far field continue to grow. In [4.6.40], when the exponent is of order unity, the excess concentration at the corresponding distance $\approx 2j_n t^{3/2}D^{1/2} / z\,F\,x^2$, which is small as compared to δc_s at $x \doteq 0$. Hence, the value of $x \approx 2(Dt)^{1/2}$ is a measure of the diffusion layer thickness. For time scales of the order of tens of seconds $(Dt)^{1/2} \approx O(10^5 \text{ nm})$, i.e. three or four orders of magnitude larger than κ^{-1}, and of course, independent of c_{salt}. With fig. 4.25 in mind, the order of magnitude of the relaxation time for the salt redistribution in the region beyond the double layer is of order

$$\tau^{\text{ff}} \approx a^2 D^{-1} \qquad\qquad\qquad [4.6.42]$$

because a is of the order of magnitude of the distances over which the excesses to the right and the deficits to the left have to diffuse to annihilate.

Finally we return to the statement, made in connection with fig. 4.23, that for $\kappa a \gg 1$ charges in a double layer leak away much easier by normal than by tangential transport. In other words, j_n should exceed j^σ. Under stationary conditions j_n is independent of x (no local charge accumulation) and equal to the bulk current density $j = K^L E$, where K^L is the bulk conductivity. As j^σ is given by [4.3.49]

$$\frac{j^\sigma}{j} \approx \frac{K^s}{K^L a} = Du \qquad\qquad\qquad [4.6.43]$$

where Du is the Dukhin number, introduced in [4.3.64] and formulated by [4.3.65] for large κa and a symmetrical (z-z) electrolyte. It is immediately seen that the required condition is met provided κa is high enough, depending on the value of ζ.

4.6d. Development of electrophoresis theories. Absence of conduction behind the slip plane

With the equations of secs. 4.6a-c the problem of computing ζ from mobilities at given κa is in principle soluble. But not in practice! The mathematics are very complicated and require a number of finesses, whereas simplifications is dangerous because the various fluxes and forces are coupled, so that approximating only one of these may offset the balance and misrepresent characteristic features. Moreover, the boundary conditions are in part determined by the composition of, and the mobilities of ions in the various parts of the double layer, for which model assumptions must be made.

At the moment of writing, the development and testing of complete theories was still in progress, although some trends emerge. Experiments with well-chosen model systems have not always kept pace with new theoretical insights (there is even a lack of good systematic mobility data for model spherical particles as a function of a, c_{salt} and σ^0). However, it is more or less established that conduction within the stagnant layer may play an important role in some systems. For the purposes of this survey we shall therefore divide the treatment into two parts:

(i) theories for "rigid particles" (particles are rigid entities up to the slip plane where the potential is ζ (this subsection);

(ii) a discussion of conduction within the stagnant layer and its inclusion in electrokinetic theories (subsecs. 4.6e and f).

The discussion requires us to go beyond the present theme of theory for electrophoresis and also consider results obtained by other types of electrokinetics.

The problem may be stated as finding the function f in [4.3.6]. For reasons discussed earlier we shall only consider dielectric particles. Figure 4.4 gave Henry's solution, in which electrophoretic retardation was accounted for, but not yet double layer polarization. If polarization is properly included, surface conduction beyond the slip plane is automatically taken into account, with $Du = Du^d$, given by [4.3.65, 66 or 67]. However, $Du^i \equiv 0$ for "rigid particles".

Early pioneering work in this direction dates back to Overbeek[1] and Booth[2]. As was later confirmed, Overbeek's and Booth's theories already contained several main features. For instance, the accumulation or deficit of electroneutral electrolyte in the far fields (secs. 3.13 and 4.6c) was already recognized and accounted for. Dating from before the computer age, these results had to be presented analytically, mostly in terms of series expansions of which only a limited number of terms could be mastered. As a result the outcome was restricted to not too high ζ. Overbeek showed that the function f is not continually rising from 1 to 1.5, as in Henry's picture (fig. 4.4), but rather passing through maxima and/or minima in the intermediate range of κa, depending on the nature of the electrolyte.

Overbeek's work was first extended by Wiersema et al.[3]. These results were an important step forward but still had some limitations because of an iterative scheme, which failed to converge at high ζ. Consequently, applicability was kept down to about $\zeta \lesssim 150$ mV and $\zeta \lesssim 25$ mV for (1-1) or (2-2) electrolytes, respectively.

Modern, rigorous theories valid for any ζ and often for arbitrary κa, stem mostly from the Ukrainian (Dukhin, Shilov and others[4]) and Australian (O'Brien, White, Hunter and others[5]) Schools. Important contributions have

[1] J.Th.G. Overbeek, Koll. Beih. **54** (1943) 287; Adv. Colloid Sci. **3** (1950) 97.
[2] F. Booth, Nature **161** (1948) 83; Proc. Roy. Soc. (London) **A203** (1950) 514.
[3] P.H. Wiersema, A.L. Loeb and J.Th. G. Overbeek, J. Colloid Interface Sci. **22** (1966) 78.
[4] See general references in sec. 4.12b and specific references below.
[5] R.W. O'Brien, L.R. White, J. Chem. Soc. Faraday Trans. II **74** (1978) 1607; R.W. O'Brien, R.J. Hunter, Can. J. Chem. **59** (1981) 1878, O'Brien in Hunter, loc. cit. and further specific references below.

been made by Fixman[1], whose theory was restricted to the case of high κa. These Schools have been working largely independently of each other. The premises are the same, namely the equations of secs 4.6a and b, but there are differences with respect to the elaboration and to assumptions or restrictions, made during the deriviation. Hence, it is not surprising that the final results for

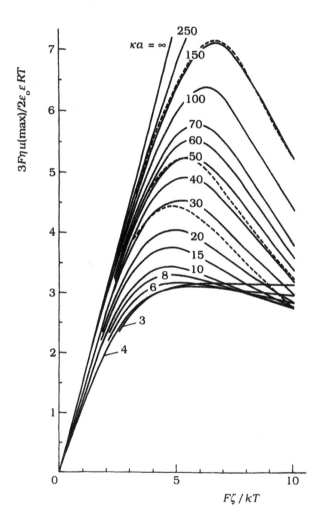

Figure 4.26. Electrophoretic mobility as a function of the potential at the slip plane, after O'Brien and White (1978). Smooth, dielectric spherical particles; (1-1) electrolyte. Conduction behind the slip plane neglected; ζ has its classical meaning as the potential of a discrete slip plane - - -, approximation [4.6.44] for high κa. (Redrawn from R.W. O'Brien, R.J. Hunter, *loc. cit.* (In the original the factor 3/2 on the ordinate is missing.).)

[1] M. Fixman, *J. Chem. Phys.* **72** (1980) 5177; **78** (1983) 1483.

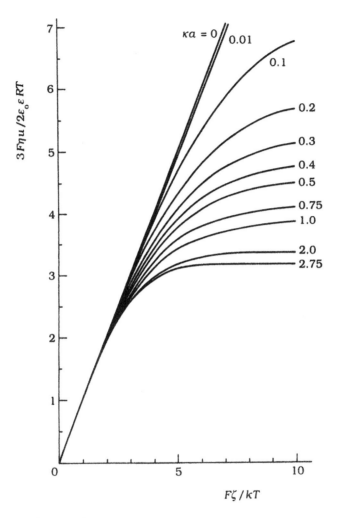

Figure 4.27. As fig. 4.26, low κa.

$u(\zeta)$ for various κa do not differ substantially between these theories. We shall first discuss the O'Brien-White results, because these authors gave extensive and rigorous numerical data, and thereafter consider results by Dukhin et al. and others. See figures 4.26-28. In order to obtain these results O'Brien and White employed the linearity of the differential equations for small perturbations (sec. 4.6b) to decompose the overall problem into two constituents: (i) thecomputationof the force needed to move the particle at velocity v in the absence of a field and (ii) computation of the force needed to keep the particle fixed in the presence of field E. The sum of these forces is the force acting on the particle when moving at velocity v under the influence of field E. This sum must be zero. In this way v can be related to the velocity in the far field and, hence, to $\textbf{\textit{E}}$. The authors also found that the result did not depend on the dielectric permittivity of the particle. The reason is that the fluxes are dominated by conduction and conductivities.

The first conclusion regarding figures 4.26 and 27 is that for $\kappa a \to \infty$ $f(\kappa a)$ approaches 1.5 (fig. 4.26) whereas for $\kappa a \to 0$ $f(\kappa a) \to 1.0$ (fig. 4.27). In other words, in the appropriate limits the Helmholtz-Smoluchowski [4.3.4] and Hückel-Onsager equations [4.3.5] are retrieved. So it is concluded that these two limiting laws also remain valid when the double layer is polarized. This is an extension of Morrison's result, quoted in connection with [4.3.20].

In the second place, the maximum in $u(\zeta)$ for $\kappa a \gtrsim 3$ is striking but theoretically well-established: it is also predicted by Dukhin's, Fixman's and, less explicitly, by Wiersema et al.'s theories. On first sight such a maximum is counterintuitive: how can a particle move slower when its electrokinetic charge is increased? It must mean that, upon increasing ζ, the driving force increases less than the braking forces. The former is proportional to σ^{ek}, the latter, force $\textbf{\textit{F}}_4$ of sec. 4.3a (i), is proportional both to σ^{ek} and to the electric field produced by the polarized double layer, which is also $\sim \sigma^{ek}$, hence the brake scales as $(\sigma^{ek})^2$, so it may prevail at high σ^{ek}, i.e. at high ζ.

According to O'Brien and White the mobility maximum depends on κa as given in fig. 4.28. With increasing counterion valency the maximum becomes lower and is observed at lower ζ (not shown).

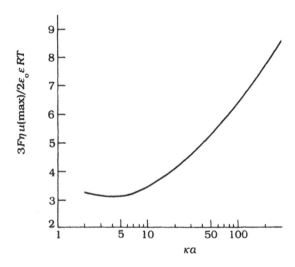

Figure 4.28. Maximum in the electrophoretic mobility according to O'Brien and White (same source as fig. 4.26).

For practical purposes analytical formulas may be useful. O'Brien and Hunter[1] derived the following formula for high κa:

$$\frac{F\eta u}{\varepsilon_0 \varepsilon RT} = \frac{F\zeta}{RT} - \frac{(F\zeta/RT) - (3\ln 2/z)\left[1 - \exp(-zF\zeta/RT)\right]}{1 + \left[\kappa a \, \exp(-zF\zeta/2RT)/2(1 + 3m/z^2)\right]} \qquad [4.6.44]$$

We come back to its derivation below. In the denominator the Dukhin number occurs in the form of [4.3.67], implying that surface conduction beyond the slip plane is accounted for. The behaviour of $u(\zeta)$ according to this equation is indicated in fig. 4.26 as dashed curves. It appears that the equation works well unless κa is very low and ζ high. For the sake of comparison we also give Dukhin and Semenikhin's equation[2]:

$$\frac{F\eta u}{\varepsilon_0 \varepsilon RT} = \frac{3}{2}\frac{F\zeta}{RT} -$$

$$6\left[\frac{\frac{F\zeta}{RT}\left(1 + \frac{3m}{z^2}\right)\sinh^2\left(\frac{zF\zeta}{4RT}\right) + \left\{\frac{2}{z}\left(1 + \frac{3m}{z^2}\right)\sinh\left(\frac{zF\zeta}{2RT}\right) - 3m\left(\frac{F\zeta}{RT}\right)\right\}\ln\cosh\left(\frac{zF\zeta}{4RT}\right)}{\kappa a + 8\left(1 + \frac{3m}{z^2}\right)\sinh^2\left(\frac{zF\zeta}{4RT}\right) - \frac{24m}{z^2}\ln\cosh\left(\frac{zF\zeta}{4RT}\right)}\right]$$

$$[4.6.45]$$

O'Brien and Hunter pointed out that, like [4.6.44], [4.6.45] was also close to the numerical results provided $\kappa a \gg 1$. An approximate analytical expression of high accuracy has also been derived by Ohshima et al.[3]. It takes about half a page for which we refer to the original literature. In the derivation [3.5.56] has been used for the $\sigma^d(\zeta)$ relationship; surface conduction in the diffuse part is again implicitly taken into account (their parameter F is our Du^d). For $\kappa a \geq 10$ their mobilities differ by less than 1% from the computer results.

Similar statements can be made about Fixman's theory[4]. It applies to high κa and differs somewhat from the preceding theories in the choice of the boundary conditions. Again about a page is needed to give the final equation and define all the parameters occurring in it. Nevertheless these expressions are not difficult, so that the theory lends itself relatively easily to extensions; an example will follow in sec. 4.6f. For $\kappa a = 50$ Fixman's results are identical to those of O'Brien and White, for $\kappa a = 10$ they differ by about 10%. Yet another

[1] R.W. O'Brien, R.J. Hunter, *Can. J. Chem.* **59** (1981) 1878.
[2] S.S. Dukhin, N.M. Semenikhin, *Koll. Zhur.* **32** (1970) 366. O'Brien and Hunter noted that the 2/z in the numerator just after the brace was erroneously written 2z in the 1974 review of Dukhin and Derjaguin (see sec. 4.12b) page 307.
[3] H. Ohshima, T.W. Healy and L.R. White, *J. Chem. Soc. Faraday Trans. II* **79** (1983) 1613.
[4] M. Fixman, *J. Chem. Phys.* **78** (1983) 1483.

elaboration for high κa by Natarajan and Schechter[1] gives essentially the same results.

Between all these theories there is no difference of principle, so it is gratifying, but not surprising that the results are so similar. Another convincing illustration[2] is that the simple equation

$$\frac{F\eta u}{\varepsilon_o \varepsilon RT} = \frac{2Du}{1+2Du}\left[\frac{2}{z}\ln 2 - \frac{2}{z}\ln\left(1+e^{zF\zeta/2RT}\right)\right] + \frac{F\zeta}{RT} \qquad [4.6.46]$$

can be derived from *Fixman's* theory but converts into the O'Brien-Hunter equation [4.6.44] for high ζ. In fact, an equation very similar to [4.6.44] has also been given by Shubin et al.[3]. The inference is that, for the "rigid particle" conditions, the mobility and the mobility maximum are irrefutably predicted.

Let us now briefly describe some steps in the derivation of [4.6.44], following some ideas by Dukhin[4,5]. The discussion is limited to large κa but contains a number of thinking steps that deserve wider attention.

The first is that, because of the local equilibrium at any θ, the ionic concentrations are related to the potential according to Boltzmann's law. In the second place, the conservation of momentum equation [4.6.5] can, for a relatively thin double layer, be decomposed into its normal and tangential parts

$$\frac{\partial p}{\partial x} + \sum_i z_i F c_i(x) \frac{\partial \psi}{\partial x} = 0 \qquad [4.6.47]$$

$$\eta \frac{\partial^2 v}{\partial x^2} - \nabla_t p - \sum_i z_i F c_i(x) \nabla_t \psi = 0 \qquad [4.6.48]$$

where, as before, ∇_t is the tangential component of the gradient and x is considered normal to the surface. We have already integrated [4.6.47], see [4.6.8]. At any x $\nabla_t p$, obtained from [4.6.8], can be substituted into [4.6.48]. Then the linear approximation (sec. 4.6b) is introduced and after some manipulations [4.6.48] is integrated twice to give

$$v_s = -\frac{1}{\eta}\sum_i \left(z_i F \nabla_t \psi^{ff} + \frac{RT}{c_i(\infty)}\nabla_t c_i^{ff}\right)\frac{1}{c_i(\infty)}\int_0^\infty x\left[c_i(x) - c_i(\infty)\right]dx \qquad [4.6.49]$$

When we introduce the *first moment* Γ_{1i} of the double layer as

[1] R. Natarajan, R.S. Schechter, *Am. Inst. Chem. Eng. J.* **33** (1987) 1110.
[2] M. Minor, A. van der Wal, pers. comm. (1995).
[3] V.E. Shubin, R.J. Hunter and R.W. O'Brien, *J. Colloid Interface Sci.* **159** (1993) 174.
[4] See Dukhin's reviews, mentioned in sec. 4.12b.
[5] O'Brien in Hunter, loc. cit. 810.

$$\Gamma_{11} = \int_0^\infty x \big[c_i(x) - c_i(\infty) \big] \, dx \qquad\qquad\qquad\qquad\qquad [4.6.50]$$

eq. [4.6.48] may also be written as

$$\boldsymbol{v}_s = -\frac{1}{\eta} \sum_i \left(z_i F \nabla_t \psi^{ff} + \frac{RT}{c_i(\infty)} \nabla_t c_i^{ff} \right) \frac{1}{c_i(\infty)} \Gamma_{11} \qquad\qquad [4.6.51]$$

For the diffuse part of the double layer Γ_{11} can simply be obtained from the Gouy-Chapman theory $c_i(x) - c_i(\infty) = c_i(\infty)[\exp\{z_i F\psi(x)/RT\} - 1]$. The integrals refer to the equilibrium double layer. Their upper limits are for simplicity set at ∞ although the far field does not contribute. As in [4.3.10], $\boldsymbol{v}_s$ is the slip velocity; it is the increase in the fluid velocity from the value at the particle surface to the same in the bulk. When the double layer is not polarized, $\nabla_t \psi \approx \boldsymbol{E}$, $\nabla_t c_i^{ff} \approx 0$ and as for a symmetrical electrolyte

$$\int_0^\infty x \big[c_+(x) - c_-(x) \big] \, dx = - \frac{\varepsilon_o \varepsilon \zeta}{zF} \qquad\qquad\qquad [4.6.52]$$

[4.6.49] reduces to the Helmholz-Smoluchowski equation. Equation [4.6.52] is obtained by substituting the Poisson equation. However, more generally eq. [4.6.49] has to be elaborated by finding ψ^{ff} and c_i^{ff}, which means that the Laplace equations [3.13.2, 19 or 20] have to be solved subject to the condition that the tangential ion flux of each ion into a portion of the double layer is just balanced by its normal outflow towards the bulk. As explained in sec. 3.13, through this latter balance $K^{\sigma d}$, and hence Du^d, enters the equations. In this way, basically any electrokinetic phenomenon can be handled.

For a spherical particle at high κa the solutions for δc and ψ^{ff} are

$$\delta c = - \frac{czFa^3}{4RTr^2} E \cos\theta \left(\frac{6Du}{1+2Du} \right) \qquad\qquad [4.6.53]$$

and

$$\psi^{ff} = - Er\cos\theta - \frac{1}{2}\left(\frac{1-Du}{1+2Du} \right) \frac{a^3}{r^2} E \cos\theta \qquad\qquad [4.6.54]$$

respectively. Far from the particle $\delta c \to 0$ but ψ continues to increase. Equation [4.6.54] has been anticipated, see [3.13.4]. When polarization is ignored it reduces to [4.3.15] with $Du = 0$. The $\cos\theta$ in [4.6.53] reflects the angle dependence of the (positive or negative, depending on the sign of z) excess of electroneutral electrolyte in the far field. At $\theta = 90°$ this excess is zero, it increases in one direction and decreases in the other, see fig. 4.25.

From [4.6.53 and 54] $\nabla_t \psi^{ff}$ and $\nabla_t c^{ff}$ can be obtained and substituted in [4.6.49]. A few more manipulations are required until eventually [4.6.44] is obtained. The difference with Dukhin's formula [4.6.45] is not one of principle but reflects relatively minor differences in approximations for which we refer to the original literature.

A phenomenological elaboration of [4.6.51], valid for high κa has been given by Zharkikh et al.[1]. It reads simply

$$u = -\frac{F}{\eta} \sum_i \frac{z_i c_i \Gamma_{ii}}{(1 + Du_i)} \tag{4.6.55}$$

and has been derived by consideration of the coupling between $\nabla_t \psi^{ff}$ and $RT \nabla_t c_i^{ff}$. Basically this stems from the identity of the Poisson equations for ψ^{ff}, c^{ff}, δc_i^{ff} and μ_i^{ff}, see [3.13.2, 19-21]. Thus, in an effective way convection and polarization are accounted for. Besides its simplicity, [4.6.55] has the advantage of specifying the individual contributions of each ionic species; Du_i is the contribution of species i to the Dukhin number.

It may be added that the terms in parentheses in [4.6.51] represent the tangential component of the gradient in the electrochemical potential, $\nabla_t \tilde{\mu}_i$. The formula that is obtained if this is substituted (but without Du) can also be found in O'Brien's work[2], illustrating again that there is analogy between the theories of different Schools.

Because the integral in [4.6.50] covers the entire double layer, [4.6.55] can be applied to the stagnant and mobile part of the double layer. Zharkikh et al.[1] applied it to a double layer which contained a thin porous surface layer that was stagnant. Extension to a Gouy-Stern layer, as considered in chapter 3, should be straightforward.

4.6e. *Experimental verification. Electrokinetic consistency tests and reconsideration of the role of inner layer conduction*

The search for experiments may, or may not, confirm the theory of sec. 4.6c. The issue can be divided into two parts:

(i) Is there experimental evidence for the maximum in $u(\zeta)$?

(ii) Is the absolute value of ζ obtained from the theory correct?

To verify the former, systems should be available where ζ can be varied over a large range (up to about 200 mV). In practice, ζ, or rather ψ^d, can only be *indirectly* enhanced by increasing σ° or ψ°. However, the extra charge that is put on the surface tends to be compensated by a charge that accumulates in the Stern layer and therefore this increase is not very effective in increasing ζ. See

[1] N.I. Zharkikh, J. Lyklema and S.S. Dukhin, *Koll. Zhur.* **56** (1994) 761.
[2] R.W. O'Brien, *J. Colloid Interface Sci.* **110** (1986) 477, eq. [3.7].

for instance figs. 3.23 and 25 where, at high y^o, σ^o and σ^i tend to become equal but with opposite signs. In addition, the viscoelectric effect also tends to level off the observed ζ-potential. In one classical example such a mobility maximum was found, viz. for silver iodide in 10^{-2} M KNO$_3$[1]. The sol was heterodisperse and the particles not spherical, hence the example is not entirely convincing. The maximum was found for $F\zeta / RT \approx 3.2$. Incidentally, the authors used the Helmholtz-Smoluchowski equation to interpret their mobilities, hence they reported a maximum of ζ(pAg) rather than as u(pAg). For polystyrene latices of given particle radius but of a widely different σ^o, Brouwer and Zsom also found such a maximum[2]. It was located at $F\zeta / RT \approx 4$ in 10^{-2} M NaCl. However, latices with other surface properties did not exhibit such a maximum, hence neither is this experiment confirming. On the other hand, for a number of latices Russell et al.[3] observed a parallel between the mobility extremum and κa (between about 10 and 50) which the theory followed. Latices are excellent model colloids in that the particles are spherical and homodisperse, but the molecular details of the surface structure ("hairiness") are not generally known and neither is the slip process, or for that matter, the location of the slip plane. Heat and other treatments may drastically modify the surface conductance. Neither are maxima in the mobility found on oxides, where σ^o can be made as high as several tens of $\mu C\,cm^{-2}$. Figure 3.63 already illustrated this.

The conclusion is that experimental verification of $u(\zeta)$ maxima is still wanting, but that there is no reason to question their existence. For the practical purpose of using the theory to obtain ζ from mobilities this maximum does not usually have many consequences, because most measurements fall on the ascending branch of fig. 4.26.

The second more critical issue, viz. whether the value of ζ, obtained from u using the black box theory of sec. 4.6d, is correct, cannot be directly answered because there is no independent way of measuring electrokinetic potentials. The only alley available is to measure different electrokinetic properties for a given material in a given solution, deduce ζ from each of these and compare the results, i.e. to carry out an *electrokinetic consistency* test. It is historically interesting that such a test, carried out for electro-osmosis in capillaries of different radii, has led to the recognition of the role played by surface conduction (sec. 4.5c). Anticipating the discussion to follow, it will appear that, at least in a number of cases, conduction behind the slip plane cannot be ignored. The trend is that for $K^{\sigma i} \neq 0$:

---

[1] S.A. Troelstra, H.R. Kruyt, *Kolloid Z.* **101** (1942) 182.
[2] W.M. Brouwer, R.L.J. Zsom, *Colloids Surf.* **24** (1987) 195.
[3] A.S. Russel, P.J. Scales, C.S. Mangelsdorf and S.M. Underwood, private comm. (1995).

(i) in electrophoresis the field lines are more strongly concentrated in the inner double layer part around the particle than was assumed in the "rigid particle" approach. At given ζ this lowers u, so that ζ's computed by the "rigid particle" theory are *underestimated*.

(ii) the conductivity of colloids (sec. 4.8) is higher than that for a rigid particle, consequently the rigid particle-ζ is *overestimated*.

(iii) low frequency dielectric dispersion is enhanced because there are more ions in the double layer that can contribute to the polarization (sec. 4.8); hence the rigid particle-ζ is *overestimated*.

(iv) streaming potentials are lowered (compare [4.3.41 and 42]) so that the rigid particle ζ's are *underestimated*.

Consistency tests should involve the determination of ζ from a phenomenon where this quantity may be underestimated and one where it may be overestimated. The list may be extended with other electrokinetic phenomena.

Besides the non-consistency of certain pairs of rigid particle-ζ's there is also direct evidence for non-zero conduction behind the shear plane in that, in a number of cases, experimental surface conductivities exceed the Bikerman value [4.3.60], plotted in fig. 4.10. More indirectly, certain observations can hardly be explained without invoking non-zero $K^{\sigma i}$. Let us give some experimental evidence.

Surface conductivities cannot be unambiguously measured experimentally because they are excess quantities. However, in a number of well-defined systems the bulk part of the total conductivity can be assessed with confidence, so that this part may be subtracted from the total to yield K^{σ}. The most typical example, already introduced in sec. 4.5c, is that of (bundles of) cylindrical capillaries for which the total conductivity can be measured. After subtraction of the bulk contribution, K^{σ} is obtained and may be compared with the Bikerman value for $K^{\sigma d}$. Alternatively, K^{σ} and K^{L} may be substituted in the denominator of [4.3.41] to obtain the streaming potential E_{str}. As long ago as 1938 Rutgers et al.[1] verified this for Jena glass and in doing so solved the then open problem, that ζ was found dependent on a, if computed via [4.3.42]. The values obtained by Rutgers et al. for K^{σ} ranged between 3 and $92 \times 10^{-9} \ \Omega^{-1}$ for c_{KCl} from ≈ 0 to 2×10^{-4} M, obviously in part exceeding the range of $K^{\sigma d}$, of up to $\approx 2 \times 10^{-9}$ S, see below [4.3.62]. Overbeek[2] has tabulated K^{σ} values for a number of glasses, dating back to the 1930s and 1940s, which ranged between $O(10^{-10} - 10^{-7} \Omega^{-1})$, confirming the above observation. A more recent study by

[1] A.J. Rutgers, Ed Verlende and Ma Moorkens, *Proc. Acad. Sci. Amsterdam* **41** (1938) 763; A.J. Rutgers, *Trans. Faraday Soc.* **36** (1940) 69.
[2] J.Th.G. Overbeek, *Electrokinetic Phenomena in Colloid Science*, Vol. I. H.R. Kruyt, Ed. Elsevier (1948) 236.

Jednacak-Biscan et al.[1] provides further confirmation: for a variety of glasses in water and in aqueous electrolytes they reported values of K^σ between 0.6 and 6.5×10^{-9} S, depending on pH; in CH_3OH and CH_3CN these values tend to be somewhat lower. Glasses are popular model substances because they are easy to draw into capillaries, but because the surfaces may be porous, they are perhaps not typical. However, evidence for substantial conduction in addition to $K^{\sigma d}$ has also been obtained on a variety of other materials, including $BaSO_4$[2], heated silica (Aerosil)[3] and latices [4,5,6], so that the phenomenon is well-established, but not necessarily general: cases have also been reported where K^σ was within the confines of the Bikerman model. When the particles are porous (as with many glasses) additional surface conductivity may stem from mobile ions in these porous layers. In the absence of such porous layers, surface conductivity in excess of $K^{\sigma d}$ can only be accounted for by tangential mobility in the Stern layer, that is: by a non-zero $K^{\sigma i}$.

A number of authors have observed that, at least in some cases, odd trends can be removed if conduction behind the slip plane is accounted for. One of these is the maximum in $u(c_{salt})$, which is often a real feature[7,8,9,10], but which is translated into a spurious $\zeta(c_{salt})$ maximum if $K^{\sigma i}$ is ignored in the conversion. Figure 4.29 gives an illustration[11]. The system is a homodisperse cationic latex with surface charge 14.2 μC cm^{-2} at pH = 6.0. It is seen that the maximum in $u(c_{salt})$ transposes into the zeta potentials if the Helmholtz-Smoluchowski (HS) equation [4.3.4] is used. Application of O'Brien-White (OBW) theory (fig. 4.26) implies an improvement. Now, together with double layer polarization, $K^{\sigma d}$ is accounted for, but the maximum fully disappears only if $K^{\sigma i}$ is also included. In this case this was done in terms of a theory by Semenikhin and Dukhin (SD) where the diffuse layer was assumed to extend down to the particle surface and to contain mobile ions everywhere[12]. A similar observation was made by

[1] J. Jednacak-Biscan, V. Mikac-Dadic, V. Pravdic and W. Haller, *J. Colloid Interface Sci.* **70** (1979) 18.

[2] D.A. Fridrikhsberg, V.Ya. Barkovskii, *Koll. Zhur.* **26** (1964) 722, transl. 617.

[3] S. Kittaka, T. Morimoto, *J. Colloid Interface Sci.* **55** (1976) 431.

[4] A. Watillon, J. Stone-Masui, *J. Electroan. Chem.* **37** (1972) 143.

[5] M.H. Wright, A.M. James, *Koll. Z.Z. Polym.* **251** (1973) 745.

[6] A.G. van der Put, B .H. Bijsterbosch, *J. Colloid Interface Sci.* **92** (1983) 499

[7] A.E.J. Meyer, W.J. van Megen and J. Lyklema, *J. Colloid Interface Sci.* **66** (1978) 99.

[8] J.R. Goff, P. Luner, *J. Colloid Interface Sci.* **99** (1984) 468.

[9] R.H. Ottewill, D.G. Rance, *Colloid Polym. Sci.* **264** (1986) 982.

[10] M. Elimelech, C.R. O'Melia, *Colloids Surf.* **44** (1990) 165.

[11] R. Hidalgo-Alvarez, J.A. Moleon, F.J. de las Nieves and B.H. Bijsterbosch, *J. Colloid Interface Sci.* **149** (1992) 23.

[12] N.M. Semenikhin, S.S. Dukhin, *Koll. Zhur.* **37** (1975) 1127, transl. 1017 The same model can also be found in V.N. Shilov, S.S. Dukhin, *Koll. Zhur.* **32** (1970) 117.

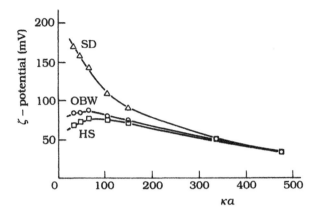

Figure 4.29. Electrokinetic potentials ζ for positively charged latex computed according to different theories. HS = Helmholtz-Smoluchowski, OBW - O'Brien-White and SD is Semenikhin-Dukhin. Further discussion, see text. (Redrawn from R. Hidalgo-Alvarez et. al. *J. Colloid Interface Sci.* **149** (1992) 23.)

Zhukov and Fridrikhsberg[1], who used experimental data by Briggs[2] (i.e. including $K^{\sigma i}$) to correct for surface conduction. Another example is fig. 4.30, referring to streaming potential measurements on a plug of negative latex. Curve a was computed without taking surface conduction into account, curve b included experimentally determined conductivities of the liquid in the plug. Similar trends were also observed with (Stöber) silicas. In this last experiment double layer overlap had to be accounted for; the outcome somewhat depends on the way in which this was done, but leaves the qualitative conclusion about $K^{\sigma i}$ unimpeded. In all these experiments the difference between corrected and rigid particle ζ's is substantial; meaning that *the major part of ζ values reported in the literature have to be revised, especially those obtained at low salt concentrations*, (upgraded in the absolute sense, that is).

Failure of electrokinetic consistency if $K^{\sigma i}$ is ignored has been observed by several researchers. In table 4.2 this is illustrated for negative polystyrene particles. The, mostly minor, spread in ζ from conductivity is again caused by uncertainties in the models applied to account for particle interaction. The difference between electrophoresis or electro-osmosis on the one hand and con-ductivity on the other is obvious; moreover $\zeta(\text{ef})$ does not systematically decrease with c_{salt}. Zukoski and Saville[3] found a similar difference for other latices. In a series of papers from Saville's group[4] it was concluded that low

[1] P.N. Zhukov, D.A. Fridrikhsberg, *Koll. Zhur.* **12** (1950) 25.

[2] D.K. Briggs, *J. Phys. Chem.* **32** (1928) 641.

[3] C.F. Zukoski, D.A. Saville, *J. Colloid Interface Sci.* **114** (1986) 32, 46.

[4] L. A. Rosen, D.A. Saville, *J. Colloid Interface Sci.* **140** (1990) 82; *Langmuir* **7** (1991) 36; *J. Colloid Interface Sci.* **149** (1992) 542; D.F. Myers, D.A. Saville, *J. Colloid Interface Sci.* **131** (1989) 461.

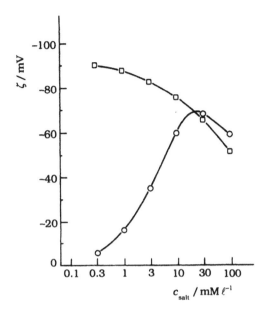

Figure 4.30. Electrokinetic potential for polystyrene plugs, computed from streaming potential measurements (a) without and (b) with inclusion of surface conduction. (Redrawn from A. van der Linde, B.H. Bijsterbosch, *Croat. Chem. Acta* **63** (1990) 455.)

Table 4.2. Electrokinetic potentials of negative polystyrene latex particles obtained from electro-osmosis, electrophoresis and conductivity. In the conversion of mobilities surface conduction behind the slip plane was ignored. (Data from A.G. van der Put, PhD. thesis, Agricultural University Wageningen, NL (1980) as elaborated by O'Brien, *J. Colloid Interface Sci.* **110** (1986) 477.)

c_{KCl} / M	κa	ζ / mV from		
		electro-osmosis	electro-phoresis	conductivity
10^{-3}	30	67	67	164-169
2×10^{-3}	42	69	72	144-146
5×10^{-3}	67	69	74	122
10^{-2}	95	62	77	107
10^{-1}	300	45	50	–

frequency dielectric dispersion gave too high ζ values if a rigid particle theory was applied. That surface conduction is a likely culprit followed from the observation that heat treatment of latices (which makes the surface "flatter" and so reduces $K^{\sigma i}$) improved the consistency. O'Brien[1] theoretically studied the

---

[1] R.W. O'Brien, *J. Colloid Interface Sci.* **110** (1986) 477.

electro-osmosis of porous plugs at $\kappa a \gg 1$. He computed ζ-potentials after accounting for double layer overlap and found these values to compare well with those from electrophoresis but to be (much) lower than the conductometric ones. O'Brien also attributed the discrepancy to additional surface conduction, that is, behind the slip plane. For latices with carboxyl groups Shubin et al.[1] studied the dielectric dispersion and measured the mobility electroacoustically. These authors also needed a finite $K^{\sigma i}$ to harmonize the ζ-potentials obtained. That low frequency dielectric dispersion $\Delta \varepsilon$ can be gigantic, requiring excessively high values of ζ or $\sigma^{ek} \approx \sigma^d$ to be explained by a rigid particle theory, is well known[2,3]. Hidalgo-Alvarez et al.[4] found for quartz in aqueous KCl that ζ from sedimentation potentials agreed well with those from streaming potentials provided the latter were computed using experimentally determined K^σ values and were corrected for double layer overlap; in this case according to a model by Levine et al.[5].

The conclusion is that surface conduction in excess of $K^{\sigma d}$ is a well-established phenomenon that has to be considered in all electrokinetic theories. This additional conductivity that we have called $K^{\sigma i}$ may be a pure Stern layer contribution, as in [4.3.54 or 57] or stem from conduction in the surface layer of the solid itself. Alternatively, polymeric hairs on the surface may give rise to a thin, hydrodynamically stagnant layer, in which ions ions may be tangentially mobile. So, $K^{\sigma i}$ is an interesting parameter because it gives specific information on numbers and/or mobilities of ions in the least accessible part of the double layer. Neglect of $K^{\sigma i}$ in the interpretation of mobilities may lead to ζ-potentials that are grossly in error. For all these reasons, systematic study for a variety of systems under different conditions (ionic composition, surface charge, porosity of the surface, the effect of polymeric hairs, etc.) with different well-defined materials continues to be highly desirable.

4.6f. Incorporation of conduction in the stagnant layer

Some theories are now available in which tangential ion transport in the stagnant layer is accounted for, or which may be modified to include this phenomenon. The quantity $K^{\sigma i}$ is system-specific, whereas $K^{\sigma d}$ is generic; therefore evaluation of the former requires insight into such properties as the distribution of charges and lateral mobilities. On the other hand, when $K^{\sigma i}$ may

[1] V.E. Shubin, R.J. Hunter and R.W. O'Brien, *J. Colloid Interface Sci.* **159** (1993) 174.

[2] J. Lyklema, M.M. Springer, V.N. Shilov and S.S. Dukhin, *J. Electroanal. Chem.*, **198** (1986) 19.

[3] D.E. Dunstan, *J. Chem. Soc. Faraday Trans.* **89** (1993) 521.

[4] R. Hidalgo-Alvarez, F.J. de las Nieves and G. Pardo, *J. Colloid Interface Sci.* **107** (1985) 295.

[5] S. Levine, J.R. Marriott, J.R. Neale and N. Epstein, *J. Colloid Interface Sci.* **52** (1975) 136.

be estimated from experiment, very useful information on the composition and dynamics of the inner layer is obtainable. For the general case of smooth inert particles, discussed in detail in chapter 3, it was concluded that the Gouy-Stern picture works well. When this is the case, $K^{\sigma i}$ and Du^i are simply given by [4.3.54] and [4.3.68], respectively. More complicated expressions have to be written if the stagnant layer corresponds to a porous sheath that allows polarization but no convection, as on the surfaces of core-shell latices or in the walls of bacterial cells. Surface heterogeneity and roughness calls for the solution of additional problems: how do the ions move tangentially? Literature models include a stagnant layer with a diffuse distribution[1], a porous surface layer[2] and detailed molecular-mechanistic pictures of the Stern layer[3,4]. Apart from a number of other differences, the last two approaches differ in that Zukoski and Saville considered type II specific adsorption, as defined in the introduction of sec. 3.6d, as compared with type I in the Mangelsdorf-White model. By both of these approaches it was possible to obtain ζ-consistency between electrophoresis and conductivity, using reasonable parameter values.

For practical purposes it would be expedient to have equations available featuring $K^{\sigma i}$ as a parameter. To this end, the required equations of sec. 4.6d could be modified. A numerical solution for large κa has been given by Kijlstra et al.[5], based on Fixman's theory. Results are presented in fig. 4.31. The reduction of the mobility by conduction behind the slip plane is obvious. For very high $K^{\sigma i}/K^{\sigma d}$ ratio the situation arises of systems with low mobility but high conductivity. In fact, mobility persists because of diffusiophoresis. Bacterial cells may be typical examples. It can be shown[6] that these curves can be well described by [4.6.46] after modification of the Dukhin number into

$$Du = \frac{\left[1 + K^{\sigma i}/K^{\sigma d} + 3m/z^2\right]}{\kappa a} \left\{ \exp\left(\frac{zF\zeta}{2RT}\right) - 1 \right\} \qquad [4.6.56]$$

Basically, this is based on the additivity of Du^i and Du^d, see [4.3.64], but the equation is rewritten in such a way that the conductivity ratio enters explicitly. For rigid particles [4.6.56] reduces to [4.3.67]. As was the case with fig. 4.26, the maximum is clearly visible. In [4.6.46], the positive and negative contributions to the mobility are clearly distinguished. The former, (the "brake") is weighed by Du. For $Du = 0$ it is absent and the Helmholtz-Smoluchowski limit is obtained.

[1] N.M. Semenikhin, S.S. Dukhin, *Koll. Zhur.* **37** (1975), 1127, transl. 1017.
[2] N.I. Zharkikh, H.P. Pendse and S.S. Dukhin, *Koll. Zhur.* **56** (1994) 573.
[3] C.F. Zukoski IV, D.A. Saville, *J. Colloid Interface Sci.* **114** (1986) 32.
[4] C.S. Mangelsdorf, L.R. White, J. Chem. Soc. Faraday Trans. **86** (1990) 2859,
[5] J. Kijlstra, H.P. van Leeuwen and J. Lyklema, *J. Chem. Soc. Faraday Trans.* **88** (1992) 3441.
[6] M. Minor, private communication.

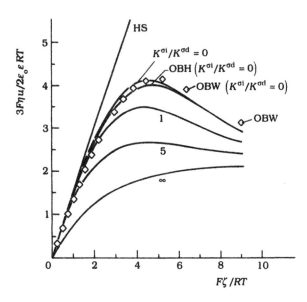

Figure 4.31. Influence of surface conduction on the electrophoretic mobility as a function of ζ. $\kappa a = 20$. Given is the ratio $K^{\sigma i}/K^d$. HS = Helmholtz-Smoluchowski, OBH = O'Brien-Hunter [4.6.44], OBW = O'Brien-White, exact numerical results for $K^{\sigma i}=0$ (fig. 4.26). (Redrawn from Kijlstra et al., loc. cit.) Further discussion in the text.

the higher Du, the lower the mobility, until for $Du \to \infty$ no maximum is found any more (this limiting case is rarely met in practice). Eventually, at very high ζ, all curves approach the same constant value. The double layer is then so strongly charged that the negative adsorption of co-ions has attained its constant maximum value, whereas the polarization of the counterionic charge around the particle obeys a Boltzmann-type of distribution where any further increase of ζ is fully compensated by additional polarization. this discussion shows that in this way a simple equation for the mobility is obtained that, for large κa, incorporates conduction in the stagnant layer. The problem is now that independent experimental information on $K^{\sigma i}$ is needed, which is hard to obtain.

Awaiting further information, table 4.3 gives indirect data, obtained in a subsequent paper by Kijlstra et al.[1]. This work concerned experiments on the electrophoretic mobility, conductivity and dielectric relaxation of (Stöber) silica and haematite (α - Fe_2O_3) sols. The reported values of $K^{\sigma i}/K^{\sigma d}$, are those needed to attain electrokinetic consistency between these three electrokinetic phenomena. The table also contains other relevant data. The values for $K^{\sigma d}$ have been obtained from the Bikerman theory (fig. 4.10), at the required electrolyte concentration, and u^i is derived from $K^{\sigma i}$, using [4.3.57]. For the

[1] J. Kijlstra, H.P. van Leeuwen and J. Lyklema, *Langmuir* **9** (1993) 1625.

Table 4.3. Surface conduction from electrokinetic consistency.

	pH	κa	$\dfrac{F\zeta}{RT}$	σ^{ek}	σ^i	$\dfrac{K^{\sigma i}}{K^{\sigma d}}$	$K^{\sigma d}$	$K^{\sigma i}$	u^i	av.
SiO_2,	7.4	19	-2.7	-0.7	0.6	1.2	3.6	4.3	7.2	
$a = 184$ nm	8.2	19	-2.9	-0.8	1.0	1.6	3.7	5.9	5.9	7.2
	8.6	19	-3.0	-0.8	1.0	1.7	5.2	8.8	8.8	
α - Fe_2O_3,	5.0	31	2.8	0.8	-3.8	5.8	3.7	21.5	5.6	
$a = 257$ nm	4.1	31	2.8	0.8	-3.9	5.9	3.7	21.8	5.6	5.6

σ^{ek} and σ^i in $\mu C\,cm^{-2}$; $K^{\sigma i}$ and $K^{\sigma d}$ in 10^{-10} S; u^i in $10^{-8}\,m^2\,V^{-1}\,s^{-1}$.

silica sol K^+ is the counterion, with a limiting mobility in bulk[1] u_{K^+} of $7.62 \times 10^{-8}\,m^2\,V^{-1}\,s^{-1}$, for haematite this role is played by Cl^-, $u_{Cl^-} = 7.91 \times 10^{-8}\,m^2\,V^{-1}\,s^{-1}$. It is gratifying, and in line with the picture developed in sec. 4.4, that u^i is close to (for K^+) or somewhat lower than (for Cl^-) the corresponding bulk values. The Stöber silica is porous, so $K^{\sigma i}$ may, to a large extent, be caused by conduction in pores. However, the haematite sample is non-porous, so that for this system $K^{\sigma i}$ is a pure Stern layer conductivity. If the difference between K^+ and Cl^- is real it would be in line with the stronger tendency of Cl^- ions to adsorb specifically, i.e. to be more resilient to tangential displacement.

In other experiments electrokinetic consistency could be achieved without consideration of conduction behind the plane of shear. Examples include streaming potentials and conduction for charged microporous membranes[2] and diffusiophoresis and electrophoresis of latices[3]. Obviously, the role of surface conduction depends on the nature of the system and on the way of measurement.

In conclusion, extensions of this type of study would be worthwhile.

4.6g. Effect of sol concentration

Most electrophoresis experiments are carried out micro-electrophoretically and hence with very dilute sols. When the sols are not so dilute, particle-particle interaction may occur and this may influence the mobility. The interaction has a two fold nature: electrostatic and hydrodynamic. Electrostatic interaction is of the range $O(\kappa^{-1})$, whereas the hydrodynamic interference is $O(a)$. For most systems it means that the latter predominates. Because of this, interaction

---

[1] Such bulk mobilities have been tabulated in table I.6.5, in sec. I.6.6a.
[2] G.B. Westermann-Clark, J.L. Anderson, *J. Electrochem. Soc.* **130** (1982) 839.
[3] J.P. Ebel, J.L. Anderson and D.C. Prieve, *Langmuir* **4** (1988) 396.

between particles undergoing electrophoresis, is much weaker than that between sedimenting particles.

According to general experience the influence of sol concentration is absent when the volume fraction remains of the order of a few percent. This may be concluded from older experiments[1,2] and from recent electroacoustic studies, discussed in sec. 4.5d. Experiments involving micro-electrophoresis are not suited to studying the volume fraction effect because the required extreme dilution may readily lead to spurious adsorption on the particles. Reed and Morrison[3] studied theoretically the hydrodynamic interactions between pairs of different particles in electrophoresis; they corrected the Helmholtz-Smoluchowski equation for various distances between the (spherical) particles and values of the electrokinetic potentials of the particles, ζ_1 and ζ_2.

A fairly detailed contribution has been given by Kozak and Davis[4], who computed the electrophoretic mobility of spherical particles of arbitrary κa in

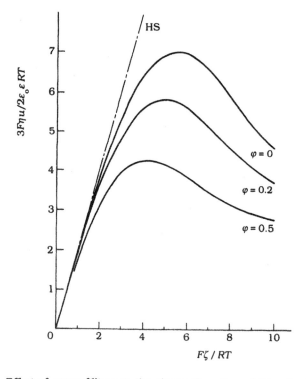

Figure 4.32. Effect of space filling on the electrophoretic mobility of spheres. $\kappa a = 50$, the volume fraction φ is indicated. HS = Helmholtz-Smoluchowski (Redrawn from Kozak and Davis, loc. cit).

[1] S.H. Maron, D. Turnbull and M.E. Elder, J. Am. Chem. Soc. **70** (1948), 582.
[2] E.A. Hauser, D.S. le Bean, J. Phys. Chem **45** (1941) 54.
[3] L.D. Reed, F.A. Morrison jr. J. Colloid Interface Sci. **54** (1976) 115.
[4] M.K. Kozak, E.J. Davis, J. Colloid Interface Sci. **129** (1989) 166.

an unconsolidated system containing many other particles, using a cell model. Electrical double layer overlap was ignored but there is of course hydrodynamic interaction and polarization was accounted for. In fig. 4.32 a result is given, valid for (z-z) electrolytes. Here, φ is the volume fraction of the particles. In the Helmholtz-Smoluchowski regime ($F\zeta/RT \leq 1.6$ or $\zeta \leq 40$ mV) there is no sol concentration effect[1], but beyond this range the reduction of the mobility may be up to 40% for $\varphi \rightarrow 0.5$, more or less proportionally, as has also been found experimentally with electroacoustics (sec. 4.5d). For $\varphi = 0$ the results are almost identical to those of O'Brien and White (fig. 4.26) and, hence, to the approximations of O'Brien and Hunter, [4.6.44] and of Ohshima et al., mentioned in connection with [4.6.45]. More recently, Kang and Sangani[2] carried out a numerical study of the same problem. These authors considered dipole moments, mobilities, conductance and resistance of randomly arranged and ordered arrays of spheres, again in the high κa limit. Conduction behind the slip plane was ignored, the analysis was numerical and also included polydispersity and the case of long aligned cylinders. Their result for $u(\zeta)$ shows the same qualitative trends as given in fig. 4.32.

For practical purposes the following equation may be useful:

$$u = \frac{\varepsilon_0 \varepsilon \zeta}{\eta}\left[1 - 1.5\,\varphi + O(\varphi^2)\right] \qquad\qquad [4.6.57]$$

It was derived by Chen and Keh[3] and applies to the Helmholtz-Smoluchowski range. The authors considered pair interactions only, but did so in some detail, also discussing differences in size, ζ-potential and alignment of the pair with respect to the applied field. Equation [4.6.57] was confirmed by Anderson[4]. So, in contradistinction to the work by Kozak and Davis, and by Shilov et al., these theories predict a finite, but minor dependence on sol concentration. Eventually experiments have to decide.

Studies like these may be considered as bridges between electrokinetics in dilute systems and in plugs (sec. 4.7).[5]

4.6h. Electrophoresis of particles with irregular shapes and charge distributions

It goes without saying that, considering the enormous problems to be solved in the simplest case of spherical particles at any κa, more difficult particles

---

[1] This result was also found by V.N. Shilov et al., using the cell model (Koll. Zhur. **43** (1981) 43, 540, 1061).

[2] S-Y Kang, A.S. Sangani, J. Colloid Interface Sci. **165** (1994) 195.

[3] S.B. Chen, H.J. Keh, Am. Inst. Chem. Eng. J. **34** (1988) 1075.

[4] J.L. Anderson, Ann. Rev. Fluid Mech. **21** (1989) 61.

[5] More information regarding secs. 4.6g and h can for instance be found in the reviews by Dukhin and Deryagin, and by Hidalgo-Alvarez, mentioned in sec. 4.11b.

pose almost untractable obstacles. This means that electrophoresis data of a number of important systems, such as clay mineral particles which have patch-wise heterogeneous surfaces, or powders of irregular shapes and sizes, are virtually non-interpretable. For clay minerals it even becomes difficult to say what the isoelectric point means: it may well be that particles of net zero charge, but with this charge patchwise distributed over the surfaces, do move in an applied field. All of this depends on the way the charge is distributed and on the way the particles move (is there a torque applied on them which gives rise to rotation?). We shall not discuss this matter in detail.

For non-spherical particles of homogeneous ζ-potential (cylinders, discs, rods ellipsoids of revolution, etc.) the induced dipole moment is different for different orientations with respect to the directions of the applied field, i.e. it acquires a tensorial nature. When the systems are sufficiently simple to require consider-ation of only two orientations (cylinders or rods) it is sufficient to distinguish only $p_{ind,//}$ and $p_{ind,\perp}$ where // and $\perp$ refer to parallel and normal orientation to the applied field, respectively. Dukhin has described the consequences for the overall polarization[1].

Regarding the electrophoretic mobilities of cylinders and rods, it is first recalled that in the absence of polarization the Henry approximation applies for $\perp$ and // orientations (dashed curves in fig. 4.4). For arbitrary orientation we have [4.3.11]. For more information on $u_\perp$ and $u_{//}$, see Van der Drift et al.[2]. The problems involve $u_\perp$ because $u_{//}$ may be simply interpreted as the Helmholtz-Smoluchowski mobility if the cylinders are long enough. Sherwood[3] extended Henry's results to include end effects for a charge cloud that was thick compared with the radius of the rod, but ignored polarization. In the limits of large $\kappa \ell$ and small $\kappa \ell$ (ℓ = length of the rod), the Henry results are retrieved. For very thick clouds (representative for double layers in non-polar media), their influence on the mobility may be ignored, a feature observed before. Harris[4] also investigated the mobility, when the double layer is very extended, considering oblate and prolate ellipsoids of revolution. He tabulates the ratios $u_\perp / u(\text{sph})$ and $u_{//} / u(\text{sph})$, where $u(\text{sph})$ is the mobility of a sphere of the same volume, as a function of the axial length ratio (the *aspect ratio*), p, defined as b/a, where b is the axis of revolution and a the maximum cross-sectional axis. In such systems the rotational ability of the particles may also count and hence it is not surprising that the magnitudes and signs of the p-effect have some correlation with those of the rotational diffusion of such non-spherical

[1] S.S. Dukhin, *Adv. Colloid Interface Sci.* **44** (1993), 1, especially sec. 3.5.
[2] W.P.J.T. van der Drift, A. de Keizer and J.Th.G. Overbeek, J. *Colloid Interface Sci.* **71** (1979) 67.
[3] J.D. Sherwood, *J. Chem. Soc. Faraday Trans. II* **78** (1982) 1091.
[4] L.B. Harris, *J. Colloid Interface Sci.* **34** (1970) 322.

particles (Perrin's analysis, see fig. I.6.19). The counterpart of this work, electrophoresis of spheroids with relatively thin double layers, has been considered by O'Brien and Ward[1]. These authors found, among other things, that v and E are not usually parallel. However, for the case of a sol with random orientation distribution the average velocity $\langle v \rangle$ is parallel to E. In that case one single mobility suffices to characterize the system electrokinetically. Figure 4.33 gives the result. It is seen that the particles move slower than according to Helmholtz-Smoluchowski (expected because of the braking effect caused by double layer polarization) but that the ratio tends to approach unity with increasing aspect ratio.

Particles of homogeneous ζ-potential do not rotate, whatever their shapes[2], provided velocity and applied field remain proportional. Therefore, a collection of non-spherical particles with a certain orientational distribution will retain this distribution upon electrophoresis. At the same time this makes the "averaging" of v over these orientations, as in [4.3.18], meaningful.

When the ζ-potential is not homogeneous, the particles tend to rotate to alignment with the field E. An illustrative example has been given by Fair and

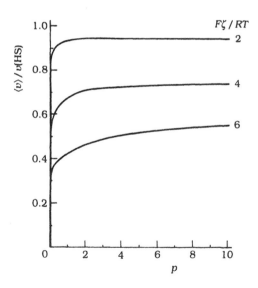

Figure 4.33. Electrophoresis of spheroids as a function of the aspect ratio p. Given is the ratio between the average velocity and the velocity according to the Helmholtz-Smoluchowski equation (Redrawn from O'Brien and Ward, loc. cit.).

[1] R.W. O'Brien, D.N. Ward, *J. Colloid Interface Sci* **121** (1988) 402.
[2] J.L. Anderson, *Ann. Rev. Fluid Mech.* **21** (1989) 61.

Anderson[1], who considered dumbbell shaped particles for which theory was derived and experiments carried out. The test particles were doublets of two latex spheres, one negative, the other positive (and different in size) that were formed from the singlets by heterocoagulation. For parallel and perpendicular orientation of the doublets with respect to E the following equations could be derived

$$v_{//} = \frac{\varepsilon_0 \varepsilon}{\eta} \left[(1 - K_{//}) \zeta_1 + K_{//} \zeta_2 \right] E \qquad\qquad [4.6.58a]$$

$$v_{\perp} = \frac{\varepsilon_0 \varepsilon}{\eta} \left[(1 - K_{\perp}) \zeta_1 + K_{\perp} \zeta_2 \right] E \qquad\qquad [4.6.58b]$$

where ζ_1 and ζ_2 are the electrokinetic potentials of components 1 and 2, respectively, and the parameters K only depend on the axial ratio, as given in table 4.4. (Note that these K's are not conductivities.) Of the two velocities $v_{//}$ was measured; its value was a weighted average between the two constituents of the doublet, agreeing well with the theoretical prediction, based on the two K and ζ-values. The perpendicular velocity cannot be measured because the particles rotate. This rotation is a novel and strong feature of heterogeneous doublets. For the angular velocity Ω the following expression was derived[1]:

$$\Omega = \frac{\varepsilon_0 \varepsilon (\zeta_2 - \zeta_1)}{(a_1 + a_2) \eta} N i \times E \qquad\qquad [4.6.59]$$

where i is the unit vector, pointing from the centre of particle 1 to that of particle 2 and N is a constant, depending on a_1/a_2, given in table 4.4. Experimentally, the time of alignment after reversal of the field could be established; it was also found in agreement with theory. In other papers theory

Table 4.4. Theoretical values for $K_{//}$, $K_{\perp}$ and N in [4.6.58 and 59].

axial ratio a_2/a_1	$K_{//}$	$K_{\perp}$	N
1	0.50	0.50	0.64
2	0.84	0.73	0.49
2.3	0.88	0.77	0.44
5	0.98	0.92	0.16

[1] M.C. Fair, J.L. Anderson, *Langmuir* **8** (1992) 2850; *Int. J. Multiphase Flow* **16** (1990) 663.

was derived for non-uniformly charged spheres with thin polarized double layers[1] and for ellipsoids[2].

Another class of "difficult" particles involves systems which are porous. Then they may no longer be classified as dielectric. Ion transport and electro-osmosis may then take place inside the particles. Miller et al.[3] found theoretically that porosity tends to increase or decrease the mobility for high and low c_{salt}, respectively. Such studies may also be interesting for the interpretation of the electrophoresis of aggregates of particles.

Generally speaking this field deserves further attention.

4.6i. Summary. Application in practice

How can the information of sec. 4.6 be used in practice? That depends on the purpose, which may well not be the same in academia and application.

For scientific purposes the understanding of charge distributions and their dynamics will be the main goals. Above, the main issues and outstanding problems have been discussed in some detail, so that a starting point for further study has been established. Perhaps the main conclusion is that *for most purposes one type of electrokinetic measurement does not usually suffice*. One should investigate at least two, but then the reward is also doubled because in addition to ζ also K^{σ} (preferably $K^{\sigma d}$ and $K^{\sigma i}$) is obtainable.

In more applied veins, ζ-potentials are not measured for their own sake, but to assess particle interaction, as is for instance required in rheology and colloid stability. These measurements have to be carried out for particles that are often ill-defined: the shapes may be known from electron microscopy, but they may be heterodisperse and heterogeneous and the surface composition may be unknown. For those cases the following summary may be helpful.

(i) Particles with very large κa: the Helmholtz-Smoluchowski equation [4.3.4] applies and remains an acceptable approximation when the particles are not spherical. The applicability limit can be inferred from fig. 4.26.

(ii) For very low κa: the Hückel-Onsager equation [4.3.5] is valid. Example: particles in media of low polarity. The applicability limit can be deduced from fig. 4.27.

(iii) When the ζ-potentials are low (which requires a measurement to know) double layer polarization is often also small so that the Henry or Ohshima approximations are acceptable (sec. 4.3a, sub iv).

(iv) In all other cases the above limiting laws are not better than semi-quantitative. Depending on ζ, κa and Du, the deviations may be a factor of 2 or more (see figures 4.26, 27, 30 and 31). The uncertainties for particle interaction

[1] Y.E. Solomentsev, Y. Pawar and J.L. Anderson, *J. Colloid Interface Sci.* **158** (1993) 1.

[2] M.C. Fair, J.L. Anderson, *J. Colloid Interface Sci.* **127** (1989) 388.

[3] N.P. Miller, J.C. Berg and R.W. O'Brien, *J. Colloid Interface Sci.* **153** (1992) 237.

may be dramatic because double layer overlap Gibbs energies scale with ζ^2. Even the i.e.p. does not necessarily coincide with the condition of $\langle\zeta\rangle = 0$, if the charge distribution is not uniform. For such "difficult" systems it is recommended to extend the electrophoresis measurements by streaming potential or conductivity experiments. Some commercial ESA or CVP apparatus also allow for this. The difference (if any) between ζ from the two techniques, using limiting formulas, gives at least an indication of the role played by surface conduction and helps to estimate ζ more precisely.

4.7 Electrokinetics in plugs and capillaries

Section 4.6 may be considered the prototype of modern electrokinetics, because all relevant features were covered: the coupling of hydrodynamic and electric fluxes and double layer polarization. However, the elaboration remained restricted to electrophoresis, which is the most familiar electrokinetic phenomenon. Other types of electrokinetics, summarized in table 4.1, basically require the same theory, although there may be considerable differences in the elaboration (what is stationary? what is moving? boundary condition?, etc.). With sec. 4.6 we consider the fundamentals sufficiently explained and illustrated and we shall therefore not repeat and apply this theory to other electrokinetic phenomena. Instead, two important extensions will now be briefly reviewed: inclusion of double layer overlap, as occurs in plugs, in the present section and measurement in alternating fields in the following.

4.7a. Conduction and flow

For the computation of streaming potentials and electro-osmotic volume flow one should know not only the conductivity of the electrolyte in the pores, K^L, but also the surface conductivity K^σ, see [4.3.24, 29 and 41]. Porous plugs, as met in practice, often contain an unknown array of irregularly stacked, heterogeneous, particles. Developing a rigorous theory for such systems is impossible for a number of obvious reasons, including the following:

(i) Convective (as in streaming potentials) and electro-osmotic flow takes place through tortuous channels of varying widths and shapes for which it is difficult to establish averages. Bulk volume flow, bulk conduction and electrokinetic quantities depend in very different ways on the local radius, i.e. with a^4, a^2 and a, respectively.

(ii) The same can be said about the bulk part of the conduction. However, surface conduction proceeds along the particle surfaces and the paths chosen by fluid flow and surface conduction are different. The former tends to be strongly dominated by the wider channels, because Poiseuille's law [I.6.4.17] predicts a proportionality to the fourth power of the radius, whereas surface conduction currents tend to avoid the wide channels. Close packing of the solid is favourable for surface conduction because that would facilitate transport of ions from one particle to the next. Such a jump possibility is particularly crucial when $K^{\sigma i} \gg K^{\sigma d}$, as may be frequently the case in practice. It becomes less relevant whent he bulk conductivity is large.

(iii) For the complex geometry of particles and liquids in plugs the electrokinetic equations of secs. 4.6a and b cannot be solved. For instance, there is not yet a solution for the Laplace equation for a random collection of identical spheres. To handle practical situations, substantial simplifications are therefore needed. Two ways are open: either very simplified models may be elaborated or experiments are coupled in such a way to other pieces of information that satisfactorily reliable data for ζ and K^{σ} are empirically estimated. Let us give illustrations of these approaches.

Theoretical geometrical idealizations can be divided into two categories: equivalent cylinder models and cell models. In the former, the plug or diaphragm is interpreted as a collection of parallel cylinders of given average radius $\langle a \rangle$. In cell models, the porous system is modelled as a collection of, usually homodisperse spherical, particles, organized into some three-dimensional array.

Relatively complete elaborations for the cylinder model have been given by for instance, Anderson and Koh[1] and Levine et al.[2]. In these two theories the solution is assumed to contain (1-1) electrolytes with $u_+ = u_-$. Both theories fail to account for conduction behind the slip plane, and both solve the electrokinetic equations, taking double layer overlap into account. Anderson and Koh assume this overlap to take place at fixed surface charge (which, because of the implicit rigid particle model of the cylinder wall, comes down to fixed $\sigma^d \approx \sigma^{ek}$), whereas Levine et al. do so for constant surface potential (essentially fixed $\psi^d \approx \zeta$). Anderson and Koh also considered capillaries of other shapes: ellipses and infinite slits. Approximations for low ζ and low σ^{ek} have been published earlier by Rice and Whitehead[3] and by Sørensen and Koefoed[4], respectively. By way of illustration we give some of Levine et al.'s results, which lend themselves more easily for analytical presentation.

The theory contains two parameters accounting for double layer overlap:

$$G(\kappa a, \zeta) = \frac{2}{F\zeta (\kappa a)^2 / RT} \int\limits_{0}^{\kappa a} \kappa r \, y(r) \, d(\kappa r) \qquad\qquad [4.7.1]$$

[1] J.L. Anderson, W. Koh, *J. Colloid Interface Sci.* **59** (1977) 149.
[2] S. Levine, J.R. Marriott, G. Neale and N. Epstein, *J. Colloid Interface Sci.* **52** (1975) 136.
[3] C.L. Rice, R. Whitehead, *J. Phys. Chem* **69** (1965) 4017.
[4] T.S. Sørensen, J. Koefoed, *J. Chem. Soc. Faraday Trans.* II **70** (1974) 665.

and

$$F_c(\kappa a, \zeta) = \cfrac{\frac{1}{2}(\kappa a)^2(1-G)}{\displaystyle\int_0^{\kappa a} \kappa r \cosh y(r)\, d(\kappa r) + \cfrac{(\varepsilon_0 \varepsilon \kappa RT)^2}{\eta K^L F^2} \int_0^{\kappa a} \kappa r \left(\cfrac{dy(r)}{d(\kappa r)}\right)^2 d(\kappa r)} \qquad [4.7.2]$$

The integrations can be carried out when the potential-distance behaviour is known. Levine et al. give plots of G and F_c. Various electrokinetic properties can be expressed in terms of these parameters. For the streaming current one may write

$$\left(\frac{I_{str}}{\Delta p}\right)_{E=0} = \frac{\varepsilon_0 \varepsilon \zeta}{\eta} \frac{1-G}{C} \qquad [4.7.3]$$

where C is the cell constant (the parameter relating the total pore conductance to the conductivity). Equation [4.7.3] extends [4.3.38]. The electrical resistance of the (bulk) liquid inside the capillary is

$$R = \frac{C F_c}{K^L (1-G)} \qquad [4.7.4]$$

and the streaming potential, which, according to the Onsager cross-relations, is equal to the electro-osmotic flux per unit of current, is given by

$$\left(\frac{E_{str}}{\Delta p}\right)_{I=0} = -\left(\frac{J_e}{I}\right)_{\Delta p=0} = \frac{\varepsilon_0 \varepsilon \zeta}{\eta K^L} F_c \qquad [4.7.5]$$

which is an extension of [4.3.42]. In sec. 4.7b we shall give an illustration.

   In addition to this theory, work by Ohshima and Kondo[1] for slit-shaped "cylinders", i.e. parallel plates, may be mentioned. The surfaces carry an ion-penetrable layer, in which the charges are uniformly distributed ("porous double layer"). Simple analytical formulas for the electro-osmotic velocity and the streaming potential are given for the case that the porous layer is thin as compared to the slit width and provided flow in that layer is suppressed. Donath and Voigt[2] considered a similar case. Cohen and Radke[3] studied streaming potentials in slits with patchwise periodical heterogeneous charge. The new feature is that flows are not parallel to the surface any more but that normal components arise. In connection with this complication, the simplifying assumption that the electrokinetic potential of such a surface is the (area-

[1] H. Ohshima, T. Kondo, J. Colloid Interface Sci. 135 (1990) 443.
[2] E. Donath, A. Voigt, J. Colloid Interface Sci. 109 (1985) 122.
[3] R.R. Cohen, C.J. Radke, J. Colloid Interface Sci. 141 (1991) 338.

weighted) average of the potentials of the patches has to be reconsidered, in agreement with conclusions in sec. 4.6h.

Cell models do not lend themselves so well to analytical representation. Examples of elaborations have been given by, for instance, Levine et al.[1]. O'Brien and Perrins[2] presented exact solutions for particles arranged in a cubic or random array. These theories consider double layer polarization but are restricted to systems containing homodisperse dielectric spheres and large κa. Large κa conditions imply absence of double layer overlap; the interaction between different particles in the plug is fully determined by the far fields, which considerably simplifies the mathematics. In cell models each particle is considered to be surrounded by an identical shell of electrolyte solution, containing the number of ions required to render particle + liquid envelope electroneutral. This last condition does not imply that $c_+ = c_-$. Levine et al. emphasized sedimentation potentials, but also gave equations for the conductivity and streaming potential. These expressions contain, as expected, the porosity of the system, but also some parameters, depending on the geometry and packing, that have to be evaluated numerically. The uncertainty about what boundary conditions to apply at the couter surface of the cell is absent in the O'Brien-Perrins model.

As an alternative to these, and other, theoretical idealizations there is the pragmatic alternative of the Briggs method, briefly discussed in sec. 4.6e. According to this procedure, K^σ is estimated by first swamping the plug with concentrated electrolyte $(K^L \gg K^\sigma)$ to determine the cell constant. This parameter is then used to compute K(plug) at low c_{salt}, so that, by subtraction of K^L from K(plug), K^σ can be assessed. That this approach has its virtues follows from experiments, also discussed in sec. 4.6e; inclusion of surface conduction in the conversion of the primary electrokinetic data leads to disappearance of the spurious $\zeta(c_{salt})$ maxima and can be achieved either by considering a model for K^σ (fig. 4.29) or by substituting experimental plug conductivities (i.e. including K^σ) in the electrokinetic equations (fig. 4.30). Obviously, one is better off for materials for which by alternative routes K^σ has been determined, see table 4.3 for haematite and silica.

In conclusion, this problem is rigorously solved only for very simple geometries. For all other cases only approximations are available. Comparison of experimental results with those from other types of electrokinetics may then be the only way to assess the quality of results obtained. In carrying out such comparisons the presumption is that ζ and K^σ are material properties, i.e.

---

[1] S.Levine, G. Neale and N. Epstein, *J. Colloid Interface Sci.* **57** (1976) 424.

[2] R.W. O'Brien, W.T. Perrins, *J. Colloid Interface Sci.* **99** (1984) 20.

determined only by the nature of the surface, its charge and the presence of adsorbates. If properly analyzed, different types of electrokinetics should therefore lead to electrokinetic and surface conduction consistency.

4.7b. *Experimental verification*

In view of our goals, we are looking for electrokinetic experiments with consolidated plugs of well-defined particles. Consideration of ill-defined materials can never lead to basic insight. Experiments with capillaries have been well understood since the 1930s. Some results for K^σ obtained with these systems have been mentioned in sec. 4.6e. Such values may be considered correct if K^σ and ζ are independent of the pore radius. Much research has been carried out on the influence of surface modification, especially by the Russian School[1].

At present it appears premature to give a table of well-established K^σ-values, because all reported data refer to one type of technique and one model. Older data for various glasses, obtained with capillaries[2], vary widely between different authors, with K^σ values ranging from those corresponding to $K^{\sigma d}$ (fig.

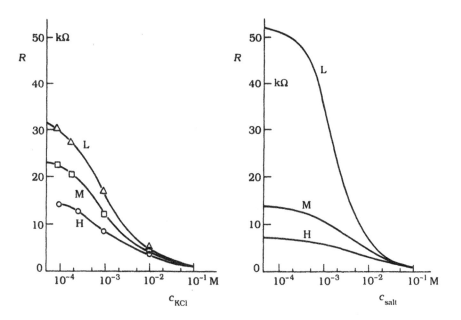

Figure 4.34. Resistance of polystyrene plugs. Left: experiments, right calculated according to [4.7.4]. Temperature 25°C. Further discussion in the text. (Redrawn from van der Put and Bijsterbosch, loc. cit.)

[1] D.A. Fridrikhsberg, K.P. Tikhomolova and M.P. Sidorova, *Croat. Chim. Acta* **52** (1979) 125; T.P. Golub, M.P. Sidorova, D.A. Fridrikhsberg, G.P. Lepnev and A.A. Belyustin, *Koll. Zhur.* **48** (1986), **428** (transl. 367); N.V. Churaev, S.P. Sergeeva, V.D. Sobolev and D.E. Ulberg, *J. Colloid Interface Sci.* **151** (1992) 490.
[2] See the tabulation by J.Th.G. Overbeek in *Colloid Science*, Vol **I**, H.R. Kruyt, Ed. Elsevier (1952) 236.

4.10) to several orders higher. Because of porous surface structures of glasses, K^σ for these systems is very sensitive to pH, mechanical and chemical pretreatment, and history (ageing). The data for (Stöber) silica and haematite in table 4.3 may also be recalled. For these systems K^σ was obtained as an adjustable parameter, needed to attain ζ-consistency.

Regarding well-defined plugs, figures 4.34 and 35 serve our purposes. They stem from work by van der Put and Bijsterbosch[1] on compacted plugs of three homodisperse polystyrene latices, called L, M and H for low, medium and high surface charge density σ°. These charges are caused by covalently bound sulfate groups; the charge densities were obtained conductometrically and amounted to 0.91, 4.00 and 9.21 $\mu C\ cm^{-2}$, respectively. The particle radii are 253, 305 and 290 nm, in this order.

The behaviour of the resistance is as expected. At low salt concentration R decreases with increasing surface charge on the latex, but at high c_{salt} there is no difference between the latices because now the bulk conductivity dominates $(R^L \ll R^\sigma)$. There is a tendency of R to level off at very low electrolyte concentration: then the particles are effectively so close to each other that they are "short-circuited", leading to a constant polarization. The theory (capillary model) applies semiquantitatively; it becomes more defective at low c_{salt} because polarization is not accounted for.

Similar comments can be made regarding the streaming potential, fig. 4.35.

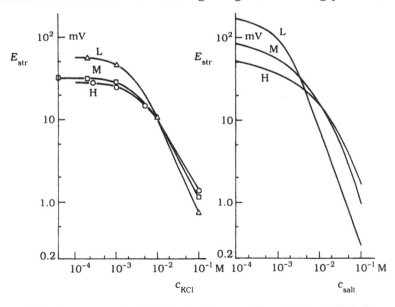

Figure 4.35. Streaming potentials for the plugs considered in fig. 4.34. Left: experiments; right eq. [4.7.5]. $\Delta p = 30$ cm Hg, temperature 25°C.

[1] A. van der Put, B.H. Bijsterbosch, *J. Colloid Interface Sci.* **92** (1983) 499.

The reliability of the experiments was verified by also carrying out electro-osmosis and checking the Saxèn relation. Note that this is a double-logarithmic plot. It is interesting to note that the authors found that agreement between theory and experiment improved if it was assumed that on the latex surface there was a "hairy" layer, in which conduction could take place, but which did not allow electro-osmosis. In fact this comes down to the desirability of accounting for conduction behind the slip plane. A similar conclusion was drawn by O'Brien and Perrins[1], who compared their cell model with these experiments. The surface conductivity of sample M was computed using the Dukhin-Semenikhin model[2,3] and amounted to about 3×10^{-9} S in 2×10^{-3} M KCl. Comparison with fig. 4.9 indicates that probably $K^{\sigma d}$ is not high enough to account for this value, so that also for latices conduction behind the slip plane is likely to occur.

4.8 Electrokinetics in alternating fields. Dielectric dispersion

Studying the response of colloidal systems under the influence of applied a.c. electric fields may be motivated by two reasons. One is of practical interest: a.c. fields may help to suppress electrode polarization and enable the implementation of certain electronic gadgets. The other is more basic: a.c. measurements allow relaxation phenomena to be studied.

Various relaxation processes may be distinguished; we introduced these in sec. I.4.4e. Some of these involve molecular processes: relaxation of dipoles (Debye relaxation) and atomic relaxation (at optical frequencies $\varepsilon \to n^2$). Such relaxations occur at high frequencies, (far) above the MHz range. Then there is the Maxwell double layer relaxation, with $\tau = O(D\kappa^2)^{-1}$ which is in the MHz range. Our present interest is focused on the low frequency relaxation of colloidal particles, because it reflects the distributions and mobilities of ions in the various parts of the double layer and in the far field. The relaxation frequency is relatively low because it is determined by the time needed to transport the excess electroneutral electrolyte in the far field, created by polarization of the particle, from one side of the particle to the other. This is a diffusion-determined process, taking place over a fairly long distance; $\tau = O(a^2/D)$ and relaxation takes place in the kHz range. In the present section only this type of relaxation will be discussed.

As electrical characterization of double layers is our present aim, particle interaction will be ignored. In practice this means that the volume fraction φ of

[1] R.W. O'Brien, W.T. Perrins, *J. Colloid Interface Sci.* **99** (1984) 20.
[2] A. van der Put, B.H. Bijsterbosch, *J. Colloid Interface Sci.* **75** (1980) 512.
[3] S.S. Dukhin, N.M. Semenikhin, *Koll. Zhur.* **32** (1970) 360 (transl. 298)

the sol has to remain below about 5% [1]. As a further restriction, only homo-disperse spherical particles will be considered.

Three parameters offer themselves as measurable quantities: the conductivity, the dielectric permittivity and the dielectric loss. The conductivity $K(\omega)$ is an ascending function of ω: at low ω the double layer polarization prevents ions to move freely around the particle, whereas at high ω polarization cannot establish itself. For low mol mass electrolytes this rise has also been established and is known as the Debye-Falkenhagen effect (sec. I.6.6c). For the dielectric permittivity $\varepsilon'(\omega)$ the trend is descending, because at low ω double layer polarization enhances ε'. These two trends are shown in fig. 4.23. The dielectric loss $\varepsilon''(\omega)$, which is related to $K(\omega)$, is a curve with its maximum around the relaxation frequency, because there maximum dissipation of electric energy in heat takes place. At very low frequency $\varepsilon''(\omega)=0$ because the double layer can follow the oscillations of the field, at high frequency $\varepsilon''(\omega)$ is again zero; then the double layer cannot polarize, so that there is nothing to dissipate.

Perhaps the most basic double layer characteristic for these phenomena is the induced dipole moment $\hat{p}_{ind}$, which can be derived from $\hat{\varepsilon}(\omega)$ and/or $\hat{K}(\omega)$. The $\wedge$ indicates that the quantities are complex: the orientation of the induced dipole lags behind the field.

For the very simplified situation that the sphere behaves electrically as a pure capacitor, and the solution as a pure resistance, the relaxation can be described by a Maxwell-Wagner mechanism, with $\tau \approx \varepsilon_0 \varepsilon / K$, see [I.6.6.32]. Although some success has been claimed by Watillon's group [2] to apply this mechanism for a model, consisting of shells with different values of ε and K, generally a more detailed double layer picture is needed. In fact, this implies starting from the transport equations of secs. 4.6a and b, generalizing these to the case of a.c. fields.

4.8a. Dielectric and conductivity relaxation. General considerations

First the phenomenology is discussed; it applies for any double layer relaxation model.

As discussed in sec. 4.5e the complex conductivity $\hat{K}(\omega)$ is measurable. It contains all the required conduction and dielectric properties and can be written as [3]

$$\hat{K}(\text{sol}, \omega) = K^L(\omega) - \omega\varepsilon_0\hat{\varepsilon}(\text{sol}, \omega) = K^L(\omega) - \omega\varepsilon_0[\varepsilon''(\text{sol}, \omega) - i\,\varepsilon'(\text{sol}, \omega)] \qquad [4.8.1]$$

[1] M.M. Springer, A. Korteweg and J. Lyklema, *J. Electroanal. Chem.* **143** (1983) 55; it is likely that interaction would reduce the permittivity increment: W.C. Vogel, H. Pauly, *J. Chem. Phys.* **89** (1988) 3823, 3830.

[2] A. Watillon, J. Stone-Masui, *J. Electroanal. Chem* **37** (1972) 143.

[3] Note that different authors use different sign conventions in writing complex quantities ($\hat{a} = b + ic$ or $\hat{a} = b - ic$) and alternating fields ($E = E_0 e^{i\omega t}$ or $E = E_0 e^{-i\omega t}$).

An alternative way of writing this is

$$\hat{K}(sol, \omega) = K^L(\omega) + \omega\,\varepsilon_0\,\varepsilon''(sol, \omega) - i\,\omega\,\varepsilon_0\,\varepsilon'(sol, \omega) \qquad\qquad [4.8.1a]$$

$$= K(sol, \omega) - i\,\omega\,\varepsilon_0\,\varepsilon'(sol, \omega) \qquad\qquad [4.8.1b]$$

Here $K^L(\omega)$ is the conductivity of the liquid. As we do not consider the high frequencies where Debye-Falkenhagen relaxation becomes prominent, this is a real quantity. Further, $\varepsilon_0\hat{\varepsilon}(\omega)$ is the complex dielectric permittivity, consisting of its real (or storage) part, $\varepsilon'(\omega)$ and its imaginary (or loss) part, $\varepsilon''(\omega)$. The fact that in the "conductivity-based" equation [4.8.1b] the dielectric loss appears without the i, and the storage part with it, is just a consequence of the formalism: had we started from $\varepsilon_0\hat{\varepsilon}(\omega)$, decomposing it into its real and imaginary parts, then $\varepsilon''(\omega)$ would have carried the i, and been interpreted as conductive loss. In fact, that is the physics of it: polarization leaks away by conduction and/or diffusion. As discussed before, the "conductivity-based" and "permittivity-based" approaches are equivalent; they are related through the Kramers-Kronig equations. Regarding the experiments, it is noted from comparison with [4.5.13] that $\hat{K}(sol)$ is just the admittance $\hat{Y}$ of the sol. The real and imaginary parts, $K(sol)$ and $\omega\,\varepsilon_0\,\varepsilon'$, respectively, can be measured.

In the further elaboration two different types of conductivity and dielectric increment can be introduced, *viz.*, with respect to ω and with respect to the volume fraction φ of the colloid. (In static experiments only the latter is relevant). Regarding the former, we can write

$$\Delta_{fr}K(\varphi, \omega) = K(\varphi, \omega) - K(\varphi, \omega_r) \qquad\qquad [4.8.2]$$

$$\Delta_{fr}\varepsilon(\varphi, \omega) = \varepsilon(\varphi, \omega) - \varepsilon(\varphi, \omega'_r) \qquad\qquad [4.8.3]$$

where ω_r is a certain reference frequency. Usually, for the conductivity $\omega_r = 0$ and for the dielectric permittivity $\omega'_r = \infty$ is chosen. See fig. 4.23 for an example.

Regarding the sol concentration effect, assuming φ to be low enough to ensure linearity, the increments can be related to the number of particles per unit volume N_p and the induced dipole moment of each particle, $\hat{p}_{ind}$:

$$\hat{K}(sol, \omega) = \hat{K}^L\left[1 + A\,N_p\,\hat{p}_{ind}(\omega)\right] \qquad\qquad [4.8.4]$$

$$\varepsilon_0\hat{\varepsilon}(sol, \omega) = i\varepsilon_0\varepsilon^L\left[1 + A\,N_p\,\hat{p}_{ind}(\omega)\right] - \frac{K^L(\omega)}{\omega}\,A\,N_p\,\hat{p}_{ind}(\omega) \qquad\qquad [4.8.5]$$

where the constant A has units $C^{-1}m^2$, i.e. it is a reciprocal polarization. The notion of polarization has been introduced in sec. I.4.5f. It is the charge passing through an interface due to polarization. Sometimes the term with $A\,\hat{p}_{\text{ind}}$ is called the dipole strength. Another useful relation, valid for dilute sols of spherical particles[1,2], is:

$$\Delta\hat{K}(\varphi,\omega) = 3\varphi\,\hat{d}_e\,\hat{K}^L(\omega) \qquad [4.8.6]$$

Here, $\hat{d}_e$ is a quantity to be defined later, see [4.8.23], and which is proportional to $\hat{p}_{\text{ind}}$, see the discussion following [4.8.29]. As in [4.8.1b], the complex conductivity $\hat{K}^L(\omega)$ of the background electrolyte can be written as $K^L - i\omega\varepsilon_0\varepsilon'^L$, where K^L and ε'^L are the static conductivity and static relative dielectric constant of the background electrolyte, respectively.

We shall now consider the increments as they follow from the equations of secs. 4.6a and b.

4.8b. More detailed theory

Starting points are the equations of secs. 4.6a and b. Most of these remain unchanged in an alternating field. This is the case for the equations of motion, the flux balance and the Poisson equation, because the potential can adjust itself instantaneously to a given charge distribution. However, ionic distributions, particularly those in the far field, require a finite time to adjust to the changing field. Therefore, these may lag behind to an extent, depending on ω. As $\hat{p}_{\text{ind}}$ depends on these distributions, it also becomes frequency-dependent. At sufficiently high ω, electro-osmosis is also suppressed.

Regarding the Stern-diffuse layer border, at low ω the situation remains as in the static case. This situation has been discussed in sec. 3.13. At high frequency, polarization of the Stern layer may to a certain extent leak away via the, more mobile, diffuse part. Otherwise stated, the Stern layer is then shunted, or short-circuited, by the diffuse part.

A number of theories have been developed by the same authors as those who have given the theoretical foundations for electrophoresis. The Dukhin-Shilov theory[3], one of the older approaches, had an analytical nature. Many of the important features can already be found there, though it was restricted to large κa. This limitation was dropped in De Lacey and White's numerical theory[4]. Later, O'Brien[5] presented semi-analytical formulas for $\zeta \leq 50$ mV and

---

[1] S.S. Dukhin, V.N. Shilov, *Dielectric Phenomena and the Double Layer in Disperse Systems and Polyelectrolytes*, Wiley (1974).

[2] E.H.B. DeLacey, L.R. White, *J. Chem. Soc. Faraday Trans. (II)* **77** (1981) 2007.

[3] S.S. Dukhin, V.N. Shilov, *Dielectric Phenomena and the Double Layer in Disperse Systems and Polyelectrolytes*, Wiley-Interscience (1974).

[4] E.H.B. De Lacey, L.R. White, *J. Chem. Soc. Faraday Trans. (II)* **77** (1981) 2007.

[5] R.W. O'Brien, *Adv. Colloid Interface Sci.* **16** (1982) 281.

Fixman[1] gave analytical expressions for high κa and $\zeta \leq 200$ mV. Chew and Sen[2] offered analytical equations valid to the first order in $(\kappa a)^{-1}$. As was the case for the electrophoretic mobility, the outcomes of these theories do not differ greatly. Fixman already compared his data with the numerical results of De Lacey and White and came to this conclusion, later elaborations confirmed it[3]. We shall now present a relatively simple analytical derivation that should be of relatively wide application.

The task is to derive an expression for $A\hat{p}_{ind}$, because from that $\hat{\varepsilon}(\omega)$ and $\hat{K}(\omega)$ are immediately obtained, using [4.8.4 and 5]. In practice $\hat{d}_e$ is computed and [4.8.6] used. The static dipole moment has been given in [3.13.4-6]. The equation from which $\hat{p}_{ind}$ is obtained is the Laplace equation.

The starting equations are, as before, those for conservation of mass, formulated as the Nernst-Planck equation, [4.6.2 or 2a], for conservation of momentum [4.6.5 or 9], the Poisson equation [4.6.13 or 12] and the condition for incompressibility, $\nabla \cdot \boldsymbol{v} = 0$. Mass conservation must apply to each ionic species, i.e. [4.6.1] should hold for any i.

We shall use the conservation of momentum equation in the form [4.6.6]. The argument for doing so is that the time derivative terms contribute only weakly to the force balance, as already pointed out by DeLacey and White. For high Reynolds number cases this simplification is not allowed, but we shall not consider this.

As the next step, we consider, as usually, small perturbations, that is, only deviations from equilibrium that are linear in $\boldsymbol{E}$ are considered. We apply [4.6.24 and 25] for $c_i(\boldsymbol{r}, t)$ and $\psi(\boldsymbol{r}, t)$, respectively. Substitution in [4.6.2] leads to

$$\frac{\boldsymbol{j}_i(\boldsymbol{r}, t)}{z_i F} = -D_i \left[\frac{z_i F}{RT} \left\{ c_i(\boldsymbol{r}, eq) \delta \nabla \psi(\boldsymbol{r}) + \delta c_i(\boldsymbol{r}) \nabla \psi(\boldsymbol{r}, eq) \right\} + \nabla \delta c_i(\boldsymbol{r}) \right] + \boldsymbol{v}(\boldsymbol{r}) c_i(\boldsymbol{r}, eq)$$

[4.8.7]

$$= -c_i(\boldsymbol{r}, eq) D_i \nabla \left[\frac{z_i F}{RT} \delta \psi(\boldsymbol{r}) + \frac{\delta c_i(\boldsymbol{r})}{c_i(\boldsymbol{r}, eq)} \right] + \boldsymbol{v}(\boldsymbol{r}) c_i(\boldsymbol{r}, eq)$$

[4.8.7a]

This equation formulates the ionic fluxes as caused by conduction, diffusion and convection. Application to the far field involves the substitution of $c_i(\infty)$ (the bulk value), for $c_i(\boldsymbol{r}, eq)$. Considering gradients of the fluxes,

$$-\frac{\nabla \cdot \boldsymbol{j}_i^{ff}}{z_i F} = \frac{\partial}{\partial t} \delta c_i^{ff} = c_i(\infty) D_i \nabla^2 \left[\frac{z_i F}{RT} \delta \psi^{ff} + \frac{\delta c_i^{ff}}{c_i(\infty)} \right]$$

[4.8.8]

[1] M. Fixman, J. Chem. Phys. **78** (1983) 1483.
[2] W.C. Chew, P.N. Sen, Proc. Electrochem. Soc. **85** (1985) 94.
[3] See, for instance, F. Carrique, L. Zurita and A.V. Delgado, J. Colloid Interface Sci. **170** (1995) 176.

The convection term dropped out because $c_i(\infty)\nabla\cdot\boldsymbol{v} = 0$. The two terms in square brackets are related through

$$\nabla^2\delta\psi = -\frac{\Sigma_i z_i F\delta c_i}{\varepsilon_0\varepsilon}$$ [4.8.9]

which is the Poisson equation for the perturbations.

In the following we shall restrict ourselves to binary symmetrical electrolytes of which the ions have equal diffusion coefficients $(D_+ = D_- = D)$. Then the concentration profiles are identical for all ions. For that situation [4.8.8] reduces to

$$\frac{\partial\delta c_s^{ff}}{\partial t} = D\nabla^2\delta c_s^{ff}$$ [4.8.10]

and [4.8.9] to

$$\nabla^2\delta\psi^{ff} = 0$$ [4.8.11]

Equation [4.8.10] is Fick's second law, which we have seen before, see [4.6.34b] and [4.8.11] is equivalent to [4.6.14]. In passing it is mentioned that for symmetrical electrolytes of unequal diffusion coefficients, this pair of equations has to be replaced by

$$\frac{\partial\delta c_s^{ff}}{\partial t} = 2\left(\frac{D_+D_-}{D_+ + D_-}\right)\nabla^2\delta c_s^{ff}$$ [4.8.12]

and

$$\nabla^2\delta\psi^{ff} = \frac{1}{Fc_s}\left(\frac{D_+ - D_-}{2zD_+D_-}\right)\frac{\partial\delta c_s^{ff}}{\partial t}$$ [4.8.13]

respectively. In this case $\delta c_+^{ff} \approx \delta c_-^{ff} = \delta c_s^{ff}$, but the difference between δc_+^{ff} and δc_-^{ff} is much smaller than the absolute values.

Returning to the case of equal diffusion coefficients, the task is now to solve [4.8.10 and 11], subject to the outer boundary conditions

$$\delta c_s^{ff}(r\to\infty) = 0$$ [4.8.14]

$$\delta\psi^{ff}(r\to\infty) = -\boldsymbol{E}(t)\cdot\boldsymbol{r}$$ [4.8.15]

and to two, more system-specific, inner boundary conditions. We need the total ionic transport from the far field into the double layer, or the other way around. To that end, we integrate [4.8.7a] with respect to r:

$$\frac{\partial}{\partial t} \int \delta c_i \, dr - D_i \int c_i(eq) \nabla_t^2 \left\{ \frac{z_i F}{RT} \delta \psi + \frac{\delta c_i}{c_i(eq)} \right\} dr$$

$$-D_i c_i(\infty) \frac{\partial}{\partial r} \left[\frac{z_i F}{RT} \delta \psi + \frac{\delta c_i}{c_i(eq)} \right] + \int \nabla \cdot (c_i(eq) \boldsymbol{v}) \, dr = 0 \qquad [4.8.16]$$

For simplicity of notation, we have dropped the (r)'s after c_i and ψ, and have written most of the equation in scalar form. Integration takes place only in radial direction. As before, ∇_t is the tangential component of ∇ and it was assumed that D_i and $c_i(eq)$ do not depend on θ. The terms $z_i F \delta \psi / RT + \delta c_i / c_i(eq)$ together constitute the perturbation in the (dimensionless) electrochemical potential of species i, $\delta \tilde{\mu}_i$. Because of the local equilibrium $\delta \tilde{\mu}_i$ will be approximately equal to $\delta \tilde{\mu}_i^{ff}$. The time derivative is again negligible, as demonstrated by Fixman[1]. The lower integration border depends on the model: it is different between the cases with and without conduction behind the slip plane, i.e. between models with and without regions of zero $\boldsymbol{v}$. In the absence of such conduction, the integration starts at the slip plane (rigid particle theory). The upper boundary is in principle infinity, but as we only need the excesses caused by the polarized particle, integration over $c_i(eq) - c_i(\infty)$ is required: this increment goes to zero for $r \to \infty$ so that the integration converges. With this in mind, [4.8.16] reduces to

$$-D_i \nabla_t^2 \left[\frac{z_i F}{RT} \delta \psi^{ff} + \frac{\delta c_i^{ff}}{c_i(\infty)} \right]_{surf} \int [c_i(eq) - c_i(\infty)] \, dr$$

$$-D_i c_i(\infty) \frac{\partial}{\partial r} \left[\frac{z_i F}{RT} \delta \psi^{ff} + \frac{\delta c_i^{ff}}{c_i(\infty)} \right]_{surf} + \int [c_i(eq) - c_i(\infty)] \nabla_t \cdot \boldsymbol{v} \, dr = 0 \qquad [4.8.17]$$

The flow term, which involves the transport of all ionic species, can be solved on the basis of the Navier-Stokes equation and can be expressed in terms of $\tilde{\mu}_i^{ff}$. Let us make one more restriction, namely that the surface potential is not very low, so that the countercharge is essentially determined by counterions; the terms with $[c_i(eq) - c_i(\infty)]$ for co-ions are then relatively small. In that case, [4.8.17] can be written as a set of uncoupled equations for each individual ionic species. For the co-ions the result is simple:

$$\frac{\partial}{\partial r} \left[\frac{zF}{RT} \delta \psi^{ff} + \frac{\delta c^{ff}}{c(\infty)} \right]_{surf} = 0 \qquad [4.8.18]$$

---

[1] M. Fixman, *J. Chem. Phys.* **72** (1980) 5177, especially 5183.

This equation expresses that the co-ion flux into (or from) the double layer is negligible; this is in line with the fact that there is no significant lateral transport of these ions. For the counterions the result can be written as

$$2Du^da\,\nabla_t^2\left[-\frac{zF}{RT}\delta\psi^{ff}+\frac{\delta c^{ff}}{c(\infty)}\right]_{surf}+\frac{\partial}{\partial r}\left[-\frac{zF}{RT}\delta\psi^{ff}+\frac{\delta c^{ff}}{c(\infty)}\right]_{surf}=0 \qquad [4.8.19]$$

In this case,

$$Du^d=\frac{1}{\kappa a}\left(1+\frac{3m}{z^2}\right)\left(e^{zF|\zeta|/2RT}-1\right) \qquad [4.8.20]$$

which is a variant of [4.3.67], to which it approaches when ζ is high. The parameter m, which reflects the electro-osmotic contributions to the surface conduction, enters [4.8.19] via the last term of [4.8.17]. Through [4.8.18 or 19] concentrations and potentials on the inner border of the far field are coupled; these boundary conditions have to be considered when solving [4.8.10 and 11] in the far field. The results are a variant of the static case. The applied field is now $E(t)=Ee^{-i\omega t}$ and in the far field the Helmholtz rather than the Laplace equation is required. For the potential and concentration perturbation in the far field we can now write[1]

$$\delta\psi^{ff}=\left[-Er\cos\theta+\frac{\hat{d}_e a^3}{r^2}E\cos\theta\right]e^{-i\omega t} \qquad [4.8.21]$$

$$\delta c^{ff}=\left[\frac{\hat{d}_c a^3}{r^2}(1+\hat{\lambda}r)e^{-\hat{\lambda}r}E\cos\theta\right]e^{-i\omega t} \qquad [4.8.22]$$

where now complex quantities enter the equation. They are defined as

$$\hat{d}_e=-\frac{1}{2}+\frac{3Du^d}{2+2Du^d(\gamma+1)} \qquad [4.8.23]$$

$$\hat{d}_c=-\frac{c(\infty)zF}{4RT}\frac{6Du\exp(\hat{\lambda}a)}{\left[1+\hat{\lambda}a+\frac{1}{2}(\hat{\lambda}a)^2\right]+2Du\left[1+\hat{\lambda}a+\frac{1}{4}(\hat{\lambda}a)^2\right]} \qquad [4.8.24]$$

$$\hat{\lambda}a=(1-i)(\omega\tau)^{1/2} \qquad [4.8.25]$$

$$\gamma=\frac{1+\hat{\lambda}a}{1+\hat{\lambda}a+\frac{1}{2}(\hat{\lambda}a)^2}=1-\frac{-i\omega\tau+(\omega\tau)^{3/2}}{(1+\sqrt{\omega\tau})(1+\omega\tau)} \qquad [4.8.26]$$

and

---

[1] There is a certain analogy with R.W. O'Brien, *J. Colloid Interface Sci.* **113** (1986) 81, where further details can be found.

$$\tau = a^2 / 2D \tag{4.8.27}$$

Equation [4.8.22] is the solution of [4.8.10]. To find the excess electrolyte profile which is encountered in physical reality, the real part of this equation has to be taken. This can be obtained by using Euler's equation, [I.A8.8]. The result is given in fig. 4.36. The curve for $\omega\tau = 0$ is the static limit. Then δc^{ff} decays as r^{-2}. With increasing frequency oscillations start to occur. Ranges where δc^{ff} differs from zero stretch till several times the particle diameter.

For $\omega \to 0$ [4.8.21 and 22] reduce to their static equivalents, [4.6.53 and 54], respectively.

The parameter $\hat{d}_e$ in [4.8.21] can be written explicitly in terms of Du and $\omega\tau$, using [4.8.26 and 25]. The resulting expression is rather involved but can be simplified without significant loss of accuracy for the low frequency field by letting $(\omega\tau)^{3/2}/(1+\omega\tau)^2 \ll 2$, leading to

$$\hat{d}_e = -\frac{1}{2}\left[\frac{1-Du^d}{1+2Du^d} - \frac{3(Du^d)^2}{(1+2Du^d)^2}\left(\frac{\omega\tau\sqrt{\omega\tau}}{(1+\omega\tau)(1+\omega\tau) - Du^d(\omega\tau)^{3/2}/(1+2Du^d)}\right.\right.$$

$$\left.\left. -\frac{i\omega\tau}{(1+\omega\tau)(1+\sqrt{\omega\tau})}\right)\right]$$

$$\tag{4.8.28}$$

For zero frequency $\hat{d}_e$ reduces to its static equivalent, d_e, which is equal to

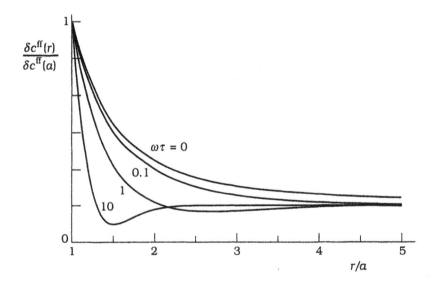

Figure 4.36. Concentration perturbations in the far field. Given is $\delta c^{\mathrm{ff}}(r)$, normalized with respect to $\delta c^{\mathrm{ff}}(r = a)$, for various values of $\omega\tau$. Conditions as in [4.8.22]. The curves are drawn in such a way that at $r = a$ the maximum is attained.

$$d_e = -\frac{1}{2}\frac{1-Du^d}{1+2Du^d} \tag{4.8.29}$$

This is identical to the coefficient in the 2nd term on the r.h.s. of [4.6.54]. According to [3.13.6] this factor is proportional to the induced dipole moment of a static double layer, $\mathbf{p}_{ind}$. In this way it can now be seen that the second term on the r.h.s. of [4.8.28] accounts for the frequency-dependent part in alternating fields. This is the part that makes the polarizability complex. It consists of a real and an imaginary part, these parts depending in a different way on $\omega\tau$: $Re(\hat{d}_e)$ increases with Du and $\omega\tau$ (except for $Du = 0$ when it is –0.5 at any $\omega\tau$), but $I_m(\hat{d}_e)$ passes through a maximum as a function of Du, the maximum being higher for higher Du. This maximum finds its origin in the inability to form electroneutral electrolyte clouds at high frequencies.

From [4.8.6 and 1b], $\hat{K}(sol, \omega) = K(\varphi, \omega) - i\omega\varepsilon_0\varepsilon'(\varphi, \omega) = \hat{K}^L + 3\varphi\hat{d}_e(K^L - i\omega\varepsilon_0\varepsilon'^L)$ from which $K(\varphi, \omega) = K^L + 3\varphi\,Re[\hat{d}_e(K^L - i\omega\varepsilon_0\varepsilon'^L)] = K^L + 3\varphi[K^L Re\,\hat{d}_e + \omega\varepsilon_0\varepsilon'^L\,Im\,\hat{d}_e]$, where in the last equation the second term is negligible as compared to the first, so that $K(\varphi, \omega) = K^L + 3\varphi\,K^L Re\,\hat{d}_e$ remains. Similarly, $\varepsilon'(\varphi, \omega) = \varepsilon'^L - 3\varphi K^L\,Im\,\hat{d}_e/\omega\varepsilon_0$. In this way, eventually the following final equations are obtained:

$$\frac{\Delta K(\omega)}{K^L} = \frac{9}{8}\varphi\left(\frac{2Du^d}{1+2Du^d}\right)^2\frac{\omega\tau\sqrt{\omega\tau}}{(1+\omega\tau)(1+\sqrt{\omega\tau}) - Du(\omega\tau)^{3/2}/(1+2Du)} \tag{4.8.30}$$

and

$$\frac{\Delta\varepsilon'(\omega)}{\varepsilon'^L} = \frac{9}{16}\varphi(\kappa a)^2\left(\frac{2Du^d}{1+2Du^d}\right)^2\frac{1}{(1+\omega\tau)(1+\sqrt{\omega\tau})} \tag{4.8.31}$$

Recall that in all these equations Du^d is given by [4.8.19]; in other approximations other expressions for Du^d will be needed.

The trends of $\Delta K(\omega)$ and $\Delta\varepsilon(\omega)$, displayed in fig. 4.23 are immediately read from these equations. $\Delta K(\omega) \to 0$ for $\omega \to 0$ and at high ω attains a plateau of

$$\frac{\Delta K(\omega\tau \gg 1)}{K^L} = \frac{9}{8}\varphi\left[\left(\frac{2Du^d}{1+2Du^d}\right)^2 \cdot \frac{1+2Du^d}{1+Du^d}\right] \tag{4.8.32}$$

whereas $\Delta\varepsilon'(\omega\tau \gg 1) \to 0$ and

$$\frac{\Delta\varepsilon'(\omega\tau \ll 1)}{\varepsilon'^L} = \frac{9}{16}\varphi(\kappa a)^2\left(\frac{2Du^d}{1+2Du^d}\right)^2 \tag{4.8.33}$$

The pairs [4.8.30 and 31] and [4.8.32 and 33] illustrate that conductivity and dielectric dispersion essentially give the same information. The equations are

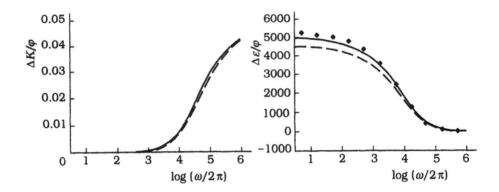

Figure 4.37. Comparison between the present theory (dashed), the Fixman theory (solid curves) and the exact numerical results by DeLacey and White (diamonds) $\kappa a = 20$, $a = 200$ nm, $F\zeta/RT = 4$. Left: excess conductivity per unit volume fractions of the colloid; right, excess dielectric loss.

relatively simple, and compare satisfactorily with other theories for large κa, as fig. 4.37 illustrates. Another advantage is that it is relatively simple to account for conduction behind the slip plane. To that end Du has to be replaced by [4.6.56] but the equations do not have to be modified.

4.8c *Some experimental illustrations*

So far, only a limited number of full dielectric relaxation spectra for well defined systems are available. Apart from the technical problems involved in the measurements (sec. 4.5e) there is the colloidal problem of synthesizing sufficiently concentrated sols with homodisperse spherical particles, preferably having different radii but fixed surface properties. Latices are popular objects because the particles are easily made homodisperse and spherical. Nevertheless they are somewhat suspect because there may be hairs on the surface, drastically affecting lateral hydrodynamic motion close to the surface. Moreover, changing the radius requires new syntheses and it is difficult to guarantee exact reproducibility of the surface structure. Inorganic particles do not have these drawbacks but it is not so easy to synthesize these as perfect spheres.

Figure 4.38 is a first illustration. The diagrams give some feeling for the quality of the data and of the fit. Regarding the latter, it appeared impossible to account for the data in terms of [4.8.31 and 30] and $Du = Du^d$, without introducing conduction behind the Stern plane. This was done numerically: by adjusting $K^{\sigma i}/K^{\sigma d}$ not only the fit of fig. 4.38 was obtained but also electrokinetic consistency. The numerical procedure gives results that are very well

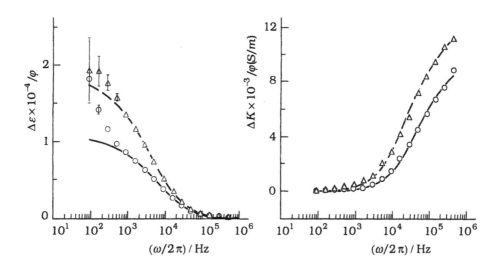

Figure 4.38. Dielectric relaxation and conductivity spectrum for haematite sols with spherical particles $a = 257$ nm, in KCl solution; (O) pH = 4.9, $\kappa a = 15$; (Δ) pH = 5, $\kappa a = 31$. Given are $\Delta\varepsilon'$ and ΔK per unit of volume fraction. (Redrawn from J. Kijlstra, H.P. van Leeuwen and J. Lyklema, *Langmuir* **9** (1993) 1625.)

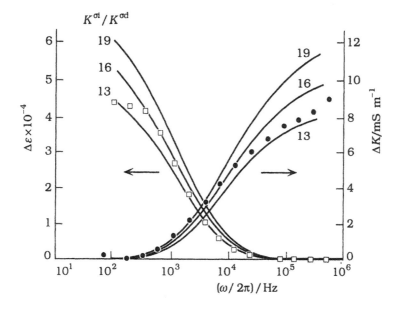

Figure 4.39. Dielectric relaxation and conductivity spectrum of a *Corynebacterium* species (strain 44016). Electrolyte, 10^{-3} M KNO_3 pH = 6.5, 25°C, φ between 0.01 and 0.1. Discussion in the text. Unpublished data by A. van der Wal et al.

represented by [4.8.31 and 30] provided instead of Du^d [4.6.56] is substituted. Figure 4.38 also shows that low-frequency dielectric increments can be gigantic.

Figure 4.39 gives another example, taken from Nature. The *Corynebacterium* considered here has more or less spherical, homodisperse cells, with diameters of 1.1 and 0.8 μm for the longer and shorter axis. Such cells are fascinating model colloids. The relaxation frequency, $\Delta\varepsilon$ and ΔK behave as expected for colloidal particles with a finite conduction behind the shear plane, which, in this case, is caused by the ions in the cell wall. As in the previous example, the data were analyzed with [4.8.30 and 31], using [4.6.56] for Du. The three curves, drawn through the measuring points, refer to this interpretation with the values of $K^{\sigma i}/K^{\sigma d}$ indicated. The conductivity of the cell wall exceeds that of the bulk by a factor of about 15. Cell walls are relatively porous structures containing large numbers of bound and mobile ions. The numbers of the mobile ions can be obtained from other sources (colloid titration, electrophoresis); together with the value of $K^{\sigma i}$ obtainable from the figure, the average ionic mobility could be estimated at about 25% of the same in bulk. Such a ratio is in the same order of magnitude as those in table 4.3.

It is concluded that by relatively simple means much interesting information can be obtained for inorganic and biological colloids. Further research along these lines looks promising.

4.9 Less familiar types of electrokinetics

The electrokinetic phenomena considered in sec. 4.3 and their counterparts discussed in sec. 4.8 are perhaps the most familiar representatives. A number of other phenomena exist, sometimes appearing in the literature under different names. We shall not treat these systematically but rather draw the attention to some relevant features.

Some of these additional types are not purely electrokinetic, in that other phenomena are "mixed". Perhaps the most familiar representatives of this group are the *primary and secondary electroviscous effects*. Both deal with the influence that tangential charge transport along a surface has on the apparent viscosity. The most basic phenomenon is the increase in resistance against convection by isolated particles due to the presence of a double layer and the ensuing electrokinetic drag. Seemingly, the viscosity of the system is increased. This is the *primary electroviscous effect*.

For low φ, the viscosity η of a sol of uncharged spherical particles can be written according to Einstein as

$$\eta = \eta^L(1 + \tfrac{5}{2}\varphi + ...)$$ [4.9.1]

where η^L is the viscosity of the solvent. The presence of the double layer leads to an apparent increase of φ or, for that matter, of η. Smoluchowski was the first to recognize this phenomenon. He gave the following expression without proof[1]

$$\eta = \eta^L \left[1 + \frac{5}{2} \varphi \left\{ 1 + \frac{1}{K^L \eta^L a^2} \left(\frac{\varepsilon_0 \varepsilon \zeta}{2\pi} \right)^2 \right\} \right]$$ [4.9.2]

Subsequently, other investigators have re-analyzed and improved this equation[2]. All of them concur that the correction is proportional to ζ^2.

It is not a very strong effect. To render it sizeable, relatively large values of φ are required but then the Einstein equation does not apply any more, because the particles begin to interact hydrodynamically. In addition, electrostatic interaction between the spheres starts to play a role, and this feature, the *secondary electroviscous effect* is also proportional to ζ^2 and tends to dominate the former. When studying the rheology of colloids, this second effect especially has to be considered.

Another phenomenon that has given rise to spurious viscosity increases is encountered in the drainage of thin liquid films from between two objects that are pressed toward each other, say two macrobodies, two colloidal particles or an air bubble against a flat surface, as occurs in flotation. When the two surfaces carry a double layer, counterions are moved out, to create a streaming potential between the centre and the periphery. According to [4.3.41], this streaming potential is proportional to $\varepsilon_0 \varepsilon \zeta / \eta (K^L + 2K^\sigma/a)$; the proportionality constant and a depending on the geometry. This streaming potential, in turn, acts as the driving force for electro-osmosis, which opposes the drainage. The resistance is again proportional to ζ^2 and is experienced as an apparent increase in viscosity. Some investigators have explained such viscosity rises as caused by liquid structuring near interfaces. Such a conclusion may be drawn only when the absence of a counter-electro osmotic flow has been established.

Relatively important is the phenomenon of *diffusiophoresis* and its counter-part *plug* or *capillary osmosis*. For both the driving force is a concentration gradient, either of an electrolyte or of a non-electrolyte. Consider for instance fig. 4.40. The presence of the gradient leads to at least an (osmotic) pressure gradient $p(\theta)$ around the particle in the double layer. Moreover, by specific adsorption it can also lead to concentration polarization and κ^{-1} may depend on θ. In this way driving forces are established to induce the particle to move.

---

[1] M. von Smoluchowski, *Kolloid-Z* **18** (1916) 190.
[2] For a review, see R.J. Hunter, *Zeta Potential in Colloid Science*, Academic Press (1981), sec. 5.2.1.

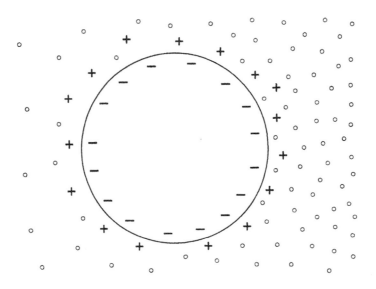

Figure 4.40. Charged colloid in a concentration gradient. The small circles indicate dissolved molecules or ions.

Diffusiophoresis is the counterpart of electrophoresis in that in the former case a concentration gradient is applied, it leads to polarization and motion; in the latter case an electric field is applied, which leads to concentration polarization and motion. When the particles are immobile, as in a porous plug, or when the concentration gradient is applied over a charged capillary the liquid starts to move. This leads to plug or capillary osmosis, the counterpart of electro-osmosis in plugs or capillaries.

We shall not consider the theory in much detail, because a variety of situations and mechanisms exist. First it is useful to reconsider [4.6.51], which is an expression for osmotic slip for high κa. The first term on the r.h.s. is the classical electro-osmotic slip, driven by a potential gradient, whereas the second term describes the slip caused by a concentration gradient, that is the phenomenon now under consideration. This is very interesting for the elaboration of the theory because instead of [4.3.19a] one may now write v(diff.phor) = $-\frac{2}{3} v_{so}$(capill.osm). When κa is not large the expressions become more difficult. An extension of this type of theory has been given by Sasidhar and Ruckenstein[1].

Let us now consider the case that the solute is an electrolyte. Mechanistically, two situations can be distinguished, depending on the diffusion coefficients of cations and anions, D_+ and D_-, respectively.

---

[1] V. Sasidhar, E. Ruckenstein, *J. Colloid Interface Sci.* **82** (1981) 430.

(i) When $D_+ \neq D_-$, say for HCl or LiCl, a *diffusion potential* is created for which we derived [I.6.7.6 or 7]. For a simple binary electrolyte this diffusion potential is, according to [I.6.7.8],

$$\Delta\psi_{\text{diff}} = -\frac{RT}{F}\left(\frac{z_- t_+ + z_+ t_-}{z_- z_+}\right)\ln\frac{c^\beta}{c^\alpha} \qquad [4.9.1]$$

where t_+ and t_- are the transport numbers of the cation and anion, respectively. For a symmetrical electrolyte $(z_+ = -z_- = z)$

$$\Delta\psi_{\text{diff}} = \frac{RT}{zF}\left(t_+ - t_-\right)\ln\frac{c^\beta}{c^\alpha} = \frac{RT}{zF}\left(\frac{D_+ - D_-}{D_+ + D_-}\right)\ln\frac{c^\beta}{c^\alpha} \qquad [4.9.2]$$

This potential drop over the gradient acts as the electric field that sets the particles into motion.

(ii) When $D_+ = D_-$, as for KCl, diffusiophoresis still occurs. The driving force is now the pressure drop around the particle, as discussed in connection with fig. 4.40.

Sometimes, the mechanisms (i) and (ii) are distinguished as *electro(diffusio-)phoresis* or *chemio(diffusio-)phoresis*, respectively. In the general case, the diffusiophoretic mobility is the sum of the two. The direction of electrodiffusiophoresis depends on the signs of ζ and $D_+ - D_-$. When these signs are identical, the particle moves *up* the gradient. Chemiodiffusiophoresis is in the direction of higher electrolyte concentrations. Here we encounter a basic difference with common diffusion which exclusively takes place *down* the gradient. However, there are cases where chemiodiffusiophoresis takes the opposite direction, depending on ζ and Du. By way of illustration we give the following equation for the (chemio-)diffusiophoretic velocity[1]

$$v_{\text{df}} = \frac{4\varepsilon_0\varepsilon}{\eta}\left(\frac{RT}{zF}\right)^2 \frac{\nabla c}{c}\frac{(1+Du)\ln\cosh(zF\zeta/4RT) - Du\,F\mid\zeta\mid/4RT}{1+2Du} \qquad [4.9.3]$$

For low Du the second term dominates; its sign is opposite to that of ∇c, whatever the sign of ζ. However, for other combinations of ζ and Du reversal of direction may occur. Equation [4.9.3] may also be used to assess how strong the phenomenon of diffusiophoresis is, as compared to electrophoresis. To compare orders of magnitudes, let us assume that the second term dominates and that $Du = 1$. The equation then reduces to

$$v_{\text{df}} = -\frac{\varepsilon_0\varepsilon\zeta}{\eta}RT\frac{\nabla c}{c} \qquad [4.9.4]$$

---

[1] E.S. Malkin, A.S. Dukhin, *Koll. Zhur.* **44** (1982) 254, (transl. 225).

which may be compared with

$$v_{ef} = -\frac{\varepsilon_0 \varepsilon \zeta}{\eta} \nabla \psi \qquad\qquad [4.9.5]$$

indicating that concentration gradients of $RT \nabla c/c$ produce a diffusiophoretic velocity that is similar to the electrophoretic velocity in a potential gradient $\nabla \psi$. Steep concentration gradients can, for instance, be created across thin membranes, or near electrodes when electrode reactions take place.

Diffusiophoresis was discovered by Deryagin et al.[1] and is now well established[2,3]. Good experimental verifications are available[4] and this can also be said of capillary osmosis[5]. The phenomenon has also received wide practical application, for instance in the deposition of colloidal particles where the driving concentration gradient is caused by surface (or electrode) reactions.

4.10 Applications

Few colloidal techniques have so deeply pervaded applied domains as electrokinetics has, electrophoresis in particular. The reason is obvious: the techniques do not require substantial investments and the results give a rapid insight into the charge of that part of the double layer that is responsible for particle interaction. Hence, electrokinetic parameters, ζ in the first place, play central roles in such stability-related phenomena as sedimentation, rheology, compaction, slurry formation and soil structure. The close analogy that is often observed between changes of ζ and the corresponding properties of the sol or suspension confirms the relevance of electrokinetics in practice. A plethora of examples serve as illustrations: thinning of suspensions by increasing ζ, increase of sediment volumes if the particles have a low ζ and the so-called *irregular series* in colloid stability: alternating stability domains upon the addition of specifically adsorbing ionic species where restabilization parallels reversal of the ζ-potential.

On closer inspection, notwithstanding the number of successful semi-quantitative interpretations of some phenomena, for quantitative studies their impact is more restricted because the experiments are rarely carried out with well-defined particles and often too simple equations are used for the $u \rightarrow \zeta$

---

[1] B.V. Deryagin, G. Sidorenko, E. Zubashchenko, and E. Kiselov, *Koll. Zhur.* **9** (1947) 335.

[2] S.S. Dukhin, B.V. Derjaguin, in *Surface and Colloid Science*, E. Matijevic, Ed., Vol. **7** Wiley (1994) chapter 3, sec. 4.

[3] J. L. Anderson, *Ann. Rev. Fluid Mech.* **21** (1989) 61.

[4] J.P. Ebel, J.L. Anderson and D.C. Prieve, *Langmuir* **4** (1988) 396.

[5] B.V. Derjaguin, S.S. Dukhin and M.M. Koptelova, *J. Colloid Interface Sci.* **38** (1972) 297.

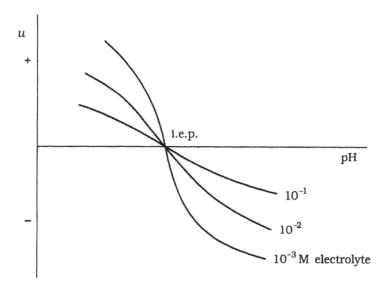

Figure 4.41. Sketch of the trend of the electrophoretic mobility of oxidic colloids as a function of pH in indifferent electrolyte. Mobilities are $O(10^{-8} m^2 v^{-1} s^{-1})$; the location of the i.e.p. depends on the nature of the oxide.

conversion, say Helmholtz-Smoluchowski under conditions where surface conduction may not be ignored. In the later Volumes we intend to come back to these applications; then the quality of the measurements and interpretations has to be considered.

Let us now briefly review some applications in the field of electrical surface characterization. As electrokinetically only a small part of the countercharge is measurable, electrokinetics is not a good method to estimate the surface charge σ^o. However, it is very helpful in locating the isoelectric point and to detect specific, or even superequivalent adsorption. Figure 4.41 gives a sketch of a typical $u(pH)$ plot for oxides; in fact, fig. 3.59 already gave a specific example. Three features deserve attention.

(i) *The isoelectric point.* When the electrolyte is indifferent and there are no other complications (like surface heterogeneity), $u(pH)$ curves at different c_{salt} intersect the pH-axis in the same point. The pristine i.e.p. should be identical to the pristine point of zero charge (p.z.c.), which is tabulated in app. 3. Specific adsorption of anions leads to a shift of u to more negative values. As more anions adsorb from $10^{-1} M$ electrolyte than from $10^{-3} M$ solutions, the shift Δ i.e.p. is larger at higher concentrations and a common intersection point at $u = 0$ is no longer observed. This is a means of establishing specific adsorption. When titration data are also available, the proof becomes stronger because upon specific adsorption p.z.c. and i.e.p. move in opposite directions, see table 3.3. Strictly speaking, isoelectric points and points of zero charge are no electric surface characteristics; they rather reflect the chemical affinity for the surface of positive and negative charge-determining ions and specifically adsorbing ions, if any. See sec. 3.8. Obviously, i.e.p.'s can be experimentally determined without any theoretical model.

(ii) *Plateaus.* With increasing distance from the i.e.p. the absolute value of u does not continue to rise but tends to level off at a level that is higher, the lower the electrolyte concentration is. Electrokinetic charges σ^{ek} for these levels seldom surpass $3\text{-}5\,\mu C\,cm^{-2}$, so the mobilities are not representative for σ^{o}, which may be higher by an order of magnitude. Hence, this plateau is not an adequate surface characteristic. Some illustrations can be found in fig. 4.13. The levelling of $u(pH)$ may be caused either by the viscoelectric effect (sec. 4.4) or by progressive ion uptake in the stagnant layer. Surface conductivity studies would be helpful to unravel these phenomena.

(iii) *slopes at the i.e.p.* The slopes $(\partial u / \partial pH)_{i.e.p.}$ decrease with increasing c_{salt}. The mobility u may be converted into ζ, using the Helmholtz-Smoluchowski equation if κa is large enough (no double layer polarization and no influence of surface conduction close to the zero point). Then, at low electrolyte concentration $\partial \zeta / \partial pH$ may approach 59 mV per pH unit at 25°, as would be the case for ψ^{o} if the Nernst equation applies. However, such a steep slope persists only close to the zero point; mostly it is much lower. Let us assume absence of specific adsorption (zeroth-order Stern theory, see fig. 3.17a) then we may write

$$\left(\frac{\partial u}{\partial pH}\right)_{iep} = \left(\frac{\partial u}{\partial \zeta}\right)_{iep} \left(\frac{\partial \zeta}{\partial \psi^{d}}\right)_{iep} \left(\frac{\partial \psi^{d}}{\partial \sigma^{d}}\right)_{iep} \left(\frac{\partial \sigma^{d}}{\partial \sigma^{o}}\right)_{iep} \qquad [4.10.1]$$

For $(\partial u / \partial \zeta)_{iep}$ the Henry equation may be used because for $\sigma^{o} \rightarrow 0$ there is negligible polarization. Again assuming that κa is large, this differential quotient becomes $\varepsilon_{0}\varepsilon / \eta$. $(\partial \zeta / \partial \psi^{d})$ is the extent by which ζ lags behind ψ^{d} because of the viscoelectric effect; at the zero point the difference between ζ and ψ^{d} is negligible (fig. 4.12), so the differential quotient is unity. Further, $\partial \psi^{d}/\partial \sigma^{d} = -1/C^{d}$ and $\partial \sigma^{d}/\partial \sigma^{o} = -1$. Hence,

$$\left(\frac{\partial u}{\partial pH}\right)_{iep} = \frac{\varepsilon_{0}\varepsilon}{\eta C^{d}} \qquad [4.10.2]$$

As C^{d} increases with increasing salt concentration, see [3.5.17], it is understood that the slope decreases in this direction.

However, [4.10.1] has another application: if on the charged surface an uncharged polymer would adsorb, the slip layer would move outward, because the water between the polymer loops becomes hydrodynamically immobilized.

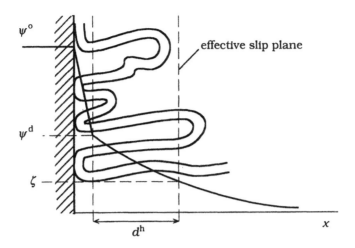

Figure 4.42. Determination of the hydrodynamic thickness of adsorbed charge-free polymer layers from the slope of the mobility curve near the isoelectric point.

The ensuing lowering of $(\partial u / \partial \text{pH})_{\text{iep}}$ has been experimentally observed[1] and is a measure of the hydrodynamic thickness of the adsorbed layer. Let us call this d^h and for the sake of simplicity interpret it as identical to the shift of a slip plane, originally situated at the outer Helmholtz plane (fig. 4.42). Let us also assume that near the p.z.c. the value of ψ^d is not much affected by the presence of segments in the Stern layer (in the plateau this approximation is poorer). Then [4.10.1] can be applied again, but $\partial \zeta / \partial \psi^d$ now follows from [3.5.22]. Carrying out the differentiation, one obtains

$$\left(\frac{\partial u}{\partial \text{pH}} \right)_{\text{iep}} = \frac{\varepsilon_0 \varepsilon}{\eta C^d} \left[\frac{\text{sech}\,(F\psi^d / 4RT)}{\text{sech}(F\zeta / 4RT)} \right]^2 e^{-\kappa d^h} \qquad [4.10.3]$$

in which ψ^d can be obtained from $\sigma^\circ = -\sigma^d$, using [3.5.13], because of the absence of specific adsorption. Hence, d^h can be obtained. (Obviously, in this simple case d^h is also immediately obtainable from [3.5.13], but the point is that the model only works well close to the p.z.c., and there $(\partial u/\partial \text{pH})$ can be better established than individual ζ values.) As from the shift of the p.z.c., which is in this case identical to the shift of the i.e.p., the occupancy of the Stern layer with train segments of the polymer can be inferred, such electrokinetic measurements are helpful to obtain experimental information about the segment

---

[1] L.K. Koopal, J. Lyklema, *Faraday Discuss. Roy. Soc. Chem.* **59** (1975) 230.

distribution of adsorbed polymers, (see further chapter 5).

Electrokinetics is of direct relevance for the interpretation of the *dynamics* of colloid particle interaction, that is interaction, considering the rates of transient charge fluxes during the brief encounters of pairs. Charge has to flow away and to a large extent this takes place laterally, so that obviously $K^{\sigma d}$ and $K^{\sigma i}$ become important variables.

In a less quantitative way, electrokinetics may be used to help characterize particles of less well-defined shapes, but which may be relevant for practice. Although the measurements defy rigorous interpretation, trends are often observed that may shed light on the surface properties of the materials studied and their modification by varying the pretreatment, adsorption of additives, changing pH, temperature or solvent. To illustrate the wide spectrum of possibilities, let us just mention a few disparate examples from the vast literature: modification of glass surfaces by Ca^{2+} uptake, to make it more corrosion-resistant, via streaming potentials[1], electrophoresis of calcium oxalate, the main component of renal stones[2], as part of a study of their aggregation, streaming potential measurements as a tool for the characterization of ultra-filtration membranes[3], glass and polymer fiber studies by streaming currents and streaming potentials[4] electrokinetics of biological cell surfaces[5], electrokinetics of fine-porous membranes in connection with reverse osmosis studies[6], of sea sediments in saline media and of mica (muscovite) as a means to characterize the surfaces that are commonly used in the surface force apparatus[7].

Electrophoresis of bubbles and drops is a story on its own. As long ago as 1861 Quincke[8] observed the electrophoresis of small air bubbles in water. Such a motion is possible only when there is a double layer at the interface, containing free ions. It is extremely difficult to keep oil-water or air-water interfaces rigorously free from adsorbed ionic species. When these are present, especially for surfactants, Marangoni effects make the surface virtually inextensible; then the drops or bubbles behave as solid spheres. Electrophoretic studies

[1] M. Nardin, E. Papirer and J. Schulz, *J. Colloid Interface Sci.* **88** (1981) 205.
[2] J. Callejas Fernández, F.J. de las Nieves, J. Salcedo Salcedo and R. Hidalgo-Alvarez, *J. Colloid Interface Sci.* **135** (1990) 154.
[3] M. Nyström, M. Lindström and E. Matthiasson, *Colloids Surf.* **36** (1989) 297.
[4] H.-J. Jacobasch, G. Bauböck and J. Schurz, *Monatsh. Chem.* **117** (1986) 1133.
[5] A. Voigt, E. Donath, in *Cell Surface Biophysics*, (Springer Series in Biophysics), R. Glaser, D. Gingell, Eds., Springer (1989) 75.
[6] B.V. Derjaguin, N.V. Churaev and G.A. Martynov, *J. Colloid Interface Sci.* **75** (1980) 419.
[7] P.J. Scales, F. Grieser and T.W. Healy, *Langmuir* **6** (1990) 582.
[8] G. Quincke, *Pogg.Ann.* **113** (1861) 513.

of such systems are relevant for flotation[1, 2]. For a recent theoretical approach regarding the electrophoresis of "clean" drops, see Baygents and Saville[3], where other references can be found. These authors concluded that for large κa conducting drops not always move in the direction expected from the sign of their surface charge. For extremely pure interfaces Marangoni effects cannot play a role; then internal circulation inside liquid drops may become important.

Proteins and peptides constitute other important classes. Often, investigators are not so much interested in ζ-potentials (their establishment and interpretation would be anyway extremely difficult) as they are in identification or, sometimes, separation. As these molecules do not exhibit strong optical contrast with the solvent, moving boundary electrophoresis presents itself as a suitable technique. Often the aqueous medium in which these molecules are dissolved is freed of convection by hydrodynamically fixing it in a gel or having it impregnated in paper. These techniques may be called *anticonvectant electrophoresis*, with *gel electrophoresis* and *paper electrophoresis* as the most prominent examples. In the latter case transport is no longer pure electrophoresis but rather a mixture with electro-osmosis and chromatography. After a certain time, mixtures of proteins have moved over certain distances, each component at its own characteristic rate; then the presence of proteins in certain spots has to be made visible, e.g. by colour reactions (for paper electrophoresis after drying). Some of the more familiar techniques belonging to this category are:

(i) *Zone electrophoresis*, an electrophoretic separation technique, conducted in a continuous buffer system. Samples are injected and move under the applied field; they separate into distinguishable zones if their mobilities are sufficiently apart.

(ii) *Isotachophoresis*[4], separation in a discontinuity between two buffer solutions (the leading and the terminating electrolyte). The sample components are introduced in small quantities in the discontinuity, a d.c. current is applied, and the various components start to move. Consecutive zones are formed and a steady state is attained where each zone moves with equal velocity, hence the name.

(iii) *Isoelectric focusing*[5]. Fractionation of amphoteric species on the basis of

[1] S. Usui, H. Sasaki, *J. Colloid Interface Sci.* **65** (1978) 36; S. Usui, H. Sasaki and H. Matsukawa, *J. Colloid Interface Sci.* **81** (1981) 80.

[2] K.P. Tikhomolova, O.V. Kokorina, *Russ. Coll. J.* **55** (1993) 467.

[3] J.C. Baygents, D.A. Saville, *J. Chem. Soc. Faraday Trans.* **87** (1991) 1883.

[4] F.M. Everaerts, J.L. Beckers and Th.P.E.M. Verheggen, *Isotachophoresis, Theory, Instrumentation and Application*, Elsevier (1976).

[5] P.G. Righetti, *Isoelectric Focussing, Theory, Methodology and Applications*, Elsevier (1983).

differences in i.e.p. Electrophoresis takes place in a pH gradient[1], and the proteins or peptides move till they are at their i.e.p. Any fortuitous displacement beyond it, say by diffusion, will lead to surface charge development, which will give the molecule a mobility, moving it back to the i.e.p. So, a stationary steady situation is eventually attained.

When these processes are carried out in capillaries, they are also named *capillary electrophoresis*, of which several variants exist.

Further information can be found in the references mentioned in sec. 4.11c.

The third category concerns autonomous applications of one of the electro-kinetic phenomena to solve a practical problem.

As the first example, *electrophoretic deposition* may be mentioned. It is an industrial process to produce thin coherent coatings on conducting surfaces, say pigments and anti-corrosion materials on metal objects in the car industry, or the deposition of phosphorescent particles in television screens. The object to be covered acts as an electrode and the particles to be deposited should be charged. These particles concentrate near the electrode; as a result of the electrode reaction and the ensuing creation of electrolyte, these concentrates coagulate, giving a coherent deposit. In addition the created concentration gradient leads to diffusiophoresis. In practice, this technique is often applied in non-aqueous media.

Electro-osmosic dewatering is a second group. Electro-osmosis can be used to squeeze water from certain materials. It has wide potential applications, although a number of conditions must be met to attain sufficient efficiency under field conditions. Dewatering of sludges is one illustration; this process is becoming increasingly urgent in view of the growing demand for disposal sites for mining slurries, tailings and other industrial wastes and the sludge left over from dredging harbours. Tree-trunks can be electro-osmotically dried so that they float better and therefore can be transported downstream on rivers in forested areas where roads are scarce. The process can also be used to dewater and stabilize soft soils.

Finally, *electrodialysis* may be mentioned, a process widely used to de-salt aqueous solutions. An electric field is applied across a stack of alternating cation-exchange and anion-exchange membranes. Ions in the electrolyte solutions between these membranes are transported till they meet a membrane of the same sign, so that electrolyte-rich and electrolyte-freed solutions are created. The process involves conduction and electro-osmosis. Obviously, irreversible thermodynamics appears very suitable to describe the various flows

---

[1] For the preparation of such gradients, see P.G. Righetti, *Immobilized pH-gradients, Theory and Methodology*, Elsevier (1990).

and fluxes. Preparing potable water or water for agricultural purposes from seawater in tropical areas is among to the applications[1].

These application examples are by no means exhaustive, but they do illustrate the richness of electrokinetics.

4.11 General references

4.11a. IUPAC recommendations

Manual of Symbols and Terminology for Physicochemical Quantities and Units; Appendix II. Definitions, Terminology and Symbols in Colloid and Surface Chemistry, part I, especially secs. 1.12 and 2.12. *Electrokinetics* (prepared for publication by D.H. Everett). *Pure Appl. Chem.* **31** (1972) 577.

Nomenclature, Symbols, Definitions and Measurements for Electrified Interfaces in Aqueous Dispersions of Solids (prepared for publication by J. Lyklema). *Pure Appl. Chem.* **63** (1991) 895.

Quantities and Units for Electrophoresis in the Clinical Laboratory (prepared for publication by R.G. Férard). *Pure Appl. Chem.* **66** (1994) 891-6.

4.11b. General books and reviews

S.S. Dukhin, B.V. Derjaguin, *Electrokinetic Phenomena*, in *Surface and Colloid Science*. E. Matijevic, Ed. Vol. **7** Wiley (1974). (Three chapters: 1. S.S. Dukhin, *Development of Notions as to the Mechanism of Electrokinetic Phenomena and the Structure of the Colloid Micelle*, p. 1; 2. S.S. Dukhin, B.V. Derjaguin, *Equilibrium Double Layer and Electrokinetic Phenomena*, p. 49; 3. B.V. Derjaguin, S.S. Dukhin, *Non-equilibrium Double Layer and Electrokinetic Phenomena*, p. 273.)

S.S. Dukhin, V.N. Shilov, *Kinetic Aspects of Electrochemistry of Disperse Systems. Part II. Induced Dipole Moment and the Non-Equilibrium Double Layer of a Colloid Particle. Adv. Colloid Interface Sci.* **13** (1980) 153. (Review on the determination and interpretation of induced dipole moments of various colloidal particles.)

S.S. Dukhin, *Non-equilibrium Surface Phenomena, Adv. Colloid Interface Sci.* **44** (1993) 1. (Review, covering a variety of electrokinetic phenomena, theory and experiments.)

---

[1] For a review on this and related matters, see N. Lakshminarayanaiah, *Transport Phenomena in Membranes*, Acad. Press (1969).

R. Hidalgo-Alvarez, *On the Conversion of Experimental Electrokinetic Data into Double layer Characteristics in Solid-liquid Interfaces*, Adv. Colloid Interface Sci. **34** (1991) 217. (Extensive review, 380 references, covering several aspects that are considered in this chapter.)

R.J. Hunter, *Zeta Potential in Colloid Science. Principles and Applications*. Academic Press (1981). (Standard book, with much information on principles, methods and applications.)

R.A. Mosher, D.A. Saville and W. Thormann, *The Dynamics of Electrophoresis*, VCH (Weinheim, D) (1992) (Theory for various forms of electrophoresis, including moving boundary, zone and isotacho-phoresis).

W.B. Russell, D.A. Saville and W.R. Schowalter, *Colloid Dispersions*. Cambridge Univ. Press (1989). (A general colloid book with much emphasis on transport phenomena, including electrokinetics.)

4.11c. Methods

A.J. Babchin, R.S. Chow and R.P. Sawatzky, *Electrokinetic Measurements by electroacoustical methods*. Adv. Colloid Interface Sci. **30** (1989) 111. (Review, theory and applications.)

Electrophoresis. Theory, Methods and Applications. M. Bier, Ed., Academic Press (1950). (Somewhat dated but still very readible anthology, with some emphasis on methods. There is a second volume, 1967.)

Electrophoresis: A Survey of Techniques and Applications, Z. Deyl, Ed., Elsevier (1979).

Capillary Electrophoresis. Theory and Practice, P.D. Grossman, J.C. Colburn, Eds., Academic Press (1992). (Emphasis on techniques and various methods of electrophoretic separation.)

Capillary Electrophoresis Technology, N.A. Guzman, Ed. Marcel Dekker (1993).

A.M. James, *Electrophoresis of Particles in Suspension* in *Surface and Colloid Science*, Vol. **II** (1979). R.J. Good, R. Stromberg, Eds., Plenum (1979), 121. (Detailed diiscussion of the micro-electrophoresis technique).

S.F.Y. Li, *Capillary Electrophoresis, Principles, Practice and Applications*, Elsevier, 1992. (Detailed information on these techniques).

Electrokinetic Separation Methods, P.G. Righetti, C.J. van Oss and J.W. Vanderhoff, Eds., Elsevier (1979). (Emphasis on anticonvectant techniques; experiments and analytical applications.)

Electrophoretic Techniques, C.F. Simpson, M. Whittaker, Eds. Academic Press (1983). (Emphasis on analytical and preparative methods.)

E.E. Uzgiris, *Laser-Doppler Methods in Electrophoresis. Progr. Coll. and Surf. Sci.* **10** (1981) 53. (Review.)

B.R. Ware, *Electrophoretic Light Scattering., Adv. Colloid Interface Sci.* **4** (1974) 1. (Review.)

5 ADSORPTION OF POLYMERS AND POLYELECTROLYTES

5.1 Introduction

This chapter concerns the adsorption of *polymers* and *polyelectrolytes*, of which the molecules consist of a large number of repeating units. Polymers and polyelectrolytes belong to the large and important category of *macromolecules*. Some of these macromolecules, such as proteins and polysaccharides, have a biological origin. In the present chapter we consider only the relatively simple classes of linear (uncharged) *homopolymers* and linear *homopolyelectrolytes*. In these molecules all building units are identical, and most of these molecules are synthetic. In several respects there are substantial quantitative and qualitative differences between uncharged polymers and polyelectrolytes; therefore we shall maintain the distinction between the two classes. More complicated macromolecules (block or random copolymers, proteins, etc.) will be treated in Volume V. There the effects of adsorbed polymers on colloid stability will also be considered.

When all chains of a polymer or polyelectrolyte are equally long, the macromolecules are called *homodisperse* or *monodisperse*. If this condition is not met the system is *heterodisperse* or *polydisperse*. Synthetic polymers and polyelectrolytes usually have a certain molar mass distribution and, hence, are to some extent polydisperse. This polydispersity has important implications for the adsorption behaviour because fractionation will occur upon adsorption.

In chapter 2 the adsorption of small molecules was treated in some detail. In such systems adsorption layers are rather thin, with a thickness of the order of the molecular diameter. In sec. 2.7d adsorption of non-ionic surfactants was considered. These molecules are oligomers and their adsorption behaviour shows already some polymer-like features. It is typical for polymers that the length of the molecules is orders of magnitude larger than their thickness. Since long sections of the absorbed chains can protrude into the solution, this may give rise to thick adsorption layers. Moreover, the polymer molecules usually have a high degree of flexibility. Unlike rigid molecules they can assume many different spatial arrangements, often called *conformations*. These conformations fluctuate in time and space, which makes it necessary to invoke statistical methods to describe their behaviour, both in solution and in the adsorbed state.

Figure 5.1 gives an impression of polymer conformations in solutions (fig.

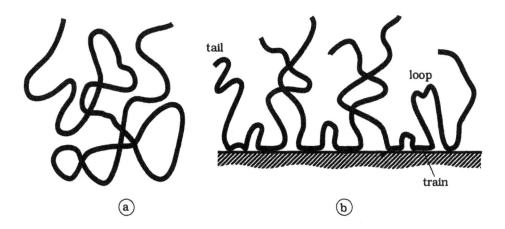

Figure 5.1. Impression of an isolated coil in solution (a) and of an adsorbed layer built up of several polymer molecules (b). In the latter case a distinction may be made between *trains, loops,* and *tails* (see sec. 5.3a).

5.1a) and next to a surface (fig. 5.1b). The average conformation in solution is more or less spherical, with a relatively high concentration of monomer units in the central region and a more dilute periphery. Because of the high number of possible conformations there is a high *conformational entropy.* On a surface (fig. 5.1b) the average conformation is different. Now there is a relatively high concentration of units on the surface because of the adsorption energy. However, long stretches of the adsorbed molecules are not in direct contact with the surface. Yet these segments, which are surrounded by solvent molecules, clearly belong to the adsorbed layer. They give this layer a thickness which far exceeds that of a monomer unit. A segment concentration gradient develops over a distance which is comparable to the dimensions of the chains in solution. The conformational entropy of the adsorbed chains is lower than in the solution but it remains much higher than that of completely flat adsorbed molecules. This is the reason why polymers quite often tend to absorb in thick layers.

In an adsorbed polymer layer, the interaction of the chains with the solvent remains very important. Therefore, before discussing the adsorption behaviour of polymers and polyelectrolytes, we review briefly some properties of these molecules in solution. In the following sections, we follow roughly the outline of a recent monograph on polymers at interfaces[1].

---

[1] G.J. Fleer, M.A. Cohen Stuart, J.M.H.M. Scheutjens, T. Cosgrove and B. Vincent, *Polymers at Interfaces,* Chapman & Hall, (1993), in this chapter further referred to as Fleer et al., loc. cit.

5.2 Polymers in solution

Long flexible polymers have a large number of internal degrees of freedom. The typical *primary structure* of such molecules is a linear chain of atoms (often carbon) connected by chemical bonds ("backbone"). Usually, every other backbone atom carries a side group. By rotation about the bonds in the backbone the molecule changes its shape, and since there are many of these bonds in a polymer, a wide spectrum of conformations is available. The rotation is hindered by the side groups, so that some of these conformations may be rather unfavourable. In some macromolecules (for instance proteins) sequences of preferred bond orientations show up as helical or folded sections in the molecules (*secondary structure*).

As we consider only flexible, linear polymers, the energy barriers associated with rotation around bonds are small with respect to the thermal motion. Such molecules have a randomly fluctuating three-dimensional *tertiary structure*, as illustrated in fig. 5.1a. The term used for such a structure is *random coil*.

In the following two sections we first consider coils in solutions which are so dilute that coil-coil interaction does not play a role. Section 5.2a deals with random-walk chains where the interaction between the units within one chain may be neglected. We denote these as *ideal chains*. Section 5.2b treats *swollen coils* in which the monomeric units repel each other due to so-called *excluded volume* effects.

5.2a. Ideal chains in dilute solutions

In the simplest possible model for an ideal chain, the bonds between atoms in the backbone are treated as vectors connecting volumeless points which do not interact. Such a model chain is depicted in fig. 5.2; the (fluctuating) distance between the end points is denoted as r. If, moreover, any orientation between two consecutive bonds is assumed to have the same probability, the conformational properties of long chains can be described by the universal *random-flight* model, first introduced by Kuhn[1]. Let the chain have N randomly oriented bonds, each of length ℓ. Such a model chain contains $N + 1$ backbone atoms. When these bonds are assumed to be fully independent of each other, the conformation resembles the trajectory of a particle diffusing under the action of a random force, for which the solution is well known[2,3,4]. The mean square displacement $\langle r^2 \rangle$ is proportional to N (i.e., the number of time "steps"), as was already discussed in sec. I.6.3d:

---

[1] W. Kuhn, *Kolloid Z.* **68** (1934) 2.
[2] See, e.g., F. Reif, *Fundamentals of Statistical and Thermal Physics*; McGraw-Hill (1965) chapters 12 and 15.
[3] P.J. Flory, *Principles of Polymer Chemistry*, Cornell University Press, Ithaca NY (1953).
[4] H. Yamakawa, *Modern Theory of Polymer Solutions*, Harper & Row, NY (1971).

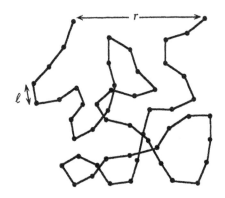

Figure 5.2. The chain conformation of fig. 5.1a, but now modelled as a random-flight chain of N bonds of length ℓ. The (fluctuating) distance between the end points is r.

$$\langle r^2 \rangle = N\ell^2 \qquad\qquad [5.2.1]$$

The quantity $\langle r^2 \rangle$ is a measure of the size of the chain; the mean *end-to-end distance* $\langle r^2 \rangle^{1/2}$ could be considered as the mean coil diameter. Another measure of the coil size is the *radius of gyration* a_g, which is the root-mean-square (r.m.s) distance of the segments from the centre of mass (see also [I.7.8.26c]). For long flexible chains $a_g = \left(\langle r^2 \rangle / 6 \right)^{1/2}$ or

$$a_g^2 = \tfrac{1}{6} N \ell^2 \qquad\qquad [5.2.2]$$

In real polymers the bonds cannot assume arbitrary directions but there are fixed valence angles between them. In addition, rotation about bonds is not entirely free, because the potential energy shows minima and maxima as a function of the rotation angle. To account for these effects, the above equations may be modified with a rigidity constant which depends on the architecture of the chain. If we introduce a *stiffness ("persistence") parameter p*, we may write

$$\langle r^2 \rangle = 6 \, pN\ell^2 \qquad\qquad [5.2.3]$$

$$a_g^2 = pN\ell^2 \qquad\qquad [5.2.4]$$

Here, p equals 1/6 for a (hypothetical) fully flexible chain, and p increases as the chain becomes less flexible (for instance, when the side groups become more bulky). Typical p values for real chains are in the range 0.5 - 4. Detailed calculations for p for various chains, based upon molecular models, are available[1,2]. In most cases a "characteristic ratio C_∞", equal to $6p$, is tabulated.

---

[1] P.J. Flory, *Statistical Mechanics of Chain Molecules*, Interscience, NY (1969).
[2] *Polymer Handbook*, J. Brandrup and E.H. Immergut, Eds, 3rd ed., Wiley, New York (1989)

Kuhn[1] was the first to point out that the dimensions of a chain with given persistence p may always be described as if it were completely flexible (see [5.1.1]) by grouping a number of monomer units together into *statistical chain elements* (s.c.e.) or *Kuhn segments*. The number a of bonds in such an s.c.e. is the larger the stiffer the chain. The basic idea is that such s.c.e.'s may be considered as orientationally independent: they are then independent subsystems as defined in sec. I.3.6. The real chain of N bonds is now modelled as an equivalent ideal chain of $N_K = N/a$ s.c.e.'s and the Kuhn length ℓ_K becomes $b\ell$ (where $a > 1$, $b > 1$). Then $\langle r^2 \rangle = N_K \ell_K^2 = (b^2/a)N\ell^2$, which equals $6a_g^2 = 6pN\ell^2$, provided that a and b satisfy $b^2/a = 6p$. In order to fully define a and b one needs another condition. Often one requires that the contour length $N_K \ell_K$ of the Kuhn chain equals the contour length $N\ell$ of the real chain. Hence, $N_K \ell_K = N\ell$, or $a = b = 6p$. A statistical chain segment typically contains several (5 to 20) backbone atoms, depending on the flexibility of the chain.

The important consequence of the above equations is that for ideal chains the dimensions are always proportional to $N^{1/2}$. When the excluded volume becomes important, as in good solvents, this does not hold because then the chains swell. These excluded volume effects are considered in the next subsection.

5.2b. Swollen chains in dilute solutions

For the ideal chains discussed above the volume of the segments and solvency effects are entirely ignored. A walk may return to its origin without any hindrance, which is unrealistic for segments which occupy volume. Because in reality segments cannot overlap there is an exclusion volume, which automatically leads to coil expansion. In addition, there may be solvency effects. In very good solvents, where the segments repel each other, the excluded volume is larger than the exclusion volume. This may become relevant for polyelectrolytes. On the other hand, in poor solvents the segments experience a net attraction (for example, due to Van der Waals forces) so that the (effective) excluded volume is small and the ideal chain model gives a reasonable description.

A convenient way to deal with the interaction between segments across solvent is through the Flory-Huggins parameter χ which was originally defined as the energy change (in units of kT) associated with the transfer of a segment from pure polymer to pure solvent (or, equivalently, a solvent molecule from solvent to polymer)[2]. This parameter, which is a measure of the excess affinity of segments for each other over that of the solvent, was introduced in sec. I.3.8c.

[1] W. Kuhn, *Kolloid Z.* **68** (1934) 2.
[2] P.J. Flory, *Principles of Polymer Chemistry*, Cornell University Press, Ithaca NY (1953).

As shown there, χ may also contain entropic contributions, in which case it is a free energy parameter. If solvent and polymer segments have the same polarity and polarizability $\chi = 0$, which is the *athermal* case (see sec. I.2.17b). Although the excess interaction in an athermal solvent is zero, so that we could call this an indifferent solvent, solvents with $\chi = 0$ (or with small χ) are commonly referred to as "good". In such a good solvent the volume exclusion implies that the walk through space has to avoid itself, which leads to a strong coil expansion (swelling). The excluded volume per segment is now of order ℓ^3. Due to this excluded volume, the chain dimensions are proportional to a higher power of N than $1/2$. Computer simulations[1] have shown that the radius of gyration for such a self-avoiding walk (SAW) is proportional to $N^{3/5}$.

For most polymer-solvent combinations $\chi > 0$ because of a net segment-segment attraction. This implies that the energy effect opposes the entropically driven dissolution of the polymer in the solvent. Such solutions are still thermodynamically stable unless χ becomes too high; see sec. 5.2e. If $\chi > 0$ the excluded volume is smaller than the real volume ($\approx \ell^3$) of a segment. Flory[1], Edwards[2], De Gennes[3] and others have shown that the excluded volume may be written as $\upsilon\ell^3$, where the dimensionless *excluded volume parameter* υ is defined as

$$\upsilon = 1 - 2\chi \qquad\qquad [5.2.5]$$

If $\chi = 1/2$ the net excluded volume (and the second virial coefficient) is zero, and the chains behave ideally ($a_g \sim N^{1/2}$). This condition is known as the Θ-point; it may be compared with the Boyle point for non-ideal gases (sec. I.2.18), where also the mutual compensation of non-zero volume and attractive forces leads to pseudo-ideal behaviour. For worse-than-Θ conditions ($\chi < 1/2$, $\upsilon < 0$) phase separation sets in (see 5.2e).

Another way to deal with the excluded volume is to introduce a linear expansion coefficient α which depends on υ (or χ) and N. Equation [5.2.4] is then modified to

$$a_g^2 = \alpha^2\, pN\ell^2 \qquad\qquad [5.2.6]$$

where $\alpha = 1$ in a Θ solvent ($\chi = 1/2$, $\upsilon = 0$) and $\alpha > 1$ in a good solvent. Flory[1] was the first to derive an analytical expression for α. For good solvents and long chains his equation takes the approximate form:

$$\alpha^5 \approx p^{-3/2}\, \upsilon\, N^{1/2} \qquad\qquad [5.2.7]$$

[1] See, e.g., D.S. McKenzie, *Phys. Rep.* **27C** (1976) 2.
[2] S.F. Edwards, *Proc. Phys. Soc.* **85** (1965) 613; **88** (1966) 265.
[3] P.G. de Gennes, *Scaling Concepts in Polymer Physics*, Cornell University Press, Ithaca, NY (1979).

which is known as the "Flory limit". Since this limit implies $\alpha \sim N^{1/10}$, [5.2.6] leads to $a_g \sim N^{3/5}$, in agreement with computer simulations. Combination of [5.2.6 and 7] gives for the radius of gyration in a good solvent:

$$a_g \approx (pv)^{1/5} N^{3/5} \ell \qquad [5.2.8]$$

The chain swelling in a good solvent, as expressed by [5.2.8], is only important for sufficiently long chains. For shorter and relatively stiff chains the segments do not meet each other and the volume exclusion plays no role; to a good approximation the chain behaves then ideally according to [5.2.4]. The cross-over chain length N_c between ideal and swollen chains is found by setting $\alpha = 1$ in [5.2.7]:

$$N_c \approx p^3 v^{-2} \qquad [5.2.9]$$

Alternatively, at given chain length N the behaviour is ideal when v is small, and the chains are swollen when v is high. The cross-over occurs at v_c, given by

$$v_c \approx p^{3/2} N^{-1/2} \qquad [5.2.10]$$

From [5.2.9] it becomes clear that N_c increases rapidly with increasing chain stiffness and decreasing solvency. In the limit of Θ-conditions ($v \to 0$), $N_c \to \infty$; even the longest chains are ideal if $v = 0$ ($\chi = 1/2$).

5.2c. Overlapping chains in a mean field

Above we discussed solutions sufficiently dilute to justify the neglect of coil-coil interactions. We now turn our attention to interacting polymer molecules. Consequently, the concentration (conveniently expressed as the average volume fraction φ) becomes an important variable. Depending on φ and v (or χ) various types of behaviour can be distinguished.

Polymer segments in a good solvent avoid each other. Similarly, polymer coils avoid each other: they are reluctant to interpenetrate. However, as the concentration is increased the molecules are forced to do so. Onset of overlap occurs when the total volume of all coils approaches the solution volume.

In this section we summarize the classical theory which assumes that, for sufficiently concentrated solutions, the distribution of segments throughout the solution is homogeneous, as if the coils could interpenetrate without any hindrance. This type of theory, which is known as a *mean-field theory* because all the segments experience, on average, the same force field, is expected to hold when the excluded volume parameter v is small. More recent theories for better solvents (higher v) will be treated in 5.2d.

Flory[1] and Huggins[2] derived a now classical mean-field expression for the configurational entropy and energy of mixing, using a lattice model. The solution, containing n_1 moles of solvent and n_2 moles of polymer, is described as a lattice of N_s sites, $N_s \varphi_1$ of which are occupied by solvent and $N_s \varphi_2$ by polymer segments, with $\varphi_1 + \varphi_2 = 1$. Each chain occupies N sites, so that $N_s = N_{Av}(n_1 + Nn_2)$. The number of polymer molecules in the lattice is $N_s \varphi_2 / N$.

According to the Flory-Huggins theory the mixing entropy $S - S^*$ is given by

$$(S - S^*)/k = -\left(n_1 \ln \varphi_1 + n_2 \ln \varphi_2\right) = -N_s \left\{ \varphi_1 \ln \varphi_1 + \frac{\varphi_2}{N} \ln \varphi_2 \right\} \qquad [5.2.11]$$

and the mixing energy by

$$(U - U^*)/kT = N_s \chi \varphi_1 \varphi_2 \qquad [5.2.12]$$

In both cases the reference state (*) is pure unmixed components: $S^* = S_1^* + S_2^*$ and $U^* = U_1^* + U_2^*$, where S_1^* and U_1^* are the entropy and energy of n_1 moles of pure solvent, and S_2^* and U_2^* refer to n_2 moles of polymer in pure amorphous polymer. The energy U in [5.2.12] becomes a Helmholtz energy if χ contains an entropic contribution. However, this Helmholtz energy only includes those entropic contributions which are proportional to n_2 (for example, those originating from orientations of the solvent molecules, as in hydrophobic bonding), and not the configurational entropy of the chains as embodied in [5.2.11].

Equations [5.2.11 and 12] may be compared with those derived for a mixture of monomeric species ($N = 1$) from regular solution theory (sec. I.2.18c). If $N = 1$, the volume fraction φ in a lattice theory is identical to the mole fraction x. Then [5.2.11 and 12] reduce to [I.2.17.12] and [I.3.8.10], respectively. Hence, the Flory-Huggins theory can be considered as a generalization of the Regular Solution model to mixtures containing species that differ in size. Basically, the only modification is the replacement of $\ln x$ by $\ln \varphi$ in the entropy of mixing. Although this may seem only a simple modification, for high N its effect is considerable. It leads to a very strong asymmetry in the phase diagram of polymer solutions.

The Helmholtz energy $F = U - TS$ is directly obtained by combining [5.2.11 and 12]. From this, the chemical potentials μ_1 of the solvent and μ_2 of the polymer are obtained by differentiation of F with respect to n_1 and n_2, respectively (analogous to [I.2.10.6]). In the following, we express all quantities in the polymer volume fraction φ, so we can drop the subscript 2: $\varphi_2 \equiv \varphi$, $\varphi_1 \equiv 1 - \varphi$. Then

[1] P.J. Flory, Principles of Polymer Chemistry, Cornell University Press, Ithaca NY (1953).
[2] M. Huggins, J. Phys. Chem. 46, 151 (1942). M. Huggins, Ann. NY Acad. Sci. 41 (1942) 1; M. Huggins, J. Am. Chem. Soc. 64 (1942) 1712.

$$\left(\mu_1 - \mu_1^*\right)/kT = \ln(1-\varphi) + \varphi\left(1 - \frac{1}{N}\right) + \chi\varphi^2 \qquad [5.2.13]$$

$$\left(\mu_2 - \mu_2^*\right)/N\,kT = \frac{1}{N}\ln\varphi - (1-\varphi)\left(1 - \frac{1}{N}\right) + \chi(1-\varphi)^2 \qquad [5.2.14]$$

where μ_1^* is the chemical potential of the pure solvent and μ_2^* that of the pure amorphous polymer. The quantity μ_2/N may be considered as the chemical potential per polymer segment.

It is sometimes convenient to expand [5.2.13] in terms of φ. Then we obtain

$$-\left(\mu_1 - \mu_1^*\right)/kT = \varphi/N + \tfrac{1}{2}\upsilon\varphi^2 + \tfrac{1}{3}\varphi^3 + \ldots \qquad [5.2.15]$$

This equation is useful to analyze the thermodynamic behaviour of mean-field polymer solutions. For example, according to [I.2.12.15] the osmotic pressure Π is simply given by

$$\Pi\upsilon_1 = -\left(\mu_1 - \mu_1^*\right) \qquad [5.2.16]$$

where $\upsilon_1 = V_{ml}/N_{Av} \approx \ell^3$ is the molecular volume of the solvent.

In very dilute solutions the first term on the r.h.s. of [5.2.15], which is proportional to the concentration (though with the very small proportionality constant N^{-1}), dominates. When in this concentration range $\upsilon < \upsilon_c$ the chains behave ideally according to [5.2.4]; this regime may be called *ideal dilute*. For better solvents ($\upsilon > \upsilon_c$) the isolated chains are swollen according to [5.2.8]; we denote this as the *swollen dilute* regime. These regimes are indicated in the "phase diagram" $\upsilon(\varphi)$ of fig. 5.3, set up following Birshtein[1] and Schaefer[2]. The value υ_c describing the cross-over between the ideal dilute and swollen dilute regimes is given by [5.2.10].

With increasing concentration the φ^2 and φ^3 terms of [5.2.15] become more important. When υ is small we can ignore the quadratic term. In the *concentrated* regime the φ^3 term is the dominant contribution to μ_1 or Π (see fig. 5.3). The transition between the ideal dilute (i) and concentrated (c) regimes occurs at a volume fraction φ^{ic} which is given by $\varphi^{ic}/N \approx (\varphi^{ic})^3$, or $\varphi^{ic} \approx N^{1/2}$. Since we are interested mainly in the functional dependencies, we ignore numerical coefficients of order unity.

When the solvency is better the quadratic term in [5.2.15] may be dominant. The transition occurs when the solvent quality is marginally good; this regime may therefore be denoted as the *marginal* (m) regime (see fig. 5.3). In very good solvents (i.e., better than marginal) the mean-field description breaks down, as

[1] T.M. Birshtein, *Vysokomol. Soedin.* **A 24** (1982) 2110.
[2] D.W. Schaefer, *Polymer* **25** (1984) 387.

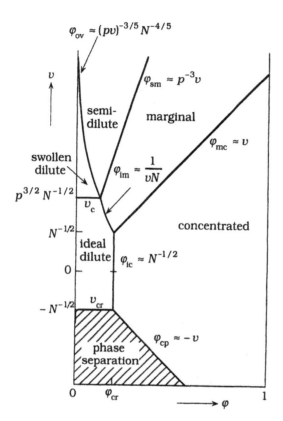

$$\varphi_{ov} \approx (pv)^{-3/5} N^{-4/5}$$

semi-dilute

$$\varphi_{sm} \approx p^{-3}v$$

marginal

$$\varphi_{mc} \approx v$$

swollen dilute

$$\varphi_{im} \approx \frac{1}{vN}$$

$$p^{3/2} N^{-1/2}$$

$$v_c$$

concentrated

$$N^{-1/2}$$

ideal dilute $\varphi_{ic} \approx N^{-1/2}$

$$0$$

$$- N^{-1/2}$$

$$v_{cr}$$

$$\varphi_{cp} \approx -v$$

phase separation

$$0 \qquad \varphi_{cr}$$

$$\longrightarrow \varphi$$

Figure 5.3. Illustration of the five stable regimes (ideal dilute, swollen dilute, semi-dilute, marginal, and concentrated) and the phase separation region (hatched) for polymer solutions. The dependencies of the various cross-overs on N, p, and v are indicated.

will be discussed in more detail in 5.2d. The cross-over concentration φ^{im} between the ideal dilute and marginal regimes is found by balancing the linear and quadratic terms of [5.2.15]: $\varphi^{im} \approx v^{-1}N^{-1}$. Similarly, the transition φ^{mc} between marginal and concentrated regimes follows from equating the quadratic and cubic terms: $\varphi^{mc} \approx v$.

In the concentrated and marginal regimes of fig 5.3 a mean-field description, which neglects any spatial fluctuations, is appropriate. In these regimes, the solution is homogeneous and there is no chain-length dependence. Neither does the persistence p of the chains play a role since the Flory-Huggins expressions do not contain the chain flexibility. This is so because the flexibility is assumed to be the same in the solution and in the reference state, so that p cancels in the entropy difference between the two states.

5.2d. Semidilute solution

When the segment repulsion is strong (good solvent) the chains in a moderately dilute (usually referred to as *semidilute*) solution do not overlap, and mean-field treatments break down because the segment concentration fluctuates in space. A model taking these fluctuations into account was developed by the French school[1] and became known as the *scaling theory* for polymers in good solvents. We first note that coil-coil interactions start to become important at the overlap concentration, which is the concentration at which the entire solution volume is occupied by coils. This overlap concentration φ^{ov} corresponds to the transition from the swollen dilute to the semidilute regimes in fig. 5.3. To a good approximation, φ^{ov} is given by the condition that the overall volume fraction in the solution corresponds to N segments (with volume $N\ell^3$) in a sphere with radius a_g: $\varphi^{ov} \approx N\ell^3/a_g^3$ or, with [5.2.6 and 7],

$$\varphi^{ov} \approx (p\upsilon)^{-3/5} N^{-4/5} \tag{5.2.17}$$

This overlap concentration is also indicated in fig. 5.3.

If we now increase the concentration beyond φ^{ov} the coils interpenetrate more and more and the concentration fluctuations, in the dilute regime taking place over distances of order a_g, become smaller. A simple picture of the semidilute solution is that of a transient network with an average mesh size ξ between interchain crossings; the quantity ξ is called the *correlation length* (see also sec. I.7.7c). A mesh volume of size ξ^3 (also called a *"blob"*) contains N_b segments which all belong to the same chain. If this subchain is sufficiently long ($N_b > N_c$, where N_c is given by [5.2.9]) it will be swollen. The overall volume fraction is given by $\varphi \approx N_b \ell^3/\xi^3$ where, analogous to [5.2.8], $\xi \approx (p\upsilon)^{1/5} N_b^{3/5} \ell$. Eliminating N_b leads to

$$\xi \approx (p\upsilon)^{-1/4} \varphi^{-3/4} \ell \tag{5.2.18}$$

which is a central result for semidilute solutions and nowadays well established by experiment[2].

With increasing φ the blobs become smaller, and ξ and N_b decrease. The upper bound of this regime is reached when $N_b \approx N_c \approx p^3 \upsilon^{-2}$, according to [5.2.9]. This leads to the relation $\xi \approx (p\upsilon)^{1/5}(p^3\upsilon^{-2})^{3/5}\ell = p^2\upsilon^{-1}\ell$ at the transition concentration φ^{sm} between the semidilute (s) and marginal (m) regimes in fig. 5.3. Substitution of this result in [5.1.17] gives $\varphi^{sm} \approx p^{-3}\upsilon$, also indicated in fig. 5.3. This equation has been used by Schaefer[3] to estimate values of φ^{sm} for several polymer-solvent systems. For most common polymer-solvent combinations υ is of order 0.01 - 0.1 so that φ^{sm} is of the order of at most 1%.

[1] P.G. de Gennes, *Scaling Concepts in Polymer Physics*, Cornell University Press, Ithaca, NY (1979).
[2] A. Lapp, C. Picot and C. Strazielle, *J. Phys. Lett.* **46** (1985) L-1031.
[3] D.W. Schaefer, *Polymer* **25** (1984) 387.

Only when the polymer is extremely flexible (p close to unity) may φ^{sm} become a few per cent in good solvents.

For $\varphi > \varphi^{sm}$ the strong density fluctuations in the semidilute regime disappear, and in the marginal regime the solution may again be described by a mean-field theory, as discussed in 5.2c.

We have now discussed all the parts of the "phase diagram" of fig. 5.3 where the solutions are thermodynamically stable. The lower left corner of this diagram corresponds to thermodynamically unstable situations; these are considered in the next section.

5.2e. Phase separation

In section 5.2b the Θ-point ($v = 0$, $\chi = 1/2$) was defined as the point where the effective segment-segment repulsion vanishes. It is therefore also the point of incipient phase separation for chains with negligible translational entropy (i.e., for $N \rightarrow \infty$). For finite chain lengths the phase separation conditions can be found by equating the chemical potentials (given by [5.2.13 and 14]) in both phases, according to the rules given in sec. I.2.19. Numerical solution of these equations gives the full curves of fig. 5.4, known as the *binodals*. For a given solvent quality, a solution of overall composition between φ^α and φ^β (fig. 5.4) separates into a dilute solution of concentration φ^α and a concentrated solution of composition φ^β. The minimum of each curve is the *critical point*, given by

$$v_{cr} = 1 - 2\chi_{cr} = 1 - \left(1 + \frac{1}{\sqrt{N}}\right)^2 \qquad [5.2.19]$$

$$\varphi_{cr} = \left(1 + \sqrt{N}\right)^{-1} \qquad [5.2.20]$$

For solvent qualities better than those corresponding to the critical point ($v > v_{cr}$, $\chi < \chi_{cr}$) the solution is stable at any composition. For $N \rightarrow \infty$, $v_{cr} = 0$ and $\chi_{cr} = 1/2$.

Apart from the binodal curves one may also define *spinodal* curves (see sec. I.2.68), found from the conditions that $d\mu/d\varphi$ (for polymer and solvent) are the same in both phases. One such spinodal (for $N = 100$) is indicated in fig. 5.4. The region inside the spinodal is unstable, that between binodal and spinodal metastable. At the critical point the binodal and spinodal curves coincide.

For long chains the critical point is given by $v_{cr} \approx -N^{-1/2}$ and $\varphi_{cr} \approx N^{-1/2}$ (again omitting numerical coefficients); this point is indicated in fig. 5.3. In this case φ^α is extremely small, corresponding to $\mu(\varphi^\alpha) \approx 0$. Hence also $\mu(\varphi^\beta) \approx 0$, where φ^β is the same as φ^{cp}, the boundary between the concentrated regime (c) and the regime where phase separation (p) occurs. From [5.2.15] we find that $\mu_1(\varphi_{cr}) \approx 0$ if $v(\varphi^{cp})^2 + (\varphi^{cp})^3 \approx 0$, or $\varphi^{cp} \approx -v$, which gives the last cross-over in

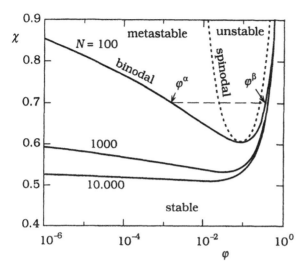

Figure 5.4. Binodals in the χ, φ plane (full curves) for different chain lengths and the spinodal for $N = 100$ (dotted curve) according to the Flory-Huggins theory.

the phase diagram of fig. 5.3. This curve $\varphi^{cp}(v)$ in fig. 5.3 corresponds to $\varphi^{\beta}(\chi)$ in fig. 5.4 (i.e., the branch to the right of the minimum).

5.2f. Polyelectrolytes in solution

We consider simple polyelectrolytes, which we define as homopolymers in which each monomer unit carries an ionizable group. Such a group may be a strong acid or base, in which case its charge is virtually independent of pH (*strong polyelectrolyte*). On the other hand, *weak polyelectrolytes* carry weakly acidic (for example carboxyl) or weakly basic (for example amino) groups; their solution behaviour depends on pH.

Examples of more complex polyelectrolytes are copolymers where only a fraction of the monomers carry charged groups, and *polyampholytes* or *amphoteric polyelectrolytes* which carry both positive and negative charges. In this context proteins may be classified as very complex polyampholytes. In the following we shall mainly survey some aspects of simple (linear) polyelectrolytes. We shall only consider a single polyelectrolyte molecule in its aqueous environment (i.e., we restrict ourselves to dilute solutions, where intermolecular interactions may be neglected; compare sec. 5.2a).

When a polyelectrolyte is dissolved in water, it acquires a certain amount of electrical charge. Due to this charge, an electrostatic repulsion builds up, the strength and range of which depend on the charge density on the molecule and on the concentration of small (counter)ions in the solution. This has two main consequences.

The first effect is that the chain becomes more rigid (it stretches) because the internal repulsion is a minimum for a straight line charge. The second effect (to be discussed below) is that the polymer coil swells due to repulsion between stretched chain sections.

The stretching is usually described in terms of the so-called worm-like chain model which pictures a chain as a thin wire with a flexibility given by its bending constant. The central quantity in this model is the *persistence length q*, which is the characteristic length along the wire over which the directional correlation between segments disappears. For very flexible chains q is only slightly larger than ℓ, whereas for stiff rods $q \to \infty$. The general expression for the mean square end-to-end distance for worm-like chains is[1]:

$$\langle R^2 \rangle = 2q^2(L/q - 1 + e^{-L/q}) \qquad\qquad [5.2.21]$$

where $L = \ell N$ is the contour length of the chain. In the limit of very stiff chain molecules, this equation predicts $\langle R^2 \rangle = L^2$, as expected. Close to this limit we may use the following expansion:

$$\langle R^2 \rangle = L^2(1 - L/3q + ...) \qquad\qquad [5.2.22]$$

For the situation of long flexible chains, the limit $q/L \ll 1$ is relevant, in which case we have $\langle R^2 \rangle = 2qL = 2q\ell N$. When this is compared to the expressions given in sec. 5.1, we see that for this latter case we can set $2q = 6p\ell = \ell_K$.

Stretching of polyelectrolytes can be readily described by allowing for an electrostatic contribution to the persistence length q, i.e., replacing the so-called *bare persistence length* q_0 of the equivalent uncharged chain by the *total persistence length* $q_t = q_0 + q_e$, where q_e accounts for the electrostatic effect. As a consequence, $q \, (= q_t)$ becomes a function of the salt concentration, q_t being small at high, and large at low ionic strength. In order to calculate the electro-static persistence length, we first have to consider the distribution of the counterions around the polyion.

Any charged body immersed in an electrolyte solution will attract counterions and repel co-ions. In the absence of specific adsorption, the charge distribution can be described by means of the Poisson-Boltzmann equation. One of the premises is that the charge on the chain be homogeneously distributed. The countercharge screens the bare charge on the macroion. For a polyelectrolyte chain, the geometry usually adopted is that of a cylinder for which the Poisson-Boltzmann distribution has been given in sec. 3.5f. For cylinders having a very small radius ("line charge") theories have been given by

---

[1] H. Yamakawa, *Modern Theory of Polymer Solutions*, Harper & Row, NY (1971).

Manning[1], Odijk[2], Fixman[3], Stigter[4] and many others. A full discussion of the polyelectrolyte double layer is beyond the purpose of this chapter. We limit ourselves to the following remarks.

Due to the non-linear nature of the Poisson-Boltzmann equation it is always necessary to distinguish between cases of low and of high potentials. It is customary, for polyelectrolytes, to express the charge density in terms of the distance ℓ_e between elementary charges on the chain. As long as the charge density on the polyion is low (i.e., ℓ_e is large), potentials in the neighbourhood of the chain will not be high. This is the Debye-Hückel regime; it is characterized by a linear relation between the charge density parameter $1/\ell_e$ and the potential at the surface of the cylinder. For high charge density, however, (line) charge and potential are no longer proportional. Moreover, most of the polyion charge is compensated by counterions that come very close to the polyion. Eventually, any increase in the polyion charge is virtually cancelled by an equal increase in countercharge in the immediate vicinity of the polyion, a situation that also occurs for other geometries, and which we denote as "strong screening". Examples are figs. 3.7 and 3.13 where, for different geometries, it is seen that strong increases in the surface potential are much less felt at some distance from the surface. As a result, the potential at a distance further away becomes rather insensitive to increases in the polyion charge, and can be seen as arising from an almost constant *effective* charge density, characterized by a parameter ℓ_{eff}. For the cylindrical case, it has been shown that this effective charge density parameter ℓ_{eff} at high polyion charges approaches the Bjerrum length ℓ_B, for monovalent ions defined by:

$$\ell_B = \frac{e^2}{4\pi \varepsilon_0 \varepsilon kT} \qquad [5.2.23]$$

This expression for ℓ_B differs by a factor of 2 from the length r_B below which ion association is assumed to take place in electrolyte solutions, according to Bjerrum's theory, see [I.5.2.30a]. The reason for this factor of 2 is that for counterion association on a polyelectrolyte only the former loses its kinetic energy, whereas for association of two small ions this occurs for both. At low polyion charge, ℓ_{eff} is of course simply given by ℓ_e.

With this result, both Odijk[5] and Skolnick and Fixman[6] were able to calculate q_e to first order. Their result may be written as

[1] G.S. Manning, *Quart. Rev. Biophys.* **11** (1978) 179.
[2] Th. Odijk, *Chem. Phys. Lett.* **100** (1983) 145.
[3] M. Fixman, *J. Chem. Phys.* **70** (1979) 4995.
[4] D. Stigter, *J. Colloid Interface Sci.* **53** (1975) 296.
[5] Th. Odijk, *J. Pol. Sci. Pol. Phys. Ed.* **15** (1977) 477.
[6] J. Skolnick and M. Fixman, *Macromolecules* **10** (1977) 944.

$$q_e = \frac{\ell_B}{4}\left(\ell_{eff}\kappa\right)^{-2}$$ [5.2.24]

This equation tells us that, since κ^2 is proportional to the molar salt concentration c_s (see, for example, [3.5.7]), the electrostatic persistence length (or chain stiffness) decreases inversely proportionally with increasing salt concentration. In fig. 5.5 we present, as an experimental example, the salt dependence of q_t for DNA, as measured by several techniques. For comparison, the theoretical dependence based on [5.2.24] is also shown.

So far, we discussed the electrostatic interactions between segments close to each other along the chain. There is also a second effect due to intramolecular repulsion between chain sections which are distant along the chain but may come spatially close in polyelectrolyte coils. It leads to swelling or, otherwise stated, to an excluded volume. As is customary with uncharged polymers, we can describe this kind of swelling in terms of the linear expansion coefficient α (see [5.1.6]). The essential difference between polyelectrolytes and neutral polymers in this respect is that stiff polyelectrolytes are locally better described by a cylindrical (rod) symmetry, whereas segments of flexible neutral polymers are usually considered as spheres. Unlike the case of spheres, the excluded volume

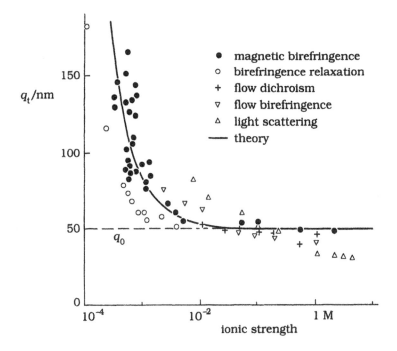

Figure 5.5. The total persistence length q_t as a function of ionic strength for DNA in univalent salt solutions and measured by a variety of techniques. The full curve gives the dependence according to $q_t = q_0 + q_e$, where q_e is given by [5.2.24].

for rods is much larger than the rod volume itself, and it would therefore be expected to be important when the chains are very stiff. One has to take into account, however, that the probability that distant chain sections will interact by excluded volume effects decreases very strongly as the chain becomes stiffer. At least three regimes can be distinguished in the swelling behaviour. We discuss these in the order of increasing electrostatic interaction, corresponding to decreasing ionic strength.

(i) *Very high salt concentrations*

Here $\kappa^{-1} \to 0$ so that only the bare excluded volume and bare persistence length play a role. The polyelectrolyte behaves like an uncharged polymer, as described in secs. 5.2a-e.

(ii) *Intermediate salt concentrations*

For long chains, the excluded volume effect is now operative. We can use an expression analogous to [5.2.8]; this expression reads

$$a_g \approx \left(q_t/\kappa\right)^{1/5} L^{3/5} \qquad\qquad [5.2.25]$$

Note that q_t increases with decreasing electrolyte concentration, whereas κ decreases, so that the expansion factor increases due to the combined effects of excluded volume and stiffening.

(iii) *Low ionic strength*

Eventually, the swelling due to the excluded volume vanishes, because of the strong increase in q_t. Then one finds

$$a_g \approx \left(q_e L\right)^{1/2} \sim \left(\kappa \ell_{eff}\right)^{-1} L^{1/2} \qquad\qquad [5.2.26]$$

Note that, in a sense, the polyelectrolyte behaves now as if it were an "ideal" (Gaussian) chain with a relatively small number of Kuhn lengths. This coil size is determined by the local stiffness, but not by long-range excluded volume. Should q_e increase even furher (approaching L), then the wormlike chain would be better described by a slightly curved rod (as expressed by [5.2.21]) than by a random-flight chain.

We note that the above description is only valid for chains that are soluble in their neutral state. If they are not, removal of the charge will lead to collapse of the coil into a dense globule and phase separation.

5.3 Some general aspects of polymer adsorption

In this section we give a brief overview of what is empirically known about polymers in the proximity of an interface. In particular, we discuss the structure and the amount adsorbed from (semi)dilute solutions. In this sense, the treatment is an extension of that in secs. 2.4e and 2.7 to long chain molecules.

5.3a. *Structure*

In sec. 5.2 we emphasised the fact that polymers have a large number of internal degrees of freedom. Adsorption must imply changes in this number. The usual description of conformations at an adsorbing interface, first proposed by Jenkel and Rumbach[1] and depicted schematically in fig. 5.1, is in terms of three types of subchains: *trains*, which have all their segments in contact with the substrate, *loops*, which have no contacts with the surface and connect two trains, and *tails* which are non-adsorbed chain ends.

Important structural information resides in the *profile c(z)*. Experimental techniques (for example, neutron scattering and neutron reflectometry, see sec. 5.6) are now available to determine this profile. For the sake of illustration we sketch a profile in fig. 5.6. For uncharged homopolymers on solid substrates the segment concentration is found to decay monotonically and to be convex with respect to the z-axis. Such a profile can only originate from a wide distribution in lengths of loops and tails. At very low coverage the chains have a tendency to "flatten" and the concentration falls steeply. At higher coverages the profile decays more gradually; this is due to an increase in the lengths of loops and tails. For long chains such a high coverage is already found at very low solution concentrations. There is a certain resemblance to the profiles for counterions and co-ions in the electric double layer (figures 3.8 and 33), but the mechanisms underlying the distributions are rather different.

It is not always possible to measure c(z) in full detail so that introduction of some average properties remains useful. One such parameter is the *train fraction* or *bound fraction p*. In terms of fig. 5.6 it is the amount of polymer within the thickness ℓ of a segment on the surface, divided by the total adsorbed amount. Spectroscopic and other techniques can often be successfully applied to determine *p*. The general trends found support the picture sketched above, i.e., *p* is close to unity when the adsorbed amount is low, and it decreases as the pseudo-plateau of the adsorption isotherm (see fig. 5.7) is approached. Reported bound fractions vary widely, depending on the system and method chosen. It should be said that bound fraction measurements leave much room for differences in interpretation; some results are still not well understood. We will briefly discuss the methods in sec. 5.6, and give some experimental data in secs. 5.7 and 5.8.

A characteristic feature of adsorbed polymers is that the width of the interfacial region can extend up to distances of the order of the radius of

---

[1] E. Jenkel and B. Rumbach, *Z. Elektrochem.* **55** (1951) 612.

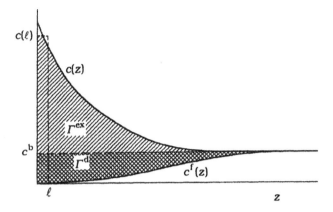

Figure 5.6. Schematic representation of a polymer concentration profile as a function of distance z from the interface. The upper curve gives the overall segment concentration $c(z)$, the lower curve the concentration $c^f(z)$ due to non-adsorbed chains. Both curves approach the bulk solution concentration c^b at large z. The hatched area is the polymer surface excess Γ^{ex}, the sum of the shaded and hatched areas represents the total adsorbed amount Γ^a. The difference between Γ^a and Γ^{ex} is denoted by Γ^d (shaded area).

gyration a_g (see sec. 5.2). This means that the thickness of adsorbed polymer layers is also informative. With the advent of accurate methods to measure diffusion coefficients of colloidal particles, or solvent fluxes through narrow pores, hydrodynamic thicknesses of adsorbed polymer layers have become available. The general finding is that for low coverage layers are thin, as are polyelectrolyte layers at low ionic strength. This is consistent with the high bound fractions found for these cases. At high coverage, on the other hand, thicknesses up to 200 nm have been reported. The thickness then increases with molecular weight roughly as M^a, where a is found to be in the range 0.4-0.8.

So far, we have discussed the structure of adsorbed layers with respect to the z direction, i.e., perpendicular to the surface. There may also be some structural variation parallel to the surface: the layer may be laterally inhomogeneous. As we have seen in sec. 5.2, dilute and semidilute polymer solutions are also inhomogeneous; density variations occur on a length scale ξ, where the correlation length ξ is equal to the coil size in dilute solutions and becomes a function of concentration in semidilute solutions. In concentrated solutions, such fluctuations are expected to be unimportant. Similar considerations hold for the adsorbed layer; at low concentrations and in good solvents we may expect fluctuations (inhomogeneities) parallel to the surface. At very low adsorbed amounts (such as occur with highly charged polyelectrolytes) any adsorbed polymer layer will clearly become inhomogeneous, since molecules are randomly scattered over the surface. However, in many practical situations the concentrations in the adsorbed layer are high enough to make a mean-field picture (in the lateral direction) appropriate.

A final remark concerns the effect of interfacial curvature. Above we have considered the surface to be flat, i.e., its radius of curvature is supposed to be large with respect to the thickness of the adsorbed layer. Some adsorption studies, however, deal with colloidal particles which are too small to fulfil this condition. Intuitively, one would expect different adsorbed amounts on a strongly convex surface, since the curvature changes the lateral repulsion. Others have argued that also the thickness could be different. Very few studies have dealt explicitly with these effects[1,2,3,4]

5.3b. Adsorbed amount

As was explained in chapter 2, adsorption can be defined in terms of a change in concentration in the interfacial region. As this change can be both negative and positive, it is customary to distinguish between *negative adsorption* or *depletion*, and *positive adsorption*. Negative adsorption leads to a layer near the interface where the concentration of polymer segments is lower than in the bulk solution. The depleted amount is difficult to study by direct experiments (especially from dilute solutions), but the thickness of depletion layers can be and has been measured[5,6]. Positive adsorption is much easier to study, because usually the difference between surface and bulk concentrations is high. In this case we have close to the surface a dense population of more or less inter-penetrating molecules. The adsorbed molecules give the interface a number of interesting properties. Therefore, this is the most widely studied situation.

In order to discuss a number of important quantities we consider the interfacial profile for the case of positive adsorption. Such a profile is sketched in fig. 5.6. It represents the polymer concentration $c(z)$ as a function of the distance z from the interface. The quantity $c(z)$ is related to the volume fraction $\varphi(z)$ through $\varphi = c/\rho$ where ρ is the partial density of the polymer in the solution. It is seen that $c(z)$ gradually drops to the value $c^b = c(\infty)$ which is the concentration of segments in the bulk of the solution. The lower curve in fig. 5.6 shows the concentration profile of segments belonging to free molecules, having no contact with the surface. The *excess* adsorbed amount Γ^{ex} (the amount of

[1] S. Alexander, *J. Phys.* (Paris) **38** (1977) 977.
[2] P. Pincus, C.J. Sandroff, and T.A. Witten jr., *J. Phys.* (France) **45** (1984) 725.
[3] M.J. Garvey, Th.F. Tadros, and B. Vincent, *J. Colloid Interface Sci.* **55** (1976) 440.
[4] G.P. van der Beek and M.A. Cohen Stuart, *J. Phys.* (France) **49** (1988) 1449.
[5] D. Ausserré, H. Hervet and F. Rondelez, *Macromolecules* **19** (1986) 85.
[6] L.T. Lee, O. Guiselin, A. Lapp, B. Farnmoux and J. Penfold, *Phys. Rev. Lett.* **67** (1991) 2838.

polymer in the interfacial region in excess of c^b) was already defined in [2.1.2]. It corresponds to the hatched area in fig. 5.6. Alternatively, one can consider the *total* amount of polymer in contact with the surface Γ^a. This is a somewhat problematic quantity, since it is not easily accessible by experiment. However, Γ^a can be split up into Γ^{ex} (hatched area) and Γ^d (shaded area). Since Γ^d vanishes for $c^b \to 0$ we then have $\Gamma^{ex} \approx \Gamma^a$, which is a good approximation for adsorption from dilute solution. For dilute solutions we shall therefore drop the distinction between Γ^a and Γ^{ex}, and simply use Γ. For the case of adsorption from a polymer melt ($\varphi^b \to 1$), the hatched area vanishes (i.e., $\Gamma^{ex} \to 0$) and only Γ^d contributes. Note that the bound fraction p, discussed above, is defined as $p = \Gamma^{tr}/\Gamma^a$, and to obtain this parameter experimentally we need to go to dilute solutions where we can simplify this to $p = \Gamma^{tr}/\Gamma$.

Adsorbed amounts are commonly presented as an adsorption isotherm, which is a plot of either Γ^{ex} or Γ^a as a function of the polymer solution concentration at a given temperature. Polymer adsorption typically leads to high-affinity (H) isotherms (see the classification in fig. 2.24). A typical example for a homodisperse polymer is shown in fig. 5.7. As a rule considerable adsorption occurs even at extremely low concentrations, typically well below 1 g m^{-3}. For somewhat higher concentrations, the isotherm shows a nearly horizontal part, the *pseudo-plateau* (often simply referred to as "plateau"), indicating saturation of the surface. For polydisperse polymers there is usually not such a well-defined pseudo-plateau, and a more rounded isotherm is found (see sec. 5.3d below).

For polymers, the plateau adsorbed amount may typically reach values of the order of a few mg m^{-2}, and it is a function of the molecular weight: higher molecular weights generally give higher plateau values. The way in which the amount adsorbed at fixed concentration varies with molecular weight depends on solvency. In not too poor solvents ($\chi < 1/2$) Γ increases with the molar mass M in the low M range but for very high molar masses the plateau value becomes independent of chain length. In Θ-solvents, however, Γ seems to increase propor-

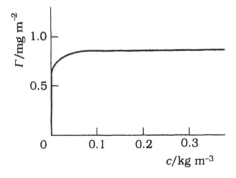

Figure 5.7. A typical high-affinty polymer adsorption isotherm.

tionally to $\log M$, without bounds. Adsorbed amounts in Θ-solvents are as a rule higher than in good solvents, especially at high M. The lower the molecular weight, the less the high-affinity character of the adsorption isotherm. One may therefore observe a more rounded isotherm when the molecular weight is sufficiently reduced.

Since adsorption from solution is an exchange process (see sec. 2.4b), the adsorbed amount depends on the difference in interaction between a polymer segment and a solvent molecule with the surface. Hence, the net adsorption (Gibbs) energy per segment depends both on the solvent and on the nature of the substrate. With solid substrates London-Van der Waals (dispersion) forces, dipolar forces, hydrogen bonds, and hydrophobic bonding are possible types of interaction. However, it is not always obvious which component, solute or solvent, will bind the strongest. For instance, poly(vinyl pyrrolidone) adsorbs from water onto hydrophilic silica[1], but polyacrylamide and the polysaccharide dextran do not[2,3]. In some cases the adsorption is driven by the poor solubility of the polymer or by the difference in surface tension between polymer and solvent. For example, poly(dimethyl siloxane), which has a lower surface tension than toluene, adsorbs at the toluene/air interface but polystyrene, having a higher surface tension, does not[4].

In polyelectrolyte adsorption electrostatic interactions play a very important role. Since these interactions are of variable range and strength, depending on the charge densities (for both the surface and the polyelectrolyte chains) and on the salt concentration, the adsorbed amount depends strongly on these variables. At low salt concentrations, one usually finds that highly charged polyelectrolytes adsorb in small amounts (a few tenth of a mg per m²), or hardly at all[5]. When salt is added, the adsorption increases in the majority of cases. However, high concentrations of salt may also lead to desorption. Weakly charged polyelectrolytes tend to give higher adsorbed amounts at low ionic strength. For a more extensive discussion we refer to sec. 5.8.

5.3c. Kinetics of adsorption

The adsorption of small molecules usually comes to equilibrium quite rapidly (see sec. 2.8). The frequency of attachment-detachment events is such that any off-equilibrium situation can relax on a time scale of milliseconds or less, so that, on the time scale of most experiments, these adsorptions can be

---

[1] M.A. Cohen Stuart, G.J. Fleer, and B.H. Bijsterbosch, *J. Colloid Interface Sci.* **90** (1982) 310.
[2] O. Griot, J.A. Kitchener, Trans. *Faraday Soc.* **61** (1965) 1026.
[3] V. Hlady, N.G. Hoogeveen, and G.J.Fleer, unpublished results.
[4] R. Ober, L. Paz, C. Taupin, P. Pincus, and S. Boileau, *Macromolecules* **16** (1983) 50.
[5] M.A. Cohen Stuart, G.J. Fleer, J. Lyklema, W. Norde, and J.M.H.M. Scheutjens, *Adv. Colloid Interface Sci.* **34** (1991) 477.

considered as completely reversible. In most cases, diffusion contants are high enough to provide rapid transport towards the surface.

With polymers, the situation is different. First of all, their diffusion coefficients are lower by one or two orders of magnitude. In addition, polymer adsorption may occur already from very dilue solutions so that considerable limitation of adsorption and desorption rates by transport in the bulk solution is a rule rather than an exception. Stirring can then be important to bring the rate to an acceptable level. For desorption, the transport rate can be shown to become effectively zero. In itself, this should not be taken as evidence that the layer behaves as "frozen"[1].

Secondly, the fact that polymers are anchored to the surface by many points of attachment severely reduces their mobility. Very slow surface processes can occur, as has been demonstrated by means of competitive adsorption studies[2,3]. Polyelectrolytes, which can form very strong ionic bonds with solid substrates, are also found to relax very slowly, or not at all. Addition of small ions (salt) can facilitate relaxation somewhat. All this implies that in judging adsorption data for polymers, one should be aware of the possibility that off-equilibrium states may persist over longer periods of time.

For polyelectrolytes a third effect is important, namely that electrostatic repulsion between the surface and the incoming molecule may strongly reduce or even entirely nullify the rate of attachment, much in the same way as an electrostatic barrier prevents colloidal particles from aggregating. Obviously, this depends on the charge present on the surface, which may originate not only from the surface itself, but also, and even more importantly, from the polyelectrolyte already adsorbed. This may explain why strongly charged polyelectrolytes can often not directly adsorb until coverage is high. However, such high coverages can be achieved indirectly by first taking away the charge from the polyelectrolyte and then, after adsorption, restoring it, for instance by changing the pH[4].

5.3d. Polydispersity

The synthesis of a polymer by a free-radical or polycondensation mechanism usually leads to a product with a rather wide distribution of molecular weights. These *polydisperse* samples must be considered as multicomponent mixtures rather than as chemically pure substances. Using analytical methods like gel permeation chromatography (GPC) it is possible to analyze such mixtures.

---

[1] J.C. Dijt, M.A. Cohen Stuart and G.J. Fleer, *Macromolecules* **25** (1992) 5416.

[2] M.A. Cohen Stuart, G.J. Fleer and B.H. Bijsterbosch, *J. Colloid Interface Sci.* **90** (1982) 321.

[3] T. Cosgrove and J.W. Fergie-Woods, *Colloids Surfaces* **25** (1987) 91.

[4] J. Meadows, P.A. Williams, M.J. Garvey, and R. Harrop, *J. Colloid Interface Sci.* **148** (1992) 160.

From early viscosity measurements on the supernatant solution in adsorption experiments it was deduced that adsorption preference occurs with respect to the molecular weight[1]. Otherwise stated, upon adsorption of a polydiperse polymer *fractionation* takes place. This polydispersity effect has been investigated in detail by several authors[2,3,4,5]. The central finding is that, at equilibrium, high molecular weight polymers adsorb preferentially over lower molecular weight ones. Although one is inclined to think that multiple anchoring gives longer molecules an advantage over shorter ones, this is not the basic reason. In reality, adsorption sites on the surface are almost equally efficiently covered by molecules of different lengths, so that the total adsorption energy per unit area in the system is not strongly affected by polydispersity. The entropy of mixing in the solution, however, decreases strongly with increasing chain length (this can be seen from the factor N^{-1} in the last term of [5.2.11]) and this favours adsorption of longer molecules. Consequently, the driving force for this fractionation depends on the concentration in solution and is larger for more dilute solutions. The thermodynamic and experimental consequences of polydispersity have been elaborated in a few keynote papers[5,6]. Below we summarize the most important effects.

The simplest example is a mixture of two homodisperse polymer samples differing only in molecular weight. The adsorption of such a mixture is conveniently discussed in terms of the total polymer mass (free and adsorbed) per unit area in the system (Γ^t), which is subdivided into the amount adsorbed on the surface (Γ) and the free molecules in the bulk solution (Γ^b). The parameter Γ is either Γ^{ex} or Γ^a; in dilute solutions, to which we shall restrict ourselves, there is no difference between Γ, Γ^a and Γ^{ex}, as discussed in sec. 5.3b. The quantity Γ^b may be defined as $c_p t_s$, where c_p is the polymer concentration and t_s the thickness of the solution layer available to the adsorbent surface. The latter parameter may also be taken as the ratio between the total solution volume and the total surface area in the system. The quantities Γ, Γ^b and Γ^t are expressed in the same units, e.g., mg/m².

Three regions can be distinguished, indicated in the schematic adsorption isotherms in fig. 5.8.

[1] J.M. Kolthoff and R.G. Gutmacher, *J. Phys. Chem.* **56** (1952) 740.
[2] R.E. Felter and L.N. Ray, *J. Colloid Interface Sci.* **32** (1970) 349.
[3] C. van der Linden and R. van Leemput, *J. Colloid Interface Sci.* **67** (1978) 63.
[4] M.A. Cohen Stuart, J.M.H.M. Scheutjens and G.J. Fleer, *J. Pol. Sci. Pol. Phys. Ed.* **18** (1980) 559.
[5] R.J. Roe, *Adhesion and Adsorption of Polymers*, L.H. Lee, Ed. in *Polymer Science and Technology*, part **12B**, Plenum N.Y. (1980) 629.
[6] J.M.H.M. Scheutjens and G.J. Fleer, in *The effect of Polymers on Dispersion Properties*, Th.F. Tadros, Ed., Academic Press, London (1982) 145.

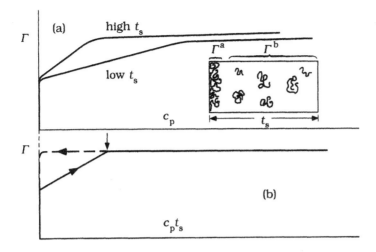

Figure 5.8. Adsorption isotherms of a mixture of two polymer fractions. The top diagram (a) shows a conventional plot of Γ^{ex} as a function of the solution concentration c_p for two different values of the thickness t_s of the solution layer. The upper curve is for adsorption carried out at the highest value of t_s. The inset illustrates the quantities Γ^a, Γ^b, and t_s. The bottom diagram (b) gives the same two isotherms replotted on a normalized scale: Γ^{ex} as a function of $\Gamma^b = c_p t_s$ (solid curve). By this normalization the two curves of the top diagram coincide. The dashed curve is a desorption isotherm obtained by dilution of the solution, i.e., by increasing t_s.

(i) When Γ^t is low, all polymer chains can find a place on the surface ($\Gamma^b = 0$) and no fractionation takes place. This is the vertical section in fig. 5.8.

(ii) As soon as saturation is approached, discrimination sets in: only the high-molecular-weight fraction is entirely adsorbed, whereas the smaller molecules have access only to the surface sites not occupied by the big ones. Consequently, low-molecular-weight polymer is left in the solution. As usually the adsorbed mass increases with increasing molecular weight of the adsorbate (see sec. 5.3b), preferential adsorption leads to an increase of the adsorbed mass. For the simple case of two fractions, Γ is linear in the solution concentration, until virtually all small molecules are displaced (see the rising part in fig. 5.8).

(iii) When the surface is saturated with the large molecules, a plateau in the adsorbed mass is reached. Upon further increase of c_p the concentration of large molecules in the solution begins to increase, but nothing changes in the adsorbate any more.

The three parts of the adsorption isotherm for a bimodal mixture are expected to be separated by rather sharp kinks, as indicated in fig. 5.8. Several examples of such mixture isotherms have been published[1,2]. One may ask what will happen when one tries to go back along such an isotherm by dilution of the supernatant, for example, from the point where, on the way up, the enrichment of the adsorbed layer with large molecules is just completed (indicated by the vertical arrow in fig. 5.8). Since the shape of the isotherm was not determined by c_p but by the *amount* of polymer added, dilution is of no consequence, i.e., Γ will remain virtually constant, as for homodisperse samples. Thus, when going up the isotherm by adding polymer and then going down by dilution, one finds different isotherms, as indicated by the arrows in fig. 5.8. This kind of *hysteresis* has sometimes been interpreted as evidence for the irreversible nature of polymer adsorption. We now see that this is not correct, since large molecules do replace smaller ones in the adsorption process, which means that desorption (of the lower molecular weights) *does* occur. In the hysteresis loop the composition of the adsorbate is different between the adsorption and desorption branches because the values of t_s are different.

An important point is that it is Γ^t (rather than the solution concentration c_p) which in a given sample determines the partitioning of polymer between surface and solution. Therefore, the adsorbed amount Γ and the amount Γ^b remaining in solution are both determined by Γ^t. A unique relation is only obtained when Γ is plotted against Γ^t or, equivalently, against $\Gamma^b = c_p t_s$. For a macroscopically flat surface, t_s equals the thickness of the solution layer in contact with the surface. More generally, t_s is the ratio of the solution volume and the available surface area, which also applies to dispersions. For an adsorbent consisting of dispersed particles, t_s is inversely proportional to the solid content in the adsorption experiment. Isotherms at different t_s, i.e., at different A/V ratio, do not coincide when plotted as Γ against c_p (fig. 5.8a), but they do when plotted as Γ versus $c_p t_s$ (fig. 5.8b). Examples will be discussed in sec. 5.8.

By generalizing the above ideas to polydisperse polymers (multicomponent mixtures), which should have many 'kinks' in their adsorption isotherms, the reader will easily understand why isotherms of polydisperse samples are usually rounded, the more so the wider the M distribution is. There are also consequences for the surface pressure, the layer thickness, and the interaction between polymer-covered surfaces.

5.3e. *Depletion*

Absence of attraction between adsorbent and polymer leads to negative adsorption. Since polymer molecules in solution resist deformation by virtue of their "elastic" conformational entropy, the centre of mass of a non-adsorbing

---

[1] L.K. Koopal, *J. Colloid Interface Sci.* **83** (1981) 116.
[2] V. Hlady, J. Lyklema and G.J. Fleer, *J. Colloid Interface Sci.* **86** (1982) 395.

polymer coil cannot come closer to the surface than a distance of the order of the correlation length which is, in dilute solution, equal to the coil radius a_g. In a semidilute solution this correlation length is given by [5.2.18]. Due to the conformational entropy loss, the region immediately adjacent to the interface has a lower segment concentration than in the solution. This phenomenon is called *depletion*. A typical volume fraction profile is sketched in fig. 5.9.

Negative adsorption is in most cases very small compared to positive adsorption and therefore not easily detected directly. A sophisticated optical technique (evanescent-wave-induced fluorescence, EWIF) has been used to prove the reduction in segment concentration close to a non-adsorbing surface[1]. Depletion also has an effect on the flux of polymer solutions through pores: since the viscosity of the liquid near the non-adsorbing surface is lower than that of the polymer solution, the flux is then higher than would be expected on the basis of the bulk viscosity[2]. Negative adsorption at a liquid-air interface leads to a measurable increase in surface tension[1].

The most conspicuous depletion effect is perhaps the destabilisation of colloidal dispersions by non-adsorbing polymer of sufficiently high concentration. This effect has been studied in much detail[3], and can be understood as a consequence of the osmotic pressure difference between the polymer solution and the pure solvent film between two particles with overlapping depletion layers. Due to the fact that the osmotic pressure *increases* with increasing concentration, whereas the depletion layer thickness (i.e., the correlation length) *decreases*, the value of the minimum in the attractive pair potential (at close contact) goes, for spherical particles, through a maximum as a function of the polymer solution concentration. As a result, both destabilisation at some critical polymer concentration and restabilisation at a higher concentration occur, and both effects have indeed been observed. With well-defined colloidal particles and polymers, very interesting phase diagrams are obtained. We intend to return to this phenomenon in Volume V.

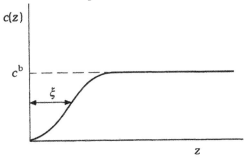

Figure 5.9. Typical interfacial concentration profile of a non-adsorbing polymer. The correlation length ξ, which is a measure of the depletion layer thickness, is indicated.

[1] C. Allain, D. Ausserré and F. Rondelez, *Phys. Rev. Lett.* **49** (1982) 1694; D. Ausserré, H. Hervet and F. Rondelez, *Phys. Rev. Lett.* **54** (1985) 1948.
[2] G. Chauvetau, M. Tirrel, and A. Omari. *J. Colloid Interface Sci.* **100** (1984) 41.
[3] B. Vincent, J. Edwards, S. Emmett and R. Croot, *Colloids Surfaces* **31** (1988) 267.

5.4 Polymer adsorption theories

5.4a. Introduction

As discussed in sec. 5.2, polymer conformations are usually modelled as walks in continuous space or on a lattice. Continuum models have the advantage of being more general, but lattice models make the counting of conformations very much simpler. It is well known that the conformations of chains in a homogeneous solution under Θ-conditions can be described as unrestricted, purely random walks. For sufficiently long chains this results in random coil conformations as illustrated in fig. 5.1a. Near an impenetrable surface such a description has to be modified because half of the space is inaccessible to polymer segments. This restriction leads to a lower conformational entropy in the surface region. If the segments have no affinity for the surface, no adsorption occurs. There is then a depletion zone near the surface, see sec. 5.3e. However, in many instances polymers do adsorb, despite these entropic restrictions. Attachment of the polymer takes place only if the adsorption energy $\Delta_{ads}U_{m1}$ of a solvent molecule (component 1) is less negative than the adsorption energy $\Delta_{ads}U_{m2}$ of a segment of the polymer (component 2). This is an illustration of the pair exchange principle discussed in secs. 2.1 and 2.4b. Following Silberberg[1] and Roe[2], we introduce a dimensionless adsorption energy parameter χ^s for the adsorption of 2 from 1. It is defined as

$$\chi^s = \left(\Delta_{ads}U_{m1} - \Delta_{ads}U_{m2}\right)/kT \qquad\qquad [5.4.1]$$

According to this definition χ^s is counted positive if there is a net segment-surface attraction.

For negative values of χ^s no adsorption occurs. Neither does adsorption take place if χ^s is zero or slightly positive because of the entropic restrictions discussed above. For long polymer chains accumulation of polymer at the surface is found only if χ^s exceeds a critical value, which is usually a few tenths, see [5.5.4] below. However, once this critical value has been surpassed, the polymer adsorbs with a high affinity because of the numerous possible contacts.

Even if $\chi^s = 0$ the statistics of polymer chains near a surface are rather complicated. In the 1960s, discussions appeared in the literature[3,4,5,6] as to

---

[1] A. Silberberg, J. Chem. Phys. **48** (1968) 2835.
[2] R.J. Roe, J. Chem. Phys. **60** (1974) 4192.
[3] R. Simha, H.L. Frisch, and F.R. Eirich, J. Phys. Chem. **57** (1953) 584; H.L. Frisch and R. Simha, J. Chem. Phys. **24** (1956) 652; **27** (1957) 702.
[4] A. Silberberg, J. Phys. Chem. **66** (1962) 1884.
[5] E.A.DiMarzio, J. Chem. Phys. **42** (1965) 2101.
[6] R.J. Roe, J. Chem. Phys. **43** (1965) 1591.

whether the surface acts as a *reflecting* or an *absorbing* boundary. In the former case a random walk hitting the surface continues its way as a reflected image of the unperturbed walk (i.e., in the absence of the surface), leading to a constant segment concentration up to the surface. In the latter case the walk is discarded ("absorbed") so that the polymer is depleted from the surface region. It is now generally accepted that the absorbing boundary condition is the proper one to use.

For the more realistic situation of segments with interactions, the walks are no longer random. There are various ways to cope with this situation. One method is to weight each step with an appropriate Boltzmann factor. A conformation is then described as a step-weighted walk. In the simplest case of adsorbing ideal chains with segments that do not "feel" each other (i.e., with zero excluded volume), the weighting factor equals $\exp(\chi^s)$ for each visit to the surface and 1 for all other steps. This case, only relevant for isolated adsorbing chains, can be solved straightforwardly (see sec. 5.5a), but even in this limit numerical methods are usually required.

In most systems of practical interest the interaction between the segments does play a role because their concentration in the surface zone is much higher than in the bulk solution. As a consequence, the Boltzmann factors in this region differ from unity even if the segments do not contact the surface. One way of tackling this problem is by Monte Carlo techniques (sec. 5.4b), another is using the concept of the so-called *self-consistent field* (SCF), to be discussed in sec. 5.5b. In the SCF method, all possible walks of a test chain in a given field are considered and from these the segment concentration profile is computed. However, this field itself depends on the local concentrations, hence some iteration procedure is required. In the final solution, the field and the concentration profile should be consistent. Usually a *mean* (averaged) *field* is used, whereby fluctuations in some directions (for example, those parallel to the surface) are neglected. If a suitable relation between field (weighting factors) and concentration profile can be formulated, the self-consistent equations may be solved. In general, such a relation contains both energetic and entropic terms.

In the next subsection (5.4b), we first give a condensed overview of the various theoretical methods which have been used to model polymers at interfaces.

5.4b. *Overview of theoretical models*

(i) *Exact enumeration.* In an exact enumeration, a computer algorithm is used to generate the complete set of translationally invariant self-avoiding conformations of a single chain on a lattice. Usually, solvent effects are not considered. At present, the exact enumeration is limited to isolated chains that have at least one segment in the surface plane. The computational problems in such an enumeration soon become prohibitive; a chain of N segments on a cubic lattice will have of the order of 5^N conformations! Exact results are known for chains up to 24 segments on various lattices[1]. As a model of polymer adsorption, the exact enumeration procedure is only of limited interest as it has been applied mainly to single walks, i.e., for extremely low coverage. However, recently some attempts have been made to consider the influence of neighbouring walks[2].

(ii) *Monte Carlo method.* An alternative method for obtaining the conformational statistics of an adsorbed chain is to use a *sampling* scheme. This means that only a representative subset of the total number of conformations is generated. Sampling methods have been described in detail in several texts[3,4], and most approaches start from the classical paper by Metropolis et al.[5]. In this procedure, known as importance sampling, the states of the system are sampled with a probability $e^{-U_i/kT}$, where U_i is the energy of conformation i, and then the accepted conformations are weighted uniformly. Lal and Clark have described the application of this method to adsorbed polymers in some detail[6], after earlier work by Skvortsov et al.[7]. The size of the system that can be simulated is restricted by computer capacity and special precautions must be introduced to avoid undesired boundary effects. This can be achieved in several ways, but the simplest is to use the idea of a periodic boundary. A better approach is to simulate several chains in one box, but this soon leads to extremely long computing times. Although in principle the Monte Carlo method gives a detailed picture on a molecular level, it cannot (yet) describe the adsorption equilibrium between adsorbed and free chains.

(iii) *Lattice models for isolated chains.* For isolated, ideal polymer chains, several authors have laid down the fundamentals of the counting of conformations[8,9,10,11]. Basically, the problem is considered as a first order Markov process. In such a process, a next step "remembers" only the position of the previous segment: immediate step reversals are allowed (backfolding).

[1] M.F. Sykes, A.J. Guttmann, M.G. Watts, and P.D. Roberts, *J. Phys.* **A 5** (1972) 653.

[2] G. M. Torrie, J.Barrett, and S.G. Whittington, *Faraday Trans.* **II 75** (1979) 369.

[3] I. Carmesin and K. Kremer, *Macromolecules* **21** (1988) 2819.

[4] K. Binder, *Colloid Polym. Sci.* **266** (1988) 871.

[5] N. Metropolis, A.W. Rosenbluth, M.N. Rosenbluth, and A.H. Teller, *J. Chem. Phys.* **24** (1953) 1087.

[6] M. Lal and A.T. Clark, *J. Chem. Soc. Faraday Trans* **II 74** (1982) 395.

[7] A.M. Skvortsov, T.M. Birshtein, and E.B. Zhulina, *Polym. Sci. USSR* **16** (1976) 2276.

[8] E.A.DiMarzio, *J. Chem. Phys.* **42** (1965) 2101.

[9] R.J. Roe, *J. Chem. Phys.* **43** (1965) 1591.

[10] R.J. Rubin, *J. Chem. Phys.* **43** (1965) 2392; *J. Res. Nat. Bur. Stand.* **B 70** (1966) 237.

[11] E.A. DiMarzio and F.L. McCrackin, *J. Chem. Phys.* **43** (1965) 539.

Consequently, a third segment (two steps from the first) may occasionally take the same position as the first segment. For isolated molecules near a surface, steps towards or within the surface layer are biased with a weighting factor $\exp(\chi^s)$, all other steps are random, i.e., their weighting factor is unity. Since the DiMarzio-Rubin single-chain model[1,2,3] forms the basis of more sophisticated theories, we return to it in sec. 5.5a.

(iv) *Lattice models for interacting chains.* In the 1970s, Roe[4] and Helfand[5] gave the first lattice treatments for interacting chains, allowing for the finite volume of the segments and for segment-solvent interactions in a self-consistent mean-field approach. These authors expressed their final equations in segment concentration profiles but did not consider individual chain conformations. The energy and entropy equations are based upon the Flory-Huggins expressions [5.2.11 and 12], but they were extended to make them applicable to a concentration gradient. In the derivation of the entropic part of the partition function, Roe assumed that the spatial distribution of the segments does not depend on the ranking number in the chain, so that anywhere end segments have the same concentration as middle segments. This boils down to a complete neglect of end effects. Roe obtained the equilibrium profiles $\{\varphi_1(z)\}$ and $\{\varphi_2(z)\}$ by minimizing $F = U - TS$ under the boundary constraints $\{\varphi_1(z)\} + \{\varphi_2(z)\} = 1$ for any layer z.

Helfand presented an alternative model in terms of so-called anisotropy factors. Under isotropic conditions these factors are unity. A value greater than 1 indicates a greater than random chance for a step from a site in layer z in a particular direction. Like Roe, Helfand neglects the ranking number dependence and his equations are formulated only in the limit of infinite chain length $(N \to \infty)$. For a full discussion of these models, we refer to the literature[4,5].

(v) *The SCF theory of Scheutjens and Fleer.* Around 1980, Scheutjens and Fleer[6] put forward a generalized lattice model for the adsorption of interacting chains. This theory can be considered as an extension of the DiMarzio-Rubin model for isolated chains. In the latter model only steps in or into the surface layer have a weighting factor different from unity. In contrast, Scheutjens and Fleer weight all the steps with an appropriate Boltzmann factor, using a mean-

---

[1] E.A.DiMarzio, *J. Chem. Phys.* **42** (1965) 2101.

[2] R.J. Rubin, *J. Chem. Phys.* **43** (1965) 2392; *J. Res. Nat. Bur. Stand.* **B 70** (1966) 237.

[3] E.A. DiMarzio and R.J. Rubin, *J. Chem. Phys.* **55** (1971) 4318.

[4] R.J. Roe, *J. Chem. Phys.* **60** (1974) 4192.

[5] E. Helfand, *J. Chem. Phys.* **63**, 2192 (1974); *Macromolecules* **9** (1976) 307; T.A. Weber and E. Helfand, *Macromolecules* **9** (1976) 311.

[6] J.M.H.M. Scheutjens and G.J. Fleer, *J. Phys. Chem.* **83** (1979) 1619; *ibid.* **84** (1980) 178.

field approach. This weighting factor accounts for the average excluded volume and energetic interactions. Whereas the DiMarzio-Rubin model gives explicit solutions but is not self-consistent, that of Scheutjens and Fleer results in a set of implicit equations which have to be solved numerically to obtain the self-consistent solution. Further details will be treated in sec. 5.5.

(vi) *Train-loop-tail models.* In these models the starting point is to consider an adsorbed molecule as an assembly of trains, loops and tails. For each of these sequences, approximate analytical expressions for the partition function are derived. From combinatory rules, the partition function of an adsorbed molecule is expressed in these functions. The Helmholtz energy is then minimized under the appropriate constraints to give the equilibrium structure in the limit of isolated adsorbed chains, whereby all segments in loops and tails have a weighting factor equal to unity. Numerical results have been given by Silberberg[1]. Hoeve et al.[2] obtained analytical approximations by neglecting the tails.

In order to extend the train-loop model to interacting chains, the Helmholtz energy of mixing of segments and solvent molecules has to be included before maximizing the partition function. Hoeve[3] evaluated this energy by assuming that, although the adsorbed layer for interacting chains is more extended than for isolated molecules, the exponential decay of the segment density (valid in the latter case) is retained. Silberberg[4] simply assumed a step function. Both authors use Flory-Huggins type expressions.

Generally speaking, the results of these two models are rather close, although typical differences do occur. A detailed comparison has been given elsewhere[5].

(vii) *Diffusion equation approach.* Edwards[6] and De Gennes[7,8] were the first to use the analogy between the paths covered by a polymer chain and a diffusing particle to describe the conformations (end point distribution) of polymer chains by a diffusion equation. In such an equation, the ranking number in the chain replaces the variable time in a diffusion process. It can be shown[9] that the diffusion equation is analogous to the recurrent relation (see [5.5.2 and 7])

[1] A. Silberberg, *J. Chem. Phys.* **48** (1968) 2835; *J. Phys. Chem* **66** (12962) 1884.
[2] C.A.J. Hoeve, E.A. DiMarzio, and P. Peyser, *J. Chem. Phys.* **42** (1965) 2558.
[3] C.A.J. Hoeve, *J. Polym Sci.* **C 30** (1970) 361; *ibid.* **34** (1971) 1.
[4] A. Silberberg, *J. Chem. Phys.* **48** (1968) 2835.
[5] G.J. Fleer and J. Lyklema, in *Adsorption from Solution at the Solid/Liquid Interface,* eds. G.D. Parfitt, C.H. Rochester, Academic Press, London (1983).
[6] S.F. Edwards, *Proc. Phys. Soc.* **85** (1965) 613; *ibid.* **88** (1966) 265.
[7] P.G. de Gennes, *Rep. Prog. Phys.* **32** (1969) 187.
[8] P.G. de Gennes, *Scaling Concepts in Polymer Physics,* Cornell University Press, Ithaca, N.Y. (1979).
[9] G.J. Fleer et al., loc. cit.

used in lattice models. The diffusion equation may be considered as the continuum version of the finite difference equation [5.5.7] valid for a lattice, where space is discretized.

Exactly as in lattice models, the walks are assumed to take place in a (self-consistent) field $U(z)$, which depends on the concentration profile $\varphi(z)$. For the relation between $U(z)$ and $\varphi(z)$ one may use the Flory-Huggins theory[1,2,3], usually in an expanded form, but other models, such as a generalized Van der Waals equation of state[4] can also be taken. The most general expression for the self-consistent mean field $U(z)$ has been given by Hong and Noolandi[3]. It has been shown[5] that this expression is the continuum analogue of the lattice version of Scheutjens and Fleer, to be discussed in sec. 5.5.

In the treatment of Hong and Noolandi[3] the ranking number dependence is fully retained and the diffusion equation has to be solved numerically. In many analytical variants[2,4,6,7,8,9] it is assumed that the end point distribution can be expanded in *eigenfunctions* weighted with the corresponding *eigenvalues*, and that the *ground-state eigenfunction*, characterized by the largest eigenvalue, dominates in the expansion. For simple forms of $U(z)$ the diffusion equation may be then be solved analytically[6,7,8]. However, it can be proven[5] that then the ranking number dependence is lost. This implies that end effects (tails) are neglected since the contribution of every segment in the chain, regardless of its ranking number, is assumed to be the same at any z.

(viii) *Square gradient.* Following the classical work of Van der Waals[10], Cahn and Hilliard[11] have shown that the Helmholtz energy in a system of non-uniform composition can to a good approximation be written as the sum of a "*local*" and a "*nonlocal*" or "*gradient*" contribution that is a function of $(d\varphi/dz)^2$, the square of the concentration gradient. The same applies to the excess Helmholtz energy. The problem is to relate the gradient term to the mixing

[1] A.K. Dolan and S.F. Edwards, *Proc. Royal Soc.* (London) **A 337** (1974) 509; *ibid.* **A 343** (1975) 427.

[2] E. Helfand and Y. Tagami, J. Chem. Phys. **56** (1972) 3592; E. Helfand, *J. Chem. Phys.* **62** (1975) 999; E. Helfand and A.M. Sapse, *J. Chem. Phys.* **62** (1975) 1327.

[3] K.M. Hong and J. Noolandi, *Macromolecules* **14** (1981) 727; *ibid.* **14** (1981) 1229.

[4] H.J. Ploehn, W.B. Russel, and C.K. Hall, *Macromolecules* **21** (1988) 1075; H.J. Ploehn, and W.B. Russel, *ibid.* **22** (1989) 266.

[5] G.J. Fleer et al,, loc. cit.

[6] I.S. Jones and P.J. Richmond, *J. Chem. Soc. Faraday* **II 73** (1977) 1062.

[7] P.R. Gerber and M.A. Moore, *Macromolecules* **10** (1977) 476.

[8] P.G. de Gennes, *Macromolecules* **14** (1981) 1637; *ibid.* **15** (1982) 492.

[9] J. Klein and P. Pincus, *Macromolecules* **15** (1982) 1129; K. Ingersent, J. Klein, and P. Pincus, *ibid.* **19** (1986) 1374.

[10] J.D. van der Waals, *Verhandel. Koninkl. Acad. Wetensch.*, Amsterdam **1** (1893) 8; *Zeit. Phys. Chem.* **13** (1894) 657; *Arch. Néerl.* **28** (1895) 121; *J. Stat. Phys.* **20** (1979) 197.

[11] J.W. Cahn and J.E. Hilliard, *J. Chem. Phys.* **28** (1958) 2.

energy and entropy of the interfacial system. So far, only ground-state expressions have been derived. This procedure has been used for the calculation of the profile of adsorbed polymer and of the interfacial tension[1].

(ix) *Scaling.* In section 5.2d it was shown that the scaling theory for polymers in a good solvent considers the solution as a transient network with a mesh size ξ, the correlation length. The central result is that in a good solvent ξ is proportional to $\varphi^{-3/4}$.

The basic ideas of the scaling theory for homogeneous polymer solutions have also been used to set up a framework for a theory describing the adsorption of polymers from good solvents. The aim is to derive power laws for $\varphi(z)$, valid in certain regimes; in most cases numerical coefficients are ignored. So far, the model has only been formulated for weak adsorption, i.e., small χ^s. We shall treat some more details of this model in the following subsection (5.4c).

5.4c. Scaling

In sec. 5.2d it was shown that in a good solvent the correlation length ξ is proportional to $\varphi^{-3/4}$. Equation [5.2.18] is often used in the simplified form

$$\xi/\ell \approx \varphi^{-3/4} \qquad\qquad [5.4.2]$$

where ℓ is the monomer length.

The concepts of the scaling theory for homogeneous solutions can also be applied to polymers in a good solvent next to and adsorbing surface. It is then possible to derive power laws for different regimes, whereby usually numerical coefficients of order unity are disregarded. The polymer is assumed to be irreversibly attached; the equilibrium with the bulk solution is ignored.

In an adsorbed layer, $\varphi(z)$ decreases with increasing distance z from the surface. Consequently, according to [5.4.2] $\xi(z)$ should increase with z until at large distance its value is $\xi^b \approx \ell(\varphi^b)^{-3/4}$. The central assumption, first introduced by De Gennes[2], is that the adsorbed layer is "*self-similar*": the correlation length $\xi(z)$ should scale as the distance z itself since z is the only length scale, or $\xi(z) \approx z$. Inserting this in [5.4.2] leads immediately to $z/\ell \approx [\varphi(z)]^{-3/4}$ or

$$\varphi(z) \approx (\ell/z)^{4/3} \qquad\qquad [5.4.3]$$

Equation [5.4.3] is valid when $\varphi(z)$ is in the semidilute regime (in between φ^{ov} and φ^{sm} (see fig. 5.3)) and when z is in between the limits $D < z < \xi^b$, where the distance D ($\gg \ell$) is determined by the adsorption energy parameter χ^s. For $z < D$

[1] E. Helfand and A.M. Sapse, *J. Chem. Phys.* **62** (1975) 1327.
[2] P.G. de Gennes, *Rep. Prog. Phys.* **32** (1969) 187; *Macromolecules* **14** (1981) 1637; *ibid.* **15** (1982) 492.

we have the "*proximal*" region dominated by the segment-surface interaction (hence, D is a function of χ^s only), in the "*central*" region ($D < z < \xi^b$) the behaviour should be universal and described by [5.4.4], and in the "*distal*" region (z around ξ^b) the profile should decay exponentially towards φ^b: $\varphi(z) = \varphi^b(1 + e^{-z/\xi^b})$. The universality of [5.4.4] for the central region implies that the dependencies on chain length and solution concentration are weak: the treatment applies only to the plateau region of the adsorption isotherm.

Originally De Gennes[1] conjectured that $\varphi(z)$ would be described by:

$$\varphi(z) \approx \varphi^s \left(\frac{4D/3}{z + 4D/3} \right)^{4/3} \tag{5.4.4}$$

in both the proximal and central regions. This form ensures the $z^{-4/3}$ power law for $z \gg D$ and gives a weaker dependence of $\varphi(z)$ on z close to the surface. The concentration profile according to this equation is pictured in fig. 5.10a. The distance D over which the proximal region extends can be identified as the "extrapolation length", obtained by extrapolating the initial tangent with slope $\varphi'(0)$ (where $\varphi' \equiv d\varphi/dz$) towards $\varphi = 0$: $D = \varphi^s/\varphi'(0)$. The surface volume fraction φ^s was assumed to be given by a similar power law:

$$\varphi^s \approx \left(\frac{\ell}{4D/3} \right)^{4/3} \tag{5.4.5}$$

From a free energy minimization, D was found to be proportional to $(\chi^s)^{-3/2}$; consequently, φ^s should scale as $(\chi^s)^2$.

Exact enumeration[2] and Monte Carlo[3] computations indicated that this

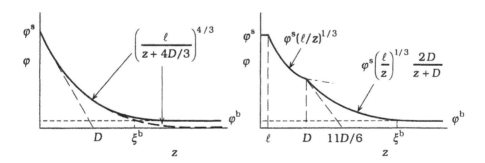

Figure 5.10a. Concentration profile of adsorbing polymer according to [5.4.4].

Figure 5.10b. Concentration profile of adsorbing polymer according to [5.4.6].

[1] P.G. de Gennes, *Rep. Prog. Phys.* **32** (1969) 187; *Macromolecules* **14** (1981) 1637; *ibid.* **15** (1982) 492.
[2] T. Ishinabe, *J. Chem. Phys.* **76** (1982) 5589.

latter result and, consequently, [5.4.4] is not correct: close to the adsorption-desorption transition φ^s should be proportional to χ^s. In order to incorporate this result, De Gennes and Pincus[1] suggested a modification of the above equations, which could be written as

$$\varphi(z) = \varphi^s \qquad\qquad\qquad 0 < z < \ell \qquad\qquad\qquad [5.4.6a]$$

$$\varphi(z) \approx \varphi^s (\ell/z)^{1/3} \qquad\qquad \ell < z < D \qquad\qquad\qquad [5.4.6b]$$

$$\varphi(z) \approx \varphi^s \left(\frac{\ell}{z}\right)^{1/3} \frac{2D}{z+D} \qquad D < z < \xi^b \qquad\qquad\qquad [5.4.6c]$$

where now

$$\varphi^s \approx \ell/D \qquad\qquad\qquad\qquad\qquad\qquad\qquad\qquad\qquad [5.4.7]$$

Unlike De Gennes and Pincus, we include a factor 2 in [5.4.6c] in order to avoid a discontinuity at $z = D$. In the proximal region, the exponent of z is not $-4/3$ but $-1/3$ (the so-called *proximal exponent*), in agreement with exact enumeration and Monte Carlo simulation[2,3]; in the central region the factor $z^{-4/3}$ in [5.4.3] is split up into a factor $z^{-1/3}$ and a factor $(z + D)^{-1}$.

A sketch of the profile according to [5.4.6] is given in fig. 5.10b. Obviously, the discontinuity at $z = D$ should be smoothed out. The distance D could still be defined through an extrapolation, in this case not from $z = 0$ but from $z = D$: $6D/5 = \varphi(D)/\varphi'(D)$. From a free energy minimization it can be derived that now $D \sim (\chi^s)^{-1}$, in agreement with [5.4.7].

This scaling picture of adsorbed layers is relevant if an appreciable fraction of the adsorbed layer has a density in the semi-dilute regime. As follows from the discussion in 5.2d, this requires a large v (very good solvent), a small persistence p (very flexible polymer) and a large ξ^b (very long polymers and $\varphi^b \leq \varphi^{ov}$). This is a rather exceptional case. It can be studied experimentally but it is not always relevant for general practice.

5.5 The self-consistent-field model of Scheutjens and Fleer

In this section we discuss in some detail the Scheutjens-Fleer (SF) model for chain molecules in a concentration gradient. This model is an extension of earlier models for *isolated* adsorbed chains, such as that of DiMarzio and

---

[3] E. Eisenriegler, K. Kremer, and K. Binder, *J. Chem. Phys.* **77** (1982) 6296.

[1] P.G. de Gennes and P. Pincus, *J. Physique Lett.* **44** (1983) L-241.

[2] T. Ishinabe, *J. Chem. Phys.* **76** (1982) 5589.

[3] E. Eisenriegler, K. Kremer, and K. Binder, *J. Chem. Phys.* **77** (1982) 6296.

Rubin[1]. The chain statistics of the latter model are included in the SF theory, and therefore we first treat these statistics in sec. 5.5a. In sec. 5.5b it will be shown how the excluded volume is incorporated so that *interacting* chains can be modelled. Then we illustrate the application of the model to systems of increasing complexity: systems with only one segment type (melt of homodisperse or polydisperse chains) in sec. 5.5c; systems of two segment types differing only in adsorption energy (5.5d); and systems with two segment types differing both in adsorption energy and in interaction with the solvent (5.5e). Some structural characteristics are dealt with in sec. 5.5f, and finally in sec. 5.5g the extension of the model to polyelectrolytes will be treated. In all cases, the discussion will remain restricted to linear polymers and polyelectrolytes. More complex systems will be considered in Volume V.

5.5a. Chain statistics in a predefined field

In the DiMarzio-Rubin model the chain statistics are treated as walks in a predefined field. This field is very simple: in the surface layer it is $-\chi^s kT$, and it is zero everywhere else. This corresponds to a weighting factor $\exp(\chi^s)$ for a step towards or within the surface layer, whereas all other steps are random, i.e., their weighting factor is unity. This model is appropriate for isolated ideal chains adjacent to a surface, since then the excluded volume plays no role. When the excluded volume becomes important, the weighting factors (and, hence, the field) are more complicated but this improvement will be deferred to sec. 5.5b.

Let us illustrate this simple model for monomers and dimers in a lattice adjoining a surface. We subdivide the space into lattice layers of spacing ℓ, the step length, and define $z = 0$ as the plane through the centre of the surface atoms. The boundary between the adsorbent and the solution is situated at $z = \ell/2$, see fig. 5.11 The first layer in the solution (also called "the surface layer") is the region $\ell/2 < z < 3\ell/2$, which will also be denoted as "layer ℓ". Generally, if z/ℓ is an integer, the region between $z - \ell/2$ and $z + \ell/2$ will be indicated as "layer z".

(i) *Segment weighting factors and the corresponding field.* Monomers that do not interact with each other may be assigned a *segment weighting factor* $G(z)$, where

$$G(z) = \begin{cases} 0 & \text{if } z \le 0 \\ e^{\chi^s} & \text{if } z = \ell \\ 1 & \text{if } z \ge 2\ell \end{cases} \qquad [5.5.1]$$

[1] E.A.DiMarzio, *J. Chem. Phys.* **42** (1965) 2101; R.J. Rubin, *J. Chem. Phys.* **43** (1965) 2392; *J. Res. Nat. Bur. Stand.* **B 70** (1966) 237; E.A. DiMarzio and F.L. McCrackin, *J. Chem. Phys.* **43** (1965) 539; E.A. DiMarzio and R.J. Rubin, *J. Chem. Phys.* **55** (1971) 4318.

Alternatively, a field $U(z)$ may be defined by $G(z) = e^{-U(z)/kT}$; in the present case $U(0) = \infty$, $U(\ell) = \chi^s kT$, and $U(z) = 0$ for $z > \ell$. The weighting factors are defined with respect to the bulk of the solution and express the preference that segments have for layer z over the solution. In this field the monomers have the same probability as in the bulk solution in all layers not contacting the surface ($z = 2\ell, 3\ell, \ldots$), and a higher probability, by a factor of $\exp(\chi^s)$, in the surface layer ($z = \ell$). Equation [5.5.1] defines the segment distribution of (non-interacting) monomers through $\varphi(z) = \varphi^b G(z)$, where $\varphi(z)$ and φ^b are the volume fractions in layer z and in the bulk solution, respectively. It leads directly to the initial linear or Henry part of the adsorption isotherm (see sec. I.2.20e):

$$\varphi(\ell) = \varphi^b \exp(\chi^s).$$

(ii) *End point distribution.* For chain molecules, consisting of more than one segment, it is easiest to first compute the end segment distribution $G(z; s)$. This quantity is defined as the relative weight (with respect to the bulk solution) that chains of s segments end in layer z. Figure 5.11a illustrates the conformations for dimers with the end segment ($s = 2$) in layer 2ℓ or in layer z (where $z \geq 3\ell$) on a simple cubic latice, in which each site has 6 neighbours (4 in the same layer and 1 in each adjacent layer).

For the sake of later generalization, we use in the following discussion $G(z)$

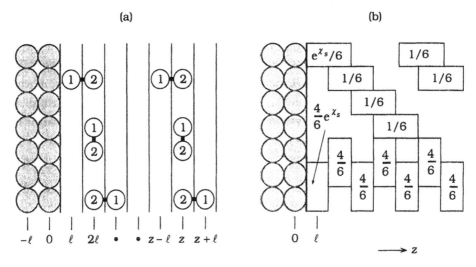

Figure 5.11. Dimer conformations in a simple cubic lattice (shown in a two-dimensional projection) near a surface (a) and their statistical weights with respect to the bulk solution (b). The centres of the solution layers are at $z = \ell$ (surface layer), 2ℓ, 3ℓ, ... In diagram (a) the segment ranking numbers ($s = 1$ or 2) are indicated. The statistical weights indicated in diagram (b) are given by $\lambda_0 G^2(z)$ for the "parallel" conformations (bottom) and $\lambda_1 G(z)G(z + \ell)$ for the "perpendicular" ones (top), where z is the layer number for the segment closest to the surface. In this example $\lambda_0 = 4/6$, $\lambda_1 = 1/6$, $G(\ell) = \exp(\chi^s)$ and $G(\ell) = 1$ for $z \geq 2\ell$.

even if, according to the simple equation [5.5.1], $G(z)$ is unity for $z \geq 2\ell$. Hence, for $z \geq 3\ell$ the weight of conformation $(z - \ell, z)$ with the first segment $(s = 1)$ at $z-\ell$ and the second $(s = 2)$ at z, is $\frac{1}{6}G(z - \ell)G(z) = \frac{1}{6}$. compare fig. 5.11a (top right). Similarly, the weights of conformations (z, z) and $(z + \ell, z)$ are $\frac{4}{6}G(z)G(z) = \frac{4}{6}$ and $\frac{1}{6}G(z + \ell)G(z) = \frac{1}{6}$, respectively. These dimer weights are indicated in fig. 5.11b. The numbers express nothing else than the fact that, of all dimers with the second segment in layer z, a fraction $\frac{1}{6}$ has its first segment in $z-\ell$ (or in $z+\ell$), and a fraction $\frac{4}{6}$ has both segments in z, just like in the bulk solution. The end segment distribution for dimers, $G(z; 2) = G(z)\left(\frac{1}{6}G(z - \ell) + \frac{4}{6}G(z) + \frac{1}{6}G(z + \ell)\right)$, is unity for $z \geq 3\ell$ since then no segment is adsorbed: the probability of ending in layer z (with $z \geq 3\ell$) is the same as in the bulk solution.

Along the same lines we find $G(2\ell; 2) = \frac{1}{6}\exp(\chi^s) + \frac{4}{6} + \frac{1}{6} = 1 + \frac{1}{6}\left(\exp(\chi^s) - 1\right)$, see the left part of fig. 5.10a and b. Due to the contribution of conformation $(\ell, 2\ell)$, with the first segment adsorbed, the probability of ending in layer 2ℓ is higher than in the bulk solution if $\chi_s > 0$. Obviously, for $\chi^s > 0$ ending in layer ℓ is even more favourable: $G(\ell, 2) = \frac{4}{6}\exp(2\chi^s) + \frac{1}{6}\exp(\chi^s)$, where the first term corresponds to conformation (ℓ, ℓ) and the second to conformation $(2\ell, \ell)$.

We can easily generalize this procedure to longer chains by relating $G(z; s+1)$ to the end segment distribution $G(z; s)$ of sequences that are one segment shorter. We have the following recurrence relation:

$$G(z; s+1) = G(z)\left\{\lambda_1 G(z - \ell; s) + \lambda_0 G(z; s) + \lambda_1 G(z + \ell; s)\right\} \qquad [5.5.2]$$

with the starting condition $G(z, 1) = G(z)$. The parameters λ_1 and $\lambda_0 = 1 - 2\lambda_1$ are the a priori probabilities of crossing to an adjacent layer or remaining in the same layer, respectively. In the simple cubic lattice discussed above, $\lambda_0 = 4\lambda_1 = \frac{4}{6}$; for a hexagonal lattice $\lambda_0 = 2\lambda_1 = \frac{6}{12}$, which is the same as the value $\frac{2}{4}$ for a tetrahedral lattice.

Through [5.5.2] the weight of chains of $s+1$ segments ending in z is expressed in that of chains of s segments, whereby segment s is necessarily in one of the adjacent lattice sites, i.e., in one of the layers $z-\ell$, z, or $z+\ell$. The factor $G(z)$ in [5.5.2] is the weighting factor for a segment in layer z and accounts in this case for segment $s+1$, which is in layer z. By applying [5.5.2] s times the end point distribution is fully and explicitly expressed in the segmental weighting factors $G(z)$ and, hence, in the field $U(z)$. In the present case, $G(z) = \exp(\chi^s)$ for $z = \ell$ and $G(z) = 1$ for $z > \ell$, corresponding to a field $U(\ell) = \chi^s kT$ and $U(z) = 0$ for $z > \ell$. Moreover, $G(z) = G(z; s) = 0$ [or $U(z) = \infty$] for $z \leq 0$. It is easily verified that the example for dimers (fig. 5.11) is a special case $(s = 1)$ of [5.5.2].

In the bulk of the solution $G(z; s + 1) = G(z; s) = G(z) = 1$ for any z, leading to a homogeneous distribution. If a chain is forced to start in a particular bulk layer

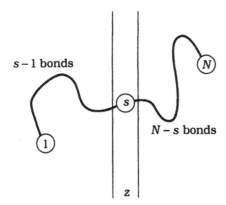

$s-1$ bonds

$N-s$ bonds

z

Figure 5.12. Graphical illustration of the composition law. The probability that segment s is in layer z is found from the joint probabilities that both the sequence 1, 2, ..., s and the sequence N, N-1, ..., s end in z. The joining segment s is included in both walks.

z' (i.e., $G(z;1)$ is unity for layer z' and zero for all other z), [5.5.2] results, for long chains, in a Gauss distribution of end segments.

(iii) *Volume fraction profile.* From [5.5.2] the distribution of end segments is obtained. Concentrations of interior segments follow by considering the probability that a segment, with ranking number s in a chain of N segments, finds itself in layer z as the joint probability that two walks, starting from different chain ends, one being s-1 steps and the other N-s steps long, both end in layer z, see fig. 5.12. The contribution $\varphi(z,s)$ of segments s to the volume fraction $\varphi(z)$ in layer z is:

$$\varphi(z,s) = \frac{C}{G(z)} G(z;s) G(z;N-s+1) \qquad [5.5.3]$$

where C is a normalization constant. The factor $G(z)$ is a correction for double counting: the joining segment s, with segment weighing factor $G(z)$, is included in both walks. For the present model (i.e., [5.5.1]) this matters only when s is in the surface layer. The total volume fraction $\varphi(z)$ is found as $\sum_s \varphi(z,s)$. Application of [5.5.3] to the bulk solution, where all G's are unity and the volume fraction $\varphi(\infty,s)$ equals φ^b/N, gives $C = \varphi^b/N$: the N chain segments contribute there equally to φ^b. Equation [5.5.3], accounting for the connectivity of the segments in the chain, is also known as the *composition law*.

(iv) *Results for isolated chains.* The main conclusions of these polymer adsorption models for isolated chains can be summarized as follows:

- There exists a critical adsorption energy χ^s_{crit}, given by

$$\chi^s_{crit} = -\ln(1-\lambda_1) \qquad [5.5.4]$$

Below χ^s_{crit} long chains do not adsorb. At $\chi^s = \chi^s_{crit}$ a sharp adsorption-desorption transition takes place, which becomes second order for $N \to \infty$. For a hexagonal or tetrahedral lattice, $\lambda_1 = \frac{1}{4}$ and $\chi^s_{crit} = 0.29$.

Equation [5.5.4] follows directly from [5.5.2]. For all segments in layer ℓ, except the first, $G(\ell; s+1) = \exp(\chi^s)\{\lambda_0 G(\ell; s) + \lambda_1 G(2\ell; s)\}$. At the critical point the end segment distribution becomes homogeneous so that all $G(z)$ and all $G(z; s)$ are unity for $s \gg 1$, or $1 = \exp(\chi^s)\{\lambda_0 + \lambda_1\} = \exp(\chi^s)\{1 - \lambda_1\}$, which is the same as [5.5.4]. This equation applies only for long chains because for small N the contribution of the end segments may not be neglected.

- At $\chi^s = \chi^s_{crit}$ tails play a very important role. Long chains are, on the average, subdivided into three roughly equal parts: two tails comprising $1/3$ of the chain length each, and a middle section of alternating loops and trains[1]. Under these conditions $\varphi(z) = \varphi^b$, except in the first layer where $\varphi(\ell) = (1 - \lambda_1)\varphi^b$.

- For χ^s only slightly above χ^s_{crit}, the importance of tails decreases very steeply. For long chains the tails become unimportant if χ^s is only a few tenths above χ^s_{crit}. The segment density profile due to loops decays exponentially with z [1]. For high χ^s nearly all segments are on the surface because the adsorption energy dominates the loss of entropy.

Most or all of these features have also been found (or extrapolated) from exact enumeration and Monte Carlo techniques.

5.5b. Incorporation of the excluded volume

In the previous section the chain statistics and the volume fraction profile were computed from a predefined field (segment weighting factors): $\varphi(z)$ was computed through [5.5.2 and 3]. The field itself did not depend on the volume fractions. This is appropriate for isolated chains but breaks down completely for higher coverages. In the isolated-chain model there is no excluded volume and according to [5.5.2 and 3] the volume fraction in the surface layer can increase without bounds, exceeding unity even for small values of χ^s and very low bulk solution concentrations. For realistic situations, the excluded volume must therefore be taken into account. Due to this excluded volume the weighting factor $G(\ell)$ in the surface layer is considerably lower than $\exp(\chi^s)$, or $-U(\ell)$ is much smaller than $\chi^s kT$ (for positive adsorption).

Scheutjens and Fleer[2] put forward a model to incorporate this excluded volume. They model the chain statistics exactly like in [5.5.2 and 3] but assume that the walks take place in a field which depends on the local concentration and the local concentration gradient according to a (generalized) Flory-Huggins

[1] R.J. Roe, *J. Chem. Phys.* **43** (1965) 1591; R.J. Roe, *J. Chem. Phys.* **44** (1966) 4264.
[2] J.M.H.M. Scheutjens and G.J. Fleer, *J. Phys. Chem.* **83** (1979) 1619; *ibid.* **84** (1980) 178.

model. Generally, this field can be written as

$$U_i(z) = U'(z) + U_i^{int}(z) \tag{5.5.5}$$

for i = 1 (solvent) and i = 2 (polymer). Here, $U'(z)$ is the *excluded volume field* which ensures that every layer is completely filled. It is the only contribution when all interactions are zero (as in a polymer melt, see sec. 5.5c). The second term on the r.h.s. of [5.5.5] is the *interaction field* representing the interaction of each segment with its surroundings. It is in fact a functional $U^{int}[\varphi(z)]$ depending not only on the function $\varphi(z)$ but also on the interaction parameters χ and χ^s. A self-consistent solution between $\varphi_i[U_i(z)]$ (according to [5.5.2 and 3]) and $U_i[\varphi_1(z), \varphi_2(z)]$ (according to [5.5.5]) must be found. The field $U'(z)$, which is the same for any type of segment in the system, is obtained from the boundary constraint $\varphi_1(z) + \varphi_2(z) = 1$ for every layer. Hence, whereas the equations in the previous section give an explicit solution (which is not self-consistent), the Scheutjens-Fleer model leads to a set of implicit (and self-consistent) equations that have to be solved numerically (for example, by iteration).

The fields in [5.5.5] are defined with respect to the bulk solution, where $U_i(\infty) = U'(\infty) = U_i^{int}(\infty) \equiv 0$. In the following sections, we will specify the form of $U_i^{int}(z)$ for systems with increasing complexity. For the moment, we assume that U_i^{int} is a known function of the concentration profile.

From the field $U_i(z)$ the segment weighting factors are obtained as

$$G_i(z) = e^{-U_i(z)/kT} \tag{5.5.6}$$

The endpoint distribution $G_i(z; s)$ for a sequence of s segments of component i can be computed from $G_i(z)$ through the recurrence relation [5.5.2]

$$G_i(z; s+1) = G_i(z)\langle G_i(z; s)\rangle \tag{5.5.7}$$

where $G_i(z; 1) = G_i(z)$. We use the abbreviated notation $\langle\ \rangle$ to denote a neighbour average over three layers. For a lattice constant $\lambda_0 = 1 - 2\lambda_1$ the neighbour average $\langle f(z)\rangle$ of a function $f(z)$ is given by

$$\langle f(z)\rangle = \lambda_1 f(z - \ell) + \lambda_0 f(z) + \lambda_1 f(z + \ell) \tag{5.5.8}$$

From $G_i(z; s)$ the distribution $\varphi_i(z, s)$ of segment s in the chain (1, 2, ..., s, ..., N_i) follows through the composition law [5.5.3]:

$$\varphi_i(z, s) = \frac{C_i}{G_i(z)} G_i(z; s) G_i(z; N_i - s + 1) \tag{5.5.9}$$

The normalization constant C_i is given by $C_i = \varphi_i^b/N_i$, where φ_i^b is the bulk solution concentration of component i and N_i its chain length. For a monomeric solvent ($N_1 = 1$), [5.5.9] reduces to

$$\varphi_i(z) = \varphi_i^b \, G_i(z) \qquad\qquad [5.5.10]$$

For chain molecules, $\varphi_i(z)$ is found by summation of [5.5.10] over s:

$$\varphi_i(z) = \sum_{s=1}^{N_i} \varphi_i(z, s) \qquad\qquad [5.5.11]$$

Full occupancy of the lattice requires

$$\sum_i \varphi_i(z) = 1 \qquad\qquad [5.5.12]$$

for each layer. If $U_i[\varphi_i(z)]$ is specified, [5.5.5-12] form a self-consistent set of equations that have to be solved numerically, usually by iteration. First, initial estimates of $\{U_i(z)\}$ for each component i are made. Then $G_i(z)$ and $G_i(z;s)$ can be computed, and from [5.5.11] the segment distributions $\{\varphi_i(z)\}$. If these obey the boundary condition [5.5.12] and give the same $U'(z) = U_i(z) - U_i^{int}(z)$ for each component i, the initial estimates represent the self-consistent solution. If not, improved estimates for $\{U_i(z)\}$ are made and the procedure is repeated until these conditions are met. A detailed description of the numerical iteration scheme is given in Appendix III of ref.[1].

The interaction energy U^{int} contains an adsorption energy part and a mixing part which may be derived from solution models. In the Flory-Huggins model, this mixing contribution will contain terms of the type $\chi\varphi_1(z)\langle\varphi_2(z)\rangle$ (compare [5.2.12]). In a concentration gradient the parameter $\varphi_2(z)$ has to be replaced by the average volume fraction $\langle\varphi_2(z)\rangle$ felt by each solvent molecule in layer z. We will not start with such expressions but will introduce a form for $U(z)$ more or less intuitively, using only relatively simple physical considerations.

5.5c. Systems with only one type of segment

In a system with only one type of segment, the volume fraction is unity in every layer: $\varphi_i(z) = \varphi^b = 1$. Mixing energy is absent. Only in the surface layer is the energy of the segments, in general, different from that in the bulk due to the adsorption energy $\Delta_{ads}U_m$ of the segments: $U^{int}(\ell) = \Delta_{ads}U_m$. The interaction with the surface is assumed to be short-range. For an air-melt interface, $\Delta_{ads}U_m$ is positive because of the cohesion energy in the liquid, whereas $\Delta_{ads}U_m$ may be negative for a melt in contact with a surface. We may write

$$U(z) = U'(z) + \delta(z - \ell)\Delta_{ads}U_m \qquad\qquad [5.5.13]$$

where δ is the Kronecker delta function.

---

[1] O.A. Evers, J.M.H.M. Scheutjens, and G.J. Fleer, *Macromolecules* **23** (1990) 5221.

(i) *Monomer liquid near a wall.* Let us first consider a monomeric liquid ($N =$ 1). According to [5.5.10], all weighting factors $G(z)$ are unity and $U(z) = 0$ for any z. For $z \geq 2\ell$, this implies $U'(z) = 0$, but in the surface layer $U'(\ell)$ is non-zero if $\Delta_{ads} U_m \neq 0$: $U'(\ell) + \Delta_{ads} U_m = 0$. It follows that the Helmholtz energy due to the excluded volume field (in this case purely entropic) in the surface layer must exactly compensate the adsorption energy in order to make $\varphi(z)$ unity also for $z = \ell$. Note that this picture neglects density variations in the surface region of a liquid. As a consequence, the surface tension γ is wholly energetic: $\gamma \ell^2 = \Delta_{ads} U_m$, where ℓ^2 is the area per surface site. In other words, the temperature dependence of γ (which has an entropic origin, compare [I.2.14.18]) is neglected. This entropic contribution to γ could be incorporated in the present model by allowing for free volume. In that case the liquid would be considered as a two-component system (monomers and "holes") and a concentration gradient of holes is allowed. In such a binary system $U^{int}(z)$ is non-zero because of the different interaction of monomers and holes. The inclusion of vacancies in the lattice also provides a way of dealing with the compressibility[1]. We do not pursue this point.

(ii) *Dimeric liquid near a wall.* As our next example we take a pure and incompressible liquid of dimers ($N = 2$). Equation [5.5.9] now reads $\varphi(z, 1) = \varphi(z, 2) = \frac{1}{2} G(z)\langle G(z)\rangle$ or, with [5.5.11], $G(z)\langle G(z)\rangle = 1$ for any z. Written out per layer:

$$\lambda_0 G(\ell) G(\ell) \quad + \quad \lambda_1 G(\ell) G(2\ell) \quad = 1$$

$$\lambda_1 G(2\ell) G(\ell) + \lambda_0 G(2\ell) G(2\ell) + \quad \lambda_1 G(2\ell) G(3\ell) \quad = 1 \qquad\qquad [5.5.14]$$

$$\lambda_1 G(3\ell) G(2\ell) + \lambda_0 G(3\ell) G(3\ell) + \quad \lambda_1 G(3\ell) G(4\ell) \quad = 1 \qquad\quad \text{etc.}$$

It is immediately clear that $G(z) = 1$ for any z does not satisfy [5.5.14]. The reason is the missing term in the first line: the conformation $(0, \ell)$ is forbidden, which is equivalent to $G(0) = 0$. If $G(z)$ would be unity for any layer, $\varphi(\ell) = G(\ell)\langle G(z)\rangle$ would be given by $\lambda_0 + \lambda_1 = 1 - \lambda_1$, which is inconsistent with the boundary constraint $\varphi(\ell) = 1$. In order to make $\varphi(\ell) = 1$, $U(\ell)$ must be negative and $G(\ell) > 1$.

The origin of the negative value for $U(\ell)$ in this case may be illustrated as follows. Suppose we want to create a surface in a dimeric bulk liquid by inserting a dividing plane (fig. 5.13). Upon inserting this plane between the layers 0 and ℓ, conformations $(0, \ell)$ or $(\ell, 0)$ crossing the new surface are forbidden. The resulting empty sites in the surface layer must be filled up, since we consider a fixed density (no holes). This is accomplished by making this layer relatively

---

[1] D.N. Theodorou, *Macromolecules* **22** (1989) 4578.

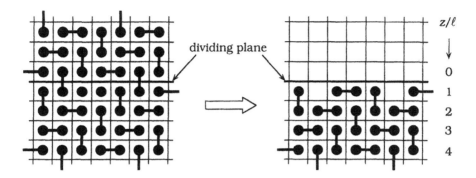

Figure 5.13. On the creation of a new surface in a dimeric liquid, conformations crossing the surface are forbidden.

more attractive: $U(\ell) < 0$, $G(\ell) > 1$. In the monomer liquid discussed above, a dividing plane does not intersect any conformation and $U(z) = 0$ for every layer.

The equilibrium values of $G(z)$ for a dimeric liquid follow from the set of M equations in M unknown $G(z)$ as given in [5.5.14]. Despite the simplicity of these equations, there is no analytical solution. The numerical solution for a simple cubic lattice ($\lambda_0 = 1 - 2\lambda_1 = 4/6$) is given in table 5.1 (left) and in fig. 5.14.

For purely space-filling reasons, the field $U(z)$ oscillates. The fact that $G(\ell) > 1$ makes the conformations (ℓ,ℓ) and $(\ell,2\ell)$ more likely: $G(\ell)\lambda_0 G(\ell) = 0.8174$ (as compared to $\lambda_0 = 0.6667$ for a "parallel" conformation in the bulk), and $2G(\ell)\lambda_1 G(2\ell) = 0.3651$ (as compared to $2\lambda_1 = 0.3333$ in bulk). In layer 2ℓ the excess due to conformation $(\ell,2\ell)$ must be compensated by less dimers in the conformations $(2\ell,2\ell)$ and $(2\ell,3\ell)$: $G(2\ell)\lambda_0 G(2\ell) = 0.6523$, $2G(2\ell)\lambda_1 G(3\ell) = 0.3301$, and $G(2\ell) < 1$, etc. Table 5.1 also gives the volume fraction $\varphi^a(z)$ due to adsorbed chains. For dimers $\varphi^a(\ell) = 0.8174 + (1/2) \times 0.3651 = 1$, $\varphi^a(2\ell) = (1/2) \times 0.3651 = 0.1826$.

The total adsorbed amount per surface site is given by

$$\theta^a = \sum_{z/\ell=0}^{M} \varphi^a(z) \tag{5.5.15}$$

and equals in this case 1.1826 segments per site.

(iii) *Polymer melt near a wall.* For long chains, [5.5.9 and 10] lead to more complicated equations for $G(z)$ than [5.5.14] because many conformations contribute to $\varphi(z)$. However, again a set of M simultaneous equations in M values of $G(z)$ is obtained. The numerical solution for $N = 10$ is given in fig. 5.14 and that for $N = 1000$ in the same figure and in table 5.1 (right). The oscillations in $U(z)$ are stronger than for dimers but have the same range (about four layers). Apparently, the "screening" length in a polymer melt is determined by the lattice

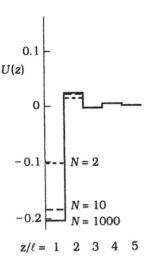

Figure 5.14. The field $U(z)$ for chain molecules in the interfacial region of the melt. The values $U(z)$ for $N = 2$ and $N = 1000$ are also included in table 5.1. Cubic lattice ($\lambda_1 = 1/6$).

spacing (Kuhn length) and not by the chain length. The range of $\varphi^a(z)$ is much larger than that of $U(z)$ and increases with increasing chain length. For long chains it is proportional to $\sqrt{N}$, reflecting the Gaussian behaviour in the melt.

An extremely large number of conformations contributes to the profile $\varphi^a(z)$. Two of them are the completely flat conformation (with all segments in layer ℓ) and the completely perpendicular one, with only the first (or last) segment

Table 5.1 The interfacial region of a melt of dimers ($N = 2$) and polymers ($N = 1000$).

	N = 2					N = 1000			
z/ℓ	$G(z)$	$U(z)/kT$	$\varphi(z)$	$\varphi^a(z)$	z/ℓ	$G(z)$	$U(z)/kT$	$\varphi(z)$	$\varphi^a(z)$
0	0	∞	0	0	0	0	∞	0	0
1	1.1073	−0.1019	1	1	1	1.2260	−0.2038	1	1
2	0.9892	0.0109	1	0.1826	2	0.9786	0.0216	1	0.9952
3	1.0011	−0.0011	1	0	3	1.0022	−0.0022	1	0.9797
4	0.9999	0.0001	1	0	4	0.9998	0.0002	1	0.9554
					5	1	0	1	0.9236
					10	1	0	1	0.6990
					20	1	0	1	0.2696
					30	1	0	1	0.0737
					40	1	0	1	0.0157
					50	1	0	1	0.0026
∞	1	0	1	0	∞	1	0	1	0

adsorbed. For $N = 1000$, the fraction "flat" is $\lambda_0^{999}G(\ell)^{1000} = 3.8 \times 10^{-88}$, a factor $G(\ell)^{1000} = 3.1 \times 10^{88}$ higher than for a parallel conformation in a layer z far from the surface. On the other hand, the adsorbed "perpendicular" conformation is only a factor $G(\ell)G(2\ell)$ ... $G(1000\ell) = 1.20$ more probable than a similar conformation wholly in the bulk.

(iv) *Mixture of two chain lengths near a wall.* For a solution of a long polymer 2 (chain length N_2) in a monomeric solvent 1 ($N_1 = 1$) in which polymer segments and solvent molecules are chemically identical, the weighting factors $G_1(z)$ and $G_2(z)$ are the same; all units feel the same field. According to [5.5.10], $\varphi_1(z)$ is given by $\varphi_1^b G(z)$. For the polymeric component the full relations [5.5.9 and 10] are needed. The M values of $G(z)$ are found from the M equations

$$\varphi_1^b G(z) + \frac{\varphi_2^b}{N_2} \sum_{s=1}^{N_2} G(z; s)\, G(z; N_2 - s + 1) = 1 \qquad [5.5.16]$$

Numerical results for a 10% solution of polymer ($N_2 = 1000$) in its own monomer are given in fig. 5.15 (solid curves). It is clear that long chains try to avoid the surface region because of the incurred loss of conformational entropy. The available space is occupied by the monomer which does not suffer from these entropical restrictions. Hence, the polymer is depleted. It can be shown that at low φ_2^b (below coil overlap) the thickness of the depletion zone is proportional to $\sqrt{N}$, see sec. 5.3e. At higher φ_2^b (as in fig. 5.15), the osmotic pressure pushes the chains closer to the surface, making the depletion layer thinner and its thickness more weakly dependent on chain length.

When the polymer (chain length N_2) is dissolved in a solvent which is not a monomer but a chain molecule with N_1 segments of the same type ($N_1 < N_2$), we have a polymer melt with a bimodal distribution of chain lengths. In [5.5.16],

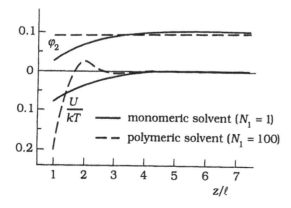

0.1
φ_2

0

0.1
$\frac{U}{kT}$ —— monomeric solvent ($N_1 = 1$)

0.2 - - - polymeric solvent ($N_1 = 100$)

1 2 3 4 5 6 7

z/ℓ

Figure 5.15. The field $U(z)$ and the polymer volume fraction $\varphi_2(z)$ for a polymer ($N_2 = 1000$) dissolved in its own monomer ($N_1 = 1$; solid curves) and in another polymer of shorter chain length ($N_1 = 100$; dashed curves). The volume fraction φ_2^b of the (long) polymer is 0.1 in both cases. Cubic lattice ($\lambda_1 = 1/6$).

the term $\varphi_1^b \, G(z)$ should now be replaced by a sum from $s = 1$ to N_1, similar to the second term in this equation.

Results for a 10% solution of long chains ($N_2 = 1000$) in a polymeric solvent ($N_1 = 100$) are also given in fig. 5.15 (dashed curves). The depletion of the long chains is much weaker than in the monomeric solvent, due to the smaller chain length difference, but it extends over longer distances. Obviously, if $N_1 = N_2$ (a homopolymer melt) there is no depletion.

It can be seen that for $N_1 = 1$ the oscillations in $U(z)$ are weak or absent. On the other hand, they are still present in the polymer mixture. Actually, $U(z)$ for $N_1 = 100$ and $N_2 = 1000$ is quite close to that for $N_1 = N_2 = 1000$ (see table 5.1 and fig. 5.13). It can be shown[1,2] that $U(z)$ for a homopolymer melt is proportional to $1 - 1/N$, which is hardly different for $N = 100$ and $N = 1000$.

5.5d. Adsorption from athermal solution

A polymer solution consists of a solvent 1 (usually monomeric) and a solute 2 (polymeric). Since in this section we consider athermal systems, there is no mixing energy and the fields only contain the adsorption energies $\Delta_{ads} U_{m1}$ and $\Delta_{ads} U_{m2}$:

$$U_1(z) = U'(z) + \delta(z-\ell)\Delta_{ads} U_{m1} \qquad\qquad [5.5.17a]$$

$$U_2(z) = U'(z) + \delta(z-\ell)\Delta_{ads} U_{m2} \qquad\qquad [5.5.17b]$$

Here, $\delta(z-\ell)$ is unity if $z = \ell$ and zero otherwise. The difference $\Delta_{ads} U_{m1} - \Delta_{ads} U_{m2}$ may be written as $\chi^s kT$, see [5.4.1].

For $z \geq \ell$ $G_1(z) = G_2(z)$, and in the surface layer $G_2(\ell)/G_1(\ell) = \exp(\chi^s)$. As compared to systems with only one type of segment, one additional variable $G_2(\ell)$ occurs and one additional relation is available so that the set of equations is fully defined. Below we give examples for the adsorption of monomers ($N_2 = 1$), dimers ($N_2 = 2$), and polymers with $N_2 = 1000$ from a monomeric solvent.

(i) *Monomers.* In this case there is a simple analytical solution because the adsorption is restricted to a monolayer. This follows directly from $G_1(z) = G_2(z) = 1$ for $z \geq 2\ell$ in combination with [5.5.10]. For the surface layer we have $\varphi_1(\ell) = \varphi_1^b \, G_1(\ell)$ and $\varphi_2(\ell) = \varphi_2^b \, G_2(\ell)$. Hence,

$$\frac{\varphi_2(\ell)}{\varphi_2^b} = \frac{\varphi_1(\ell)}{\varphi_1^b} \, e^{\chi^s} \qquad\qquad [5.5.18]$$

which, after substitution of $\varphi_1(\ell) = 1 - \varphi_2(\ell)$, is nothing else than the Langmuir

[1] E. Helfand, *J. Chem. Phys.* **63**, 2192 (1974); *Macromolecules* **9** (1976) 307; T.A. Weber and E. Helfand, *Macromolecules* **9** (1976) 311.
[2] D.N. Theodorou, *Macromolecules* **22** (1989) 4578.

adsorption isotherm equation [2.4.16] with $K = \exp(\chi^s)$. Hence, the Scheutjens-Fleer model reduces to the correct limit for monomers.

(ii) *Dimers.* For dimers (component 2) adsorbing from a monomeric solvent 1, [5.5.9 and 10] give $\varphi_1(z) = \varphi_1^b G_1(z)$ and $\varphi_2(z) = \varphi_2^b\, G_2(z)\langle G_2(z)\rangle$. Because, according to [5.5.6 and 17], $G_2(z) = G_1(z)\exp(\chi^s)\delta(z - \ell)$ we find upon elimination of $G_1(z)$:

$$G_2(z) = \frac{\varphi_1(z)}{\varphi_1^b}\, e^{\chi^s}\, \delta(z - \ell) \qquad\qquad [5.5.19]$$

Substituting $G_2(z)$ into $\varphi_2(z)$ we immediately obtain the dimer adsorption isotherm equations:

$$\frac{\varphi_2(\ell)}{\varphi_2^b} = \frac{\varphi_1(\ell)}{(\varphi_1^b)^2}\left\{\lambda_0\varphi_1(\ell)\,e^{2\chi^s} + \lambda_1\varphi_1(2\ell)\,e^{\chi^s}\right\} \qquad [5.5.20a]$$

$$\frac{\varphi_2(2\ell)}{\varphi_2^b} = \frac{\varphi_1(2\ell)}{(\varphi_1^b)^2}\left\{\lambda_1\varphi_1(\ell)\,e^{\chi^s} + \lambda_0\varphi_1(2\ell) + \lambda_1\varphi_1^b\right\} \qquad [5.5.20b]$$

$$\frac{\varphi_2(z)}{\varphi_2^b} = \frac{\varphi_1(z)}{\varphi_1^b} = 1 \qquad\qquad (z > 2\ell) \qquad [5.5.20c]$$

With $\varphi_1(\ell) = 1 - \varphi_2(\ell)$ and $\varphi_1(2\ell) = 1 - \varphi_2(2\ell)$, [5.5.20a and b] constitute a set of two (quadratic) equations in the two unknowns $\varphi_2(\ell)$ and $\varphi_2(2\ell)$. These equations may be considered as the dimeric analogue of the Langmuir equation. In these equations, one recognizes in the factor $\exp(2\chi^s)$ the contribution of the "parallel" adsorbed conformation (ℓ,ℓ), and in the terms $\exp(\chi^s)$ those of the "perpendicular" adsorbed conformations $(\ell,2\ell)$ and $(2\ell,\ell)$. The last two terms on the r.h.s. of [5.5.20b] account for the (non-adsorbed) conformations $(2\ell,2\ell)$ and $(2\ell,3\ell)$, respectively.

In a similar way one could write down the expressions for trimers, tetramers, etc., adsorbing from an athermal mixture. The equations become more and more tedious, and we will not give them here. For trimers, one obtains three cubic equations in three unknown volume fractions $\varphi_2(z)$, for tetramers four equations of fourth degree in four values of $\varphi_2(z)$, etc.

(iii) *Polymers.* For long chains the recurrence relations become rather complicated but can be handled easily in a matrix formalism. Results for 10% polymer ($N_2 = 1000$) in a solvent ($N_2 = 1$) in which $\chi^s = 1$ are given in fig. 5.16 (solid curves). The only difference with fig. 5.15 is that now $U_2(\ell)$ is 1 kT more negative than $U_1(\ell)$. For $z \geq 2\ell$ we still have $U_1(z) = U_2(z)$ As a result, the depletion in fig. 5.15 is changed to a strong preferential adsorption. The adsorption energy overcompensates the conformational entropy loss and $\varphi_2(z)$

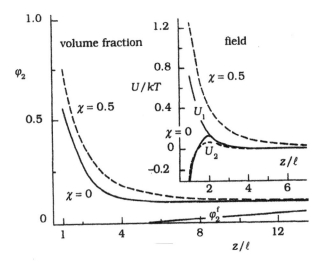

Figure 5.16. Volume fraction profiles $\varphi_2(z)$ (main figure) and the fields $U_1(z)$ for the solvent and $U_2(z)$ for the polymer (inset), for adsorbing polymer (N_2 = 1000) from a solution with $\varphi_2^b = 0.1$. Solid curves are for athermal solutions ($\chi = 0$), dashed curves for Θ-solvents ($\chi = 0.5$). Cubic lattice ($\lambda_1 = 1/6$), $\chi_s = 1$.

is greater than φ_2^b over several layers. Due to this accumulation of segments, $U_1(z)$ is positive for any z (the solvent is pushed away by segments), in contrast to fig. 5.15 where $U(z)$ is negative because here $\varphi_1(z) > \varphi_1^b$.

The volume fraction $\varphi_2(z)$ may be subdivided in $\varphi_2^a(z)$ due to adsorbed chains and $\varphi_2^f(z)$ due to free, non-adsorbed chains, compare fig. 5.6. The latter is obtained by setting $G^f(\ell) = 0$, see sec. 5.5f. The contribution of $\varphi_2^f(z)$ is rather small in the first eight layers. However, $\varphi_2^f(z)/\varphi_2^b = 0.363$ for $z = 10\ell$, 0.821 for $z = 20\ell$ and, obviously, approaches unity for $z \to \infty$.

5.5e. Adsorption with lateral interaction

In general two segment types do not mix athermally and the field should contain a mixing energy contribution. To account for that, the Flory-Huggins solvency parameter χ may be used. As explained in sec. 5.2b, the energy associated with the transfer of one segment 1 from pure liquid 1 ($\varphi_2 = 0$) to pure liquid 2 ($\varphi_2 = 1$) equals, by definition, χkT. Upon transfer of a segment 1 from the bulk solution (φ_2^b) to layer z, where the concentration is $\varphi_2(z)$, the energy change is $\chi kT(\varphi_2(z) - \varphi_2^b)$. However, in the concentration gradient around layer z only a fraction λ_0 of the contacts is within layer z; a fraction λ_1 is with $z-\ell$ and a fraction λ_1 with $z+\ell$. Hence, $\varphi_2(z)$ has to be replaced by $\langle \varphi_2(z) \rangle$, see [5.5.8]. Therefore:

$$U_1(z) = U'(z) + \delta(z-\ell)\Delta_{ads}U_{m1} + \chi kT\left\{\langle \varphi_2(z) \rangle - \varphi_2^b\right\} \qquad [5.5.21a]$$

$$U_2(z) = U'(z) + \delta(z-\ell)\Delta_{ads}U_{m2} + \chi kT\left\{\langle \varphi_1(z) \rangle - \varphi_1^b\right\} \qquad [5.5.21b]$$

Due to the mixing energy, $U_1(z)$ and $U_2(z)$ are now different for any z, and not just for $z = \ell$ as in athermal conditions $(\chi = 0)$. We now have two variables $G_1(z)$ and $G_2(z)$ per layer and one additional relation for $G_1(z)/G_2(z)$ for each z. In this ratio the parameter $U'(z)$ cancels. If the solvent is monomeric, the excluded volume term $U'(z)$ can always be expressed in the solvent profile, by substituting $U_1(z) = kT \ln G_1(z) = -kT[\varphi_1(z)/\varphi_1^b]$ in [5.5.21a]. The second equality follows from [5.5.10].

We give examples for monomers $(N_2 = 1)$ and polymers $(N_2 = 1000)$ from a monomeric solvent $(N_1 = 1)$.

(i) *Monomers.* For $\chi \neq 0$ the Langmuir equation [5.5.18] does not apply. The volume fraction $\varphi_2(\ell)$ in the surface layer does not only depend on the bulk solution concentration φ_2^b but also on $\varphi_2(2\ell)$ which is larger than φ_2^b if the solvency is poor (i.e., if $\chi \gg 0$). Hence, multilayers may be formed even for a mixture of monomers. We still have $\varphi_2(z) = \varphi_2^b G_2(z)$ and $\varphi_1(z) = \varphi_1^b G_1(z)$ as for monomers (sec. 5.5d), but now $G_1(z) \neq 1$ for several layers. From [5.5.6 and 21] we find an expression for $G_1(z)/G_2(z)$. With $\langle \varphi_1(z)\rangle + \langle \varphi_2(z)\rangle = 1 - \lambda_1 \delta(z - \ell)$ we obtain

$$\frac{\varphi_2(z)}{\varphi_2^b} = \frac{\varphi_1(z)}{\varphi_1^b} \, e^{(\chi^s + \lambda_1 \chi)\delta(z - \ell)} e^{2\chi\{\langle\varphi_2(z)\rangle - \varphi_2^b\}}$$ [5.5.22]

for any layer. If $\chi = 0$, this equation reduces to the Langmuir equation [5.5.18] for monolayer adsorption. If $\chi \neq 0$, a concentration gradient over several layers develops and a numerical solution is required. Equation [5.5.22] may be compared to common approximations, such as the BET equation [1.5.47] and the FFG equation [A.1.5a]. The former may be considered as the limiting case $\lambda_0 \equiv 0$ (only interaction between layers, no lateral interaction), whereas the FFG model corresponds to $\lambda_0 \equiv 1$ (lateral interaction within a layer, no interaction between the layers).

There is also a strong analogy between [5.5.22] and [2.4.21]. The latter equation was derived for monolayer adsorption with lateral interaction. The constant K in [2.4.21] is again $\exp(\chi^s)$. When the multilayer character of [5.5.22] is ignored by forbidding contact between the first (surface) and second layers, the term $2\chi\langle\varphi_2(\ell)\rangle$ equals $2\chi\lambda_0\theta$, which is the same as in the l.h.s. exponent of [2.4.21] provided χ^σ is identified as $\lambda_0\chi$. This is a very reasonable result because the lateral contacts form only a fraction λ_0 of the total number of contacts. The term $\lambda_1\chi$ in the exponent of [5.5.22] is replaced by the difference $\chi^L - \chi^\sigma \approx 2\lambda_1\chi$ in [2.4.21].

(ii) *Polymers.* For long polymers, there are no simple analytical equations, and a numerical solution is required. The dashed curves in fig. 5.16 give $\varphi_2(z)$ and the fields $U_1(z)$ and $U_2(z)$ for adsorption from a Θ-solvent $(\chi = 0.5)$. The volume fraction and adsorbed amount of the polymer are higher than for an athermal solvent because the poor solvency favours accumulation of segments. As a result, $U_1(z)$ is considerably more positive than for $\chi = 0$. The field $U_2(z)$ is slightly more negative, except in the surface layer $(z = \ell)$: in this layer the negative mixing energy contribution is overcompensated by the crowding effect due to the high segment concentrations.

In sec. 5.7 several other results for polymers adsorbing from solution will be discussed.

5.5f. *Structure of the adsorbate*

So far, we have considered only the overall properties of the adsorbed layer, without making a distinction between adsorbed and free chains, or between trains, loops, and tails. Since in the Scheutjens-Fleer model the contribution of all possible conformations is taken into account, it is possible to extract a wealth of structural information. In this section we will not consider the full details; for these we refer to the literature[1,2]. We restrict ourselves to showing how the volume fractions of adsorbed and non-adsorbed chains, such as presented in fig. 5.6, can be individually computed.

Adsorbed chains have at least one segment adsorbed (i.e., they have segments in layer ℓ), whereas free chains visit only $z = 2\ell, 3\ell, \ldots$. All the information is contained in the $N_i \times M$ weighting factors $G_i(z; s)$, expressing the probability that a walk from segment 1 to segment s ends at z. A part $G_i^a(z; s)$ of these walks visits the surface layer, the other part $G_i^f(z; s)$ does not:

$$G_i(z; s) = G_i^a(z; s) + G_i^f(z; s) \qquad\qquad [5.5.23]$$

The only restriction for free chains is that $z = \ell$ is forbidden. Hence,

$$G_i^f(\ell; s) = 0 \qquad\qquad [5.5.24a]$$

$$G_i^f(z; s+1) = G_i(z) \langle G_i^f(z; s)\rangle \qquad (z/\ell = 2, 3, \ldots) \qquad [5.5.24b]$$

The second relation is a special case of [5.5.7]: if, for the layers $z \geq 2\ell$, a free walk of s segments is extended by one segment that segment contributes a weighting factor $G_i(z)$.

In a similar way, the recurrence relation for $G_i^a(z; s)$ may be written as:

$$G_i^a(\ell; s) = G_i(\ell; s) \qquad\qquad [5.5.25a]$$

$$G_i^a(z; s+1) = G_i(z) \langle G_i^a(z; s)\rangle \qquad (z/\ell = 2, 3, \ldots) \qquad [5.5.25b]$$

---

[1] J.M.H.M. Scheutjens and G.J. Fleer, *J. Phys. Chem.* **84** (1980) 178.
[2] G.J. Fleer et al., loc. cit.

Adding [5.5.24b and 25b] gives, with [5.5.23], the former result as expressed in [5.5.7].

In a free chain both the chain parts 1, 2, ..., s and N, $N-1$, ..., s must be non-adsorbed. Analogously to [5.5.9 and 11], the volume fraction due to free chains is found as

$$\varphi_i^f(z) = \frac{C_i}{G_i(z)} \sum_{s=1}^{N_i} G_i^f(z; s) \, G_i^f(z; N_i - s + 1) \qquad [5.5.26]$$

where $C_i = \varphi_i^b/N_i$. Obviously, we can write

$$\varphi_i^a(z) = \varphi_i(z) - \varphi_i^f(z) \qquad [5.5.27]$$

The amount of chains contacting the surface is expressed in the adsorbed amount θ_i^a of component i, which is found by summing [5.5.27] over all all z, just as in [5.5.15]. This amount corresponds to the sum of hatched and shaded area in fig. 5.6. The excess amount indicated in this figure is obtained analogously:

$$\theta_i^{ex} = \sum_z \left(\varphi_i(z) - \varphi_i^b(z) \right) \qquad [5.5.28]$$

5.5g. Polyelectrolytes

For polyelectrolytes, the way in which the conformation statistics are computed from the field is the same as discussed in secs. 5.5a-f. The new feature originates from the electrostatic interactions, which affect the field. We make a distinction between strong polyelectrolyte segments which have a fixed (pH-independent) charge and weak groups that adjust their charge to the (local) pH. For the former case the situation is relatively simple: we may extend the field as given in [5.5.5] with an electrostatic term U_i^{el}:

$$U_i(z) = U'(z) + U_i^{int}(z) + U_i^{el}(z) \qquad [5.5.29]$$

where U_i^{int} refers to contact interactions: it is still given by [5.5.21]. The electrostatic contribution U_i^{el} for each species i (solvent, small ion, polyelectrolyte) depends on the valency z_i and on the local electrostatic potential $\psi(z)$:

$$U_i^{el}(z) \equiv z_i e \psi(z) \qquad [5.5.30]$$

where e is the elementary charge and z_i the valency of species i. Note that z without subscript is still used as the distance from the surface. The valency z_i is zero for the solvent, +1 for cations like H_3O^+, Na^+, or quaternary ammonium groups, and -1 for anions such as OH^-, Cl^- or sulfonate groups. Hence, several components are distinguished: a polyelectrolyte solution is a multicomponent mixture. For example, in an aqueous solution of sodium polystyrene sulfonate and NaCl the solution contains four species: H_2O, Na^+, Cl^-, and the polyanion. In the spirit of the lattice model all species, i.e., (clusters of) water molecules, hydrated ions and segments are assigned the same volume. In principle, we would need three χ_s parameters and six χ parameters for the four species. Usually, the small ions are considered to have the same non-electrostatic properties as a solvent entity, so that only χ_s and χ for the polyanion are non-zero. However, specific adsorption of a particular type of ion could be modelled easily by assigning a non-zero χ^s to that ion.

The problem is to relate $\psi(z)$ to the surface potential $\psi^0 = \psi(0)$ [or the surface charge density $\sigma^0 = \sigma(0)$] and the volume fraction profiles of the components. Early versions[1,2] of a polyelectrolyte adsorption model neglected the volume of the small ions and solved (numerically) the Poisson-Boltzmann equation [3.5.6]. A more sophisticated, yet simpler, approach was proposed by Böhmer et al.[3], who accounted for the ion volume by adopting a multilayer Stern model, see fig. 5.17. This is a straightforward extension of the monolayer Stern model discussed in sec. 3.6c. The charges of the ions and the segments are assumed to be located on planes in the centres of the lattice layers. The lattice is thus con-

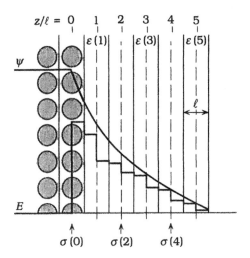

Figure 5.17. Electric potential profile $\psi(z)$ and field strength profile $E(z)$ in the multilayer Stern model of Böhmer et al.[3].

[1] H.A. van der Schee and J. Lyklema, *J. Phys. Chem.* **88** (1984) 6661; J. Papenhuijzen, H.A. van der Schee, and G.J. Fleer, *J. Colloid Interface Sci.* **104** (1985) 540.
[2] O.A. Evers, G.J. Fleer, J.M.H.M. Scheutjens, and J. Lyklema, *J. Colloid Interface Sci.* **111** (1986) 446.
[3] M.R. Böhmer, O.A. Evers, and J.M.H.M. Scheutjens, *Macromolecules* **23** (1990) 2288.

sidered as an assembly of Stern layers (see fig. 5.17). The plane charge densities
for $z/\ell = 1, 2, ...$ are given by

$$\sigma(z) = \sum_i z_i e \varphi_i(z)/\ell^2 \qquad\qquad [5.5.31]$$

Equation [5.5.31] expresses $\sigma(z)$ in the volume fraction profiles. From $\sigma(z)$ the
dielectric displacement $D(z)$, which is the product of the dielectric permittivity
$\varepsilon_0 \varepsilon(z)$ and the electric field strength $E(z)$, follows from standard electrostatics
(see [I.4.5.12 and 16]:

$$D(z) = \varepsilon_0 \varepsilon(z) E(z) = \sum_{z'=0}^{z} \sigma(z') \qquad\qquad [5.5.32]$$

which is valid provided $E(z) = 0$ for $z < 0$. In [5.5.32] $\sigma(0)$ is equal to σ^0. For the
case of constant potential $\sigma(0)$ is computed from the given value of ψ^0.

The field strength $E(z)$ changes discontinuously at the mid-planes $(z = 0, \ell, 2\ell,$
...) because of the presence of charges. However, also at the layer boundaries $(z =$
$\ell/2, 3\ell/2, ...)$ there are (small) discontinuities in $E(z)$ because a different
composition in each layer leads to another $\varepsilon(z)$. As a first approximation, $\varepsilon(z)$ is
taken to be a linear combination of the permittivities ε_i of the various pure
species:

$$\varepsilon(z) = \sum_i \varepsilon_i \varphi_i(z) \qquad\qquad [5.5.33]$$

From [5.5.31-33] the field strength $E(z)$ can be directly computed from the composition profiles $\{\varphi_i(z)\}$ and given σ^0 or ψ^0. The potential $\psi(z)$ varies linearly
across each half-layer, see fig. 5.17. Its value on each plane $(z = \ell, 2\ell, ...)$ is
related to that on the previous plane $(z = 0, \ell, ...)$ by

$$\psi(z + \ell) = \psi(z) - \frac{1}{2}\ell \left\{ E(z) + E(z + \tfrac{1}{2}\ell) \right\} \qquad\qquad [5.5.34]$$

Equations [5.5.32 and 34] are discrete representations of the Poisson equation
[3.6.23]. If the charge is not fixed on the midplanes but smeared out over the
layer, the space charge density $\rho(z)$ is equal to $\sigma(z)/\ell$. Equation [5.5.32] is then
equivalent to $\varepsilon_0 \varepsilon\, dE/d(z/\ell) = \sigma$ or $dE/dz = \rho/\varepsilon$. Equation [5.5.34] is the discrete
version of $E = -d\psi/dz$. Together these expressions lead to [3.6.23].

From [5.5.32-34] $\psi(z)$ can be obtained from the volume fraction profiles;
through [5.5.29 and 30] the fields $U_i(z)$ and the weighting factors $G_i(z)$ are
found. These serve as the basis for the computation of the end point
distributions $G_i(z; s)$ and, through the composition law [5.5.3], of the volume
fractions. Again, the final solution requires a numerical iteration.

The multilayer Stern model describes the electrostatic contributions quite

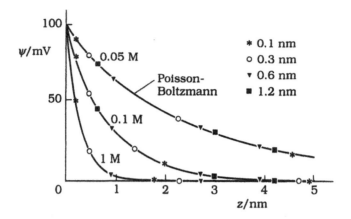

Figure 5.18. Potential decay in the electrical double layer at three salt concentrations as calculated with the PB equation (solid curves) and with the multilayer Stern model at various lattice spacings (symbols, as indicated in the legend).

well. For a charged surface in contact with a solution containing only small ions the results are identical to those of the well-known Gouy-Chapman theory. An illustration is given in fig. 5.18 for the case of a rather low ψ^o (100 mV) and three different salt concentrations. This figure may be compared with fig. 3.7. The solid curves in fig. 5.18 were computed from the PB equation, the symbols from the multilayer Stern model at four different lattice spacings. At these low potentials the ion concentrations are low everywhere and both models give identical results. Also the lattice spacing has no effect. For higher values of ψ^o (well above 200 mV) deviations show up[1], which is to be expected because in the Gouy-Chapman model the volume of the ions is neglected.

The above model for the adsorption of strong polyelectrolytes could also be applied to randomly charged copolymers with a fraction α of strongly charged groups. The only modification would then be to replace the valency z_p of the polymer segments by $z_p\alpha$. The model does not work for weak polyelectrolytes in a concentration gradient. Then the segment charge is determined by the dissociation equilibrium involving the *local* concentrations of ions. Hence, for weak groups the degree of dissociation α is not a constant, but it becomes a function $\alpha(z)$ of the distance from the surface: the segment charge adjusts itself to the local environment. Moreover, for weak polyelectrolytes the charged groups can distribute themselves in several ways along the chain: the associated entropy has to be accounted for. This is a very complex problem indeed. It has

---

[1] M.R. Böhmer, O.A. Evers, and J.M.H.M. Scheutjens, *Macromolecules* **23** (1990) 2288.

been shown[1] that in this case the weak polyelectrolyte can be considered as a homopolymer with an average weighting factor G_p which is given by

$$G_p(z) = \alpha^b G_c(z) + \left(1 - \alpha^b\right) G_u(z) \qquad [5.5.35]$$

where $G_c(z) = \exp[-U'(z) - U^{int}(z) - U^{el}(z)]$ is the weighting factor of charged segments, and $G_u(z) = \exp[-U'(z) - U^{int}(z)]$ that of uncharged ones. In the calculation of $G_c(z) = \exp[-U(z)/kT]$, the valency z_c has to be taken as -1 for carboxylic groups, and as $+1$ for amino groups. The parameter $\alpha^b = \alpha(\infty)$ is the degree of dissociation in the bulk solution; it follows directly from the bulk pH and the dissociation constant through the classical acid/base dissociation equilibrium.

As stated above, $\alpha(z)$ will in general be different from α^b. This parameter can be calculated from the weighting factors through[1]:

$$\alpha(z) = \alpha^b \, G_c(z) / G_p(z) \qquad [5.5.36]$$

In sec. 5.8 we shall consider a few examples of polyelectrolyte adsorption as calculated with the multilayer Stern model.

5.6 Experimental methods

In this section we discuss briefly some experimental methods for investigating adsorbed polymers. Determination of adsorbed amounts is for polymers not much different from that discussed in chapter 2. We therefore concentrate on three aspects which are typical for polymers. These are the (relative) number of segments in contact with the surface (i.e., the trains) (sec. 5.6a), the extension (thickness) of the adsorbed layer (sec. 5.6b), and the volume fraction profile normal to the surface (sec. 5.6c).

5.6a. Bound fraction and train density

For the determination of the fraction of segments in trains or, equivalently, the *bound fraction p*, one relies on features which allow differentiation between free (non-adsorbed) segments and segments in contact with the surface. In some cases, it is also possible to discriminate between parts of the solid substrate covered with polymer segments, and bare surface parts. One can then also determine the *train density*.

(i) *Infrared spectroscopy.* Small changes in the force constant of a chemical bond due to physical interactions (such as dipole-dipole interactions or hydrogen bonding with the surface) give rise to shifts of the vibrational

[1] R. Israëls, J.M.H.M. Scheutjens, and G.J. Fleer, *Macromolecules* **26** (1993) 5405; R. Israëls, F.A.M. Leermakers, and G.J. Fleer, *ibid*, **27** (1994) 3087.

frequencies. These can be monitored by infrared spectroscopy, provided the "bound mode" and the "free mode" can be resolved. The carbonyl stretching vibration is a good probe, as it is strong and usually well separated from other spectral features. A problem is that IR absorption by the solvent may severely limit the range of spectral bands that can be used. Preferably, a solvent with an appropriate spectral window should be chosen. Some H-bonding solvents, especially water, are notoriously difficult in this respect. A good alternative is then to use ATR (attenuated total reflection) where the IR excitation occurs in an evanescent wave in the immediate vicinity of the surface. The penetration depth of such an evanescent wave is given by [I.7.10.12]. The evanescent IR beam samples the adsorbed layer while keeping the effective path length through the solvent short.

Pioneers in measuring polymer bound fractions by IR were Fontana and Thomas[1], who studied poly(alkyl methacrylates) in n-dodecane on silica. Upon adsorption, part of the carbonyl stretching band (originally at 1736 cm^{-1}) shifted to a lower frequency (1714 cm^{-1}) due to formation of bound carbonyls. Fontana and Thomas also observed shifts in the vibration of silanol groups from the substrate and used these to obtain the train density. The IR technique has been exploited by several authors using different polymer systems, mostly with non-aqueous media and silica or titania as the substrate[2,3]. The ATR approach has been used in several more recent studies[4,5,6].

(ii) *Magnetic resonance.* The use of magnetic resonance techniques (nuclear magnetic resonance, NMR, and electron paramagnetic resonance, EPR) is mainly based on the difference in mobility between free and bound segments. In NMR, the probe is an excited nuclear spin, in EPR it is an electron spin. Both are sensitive to fluctuating electric and magnetic fields in their surroundings. These fluctuations lead to broadening of the spectrum, and to faster relaxation of the excited state. Provided the resonance signal of the polymer can be resolved in a mobile contribution (supposed to correspond to loops and tails) and an immobile one (trains), a bound fraction can be deduced[7,8].

The problem is how such a decomposition can be achieved. This is not straightforward. In NMR, one can use special sequences of radiofrequency

[1] B. Fontana and J. Thomas, *J. Phys. Chem.* **65** (1962) 480.
[2] I.D. Robb, *Comprehensive Polymer Sci.* **2** (1989) 733.
[3] I.D. Robb and J.C. Day, *Polymer* **21** (1980) 408.
[4] K.K. Chittur, D.J. Fink, R.L. Leininger, and T.B. Hutson, *J. Colloid Interface Sci.* **111** (1986) 419.
[5] D.J. Kuzmenka and S. Granick, *Colloids Surfaces* **31** (1988) 105.
[6] M. Hair, C. Tripp, and D. Guzonas, *ACS Symposium SER.* **447** (1990) 237.
[7] T. Miyamoto and H.J. Cantow, *Makromol. Chemie* **162** (1972) 43.
[8] A. Diaz-Barrios and G.J. Howard, *Makromol. Chemie* **182** (1981) 108.

pulses[1] designed to excite only a particular fraction of the resonating spins, for example, the mobile part. In this way one can then eliminate the "solid" signal, i.e., that due to the immobile segments. The pulses stimulate the system to produce so-called *echoes* (emitted pulses after a certain delay time). The idea is to choose the pulse sequence such as to first produce a "solid plus liquid" echo which contains both the mobile and immobile components, and then a "liquid" echo which only contains the mobile part. The bound fraction p is then found from the intensity ratio of these two echoes, which equals $1/(1-p)$.

An elegant alternative is to use the NMR relaxation rate of the *solvent*. The method relies on the fact that, when there is rapid exchange between solvent molecules on the solid surface and in the bulk solution, the contributions of these two populations to the (average) relaxation rate are additive. In a colloidal dispersion, mobile solvent in the bulk (b) and less mobile adsorbed solvent (a) have widely different relaxation rates, indicated by T_b and T_a, respectively. The average solvent relaxation time $\langle T \rangle$ is then given by[2]

$$1/\langle T \rangle = (1 - f_a)/T_b + f_a/T_a \qquad\qquad [5.6.1]$$

where f_a is the fraction of time a solvent molecule spends at the surface. This fraction is modified by the presence of adsorbed polymer segments. Both decreases in f_a (if solvent is displaced from the surface by train segments) and increases (if solvent remains attached to immobilized segments) are possible. From the data one can extract the *specific relaxation rate* of the solvent, i.e., the enhancement in relaxation rate per unit area caused by the adsorbing polymer. This quantity is found[1,3] to be essentially linear in Γ^{tr}. The (reasonable) assumption that $p \to 1$ as $\Gamma \to 0$ provides a calibration for Γ^{tr} so that p can be calculated.

The application of EPR to polymer adsorption was introduced by Robb and co-workers[4,5] and also employed by Kobayashi et al.[6]. The method is based on a mobility criterion similar to that for NMR. The decomposition of an EPR spectrum into contributions that can be associated with loops/tails and with trains is usually achieved by matching the spectrum to a weighted sum of two spectra of the same polymer, one when it is dissolved in a viscous fluid such as glycerol (immobile), and one in a low-viscosity solvent (mobile). The bound

---

[1] K.G. Barnett, T. Cosgrove, M.A. Cohen Stuart, D.S. Sissons, and B. Vincent, *Macromolecules* **14** (1981) 1018.
[2] G.P. van der Beek, M.A. Cohen Stuart, and T. Cosgrove, *Langmuir* **7** (1991) 327.
[3] T. Cosgrove, T. Obey, and M. Taylor, *Colloids Surfaces* **64** (1992) 311.
[4] K.K. Fox, I.D. Robb, and R. Smith, *J. Chem. Soc. Faraday I* **70** (1974) 1186.
[5] I.D. Robb and R. Smith, *Polymer* **18** (1977) 500.
[6] K. Kobayashi, H. Yajima, Y. Imamura, and R. Endo, *Bull. Chem. Soc. Japan* **63** (1990) 1813.

fractions obtained with EPR seem to be consistent with those from NMR[3,4,5]. Kobayashi et al.[1] have decomposed their spectra into *three* components with low, intermediate and high mobility, respectively.

(iii) *Calorimetry.* Since the formation of segment/surface bonds will, in general, involve an enthalpy change which, in favourable cases, is proportional to the number of bonds formed, calorimetry can in principle be used to measure the number of these bonds. If it is assumed that the enthalpy of adsorption per unit of surface area, ΔH, is proportional to the coverage in trains, we can write:

$$\Delta H = \Delta H^{mon} \, \Gamma^{tr}/\Gamma^{mon} \qquad\qquad [5.6.2]$$

where ΔH^{mon} is the enthalpy for a surface entirely covered with a monolayer of trains and Γ^{mon} is the monolayer capacity (mass per unit area). Two methods have been proposed to obtain ΔH^{mon}, namely (i) measurement of the immersion enthalpy for a monomeric analogue of the polymer (compare fig. 2.27), and (ii) using the assumption that $p \to 1$ as $\Gamma \to 0$, which was also used above. Both methods were investigated by Cohen Stuart et al.[2] and found to be consistent for low-molar-mass polymer. When higher molecular weights were studied, however, Γ^{tr} seemed to decrease with increasing molar mass, which is inconsistent with equilibrium conformations. Denoyel et al.[3] carried out adsorption enthalpy measurements on a highly exothermic system in which ionic bonds between polymer and surface were formed. These authors also found a decrease of the enthalpy with increasing M at low coverage. They suggested that the decrease in Γ^{tr} was due to the topological constraints encountered by a polymer molecule when it tries to maximize its number of segment/surface contacts. If one accepts this argument, the calorimetric values for Γ^{tr} and p would not be representative for equilibrium. The same group later reported measurements on the adsorption of polyethylene oxide on silica from water[4], which are more in line with theoretical predictions.

(iv) *Electrochemical titrations.* Train segments change the composition of the Stern layer, and thereby modify the electrical properties of this layer and, consequently, those of the electrical double layer as a whole. The property of interest is the *differential capacitance C* per unit area. For a Gouy-Stern double layer without specific adsorption, the capacitance can be split into two parts: a contribution C^d of the diffuse layer, and a part C^s for the Stern layer, see

[1] K. Kobayashi, H. Yajima, Y. Imamura, and R. Endo, *Bull. Chem. Soc. Japan* **63** (1990) 1813.

[2] M.A. Cohen Stuart, G.J. Fleer, and B.H. Bijsterbosch, *J. Colloid Interface Sci.* **90** (1982) 321.

[3] R. Denoyel, G. Durand, F. Lafuma, and R. Audebert, *J. Colloid Interface Sci.* **139** (1990) 281.

[4] P. Trens and R. Denoyel, *Langmuir* **9** (1993) 519.

[3.6.25]. In sec. 3.6 it was shown that, to a first approximation, C^s does not depend on the salt concentration, whereas C^d does. At salt concentrations of about 0.1 mol dm^{-3} C^d becomes so large that $C \approx C^s$. Under these conditions any change in C upon adsorption of polymer must reflect the presence of train segments. The amount Γ^{tr} adsorbed in trains is then obtained by comparing the Stern capacitance in the absence (C^{s0}) and in the presence (C^s) of adsorbed polymer, provided the capacitance for a surface fully covered with a monolayer of segments (denoted by C^{sm}) is known. Then

$$\frac{\Gamma^{tr}}{\Gamma^m} = \frac{C^s - C^{s0}}{C^{sm} - C^{s0}} \qquad\qquad [5.6.3]$$

and the bound fraction p is found as Γ^{tr}/Γ^a [1].

5.6b Layer thickness

Various methods have been proposed to measure the thickness of an adsorbed polymer layer. Depending on the method, a different property of the layer is determined. For example, hydrodynamic and electrokinetic techniques probe the extension of the tails and give a thickness which may exceed considerably the average thickness as obtained from ellipsometry or from the reflected or scattered intensity of visible light, of X-rays, or of neutron radiation. In this section we can touch upon just a few aspects of the various techniques.

(i) *Hydrodynamic thickness.* Hydrodynamic methods are based on the fact that the adsorbed layer impedes the flow of solvent along the surface. For solvent flow in a capillary or in a porous plug this leads to smaller fluxes; for particles diffusing in a liquid medium it implies a lowering of the diffusion coefficient. These two phenomena have the same origin, although they require very different experimental approaches.

It should be realized that hydrodynamic techniques measure an *effective* thickness obtained by comparing the tangential flow along a polymer-covered surface with that along a bare surface. The effective thickness thus found is usually called the *hydrodynamic layer thickness*. The exact shape of the flow velocity profile is important, and this depends on the shape of the surface in question, and on its orientation with respect to the flow field. Detailed data interpretation is therefore only possible if simple geometries are chosen such as smooth cylindrical channels or spherical colloidal particles.

In order to establish the relation between the segment density profile and the measured thickness, a model of the flow inside the layer is needed. We will not

---

[1] L.K. Koopal and J. Lyklema, *Discuss. Faraday Soc.* **59** (1975) 230; *J. Electroanal. Chem.* **100** (1979) 895.

elaborate any of these models here, but note that all treatments find that the solvent flow velocity inside adsorbed layers tends to be considerably reduced due to the presence of the polymer, even when the segment density is rather low. As a result, the hydrodynamic layer thickness is mainly determined by the dilute periphery of adsorbed layers where tails dominate the segment density.

The method of *capillary flow* measures the increase in resistance for solvent flow through a capillary (or a porous plug) due to an adsorbed polymer layer. This increase can be translated into a smaller effective capillary (or pore) radius through the Hagen-Poiseuille law [I.6.4.18]. The hydrodynamic radius d^h is supposed to be given by the difference between the "covered" and the "bare" radius. In such experiments the observed hydrodynamic thickness sometimes turns out to be flow-rate dependent. In such cases an extrapolation to zero flow rate needs to be carried out.

For adsorption onto colloidal particles two approaches are possible. The first is to determine viscometrically the increase in the effective particle volume fraction upon adsorption, and the second is to measure the decrease in the particle diffusion coefficient.

The effective volume fraction can be obtained from the *intrinsic viscosity* $[\eta]$ of the dispersion. According to Einstein's viscosity relation, the product of $[\eta]$ and the mass concentration c is proportional to the particle volume fraction, which is bigger for covered particles than for bare ones. The ratio $[\eta]/[\eta]_0$, where $[\eta]_0$ is the intrinsic viscosity of the dispersion of bare particles, is therefore equal to the ratio between the volume of a covered particle and that of a bare one, i.e., it equals $(1 + d^h/a)^3$. This method has been used by several authors[1,2,3], especially in older work.

Diffusion coeffficients can be readily measured by means of *quasi-elastic* or *dynamic* light scattering, also called *photon correlation spectroscopy*. For a description of the technique, we refer to sec. I.7.8. In a dilute dispersion of spherical particles of bare radius a, the diffusion coefficient D can be directly related to d^h and a by the Stokes-Einstein equation [I.6.3.32]:

$$D = \frac{kT}{6\pi\eta(a + d^h)} \qquad\qquad [5.6.4]$$

where η is the viscosity of the bulk medium. The method gives accurate results with dispersions, provided that the particle size distribution is narrow and that the dispersion is stable. Flocculation of any kind will give very misleading

[1] M.J. Garvey, Th.F. Tadros, and B. Vincent, *J. Colloid Interface Sci.* **49** (1974) 57.
[2] M.D.Croucher and T.H. Milkie, in *The Effect of Polymers on Dispersion Properties*, Ed. Th.F. Tadros, Academic Press, London (1982) 101.
[3] A. Doroszkowski and R. Lambourne, *J. Colloid Sci.* **26** (1968) 214.

results and must be avoided. Details of the experimental equipment required to carry out these experiments have been described in the literature[1], as are applications to adsorbed polymer layers[2,3].

(ii) *Electrokinetic methods.* The presence of an electrical double layer offers another possible way to measure the thickness of an adsorbed polymer layer. The idea is to use the fact that such a layer, even though it may be highly porous, is capable of strongly reducing the flow of charge along the wall because the adsorbed polymer impedes the flow of solvent. This will then show up as a reduction in the electrokinetic quantities, such as electrophoretic mobility, electro-osmotic flow, streaming current and streaming potential. By analogy with the hydrodynamic layer thickness one can define an *electrokinetic thickness* d^{ek} in terms of the effective shift in the plane of shear that separates mobile and immobile charges. For example, in a single capillary one may determine the streaming potentials $E_{str,0}$ and E_{str} corresponding to a bare and a covered surface, respectively. From this, d^{ek} follows through[4,5]

$$d^{ek} = \kappa^{-1} \ln\left(\frac{E_{str,0}}{E_{str}}\right) \qquad [5.6.5]$$

where κ is the reciprocal Debye length, introduced in [3.5.7].

One would perhaps expect that the hydrodynamic and electrokinetic methods measure the same thickness. This is in general not true, however. Contributions to the electrokinetic flux are located exclusively in the electrical double layer (i.e., up to distance of order κ^{-1}) where the hydrodynamic flux is strongly impeded by the adsorbed polymer. Also, since the extension of the electrical double layer is variable with c_{salt}, the location d^{ek} of the electrokinetic slip plane is not fixed but it depends on the ionic strength. The theory for this dependency has been worked out for electro-osmosis and electrophoresis[6,7] and for streaming current[4]. A strong dependency of d^{ek} on the ionic strength is predicted for the case of double layers not extending beyond the hydrodynamic thickness $(\kappa^{-1} < d^h)$. The dependence of d^{ek} on κ disappears for very thick

[1] K.S. Schmitz, *Dynamic Light Scattering by Macromolecules*, Academic Press, New York (1990).
[2] M.A. Cohen Stuart, F.H.W.H. Waajen, T. Cosgrove, T.L. Crowley, and B. Vincent, *Macromolecules* **17** (1984) 1825.
[3] Th. van den Boomgaard, T. A. King, Th. F. Tadros, H. Tang, and B. Vincent, J. *Colloid Interface Sci.* **66** (1978) 68.
[4] M.A. Cohen Stuart, F.H.W.H. Waajen, and S.S. Dukhin, *Colloid Polym. J.* **262** (1984) 423.
[5] M.A. Cohen Stuart and J.W. Mulder, *Colloids Surfaces* **15** (1985) 49.
[6] R. Varoqui, *Nouveau J. Chimie* **6** (1982) 187.
[7] P.G. de Gennes, *C.R. Acad. Sci. Paris* **197** (1983) 883.

double layers ($\kappa^{-1} \gg d^h$), known as the "hydrodynamic limit") and in this case $d^{ek} = d^h$. It has been demonstrated that layer thicknesses of neutral, water-soluble polymers can be rapidly and sensitively measured[1]. Kinetic studies are also possible[2].

(iii) *Reflectometry.* For a single planar interface the reflectivity is given by the Fresnel equations[3]; see sec. I.7.10a. For an interface covered with one or more thin films of different refractive index multiple reflections occur, leading to interference. Since the optical path length within the film depends on its thickness it should be possible to extract this thickness from reflectivity data. Formalisms have been developed to deal with the combined effect of multiple reflections[4]. For typical adsorbed polymer layers, which are relatively thin, short wavelengths would be needed to obtain sufficient interference. Both X-rays and neutron beams are, in principle, suitable, but visible light is less so.

(iv) *Ellipsometry.* With ellipsometry one measures not only reflected intensities, but also the changes in polarization upon reflection. The technique has been discussed in sec. I.7.10b.

Ellipsometric data are usually interpreted in terms of one equivalent homogeneous film with a refractive index n^{ell} and an *ellipsometric thickness* d^{ell}. This homogeneous layer is defined as a layer that gives the same reflected intensity and phase shift as the actual polymer layer with a z-dependent concentration. The parameters n^{ell} and d^{ell} can be extracted from the experimental data using the Drude equations (sec. I.7.10b); usually, a numerical iteration is required.

As with other thicknesses, the question arises what kind of average is determined in ellipsometry. This problem has been studied both by numerical[5] and analytical[6] calculations, with slightly different results. The calculations show that d^{ell} is much smaller than d^h because it is mainly determined by the loops, where the density of the layer is large enough to affect the refractive index. For films which are thin compared to the wavelength of the light ellipsometry is rather insensitive to the film thickness. In such cases (which are typical for adsorbed polymers) only the product of film index and thickness (i.e.,

[1] M.A. Cohen Stuart and J.W. Mulder, *Colloids Surfaces* **15** (1985) 49.

[2] J.C. Dijt, M.A. Cohen Stuart, and G.J. Fleer, *Macromolecules* **25** (1992) 5416.

[3] See, e.g., R.M.A. Azzam and N.M. Bashara, *Ellipsometry and Polarised Light*, North Holland Amsterdam (1989).

[4] F. Abelès, in *Ellipsometry in the measurement of Surfaces and Thin Films*, E. Passaglia, R.R. Stromberg, and J.Kruger, Eds., *Nat. Bureau. Stand. Misc. Publ.* **256** (1964) 41.

[5] F.L. McCrackin and J.P. Colson, in *Ellipsometry in the measurement of Surfaces and Thin Films*, E. Passaglia, R.R. Stromberg, and J. Kruger, Eds., *Nat. Bureau. Stand. Misc. Publ.* **256** (1964) 61.

[6] J.C. Charmet and P.G.de Gennes, *J. Phys.* (Colloque C10) **44** (1983) 27.

the adsorbed amount) can be determined accurately. The sensitivity for the thickness can be enhanced, however, by using substrates with a top layer with appropriate thickness and index (for example, silicon with an oxide layer).

(v) *Disjoining pressure methods.* When two surfaces covered by adsorbed polymer layers approach each other, usually a repulsive force is experienced when the separation h between the surfaces is of the order of twice the adsorbed layer thickness. Hence, measurement of the separation at which repulsion is first detected provides a method for determining a kind of layer thickness. The value so obtained may be termed the "steric" thickness d^{st}. Obviously, it will depend on the sensitivity of the force apparatus used. Because in most case the steric force increases rather steeply with decreasing separation between the surfaces, measurement of d^{st} is nevertheless meaningful.

One limitation of this method is that no other repulsive forces (for example, electrostatic) between the two surfaces should be present. Thus, this method is unsuitable for determining d^{st} for adsorbed polyelectrolytes.

Two basic approaches are possible: one involves compression of a dispersion of (monodisperse) particles carrying an adsorbed polymer layer and monitoring the pressure as a function of the volume fraction. In the second approach the force between two macroscopic surfaces with adsorbed polymer layers is measured as a function of the surface separation.

In the compression experiments, the distance r between the centres of the particles can be measured directly by scattering techniques, or it can be inferred from the volume fraction. In the latter case, an assumption is needed about the packing: the calculation of r is then a simple geometrical problem. When repulsion is first detected $r = 2(a + d^{st})$, where a is the bare particle radius. The packing may be two- or three-dimensional. For the two-dimensional case, a Langmuir surface balance has been used by Doroszkowski and Lambourne[1,2] and by Garvey et al.[3]. Experiments with three-dimensional compression were done by Cairns et al.[4] and by Homola and Robertson[5]. The necessary pressure can be applied externally[1,2], by centrifugation[6], or by osmosis[7].

Succesful studies with macroscopic surfaces have mostly been carried out with the force-balance apparatus designed by Israelachvili and co-workers[8],

[1] A. Doroszkowski and R. Lambourne, *J. Polymer Sci.* **34** (1971) 253.
[2] A. Doroszkowski and R. Lambourne, *J. Colloid Interface Sci.* **43** (1973) 97.
[3] M.J. Garvey, D. Mitchell, and A.L. Smith, *Colloid Polymer Sci.* **257** (1976) 45.
[4] R.J.R. Cairns, R.H. Ottewill, D.W.J. Osmond, and I. Wagstaff, *J. Colloid Interface Sci.* **54** (1976) 45.
[5] A. Homola and A.A. Robertson, *J. Colloid Interface Sci.* **54** (1976) 286.
[6] M.J. Garvey, Th.F. Tadros, and B. Vincent, *J. Colloid Interface Sci.* **55** (1976) 440.
[7] V.A. Parsegian, N. Fuller, and R.P. Rand, *Proc. Natl. Acad. Sci.* **76** (1979) 2750.
[8] J.N. Israelachvili, *J. Colloid Interface Sci.* **44** (1973) 255; J.N. Israelachvili and G.E. Adams, *J. Chem. Soc. Faraday Trans. I* **74** (1978) 975.

using molecularly smooth mica surfaces. The first experiments with adsorbed polymer layers were done by Klein et al.[1,2]. Sonntag et al.[3,4] have employed a similar method, but with crossed quartz filaments rather than mica. Their technique is inherently less accurate than that of Israelachvili. Recently, Horn et al.[5] have used thin quartz sheets in an Israelachvili-type apparatus. Clearly, the use of surfaces other than mica is to be welcomed.

5.6c Volume fraction profile

The determination of the detailed structure of adsorbed polymer layers, in terms of the volume fraction profile, constitues a major challenge to investigators. Typical length scales are in the nanometer range, and the amount of adsorbed material per unit of volume in a typical experiment is rather low. It is now generally accepted that the layer structure of adsorbed polymers is best probed by neutron radiation, either in scattering or in reflection mode. In the (small angle) scattering mode, one uses colloidal particles covered with adsorbed polymer; in the reflection mode a macroscopically flat surface, typically a silicon crystal, is employed.

The basic theory of scattering and reflection can be found in secs. I.7.9 and 10 and will not be repeated here. The expressions given in that chapter apply equally well to neutron radiation, with the proviso that the material property n, the *refractive index*, is replaced by its analogue μ_n for neutron radiation[6]:

$$\mu_n = 1 - \lambda^2 \rho / 2\pi \qquad\qquad [5.6.6]$$

Here, λ is the wavelength, and ρ is the so-called *scattering length density*, which can be simply calculated by summing the product of the nuclear *scattering length* and the atomic number density over all kinds of nuclei in the material. This scattering length can take both negative and positive values. Note that for neutrons the refractive index can become negative, which is never the case for electromagnetic radiation. As a consequence, total reflection may occur even when the beam comes in through vacuum.

An important option of neutron scattering and reflectometry is the possibility of contrast matching, whereby a certain component (for example, the particles) can be made "invisible" by modifying ρ, just as one can suppress

[1] J. Klein, Y. Almog and P. Luckham, *ACS Symp. Ser.* **240** (1984) 99.
[2] J. Klein, *J. Chem. Soc. Faraday Trans. I* **79** (1983) 99.
[3] L. Knapschinski, W. Katz, B. Elinke, and H. Sonntag, *Colloid Polym. Sci.* **260** (1982) 1153.
[4] Th. Götze and H. Sonntag, *Colloids Surfaces* **25** (1987) 77.
[5] R. Horn, D.T. Smith, and W. Haller, *Chem. Phys. Lett.* **162** (1989) 404.
[6] *Neutron, X-ray and Light Scattering*, Eds. P. Linder and Th. Zemb, North Holland (1991).

scattering of light by matching the refractive indices of the particles and the medium. The deuterium nucleus is very interesting in this respect because its scattering properties are greatly different from those of hydrogen. It is thus possible to change μ_n strongly without affecting the chemistry, and this feature can be used advantageously to get the best sensitivity for a particular system. By varying the ratio of ^1H to ^2D in either solvent or polymer, one can systematically vary the contrast, which can be used to obtain additional information. Details about the technique can be found in a collection of specialist papers[1]. Pioneering scattering experiments on adsorbed polymer layers have been done by Cosgrove et al.[2,3,4] and by Auvray and coworkers[5]. A general problem with scattering techniques is that the calculated profiles are not always unique: different profiles might give the same scattering function. In this respect contrast variation is a very useful tool.

5.7 Theoretical and experimental results for uncharged polymers

In this section we give a selection of theoretical and experimental results for homopolymer adsorption. For a meaningful comparison between theory and experiment it is mandatory that the experimental system is well defined, with as many parameters known as possible (chain length and chain-length distribution, solvency, adsorbent properties, etc.). Wherever feasible, we shall discuss theoretical predictions in combination with experimental data. However, this correspondence cannot be maintained in all cases: there are useful theoretical predictions that, as yet, cannot be checked experimentally (for example, the relative contributions of loops and tails), whereas for some measurable quantities no quantitative theory has yet been developed (for example, most kinetic data).

We divide this section into three main themes: structure, adsorbed amount, and polydispersity. In sec. 5.7a we discuss the structure of the adsorbed layer, paying attention to the volume fraction profile, the composition in terms of loops and tails, the bound fraction, and the layer thickness. Section 5.7b deals with the adsorbed amount as a function of the polymer concentration, the chain length, the solvency and the surface affinity. These two sections are mainly concerned with monodisperse polymer. Polydispersity, where competition between different chain lengths shows up, is treated in sec. 5.7c.

---

[1] *Neutron, X-ray and Light Scattering*, Eds. P. Linder and Th. Zemb, North Holland (1991).

[2] T. L. Crowley, PhD Thesis Oxford (1984).

[3] T. Cosgrove, T.G. Heath, K. Ryan, and T. L. Crowley, *Macromolecules* **20** (1987) 2879.

[4] T. Cosgrove, T.L. Crowley, and B. Vincent, in *Structure of the Interfacial Region*, R.H. Ottewill and C.H. Rochester, Eds., Academic Press (1982) 287.

[5] L. Auvray and J.P. Cotton, *Macromolecules* **20** (1987) 202.

5.7a. *Structure of the adsorbed layer*

(i) *Volume fraction profiles.* We start with theoretical volume fraction profiles for the adsorption of long homodisperse polymer chains and the contribution of trains (tr), loops (lp), and tails (tl) to these profiles. Figure 5.19 gives, for $N = 5000$ and $\varphi^b = 10^{-3}$, the dependence of φ, φ^{lp} and φ^{tl} on the distance z from the

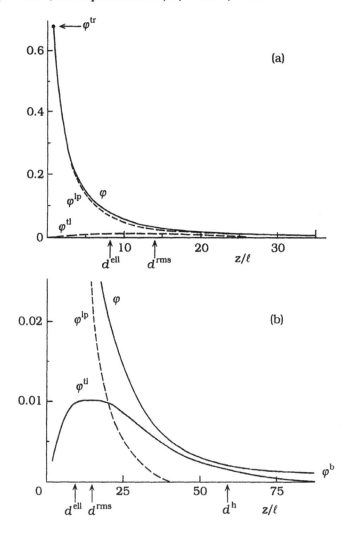

Figure 5.19. Volume fraction profile $\varphi(z)$ calculated with the Scheutjens-Fleer (SF) theory. Parameters: $N = 5000$, $\varphi^b = 10^{-3}$, $\chi^s = 1$, $\chi = 0.5$, hexagonal lattice.
(a) Entire profile φ (solid curve) and the contributions φ^{lp} due to loops and φ^{tl} due to tails (dashed curves); at $z/\ell = 1$ the volume fraction φ^{tr} in trains is indicated. The r.m.s thickness d^{rms} and the ellipsometric thickness d^{ell} are indicated by arrows.
(b) Expanded view of the lower part of diagram (a) for a larger range of z, in order to show the (dilute) volume fraction profiles of tails and loops more clearly. In addition to d^{rms} and d^{ell}, the hydrodynamic thickness d^h, which was outside the range of diagram (a), is indicated.

surface, with linear scales for φ and z. The train density is given by $\varphi^{tr} = \varphi(\ell)$. In the lower diagram the dilute region of the profile ($\varphi < 0.025$) is redrawn on a larger scale, in order to show the contribution of tails in more detail. The figure was calculated using the model of Scheutjens and Fleer (SF)[1]. For the given interaction parameters ($\chi^s = 1$, $\chi = 0.5$) 25% of the segments are in trains, 61% in loops and 14% in tails. In the inner (concentrated) region of the adsorbed layer, beyond the train layer, loops dominate, whereas in the outer (dilute) region only tails contribute to the profile. The tail volume fraction is at most about 1%: the tail region is very dilute throughout. The difference between $\varphi(z)$ and $\varphi^{tl}(z)$ at large z ($\geq 50\ell$) is due to free (non-adsorbed) polymer; this difference is always smaller than the bulk solution concentration φ^b (which in this example is 0.001).

The arrows along the abscissa in fig. 5.19 indicate various layer thicknesses: the ellipsometric thickness d^{ell}, the root-mean-square thickness d^{rms}, and the hydrodynamic thickness d^h. For a definition of these quantities we refer to sec. 5.6. From fig. 5.19 it is clear that d^{ell} and d^{rms} are mainly determined by the loops, whereas d^h measures the extension of the tails. We return to this point below.

Figure 5.20 gives a similar plot for the profile of a longer chain ($N = 10000$) at a lower solution concentration ($\varphi^b = 10^{-6}$). In this case the volume fractions are plotted on a logarithmic scale. It can be seen that the loop contribution φ^{lp} decays exponentially and that φ^{tl} passes through a maximum, followed by an

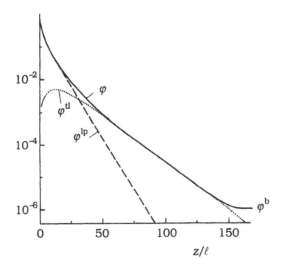

Figure 5.20. Semi-logarithmic segment volume fraction profile $\varphi(z)$ as a function of the distance z from the surface (expressed in units of the lattice layer thickness ℓ). The contributions φ^{lp} due to loops and φ^{tl} due to tails are also indicated. Parameters: $N = 10000$, $\chi^s = 1$, $\chi = 0.5$, $\varphi^b = 10^{-6}$, hexagonal lattice. Redrawn from ref.[2].

[1] J.M.H.M. Scheutjens and G.J. Fleer, *J. Phys. Chem.* **83** (1979) 1619; *ibid.* **84** (1980) 178.

[2] J.M.H.M. Scheutjens, G.J. Fleer and M.A. Cohen Stuart, *Colloids Surfaces* **21** (**1986**) **285**.

exponential decrease with a decay length which is about twice that of φ^{lp}. Because of the lower solution concentration, the tail segment fraction in fig. 5.20 is lower (6%) than in fig. 5.19 (14%). Nevertheless, the outer region of the adsorbed layer is compl etely dominated by the tails; in this case ($\varphi^b = 10^{-6}$) the contribution of free polymer (i.e., the difference between φ and φ^u) is negligible over a wide range of z.

(ii) *Loops, tails, and trains.* Figure 5.21 gives some insight into the fractions of trains, loops, and tails in the adsorbed layer as a function of chain length, for two bulk solution concentrations. In these diagrams, the bound fraction p can be read from the bottom, the tail fraction from the top, and the loop fraction is the remaining middle part.

The train fraction is unity for monomers ($N = 1$) and it decreases as the chains become longer, mainly because of the formation of more loops (see also fig 5.22). At constant φ^b the loop fraction is a monotonically increasing function of the chain length. At low φ^b the tail fraction has a maximum for oligomers (around $N \approx 4$) because such very short chains cannot easily form loops. For slightly longer chains the contribution of loops increases at the expense of the tail fraction. However, for chains longer than about 20 segments the tail fraction remains nearly constant, indicating that the length of the tails is, for long chains, proportional to the chain length. The contribution of tails becomes more important at a higher solution concentration. This increase of the tail fraction with φ^b is accompanied by a decrease of the loop contribution and, to a smaller extent, of the train fraction. It is noteworthy that for very long chains the importance of tails increases again. This indicates that the tail

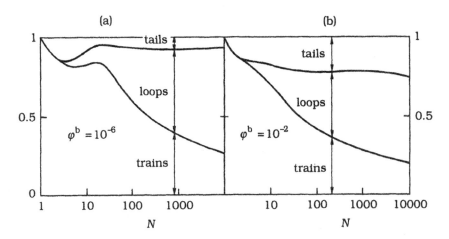

Figure 5.21. The relative contribution of trains, loops, and tails in an adsorbed layer for $\varphi^b = 10^{-6}$ (a) and $\varphi^b = 10^{-2}$ (b) as a function of chain length N, calculated with the SF model. Parameters: $\chi^s = 1$, $\chi = 0.5$, hexagonal lattice.

fraction in saturated layers of long chains at finite solution concentrations remains significant, whereas it vanishes for isolated chains.

(iii) *Bound fraction and train density*. Above we discussed some aspects of trains from a theoretical point of view. Now we make a comparison with experimental results. As mentioned in sec. 5.6 there are several ways to determine the number of segments in contact with the surface.

Figure 5.22 gives results for PVP/water/SiO$_2$ as studied by EPR[1] (using spin labels attached to PVP) and NMR[2] (using the proton relaxation rate of the polymer). The data are presented both as Γ^{tr}/Γ and as $p(\Gamma)$. These three quantities are related through $p = \Gamma^{tr}/\Gamma$. The top curve in fig. 5.22, where the filled symbols were obtained by NMR and the open circles by EPR, has the same general shape as that predicted by theory: a linear increase for an under-

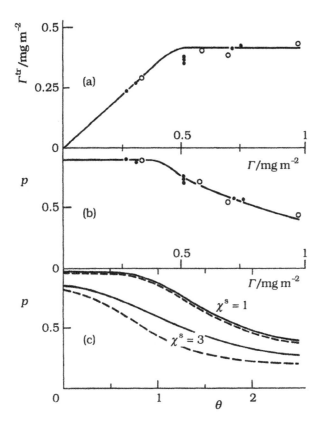

Figure 5.22. Adsorbed amount Γ^{tr} in trains (a) and the bound fraction p as a function of coverage (b) as measured by NMR and ESR, and $p(\theta)$ according to SF theory (c). The experimental data are for the system PVP (900K)/ H$_2$O/SiO$_2$, taken from refs.[1] (ESR) and [2] (NMR). The parameters for the theoretical diagram are $N = 1000$, hexagonal lattice, two values of χ^s (indicated), and $\chi = 0$ (dashed curves) or 0.5 (solid curves). The latter diagram was redrawn from ref.[3].

[1] I.D. Robb and R. Smith, *Eur. Polym. Sci.* **10** (1974) 1008.
[2] K.G. Barnett, T. Cosgrove, B. Vincent, B.S. Sissons, and M.A. Cohen Stuart, *Macromolecules* **14** (1981) 1018.
[3] J.M.H.M. Scheutjens and G.J. Fleer, *J. Phys. Chem.* **83** (1979) 1619.

saturated surface (where the conformations are flat), followed by a nearly constant level where the surface becomes saturated (though not all surface sites are occupied because of entropic restrictions).

The experimental $p(\Gamma)$ curve in fig. 5.22b may also be compared with its theoretical counterpart, shown in fig. 5.22c. This diagram gives SF predictions for two surface affinities (χ^s = 1 and 3) and two solvencies: χ = 0.5 (solid curves) and χ = 0 (dashed). The curves in the figure are for N = 1000, but it has been shown[1] that the chain-length dependence is only very weak, so this picture is representative. There is an excellent qualitative agreement between figs. 5.22b and c. Quantitatively, the correspondence is also quite reasonable. The Γ axis in fig. 5.22b may be transformed into the number of equivalent monolayers through $\theta = \Gamma/\Gamma^{mon}$, where the monolayer capacity Γ^{mon} is around 0.45 mg/m^2; hence, Γ = 1 mg/m^2 corresponds to about 2.3 monolayers. The data suggest that χ^s for PVP from water on to silica is reasonably high, for which there is independent experimental evidence: a value χ^s = 4 has been reported for this system[2].

(iv) *Layer thickness.* An adsorbing polymer does not produce a uniform layer but gives rise to a strongly varying concentration profile, as shown in figs 5.19

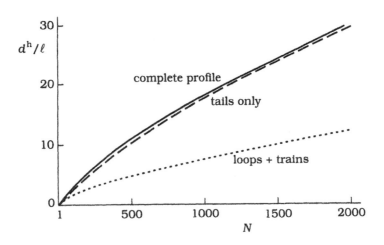

Figure 5.23. Theoretical hydrodynamic layer thickness as a function of chain length. The figure gives d^h as computed[3] from the complete volume fraction profile (solid curve), from the loops and trains only (dotted curve), and from the tails only (dashed curve). Parameters: χ^s = 1, χ = 0.5, φ^b = 10^{-3}, hexagonal lattice. Redrawn from ref.[3].

[1] J.M.H.M. Scheutjens and G.J. Fleer, *J. Phys. Chem.* **83** (1979) 1619.
[2] M.A. Cohen Stuart, G.J. Fleer, and B.H. Bijsterbosch, *J. Colloid Interface Sci.* **90** (1982) 321.
[3] M.A. Cohen Stuart, F.H.W.H. Waajen, T. Cosgrove, B. Vincent, and T.L. Crowley, *Macromolecules* **17** (1984) 1825.

and 20. As a result, any method of estimating the layer thickness provides some average whereby the contributions of loops and tails may differ greatly. As indicated in the discussion of fig. 5.19 some averages, such as the ellipsometric thickness d^{ell} and the root-mean-square thickness d^{rms}, are representative for the loop contribution, whereas the hydrodynamic thickness d^h is mainly determined by the tails.

In order to illustrate this latter feature we present in fig. 5.23 theoretical predictions of d^h as a function of chain length. The solid curve in fig. 5.23 gives the thickness obtained from the complete profile. The dashed curve is the result if only tail segments were present (i.e., assuming that loop and train segments would have zero friction), and the dotted curve represents the thickness due to loops and trains only (omitting the tail fraction). It is clear that the tail segments (which contribute by only around 10% to the adsorbed amount) nearly completely determine the hydrodynamic thickness. If these tail segments are omitted, the remaining 90% of loop and train segments give rise to a hydrodynamic thickness, but this is lower than d^h by a factor of 2.5. Apparently, the segments in the periphery of the adsorbed layer are the dominant factor in perturbing the liquid flow, thereby hydrodynamically screening the inner segments. Indeed, the hydrodynamic thickness is a sensitive parameter to probe the extension of the tail segments.

When the hydrodynamic thickness is plotted as a function of the adsorbed amount Γ, a steep increase is found as Γ approaches saturation. This is schematically illustrated in fig. 5.24, both for a good solvent and for a poor solvent. In the good solvent saturation corresponds to a relatively low Γ, whereas in a poor solvent higher adsorbed amounts can be accomodated. The key feature of the two drawn curves in fig. 5.24 is that a small increase in Γ on a crowded surface (saturation) leads to a considerable increase of the hydrodynamic layer thickness because the additional segments are forced to form tails. Another important aspect is that d^h at given Γ and solvency hardly depends on the chain length[1,2]. However, in order to obtain the same Γ for short and long chains, quite different bulk concentrations are usually required.

The shapes of the $d^h(\Gamma)$ curves in fig. 5.24 have a number of consequences that are important in the interpretation of experiments. If, for a given polymer, d^h is measured as a function of the adsorbed amount Γ, there should (in dilute solutions) be no effect of the molar mass M. However, in the steep part of the $d^h(\Gamma)$ curve, which is in the (pseudo)plateau region of the adsorption isotherm, Γ

---

[1] M.A. Cohen Stuart, F.H.W.H. Waajen, T. Cosgrove, B. Vincent, and T.L. Crowley, *Macromolecules* **17** (1984) 1825.
[2] G.J. Fleer et al., loc. cit.

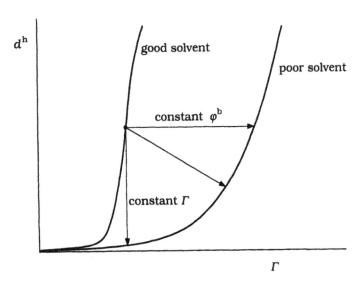

Figure 5.24. The hydrodynamic layer thickness d^h as a function of the adsorbed amount Γ (schematic), and the effect of changes in the solvency under various experimental conditions, as symbolized by the arrows. For explanation see text.

has to be known very accurately since a small variation in Γ leads to a large change in d^h.

If, on the other hand, d^h is measured *at constant* φ^b, a clear molecular weight dependence might be found[1,2,3,4]. In poor solvents Γ increases with M (see secs. 5.3 and 5.7b), leading to a higher hydrodynamic thickness. In good solvents the dependence of Γ on M is much weaker, but since d^h now increases much more steeply with Γ the same conclusion applies. However, plots of log d^h against log M are only meaningful if φ^b is constant (and well controlled!) because the lengths of the tails and, consequently, d^h depend strongly on φ^b, even in the plateau region of the adsorption isotherm. The latter aspect has often been disregarded. It is clearly not sufficient to state that "measurements were made in the plateau region of the isotherm"; the precise value of φ^b should be specified.

Another interesting aspect is the variation of d^h when the solvency is changed in a closed system, for example by temperature variation. Let us assume that the temperature change is such that the solvency is decreased (higher χ). Then more polymer adsorbs, thereby decreasing φ^b. Figure 5.24 illustrates various possibilities. If the ratio between the surface area and the volume is

[1] G.P. van der Beek and M.A. Cohen Stuart, *J. Phys.* (Les Ulis, Fr.) **49** (1988) 1449.
[2] E. Killmann, Th. Wild, N. Gütling, and H. Maier, *Colloids Surfaces* **18** (1986) 241.
[3] M. Kawaguchi, M.M. Mikura, and A. Takahashi, *Macromolecules* **17** (1984) 2063.
[4] T. Kato, N. Nakamura, M. Kawaguchi, and A. Takahashi, *Polymer J.* **13** (1981) 1037.

small, most of the polymer is in the solution and Γ may increase without an appreciable change in φ^b. Now d^h remains virtually constant, as indicated by the horizontal arrow in fig. 5.24. The increase of Γ in this case comes mainly from the loops, without affecting the tails. However, in a dispersion of high surface area in contact with a dilute polymer solution, the solution is readily depleted upon decreasing the solvency, and Γ (or θ) is approximately constant: d^h drops steeply, as shown by the vertical arrow in fig. 5.24. In this case tails are transformed into loops. In practical situations an intermediate behaviour might be found with decreasing φ^b and decreasing d^h, as illustrated by the middle arrow in the figure. The slope of this arrow depends on the ratio between the total amounts of polymer on the surface and in the solution.

   In fig. 5.25 we show some experimental results for the hydrodynamic thickness as a function of the adsorbed amount. This figure gives two examples for poly(ethylene oxide) (PEO) from water on polystyrene latex (PSL)[1] or on silica (SiO_2)[2]. The PSLdata were obtained by PCS, the SiO_2 results from the change in the dispersion viscosity. The curves show the same general dependence $d^h(\Gamma)$ as predicted by theory, with a low d^h at low adsorption and a steep increase at high Γ, compare fig. 5.24. The data points of fig. 5.25a apply to seven different molecular weights (ranging from 25 K to 1300 K) in the semi-plateau region of the adsorption isotherm, those of fig. 5.25b to three samples with varying coverage. In both cases d^h is a unique function of Γ, regardless of M. On both

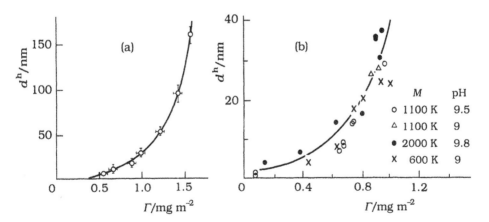

Figure 5.25. The hydrodynamic layer thickness d^h as a function of adsorbed amount Γ for (a) PEO/H_2O/PSL and (b) PEO/H_2O/SiO_2. In both cases a range of molecular weights was used. The data in diagram (a) were obtained by PCS[1], those in diagram (b) by viscosity measurements on dispersions[2].

[1] M.A. Cohen Stuart, F.H.W.H. Waajen, T. Cosgrove, B. Vincent, and T.L. Crowley, *Macromolecules* **17** (1984) 1825.
[2] F. Lafuma, K. Wong, and B. Cabane. *J. Colloid Interface Sci.* **143** (1991) 9.

surfaces d^h has about the same magnitude (at equal coverage). This features are fully consistent with theoretical predictions as discussed above.

5.7b. Adsorbed amount

The adsorbed amount is one of the basic parameters in polymer adsorption. Yet its measurement is not without pitfalls and examples of good experiments on well-defined systems over a wide range of parameters are scarce. As mentioned above, the adsorbed amount is mainly determined by the loops and the trains.

Adsorbed amounts are usually presented in the form of an adsorption isotherm. Figure 5.26 gives a set of theoretical (top) and experimental (bottom) isotherms, with linear scales for both the adsorbed amount and the equilibrium concentration. The theoretical curves in fig. 5.26a show the typical high-affinity type isotherms with a nearly horizontal plateau over a wide range of concentrations. In Θ-solvents this (pseudo)plateau depends strongly on the chain length. Oligomers ($N = 1$, 10, 20) adsorb weakly in this concentration range (at the given

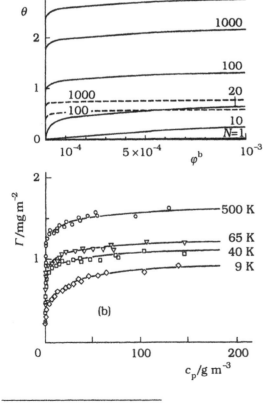

Figure 5.26. Adsorption isotherms for various molecular weights. Diagram (a) shows theoretical data from SF theory; the chain length N for each curve is indicated. Solid curves apply to a Θ-solvent ($\chi = 0.5$) and dashed curves to a good solvent ($\chi = 0$). Parameters: $\chi^s = 1$, hexagonal lattice. Diagram (b) shows experimental results for the system dextran/H_2O/ silver iodide, and is redrawn from ref.[1].

---

[1] V. Hlady, J. Lyklema, and G.J. Fleer, *J. Colloid Interface Sci.* **87** (1982) 395.

value of χ^s) and show more rounded (Langmuir-like) isotherms. In better solvents (dashed curves) the adsorbed amount is much less and, moreover, more weakly dependent on chain length than in poor solvents. The molecular weight effect is more extensively discussed in relation to fig. 5.28, below. The higher adsorption in poor solvents is caused by the fact that the weaker effective repulsion between the segments makes the accumulation at the surface easier.

The experimental isotherms of fig. 5.26b, for dextran fractions from water on to silver iodide[1], show the same general features as the theoretical ones, with the adsorption increasing with molecular weight. However, the isotherms are more rounded than the theoretical ones. This is caused by the polydispersity of the dextran fractions, a feature to be discussed in more detail in sec. 5.7c.

In order to explore the $\theta(\varphi^b)$ dependence over a wider range of parameter values, the curves for $N = 100$ in fig. 5.26a are replotted in fig. 5.27 with logarithmic scales for θ and φ^b. The usual experimental range for θ and φ^b is within the box in the upper right corner of fig. 5.27; the data shown in fig. 5.26 also lie within this box. The curve was calculated by fixing the chemical potential, which can be chosen very low but then corresponds to an extremely low (in fact fictitious) concentration. The parts of the curves outside this box are not easily

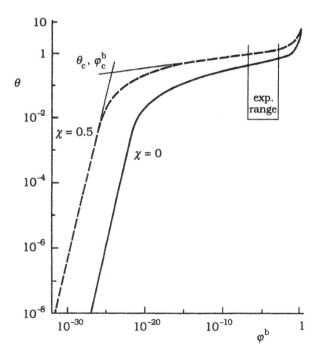

Figure 5.27. Double-logarithmic plot of theoretical adsorption isotherms over a wide range of bulk concentrations. Solid curve: θ-solvent ($\chi = 0.5$); dashed curve: good solvent ($\chi = 0$). Parameters: chain length $N = 100$, $\chi^s = 1$, hexagonal lattice. The cross-over parameters θ_c and φ_c^b, characterizing the transition between the Henry regime and the pseudo-plateau region are indicated. The usual experimental range is within the box in the top right part of the diagram. Redrawn from ref.[2].

[1] V. Hlady, J. Lyklema, and G.J. Fleer, *J. Colloid Interface Sci.* **87** (1982) 395.
[2] J.M.H.M. Scheutjens and G.J. Fleer, in *The Effect of Polymers on Dispersion Properties,* Th.F. Tadros, Ed., Academic Press, London (1982) 145.

accessible in an experiment; nevertheless they show an interesting physical picture.

At extremely low concentrations (below $\varphi^b \approx 10^{-25}$ for $N = 100$ and $\chi = 0.5$), the adsorbed amount θ is proportional to φ^b (the slope in the double-logarithmic plot is unity), which corresponds to an unsaturated surface. This is the Henry region of the isotherm, described by $\theta \approx \varphi^b \exp\{pN(\chi^s + \lambda_1\chi)\}$ (compare the low-φ^b limit of [5.5.22], which reads $\theta \approx \varphi^b \exp\{\chi^s + \lambda_1\chi\}$. As φ^b increases, the surface becomes saturated, the chains have to compete for surface sites, and a pseudo-plateau is found which extends over many decades of φ^b. In this pseudo-plateau region, θ increases typically by a few per cent (or less) per decade in φ^b. This region extends across the experimental range. Beyond that ($\varphi^b \gtrsim 0.01$) the concentrated regime is found where θ increases more strongly; as φ^b approaches unity (i.e. in a polymer melt) the effect of χ disappears and θ becomes proportional to $N^{1/2}$, which is typical for Gaussian chains. The difference between $\chi = 0$ and $\chi = 0.5$ in fig. 5.27 is in the Henry region only determined by the effective adsorption energy $\chi_s + \lambda_1\chi$. In the pseudo-plateau regime crowding effects play the dominant role.

The molecular weight dependence of the adsorbed amount in the (pseudo)-plateau region of the adsorption isotherms is shown in fig. 5.28. The solid curves give SF predictions $\theta(N)$ for three adsorption energies ($\chi^s = 0.6$, 1 and 3) in a Θ-

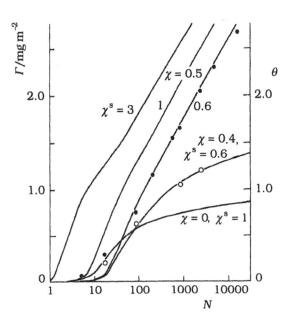

Figure 5.28. Adsorbed amount as a function of chain length on a semi-logarithmic scale. Solid curves are theoretical results [i.e., $\theta(N)$, r.h.s. scale] obtained with SF theory, for $\varphi^b = 10^{-3}$, and various values of χ^s and χ (as indicated). Points represent experimental results [i.e., $\Gamma(M)$, l.h.s. scale] for the systems PS/cyclohexane(35°C)/SiO$_2$ (filled circles) and PS/CCl$_4$/SiO$_2$ (open circles) at a concentration ≈ 1 g/m^3, taken from ref.[1]. Redrawn from ref.[2].

[1] C. Vander Linden and R. van Leemput, *J. Colloid Interface Sci.* **67** (1978) 48.
[2] J.M.H.M. Scheutjens and G.J. Fleer, in *The Effect of Polymers on Dispersion Properties*, Th.F. Tadros, Ed., Academic Press, London (1982) 145.

solvent (χ = 0.5) and two curves for better solvents (χ = 0.4, 0); for θ the r.h.s. ordinate axis (in equivalent monolayers) applies. In a good solvent (χ = 0), θ is low and, for long chains, nearly independent of chain length. In a poorer solvent (χ = 0.4) the adsorbed amount is higher because of the mutual attraction of the segments and it is more strongly dependent on N. In a Θ-solvent (χ = 0.5) θ increases linearly with log N (for $N \geq$ 100). This linear dependence applies for any χ^s. The effect of the adsorption energy is to increase the adsorbed amount (or equivalently, to shift the linear sections to the left). This increase is most pronounced in the unsaturated region. For the highest χ^s shown in fig. 5.28 (χ^s = 3) the surface layer is (for $N \geq$ 100), nearly fully occupied ($\varphi^{tr} \approx$ 1); if χ^s is increased even more the results are hardly affected, except in the oligomer range.

There is now clear experimental evidence corroborating the trends predicted by the SF model. Figure 5.28 shows also some experimental data[1] for homodisperse PS fractions adsorbing on SiO_2 from the Θ-solvent cyclohexane (filled circles) or the good solvent CCl_4 (open circles). For the latter solvent χ = 0.4[2]. The experimental adsorbed amount Γ is expressed in mg/m^2 (l.h.s. ordinate axis), the experimental molecular weight is converted to the number of segments by assuming (arbitrarily) that one segment corresponds to one monomer unit. The general trends are fully confirmed. The theoretical curves would quantitatively agree with the experimental points if χ^s would be 0.6 (in both solvents) and if one equivalent monolayer would contain 1 mg/m^2. These assumptions are presumably not warranted; a more likely value of χ^s in the Θ-solvent is about 2[3]. The experimental filled points could be made to coincide with the theoretical curve for χ^s = 2 (not shown) by taking $\approx$5 monomer units per theoretical segment, which would shift the points about 0.7 decade to the left. For this choice the logarithmic dependence would be also maintained. As yet, there is no independent evidence for the proper scaling factors (monolayers $\leftrightarrow$ mg/m^2; number of segments $\leftrightarrow$ degree of polymerization). Despite these uncertainties, there is at least semi-quantitative agreement between theory and experiment as to the molecular weight dependence of the adsorbed amount.

5.7c. Polydisperse polymer

In most situations the experimental system is more complicated than one (homodisperse) polymer adsorbing from a single solvent. In multicomponent systems preferential adsorption always plays a role. A common example is the adsorption of a polydisperse polymer, where usually long chains adsorb preferentially over short ones, even if the adsorption energy per segment is the same.

[1] C. Vander Linden and R. van Leemput, J. Colloid Interface Sci. **67** (1978) 48.
[2] G.M. Bristow and W.F. Watson, Trans. Faraday Soc. **54** (1958) 1742.
[3] G.P. van der Beek, M.A. Cohen Stuart, G.J. Fleer, and J.E. Hofman, Macromolecules **24** (1991) 6600.

This fractionation was discussed qualitatively in sec. 5.3d. Here, we consider some quantitative results.

In dilute solutions of polydisperse polymers, the long chains are found preferentially on the surface because they lose less translational entropy (per unit mass) in the solution, while they gain approximately the same (total) adsorption energy per unit area. This kind of preference shows up most clearly with a bimodal distribution of chain lengths. When the surface becomes saturated with long molecules the adsorption isotherm reaches a pseudo-plateau. For the mixture this plateau is slightly below the isotherm for a pure sample of long molecules. This is due to the (much) lower solution concentration of long chains in the mixture: the total solution concentration is predominantly made up of the short chains which have lost the competition for surface sites.

An experimental example of this kind is shown in fig. 5.29, for a mixture of two dextran fractions (of equal mass fractions w_a and w_b) adsorbing from water on to silver iodide[1]. The top figure gives $\Gamma(c_p)$ adsorption isotherms at three different values of t_s; as explained in sec. 5.3d this parameter is the ratio V/A between the solution volume V and the surface area A in the system. The isotherms consist of three linear sections: in the initial steep rise the surface is unsaturated and all molecules can find a place on the surface, in the linearly increasing branch the small molecules are gradually displaced by big ones, and in the pseudo-plateau only the long chains are found on the surface.

For a surface fully covered by molecules a and b we have $\varphi^{tr} = p_a\theta_a + p_b\theta_b$. For saturated adsorption layers of pure a and b $\varphi^{tr} = p_a\theta_a^m$ and $\varphi_{tr} = p_b\theta_b^m$, respectively, where θ_a^m and θ_b^m are the surface coverages in the pseudo-plateau of the individual adsorption isotherms. Assuming the chains to have the same conformation (i.e., the same bound fraction p_a or p_b) in mixed and pure adsorbate layers, and taking also φ^{tr} to be the same (which is very reasonable), we find $\theta_a/\theta_a^m + \theta_b/\theta_b^m = 1$. In terms of adsorbed mass, instead of monolayers:

$$\Gamma_a/\Gamma_a^m + \Gamma_b/\Gamma_b^m = 1 \qquad\qquad [5.7.1]$$

where Γ_a^m and Γ_b^m correspond to the plateau levels of the individual isotherms. Applying this equation to the first kink in the isotherms of fig. 5.29 (where $\Gamma_a = \Gamma_b = \frac{1}{2}\Gamma$), we obtain in this point $1/\Gamma = \frac{1}{2}(1/\Gamma_a^m + 1/\Gamma_b^m)$, which is the average of the reciprocal values of Γ^m. For the second kink we have $\Gamma_a = 0$ (if a denotes the small molecules) and $\Gamma = \Gamma_b^m$. The equilibrium concentration at this point is found from the mass balance of molecules a (which are all in the solution): $\Gamma = c_p V/A = c_p t_s$. At this point the amount on the surface (i.e.,

---

[1] V. Hlady, J. Lyklema, and G.J. Fleer, J. Colloid Interface Sci. 87 (1982) 395.

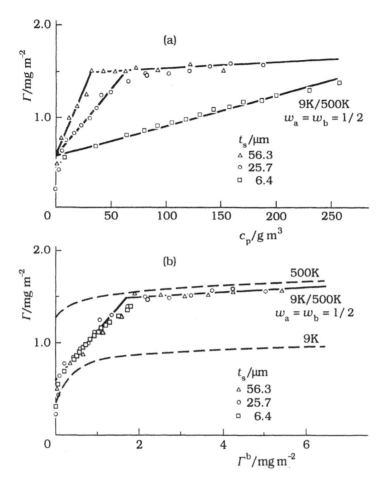

Figure 5.29. Adsorption isotherms for a binary mixture of dextran samples (9 K and 500 K in equal mass fractions) adsorbed on AgI from aqueous solution, for three values of t_s (indicated)[1]. In the top figure (a) the isotherms are plotted in the conventional way (Γ versus c_p), in the lower diagram (b) they are plotted as $\Gamma(\Gamma^b)$, where $\Gamma^b = c_p t_s$ is the amount of polymer (per unit area) in the bulk solution. In (b) the adsorption isotherms of the individual fractions are given as the dashed curves.

molecules b) equals that in solution (i.e., molecules a) since, in this case, the weight fractions of a and b are the same.

The quantity $c_p t_s$ is the amount of polymer in the bulk solution per unit area; its dimensions (mass/unit area) are the same as for Γ. We denote this bulk solution amount (per unit area) as Γ^b. The consequence of the above considerations is that an adsorption isotherm $\Gamma(\Gamma^b)$ should become independent of t_s, see also sec. 5.3d. Figure 5.29b illustrates that the three curves $\Gamma(c_p)$ merge perfectly

---

[1] V. Hlady, J. Lyklema, and G.J. Fleer, *J. Colloid Interface Sci.* **87** (1982) 395.

into one $\Gamma(\Gamma^b)$ isotherm. For a polydisperse system the quantity Γ^b is more fundamental than the quantity c_p, since it eliminates the dependence on the size of the system and accounts automatically for the mass balance. For comparison, fig. 5.29b also shows the individual isotherms of the two fractions; these are the same as in fig. 5.26b. The plateau section of the mixture isotherm lies below that of the high M fraction, because of the lower solution concentration of long chains in the mixture.

If a mixture contains more than two fractions the treatment can easily be generalized. For example, a trimodal mixture gives four linear isotherm sections[1]. The increasing part of fig. 5.29 is replaced by two sections of different slope: in the first the lowest molecular weight is displaced from the surface, in the second the middle fraction. For a continuous distribution of molecular weights a rounded isotherm is obtained: the adsorption fractionation varies smoothly with the polymer dose.

The fractionation effect occurring in polydisperse samples is shown more explicitly in fig. 5.30. The theoretical computations (left) were performed with the SF model, assuming a Schultz-Flory distribution with $M_w/M_n = 2.05$[2,3]. The experiments apply to polydisperse PS adsorbing from CCl_4 on SiO_2 (right)[4]. In both cases a rather sharp fractionation occurs, with the long chains (the "winners") on the surface and the short ones (the "losers") in the solution. It

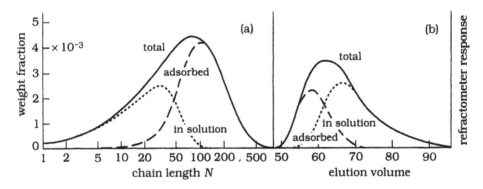

Figure 5.30. Adsorption fractionation of a polydisperse sample as calculated from the SF model (a) and as measured by gel permeation chromatography (b)[1]. The calculations apply to a Schultz-Flory distribution with $M_w/M_n = 2.05$[2,3], the experiments to the system $PS/CCl_4/SiO_2$[4]. Note that a higher elution volume corresponds to a lower molecular weight, and that it is linear in log M. Both in theory and in experiment long chains adsorb preferentially.

[1] V. Hlady, J. Lyklema, and G.J. Fleer, *J. Colloid Interface Sci.* **87** (1982) 395.
[2] G.J. Fleer and J.M.H.M. Scheutjens, in *Coagulation and Flocculation: Theory and Applications*, B. Dobias, Ed., *Surfactant Science Series* Vol **47**, Marcel Dekker (1993) 209.
[3] S.P.F.M. Roefs, J.M.H.M. Scheutjens, and F.A.M. Leermakers, *Macromolecules* **27** (1994) 4810.
[4] C. Vander Linden and R. van Leemput, *J. Colloid Interface Sci.* **67** (1978) 63.

should be realized that the experimental elution volume decreases with increasing molecular weight and that it is linear in log M; therefore figs. 5.30a and b are approximately mirror images of each other. The similarity in the shape of the curves is striking. The data in fig. 5.30 apply to one specified value of t_s. As this parameter increases, the "solution" peak increases and the "adsorbed" peak decreases; both peaks shift to higher chain lengths. The result is that the average M in the adsorbate becomes higher, whereas that in the solution approaches the sample average[1].

5.7d. Polydispersity and (ir)reversibility

The last issue to be dealt with is the apparent irreversibility of the adsorption. One quite often encounters the opinion, especially in the older literature, that polymer adsorption would be an irreversible phenomenon. These ideas are based on the hysteresis found in the adsorption isotherms: desorption isotherms (obtained by dilution with solvent) do not coincide with adsorption isotherms (obtained by adding more polymer at given amount of solvent). Qualitatively, this was already discussed in sec. 5.3d. An experimental example is given in fig. 5.31, for the adsorption of a polydisperse rubber from heptane on two types of carbon black (differing in specific surface area)[2]. The desorption isotherms are found to lie considerably above the adsorption isotherms, the extent of desorption being very small.

Although these hysteresis effects are real, there is no reason to conclude that the adsorption of polymers is irreversible. The situations for a point on the desorption isotherm and on the adsorption isotherm are different in one important aspect: t_s is higher in the former case because the solution has been diluted with solvent. This solvent addition lowers the concentration in the solution (which contains only the "losers") but does not affect the "winners" on the surface. The latter would desorb only by excessive dilution (below φ_c^b, see fig. 5.27), which cannot be done under the usual experimental conditions (unless the chains are short and the adsorption weak). As a matter of fact, a "desorption isotherm" as given in fig. 5.31 (whereby t_s is continuously increased) corresponds to going from some point on the plateau of an adsorption isotherms at low t_s to a similar point (with nearly the same Γ) on an adsorption isotherm at higher t_s: one crosses a set of adsorption isotherms to the left at essentially constant Γ. This does not reflect irreversibility, but simply a change in the mass balance.

In the same way it can be shown that desorption experiments by repeated solvent exchanges (using, for example, labelled polymer adsorbed on a metal

---

[1] V. Hlady, J. Lyklema, and G.J. Fleer, *J. Colloid Interface Sci.* **87** (1982) 395.
[2] G. Kraus and J. Dugone, *Ind. Eng. Chem.* **47** (1955) 1809.

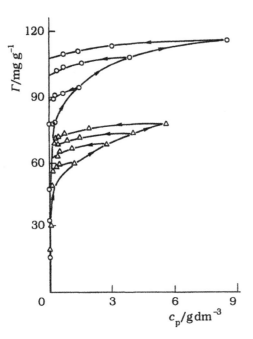

Figure 5.31. Adsorption-desorption hysteresis as observed for rubber adsorbing from n-hexane on to two different types of carbon black. Arrows indicate ascending (increasing c_p) and descending (decreasing c_p) branches. Redrawn from ref.[1].

slide[2]) are bound to be unsuccessful. Even if the polymer would be strictly homodisperse, each solvent exchange would correspond to a (small) change to the left along an isotherm as given in fig. 5.27. If, after several dilutions, the solution concentration attains the range of 10^{-10} or lower, further solvent exchange has hardly an effect any more because the desorbed amount needed to make up a new equilibrium solution is minute indeed. We conclude that the lack of desorption upon dilution may not be invoked to prove "irreversibility" of the adsorption, whether the polymer sample is monodisperse or not.

5.8 Theoretical and experimental results for polyelectrolytes

5.8a. General trends

The interaction parameters governing the adsorption of uncharged polymers are the adsorption energy (modelled by the parameter χ^s) and the solvency (described by the parameter χ). These two parameters refer to short-range "chemical" or "specific" interactions. For polyelectrolytes the long-range electrostatic interaction is the additional feature. It depends on three main factors: the surface charge density σ^0, the polyelectrolyte charge, and the ionic strength. We

---

[1] G. Kraus and J. Dugone, *Ind. Eng. Chem.* **47** (1955) 1809.
[2] W.H. Grant, L.E. Smith, and R.R. Stromberg, *Faraday Discussions Chem. Soc.* **59** (1975) 209.

may express the polyelectrolyte charge through the monomer charge $q_m = \pm t\alpha e$, where α is the degree of dissociation of the segments. Usually, we will consider only monovalent salt ions, in which case the ionic strength equals the electrolyte concentration c_s. The effect of pH, if any, manifests itself in the value of σ^0 and/or α.

The effects of the parameters σ^0. q_m. c_s. χ, and χ^s on the adsorbed amount of polyelectrolytes are qualitatively illustrated in fig. 5.32. In this figure the excess adsorbed amount θ^{ex} is plotted as a function of log c_s for different values of σ^0, α, and χ^s. Note that, as for uncharged polymers, χ and χ^s contain only short-range interactions. Figure 5.32 shows the wide variety of trends encountered in polyelectrolyte adsorption. The curves are most easily interpreted by first considering the limiting cases of very low and very high c_s.

In the low-salt limit the electrostatics dominate: depletion occurs when σ^0 and q_m have the same sign (curve 2), only weak adsorption is found on an uncharged surface even when $\chi^s \gg \chi^s_{crit}$ (curve 1), and *charge compensation* (or, equivalently, exchange of small ions by the polyions) takes place when σ^0/q_m is negative (curves 3-5). When charge compensation is the only effect, the polyions are bound in the electric double layer and the sum of the surface charge ($\sigma^0 \ell^2$ per site) and the polyelectrolyte charge ($q_m \theta^{ex}$ per site) is zero, so that $\theta^{ex} \approx -\sigma^0 \ell^2/q_m$. At low c_s the co-ions hardly contribute to the charge balance. Molecular weight effects are nearly absent in this case, because only the adsorbed charge matters.

In the limit of high c_s the electrostatic contributions are highly screened so that the chemical interactions χ^s and χ become important: no adsorption occurs

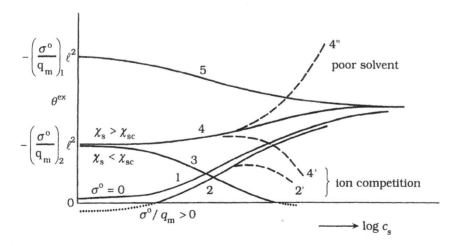

Figure 5.32. Schematic overview of polyelectrolyte adsorption behaviour. The adsorbed amount is plotted as a function of the salt concentration c_s, for various combinations of the segment charge q_m, the surface charge density σ^0, and the chemical parameters χ and χ^s. The curves are numbered to facilitate the discussion (see text).

if $\chi^s < \chi^s_{crit}$ (curve 3), a given adsorption level is approached if $\chi^s > \chi^s_{crit}$ and if χ is not too high (curves 1, 2, 4 and 5), and very high adsorption (leading to phase separation) may be observed if the solvent is very poor for the screened polyelectrolyte (as exemplified by curve 4"). In this limit molecular weight effects are important, especially if χ is not too low. For uncharged polymers, these chain-length effects were extensively discussed in sec. 5.7b.

Curve 1 refers to polyelectrolyte adsorption on an uncharged surface. In this case the electrostatic contribution opposes the adsorption because the strong mutual repulsion between segments (at low c_s) prevents their accumulation in the surface region. This mutual repulsion could be viewed upon as contributing a negative increment to the solvency parameter χ: electrostatics make the solvent better. The *effective* solvency parameter χ^{eff} (i.e., χ plus a negative electrostatic term) is negative at low c_s and not too small α. The low adsorption in this case is analogous to that of uncharged polymers from extremely good solvents, where the adsorbed amount is also low. Since salt addition weakens the electrostatic interactions, it promotes the adsorption in a similar way as a strong decrease in solvency would do. At very high ionic strengths, χ^{eff} approaches χ and the adsorption of polyelectrolytes tends towards that of uncharged polymers.

Curve 2 describes the case where the charges on the polyelectrolyte and the surface have the same sign and where $\chi^s > \chi^s_{crit}$. Not only is χ^{eff} now affected by the salt concentration, but also the *effective* adsorption energy parameter $\chi^{s,eff}$ which includes the electrostatic interaction between segments and surface. The chemical adsorption energy of a segment is $-\chi^s kT$ (see [5.4.1]), the electrostatic energy of an adsorbed segment is $q_m \psi^i$, where ψ^i is the Stern potential (see sec. 3.6c). Hence, $\chi^{s,eff}$ could be written as $\chi^s - q_m \psi^i/kT$. When $q_m \psi^i$ is positive, as for curve 2, $\chi^{s,eff}$ is smaller than χ^s. At low c_s the repulsion between segments and surface is so strong that $\chi^{s,eff} < \chi^s_{crit}$, resulting in depletion ($\theta^a = 0$ but $\theta^{ex} < 0$, as indicated by the dotted part of the curve). With increasing c_s this depletion is changed to adsorption: $\chi^{s,eff} \approx \chi^s_{crit}$ at the intersection of curve 2 with the abscissa axis. For still higher c_s the adsorption increases in a similar way as for an uncharged surface (curve 1), approaching nearly the same value because the electrostatic contribution to $\chi^{s,eff}$ vanishes.

Curve 3 is typical for *purely electrostatic adsorption*: opposite charges for polyelectrolyte and surface and no chemical affinity between the two ($\chi^s \approx 0$). The polyvalent ions assume the role of the counterions and are at low c_s kept in the double layer by electrostatic forces. With increasing concentration of added salt the number of small counterions increases and the polyvalent ions desorb by an ion exchange process. At very high ionic strength θ^{ex} becomes even negative: then depletion takes place, as for uncharged polymers with $\chi^s < \chi^s_{crit}$.

If σ^o/q_m is negative and χ^s positive, there is a non-zero adsorption at high ionic strengths, where the electrostatic contributions are screened. In this case the electrostatic contribution to $\chi^{\text{s,eff}}$ is positive (in contrast to curve 2) and that of χ^{eff} is negative (as always). The effect of increasing the salt concentration is a gradual transition from mainly electrostatic to mainly chemical attraction. If $-\sigma^{\circ}/q_m$ is small, salt promotes the adsorption (curve 4), if $-\sigma^{\circ}/q_m$ is high (low α) it opposes it. With increasing salt content $\chi^{\text{s,eff}}$ decreases and χ^{eff} increases. The former effect tends to lower the adsorption, the latter to increase it. Depending on the balance between these two effects, we might find an *increase* or a *decrease* of θ^{ex} with c_s. An increase is found for highly charged polyelectrolytes, where the mutual repulsion, as embodied in χ^{eff}, dominates; see curve 4. A decrease of θ^{ex} occurs for low α, where at low c_s the electrostatic component of $\chi^{\text{s,eff}}$ is the most important, as in curve 5.

Finally, we consider the dashed curves in fig. 5.32. The curve labelled 4" represents a common situation where the solubility of the screened polyelectrolyte is poor ($\chi > 0.5$). At low c_s the electrostatic contribution makes χ^{eff} small and the polymer soluble, but as this contribution is screened by salt the polyelectrolyte might precipitate at very high ionic strengths. Before this precipitation occurs, the adsorption can become very high since the surface acts as a nucleus for phase separation (through χ^s). Clearly, this effect is not restricted to curve 4; it is indicated in fig. 5.32 only once in order not to overcrowd the figure.

The dashed curves 2' and 4' are typical for specifically adsorbing (counter)ions which may displace the polyelectrolyte if their concentration is sufficiently high. As indicated in fig. 5.32 the adsorbed amount may now pass through a maximum as a function of the ionic strength. Although this displacement at high c_s is shown only for the curves 2' and 4' it may, obviously, also occur in other situations.

In the following subsections, we shall discuss in a more quantitative way a few of the situations referred to above and, where possible, present some experimental evidence.

5.8b Polyelectrolytes on uncharged surfaces

For polyelectrolytes on uncharged surfaces there is no electrostatic contribution to the adsorption energy. The only electrostatic effect is the mutual repulsion between the segments, leading to a χ^{eff} which is lower than χ. Since salt screens the electrostatic interactions, χ^{eff} and θ^{ex} increase with increasing c_s, as illustrated qualitatively in curve 1 of fig. 5.32.

A quantitative example, calculated using the multilayer Stern model[1], is presented in fig. 5.33. The lower curve gives the dependence of θ^{ex} on $\sqrt{c_s}$ for an uncharged surface ($\sigma^{\circ} = 0$); the two curves for non-zero values of σ° will be

---

[1] M.R. Böhmer, O.A. Evers, and J.M.H.M. Scheutjens, *Macromolecules* **23** (1990) 2288.

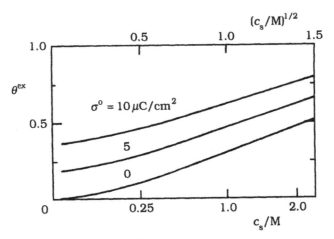

Figure 5.33. Adsorbed amount θ^{ex} of a strong negative polyelectrolyte as a function of the salt concentration c_s on an uncharged surface ($\sigma^o = 0$) and on a positively charged surface at two values of the surface charge density σ^o (indicated) [1]. Note that the abscissa scale is not linear in c_s but in $\sqrt{c_s}$. Parameters: $N = 500$, $\varphi^b = 10^{-4}$, $\chi^s = 1$, $\chi = 0.5$, $\ell = 0.6$ nm, hexagonal lattice. Ions have the same non-electrostatic interactions as the solvent. The dielectric permittivity was taken as $20\varepsilon_o$ for the polymer and $80\varepsilon_o$ for the other components.

discussed in sec. 5.8d. On an uncharged surface, the amount adsorbed at low c_s is very low despite the chemical adsorption energy ($\chi^s = 1$) of the segments. Only at higher ionic strengths does the decreased segment repulsion allow more adsorption. For $c_s \geq 0.5$ M θ^{ex} is approximately linear in $\sqrt{c_s}$. Even at very high salt concentrations (several moles/l) there is still an appreciable salt effect. Under those conditions, double layers around colloidal particles are considered to be fully compressed. However, for interactions between two such particles only long-range interactions play a role, whereas for the present case interactions between segments in close proximity are relevant; these are still affected by salt, even if the Debye length κ^{-1} is well below 1 nm. As a result, highly charged polyelectrolytes at (very) high salt content do not quite reach the limiting behaviour of the corresponding uncharged polymer although, clearly, the trends converge.

It is illustrative to consider the volume fraction profiles in the adsorbed layer at various ionic strengths. Figure 5.34 gives theoretical profiles $\varphi(z)$ (on a semi-logarithmic scale) of an uncharged polymer (top curve) and of a strong polyelectrolyte at four salt concentrations c_s [1,2]. For the neutral polymer there

[1] G.J. Fleer, in *Food Polymers, Gels and Colloids* (Proceedings International Symposium Royal Society of Chemistry, Norwich, U.K., May 1990), E. Dickinson, Ed., The Royal Society of Chemistry (1991) 34.
[2] J. Marra, H.A. van der Schee, G.J. Fleer, and J. Lyklema, in *Adsorption from Solution*, R.H. Ottewill, C.H. Rochester, and A.L. Smith, Eds., Academic Press (1983) 245.

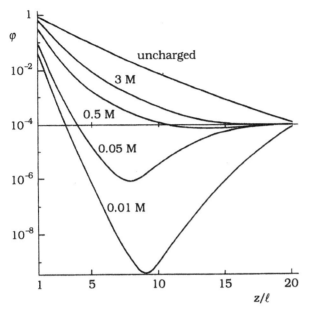

Figure 5.34. Semi-logarithmic volume fraction profiles of a strong polyelectrolyte adsorbing on an uncharged surface, at various salt concentrations (indicated). For comparison, the profile of an uncharged polymer is also given. Parameters: $N = 2000$, $\varphi^b = 10^{-4}$, $\chi^s = 2$, $\chi = 0.5$, $\ell = 0.71$ nm, hexagonal lattice, $\varepsilon = 80$ for all components. Redrawn from ref.[1].

is a gradual (more or less exponential) decay towards φ^b, analogous to fig. 5.20. For the polyelectrolyte $\varphi(z)$ is much lower everywhere, and for low ionic strength the profile shows a *minimum* at some distance from the surface. This minimum originates from the high and long-range electric potential generated by the adsorbed molecules; these repel other (non-adsorbed) chains which are, consequently, depleted. The adsorbed chains lie essentially flat, with only very short loops and hardly any tails; the adsorbate finds itself almost exclusively in the first two layers. Under these conditions, the chain-length dependence is very weak. With increasing c_s the adsorbed amount (and the chain-length dependence) increases and the minimum in the profiles becomes shallower. On the logarithmic scale for φ the minima are rather pronounced. However, the absolute concentrations are very low, which makes a direct experimental verification of the small negative adsorption (on top of the positive adsorption) nearly impossible.

The experimental counterpart of fig. 5.33 ($\sigma^o = 0$) is shown in fig. 5.35, for poly(styrene sulfonate) (PSS) on essentially uncharged surfaces[1,2]. The curves have the same shape as that for $\sigma^o = 0$ in fig. 5.33, showing a weak initial increase followed by a linear section above $c_s \approx 0.5$ M. The slope of this part

[1] J. Marra, H.A. van der Schee, G.J. Fleer, and J. Lyklema, in *Adsorption from Solution*, R.H. Ottewill, C.H. Rochester, and A.L. Smith, Eds., Academic Press (1983) 245.
[2] J. Papenhuijzen, G.J. Fleer, and B.H. Bijsterbosch, *J. Colloid Interface Sci.* **104** (1985) 530.

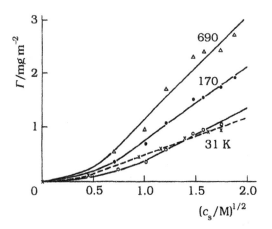

Figure 5.35. Adsorbed amount of poly(styrene sulfonate) of three different molar masses as a function of the square root of the NaCl concentration. The adsorbent is crystalline polyoxymethylene[1] (solid curves) or silica at pH = 2[2] (dashed curve).

increases with molecular weight, as expected for uncharged polymers in poor olvents. At high ionic strengths polyelectrolyte adsorption tends towards that of uncharged polymers, which form loops and tails. Consequently, the chain-length dependence, which is weak at low c_s, becomes stronger with increasing salt concentration. The adsorption is nearly equal for two different kinds of substrate (silica and polyoxymethylene), as shown by the two curves for the lowest molecular weight (31 K). One might infer that χ^s is about the same for both substrates. However, even for different χ^s such a behaviour is not unexpected since at sufficiently high χ^s the adsorbed amount in concentrated salt solutions is high: the precise value of χ^s is then not important. At low c_s the electrostatic effects (χ^{eff}) dominate the adsorption characteristics.

5.8c Polyelectrolytes on surfaces with the same charge sign

As pointed out in sec. 5.8a, there is for positive σ^o/q_m a negative electrostatic contribution to $\chi^{s,eff}$, which at low c_s may be high enough to prevent adsorption: see curve 2 in fig. 5.32. If, at given c_s, the ratio σ^o/q_m is increased from negative values to zero in the point of zero charge (p.z.c.) and then to positive values, one expects to find adsorption of the polyelectrolyte at a certain critical surface charge density σ^o_{crit}. A theoretical example is given in fig. 5.36[1]. This figure shows the adsorbed amount of a weak polyacid with $\chi^s = 1$ as a function of σ^o for three values of the pH. At pH = 2 the polymer is virtually uncharged and θ^{ex} is (nearly) independent of the surface charge. At higher pH, the polymer is

[1] J. Papenhuijzen, G.J. Fleer, and B.H. Bijsterbosch, *J. Colloid Interface Sci.* **104** (1985) 530.
[2] J. Marra, H.A. van der Schee, G.J. Fleer, and J. Lyklema, in *Adsorption from Solution*, R.H. Ottewill, C.H. Rochester, and A.L. Smith, Eds., Academic Press (1983) 245.

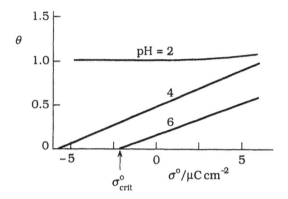

Figure 5.36. Adsorption of a weak polyacid as a function of the surface charge density, for three different values of the pH[1]. Parameters: $pK = 4$, $c_s = 0.1$ M, $\ell = 1$ nm; other parameters as in fig. 5.33.

charged and θ^{ex} increases linearly with σ^o (i.e., with increasing $\chi^{s,eff}$). The polymer and the surface are both negative if $\sigma^o < 0$. Yet, some adsorption does occur as long as $\sigma^o > \sigma^o_{crit}$. At pH = 4, $\chi^{s,eff}$ at given negative σ^o is higher than at pH = 6, because of the weaker repulsion between the segments and the surface. Consequently, the critical value σ^o_{crit} is more negative at pH = 4: a weakly charged polymer adsorbs more easily on a negative surface than a highly charged one.

There are only a few experimental studies where polyelectrolytes are adsorbed on surfaces with the same charge sign. One example is given in fig. 5.37, applying to poly(methacrylic acid) (PMA) on AgI[2]. In this system the polymer charge and the surface charge can be varied independently by changing the pH (affecting only the polymer charge) and the pAg (determining the surface charge), respectively. Figure 5.37 is analogous to fig. 5.36: once adsorption occurs Γ is linear in ψ^o (which around the p.z.c. is more or less proportional to σ^o, according to fig. 3.28). Some adsorption is still found when polymer and surface are both negative, indicating specific adsorption ($\chi^s > \chi^s_{crit}$).

From the transition point, one could try to estimate χ^s from the condition $\chi^{s,eff} = \chi^s_{crit}$ in this point. To that end, an explicit relation between χ^s, $\chi^{s,eff}$, α, and σ^o is needed. Such an attempt was made by Fleer et al.[3]. They concluded that in this system χ^s is only slightly above χ^s_{crit}. This points to a weak chemical affinity and agrees with the observation that on an uncharged surface (pAg = pAg°) the adsorbed amount is well below one monolayer, despite the fact that the polymer is only weakly charged.

[1] O.A. Evers, G.J. Fleer, J.M.H.M. Scheutjens, and J. Lyklema, *J. Colloid Interface Sci.* **111** (1986) 446.
[2] M.A. Cohen Stuart, *J. Phys. France* **49** (1988) 1001.
[3] G.J. Fleer et al., loc. cit.

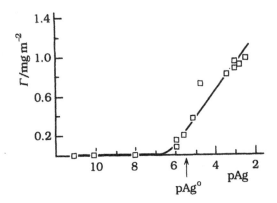

Figure 5.37. Adsorption of poly(methacrylic acid) from 0.0165 M KNO$_3$ on AgI, at a constant pH of 3, as a function of pAg (= $-$log [Ag$^+$]). The surface is negatively charged for pAg > pAg^o and positive for pAg > pAg^o. Redrawn from ref.[1].

Also for other systems similar trends were observed. For example, Cosgrove et al.[2] found that negative poly(styrene sulfonate) does not adsorb on negatively charged PS latex without added salt, whereas adsorption does occur when salt is added.

5.8d Polyelectrolytes on surfaces with opposite charge signs

In the examples discussed in subsections 5.8a and b we considered situations where the driving force for polyelectrolyte adsorption is a positive chemical χ^s, and where the electrostatics counteract this adsorption. Now we turn our attention to cases of electrostatic attraction between segments and surface ($\chi^{s,\text{eff}}$ > χ^s). We start with purely electrostatic adsorption (χ^s = 0), and then deal briefly with a non-zero χ^s.

(i) *Purely electrostatic adsorption.* In sec. 5.8a we defined purely electrostatic adsorption as an adsorption where there is only one driving force, which is of an electrostatic nature: $\chi^s \leq 0$ but $\chi^{s,\text{eff}} > 0$. Curve 3 in fig. 5.32 is representative for this situation. Again there is a balance of forces: the electrostatics drive the adsorption process (through the electrostatic part of $\chi^{s,\text{eff}}$) but at the same time limit it because χ^{eff} is low or negative due to the mutual repulsion between charges on the polyelectrolyte chains. Salt weakens both the driving and opposing forces, but not to the same extent. One can regard the adsorption as ion exchange: the polyelectrolyte (macro-ion) replaces the small counterions in the electrical double layer (compare sec. 3.5d, where the structure of electric double layers in electrolyte mixtures is discussed). Upon liberation of the small counterions the entropy in the system is strongly increased, in particular when the salt concentration is low. Hence, for highly charged macroions we expect

---

[1] M.A. Cohen Stuart, *J. Phys. France* **49** (1988) 1001.
[2] T. Cosgrove, T.M. Obey, and B. Vincent, *J. Colloid Interface Sci.* **111** (1986) 409.

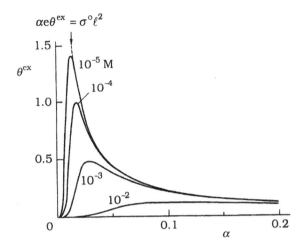

Figure 5.38. The adsorbed amount θ^{ex} of a polyelectrolyte on a surface of opposite charge sign, as a function of the fractional charge α per segment, for various salt concentrations (indicated). The dashed curve represents the charge compensation limit $\alpha\theta^{ex}$ = constant. Parameters: N = 100, χ^s = 0, χ = 0.5, φ^b = 10^{-3}, σ^o = 1 $\mu C/cm^2$, ℓ = 0.6 nm, hexagonal lattice. Redrawn from ref.[1].

charge compensation: $\theta^{ex} \approx - \sigma^o\ell^2/q_m$. For weakly charged polyelectrolytes this ion exchange is not complete; when $\alpha = 0$ it obviously does not occur at all. Hence, the adsorbed amount as a function of the polymer charge displays a maximum.

Figure 5.38, taken from ref.[1], shows this dependence of the adsorbed amount on the polyelectrolyte charge for four different salt concentrations. For high α and not too high c_s all curves approach the charge compensation limit $\alpha\theta^{ex}$ = $\sigma^o\ell^2/e$ = constant, which is the equation for a hyperbola. This hyperbola is indicated in fig. 5.38. At c_s = 10^{-5} M the polyion compensates the surface charge down to $\alpha \approx 0.015$ at this polyelectrolyte concentration (φ^b= 10^{-3}), corresponding to 10^{-5} M polyelectrolyte. The total valency $N\alpha$ of an entire polyion chain is then 1.5, which is only 50% more than of the small counterions. At still lower α the small counterions take over and the polyelectrolyte desorbs. Hence, the $\theta^{ex}(\alpha)$ curves show a maximum, as predicted earlier by Hesselink[2]. The effect of a higher salt concentration is to increase the competition of counterions, leading to a lower maximum which is situated at higher α. For several systems with purely electrostatic adsorption experimental data are now available. They all show the expected maximum as a function of α [3,4,5].

(ii) *Electrosorption with positive χ^s.* When, in addition to electrostatic long-range attraction between polyelectrolyte and surface, there is a chemical or

[1] H.G.M. van de Steeg, M.A. Cohen Stuart, A. de Keizer, and B.H. Bijsterbosch, *Langmuir* **8** (1992) 2538.
[2] F.Th. Hesselink, *J. Colloid Interface Sci.* **60** (1977) 448.
[3] T.K. Wang and R. Audebert, *J. Colloid Interface Sci.* **121** (1988) 32.
[4] G. Durand-Piana, F. Lafuma, and R. Audebert, *J. Colloid Interface Sci.* **119** (1987) 474.
[5] S. Vaslin-Reimann, F. Lafuma, and R. Audebert, *Colloid Polym. Sci.* **268** (1990) 476.

specific short-range attraction (χ^s), these two effects become synergistic. In this case $\chi^{s,\text{eff}}$ contains two contributions (electrostatic and chemical) both promoting the adsorption, just as in the case of a Stern layer with specific adsorption. As for polyelectrolytes χ^{eff} is always smaller than χ, this solvency effect opposes accumulation near the surface. Salt screens the electrostatic contributions of both $\chi^{s,\text{eff}}$ and χ^{eff}, resulting in an *increased* adsorption for relatively highly charged polyelectrolytes (curve 4 of fig. 5.32) and in a *lower* adsorption of the polyelectrolyte charge when the charge density is small (curve 5).

An example of the former case was already shown in fig. 5.33, where two curves for a fully charged polyelectrolyte $(\alpha = 1)$ with $\chi^s = 1$ on oppositely charged surfaces are given. As compared to the adsorption on an uncharged surface, the additionally adsorbed amount due to electrostatic attraction is found to be nearly proportional to σ^0. At low c_s (high $\chi^{s,\text{eff}}$, low χ^{eff} the adsorbed amount corresponds to charge neutralization. Upon addition of salt the adsorption becomes superequivalent due to the screening of the mutual repulsion between segments. Apparently, weakening of the opposing effect (χ^{eff}) is then more important than that of the promoting effect $(\chi^{s,\text{eff}})$. At high c_s a linear increase of θ^{ex} with σ^0 is still found, but the proportionality constant is slightly smaller, reflecting that the segment/surface attraction is only weakly screened.

For weakly charged polyelectrolytes, where the electrostatics are less important, specific adsorption $(\chi^s > 0)$ leads to an increase of the adsorbed amount. A maximum in $\theta^{\text{ex}}(\alpha)$, as shown in fig. 5.38, may still be present. The effect of salt on the two sides of the maximum is quite different. Figure 5.39a illustrates this for $\chi^s = 0.6$ and various charge densities of the polyelectrolyte[1]. The trends indicated in fig. 5.32 are fully corroborated. At low α the decrease of $\chi^{s,\text{eff}}$ with increasing c_s outweighs the increase of χ^{eff} (hence θ^{ex} decreases), at high α the opposite situation is found (as in fig. 5.33). In concentrated salt solutions the adsorbed amount converges to that of the uncharged polymer, irrespective of the charge density on the polyelectrolyte.

An experimental analogue of fig. 5.39a is shown in fig. 5.39b[2]. The polymer is a cationic polyacrylamide and the adsorbent is a montmorillonite, onto which the uncharged polymer also adsorbs, as can be deduced from the curve for $\alpha = 0$. The qualitative agreement between experiment and theory is excellent: for a weakly charged polymer $(\alpha = 0.01)$ Γ decreases with c_s, for higher polyelectrolyte charges Γ is constant over a wide range of c_s or it increases slightly. Note that a polymer with high α (requiring only a low adsorption for charge compensation) adsorbs less than the uncharged polymer, whereas for low α (needing

[1] H.G.M. van de Steeg, M.A. Cohen Stuart, A. de Keizer, and B.H. Bijsterbosch, *Langmuir* **8** (1992) 2538.
[2] G. Durand, F. Lafuma, and R. Audebert, *Progr. Colloid Polym. Sci.* **266** (1988) 278.

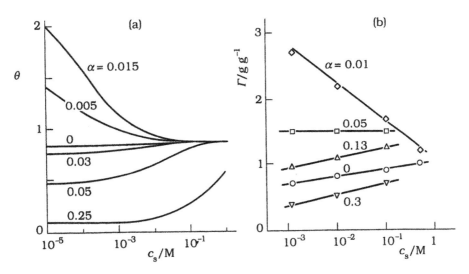

Figure 5.39. Theoretical (a) and experimental (b) amount adsorbed as a function of the salt concentration for polyelectrolytes of varying charge density α, as indicated. The theoretical figure[1] was calculated for χ^s = 0.6 and a surface charge density σ^0 of 1 $\mu C/cm^2$; other parameters as in fig. 5.38. The experimental data are for a cationic polyacrylamide adsorbing on montmorillonite particles[2].

more polymer to compensate the surface charge) the adsorption is higher. These features show up both in the theoretical model and in the experiment. Several more examples where the adsorbed amount decreases with increasing ionic strength are available in the literature[3,4,5,6]. In all these cases there is a different charge sign of polymer and surface, and electrosorption effects as discussed above (i.e., ion exchange and a decreasing $\chi^{s,\text{eff}}$) are likely to play a role.

Figure 5.40 shows an experimental corroboration of the importance of charge compensation. The figure gives the ratio ϱ_{ads} between the adsorbed polyelectrolyte charge and the surface charge (defined by $\varrho_{\text{ads}} = -q_m \theta^{\text{ex}}/\sigma^0 \ell^2$) as a function of ϱ_{add}, the added polyelectrolyte charge per surface charge, for cationic polyacrylamides of different molecular weights (a) and different polyelectrolyte charge densities (b) on an anionic PS latex. If the adsorption would be solely driven by charge compensation, one would expect ϱ_{ads} in salt-free solutions to be unity, as indicated by the dashed line. The experimental ratio at high polymer doses is 20-30% higher and it is nearly independent of chain length and polymer

[1] H.G.M. van de Steeg, M.A. Cohen Stuart, A. de Keizer, and B.H. Bijsterbosch, *Langmuir* **8** (1992) 2538.

[2] G. Durand, F. Lafuma, and R. Audebert, *Progr. Colloid Polym. Sci.* **266** (1988) 278.

[3] R.E. Frockling, A. Bantjes, and Z. Kolar, *J. Colloid Interface Sci.* **62** (1977) 35.

[4] R.J. Davis, L.R. Dix, and C. Toprakcioglu, *J. Colloid Interface Sci.* **129** (1989) 145.

[5] E.R. Hendrickson and R.D. Neuman, *J. Colloid Interface Sci.* **110** (1986) 243.

[6] R.H. Pelton, *J. Colloid Interface Sci.* **111** (1986) 475.

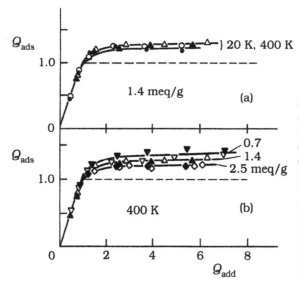

Figure 5.40. Charge due to adsorbed polyelectrolyte per unit of surface charge (θ_{ads}) as a function of the total (added) amount of polyelectrolyte charge per unit surface charge (θ_{add}), for cationic polyacrylamides of different molecular weights (expressed in K = kg/mole) (a) and of different polymer charge densities (b) adsorbed from salt-free solutions on anionic polystyrene latex. Redrawn from ref.[1].

cationicity. This points to a chemical contribution (χ^s) to the adsorption process, which leads to some superequivalent adsorption $(\theta_{ads} > 1)$.

5.9 Applications

The adsorption of polymers has numerous technological applications. In the majority of these applications, the adsorption of the polymer occurs on the surface of colloidal particles, and the purpose is to modify the properties of colloidal dispersions.

One important role of polymers is to form a protective sheath around colloidal particles to stabilize them against aggregation. Perhaps the oldest known application is the use of water-soluble gums in the stabilization of colloidal soot particles for the preparation of ink, already mentioned in sec. I.1.1. More recent examples are the stabilization of pigments in paints, of magnetic particles in electronic applications, of silver bromide particles in photographic "emulsions", of oxide particles in the "green" (wet) phase which is the precursor of a ceramic product or an inorganic coating, and of clays in drilling fluids and dispersions for borehole treatment as employed in the oil industry. In many industrial applications polymers and polyelectrolytes are used to control the rheology of concentrated dispersions. Both adsorption and depletion play a role in this respect.

[1] H. Tanaka, L. Ödberg, L. Wågberg, and T. Lindström, *J. Colloid Interface Sci.* **134** (1990) 219.

The purpose of adding a polymer is not always to stabilize colloidal particles; in other cases one aims at imparting an effective attraction which will bring the particles together in a floc or a network. The interparticle attraction is then caused by the formation of so-called polymer bridges, strands of polymer that are adsorbed simultaneously onto two or more particles, in this way bridging the gap between them. In these applications kinetic aspects, which have not received much attention in the present chapter, may be crucial. We mention a number of examples of this type of destabilization.

An interesting illustration is found in the production of aluminium from the mineral bauxite. After grinding the ore, the Al_2O_3 is dissolved by alkali to give aluminate. The insoluble ore particles left behind are flocculated by polymer and thus separated. Then the aluminate is acidified again, and the colloidal Al_2O_3 is concentrated by flocculants and further used for the production of aluminium.

Waste water, containing a variety of colloids, is often treated by polymeric flocculants such as polyacrylamides to precipitate the colloidal material and make it better filterable. Quite often these flocculants are high-molecular-weight weakly charged polyelectrolytes. Bacterial suspensions resulting from biological waste water treatment must also be flocculated before they can be properly separated from the purified effluent by sedimentation or filtration.

Other applications are found in powder technology. The cohesion in pellets formed out of dry powders can be improved by treating the powder with an appropriate polymer solution. This process is used on an enormous scale to prepare iron ore in a form suitable for blast furnaces, but also for the preparation of pharmaceutical specialty products. Polymers are applied at a large scale in paper-making where they help to strengthen the network of cellulosic fibres and to trap different kinds of particles in this network. The building of a network is also the purpose of adding carbon black to rubber, which improves its resilience and abrasive resistance.

An interesting application of polymer adsorption is the control of crystal growth. Polymers are added to ice cream to help reducing the size of the ice crystals, thus improving the smoothness of this delicacy. Crystal growth is also undesirable in diesel fuel stored at low temperatures, since otherwise pumps and pipes may get plugged. Again polymers, now called flow-point depressants, do the job by keeping the crystals small.

Adhesives form an important class of materials entirely consisting of poly-mers. In this application, both the bulk mechanical properties of the polymers and their interfacial behaviour determine the performance. Strong adsorption is required for proper adhesive joints. Small amounts of low-molecular-weight impurities may adversely affect the segment adsorption energy. A fundamental understanding of such competitive adsorption phenomena is mandatory for improving these products.

Another important category of applications is surface modification. For medical applications, properties of surfaces that may come in contact with body fluids or with living cells are very important. Quite often, undesired adsorption of proteins occurs on synthetic materials. Coating the surface of the materials with polymers, in particular with poly(ethylene oxide), often solves the problem. Wettability is the issue for many separation membranes, often produced from rather hydrophobic materials. Adsorbed polymers are used when good wetting by water is wanted. A last example is lubrication, where adsorbing polymers are very effective in reducing wear of hard surfaces exposed to strong shear.

Although we have restricted our discussion of adsorption to well-defined homopolymers (both charged and uncharged), most polymers used in technology are less well defined. In many applications commercial copolymers are used, which are not only heterogeneous in molar mass (polydisperse), but also heterogeneous in their chemical composition. Getting control over the copolymer composition at reasonable production price is an important challenge for the chemical industry.

For these copolymers, a distinction has to be made between *random* and *block* copolymers. These two classes may differ considerably in their interfacial behaviour. Most of the trends discussed for homopolymer adsorption apply also to random copolymers, so that the insights presented in this chapter can still serve as a guide when dealing with practical systems in which, so far, mostly random copolymers are used. Block copolymers are potentially very interesting for many specific applications but are not yet applied on a large scale. We return to these more sophisticated systems in Volume V.

Not only for application purposes, but also from a more fundamental point of view adsorption of polymers and polyelectrolytes is very important. Therefore, these issues will come back several times in the forthcoming Volumes.

5.10 General references

5.10a. Polymers and polyelectrolytes in solution

This field is well developed. Numerous textbooks in this area have appeared, but there are a few classical texts which cover most of the field.

P.J. Flory, *Principles of Polymer Chemistry*, Cornell University Press, Ithaca NY (1953). (The "bible" for polymer chemistry and physics, including polymer statistics and polymers in solution. Old, but still valuable.)

C. Tanford, *Physical Chemistry of Macromolecules*, Wiley, New York (1961). (Classical, like Flory's book, but with more emphasis on proteins.)

P.J. Flory, *Statistical Mechanics of Chain Molecules*, Interscience, NY (1969). (A more refined and rather mathematical treatment of chain statistics, starting from an atomic level.)

H. Yamakawa, *Modern Theory of Polymer Solutions*, Harper & Row, NY (1971). (Thorough statistical thermodynamic treatment of polymer solution properties.)

P.G. de Gennes, *Scaling Concepts in Polymer Physics*, Cornell Univ. Press, Ithaca, NY (1979). (Scaling description for polymers in good solvents, emphasizing the "blob" structure as compared to homogeneous "mean-field" solutions.)

5.10b. *Polymers and polyelectrolytes at interfaces*
This area is much newer and still developing. Only very few textbooks are available.

Yu. S. Lipatov and L.M. Sergeeva, *Adsorption of Polymers*, Wiley, New York (1974). (A rather uncritical compilation of older data.)

P.G. de Gennes, *Scaling Concepts in Polymer Physics*, Cornell University Press, Ithaca, NY (1979). (This text, mentioned above, also pays some attention to polymers adsorbing from good solvents.)

D.H. Napper, *Polymeric Stabilissation of Colloidal Dispersions*, Acad. Press (1983). (Gives much attention to steric repulsion between polymer layers, but adsorption as such is not considered in detail.

G.J. Fleer, M.A. Cohen Stuart, J.M.H.M. Scheutjens, T. Cosgrove and B. Vincent, *Polymers at Interfaces*, Chapman & Hall (1993). (The only comprehensive text so far on polymer adsorption, with emphasis on both theoretical developments and comparison with experiment. Chapter 5 of this Volume is a condensed version of the homopolymer and homopolyelectrolyte sections of this book.)

Apart from these textbooks, only reviews are available. The most relevant ones are (in chronological order):

E. Dickinson and M. Lal, *Adv. Mol. Relax. Interact. Proc.* **17** (1980) 1.

A. Takahashi and M. Kawaguchi, *Adv. Polym. Sci.* **46** (1982) 3.

G.J. Fleer and J. Lyklema, in *Adsorption from Solution at the Solid/Liquid Interface*, G.D. Parfitt and C.H. Rochester, Eds., Academic Press London (1983) 153.

A. Silberberg, in *Encyclopedia of Polymer Science and Engineering* Vol I, Wiley, New York (1985) 577.

M.A. Cohen Stuart, T. Cosgrove, and B. Vincent, *Adv. Colloid Interface Sci.* **24** (1986) 143.

G.J. Fleer, in *Reagents in Mineral Technology* (Surfactant Science Series Vol. **27**), P. Somasundaran and B.R. Moudgil, Eds., Marcel Dekker, New York (1987) 105.

I.D. Robb, *Comprehensive Polymer Sci.* **2** (1989) 733.

H. J. Ploehn and W.B. Russel, *Adv. Chem. Eng.* **15** (1990) 137.

M.A. Cohen Stuart, G.J. Fleer, J. Lyklema, W. Norde, and J.M.H.M. Scheutjens, *Adv. Colloid Interface Sci.* **34** (1991) 477.

M. Kawaguchi and A. Takahashi, *Adv. Colloid Interface Sci.* **37** (1992) 219.

G.J. Fleer and J.M.H.M. Scheutjens, in *Coagulation and Flocculation: Theory and Applications*, B. Dobias, Ed., *Surfactant Science Series* Vol **47**, Marcel Dekker (1993) 209.

APPENDIX 1

Survey of adsorption isotherms and two-dimensional equations of state for homogeneous, nonporous surfaces. Monolayer regime.

For (sub-) monolayer coverage θ = adsorbed fraction, w = pair interaction parameter, z coordination number, a_m = molecular cross section or site area, a^σ = surface Van der Waals constant, $\beta = \left[1 - 4\theta(1-\theta)\{1-\exp(-w/kT)\}\right]^{1/2}$.

The constants K depend on the nature and units into which the equilibrium concentration (mol fraction, relative pressure, concentration, ...) is expressed. The equations below are formulated for gas adsorption.

Linear initial part of most isotherms (Henry)

$$N/A = K_H p \tag{A1.1a}$$

$$\pi A = kT N \tag{A1.1b}$$

(ideal two-dimensional gas)

Localized without lateral interaction (Langmuir)

$$\frac{\theta}{1-\theta} = K_L p \tag{A1.2a}$$

$$\pi a_m = -kT \ln(1-\theta) \tag{A1.2b}$$

Mobile without lateral interaction (Volmer)

$$\frac{\theta}{1-\theta} e^{\theta/(1-\theta)} = K_V p \tag{A1.3a}$$

$$\pi a_m = kT \theta/(1-\theta) \tag{A1.3b}$$

Localized with lateral interaction

virial expansion

$$\theta = Kp + \left[z(e^{-w/kT}-1)-1\right](Kp)^2 + ... \tag{A1.4a}$$

$$\pi a_m = kT\theta + \underbrace{\tfrac{1}{2}kT\left[1 - z(e^{-w/kT}-1)\right]}_{kT B_2^\sigma(T)/a_m} \theta^2 + ... \tag{A1.4b}$$

Bragg-Williams approximation *(Frumkin-Fowler-Guggenheim, FFG)*

$$\frac{\theta}{1-\theta} e^{zw\theta/kT} = K_L p \tag{A1.5a}$$

A1.2

$$\pi a_m = -kT \ln(1-\theta) + \tfrac{1}{2} z w \theta^2 \qquad\qquad\qquad [\text{A1.5b}]$$

Quasi-chemical approximation

$$\frac{\theta}{1-\theta}\left[\frac{(\beta-1+2\theta)(1-\theta)}{(\beta+1-2\theta)\theta}\right]^{z/2} e^{w/kT} = K_L p \qquad [\text{A1.6a}]$$

$$\pi a_m = -kT \ln(1-\theta) - \tfrac{1}{2} z k T \ln\left[\frac{\beta+1-2\theta}{(\beta+1)(1-\theta)}\right] \qquad [\text{A1.6b}]$$

Mobile with lateral interaction

$$\frac{\theta}{1-\theta} e^{\theta/(1-\theta)} e^{-2a^\sigma \theta / a_m k T} = K_V p \qquad\qquad [\text{A1.7a}]$$

$$\left(\pi + \frac{a^\sigma N^2}{A^2}\right)\left(A - N a_m\right) = N k T \qquad\qquad [\text{A1.7b}]$$

(Hill-de Boer or two-dimensional van der Waals equation), also written as

$$\pi a_m = kT \frac{\theta}{1-\theta} - \frac{a^\sigma}{a_m} \theta^2 \qquad\qquad\qquad [\text{A1.7c}]$$

APPENDIX 2

Hyperbolic functions

Hyperbolic functions are combinations of positive and negative exponentials. They resemble goniometric functions and derive their names from the fact that they describe the coordinates of points on rectangular hyperbolas. They are often encountered in diffuse double layer theory .

The definitions and some important properties are summarized below.

a. Defining equations

$$\sinh x = \tfrac{1}{2}(e^x - e^{-x})$$ [A2.1]

$$\cosh x = \tfrac{1}{2}(e^x + e^{-x})$$ [A2.2]

$$\tanh x = \frac{\sinh x}{\cosh x} = \frac{e^x - e^{-x}}{e^x + e^{-x}}$$ [A2.3]

$$\coth x = \frac{\cosh x}{\sinh x} = \frac{e^x + e^{-x}}{e^x - e^{-x}}$$ [A2.4]

$$\operatorname{cosech} x = \frac{1}{\sinh x} = \frac{2}{e^x - e^{-x}}$$ [A2.5]

$$\operatorname{sech} x = \frac{1}{\cosh x} = \frac{2}{e^x + e^{-x}}$$ [A2.6]

The cosh and sech are even functions (cosh x = cosh $(-x)$, etc.) and always positive; all others are uneven (sinh x = $-$sinh $(-x)$, etc.), and may be positive or negative. Trends are sketched in fig. A2.1.

b. Series expansions

$$\sinh x = x + \frac{x^3}{3!} + \frac{x^5}{5!} + \dots$$ [A2.7]

$$\cosh x = 1 + \frac{x^2}{2!} + \frac{x^4}{4!} + \dots$$ [A2.8]

$$\tanh x = x - \frac{x^3}{3} + \frac{2x^5}{15} - \dots$$ $(x^2 < \pi^2/4)$ [A2.9]

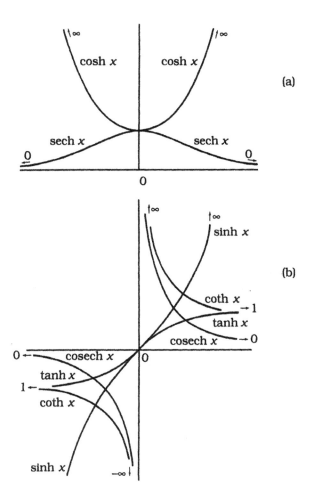

Figure A2.1. Hyperbolic functions. (a) even; (b) uneven functions. Limits are indicated.

c. Inverse functions

Note: $\sinh^{-1} x$ is also written as arc sinh x, etc.

$$\sinh^{-1} x = \ln\left(x + \sqrt{x^2 + 1}\right) = \int \frac{dx}{\sqrt{x^2 + 1}} = \cosh^{-1} \sqrt{x^2 + 1} \qquad \text{[A2.10]}$$

$$\cosh^{-1} x = \ln\left(x + \sqrt{x^2 - 1}\right) = \int \frac{dx}{\sqrt{x^2 - 1}} = \sinh^{-1} \sqrt{x^2 - 1} \qquad \text{[A2.11]}$$

$$\tanh^{-1} x = \tfrac{1}{2}\ln\left(\frac{1 + x}{1 - x}\right) = \int \frac{dx}{1 - x^2} \qquad (|x| < 1) \qquad \text{[A2.12]}$$

$$\coth^{-1} x = \tfrac{1}{2}\ln\left(\frac{1+x}{x-1}\right) = \int \frac{dx}{1-x^2} \qquad (x>1 \text{ or } x<1) \qquad [\text{A2.13}]$$

$$\operatorname{sech}^{-1} x = \ln\left(\frac{1}{x} + \sqrt{\frac{1}{x^2}-1}\right) = -\int \frac{dx}{x\sqrt{1-x^2}} \qquad\qquad [\text{A2.14}]$$

$$\operatorname{cosech}^{-1} x = \ln\left(\frac{1}{x} + \sqrt{\frac{1}{x^2}+1}\right) = -\int \frac{dx}{x\sqrt{x^2+1}} \qquad\qquad [\text{A2.15}]$$

$$\sinh^{-1} x = x - \frac{1}{2}\frac{x^3}{3} + \frac{1.3}{2.4}\frac{x^5}{5} - \frac{1.3.5}{2.4.6}\frac{x^7}{7} + \ldots \qquad (x^2<1) \qquad [\text{A2.16}]$$

$$\sinh^{-1} x = \ln 2x + \frac{1}{2}\frac{1}{2x^2} - \frac{1.3}{2.4}\frac{1}{4x^4} + \frac{1.3.5}{2.4.6}\frac{1}{6x^2} - \ldots \qquad (x^2>1) \qquad [\text{A2.17}]$$

$$\cosh^{-1} x = \ln 2x - \frac{1}{2}\frac{1}{2x^2} - \frac{1.3}{2.4}\frac{1}{4x^4} - \frac{1.3.5}{2.4.6}\frac{1}{6x^6} - \ldots \qquad (x^2>1) \qquad [\text{A2.18}]$$

$$\tanh^{-1} x = x + \frac{x^3}{3} + \frac{x^5}{5} + \frac{x^7}{7} + \ldots \qquad (x^2<1) \qquad [\text{A2.19}]$$

d. Integrals

$$\int \sinh x \, dx = \cosh x \qquad\qquad [\text{A2.20}]$$

$$\int \cosh x \, dx = \sinh x \qquad\qquad [\text{A2.21}]$$

$$\int \tanh x \, dx = \ln \cosh x \qquad\qquad [\text{A2.22}]$$

$$\int \coth x \, dx = \ln \sinh x \qquad\qquad [\text{A2.23}]$$

$$\int \operatorname{sech} x \, dx = \tan^{-1}(\sinh x) \qquad\qquad [\text{A2.24}]$$

$$\int \operatorname{cosech} x \, dx = \ln(\tanh(x/2)) \qquad\qquad [\text{A2.25}]$$

e. Relations between the functions

$$\sinh x = \frac{\tanh x}{\sqrt{1-\tanh^2 x}} \qquad\qquad [\text{A2.26}]$$

$$\cosh x = \frac{1}{\sqrt{1-\tanh^2 x}} \qquad\qquad [\text{A2.27}]$$

A2.4

$$\cosh^2 x - \sinh^2 x = 1 \qquad \text{[A2.28]}$$

$$\tanh x = \pm\sqrt{1 - \operatorname{sech}^2 x} \qquad \text{[A2.29]}$$

$$\coth x = \sqrt{\operatorname{cosech}^2 x + 1} \qquad \text{[A2.30]}$$

$$\operatorname{sech} x = \sqrt{1 - \tanh^2 x} \qquad \text{[A2.31]}$$

$$\operatorname{cosech} x = \pm\sqrt{\coth^2 x - 1} \qquad \text{[A2.32]}$$

f. Relations between functions involving x/2

$$\sinh x = \frac{2\tanh(x/2)}{1 - \tanh^2(x/2)} \qquad \text{[A2.33]}$$

$$\sinh x = 2\sinh(x/2)\cosh(x/2) \qquad \text{[A2.34]}$$

$$\cosh x = \frac{1 + \tanh^2(x/2)}{1 - \tanh^2(x/2)} \qquad \text{[A2.35]}$$

$$\cosh x = \cosh^2(x/2) + \sinh^2(x/2) \qquad \text{[A2.36]}$$

$$\cosh x = 2\cosh^2(x/2) - 1 \qquad \text{[A2.37]}$$

$$\cosh x = 1 + 2\sinh^2(x/2) \qquad \text{[A2.38]}$$

$$\tanh x = \frac{2\tanh(x/2)}{1 + \tanh^2(x/2)} \qquad \text{[A2.39]}$$

$$\sinh(x/2) = \sqrt{\tfrac{1}{2}(\cosh x - 1)} \qquad \text{[A2.40]}$$

$$\cosh(x/2) = \sqrt{\tfrac{1}{2}(\cosh x + 1)} \qquad \text{[A2.41]}$$

APPENDIX 3

Pristine points of zero charge

a. Metals (Room temperature)

Metal[1]	Electrolyte	Method	p.p.z.c./V (NHE[2])	Ref.
mercury	NaF	electrocapill. curve or cap. minimum	-0.193	1
bismuth, polycryst.	KF	cap. minimum	-0.39	2
silver (111)	KF	cap. minimum	-0.46	3,4
silver (100)	NaF	cap. minimum	-0.61	3
silver (110)	NaF	cap. minimum	-0.77	3
gold (110)	NaF	cap. minimum	0.19	5
cadmium	NaF	cap. minimum	-0.75	6
lead	NaF	cap. minimum	-0.65	7
tin	K_2SO_4	cap. minimum	-0.38	8
thallium	NaF	cap. minimum	-0.71	9

b. Oxides

Most data are from acid-base titrations at various salt concentrations, leading to a common intersection point. Data obtained for only two salt concentrations are, as a rule, avoided. This is also the policy for curves where the sample conditions and/or the reversibility of the titrations are insufficiently controlled or which are suspect for other reasons. Isoelectric points or points of zero charge obtained by other methods are sometimes recorded (in *italics*) if titration data are not available, or if such points are interesting for other reasons. When the temperature is known $\Delta_{cf}G^\circ_m = -2.303\ RT(\text{pH}^\circ - \text{pOH}^\circ)/2$, see [3.8.16], is also given. It is abbreviated to ΔG°_m. Although sometimes two decimals are given, the reproducibility is often not better than 0.1 pH unit. Older reviews tend to be somewhat less critical with respect to method and/or definition of the

---

[1] Most data: values recommended by A.N. Frumkin, O.A. Petrii and B.B. Damaskin, in *Comprehensive Treatise of Electrochemistry*, (J. O'M Bockris, B.E. Conway and E. Yeager (Eds.), Vol. 1, Plenum Press (1980) 228. For other tabulations on metals see R.S. Perkins, T.N. Andersen, *Potentials of Zero Charge on Electrodes*, in *Modern Aspects of Electrochemistry*, Vol. 5, J. O'M Bockris, B.E. Conway, Eds. Plenum Press (1969) (401 references); L. Campanella, *J. Electroanal. Chem.* 28 (1970) 228.
[2] NHE = with respect to normal hydrogen electrode. Subtract 0.244 V to obtain values with respect to saturated calomel electrode (SCE).
[3] References at the end of table c.

sample[27,28]. Differences between different authors are often attributable to differences in preparation; details can be found in the references.

Abbreviations: titr. titration; el. phor. = electrophoresis; el. cap. = electric capacitance; str. pot. = streaming potential; stab. = colloid stability; comm. = commercial product; chromatogr. = chromatographic; imm. = immersion; susp. eff. = suspension effect, prec. = precipitated.

Substance	T	Electrolyte	Method	pH^o	ΔG_m^o	ref.
SiO₂						
pyrogenic		NaCl, KCl,	titr.	<1.8		24
(Cab-O-Sil)		LiCl, CsCl		(if any)		
precipitated, BDH	25	LiCl, KCl,	titr.	~ 2.8	24	59
		CsCl				
	25	KCl, 10⁻³	str. pot.	2.1	28	59
oxid. Si electr.		NaCl	el.cap.	2.2		64
Aerosil OX 50		KCl, etc.	el. phor.	2.5		68
Min-U-Sil	25	NaCl	el. phor.	2.3	26.8	69
homodisperse sol	25	KCl	el. phor.	3.4	20.5	70
"Ludox"		NaCl	titr.	≤3.5		75
TiO₂						
rutile, synth.		KNO₃	i.e.p.=p.z.c.	5.8		17
		various	stab.	5.9		18
other sample[18]		various	stab.	4.6		18
other sample		various	stab.	3.9		18
other sample		various	stab.	2.9		18
		KNO₃	p.z.c.=i.e.p.	5.9		19
	10	KNO₃	titr.=i.e.p.	5.63	8.88	20
	20	KNO₃	titr.	5.52	8.77	20
	30	KNO₃	titr.	5.41	8.75	20
	40	KNO₃	titr.	5.33	8.62	20
	50	KNO₃	titr.	5.24	8.61	20
	60	KNO₃	titr.	5.15	8.66	20
	70	KNO₃	titr.	5.07	8.71	20
	25	KNO₃	titr.	5.8	6.8	66
	25	KNO₃	titr.	5.70	7.41	26
	25	KNO₃	titr.	5.5	8.6	36
	25	NaCl	titr.	6.00	5.70	32
	50	NaCl	titr.	5.67	5.95	32
	75	NaCl	titr.	5.50	5.62	32
	95	NaCl	titr.	5.35	5.67	32
	25	KNO₃	el. phor.	5.9	6.3	37
	21	KNO₃	titr.=i.e.p.	5.9	6.6	38

Substance	T	Electrolyte	Method	pH°	ΔG_m^o	ref.
TiO₂ (cont.)						
aged, homodisperse powders	25	KCl	titr.	5.2	10.3	39
anatase	25	NaCl	titr.	5.90	6.27	32
anatase	25	NaClO₄, 3M	indirect	*6.39*		34
		NaCl	titr.	5.9		35
anatase (BDH)	20	NaNO₃	imm.	*5.96*		80
anatase (Degussa)		KCl	titr.	5.3		63
" "		LiCl	titr.	5.5		63
" "		NaCl	titr.=i.e.p.	6.2		79
" "	20	KCl, KNO₃	titr.	6.6	2.7	73
anatase (Merck)	25	NaCl	titr.	6.0	5.7	67
" "	20	KCl, KNO₃	titr.	6.6	2.7	73
MnO₂						
α-(synthetic)		NaNO₃ 10⁻³	stab., i.e.p.	*4.5±0.5*		22
β-(synthetic)		NaNO₃ 10⁻³-10⁻¹	stab., i.e.p.	*7.3±0.2*		22
δ-(synthetic)		NaNO₃ 10⁻³	stab.	*1.5±0.5*		22
δ-(synthetic)		KNO₃ 10⁻⁴	el. phor., ion ads.	*2.2*		31
δ-(synthetic)	30±2	NaCl	titr.	*2.25-3.0*		71
γ-(synthetic)		NaNO₃ 10⁻³	stab., i.e.p.	*5.5-5.6*		22
γ-(int. sample)	25	NaNO₃	titr.	*4.8*	12.6	74
Mn (II)-manganite			stab.	*1.8±0.5*		22
Fe₃O₄						
magnetite, synth.	35	KNO₃	titr.	6.30	3.19	25
" "	44	KNO₃	titr.	6.05	4.02	25
" "	53	KNO₃	titr.	5.89	4.39	25
" "	60	KNO₃	titr.	5.79	4.58	25
" "	80	KNO₃	titr.	5.50	5.36	25
" "	90	KNO₃	titr.	5.41	5.48	25
" "	30	KNO₃	titr.	6.80	0.68	42, 43
" "	50	KNO₃	titr.	6.45	1.12	43
" "	80	KNO₃	titr.	6.00	1.98	43
" "	25	KNO₃	i.e.p.	*6.9*	*0.6*	43

Substance	T	Electrolyte	Method	pH°	ΔG_m^o	ref.
α-Fe₂O₃						
(haematite, synth.*)			titr.	8.5		44
" "			titr.	8.3-8.4		45
" "	20	KCl	titr.	8.5	–8.0	46
" "				8.6-9.3		47
" "				9.0		48
" "	5	KNO₃	titr.	9.1	–9.2	20
" "	20	KNO₃	titr.	8.7	–9.1	20
" "	40	KNO₃	titr.	8.3	–9.2	20
" "	60	KNO₃	titr.	7.9	–8.9	20
(homodisperse sol)	20	KNO₃	titr. = i.e.p.	9.1	–11.3	19
" "	25	NaNO₃	titr.	7.2	–1.2	50
" "	25	KCl	titr.	9.5	–14.3	49
α–FeOOH						
(goethite, synth.)	25	KNO₃	titr.	7.5	–2.9	40
" "		KNO₃	titr.	9.3		41
" "		NaClO₄	el. phor.	9.0±0.2		55
FeOOH (others)						
β-(akaganeite, synth.)		KNO₃	titr. = i.e.p.	7.15		72
" " "	25	KNO₃	titr.	7.10		78
(amorph., hydr.)		KNO₃	titr. = i.e.p.	7.05		72
" "	25	KNO₃	titr.	7.25		78
Al₂O₃						
(chromatogr.)	30	KNO₃	titr.	9.02	–12.21	25
(chromatogr.)	44	KNO₃	titr.	8.87	–13.11	25
(chromatogr.)	60	KNO₃	titr.	8.67	–13.79	25
(chromatogr.)	80	KNO₃	titr.	8.47	–14.72	25
(chromatogr.)	90	KNO₃	titr.	8.31	–14.69	25
γ- (comm. Linde)	25	KNO₃	el. phor.	8.9	–10.9	37
γ- (comm. Linde)	23	KNO₃	imm.	8.8	–10.0	77
γ- (comm. Linde)	43	KNO₃	imm.	8.1	–8.3	77
γ- (comm. Alon)	25	NaCl	titr.	8.5	–8.6	51
γ- (comm. Alon)	25±2	NaClO₄ 10⁻³ M	titr.	9.0		52
γ- (comm. Alon)	25±2	NaClO₄ 10⁻¹ M	titr.	9.7		52

* Natural haematites tend to have much lower zero points, with considerable scatter.

Substance	T	Electrolyte	Method	pH°	ΔG_m^o	ref.
Al_2O_3 (cont.)						
γ- (comm. Alon)		NaCl	titr. = el. phor.	8.1		56
γ- (comm. Degussa)		KCl	titr.	9.1		63
(prec.)	25	NaCl	titr. = el. phor.	8.2-8.4	~7.4	53
α-(corundun)	25		imm.	8.8	−10.3	85
α-(comm. Linde A)	25±0.1	KCl, KNO_3	titr. = el. phor.	9.1	12.0	61
(comm. POCh. Poland)		NaCl	titr.	8.5		75
pseudoboehmite		$NaClO_4$ 10^{-2} M	el. phor.	9.2	−12.6	54
	25°	–	imm.	8.8	−10.3	85
$AlOOH$						
boehmite		$NaClO_4$	el. phor.	10.4		55
comm.(El. Space Ind.)		KNO_3	titr.	8.5		62
ZrO_2						
comm. baddelyte	30	KNO_3	titr.	6.4	3.0	42
synth.		KNO_3	titr.	8.0		57
synth., hydrous Zr(IV) oxide	31	KNO_3	titr.	6.0-6.4		65
synth., calcined at 450°C	25	KNO_3	titr.= el.phor.	6.0	5.7	88
ibid, 650°C	25	KNO_3	"	5.1	10.8	88
ibid, 850°C	25	KNO_3	"	7.6-8.1	−3.4 - −6.4	88
Chromium oxides						
$Cr(OH)_3$ (sol. synth.)	25	$NaClO_4$	el. phor.	8.4	−8.0	29
α-chromia, (synth.)	25	KNO_3	salt titr.	6.35	3.70	54
La_2O_3 (synth.)	25	NaCl	titr. = el. phor.	9.6		58
β-PbO_2 (electr.)	25±3	KNO_3	titr.	9.2		30
CeO_2		KNO_3	titr.	8.1		87

Substance	T	Electrolyte	Method	pH°	ΔG_m^o	ref.
SnO₂						
natural		KNO$_3$ 10^{-3}	imm.	5.5		21
cassiterite		KNO$_3$ 10^{-1}	imm.	5.4		21
RuO₂ (synth.)	20	KNO$_3$	titr.	5.75	7.48	33
ZnO, diff. synth.		NaNO$_3$, div.	titr.	8.5-9.5		23
Nickel oxides						
Ni(OH)$_2$	25	KNO$_3$	titr.	11.20	-23.99	86
	40	KNO$_3$	titr.	11.02	-25.50	86
	60	KNO$_3$	titr.	10.70	-26.74	86
	80	KNO$_3$	titr.	10.50	-28.45	86
NiO	25	KNO$_3$	titr.	11.30	-24.56	86
	80	KNO$_3$	titr.	10.50	-28.45	86
Cobalt oxides						
Co$_3$O$_4$	25	KNO$_3$	titr.	11.35	-24.85	86
	40	KNO$_3$	titr.	10.84	-24.42	86
	80	KNO$_3$	titr.	10.38	-27.64	86
Co(OH)$_2$	25	KNO$_3$	titr.	11.42	-25.25	86
	60	KNO$_3$	titr.	11.00	-28.66	86
	80	KNO$_3$	titr.	10.72	-29.94	86

c. *Other materials*

For silver halides (AgX) $\Delta_{cf}G^o_m = -\dfrac{2.303}{2} RT(pAg^o - pX^o)$, see [3.8.10].
Abbreviations as in table b.

Material	T	Electrolyte	Method	p.p.z.c. (pAg^o)	ΔG^o_m	ref.
AgI		KNO$_3$	titr.	5.63-5.65		10, 11
AgI	10	KNO$_3$+10^{-3} K-bipht.	titr.	5.88	13.34	12
AgI	20	" " "	titr.	5.50	13.19	12
AgI	30	" " "	titr.	5.63	11.84	12
AgI	40	" " "	titr.	5.58	10.85	12
AgI	50	" " "	titr.	5.53	9.93	12
AgI	60	" " "	titr.	5.46	9.09	12
AgI	70	" " "	titr.	5.42	8.14	12
AgI	80	" " "	titr.	5.37	7.10	12
AgBr				5.4		13
AgBr			titr.	5.6-5.9		14
AgBr			susp. eff.	5.4		15
AgCl				4.6		13
AgCl			susp. eff.	4.6		15
Ag$_2$S, pH 4.7			titr., cap. min.	10.2		16
				(pH^o)		
Ca$_{10}$(OH)$_2$ (PO$_4$)$_6$ (hydroxyapatite)	20	KCl, KNO$_3$	titr.	8.5		60
Ca$_{10}$(OH)$_2$ (PO$_4$)$_6$ (synth.)		KCl, KNO$_3$				
Ca$_{10}$F$_2$ (PO$_4$)$_6$ (fluorapatite synth.)	20	KCl, KNO$_3$	titr.	6.8		60

Literature and further notes

1. D.C. Grahame, *Chem. Revs.* **41** (1947) 441.

2. U. Palm, B.V. Damaskin, *Itogi Nauki i Tekhn. Elektrokhim.* **12** (1977) 99.

3. A. Valette, *C.R. Acad. Sci.* **C273** (1971) 320.

4. E. Sevastyanov, T. Vitanov and A. Popov, *Elektrokhim.* **8** (1972) 412.

5. A. Hamelin, J. Lecoueur, *Coll. Czech. Chem. Commun.* **36** (1971) 714.

6. D. Leikis, V. Panin and K. Rybalka, *J. Electroanal. Chem.* **40** (1972) 9.

7. K. Rybalka, D. Leikis, *Elektrokhim.* **3** (1967) 383.

8. T. Ehrlich, Yu-kukk and V. Past, *Ucheb. Zap. Tartu Gos. Univ.* **289** (1971) 9.

9. I. Dagaeva, D. Leikis and E. Sevastyanov, *Elektrokhim.* **3** (1967) 301.

10. J.A.W. van Laar, Ph.D. Thesis, State Univ. of Utrecht, 1957. See also J.Th.G. Overbeek in *Colloid Science*, H.R. Kruyt, Ed., Elsevier (1952) Vol. **1**, 160.

11. B.H. Bijsterbosch, J. Lyklema, *Adv. Colloid Interface Sci.* **9** (1978) 147.

12. A.C.M. Korteweg, J. Lyklema (partly published in ref. 11) fig. 12 and in J. Lyklema, *Discuss. Faraday Soc.* **42** (1966) 81.

13. B.B. Owen, S.R. Brinkley, *J. Am. Chem. Soc.* **60** (1938) 2233.

14. A. Basiński, *Rec. Trav. Chim.* **60** (1941) 267.

15. E.P. Honig, J.H.Th. Hengst, *J. Colloid Interface Sci.* **29** (1969) 510.

16. I. Iwasaki, P.L. de Bruyn, *J. Phys. Chem.* **62** (1958) 594. Protons also play a role because H_2S is a weak acid.

17. R.O. James, P.J. Stiglich and T.W. Healy, in *Adsorption from Aqueous Solution*, P.H. Tewari, Ed., Plenum (1981) 19.

18. F. Dumont, J. Warlus and A. Watillon, *J. Colloid Interface Sci.* **138** (1990) 543.

19. A.W. Gibb, L.K. Koopal, *J. Colloid Interface Sci.* **134** (1990) 122.

20. L.G.J. Fokkink, A. de Keizer and J. Lyklema, *J. Colloid Interface Sci.* **127** (1989) 116.

21. S.A. Ahmed, D. Maksimov, *J. Colloid Interface Sci.* **29** (1969) 97.

22. T.W. Healy, A.P. Herring and D.W. Fuerstenau, *J. Colloid Interface Sci.* **21** (1966) 435.

23. L. Blok, P.L. de Bruyn, *J. Colloid Interface Sci.* **32** (1970) 518.

24. R.P. Abendroth, *J. Colloid Interface Sci.* **34** (1970) 591.

25. P.H. Tewari, A.W. McLean, *J. Colloid Interface Sci.* **40** (1972) 267.

26. D.E. Yates, (1975), as quoted by J.A. Davis, R.O. James and J.O. Leckie, *J. Colloid Interface Sci.* **63** (1978) 480.

27. G.A. Parks, *Chem. Revs.* **65** (1965) 177.

28. T.W. Healy, D.W. Fuerstenau *J. Colloid Interface Sci.* **20** (1965) 376.

29. R. Sprycha, E. Matijevic, *Langmuir* **5** (1989) 479. (As the i.e.p. was independent of the $NaClO_4$ concentration, the surface is probably pristine and p.z.c. = i.e.p.)

30. N. Munichandraiah, *J. Electroanal. Chem.* **266** (1989).

31. J.W. Murray, *J. Colloid Interface Sci.* **46** (1974) 357. From electrophoresis and ion adsorption (Na^+, K^+). Around the p.z.c. Mn^{2+} ions also play a role.

32. Y.G. Bérubé, P.L. de Bruyn, *J. Colloid Interface Sci.* **27** (1968) 305.

33. J.M. Kleijn, J. Lyklema, *Colloid Polym. Sci.* **265** (1987) 1105.

34. P.W. Schindler, H. Gamsjäger, *Kolloid Z. Z. Polymere* **250** (1972) 759. (Indirect, pK_a and pK_b from titration analysis, then pH^o from [3.6.23b].)

35. J.T. Webb, P.D. Bhatnagar and D.G. Williams, *J. Colloid Interface Sci.* **49** (1974) 346.

36. R.M. Cornell, A.M. Posner and J.P. Quirk, *J. Colloid Interface Sci.* **53** (1975) 6.

37. G.R. Wiese, T.W. Healy, *J. Colloid Interface Sci.* **51** (1975) 427.

38. H.M. Jang, D.W. Fuerstenau, *Colloids Surf.* **21** (1986) 235.

39. E.A. Barringer, H.K. Bowen, *Langmuir* **1** (1985) 420 (i.e.p. = 5.5).

40. D.E. Yates, T.W. Healy, *J. Colloid Interface Sci.* **52** (1975) 222.

41. T.D. Evans, J.R. Leal and P.W. Arnold, *J. Electroanal. Chem.* **105** (1979) 161. For less pure samples pH° = 8.75.

42. A.E. Regazzoni, M.A. Blesa and A.J.G. Maroto, *J. Colloid Interface Sci.* **91** (1983) 500. (I.e.p. ZrO_2 = 6.51; i.e.p. Fe_2O_3 = 6.9, both at 25°; ibid on p. 560.)

43. M.A. Blesa, N.M. Figliolia, A.J.G. Maroto and A.E. Regazzoni, *J. Colloid Interface Sci.* **101** (1984) 410.

44. G.A. Parks, P.L. de Bruyn, *J. Phys. Chem.* **66** (1962) 967.

45. G.Y. Onoda, P.L. de Bruyn, *Surface Sci.* **4** (1966) 48.

46. A. Breeuwsma, J. Lyklema, *J. Colloid Interface Sci.* **43** (1973) 437.

47. R.J. Atkinson, A.M. Posner and J.P. Quirk, *J. Phys. Chem.* **71** (1967) 550.

48. R.H. Yoon, T. Salman and G. Donnay, *J. Colloid Interface Sci.* **70** (1979) 483.

49. N.H.G. Penners, L.K. Koopal and J. Lyklema, *Colloids Surf.* **21** (1986) 457.

50. P. Hesleiter, D. Babic, N. Kallay and E. Matijevic, *Langmuir* **3** (1987) 815.

51. C-P. Huang, W. Stumm *J. Colloid Interface Sci.* **43** (1973) 409.

52. A.R. Bowers, C-P. Huang, *J. Colloid Interface Sci.* **110** (1986) 575.

53. E. Rakotonarivo, J.Y. Bottero, F. Thomas, J.E. Poirier and J.M. Cases, *Colloids Surf.* **33** (1988) 191.

54. R.S. Alwitt, *J. Colloid Interface Sci.* **40** (1972) 195. (Pseudoboehmite is a poorly crystallized modification of boehmite, containing extra water: 1.5-2.5 mol H_2O/mol Al_2O_3.)

55. W.F. Bleam, M.S. McBride, *J. Colloid Interface Sci.* **103** (1985) 124.

56. R. Sprycha, *J. Colloid Interface Sci.* **127** (1989) 1, 12.

57. S. Ardizzone, G. Bassi and G. Liborio, *Colloids Surf.* **51** (1990) 207. (Calcination between 200° and 850°C did not change pH°.) See also S. Ardizzone, G. Bassi, *J. Electroanal. Chem.* **300** (1991) 585.

58. S.K. Roy, P.K. Sengupta, *J. Colloid Interface Sci.* **125** (1988) 340.

59. Th.F. Tadros, J. Lyklema, *J. Electroanal. Chem.* **17** (1968) 267.

60. L.C. Bell, A.M. Posner and J.P. Quirk, *J. Colloid Interface Sci.* **42** (1973) 250.

61. J.A. Yopps, D.W. Fuerstenau, *J. Colloid Interface Sci.* **19** (1964) 6.

62. R. Wood, D. Fornasiero and J. Ralston, *Colloids Surf.* **51** (1990) 389 (i.e.p. = 9.1).

63. M. Tschapek, C. Wasowski and R.M. Torres Sanchez, *J. Electroanal. Chem.* **74** (1976) 167.

64. (Cited in) L. Bousse, P. Bergveld, *J. Electroanal. Chem.* **152** (1983) 25.

65. F.S. Mandel, H.G. Spencer, *J. Colloid Interface Sci.* **77** (1980) 577. (pH° depends on method of preparation.)

66. D.E. Yates, T.W. Healy, *J. Chem. Soc. Faraday Trans. (I)* **76** (1980) 9. (i.e.p. = 5.85 ± 0.1; G.R. Wiese, T.W. Healy, *J. Colloid Interface Sci.* **57** (1975) 434.)

67. R. Sprycha, *J. Colloid Interface Sci.* **110** (1986) 278.

68. H. Sonntag, V. Itschenskij and R. Koleznikova, *Croat. Chem. Acta* **60** (1987) 383. (No positive charges could be detected by titration.)

69. H.Ll. Michael, D.J.A. Williams, *J. Electroanal. Chem.* **179** (1984) 131. (No positive charges detectable by titration .)

70. M. Kosmulski, E. Matijevic, *Colloid Polym. Sci.* **270** (1992) 1046.

71. S.B. Kanungo, D.M. Mahapatra, *J. Colloid Interface Sci.* **131** (1989) 103. (pH° depends on method of preparation.)

72. S.B. Kanungo, D.M. Mahapatra, *Coll. Surf.* **42** (1989) 173.

73. M.J.G. Janssen, H.N. Stein, *J. Colloid Interface Sci.* **109** (1986) 508. (The Degussa anatase contained 20% rutile.)

74. H. Tamura, T. Oda, M. Nagayama and R. Furuichi, *J. Electrochem. Soc.* **136** (1989) 2782. (International common sample no. 22 (IC 22), ex. IC. Sample Office, Cleveland Ohio, USA.)

75. G.H. Bolt, *J. Phys. Chem.* **61** (1957) 1166. (The ludox might have contained some Al.)

76. W. Janusz, J. Szczypa, *Mater. Sci. Forum* **25-26** (1988) 427.

77. D.A. Griffiths, D.W. Fuerstenau, *J. Colloid Interface Sci.* **80** (1981) 271.

78. S.B. Kanungo, *J. Colloid Interface Sci.* **162** (1993) 86. (I.e.p.'s of amorphous FeOOH and akagenite 7.15 and 7.20, respectively.)

79. A. Foissy, A.M.' Pandon, J.M. Lamarche and N. Jaffrezic-Renault, *Colloids Surf.* **5** (1982) 363 (Degussa anatase P25, contains 5% of rutile).

80. N. Kallay, S. Žalac and G. Štefanić, *Langmuir* **9** (1993) 3457.

81. M. Kosmulski, J. Matysiak, J. Szczypa, *J. Colloid Interface Sci.* **164** (1994) 280.

82. J. Szcypa, J. Matysiak, M. Kosmulski, in press.

83. N.M. Figliolia, Ph.D. thesis, Universidad de Buenos Aires, Argentina, 1987.

84. W.A. Zeltner, E. Yost, M.L. Macheski, M.I. Tejedor-Tejedor and M.A. Anderson, in *Geochemical Processes at Mineral Surfaces*, ACS Symp. Ser. Vol. **323**, p. 142, American Chemical Society, Washington DC, 1986.

85. D.A. Griffiths and D.W. Fuerstenau, *J. Colloid Interface Sci.* **80** (1981) 271.

86. P.H. Tewari and A.B. Campbell, *J. Colloid Interface Sci.* **55** (1976) 531.

87. L.A. de Faria and S. Trasatti, *J. Colloid Interface Sci.* **167** (1994) 352.

88. M. Prica, *Ph.D. Thesis*, University of Melbourne, Austr. (1993).

CUMULATIVE SUBJECT INDEX OF VOLUMES I (FUNDAMENTALS) AND II (SOLID-FLUID INTERFACES)

In this index bold face print refers to chapters or sections; app., and fig. mean appendix, and figure, respectively. The roman numeral I refers to Volume I. When a subject is referred to a chapter or section, specific pages of that chapter or section are usually not repeated. Sometimes a reference is made even if the entry is not explicitly mentioned on the page indicated. Entries in square brackets [..] refer to equations. The following abbreviations are used: (intr.) = introduced; (def.) = definition of the entry, ff = and following page(s). Combinations are mostly listed under the main term (example: for negative adsorption, see adsorption, negative), except where only the combination as such makes sense or is commonly used (example: capillary rise). Entries with 'surface' are often also found under 'interface' except where one of the two is uncommon. Entries to incidentally mentioned subjects are avoided. For the spelling of non-English names, see the preface to this volume.

adsorption (continued),

    superequivalent; 3.62(def.), see further: specific adsorption

    t-plot; figs. 1.27-28, 1.88, fig. 1.34

    (statistical) thermodynamics; I.**2.20e**, I.**2.22**, **1.3**, **2.3**, **3.6d-e**, **3.12**, **5.5**

    α-plot; 1.90

    (see also: adsorption isotherm (equation), calorimetry; for specific examples
        see under the chemical name of the adsorbate)

adsorption isosters; **1.3d**, fig. 1.9, 1.37, fig. 1.12f, 2.26

adsorption isotherm (equation); I.1.17ff(intr.), fig. I.1.12, 1.3ff, fig. 1.12, **1.5**,
    app. **1**

    Brunauer-Emmet-Teller (BET); I.**3.5f**, **1.5f**, [1.5.47], [1.5.50], fig;. 1.24a

    classification,

        adsorption from dilute solution; **2.7b**, fig. 2.24

        gas adsorption; **1.4b**

        surface excess; **2.3c**, fig. 2.8, **2.4**,

    composite; I.2.85

    (see surface excess)

    Dubinin-Radushkevich; [1.5.56]

    electrosorption; **3.12b**, **5.5g**

    Frenkel-Halsey-Hill; [1.5.55]

    Freundlich; I.1.19, 1.2, [1.7.7] (generalized), fig. 2.24c

    Frumkin-Fowler-Guggenheim (FFG); I.**3.8d**, **1.5e**, fig. 1.19, 1.64, fig. 1.43,
        [A1.5a], 2.65, 3.195

        for surface excess isotherm; **2.4d**

        for specific adsorption of ions; **3.6d**

    Harkins-Jura; [1.5.57]

    Henry; I.1.19, I.2.73, I.6.65, 1.2, [A1.1a], fig. 2.24a

        heterogeneous surface; fig. 1.43

    high affinity; I.1.19, fig. 2.24d, figs. 5.7-8, fig. 5.26, fig. 5.29, fig. 5.31

    Hill-De Boer = Van der Waals; see there

    individual; see partial

    Langmuir; I.1.20, I.2.74, I.**3.6d**, fig. I.3.2, I.3.46, 1.2, 1.28, **1.4a**, **1.5a**, **1.5b**,
        figs. 1.14-17, **1.5d**, **1.5e**, [1.7.7] (generalized), [A1.2a], fig. 2.24b, 2.86,
        3.196

        (binary mixture; **2.4b**, **2.4c**, fig. 2.11

    local; 1.104, fig. 1.43, 1.108

    one-dimensional; I.**3.8a**

    Ostwald-Kipling; **2.3b**, [2.3.6], [2.6.1]

    partial (= individual); fig. 2.9, figs. 2.11-14

    partially mobile; **1.5d**, fig. 1.18

SUBJECT INDEX

diffusion (coefficient) (continued),
  of colloids
    from dynamic light scattering; I.**7.8b**, I.**7.8c**, I.**7.8d**, I.**7.15**
  rotational; I.5.44, I.6.20, I.6.53, I.6.70ff, I.**7.8c**, I.7.59
  self; I.5.44, I.6.53, I.**7.15**, I.app. **11c**
  semi-infinite; I.6.59
  thermal; I.7.44, I.7.48
  to/from (almost) flat surface; I.**6.5d**, **2.8**, **4.6c**, **4.8b**
diffusion equation (theory for polymer adsorption); 5.32ff
diffusion impedance; 3.96
diffusion layer; I.6.63, I.6.68
diffusion potentials; I.**5.5d**, 4.125
  see also: potential difference
diffusion relaxation; **3.13**, 3.219, **4.6**, **4.8**
diffusiophoresis; I.6.91, 3.214, **4.9**
dilatometry (and surface excesses); 2.7
dipole field; I.**4.4b**, **3.13**, **4.6**, **4.8**
dipole moment; I.**4.4b**(def.), I.**7.3b**
  data for molecules; table I.4.1
  of colloids; **3.13**
dipoles; I.**4.4b**
  ideal (point dipole); I.4.20
  induced; I.4.22, I.4.27, I.7.18, I.7.93ff, **3.13**, **4.6**
  of colloids; **3.13**, fig. 4.1, **4.8**
  oscillation; I.**7.3b**, **4.8**
  permanent; I.4.22, I.4.27, I.7.17
Dirac delta function; I.7.40(def.)
disjoining pressure; I.4.6(def.), 1.22, fig. 1.37, 1.95ff, 1.101, [2.2.1], 5.65
  (see also: films, interactions)
dispersion,
  of colloids; I.1.5
  dielectric; see there
  of refractive index; I.4.37, I.7.13
dispersion forces; see Van der Waals forces
displacement ,
  of particles; I.6.18ff, I.6.30ff
  dielectric; see there
dissipation; I.2.7, I.2.22, I.4.3, I.4.34, I.6.9, I.6.13, I.6.35ff, I.**7.2c**, I.7.14

ISBN 0-12-460524-9